ESTUARINE INDICATORS

 *Marine Science Series*

The CRC Marine Science Series is dedicated to providing state-of-the-art coverage of important topics in marine biology, marine chemistry, marine geology, and physical oceanography. The series includes volumes that focus on the synthesis of recent advances in marine science.

CRC MARINE SCIENCE SERIES

SERIES EDITOR

Michael J. Kennish, Ph.D.

PUBLISHED TITLES

Artificial Reef Evaluation with Application to Natural Marine Habitats, William Seaman, Jr.

The Biology of Sea Turtles, Volume I, Peter L. Lutz and John A. Musick

Chemical Oceanography, Second Edition, Frank J. Millero

Coastal Ecosystem Processes, Daniel M. Alongi

Ecology of Estuaries: Anthropogenic Effects, Michael J. Kennish

Ecology of Marine Bivalves: An Ecosystem Approach, Richard F. Dame

Ecology of Marine Invertebrate Larvae, Larry McEdward

Ecology of Seashores, George A. Knox

Environmental Oceanography, Second Edition, Tom Beer

Estuarine Research, Monitoring, and Resource Protection, Michael J. Kennish

Estuary Restoration and Maintenance: The National Estuary Program, Michael J. Kennish

Eutrophication Processes in Coastal Systems: Origin and Succession of Plankton Blooms and Effects on Secondary Production in Gulf Coast Estuaries, Robert J. Livingston

Handbook of Marine Mineral Deposits, David S. Cronan

Handbook for Restoring Tidal Wetlands, Joy B. Zedler

Intertidal Deposits: River Mouths, Tidal Flats, and Coastal Lagoons, Doeke Eisma

Marine Chemical Ecology, James B. McClintock and Bill J. Baker

Morphodynamics of Inner Continental Shelves, L. Donelson Wright

Ocean Pollution: Effects on Living Resources and Humans, Carl J. Sindermann

Physical Oceanographic Processes of the Great Barrier Reef, Eric Wolanski

The Physiology of Fishes, Second Edition, David H. Evans

Pollution Impacts on Marine Biotic Communities, Michael J. Kennish

Practical Handbook of Estuarine and Marine Pollution, Michael J. Kennish

Practical Handbook of Marine Science, Third Edition, Michael J. Kennish

Seagrasses: Monitoring, Ecology, Physiology, and Management, Stephen A. Bortone

Trophic Organization in Coastal Systems, Robert J. Livingston

ESTUARINE INDICATORS

Edited by

Stephen A. Bortone

CRC PRESS

Boca Raton   London   New York   Washington, D.C.

Cover Art: Maggie May, Marine Laboratory, Sanibel-Captiva Conservation Foundation, Sanibel, Florida.

Library of Congress Cataloging-in-Publication Data

Catalog record is available from the Library of Congress

Visit the CRC Press Web site at www.crcpress.com

© 2005 by CRC Press

No claim to original U.S. Government works
International Standard Book Number 0-8493-2822-5
Library of Congress Card Number
Printed in the United States of America   2 3 4 5 6 7 8 9 0
Printed on acid-free paper

Durbin Tabb was a pioneer in establishing a long-standing database on the estuarine-dependent, spotted seatrout. Gustavo Antonini was instrumental in bringing the historical configurations of estuaries to bear on our current understanding of estuarine processes. Rich Novak explained estuaries to a new generation of citizens who will have a voice in the fate of estuaries. Last, Dave Lindquist admirably displayed dedication, good humor, and courage in the practice of his science.

Although they are no longer with us, this volume is dedicated to them for their efforts in helping to elevate our understanding of estuaries.

Preface

Our current level of long-term and comparative information on estuaries in many cases prohibits objective determination of the status and trends among these ecosystems. The way to resolve this situation is to develop and evaluate estuarine environmental indicators that will permit objective and meaningful evaluation of estuaries. However, this effort far exceeds the ability of one or a few well-intentioned scientists. Only the collective wisdom of the larger scientific community has the potential to make considerable strides in the direction of developing meaningful estuarine indicators. It is within this process that the idea for an Estuarine Indicators Workshop was born. The workshop, held on 29–31 October 2003 on Sanibel Island, Florida, served to bring together many of the world's leading estuarine scientists for the express purpose of presenting their views on estuarine indicators. Oral presentations were organized to address several features of estuarine indicators. These included the theory behind environmental indicators and the presumed attributes of effective estuarine indicators; the methods and protocols of indicator development and evaluation; a presentation of effective and failed examples of estuarine indicators; and a discussion that led contributors to speculate on the future direction of this dynamic field.

The workshop was an initial step toward resolving the issues associated with the development of successful estuarine indicators. A second step is the refinement of the ideas presented in the workshop in the form of this edited volume. It is hoped that future efforts will build upon the earnest efforts of the collective body of wisdom that resulted from these efforts.

Acknowledgments

This book is based largely on the chapter authors' contributions presented at the Estuarine Indicators Workshop. Sponsorship for the workshop was essential in bringing these estuarine experts together. Accordingly, I thank the South Florida Water Management District, especially Tomma Barnes; the Florida Department of Environmental Protection, specifically Eric Livingston and Pat Fricano; and the Charlotte Harbor National Estuary Program with special thanks to Lisa Beever, Catherine Corbett, and Maran Brainard Hilgendorf. Thanks also to Rob Jess, Susan White, Kevin Godsea, Cindy Anderson, and the entire staff at the J.N. "Ding" Darling National Wildlife Refuge, site of the workshop on Sanibel Island, Florida. Additionally, the staff and associates of the Sanibel-Captiva Conservation Foundation helped in many aspects of the logistics associated with hosting the workshop. Specifically, I thank Marti Bryant, Cheryl Giattini, and Erick Lindblad.

Technical reviews of the manuscripts were conducted on all the chapters. I gratefully acknowledge the following individuals for generously donating their expertise to this endeavor: Tomma Barnes (South Florida Water Management District), John W. Burns (Everglades Partners Joint Venture, U.S. Army Corps of Engineers). Dan Childers (Florida International University), Sherri Cooper (Bryn Athyn College), Jaime Greenawalt (Sanibel-Captiva Conservation Foundation), Holly Greening (Tampa Bay Estuary Program), John Hackney (University of North Carolina at Wilmington), Megan Tinsley (Sannibel-Captiva Conservation Foundation), Michael Hannan (Sannibel-Captiva Conservation Foundation), Steve Jordan (U.S. Environmental Protection Agency–Gulf Breeze, Florida), Ken Portier (University of Florida), Eric Milbrandt (Sanibel-Captiva Conservation Foundation), Chris Onuf (National Wetlands Research Center, U.S. Geological Survey), Ken Portier (University of Florida), Chet Rakocinski (University of Southern Mississippi), Steve W. Ross (University of North Carolina at Wilmington), Stanley Rice (University of Tampa), Joel Trexler (Florida International University), and Kendra Willet (J.N. "Ding" Darling National Wildlife Refuge). Thanks to John Sulzycki, Pat Roberson, Donna Coggshall, and Christine Andreasen for their direction and help in the production of this volume.

Last, a special thanks to the contributors. Their willingness to exchange information and ideas cooperatively captured the essence of scientific exchange. It is through this process and their efforts that this book is possible.

The Editor

Stephen A. Bortone, Ph.D., is Director of the Marine Laboratory at the Sanibel-Captiva Conservation Foundation in Sanibel, Florida. He holds an administrative appointment to the Graduate Faculty at the University of South Alabama, a courtesy faculty appointment at the Florida Gulf Coast University, and Research Professor status at the Florida Atlantic University and its Florida Center for Environmental Studies. Previously, he was Professor of Biology at the University of West Florida, where he served as Director for the Institute for Coastal and Estuarine Research. He also served as Director of Environmental Science at the Conservancy of Southwest Florida. Dr. Bortone received his B.S. from Albright College in Reading, Pennsylvania; his M.S. from Florida State University, Tallahassee; and his Ph.D. from the University of North Carolina, Chapel Hill.

For the past 37 years, Dr. Bortone has conducted research on the life history of estuarine organisms, especially fishes and seagrasses, chiefly in the southeastern United States and in the Gulf of Mexico. He has published more than 140 scientific articles on the broadest aspects of biology, including such diverse fields as anatomy, behavior, biogeography, ecology, endocrinology, evolution, histology, oceanography, physiology, reproductive biology, sociobiology, systematics, and taxonomy.

In conducting his research and teaching activities, Dr. Bortone has traveled widely. He has served as Visiting Scientist at The Johannes Gutenberg University (Mainz, Germany) and conducted extensive field surveys with colleagues from La Laguna University in the Canary Islands. He was a Mary Ball Washington Scholar at University College Dublin, Ireland. He has received numerous teaching and research awards, including the title "Fellow" from the American Institute of Fishery Research Biologists.

Dr. Bortone has served as scientific editor and reviewer for numerous organizations, such as the National Science Foundation, the U.S. Environmental Protection Agency, the National Marine Fisheries Service, and the U.S. Fish and Wildlife Service, and several journals, including the *Bulletin of Marine Science, Copeia, Estuaries, Marine Biology,* and *Transactions of the American Fisheries Society.*

Contributors

S. Marshall Adams   Environmental Sciences Division, Oak Ridge National Laboratory, Oak Ridge, Tennessee

Tomma Barnes   South Florida Water Management District, Fort Myers, Florida

Brian Bendis   AMJ Equipment Corporation, Lakeland, Florida

Marcia R. Berman   Virginia Institute of Marine Science, The College of William and Mary, Gloucester Point, Virginia

Patrick D. Biber   Institute of Marine Science, University of North Carolina at Chapel Hill, Morehead City, North Carolina

Donna Marie Bilkovic   Virginia Institute of Marine Science, The College of William and Mary, Gloucester Point, Virginia

Stephen A. Bortone   Marine Laboratory, Sanibel-Captiva Conservation Foundation, Sanibel, Florida

David R. Breininger   Dynamac Corporation, Kennedy Space Center, Florida

Marius Brouwer   Department of Coastal Sciences, University of Southern Mississippi, Ocean Springs, Mississippi

Nancy J. Brown-Peterson   Department of Coastal Sciences, University of Southern Mississippi, Ocean Springs, Mississippi

Billy D. Causey   Florida Keys National Marine Sanctuary, Marathon, Florida

Catherine A. Corbett   Charlotte Harbor National Estuary Program, Fort Myers, Florida

Nancy Denslow   Department of Biochemistry and Molecular Biology and Center for Biotechnology, University of Florida, Gainesville, Florida

Thomas L. Dix   Environmental Protection Commission of Hillsborough County, Tampa, Florida

Peter H. Doering   South Florida Water Management District, West Palm Beach, Florida

William A. Dunson   Pennsylvania State University (Emeritus Professor), Englewood, Florida

Michael J. Durako   Center for Marine Science, University of North Carolina at Wilmington, Wilmington, North Carolina

Julianne Dyble   Institute of Marine Science, University of North Carolina at Chapel Hill, Morehead City, North Carolina

John Edinger   J. E. Edinger Associates, Inc., Wayne, Pennsylvania

Anne-Marie Eklund   Southeast Fisheries Science Center, NOAA–Fisheries, Miami, Florida

Dana Fike   Florida Department of Environmental Protection, Port St. Lucie, Florida

Peter C. Frederick   Department of Wildlife Ecology and Conservation, University of Florida, Gainesville, Florida

Russel Frydenborg   Bureau of Laboratories, Florida Department of Environmental Protection, Tallahassee, Florida

Evelyn E. Gaiser   Department of Biology and Southeast Environmental Research Center, Florida International University, Miami, Florida

Charles L. Gallegos   Smithsonian Environmental Research Center, Edgewater, Maryland

Barbara K. Goetting   Environmental Protection Commission of Hillsborough County, Tampa, Florida

Stephen A. Grabe   Environmental Protection Commission of Hillsborough County, Tampa, Florida

Gregory A. Graves   Florida Department of Environmental Protection, Port St. Lucie, Florida

Jaime M. Greenawalt   Marine Laboratory, Sanibel-Captiva Conservation Foundation, Sanibel, Florida

John W. Hackney   NOAA/National Ocean Service, Center for Coastal Fisheries and Habitat Research, Beaufort, NC

M. Jawed Hameedi   Center for Coastal Monitoring and Assessment, National Centers for Coastal Ocean Science–NOAA, Silver Spring, Maryland

Kirk J. Havens   Virginia Institute of Marine Science, The College of William and Mary, Gloucester Point, Virginia

Ryan F. Hechinger   Marine Science Institute and Department of Ecology, Evolution and Marine Biology, University of California, Santa Barbara, California

Carl H. Hershner   Virginia Institute of Marine Science, The College of William and Mary, Gloucester Point, Virginia

Christina M. Holden   Environmental Protection Commission of Hillsborough County, Tampa, Florida

Xiaohong Huang   J. E. Edinger Associates, Inc., Wayne, Pennsylvania

Jon Hubertz   Florida Regional Office, J. E. Edinger Associates, Inc., Punta Gorda, Florida

Melody J. Hunt   Coastal Ecosystems Division, South Florida Water Management District, West Palm Beach, Florida

Todd C. Huspeni   Department of Biology, University of Wisconsin–Stevens Point, Stevens Point, Wisconsin

Tim Jones   Rookery Bay National Estuarine Research Reserve, Florida Department of Environmental Protection, Naples, Florida

Stephen J. Jordan   Gulf Ecology Division, U.S. Environmental Protection Agency, Gulf Breeze, Florida

David J. Karlen   Environmental Protection Commission of Hillsborough County, Tampa, Florida

Brian D. Keller   Florida Keys National Marine Sanctuary, Marathon, Florida

Carrie Kelly   Florida Department of Environmental Protection, Port St. Lucie, Florida

W. Judson Kenworthy   Center for Coastal Fisheries and Habitat Research, National Centers for Coastal Ocean Research, Beaufort, North Carolina

Venkat Kolluru   J. E. Edinger Associates, Inc., Wayne, Pennsylvania

Kevin D. Lafferty   U.S. Geological Survey, Western Region & Marine Science Institute, University of California, Santa Barbara, California

Patrick Larkin   ECOArray LLC, Alachua, Florida

Joe E. Lepo   CEDB-Biology, University of West Florida, Pensacola, Florida

Michael A. Lewis   Gulf Ecology Division, U.S. Environmental Protection Agency, Gulf Breeze, Florida

Edward R. Long   ERL Environmental, Salem, Oregon

Kevin A. Madley   Florida Fish and Wildlife Conservation Commission–Fish and Wildlife Research Institute, St. Petersburg, Florida

Steve Manning   Department of Coastal Sciences, The University of Southern Mississippi, Ocean Springs, Mississippi

Sara E. Markham   Environmental Protection Commission of Hillsborough County, Tampa, Florida

Frank K. Marshall III   Environmental Consulting & Technology, Inc., New Smyrna Beach, Florida

Frank J. Mazzotti   Fort Lauderdale Research and Education Center, University of Florida, Fort Lauderdale, Florida

Ellen McCarron Division of Water Resource Management, Florida Department of Environmental Protection, Tallahassee, Florida

Vicki McGee Rookery Bay National Estuarine Research Reserve, Florida Department of Environmental Protection, Naples, Florida

Eric C. Milbrandt Marine Laboratory, Sanibel-Captiva Conservation Foundation, Sanibel, Florida

David F. Millie Florida Institute of Oceanography, University of South Florida, St. Petersburg, Florida

Pia H. Moisander Institute of Marine Sciences, University of North Carolina at Chapel Hill, Morehead City, North Carolina

James T. Morris Department of Biological Sciences, University of South Carolina, Columbia, South Carolina

Andreas Nocker CEDB-Biology, University of West Florida, Pensacola, Florida

Patrick O'Donnell Rookery Bay National Estuarine Research Reserve, Florida Department of Environmental Protection, Naples, Florida

Judith A. Ott Florida Department of Environmental Protection–Charlotte Harbor Aquatic Preserves, Punta Gorda, Florida

Hans W. Paerl Institute of Marine Science, University of North Carolina at Chapel Hill, Morehead City, North Carolina

Michael F. Piehler Institute of Marine Science, University of North Carolina at Chapel Hill, Morehead City, North Carolina

James L. Pinckney Department of Oceanography, Texas A&M University, College Station, Texas

Chet F. Rakocinski Department of Coastal Sciences, Gulf Coast Research Laboratory, University of Southern Mississippi, Ocean Springs, Mississippi

Kenneth Rose Coastal Fisheries Institute and Department of Oceanography and Coastal Sciences, Louisiana State University, Baton Rouge, Louisiana

Michael Ross Southeast Environmental Research Center, Florida International University, Miami, Florida

Pablo Ruiz Southeast Environmental Research Center, Florida International University, Miami, Florida

Gitta Schmitt Florida Department of Environmental Protection, Port St. Lucie, Florida

Michael Shirley Rookery Bay National Estuarine Research Reserve, Florida Department of Environmental Protection, Naples, Florida

Gail M. Sloane   Florida Department of Environmental Protection, Tallahassee, Florida

Lisa M. Smith   Gulf Ecology Division, U.S. Environmental Protection Agency, Gulf Breeze, Florida

Richard A. Snyder   CEDB-Biology, University of West Florida, Pensacola, Florida

David M. Stanhope   Virginia Institute of Marine Science, The College of William and Mary, Gloucester Point, Virginia

Eric D. Stolen   Dynamac Corporation, Kennedy Space Center, Florida

Mark Thompson   Department of Environmental Protection, Port St. Lucie, Florida

Franco Tobias   Southeast Environmental Research Center, Florida International University, Miami, Florida

David A. Tomasko   Southwest Florida Water Management District, Brooksville, Florida

Leigh G. Torres   Duke University Marine Laboratory, Nicholas School of the Environment and Earth Sciences, Beaufort, North Carolina

Louis A. Toth   Vegetation Management Division, South Florida Water Management District, West Palm Beach, Florida

Jillian Tyrrell   Department of Environmental Protection, Port St. Lucie, Florida

Dean Urban   Nicholas School of the Environment and Earth Sciences, Duke University, Durham, North Carolina

Lexia M. Valdes   Institute of Marine Science, University of North Carolina at Chapel Hill, Morehead City, North Carolina

Anna Wachnicka   Department of Earth Sciences and Southeast Environmental Research Center, Florida International University, Miami, Florida

Kathy Worley   The Conservancy of Southwest Florida, Naples, Florida

Glenn A. Zapfe   NOAA/NMFS, Southeast Fisheries Science Center, Pascagoula, Mississippi

Contents

1

The Quest for the "Perfect" Estuarine Indicator: An Introduction

Stephen A. Bortone

CONTENTS

Environmental Indicators

The hallmark of the condition of our environment in recent years is *change*. Detecting the status and trends of our environment has become one of the central themes of modern ecology. Paramount in this effort is that methods of environmental assessment and analysis should adhere to the scientific method regardless of the questions being asked and the habitat being examined.

Aiding the meaningful assessment of environmental change is the implementation and development of insightful environmental indicators. Indicators such as species community features, their biological attributes, or other innovative metrics of abiotic features have promise in assessing trends in environmental conditions. However, scientists must be mindful in the development of these indicators to avoid the circularly reasoned, tautological "trap" of using a biological parameter to predict or classify an environmental condition and, subsequently, using the same environmental condition to classify the same biological parameter. With this caveat, effective environmental indicators have an advantage in assessing environmental change in that they are often directly related to the problem being evaluated and thus are ecologically meaningful.

Recently, a large-body of information has been developed, directed toward the development of environmental indicators. Notable among these are indicators for streams, lakes, and ponds as well as terrestrial biotopes. Methods and protocols are likely to continue to improve, becoming more precise and accurate. Consequently, today we can list many environmental indicators that have proven their utility in being able to describe and assess environmental change objectively and efficiently.

Estuarine Indicators

Missing from the euphemistic "Manual of Environmental Science" is a consensus among scientists regarding the most effective and meaningful methodologies and protocols needed to accurately and precisely assess the status and trends within and between estuarine biotopes. This is due, in large part, to the relative infancy of estuarine status and trend assessment coupled with the inexorable truth that estuaries, by their very nature, are places of extraordinary (both predictable and unpredictable) natural change in both time and space, each at very broad scales.

Each estuary varies independently relative to stressors with regard to space and time. Although this is true for all environments, the scale, frequency, and duration of change is unique to estuarine ecosystems. Moreover, generalizations regarding estuaries have been difficult because many of the stressors are individually unique in their scale of impact to each estuary. Many of the factors such as the salinity regime are influenced by tide, which is estuary specific because of shape and latitude parameters and unique to each estuary. This feature alone makes predictions and generalizations regarding estuarine ecosystem responses to stressors of special significance. Large-scale estuarine changes are often dominated by both predictable and unpredictable factors. The predictable factors include tides, seasons, circadian adaptations, and human development while the unpredictable factors include storms, accidents, and the yet-to-be-understood coupling of events.

Goal of an Effective Estuarine Indicator

It is likely that more than a few carefully thought-out and tested estuarine indicators will be deemed sufficient to satisfy the demands for information required for estuarine assessment. This is because of the varied questions that will most assuredly be asked from a variety of perspectives. Moreover, the complete suite of estuarine indicators should allow comparisons between estuaries and comparisons within estuaries relative to both space and time.

As proposed in many of the chapters that follow in this volume, estuarine indicators can be abiotic or biotic. Abiotic indicators will have measuring and scale problems associated with the application of their specific protocols to the specific questions being asked. Appropriate methodologies, verification of protocols, and meaningful measures are areas explored here. Clearly, progress is being made in the refinement of abiotic measures of estuarine condition. As a personal aside, it is interesting to note that researchers are beginning to realize the folly of measuring factors such as temperature, salinity, and dissolved oxygen (along with a plethora of other water quality components) to levels of accuracy that far exceed the natural variation in the system. Gathering data at levels that far exceed the space and time variation known for the system is not only inefficient, but can be misleading.

Ideal estuarine bioindicators (usually in the form of measured responses of a species, its population, or its community) should be gathered from species that are broadly distributed between estuaries and regions. Concomitantly, individual species, if selected for examination, should display limited movement between estuaries, thus assuring that responses within a system are the result of factors predominately from that same system. In addition, individual response variables should show little variation relative to latitude. Most importantly, the measured biological responses should be ascribable to environmental attributes so that the responses reflect environmental features of the estuary being evaluated. Ideally, several biological responses should be measured for each environmental factor, and the biological response variables should be consistent for similar environmental conditions.

The chapters included here (largely derived from the Estuarine Indicators Workshop held on Sanibel Island, Florida in October 2003) represent a broad range of estuarine indicators. Some of the chapters offer presentation on the application and effectiveness of estuarine indicators currently used by research scientists. Other chapters present documented arguments for the future consideration of indicators not previously considered nor generally accepted as estuarine indicators. Still other chapters offer insight into the overall role that estuarine indicators play in estuarine management decisions, now and in the future. The chapters are arranged to lead the reader to fully appreciate the need, problems, complexity, breadth, and application of estuarine indicators. Although each chapter contains elements of each of these features, the particular organization begins with an overall introduction to the multifaceted nature of estuarine indicators, followed by a series of chapters that demonstrate the range and complexity of estuarine indicators, including biotic and abiotic indicators. The diverse array of biotic indicators is arranged, more or less, in a phylogenetically hierarchical order and includes indicators that are molecular, species-based, populational, and community oriented. Last, a series of chapters offers glimpses of larger-scale applications and considerations of estuarine indicators culminating in demonstrations of their utility in the management of estuarine ecosystems.

When addressing the general public, scientists are often asked, "How's the estuary doing?" While the public is often perplexed by answers that involve far too many qualifiers, it can be assured that the assessment process is now proceeding with a degree of rigor and direction that is unprecedented in estuarine science. With the knowledge base offered here, the scientific community will at least have the "tools" necessary to calibrate the environmental barometer that measures estuarine condition.

2

Using Multiple Response Bioindicators to Assess the Health of Estuarine Ecosystems: An Operational Framework

S. Marshall Adams

CONTENTS

Introduction

Estuaries are complex ecosystems that are controlled and regulated by a variety of physicochemical and biological processes. In addition, estuarine organisms experience a variety of natural and anthropogenic stressors, both of which vary spatially and temporally. High variability of environmental factors combined with synergistic and cumulative interactions of these factors complicates the interpretation and evaluation of the effects of stressors on estuarine biota. Because of their complexity, estuaries present unique challenges relative to understanding the effects of stressors and the underlying causes of these effects on biological components of estuarine ecosystems. Understanding the relationships between environmental stressors, causal mechanisms of stress, and biological effects is critical for achieving effective management and regulation of estuarine resources. As reflected in the chapters in this book, many studies focus on the structural aspects of estuarine systems, such as identification and description of organisms, populations, and communities. Several chapters describe the occurrence, distribution, and abundance of estuarine biota relative to spatial and temporal patterns of various influential or controlling physicochemical and biological factors, such as salinity, nutrients, habitat availability, and food availability. Few studies, however, have focused on understanding the mechanisms or the functional processes responsible for observed changes in biological components of estuaries.

    Biological indicators have been used in a limited number of studies to assess the health of aquatic systems and to help identify underlying processes or mechanisms responsible for observed changes in these systems. Bioindicators have traditionally been considered structural entities of ecosystems, which are used as sentinels of overall condition or health. Within this context, a bioindicator can be defined as a particular species, population, or community, which serves as an early-warning indicator that reflects the "health" status of an aquatic system (Van Gestel and Van Brummelen, 1996). Recently, however, bioindicators have been applied more within a functional context to include biological responses at the organism level and above. Bioindicators can thus be considered, not only as indicators of ecosystem status, but also as processes or components of organisms, populations, or communities that provide various degrees or levels of information regarding the functional status of aquatic systems (Engle and Vaughan, 1996; Adams, 2002). In contrast to bioindicators, biomarkers are operationally considered to be indicators of exposure to environmental stressors, which are usually expressed at the suborganismal levels of biological organization including the biomolecular, biochemical, and physiological levels (Adams, 1990; McCarthy and Shugart, 1990; Huggett et al., 1992).

Given this background the objectives of this chapter are then to (1) provide a framework or a basis by which bioindicators can be used to assess the effects of environmental stressors on ecological components of estuaries, (2) demonstrate how bioindicators can be used to help identify the processes or causal mechanisms responsible for these effects, and (3) provide some basic guidance for use and application of bioindicators within the framework of effective environmental management of estuarine ecosystems.

Characteristics of Bioindicators

The underlying concept of using multiple response indicators (bioindicators) to assess the health of estuarine ecosystems is that the effects of environmental stressors are manifested at lower levels of biological organization before they are realized at higher levels of organization such as at the population and community levels (Adams, 1990; Figure 2.1). Sublethal stress is generally expressed first at the

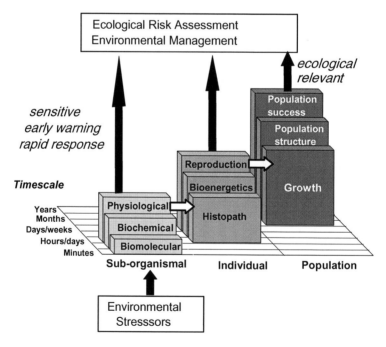

FIGURE 2.1 Hierarchical responses of organisms along a time–response scale to environmental stressors illustrating that lower-level responses serve as rapid, sensitive, and early-warning indicators of stress while organism-, population-, and community-level responses reflect the ecological significance of environmental stress. This multivariate bioindicator approach is used to help establish causal relationships between environmental stressors and effects.

molecular and biochemical levels through interference with genetic material, enzymes, or cell membranes. Such changes induce a series of structural and functional responses at increasingly higher levels of biological organization (Figure 2.1). Induced biological responses at these lower levels can impair, for example, integrated processes related to hormonal regulation, metabolism, bioenergetics, and immunocompetence. These effects, in turn, may eventually affect the organism's ability to survive, grow, or reproduce (Figure 2.1). Ultimately, irreversible and detrimental effects may be observed at the population or community levels. For effects to be realized at increasing higher levels of organization, however, the stressor(s) must be of sufficient magnitude and or duration to overwhelm the normal homeostatic capacity of specific biological systems (Schlenk et al., 1996a). For example, when the capacity of protein systems such as the HSP70 stress proteins is exceeded, pathological lesions can develop in tissues and organs such as the liver, gill, or kidney. Consequently, structural damage to liver tissue can impair the ability of this organ to produce vitellogenin, a critical component of egg development and, therefore, ultimately compromise reproductive success. In the sequence of biological organization from molecules and cells to populations and communities, each level of organization finds its explanation of mechanism in the levels below and its significance in the levels above (Bartholomew, 1964).

Exposure to environmental stressors, at increased frequencies and/or durations, results in a progressive deterioration in organism health that may ultimately compromise the success of populations and communities. The first stage of a stress response in an organism involves departures from the healthy state, which are associated with the initiation of a compensatory stress response resulting in little or no loss of functional ability. With increased environmental challenge, the survival potential of organisms is reduced because of the loss of compensatory reserves. Once these compensatory reserves have been depleted, the ability of organisms to mount a successful response to additional challenges is severely compromised, resulting in increased disability and disease. Disabilities such as pathologies and disease are usually not detected until after the loss of compensation, whereas impairments (i.e., biochemical, physiological), because of their sensitivity, can be detected much earlier and can be reversible and even curable (Figure 2.1). Therefore, biochemical, physiological, and behavioral responses can provide sensitive and early warning indicators of injuries or disabilities to biota (Depledge, 1989).

In general, biomarkers are used to indicate exposure to environmental stressors, while bioindicators, because of their integrative nature, reflect the effects of exposure to stressors at higher levels of biological organization. The main attributes of biomarkers and bioindicators that are important for consideration in assessing the health of estuarine systems are listed in Table 2.1. Because biomarkers are stressor-sensitive and rapidly responding end points, they can be used to identify the mechanistic basis of possible causal relationships between stressors and effects (Figure 2.1). Biomarkers can also be used to help identify the source of a stressor or determine if organisms have been exposed to a specific stressor (such as contaminants) or a group of similar stressors. Conversely, bioindicators have limited ability for helping to establish causal relationships between stressors and effects because their sensitivity and specificity to stressors is low and they tend to integrate effects of multiple stressors over larger spatial and temporal scales (Adams, 1990, 2002). The advantage of bioindicators, however, is their relatively low response

TABLE 2.1

Major Features of Biomarkers and Bioindicators Relative to Their Advantages and Limitations for Use in Assessing the Health of Estuarine Ecosystems

Major Features	Biomarkers	Bioindicators
Types of response	Subcellular, cellular	Individual–community
Primary indicators of	Exposure	Effects
Sensitivity to stressors	High	Low
Relationship to cause	High	Low
Response variability	High	Low
Specificity to stressors	Moderate–high	Low–moderate
Timescale of response	Short	Long
Ecological relevance	Low	High

variability and high ecological relevance or significance (Table 2.1, Figure 2.1). The complexity of estuarine systems, with their inherently high variability and presence of multiple stressors, suggests that no single measure (or perhaps not even a few measures) is adequate for assessing the health of these systems. Instead, an appropriate suite of end points is required not only to determine the biological significance of stress, but also to understand its underlying cause (Hodson, 1990; Attrill and Depledge, 1997). The basic concept of using a variety and suite of biomarkers and bioindicators to understand the mechanistic basis of stress and the ecological significance of stress is shown in Figure 2.1.

Application of Bioindicators in Estuarine Ecosystems

As an operational framework for using multivariate bioindicators to assess the health status of estuarine systems and to help diagnose causes of stress, several strategies or approaches should be considered: (1) direct vs. indirect effects of stressors, (2) temporal response scaling, (3) establishment of causal relationships between stressors and effects, (4) environmental profiling and diagnosis, and (5) integrated effects assessment.

Direct vs. Indirect Effects

Responses of organisms to environmental stressors are the integrated result of both direct and indirect pathways or effects. Direct pathways operate primarily through metabolic processes that are initiated at the lower levels of biological organization and are propagated upward through increasingly higher levels of organization (Figure 2.2). Indirect pathways, however, operate mainly through effects on the food chain, on habitat availability, or through behavioral modification of organisms (Figure 2.2). The effects of multiple stressors acting through direct mechanisms occur initially at the molecular or subcellular level and can be expressed, for example, as changes in biomolecular, biochemical, and physiological

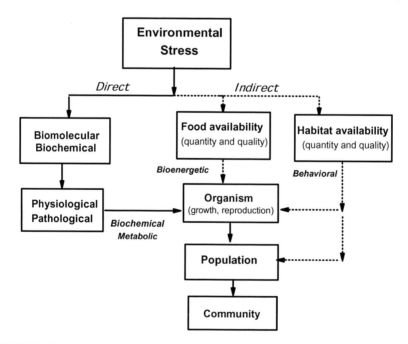

FIGURE 2.2 Relationships between environmental stress and direct and indirect effects on biological systems. Direct pathways affect organisms primarily through biochemical and metabolic processes, and indirect pathways influence biota through effects on food and habitat availability.

TABLE 2.2

Major Categories of Response Indicators to Environmental Stressors Representing Direct Indicators of Exposure (Biomarkers), Direct Indicators of Effects (Bioindicators), and Indirect Indicators of Exposure/Effects That Can Be Used to Help Identify Causes of Effects Due to Environmental Stressors in Estuarine Systems

Direct Indicators of Exposure (biomarkers)	Direct Indicators of Effects (bioindicators)	Indirect Indicators of Exposure/Effects
Detoxification enzymes	Lipid metabolism	Nutrition
DNA damage	Organ dysfunction enzymes	Lipid pools
Antioxidant enzymes	Immunocompetence	Growth
Selected serum chemistries	Selected histopathologies	Reproduction
Stress proteins	Steroid hormones	Behavior
Osmoregulatory responses	Metabolism/respiration	Bioenergetic processes

components or processes such as DNA integrity, enzyme activity, metabolism, and respiration, respectively. Responses at these lower levels can be propagated upward through increasing levels of biological complexity, ultimately affecting higher-level metabolic processes such as lipid dynamics, immunocompetence, and hormonal regulation (Larsson et al., 1985). Ultimately, these effects may be manifested as changes at the organism, population, and community levels. Environmental stressors may also have an impact on organisms indirectly through the food chain by influencing the quality (energy and protein content) and quantity (biomass) of energy available to consumers. In addition, stressors can indirectly impair the health of estuarine biota by affecting the quality and quantity of the habitat, resulting in altered behavior related to reproduction, feeding, or habitat selection (Reynolds and Casterlin, 1980; Little, 2002). The more ecologically relevant parameters of aquatic systems, such as growth, reproduction, and population-level attributes, can therefore be affected by both direct and indirect pathways, which involve the integrated effects of metabolic impairment, energy availability, and behavioral alterations (Adams, 1990, 2002).

An approach for helping to assess the relative influence of direct and indirect pathways on the health of estuarine biota is to measure a selected suite of bioindicators representing different types or categories of response variables (Table 2.2). Examples of the types and categories of indicators that could be measured are (1) direct indicators of exposure to stressors including biomarkers of exposure, (2) direct indicators of effects that include bioindicators of metabolic and bioenergetic impairment and dysfunction, and (3) indirect indicators of effects including nutrition and feeding indices, growth, reproduction, behavior, and various measures of lipid pools within the organism (Table 2.2). Identification of the pathway primarily responsible for any particular observed effect can therefore be qualitatively assessed based on the relative proportion of responses that can be measured (compared to a measured change from a reference or standard condition) in each of the three categories above. For example, if stressors affect organisms primarily through indirect pathways such as the food chain, we would expect to see corresponding effects on nutrition and feeding indices, growth, and various lipid pools within the organism. Oppositely, if impacts occur primarily through direct metabolic stress pathways, we might expect to see a higher proportion of effects expressed as metabolic responses, such as changes in molecular function, enzyme concentration or activity, stress proteins (concentration or induction), and osmogulatory ability.

Temporal Response Scaling

In addition to stressors operating through direct and indirect pathways, estuarine organisms are typically subjected to two general types of environmental disturbances: (1) chronic, sustained, or long-term stressors, and (2) periodic, pulsed, or short-term stressors (Figure 2.3). In estuarine systems, these pulsed or short-term stressors are usually superimposed, on a periodic basis, over the longer-term sustained types of stressors. These longer-term stressors can cause (1) gradual modifications in the quality and quantity of important habitats such as seagrass, mangrove, and salt marsh systems; (2) insidious changes in water quality due to contaminant and sediment loading; and (3) subtle changes in the eutrophic status

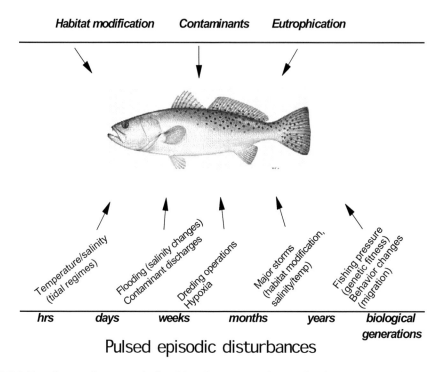

FIGURE 2.3 Estuarine organisms are typically subjected to two general types of environmental stress, including chronic, sustained, or long-term stressors and pulsed or short-term stressors, which are superimposed on a periodic basis over the chronic stressors. The pulsed stressors can occur within a wide range of timescales from hours to biological generations.

of estuaries including associated zones of hypoxia and blooms of toxic algae. Superimposed on these sustained and chronic changes are the more pulsed or episodic events, which can vary over the short term from minutes to hours to longer-term scales spanning years or biological generations. Although ecologically relevant indicators such as population- and community-level variables provide integrated responses to environmental stressors, they are characterized by relatively long response times. Consequently, responses of these higher-level indicators do not typically occur within the same time frame as the stressors that originally caused the observed change in the biological receptor of interest. Therefore, bioindicators at higher levels of organization provide useful information about the health status of estuarine ecosystems, but because of their relatively slow response times, low sensitivity, and low specificity to stressors (see Table 2.1), they have limited use as early-warning sentinels of ecosystem health. For example, by the time a stressor interacts with a biological receptor at the lower levels of biological organization and a change is ultimately manifested and observed at the population, community, or ecosystem level, damage to an ecological system has already occurred. Such a long time lag between initiation of a stress response at lower levels of organization and an observed change at higher levels of organization minimizes the probability that proactive mitigation, including restoration and recovery of these impaired systems, can be effectively achieved.

As environmental stressors and their associated effects can occur over wide temporal scales ranging from minutes to biological generations, in order to capture and identify biological effects that occur within the corresponding timeframes of stressors, it is important to apply a battery or a suite of bioindicators that also represent a wide range of response times to these stressors. Bioindicators that respond over a wide range of timescales would, in many cases, represent and effectively capture those particular responses that also occur over a wide range of specificities and sensitivities to stressors. For example, biomolecular and biochemical responses generally occur within a short period (minutes–hours–days) following exposure to a stressor, and they are also relatively specific and sensitive to stressors. Conversely, organism-, population-, and community-level responses are manifested over longer timescales of weeks, months, and years and generally have much lower specificity and sensitivity to stressors.

In estuarine systems, chronic low-level disturbances can include subtle changes over time in habitat quality and quantity, contaminant loading, and eutrophication (Figure 2.3). In the case of these chronic low-level disturbances, which are sustained over long periods, suites of bioindicators can be applied to evaluate the health status of estuarine biota, but causality is more difficult to assess because of the long-term and integrated nature of these responses. However, cause and effect due to disturbances that occur on shorter timescales are more straightforward to identify and diagnose because some disturbances in estuarine systems occur within the same timescales of the biomarker and bioindicator responses of interest. In estuaries, short-term or episodic types of disturbances can occur over a range of timescales from hours, weeks, or months. In the case of disturbances that take place on shorter timescales, such as changes in salinity or temperature, biomarkers and bioindicators that respond within similar or corresponding timescales can be used to help identify or assess the cause. For example, the stress proteins or HSP70 proteins, physiological measures related to osmogulation, such as serum electrolytes, and biochemical measures, such as cortisol and glucose, can be used to help assess causal relationships based on these shorter-term response scales. At scales of days to weeks, causal relationships related to other periodic disturbances such as flooding (i.e., salinity changes), contaminant discharges, dredging operations, and hypoxia could be assessed by using a suite of biomarkers and bioindicators with intermediate response times such as selected histopathological markers, bioenergetic and lipid indicators, individual health and condition indices, and immunological parameters (Table 2.2).

Establishment of Causality

Identification of the causal relationships or mechanisms that occur between environmental stressors and impairment of estuarine systems is a difficult task. A causal relationship is assumed to exist whenever evidence indicates that certain environmental factors increase the probability of the occurrence of injury to a system and when a reduction in one or more of these factors decreases the frequency of that injury to a biological resource (Fox, 1991). A cause can therefore be defined as a stressor that occurs at an intensity, duration, and frequency of exposure that results in a detectable change in the integrity of ecosystems or components of ecosystems such as key biota. The ability to establish causality between environmental stressors and ecological effects is particularly important in the environmental policy and regulatory arena because of the critical decisions that often must be made regarding remediation, legacy of contaminated sites, atmospheric deposition, and other environmental compliance and regulatory issues. Definitive evidence of causality could reduce the uncertainty of such decisions, resulting in less costly environmental policies and streamlining of regulatory and compliance procedures. Investigative procedures that successfully identify causal stressors could result in appropriate corrective action measures through habitat restoration, point- and non-point-source controls, and invasive species control (U.S. EPA, 2000.

Establishing definitive causal relationships between stressors and observed effects in estuarine ecosystems continues to be a challenge because of the complex nature of these systems, the many biotic and abiotic factors that can influence or modify responses of biological systems to stressors (McCarty and Munkittrick, 1996; Wolfe, 1996), the orders of magnitude involved in extrapolation over both spatial and temporal scales (Holdway, 1996), compensatory mechanisms that operate in natural populations (Power, 1997), and the many possible modes and pathways by which stressors can disrupt and destabilize ecosystems. For example, not only can stressors affect biological systems directly, but indirect effects can also occur as a result of such factors as habitat and food availability, predator–prey interactions, and competition (McCarty and Munkittrick, 1996; Adams et al., 1998). In addition, time lags between the initial cause and effect can be long (Vallentyne, 1999) and interdependence among disturbance events, ecosystem properties, and biological invasions often make causal relationships difficult to discern (Bart and Hartman, 2000).

High variability of environmental factors, combined with synergistic and cumulative interactions of these factors in estuaries, also complicates our ability to establish definitive causal links between stressors and effects. Cumulative and synergistic effects of stressors manifest themselves at a variety of spatial, temporal, and organizational scales, which also makes establishment of causality in field situations particularly problematic. Culp et al. (2000) have identified three categories of impacts in aquatic ecosystems that complicate the establishment of causality:

1. Incremental impacts. The total effect of successive stressor events whose combined effect exceeds a critical ecological threshold thereby compromising ecosystem integrity
2. Multiple source impacts. Impacts that occur when sources of stressors and their effects overlap spatially
3. Multiple stressor impacts. Impacts that include situations where different classes of stressors interact in an additive fashion preventing *a priori* prediction of biotic response

Given all these complicating factors in assessing causality, Nacci et al. (2000) have stated that relationships between stressors and ecological effects may be observable only under situations when compensatory processes of organisms have been overwhelmed and specific injuries are observed (also see discussion related to this in the Bioindicators Characterization section).

Studies that integrate responses across levels of organization are especially valuable in helping to establish causality because they can aid in identifying mechanistic linkages between lower-level responses (biomarkers) and population and community responses (bioindicators) (see Figure 2.1). Because protection and management of biological resources generally require that effects at higher levels of biological organization (e.g., populations, communities) be utilized in the ecological risk assessment process (U.S. EPA, 1998; Power, 2002), it is important to establish linkages and relationships between lower-level responses (i.e., biomarkers) and higher-level responses (McCarty and Munkittrick, 1996) using a suite and variety of indicators that reflect a range of sensitivities, specificities, and response timescales to environmental stressors (Adams et al., 2002).

Environmental Profiling and Diagnosis

Important to the assessment and diagnosis of estuarine health is the use and application of bioindicators in separating or partitioning out the effects of natural environmental factors from those effects due to anthropogenic stressors. Estuaries can be affected by a variety of anthropogenic activities including point-source discharges of domestic and industrial pollutants, atmospheric deposition, agricultural practices, land-use activities including urban development, modification of water flow regimes, dredging practices, and physical habitat alteration. Many of these activities are related to specific stressors such as contaminants, nutrients, or sediments, which characterize that particular activity. These sets of stressors, which are unique or characteristic to each type of activity, can be used to help distinguish these activities from each other. For example, point-source discharges from paper mill operations are typically characterized by chlorophenolic and resin acid compounds, dioxin-type contaminants, and high nutrient loading. In contrast, non-point-source agricultural activities contribute pesticides, nutrients, and sediment to receiving estuaries, which typically results in specific or predictable types of biological responses.

Because certain types of stressors cause predictable responses in biological systems, we can use environmental "profiling" or diagnosis methods to help partition or separate out the effects of natural vs. anthropogenic stressors on estuarine biota. The use of multiple response bioindicators for environmental diagnosis or profiling is conceptually similar to approaches used by the medical profession to diagnose the health of human patients. In human subjects, a variety and suite of medical procedures are performed such as chemical profiling of blood and urine, and the results are compared with standardized norms for diagnosis of pathology and disease. In humans, diagnosis of health is relatively straightforward because the individual is the ultimate end point of interest. In considering ecosystem health, however, the end points of interest are typically populations, communities, and ecosystems, and diagnosing the causes of effects at these higher levels of biological organization is particularly problematic because of

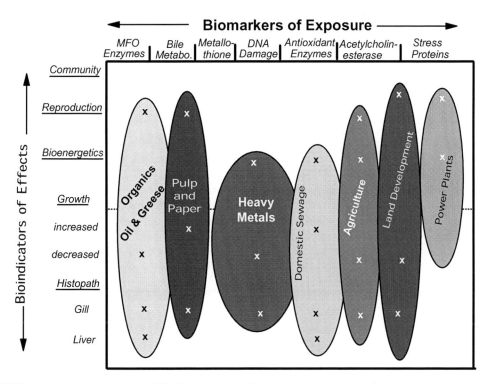

FIGURE 2.4 (Color figure follows p. 266.) Biomarker–bioindicator response profiles characteristic of several major types of anthropogenic activities. Use of such characteristic profiles can help diagnose and identify sources of stress in estuarine systems affected by multiple stressors.

the interacting effects of biotic and abiotic factors, high temporal and spatial variability, and compensatory mechanisms that operate in ecological systems.

To demonstrate the use of a diagnostic approach that can help identify and differentiate among sources of stress potentially responsible for biological effects in aquatic systems, exposure–response profiles were constructed for various stressor exposure–effects relationships corresponding to each major type of anthropogenic activity shown in Figure 2.4. The principal types of stressors associated with each activity shown in Figure 2.4 were first determined and then matched with the types of biomarker responses characteristic of that specific activity. Because certain stressors, and in particular various types of contaminants, are associated with specific responses at the biomolecular, biochemical, or physiological levels, this analysis matched each major type of stressor or activity to its corresponding responses at these lower levels of biological organization. Once the exposure responses (biomarkers) associated with each anthropogenic activity were identified, a stress exposure–biological effects profile was generated for each of these activities by plotting exposure biomarkers on one axis (*x* axis) and bioindicator responses or effects on the other axis (*y* axis) (Figure 2.4). A literature review was then conducted to identify which major types of biological responses at the higher levels of organization (bioindicators) were typically associated with each major type of anthropogenic activity. Cross marks in Figure 2.4 within the exposure–effects profile (ellipse) for each activity indicate those specific bioindicators of effects on the *y* axis that are associated with the various biomarkers of exposure on the *x* axis. For example, the principal biomarkers of exposure for petrochemical and pulp and paper activities are induction of the P450-detoxification enzymes and production of aromatic and chlorophenolic biliary metabolites, respectively. For bioindicator responses corresponding to these two activities, both petrochemical and paper mills have been reported to cause various gill lesions in fish and impaired reproductive function in aquatic organisms. Growth, however, typically decreases under petrochemical exposure and actually increases in aquatic systems receiving paper mill effluents, due primarily to nutrient enrichment and increased productivity of receiving waters. In addition, organisms inhabiting systems affected by polycyclic aromatic hydrocarbon (PAH) compounds typically have relatively high incidences of liver tumors, a situation not normally observed in aquatic systems receiving paper mill effluents. Thus, even though this environmental diagnosis and profiling approach is far from the level of sophistication enjoyed by the medical profession for diagnosing disease in humans, if properly tested and applied, it could be a useful management tool in helping to assess the health of estuarine systems.

Integrated Effects Assessment

As emphasized above, estuaries are complex systems composed of many interacting factors that complicate the understanding of how environmental stressors act and ultimately affect biota. The complexity of estuarine systems dictates that an integrated or holistic approach should be taken in evaluating the effects of stressors on important biological components of these systems. Most field studies typically measure one or only a few variables. For the purpose of assessing the effects of stressors on the health of estuarine systems, only a few studies have attempted to assess effects of multiple stressors on one species, assess effects of a single stressor on multiple biological end points, or evaluate the effects of multiple stressors on multiple biological end points. Whereas single-variable responses may reflect specific structural or functional attributes of an organism (usually at one particular level of biological organization), single responses, in themselves, do not usually provide an integrated measure of organism or ecosystem health (Adams et al., 1994). If single-variable measurements are used separately to evaluate the effects of stress, then the interrelationships among variables may not be properly considered in assessing responses of organisms to stressors. Quantitative approaches that use integrated multivariate analysis are useful to aid understanding of the interrelationships and associations that exist among multiple response variables. Therefore, multivariate approaches more accurately reflect the myriad of interactions that occur between biota and the environment than single-variable approaches (Capuzzo, 1985; Smith, 2002). For both field and laboratory studies that incorporate multiple response variables in the experimental design, it is important to consider the response variables jointly within a multivariate context and to analyze the data with multivariate procedures that reveal the integrative or holistic nature of the responses.

Canonical discriminant analysis is a quantitative statistical procedure that can be used for analyzing multivariate data sets that are composed of a large number of biomarker and bioindicator response variables. The canonical discriminant analysis procedure facilitates the graphical comparison of holistic responses among data sets because the differences among means can be visualized on a reduced number of axes (Adams et al., 1994; Rencher, 2002). Comparisons of means can also be made by plotting circular confidence regions around each estimated mean (Schott, 1990). One particularly useful application of canonical discriminant analysis has been to measure several response variables in organisms collected from both stressed sites and reference areas and to graphically compare the integrated canonical mean response of organisms among sample sites. To illustrate how this multivariate approach can be used to assess the integrated effects of stressors on organisms, an example of a river contaminated by discharges from a pulp and paper mill will be used. From the discharge point of the paper mill, the river has a distinct spatial gradient in contaminant loading as evidenced by decreasing downstream levels of contaminants, including dioxin, in sediment and biota. At each of three sites located at increasing distances below the paper mill and at each of three reference sites, 15 individual redbreast sunfish (*Lepomis auritus*) of both sexes were collected (Adams et al., 1996). For each fish, we measured a variety of biochemical, physiological, histopathological, general condition indices, nutrition indicators, and reproductive variables. This suite of bioindicators measured represents six different functional response groups including biochemical markers of exposure, organ dysfunction, tissue and organ damage, overall condition, nutrition/bioenergetics, and reproductive integrity. Population- and community-level surveys were also conducted at each site including relative abundance, size distribution, sex ratios, community diversity, and the index of biotic integrity (Adams et al., 1996).

The integrated health responses of sunfish at these six sites are shown in Figure 2.5 as three-dimensional displays that provide a basis for comparing the overall health status of fish among sites. This analysis uses the individual bioindicators jointly within a multivariate context and takes into account the interrelationships and associations that exist among the individual bioindicator responses. Each ellipse in

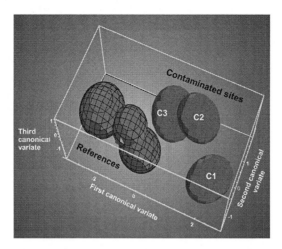

FIGURE 2.5 (Color figure follows p. 266.) Integrated health responses of sunfish sampled from three sites in a contaminated river and three reference sites. Boundaries of each ellipse are based on the 95% confidence radii of the integrated site means.

Figure 2.5 represents the integrated health response of all fish collected at a particular site based on using, within the analysis, all of the biomarkers and bioindicators from the six functional response groups. The boundaries of each ellipse are set based on the 95% confidence radii of the canonical mean for each ellipse. These integrated site responses are shown as three-dimensional functions because the first three canonical variables account for 97% of the total variation (discrimination) among sites while two variables explain 81% of the total variation. Two distinct patterns in integrated site responses are evident in Figure 2.5. Because the ellipses represented by the three reference sites overlap with each other and do not overlap the contaminated sites, the integrated health status of fish at the three reference sites is considered similar but distinctly different from fish at the three contaminated sites. In addition, there is an obvious downstream gradient in fish health in the contaminated system with the site nearest the discharge (site C1) most dissimilar to the references (i.e., the poorest health) and site C3, the greatest distance from the outfall (i.e., the best health), most similar to the reference. A measure of the linear statistical distances between sites (the Mahalanobis distance) indicates that the three reference sites are not statistically different from each other, whereas each of the contaminated sites is statistically different from each of the references. Fish at contaminated sites C2 and C3 are more similar to each other than either is to fish from site C1, the most polluted site. Interestingly, these integrated health responses at the individual organism level follow the same spatial pattern in the river as the population- and community-level responses (Adams et al., 1996), illustrating that these integrated individual responses are good indicators of higher-level effects. At least for this example, the integrated health response at the individual organism level serves as a predictive model of effects occurring at higher levels of biological organization (i.e., at the population and community level).

The individual variables providing the greatest amount of discrimination among these integrated site health responses are a biochemical response or a detoxification enzyme (canonical axis 1), size (growth) of age 2 sunfish (axis 2), and a lipid metabolism indicator (axis 3). The first canonical variable or axis accounted for 59% of the variability in discriminating among sites, growth of age 2 sunfish an additional 22% (axis 2) to this discriminatory ability, and triglycerides (axis 3) another 16%. Thus, these three variables accounted for 97% of the variation in discriminating the integrated health status of fish among sites. These three variables represent responses from three different levels of biological function including the biochemical level (detoxification enzyme), individual-population level (growth), and lipid dynamics (triglycerides). The results of this multivariate analysis demonstrate that when evaluating the effects of environmental stressors on the health of biota, it may be advantageous to use bioindicators that represent multiple biological functions that reflect different sensitivities, specificities, and response time (response scales) to stressors. In assessing the condition of organisms in aquatic environments, individual variables are not generally adequate to reliably predict changes at the population or community level (Capuzzo, 1985), and the exclusive use of only one indicator may lead to invalid conclusions regarding organism health (Schlenk et al., 1996b). Therefore, inclusion of a suite of bioindicator variables in bioassessment programs is important for detecting large-scale disturbances in organism health due to environmental stressors (Goksoyr et al., 1991; Balk et al., 1993). The number of indicators measured, however, is not as important as the nature of these variables and what they reflect functionally about stress responses in aquatic systems. In addition, comparison of the integrated health responses at the individual organism level to the population- and community-level indicators established, at least in this case, that organism-level health is a reliable indicator of effects at higher levels of biological organization.

Conclusions and Synthesis

As a framework for using multivariate bioindicators, there are several strategies or approaches that should be considered when designing field studies for the purpose of assessing the health of estuaries, determining the effects of stressors on biological components of estuaries, and identifying the cause(s) of observed effects. These approaches are (1) separating out the effects of direct vs. indirect pathways on biological components of estuaries, (2) using bioindicators that are reflective of different temporal response scales to environmental stressors, (3) identifying and establishing causal factors or mechanisms responsible for effects on estuarine resources, (4) environmental profiling and diagnosing natural vs. anthropogenic effects on estuarine biota, and (5) employing integrated effects assessment. Application of some or all of these approaches is important for understanding, assessing, and evaluating the effects of environmental stressors on estuarine resources so that more reliable decisions can be made regarding the management and protection of estuaries. Consideration of these integrative approaches as part of an overall environmental monitoring and assessment framework is particularly important in the face of increasing coastal zone development and the increasing vulnerability of estuarine systems to environmental disturbances.

References

Adams, S. M. 1990. Status and use of bioindicators for evaluating effects of chronic stress on fish. *American Fisheries Society Symposium* 8:1–8.

Adams, S. M. 2001. Biomarker/bioindicator response profiles of organisms can help differentiate between sources of anthropogenic stressors in aquatic ecosystems. *Biomarkers* 6:33–44.

Adams, S. M. 2002. Biological indicators of aquatic ecosystem stress: introduction and overview. In *Biological Indicators of Aquatic Ecosystem Stress*, Adams, S. M. (ed.). American Fisheries Society, Bethesda, MD, pp. 1–11.

Adams, S. M., W. R. Hill, M. J. Peterson, M. G. Ryon, J. G. Smith, and A. J. Stewart. 2002. Assessing recovery from disturbance in a stream ecosystem: application of multiple chemical and biological endpoints. *Ecological Applications* 12:1510–1527.

Adams, S. M., K. D. Ham, and J. J. Beauchamp. 1994. Application of canonical variate analysis in the evaluation and presentation of multivariate biological response data. *Environmental Toxicology and Chemistry* 13:1673–1683.

Adams, S. M., K. D. Ham, M. S. Greeley, R. F. LeHew, D. E. Hinton, and C. F. Saylor. 1996. Downstream gradients in bioindicator responses: point source contaminant effects on fish health. *Canadian Journal of Fisheries and Aquatic Science* 53:2177–2187.

Adams, S. M., K. D. Ham, and R. F. LeHew. 1998. A framework for evaluating organism responses to multiple stressors: mechanisms of effect and importance of modifying ecological factors. In *Multiple Stresses in Ecosystems*, J. J. Cech, B.W. Wilson, and D.G. Crosby (eds.). Lewis Publishers, Boca Raton, FL.

Attrill, M. J. and M. H. Depledge. 1997. Community and population indicators of ecosystem health: targeting links between levels of biological organisation. *Aquatic Toxicology* 38:183–197.

Balk, L., L. Forlin, M. Soderstrom, and A. Larsson. 1993. Indications of regional and large-scale biological effects caused by bleached pulp mill effluents. *Chemosphere* 27:631–650.

Bart, D. and J. M. Hartman. 2000. Environmental determinants of *Phragmites australis* expansion in a New Jersey salt marsh: an experimental approach. *Oikos* 89:59–69.

Bartholomew, G.A. 1964. The roles of physiology and behavior in the maintenance of homeostasis in the desert environment. *Symposia of the Society for Experimental Biology* 18:118–124.

Capuzzo, J. M. 1985. Biological effects of petroleum hydrocarbons on marine organisms: integration of experimental results and predictions of impacts. *Marine Environmental Research* 17:272–276.

Culp, J. M., K. J. Cash, and F. J. Wrona. 2000. Cumulative effects assessment for the Northern River Basins Study. *Journal of Aquatic Ecosystem Stress and Recovery* 8:87–94.

Depledge, M. 1989. The rational basis for detection of the early effects of marine pollutants using physiological indicators. *Ambio* 18:301–302.

Engle, D. W. and D. S. Vaughan. 1996. Biomarkers, natural variability, and risk assessment: can they coexist? *Human and Ecological Risk Assessment* 2:257–262.

Fox, G. A. 1991. Practical causal inference for ecoepidemiologists. *Journal of Toxicology and Environmental Health* 33:359–373.

Goksoyr, A. and 11 co-authors. 1991. Environmental contaminants and biochemical responses in flatfish from the Hvaler Archipelago in Norway. *Archives of Environmental Contamination and Toxicology* 21:486–496.

Hodson, P. V. 1990. Indicators of ecosystem health at the species level and the example of selenium effects on fish. *Environmental Monitoring and Assessment* 15:241–254.

Holdway, D. A. 1996. The role of biomarkers in risk assessment. *Human and Ecological Risk Assessment* 2:263–267.

Huggett, R. J., R. A. Kimerle, P. M. Mehrle, and H. L. Bergman (eds.). 1992. *Biomarkers: Biochemical, Physiological, and Histological Markers of Anthropogenic Stress.* Lewis Publishers, Boca Raton, FL, 347 pp.

Larsson, A., C. Haux, and M. Sjobeck. 1985. Fish physiology and metal pollution: results and experiences from laboratory and field studies. *Ecotoxicology and Environmental Safety* 9:250–281.

Little, E.E. 2002. Behavioral measures of environmental stressors in fish. In *Biological Indicators of Aquatic Ecosystem Stress,* S. M. Adams (ed.). American Fisheries Society, Bethesda, MD, pp. 431–472.

McCarthy, J. F. and L. R. Shugart (eds.). 1990. *Biomarkers of Environmental Contamination.* Lewis Publishers, Boca Raton, FL, 457 pp.

McCarty, L. S. and K. R. Munkittrick. 1996. Environmental biomarkers in aquatic toxicology: fiction, fantasy, or functional? *Human and Ecological Risk Assessment* 2:268–274.

Nacci, D., J. Serbst, T. R. Gleason, S. Cayula, W. R. Munns, and R. K. Johnson. 2000. Biological responses of the sea urchin, *Arbacia punctuata,* to lead contamination for an estuarine ecological risk assessment. *Journal of Aquatic Ecosystem Stress and Recovery* 7:187–199.

Power, M. 1997. Assessing the effects of environmental stress on fish populations. *Aquatic Toxicology* 39:151–169.

Power, M. 2002. Assessing fish population responses to stress. In *Biological Indicators of Aquatic Ecosystem Stress,* S. M. Adams (ed.). American Fisheries Society, Bethesda, MD, pp. 379–430.

Rencher, A. 2002. *Methods of Multivariate Analysis,* 2nd ed. John Wiley & Sons, New York.

Reynolds, W. W. and M. E. Casterlin. 1980. The role of behavior in biomonitoring of fishes: laboratory studies. In *Biological Monitoring of Fish,* C. H. Hocutt and J. R. Stauffer (eds.). Lexington Books, Lexington, MA, pp. 59–82.

Schlenk, D., E. J. Perkins, G. Hamilton, Y. S. Zhang, and W. Layher. 1996a. Correlation of hepatic biomarkers with whole animal and population-community metrics. *Canadian Journal of Fisheries and Aquatic Science* 53:2299–2309.

Schlenk, D., E. J. Perkins, W. G. Layher, and Y. S. Zhang. 1996b. Correlating metrics of fish health with cellular indicators of stress in an Arkansas bayou. *Marine Environmental Research* 42:247–251.

Schott, J. R. 1990. Canonical mean projections and confidence regions in canonical variate analysis. *Biometrika* 77:587–596.

Smith, E. P. 2002. Statistical considerations in the development, evaluation, and use of biomarkers in environmental studies. In *Biological Indicators of Aquatic Ecosystem Stress,* Adams, S. M. (ed.). American Fisheries Society, Bethesda, MD, pp. 565–590.

U.S. Environmental Protection Agency (U.S. EPA). 1998. Guidelines for ecological risk assessment. Office of Research and Development, Risk Assessment Forum, Washington, D.C. EPA-630-R-95002F.

U.S. Environmental Protection Agency (U.S. EPA). 2000. Stressor identification guidance document. Office of Water, Washington, D.C. EPA-22-B-00-025

Vallentyne, J. R. 1999. Extending causality in the Great Lakes basin ecosystem. *Aquatic Ecosystem Health Management* 2:229–237.

Van Gestel, C. A. M. and T. C. Van Brummelen. 1996. Incorporation of the biomarker concept in ecotoxicology calls for a redefinition of terms. *Ecotoxicology* 5:217–222.

3

Physical Processes Affecting Estuarine Health

Jon Hubertz, Xiaohong Huang, Venkat Kolluru, and John Edinger

CONTENTS

Introduction

An estuary can be described as an arm of the sea, where the tide meets and mixes with fresh water run-off from the land, rivers, and rain from the atmosphere. It is a semienclosed region, bordering a sea, with boundaries and bathymetry that influence the circulation of the estuary volume. The tide, wind, freshwater inflows, density stratification, and the Earth's rotational effects can act at various levels to control the circulation. As human population pressures mount in these coastal areas, stewards need to understand the physics of their estuary to avoid the negative affects of this growth and mitigate them if possible.

What indicates the health of an estuary? Can the relative change in an indicator be measured to decide if the estuary is degrading, remaining the same, or improving? These are key questions, but difficult to answer. Indicators can range in scales of size and time. Microorganisms can be stressors at the bottom of the biological, chemical web or large-scale physical processes can stress the whole ecosystem process. Short-term variability of a potential stress can be an early warning of future problems, or long-term monitoring can provide a history of an ecosystem and hints of long-term climate change. There is not a single indicator of the health of an estuary. There is a complex interaction of physical, chemical, biological, and geological processes that act on multiple space- and timescales affecting an ecosystem. This chapter addresses the physical processes.

0-8493-2822-5/05/$0.00+$1.50
© 2005 by CRC Press

Examining the basic physical process in an estuary, namely, the movement of water in the system, can be a beginning to understanding an important process affecting the movement (or lack thereof) of many chemical and biological indicators. Without physical processes acting, the definition of an estuary would have to be changed. If there were no tide and wind, two important physical processes, there would be greatly reduced mixing of water between the enclosed area and the sea. Omit rainfall, river inflow, insolation, evaporation, water density differences, and any currents, and the water volume would become stagnant. Such a water body, no longer an estuary, would be an unhealthy habitat.

Humans cannot change most physical processes (e.g., tide, wind, rain, insolation, etc.) but can influence others. One example is the controlled release of fresh water into an estuary. Another is modifying the opening between the sea and interior water body; for example, dredging a pass or inlet. Both of these processes affect the mixing of salt and fresh water as well as other constituents and their residence time in a water body. In addition, the demand for water from rivers entering a system as well as runoff from the land and into rivers of undesirable material are potential anthropogenic changes that could affect the estuary.

How does one go about studying these effects and defining the movement and mixing of water in an estuary in space and time? It is not feasible to assess an estuary in space and time to define the movement of water in the past, present, and future. Technically and logistically, it is feasible to make measurements at a limited number of points for a limited time. Such a database is a necessary and sufficient condition for the feasible and practical application of current numerical models as tools to study these effects and others. One of these numerical modeling systems at present applied to estuaries is the Generalized Environmental Modeling System of Surfacewaters (GEMSS) as described in Kolluru (2003).

There are about six other models currently being applied to coastal problems around the United States (Estuarine Research Federation annual conference, 2003, Seattle, Washington), and others internationally. The trend is to address not only the hydrodynamics, but also biological and chemical water quality processes and sediment dynamics using a modular modeling system. Some investigators propose one general open-source model for all applications to ensure consistency in calculations and output display, while others prefer the familiarity and confidence in their own system. Regardless, there is need for a simplified comparison of results among models, to show differences in application procedures, for comparison to measurements, and to research the best formulation of processes.

Background

Physical Processes

Examining the basic physical process in an estuary, i.e., the movement of water in the system, can initiate the understanding of an important process affecting many chemical and biological indicators. There are a number of physical processes that influence the movement of water in an estuary. Probably the two most important are the tidal signal at the mouth of the estuary and the rate of volume inflow from rivers usually located at the head of the estuary. Other physical processes are wind and atmospheric pressure, water runoff from the land boundaries, the land boundaries of the estuary, bathymetry of the estuary, the Coriolis effect due to the rotation of the Earth, buoyancy or density effects related to salinity and temperature, rain and evaporation as affecting salinity and temperature, and insolation affecting temperature and dependent on season and latitude. All of the above processes have implications for the movement of pollutants, sediment, phytoplankton, zooplankton, dissolved nutrients, salt, heat, and anything else at the mercy of currents.

Types of Currents

Tidal

Tidal currents are probably the most recognized and have the most interaction within an estuary. Tidal currents in Charlotte Harbor are semidiurnal in nature, "semi" meaning half and "diurnal" meaning daily,

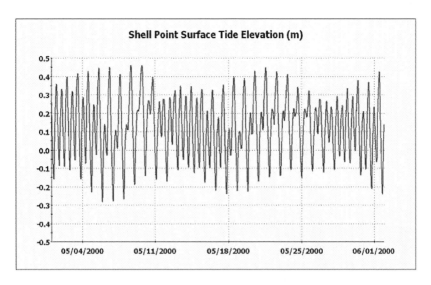

FIGURE 3.1 Tide elevation for Shell Point (1 month).

which translates into a high and low tide in a half day or two highs and two lows each day. These highs and lows are the crests and troughs of a shallow water wave, which is a wave where a representative particle or small volume of water follows a path related to the wave velocity that is dependent on the water depth. Thus, tidal waves will vary not only due to the height of the tide (making the water deeper), but also by the varying depth of the estuary. Tidal currents between the sea and estuary will vary due to the relative difference in water elevations between the two and the entrance geometry.

Tidal heights and currents also vary due to the forces that cause them, namely, the positions of the sun, moon, and planets with respect to the Earth. Thus, tidal heights and currents are a function of time as well as space. The tide may also have a longer period of oscillation, on the order of weeks, rather than the commonly seen daily oscillation. This can cause a phenomenon called "tidal pumping," where more water will enter an estuary during weeks when the tide is higher than those weeks when it is lower. Figure 3.1 depicts an example of the tide in this region obtained from about a month of measurements at Shell Point near the entrance of the Caloosahatchee River (see Figure 3.4 below for location). This variation can cause nonzero mean or residual tidal currents over a time period.

In addition to the gravitational force affecting tides, the rotation of the Earth also affects tidal currents. In the Northern Hemisphere, the path a water particle takes will tend to the right of its direction (to the left in the Southern Hemisphere) for an observer on Earth. This deflecting force due to the Earth's rotation is termed the Coriolis effect. Thus, water entering a harbor (in the Northern Hemisphere) on a flood (rising) tide will be deflected to the right, and to the left on an ebb (lowering) tide.

This introduces spatial as well as temporal gradients in the flow that can affect the residual current at a point and the transport of constituents through the estuary. Discounting other forces such as the wind, inflows, density effects, etc., the tide results in a flushing or residence time for locations within a particular estuary. These times will vary depending on the physical processes acting, for example, increased or decreased river and runoff inflows, variations in the tide and winds, changes to boundaries and bathymetry, etc. Flushing times can also vary depending on the percentage reduction of a constituent assumed in the calculation. For example, flushing times will be longer for a constituent concentration reduction of 50% vs. 30% since it simply takes longer to reduce a concentration when all other factors are the same.

The simple idea of a high and low tide and flood and ebb currents quickly becomes more complicated because of all the possible interactions when describing tidal currents at a particular place and time in an estuary or entire ecosystem.

River Inflows

Currents issuing from river mouths are usually well defined and become more diffuse as they interact with the tidal currents, and the usually less-dense water from the river mixes in the surface layers above the more dense water originating from the sea. These currents may vary naturally, due to the climate, and artificially, due to human intervention on river control structures. Unchanneled surface runoff can also add to a lower-density surface layer. Momentum carries this surface layer toward the mouth of the estuary. Non-natural heat sources (power plants) and solar radiation can affect the density of river inputs as well. In the Charlotte Harbor region, warm water from power plant discharges provides a winter habitat for endangered West Indian manatees.

Wind-Induced Currents

The wind can be a major factor in moving water into and out of an estuary especially if the wind direction is aligned with the mouth of the estuary. The wind acts on the surface layer and induces currents through the depth, which are acted upon by the Coriolis effect. These currents will also interact with the tidal currents. Atmospheric pressure effects associated with a weather system and winds will also affect the elevation of the water. Generally, the water surface will rise a centimeter for every drop of a millimeter in surface atmospheric pressure. The climatic passage of cold fronts in the winter through the North–South-oriented Charlotte Harbor results in lower water levels due to northerly winds pushing water out of the estuary. If tropical storms occur in the summer, winds may push water into the estuary (depending on their path) and result in increased levels or the storm surge effect.

Other Physical Processes

In addition to the above processes affecting currents and water levels, the health of an estuary can be affected by the turbidity and color of the water. Decaying vegetative matter runs off into Charlotte Harbor, giving the water a tea-stained appearance. This, in turn, affects the light transmitted through the water column and, subsequently, inhibits the growth of submerged aquatic vegetation and affects oxygen concentration, etc. Light or insolation is affected by the weather and season and, in turn, affects the temperature of the water. Temperature and salinity affect the water density and thus can inhibit vertical mixing, leading to stagnant areas where chemical and biological processes have a shorter reaction time than mixing of the water column. These processes along with currents and water levels quickly lead to the interaction of multiple factors requiring knowledge of their boundary values in time to describe how a system functions physically.

Approach

Taken together, all these physical effects produce a complicated, interacting water level and current field in three dimensions and time, which ultimately determines the concentration of constituents in the water. The result is measurable and could possibly be broken down to determine causes. Such a current measurement at a point (Eulerian flow) neglects what goes on at neighboring points, and measurement of a path between two points (Lagrangian) is an integration of effects between the points and hence neglects history.

The desirable solution is a history of all processes at all points of the system due to all effects. This can be estimated using present numerical modeling technology. Not only is the complete surface level, levels of density surfaces, currents, and the general circulation specified in space and time, other effects such as mixing of constituents and description of biological and chemical interactions are possible. This requires boundary conditions of tidal information at seaward points, wind stress over the estuary, and inflow conditions. Other boundary conditions such as salinity, temperature of water input, solar insolation, evaporation, and rainfall are required for salinity and temperature mixing problems. Additional boundary conditions may be required for chemical and biological water quality simulations. The numerical model acts as an "intelligent" interpolator using conditions on the boundaries of the system and formulations of physical, chemical, biological, and geological interactions to provide a picture in time and space of the ecosystem.

In addition, there is the challenge of model calibration to the estuary for a length of time and verification of model results for a separate length of time. This is a necessary and costly step since it requires *in situ* Eulerian and or Lagrangian current measurements with which to compare model results. Enough such measurements should be made throughout the estuary to describe the general circulation in the horizontal and vertical dimensions. They should also be recorded with sufficient duration to describe the major periods of motion. For a large system, this is not practical, and one uses what is available to provide some confidence the system is being described properly by the model for a given purpose.

A numerical hydrodynamic, water quality model of an estuary should be considered an enduring tool that becomes more refined as different measurements and projects provide more data for comparison. It will become more useful and a standard as more groups become familiar with it, use it, and interactively evaluate it. Once a model is configured for a location, initially calibrated, and verified with available data, it is an efficient tool for study. It is possible that a number of different models can be applied to an ecosystem. This introduces the possibility of obtaining different results and may require alternative explanations. The situation of using more than one model is an argument for one standard model system, but it also presents an argument for a variety of systems, the results of which may indicate more accurate modeling approaches and formulations.

Numerical Modeling System — GEMSS

The GEMSS program application discussed in this study is an example of a generalized system for coastal work being used in the United States and abroad. GEMSS is embedded in a geographic information system (GIS), and it is designed in a modular fashion for easy coupling of existing as well as other user-defined models. Currently, the system has the following modules supported by a graphical user interface (GUI) and static and animated displays. The modules are as follows:

HDM: 2-D and 3-D Hydrodynamic Module
WQM: Water Quality Module
STM: Sediment Transport Module
PTM: Particle Tracking Module
OCSM: Oil and Chemical Spill Module
1DM: 1-D Hydrodynamic Module

For estuary water bodies, GEMSS-HDM is the main frame for hydrodynamic studies. The GEMSS-HDM module uses GLLVHT (generalized, longitudinal-lateral-vertical hydrodynamic and transport), which is a state-of-the-art, 3-D numerical model that computes time-varying velocities, water surface elevations, and water quality constituent concentrations in water bodies. The hydrodynamic and transport relationships used in GLLVHT are developed from the horizontal momentum balance, continuity, constituent transport, and the equation of state. The basic relationships are given in Edinger and Buchak (1980, 1994, and 1995).

The computations are done on a horizontal curvilinear and vertical grid that represents the water body bounded by its water surface, shoreline, and bottom. A grid generation, change, and display system is included in the system as grid generator and editor. 2-D and 3-D postprocessing viewers and additional tools that include meteorological and time-varying data generators are included in the GEMSS software to support all-purpose modeling. The GEMSS is operated through a GUI with an extensive help file for support. The GEMSS software main window is shown in Figure 3.2.

The water surface elevations are computed simultaneously with the velocity components. The water quality constituent concentrations are computed from the velocity components and elevations. Included in the computations are boundary condition formulations for friction, wind shear, turbulence, inflow, outflow, surface heat exchange, water quality, and chemical and sediment kinetics. The model can be used to analyze water body system dynamics and predict and "hindcast" the impacts of actual events or possible design or management alternatives. The model also uses a nonhydrostatic condition in the vertical direction to simulate high discharge flows.

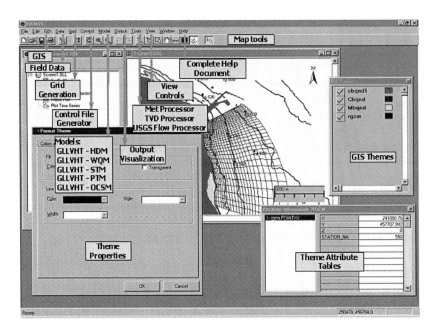

FIGURE 3.2 GEMSS software main window.

The flow and constituent fields are discretized in time, and the computation marches forward in time using the most efficient variable time step while computing the dependent variables throughout the grid at each of these steps. To march the calculations through time, boundary-condition data consisting of meteorological data, inflow rates, temperatures, constituent concentrations, and outflow rates are required.

The model was built to accept a large number of transport constituents and constituent relationships depending on the water quality model, sediment transport model, oil and chemicals model, and toxic model being used. The list of transport variables available in GEMSS-HDM to analyze flushing, entrainment, cooling water discharges, boundary exchange, etc. is contained in ENTRIX (2001): temperature, salinity, excess temperature (due to inflow heat source), instantaneous tracer dye, and continuous tracer dye.

The inputs to all these GEMSS embedded modules are obtained from their respective control files and the outputs are designed for easy uploading into existing 2-D and 3-D display modules. Output can be displayed as fixed or time-varying plots of scalar and vector quantities in one or two dimensions. The unique design of the GEMSS suite of models gives the user the power of writing adaptation routines to introduce different initial conditions, time-variant boundary conditions, replace existing algorithms for source and sink computations related to water quality, sediment transport, etc. and nonstandard features to customize the output. In this scheme, the main kernel of GLLVHT-HDM behaves like a black box.

Applications

The GEMSS has the potential to be a standard hydrodynamic water quality numerical modeling tool for any estuary. It has local support through an extensive "Help" file and a GUI for input and analysis of results, and it can be run using different types of operating systems including those found on personal computers. It has been developed for more than a decade and widely applied in coastal regions, rivers, reservoirs, lakes, and estuaries. GEMSS is currently used for studies on the Charles River in Boston, Puget Sound in Washington State, Indian River/Rehoboth Bay Delaware, the Arabian Gulf (Doha, State of Qatar), and South Korea.

The GEMSS program is applied here to a case of salinity intrusion up the Caloosahatchee River due to a fixed tidal amplitude and period from the Gulf of Mexico and no freshwater input to the river. To

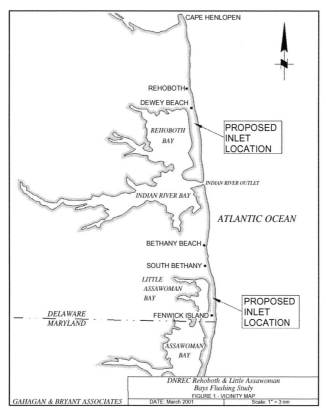

FIGURE 3.3 The vicinity map for Indian River/Rehoboth.

further demonstrate the importance of the residence time/flushing time analysis, another GEMSS study on Indian River/Rehoboth Bay Delaware Inland Bay, another estuary area from the U.S. EPA National Estuaries Program (NEP) list, is also presented to predict the effect of a proposed inlet on the change in residence time. The interlocked Delaware Inland Bay System includes two main water bodies: Indian River and Rehoboth Bay. Both water bodies are shown in Figure 3.3. The Indian River is connected to the Atlantic Ocean on the east via an inlet and to Little Assawoman Bay via a canal. Rehoboth Bay is connected to Delaware Bay to the north via a canal and to Indian River Bay to the south. The western portion of Indian River Bay, referred to as the Indian River, terminates at Millsboro Dam.

Results

The Salinity Mixing Problem

The health of a region of the Caloosahatchee River estuary is described by the range in salinity that submerged aquatic vegetation (SAV) can tolerate in this region (South Florida Water Management District, 2003). SAV tolerance is directly related to minimum flows and levels (MFL) mixing between fresh water flowing into the region (river and runoff) and the propagation of salt water from the Gulf into the region. Figure 3.4 shows the GEMSS grid used to investigate the salinity in the region between Fort Myers and Beautiful Island.

As an example of mixing, a constant tide of 1-m range, 12.4-h period was input at Shell Point where the salinity was a constant 32 parts per thousand (ppt). As time progressed through a year, salinity rose along the river from an initial value of zero everywhere. Salinity values were tracked in time at five locations along the river. An example of the response near Fort Myers is shown in Figure 3.5.

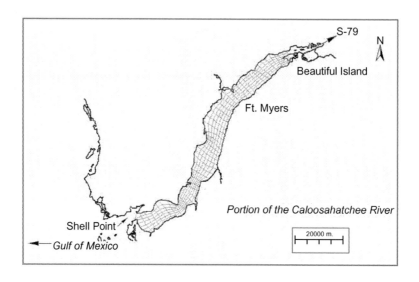

FIGURE 3.4 GEMSS grid for the Caloosahatchee River.

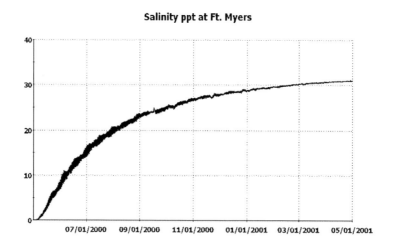

FIGURE 3.5 Salinity time series plot for Fort Myers.

The salinity at Fort Myers increases gradually due to water from the Gulf mixing with the river and approaches a steady state after about 6 months. This simulation shows how long it takes for salinity to reach a value given constant tidal input and zero river inflow. One could easily simulate time-varying tidal and salinity values at Shell Point and freshwater discharges at the S-79 lock (up river) as well as rainfall and other freshwater input along the river. A variety of postprocessing visual displays is available to help analyze results. The mixing gradient in rivers is important information to help manage flows and water levels. Using an approach like this, an atlas of salinities along a river could be developed for various scenarios of controlled fresh water input and climatic rainfall conditions. Such a document could help managers plan for the best management practices.

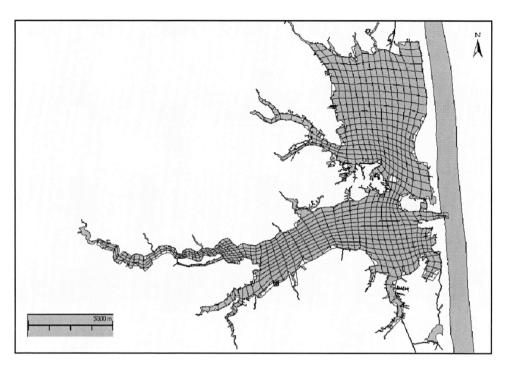

FIGURE 3.6 3-D hydrodynamic, transport, and water quality model grid for Indian River, Rehoboth Bay.

Residence/Flushing Time Analysis in Indian River/Rehoboth Bay, Delaware

A grid of Indian River/Rehoboth Bay was developed for a hydrodynamic calibration study. It is shown in Figure 3.6. The proposed inlet location in the Delaware Seashore State Park is shown in Figure 3.3. The width of the inlet is approximately 122 m (400 ft) with a depth of 6 m (20 ft) extending to a distance of approximately 1.5 km (4920 ft) into Rehoboth Bay. Using these dimensions, a new grid was created by altering the grid lines around the new proposed inlet to match its dimensions inside the Rehoboth Bay region. The final grid in the vicinity of the inlet is shown in Figure 3.7.

The contour plot of computed flushing time without the proposed inlet is shown in Figure 3.8. It can be seen that the flushing time in the immediate vicinity of the Indian River Inlet and the Indian River Bay was 1 to 3 days. Along the southern shoreline of Indian River Bay, it was as long as 20 to 30 days. Upstream in the Indian River, flushing times vary from 20 to 40 days. In Rehoboth Bay, the existing flushing times ranged from 1 to 2 days at its connection with the Indian River Bay, to 30 to 35 days in the vicinity of Dewey Beach. Flushing was as long as 40 days at the ends of the tributaries joining the southwestern corner of Rehoboth Bay.

The contour plot of change in flushing time with and without the proposed inlet is shown in Figure 3.9. The flushing time in the vicinity of the proposed inlet reduced by 25 to 30 days from the "without" case. This improvement extends across the western shore of Rehoboth Bay where the flushing is reduced by 12 to 24 days. For the southwestern end of Rehoboth Bay, the flushing time is reduced by up to 12 days from the base case. The rest of the Indian River and the region that connects the Indian River Bay with the Rehoboth Bay either remained same or increased by 5 days.

Beyond Physical Processes — Water Quality Modeling

Modeling water quality in the different water bodies will allow investigators to determine what inputs and processes control different water quality variables such as dissolved oxygen, nutrient levels, algal densities, and the onset of algal blooms. A series of water quality models has been developed for coupling

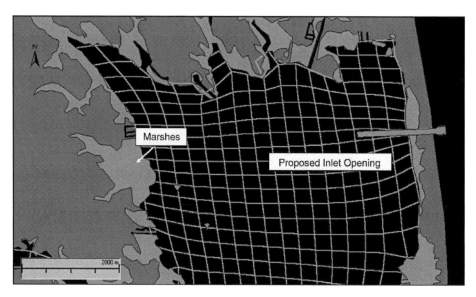

FIGURE 3.7 Numerical grid in the vicinity of a proposed inlet.

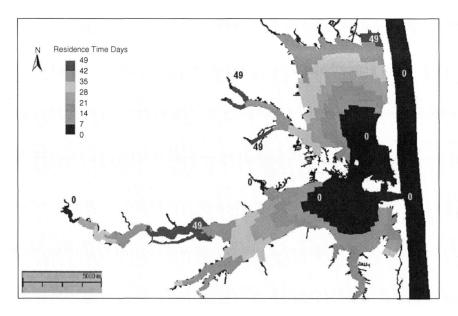

FIGURE 3.8 Residence time in days for the existing conditions inside the Indian River and Rehoboth Bay.

with the computationally efficient HDM module of GEMSS with different levels of usage (ENTRIX, 2001). The water quality variables and processes included in GEMSS-WQM are shown in Figure 3.10.

For nutrients, Figure 3.10 shows that organic carbon (expressed here as CBOD, carbonaceous biochemical oxygen demand) decays or oxidizes to some simpler form (expressed here as CO_2), and in the process consumes dissolved oxygen (DO). The organic nitrogen (ON) first mineralizes to ammonia (NH_3) that in turn nitrifies to nitrate (NO_3^-) and consumes DO in the process. The organic phosphorus (OP) mineralizes to phosphate (PO_4).

The role of phytoplankton (PHYT in Figure 3.10) in eutrophication shows that phytoplankton recycle nutrients to their dissolved and particulate organic forms by excretion and death, and to ammonia and

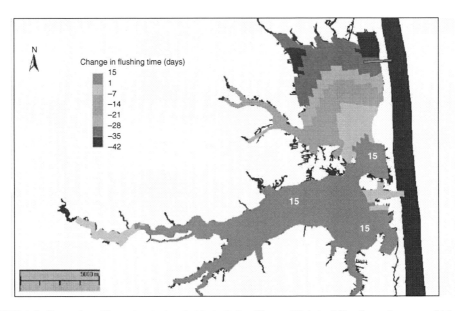

FIGURE 3.9 Change in residence time in days inside the Indian River and Rehoboth Bay due to the proposed inlet opening in Rehoboth Bay.

phosphate by respiration. Phytoplankton produce DO by photosynthesis when sufficient light is available, and consumes DO by respiration. The levels of ammonia, nitrate, and phosphate and the light level at different locations within the water column determine the rate of growth of phytoplankton, ON, and OP. Additional processes are settling of the particulate forms of organic carbon, nitrogen, and phosphorus and of dead phytoplankton and exchanges with the sediment. These processes are sediment oxygen demand and the release of ammonia and phosphate from the sediment.

A major process controlling phytoplankton densities and recycling nutrients is zooplankton grazing. Grazing recycles nutrients obtained from phytoplankton to particulate organic carbon. Different formulations of zooplankton grazing have been developed (Edinger et al., 2003a) and are included in GEMSS-WQM. Examination of phytoplankton densities and blooms requires the ability to identify the sources of "seed" for them. The seed is thought to be from cyst and spore forms residing in the sediment or attached to other plant life, which eventually enters the water column. Quantitative formulations of the generation of phytoplankton from cysts and spore forms are being studied for eventual inclusion in GEMSS-WQM (Edinger et al., 2003b).

Conclusions

- Physical processes result in the forces moving water in time and the three dimensions of an estuary. Given a state-of-the-art hydrodynamic modeling system, measurements in space and time with which to check calculations, and the proper forcing data, the movement of water in an estuary system can be accurately estimated in time (past, present, and future) and in three dimensions.

- The reaction times of biological and chemical constituents and the mixing/flushing times of an estuary may affect indicators of the health of an estuary.

- Combining hydrodynamic/transport capabilities with water quality (biological/chemical) calculations in a modeling system such as GEMSS allows estimation of the effectiveness of both conservative and nonconservative constituents in acting as indicators of the health of an estuary.

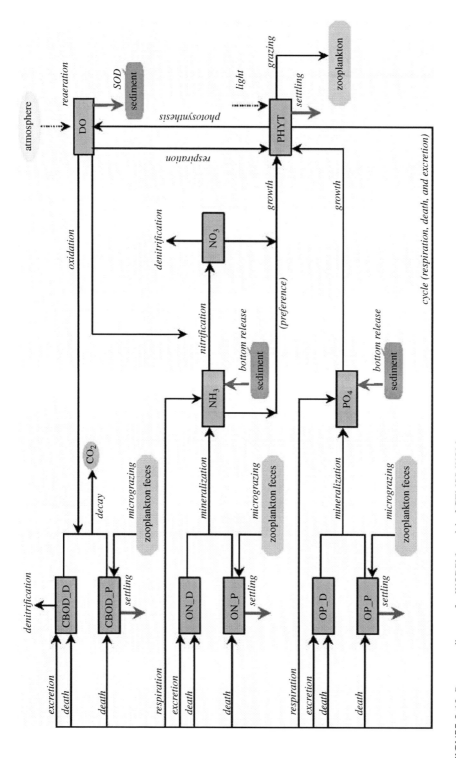

FIGURE 3.10 Processes diagram for WQDPM model of GEMSS-WQM.

References

Edinger, J. E. and E. M. Buchak. 1980. Numerical hydrodynamics of estuaries. In *Estuarine and Wetland Processes with Emphasis on Modeling*, P. Hamilton and K. McDonald (eds.). Plenum Press, New York, pp. 115–146

Edinger, J. E. and E. M. Buchak. 1995. Numerical intermediate and far field dilution modelling. *Water, Air, and Soil Pollution* 83:147–160.

Edinger, J. E., E. M. Buchak, and M. D. McGurk. 1994. Analyzing larval distributions using hydrodynamic and transport modeling. In *Estuarine and Coastal Modeling*, Vol. III. American Society of Civil Engineers, New York.

Edinger, J. E., S. Dierks, and V. S. Kolluru. 2003a. Density dependent grazing in estuarine water quality models. *Water, Air, and Soil Pollution* 147:163–182.

Edinger, J. E., C. Boatman, and V. S. Kolluru. 2003b. Influence of multi algal groups in the calibration of water quality model. Paper presented at 8th Annual ASCE Estuarine and Coastal Modeling Conference, San Diego, CA.

ENTRIX, 2001. Hydrodynamic and water quality modeling and feasibility analysis of Indian River, Rehoboth Bay and Little Assawoman Bay. Delaware Department of Natural Resources and Environmental Control, Dover.

South Florida Water Management District. 2003. Technical documentation to support development of minimum flows and levels for the Caloosahatchee River and Estuary. South Florida Water Management District, West Palm Beach.

Weisberg, R. H. and L. Zheng. 2003. How estuaries work: a Charlotte Harbor example. *Journal of Marine Research* 61:635–657.

4

Using Statistical Models to Simulate Salinity Variability in Estuaries

Frank E. Marshall III

CONTENTS

Introduction

Models to simulate the variability of salinity in estuaries are important as tools to characterize the current salinity situation, and to estimate the salinity regime that may result from a change in the watershed hydrology. The salinity regime of Florida Bay is a primary factor in the composition of the estuarine ecosystems in Everglades National Park. Because of this, salinity data have been collected in the bay for a relatively long time, in some places since the 1930s. Over the past decade, continuous measurements have been collected.

Because of the global significance of the Everglades and Florida Bay ecosystems, there is a significant commitment of monetary resources to restore (to the extent possible) the modified hydrologic cycle of south Florida. To curtail flooding and encourage urban and agricultural development, the timing and distribution of freshwater delivery to Florida Bay from the Everglades was altered during the past century, thereby altering the natural salinity regime of Florida Bay. The effects of those alterations on the diverse ecosystems within Everglades National Park are still being evaluated. However, it is accepted that the overall health of the biota of the region including Everglades National Park will benefit from hydrologic modifications that restore the historic hydrologic cycle as much as possible.

There have been a number of attempts to develop models of the salinity regime in Florida Bay. Numerical modeling of salinity in Florida Bay on a three-dimensional basis has proved to be difficult (USACOE and South Florida Water Management District, 2002). In 2002, the Committee on Restoration of the Greater Everglades Ecosystem (CROGEE) concluded that an estimate of the Comprehensive Everglades Restoration Plan (CERP) influence on Florida Bay can be inferred from statistical models and time-series analysis, including the coupling of statistical salinity models with the output of the South Florida Water Management District watershed models (CROGEE, 2002).

0-8493-2822-5/05/$0.00+$1.50
© 2005 by CRC Press

The first known attempt to use statistical methods to make a connection between the salinity of Florida Bay and the upstream hydrology is attributed to Tabb (1967). Tabb (1967) based his prediction method on the close relationship that had been observed between salinity and the elevation of the water level at two monitoring wells in Everglades National Park (P35 and P38) and one well in Homestead. Other early efforts at statistical analysis and the development of simple statistical models included Scully (1986) and Cosby (1993). Nuttle (1997) was the first to use more-sophisticated statistical techniques. Marshall (2000) evaluated the linear regression salinity performance measures that had been developed for the RECOVER program by the South Florida Water Management District and recommended applying time series statistical methods for salinity modeling. Marshall (2003) reported the early activities in the development of multivariate linear regression (MLR) models using time series cross-correlation techniques.

With the exception of Marshall (2003), all of the previous studies have used data gathered monthly. The present study concentrates on the development of statistical models using daily average time series data for the evaluation of water management scenarios. Daily variability in salinity is important to biologists, because daily variability captures the "flashiness" of the salinity regime. It is known from anecdotal information that the nearshore areas of Florida Bay historically experienced periods of rapid fluctuation in salinity, in addition to long periods of high and low salinity. The open-water areas likely were rarely fresh, and periods of hypersalinity occurred, on a recurring basis. Because of large-scale diversions of fresh water, it is thought that the periods of hypersalinity in the open bay now last longer compared to pre-drainage conditions. Alternatively, in the northeast near-shore embayments where diverted water is discharged, freshwater conditions may remain longer and the average salinity may be less than in the past. The salinity models described below were developed as an analytical tool for the evaluation of water delivery alternatives to Florida Bay. They were used with long-term simulations of upstream water levels by the South Florida Water Management District 2X2 model and observed data to evaluate the impact of modifications to the water management system on the salinity regime in Florida Bay.

Methods and Materials

Study Area and Data

The study area encompasses northeastern, north central, and, to a limited extent, northwestern Florida Bay and the Everglades within Everglades National Park. This modeling effort utilized data that were collected at 10- to 60-min increments and averaged to daily and monthly values. The backbone of the data set for this study is the Everglades National Park Marine Monitoring Network (MMN) salinity data in Florida Bay and South Florida Water Management District's Everglades water levels. Details about these data can be found in Everglades National Park (1997a,b), Smith (1997, 1998, 1999, 2001), and South Florida Water Management District (2003). To these data other time-series data were added, including wind data from the National Weather Service, and water level data collected at Key West from the National Ocean Service. Wind data from Key West and Miami were used as these locations had the longest continuous records for wind and were considered to be representative of the regional wind patterns. Sea level data from Key West were considered to be representative of the average effect of oceanic water level influences and, to some extent, the average water level patterns within Florida Bay.

The locations of each of the monitoring stations where salinity and Everglades water-level data were collected are presented in Figure 4.1. The near-shore salinity stations are identified in one or more restoration programs as locations where performance measures will be established for evaluation of alternative water-delivery schemes. The other three salinity stations (Duck Key, Butternut Key, and Whipray Basin) were chosen because they are representative of different open-water conditions. Stage (water level) stations were selected to be representative of the water-level conditions in the Everglades, including Shark River Slough, Taylor Slough, the eastern panhandle, and other areas.

Because the data record for one of the water-level monitoring stations (E146) begins on 24 March 1994, the period of data used for these modeling activities begins on this date. The period of record extends through 31 October 2002, which means that there are 3143 daily values in a record with no

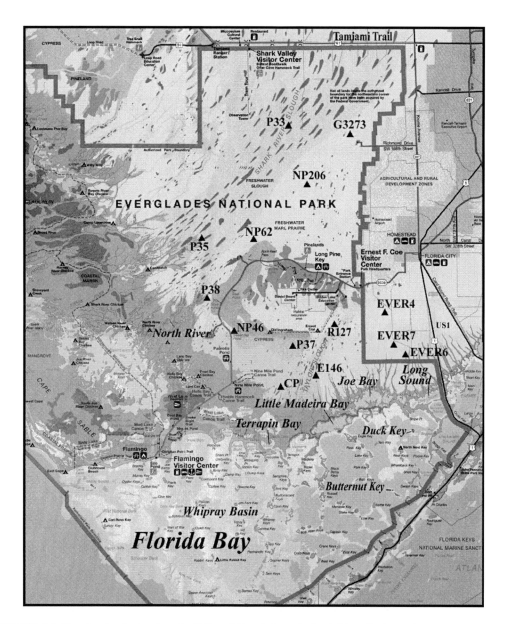

FIGURE 4.1 Map showing the location of salinity and Everglades water-level data collection stations.

missing data. In reality, most data sets contained some missing values. The model development period was 24 March 1995 through 31 October 2002. The verification period was 24 March 1994 through 23 March 1995. Information on the parameters that were used for the modeling activities is presented in Table 4.1.

Model Development

With the addition of the hydraulic gradient parameters, there were 25 independent variables that were subjected to a correlation analysis with the five near-shore salinity variables using a technique from SAS© SARIMA (SAS Institute, Inc., 1989, 1993) time-series model development procedures. This procedure (Brockwell and Davis, 1996; Marshall, 2003) uses pre-whitened time series, which have been filtered to explain as much of the variation in the parameter as possible using autoregressive and moving

TABLE 4.1

Summary of Information about the Dependent and Independent Variables Used in Model Development and Verification, and in Simulations

Variable Name	Variable Type	Variable Measure	Units	Data Source	Location
Little Madeira Bay	Dependent	Salinity	psu	ENP	Northeast Florida Bay
North River	Dependent	Salinity	psu	ENP	Northwest Florida Bay
Terrapin Bay	Dependent	Salinity	psu	ENP	Northeast Florida Bay
Long Sound	Dependent	Salinity	psu	ENP	Northeast Florida Bay
Joe Bay	Dependent	Salinity	psu	ENP	Northeast Florida Bay
Whipray Basin	Dependent	Salinity	psu	ENP	Open-water Florida Bay
Duck Key	Dependent	Salinity	psu	ENP	Open-water Florida Bay
Butternut Key	Dependent	Salinity	psu	ENP	Open-water Florida Bay
Cp	Independent	Water level	ft, NGVD 29	SFWMD	Craighead Pond
E146	Independent	Water level	ft, NGVD 29	SFWMD	Taylor Slough
Ever4	Independent	Water level	ft, NGVD 29	SFWMD	South of Florida City
Ever6	Independent	Water level	ft, NGVD 29	SFWMD	South of Florida City
Ever7	Independent	Water level	ft, NGVD 29	SFWMD	South of Florida City
G3273	Independent	Water level	ft, NGVD 29	SFWMD	East of Shark River Slough
NP206	Independent	Water level	ft, NGVD 29	SFWMD	East of Shark River Slough
NP46	Independent	Water level	ft, NGVD 29	SFWMD	Rocky Glades
NP62	Independent	Water level	ft, NGVD 29	SFWMD	East of Shark River Slough
P33	Independent	Water level	ft, NGVD 29	SFWMD	Shark River Slough
P35	Independent	Water level	ft, NGVD 29	SFWMD	Shark River Slough
P37	Independent	Water level	ft, NGVD 29	SFWMD	Taylor Slough
P38	Independent	Water level	ft, NGVD 29	SFWMD	Shark River Slough
R127	Independent	Water level	ft, NGVD 29	SFWMD	Taylor Slough
uwndkw	Independent	Wind vector	N/A	NWS	Key West
vwndkw	Independent	Wind vector	N/A	NWS	Key West
uwndmia	Independent	Wind vector	N/A	NWS	Miami
vwndmia	Independent	Wind vector	N/A	NWS	Miami
Kwwatlev	Independent	Tide elevation	ft, NGVD 29	NOS	Key West

Abbreviations: ENP = Everglades National Park, SFWMD = South Florida Water Management District, NOS = National Ocean Service, NWS = National Weather Service.

average terms. After pre-whitening, the remaining series is similar to a white noise sequence, containing an approximation of the non-explained variation in the parameter. The correlation coefficient between each independent variable in pre-whitened form and each dependent salinity variable (also pre-whitened) is then examined in a graphical format for a number of lagged values of the independent variable. For this analysis, the graphical format allowed 50 daily lags to be examined easily. All lags with a significant correlation coefficient (two standard deviations above or below the average correlation coefficient for all lags) were then identified as candidate lagged-model parameters for model development. Unlagged values of all independent variables, including hydraulic gradient variables, were also considered as candidate model parameters.

Following the correlation analysis, a modified stepwise regression technique was used to determine the significant variables in the five near-shore location MLR models (Kashigan, 1991). Because there was a high degree of cross-correlation between the stage variables, allowing the SAS stepwise regression procedure to automatically select the significant variables sometimes resulted in MLR models with terms that do not seem to make physical sense, such as a stage parameter with a positive sign seemingly indicating that increasing water levels in the Everglades results in increasing salinity in Florida Bay. Therefore, to make sure that the MLR models contain only terms that make physical sense, the level of significance for inclusion in the model was raised to 0.999, a high threshold, and all terms in an MLR model were tested to be sure that they were physically defensible. If not, they were eliminated from the

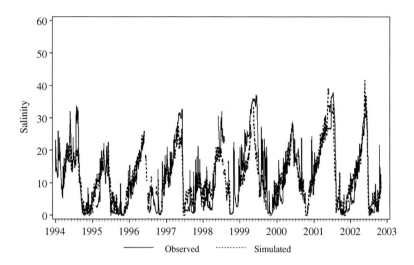

FIGURE 4.2 Joe Bay salinity model development plot. Calibration is 24 March 1995 to 31 October 2002; verification is 24 March 1994 to 23 March 1995.

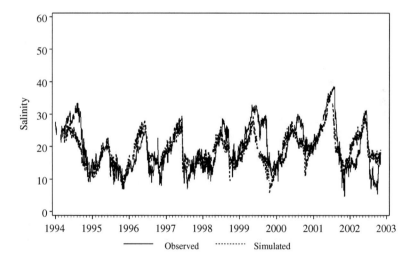

FIGURE 4.3 Little Madeira Bay salinity model development plot. Calibration is 24 March 1995 to 31 October 2002; verification is 24 March 1994 to 3 March 1995.

list of candidate parameters, and the stepwise model development procedure was rerun. The results of this selective stepwise procedure were relatively simple MLR models with a high level of significance for all model parameters, but with a lower model coefficient of determination (R^2) value than a model with parameters that were selected by the SAS stepwise automated procedure. However, this modification to model development did not hinder greatly the simulative capabilities of the MLR models. This process yielded MLR salinity models for all five near-shore locations (Joe Bay, Little Madeira Bay, Terrapin Bay, Long Sound, and North River; Figure 4.2 through Figure 4.6).

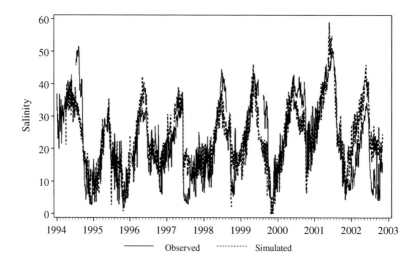

FIGURE 4.4 Terrapin Bay salinity model development plot. Calibration is 24 March 1995 to 31 October 2002; verification is 24 March 1994 to 3 March 1995.

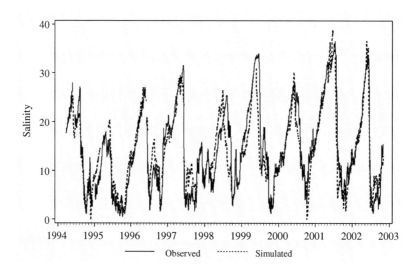

FIGURE 4.5 Long Sound salinity model development plot. Calibration is 24 March 1995 to 31 October 2002; verification is 24 March 1994 to 3 March 1995.

To simulate salinity at the open bay stations (Whipray Basin, Duck Key, and Butternut Key) a similar approach was used. However, instead of using Everglades water levels with wind and tide, salinity measurements at Joe Bay, Little Madeira Bay, and Terrapin Bay were used with wind and tide, at the same 0.999 level of significance. This yielded three MLR salinity models for the open-water stations (Figure 4.7 through Figure 4.9).

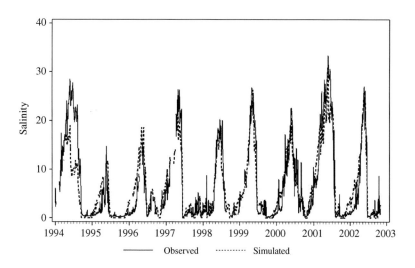

FIGURE 4.6 North River salinity model development plot. Calibration is 24 March 1995 to 31 October 2002; verification is 24 March 1994 to 3 March 1995.

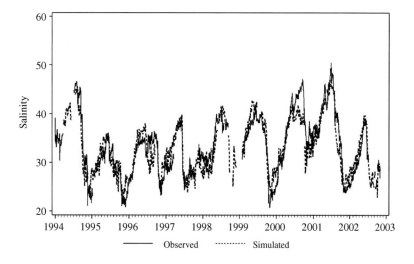

FIGURE 4.7 Whipray Basin salinity model development plot. Calibration is 24 March 1995 to 31 October 2002; verification is 24 March 1994 to 3 March 1995.

Model Runs for Operational Evaluations

The salinity models developed from the independent variables in the data set were produced for a Congressional report on certain modifications that had been made to the operations of the water management system in the Everglades, called the Interim Operations Plan (IOP). Specifically, for the IOP operations, fresh water that was being diverted into Shark River Slough had created high water-level conditions that were threatening the nesting habitat of the Cape Sable Seaside Sparrow. To alleviate these conditions, some of the water was re-diverted into the C-111 Canal system, for dispersal to Taylor Slough

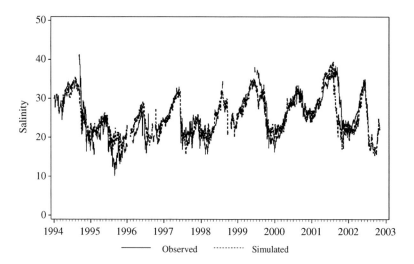

FIGURE 4.8 Duck Key salinity model development plot. Calibration is 24 March 1995 to 31 October 2002; verification is 24 March 1994 to 3 March 1995.

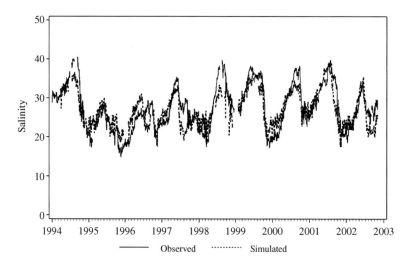

FIGURE 4.9 Butternut Key salinity model development plot. Calibration is 24 March 1995 to 31 October 2002; verification is 24 March 1994 to 3 March 1995.

and northeastern Florida Bay. At question was the benefit or impact to the Everglades and Florida Bay ecosystems, and salinity impacts were a concern. Because the salinity regime in Florida Bay responds to changes in water deliveries over a long period of time, simulations of long-term salinity variations were made by coupling the MLR salinity models with water levels simulated for 31 years by the South Florida Water Management Model (SFWMM), also known as the 2X2 model for the size of the grid cells in the model (2 miles by 2 miles). The 2X2 model domain encompasses most of the freshwater areas of the Everglades but does not extend into the mangrove zone.

The five near-shore models were used to make the various IOP evaluation runs using 2X2 model output for the input water levels, along with the real wind and tide data. For this case study, the two operational scenarios that are compared are the Base95 and IOP scenarios. The South Florida Water Management District produced model runs for both scenarios that supplied Everglades water levels in the model grid cells where each of the 14 monitoring stations were located. The Base95 run represents the operational conditions as they were in 1995, before the IOP operational changes. The 2X2 model used 31 years of historic rainfall to produce a time series of water levels as if the system had been operating for that period with the 1995 schedules in effect at the water management system structures such as weirs, gates, pumps, reservoirs, etc. In this manner, wet and dry period responses can be simulated, such as the severe drought that the region experienced in the late 1980s. Similarly, the IOP 2X2 model run simulated 31 years of operations with the IOP modifications to water delivery schedules in effect at the hydraulic structures.

As part of the preliminary evaluation of uncertainty, comparisons were made between the 2X2 model water-level output and the observed data. Evaluations of data plots showed that, at most stations, there was a systematic variation — systematically greater for some stations and systematically less for others compared to the observed data. Therefore, the average value for the observed data and the average value for the 2X2 model data were computed, and the 2X2 model output was adjusted by adding or subtracting the difference in the mean values. In previous work with 2X2 model output (Marshall, 2003), this step was found to improve the simulative capabilities of the 2X2 model output used as input to MLR salinity models.

The adjusted 2X2 model output was then used with historic wind and tide to simulate salinity at the near-shore stations for both Base95 and IOP operational conditions. To simulate the salinity at the open-water stations, these modeled salinity time series values were then transferred to the Whipray Basin, Duck Key, and Butternut Key salinity models and used with historic wind and tide data. The products are simulated salinities for a 31-year period for both operational scenarios that can be evaluated in a variety of ways.

Results

The salinity models that were developed using the techniques described above are as follows:

JOE BAY = 37.1 − 3.1CP − 3.5EVER6[lag6] − 10.5E146[lag6] − 0.19uwndkw − 0.09uwndkw[lag2] − 0.1vwndkw − 0.16vwndmia[lag1]; R^2 = 0.74

LITTLE MADEIRA BAY = 66.4 − 3.6CP[lag2] − 6.3P33[lag2] − 0.83srsdiff2 − 0.21uwndkw + 0.15uwndmia − 0.14vwndmia[lag1] + 0.8kwwatlev[lag2]; R^2 = 0.56

TERRAPIN BAY = 106.9 − 6.3CP[lag1] − 11.1P33[lag2] − 0.45uwndkw − 0.23uwndkw[lag1] − 0.2uwndkw[lag2] − 0.14vwndkw[lag2] + 0.46uwndmia + 1.9kwwatlev[lag2]; R^2 = 0.76

LONG SOUND = 42.2 − 9.5CP[lag4] − 5.2EVER7[lag2] − 1.7EVER6[lag2] −0.04vwndmia[lag1]; R^2 = 0.80

NORTH RIVER = 36.7 − 4.3CP − 3.8CP[lag3] − 3.4NP206[lag3] + 0.6kwwatlev[lag2]; R^2 = 0.86

WHIPRAY BASIN = 21.1 + 0.24ltmad[lag3] + 0.2terbay + 0.15terbay[lag3] − 0.04vwndkw[lag2] − 0.5kwwatlev[lag2]; R^2 = 0.80

DUCK KEY = 10.2 + 0.3ltmad[lag1] + 0.4ltmad[lag3] + 0.10uwndkw[lag1] + 0.13vwndkw[lag2] + 0.5kwwatlev; R^2 = 0.70

BUTTERNUT KEY = 15.4 + 0.14ltmad[lag1] + 0.44ltmad[lag3] + 0.03terbay[lag3] − 0.08uwndkw − 0.10uwndkw[lag2] + 0.4kwwatlev; R^2 = 0.65

As an evaluation of the performance of the models, salinity simulations produced using the above MLR models with Base95 (2X2 model) water levels and historic wind and tide were compared to

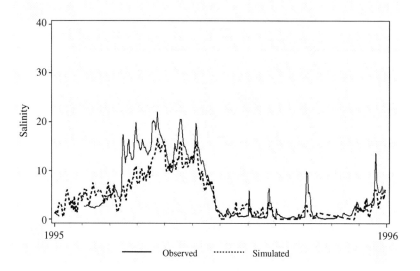

FIGURE 4.10 Comparison of simulated salinity from Base95 2X2 model run to observed salinity at Joe Bay.

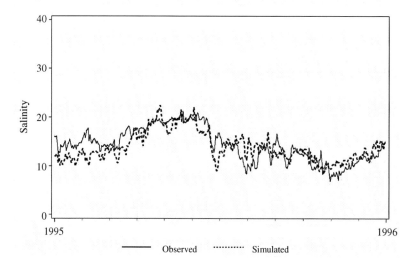

FIGURE 4.11 Comparison of simulated salinity from Base95 2X2 model run to observed salinity at Little Madeira Bay.

observed salinity data for calendar year 1995, on the premise that the Base95 2X2 model output for Everglades water level should be similar to the observed water level for the base year (1995). Salinity estimated by the MLR models driven with this input should reasonably simulate observed salinity if the models are working well in explaining salinity variation. Figure 4.10 through Figure 4.17 show that the simulated salinity values are similar to the observed salinity at all eight stations for this period.

The above models were then used as described above to produce 31-year simulations of salinity at each of the eight stations for Base95 and IOP operations using adjusted 2X2 model water levels, and historic wind and sea level data for the period 1 January 1965 through 31 December 1995. A plot of the Base95 Little Madeira Bay simulation is presented as Figure 4.18 as an example of the 31-year simulations. Similar plots were produced for all eight stations, for both operational scenarios.

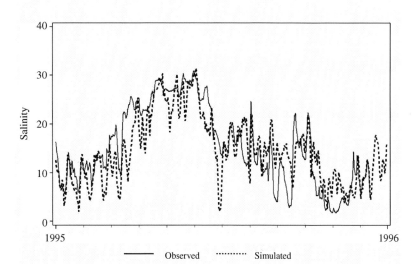

FIGURE 4.12 Comparison of simulated salinity from Base95 2X2 model run to observed salinity at Terrapin Bay.

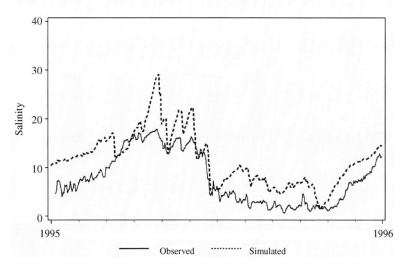

FIGURE 4.13 Comparison of simulated salinity from Base95 2X2 model run to observed salinity at Long Sound.

Time series simulations of operational scenarios can be compared in a number of different ways. The average values from each of the 31-year runs that were made using the 2X2 model output are provided in Table 4.2. A cursory evaluation of the results at the near-shore stations shows that, in general, the long-term average salinity is reduced for IOP runs compared to the Base95 scenario for Joe Bay and, to a limited extent, at Long Sound. Average salinity increased for IOP runs over Base95 simulations for the near-shore stations of Little Madeira Bay, Terrapin Bay, and North River. This result is reasonable when it is considered that the IOP and ISOP operations sent more water through the C-111 Canal system to discharge through the degraded berm into the mangrove zone of the eastern panhandle. The effect of changes to water level in Shark River Slough by IOP operations can be seen in the results from the North River runs, where average salinity is greater for IOP runs. In the open water, all three stations

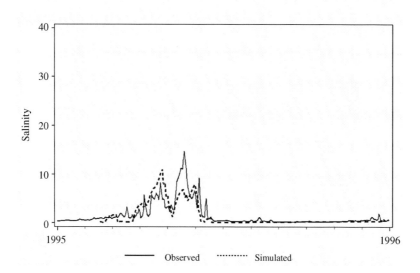

FIGURE 4.14 Comparison of simulated salinity from Base95 2X2 model run to observed salinity at North River.

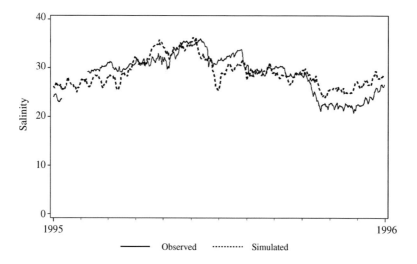

FIGURE 4.15 Comparison of simulated salinity from Base95 2X2 model run to observed salinity at Whipray Basin.

(Whipray Basin, Duck Key, and Butternut Key) showed an increase in average salinity for IOP operations compared to Base95 conditions (see Table 4.2). The relative increase was greater at Whipray Basin than at either Duck or Butternut Keys. Although the values are not presented, the standard deviation computed for the 31-year simulated series compares well to the standard deviation of the real data at all stations, even though the number of values for the simulated series (11,322) is much greater than the number of values in the real time series (2469–3062).

When the 31-year model run average values for Base95 and IOP presented in Table 4.2 were evaluated statistically using the average and standard deviation values there was no statistically significant difference between the two operational scenarios. However, when wet/dry season averages and monthly averages were compared, statistically significant differences were noted, as described below.

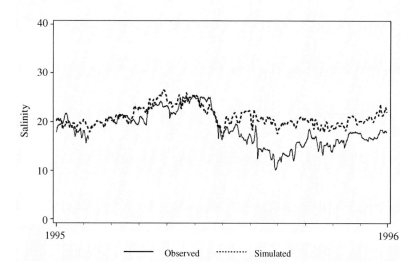

FIGURE 4.16 Comparison of simulated salinity from Base95 2X2 model run to observed salinity at Duck Key.

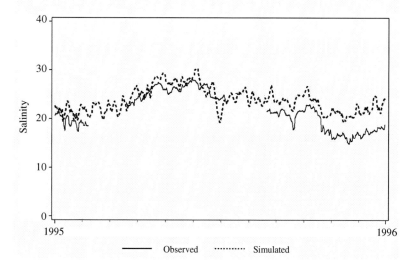

FIGURE 4.17 Comparison of simulated salinity from Base95 2X2 model run to observed salinity at Butternut Key.

For this IOP analysis, Everglades National Park has defined the dry season as 1 November through 31 May of each year (7 months). The wet season is defined as the period that begins on 1 June and ends on 31 October of each year (5 months). Table 4.3 presents dry season statistics and Table 4.4 presents wet season statistics, including 80% confidence intervals. The 80% confidence intervals were used because this was the highest level at which confidence intervals for some of the stations began to exhibit separation. At the 85% level all of the confidence intervals overlapped for all of the evaluations.

The only dry season confidence interval in Table 4.3 that does not overlap is the interval for Whipray Basin, although the end points of the intervals are the same. However, the confidence interval overlap for the dry season for Little Madeira Bay, Terrapin Bay, and Duck Key is small, meaning Base95 and IOP average values were significantly different at a slightly lower level, at approximately 75%

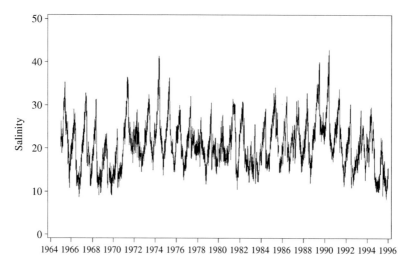

FIGURE 4.18 Simulated salinity for Little Madeira Bay for the Base95 operational scenario.

TABLE 4.2

Comparison of Overall Average Values of Salinity Produced by MLR Models for the Indicated Operational Scenario Simulated from 2X2 Model Output and Historic Wind and Tide Data for 31-Year Simulations

Operation	Joe Bay	Little Madeira Bay	Terrapin Bay	North River	Long Sound	Whipray Basin	Duck Key	Butternut Key
IOP	12.48	23.45	32.15	9.34	19.07	37.17	28.26	29.77
Base95	13.08	20.76	27.19	7.35	19.93	34.89	26.26	28.01

TABLE 4.3

Comparison of Average, Standard Deviation, and Confidence Intervals (80% significance level) for Dry Season Base95 and IOP Operations for 31-Year Simulations

	Operation	Average	Std. Dev.	80% Lower	80% Upper
Joe Bay	Base95	33.5	4.1	32.5	34.5
	IOP	35	4.6	33.8	36.2
Little Madeira Bay	Base95	22.0	5.8	20.5	23.5
	IOP	25.2	7.2	23.4	27.1
Terrapin Bay	Base95	28.9	11.5	26.0	31.8
	IOP	35.0	13.7	31.5	38.5
North River	Base95	10.8	9.3	8.4	13.2
	IOP	13.4	9.5	10.9	15.8
Long Sound	Base95	24.6	11.9	21.6	27.7
	IOP	24.1	11.4	21.1	27.0
Whipray Basin	Base95	35.8	5	34.5	37.1
	IOP	38.6	6	37.1	40.1
Duck Key	Base95	27.1	4.7	25.9	28.3
	IOP	29.5	5.7	28.0	31.0
Butternut Key	Base95	28.3	4.8	27.1	29.5
	IOP	30.3	5.6	28.9	31.7

TABLE 4.4

Comparison of Average, Standard Deviation, and Confidence Intervals (80% significance level) for Wet Season Base95 and IOP Operations for 31-Year Simulations

	Operation	Average	Std. Dev.	80% Lower	80% Upper
Joe Bay	Base95	7.9	9.2	5.5	10.3
	IOP	7.19	8.9	4.9	9.5
Little Madeira Bay	Base95	17.9	5.8	16.4	19.4
	IOP	19.7	6.5	18.0	21.3
Terrapin Bay	Base95	23.2	9.9	20.7	25.8
	IOP	26.4	11.1	23.5	29.2
North River	Base95	3.2	5.8	1.8	4.7
	IOP	3.9	6.1	2.3	5.4
Long Sound	Base95	13.4	8.0	11.3	15.4
	IOP	12.2	8.0	10.1	14.2
Whipray Basin	Base95	23.4	5.9	21.9	24.9
	IOP	24.7	6.4	23.1	26.3
Duck Key	Base95	26	6.2	24.4	27.6
	IOP	27.1	6.7	25.4	28.8
Butternut Key	Base95	23.3	9.9	20.8	25.8
	IOP	26.4	11.1	23.6	29.2

significance level. Typically, for statistical evaluations the lowest level of significance utilized is 80%. For the wet season (Table 4.4), all of the confidence intervals overlap by a relatively large margin, meaning there was no statistically significant difference between the wet season average values.

To investigate further the potential of a dry season effect of IOP operations, monthly average values were computed for each month over the 31-year simulation period. Table 4.5 presents a summary of this comparison. The average values are the average for the month of the 31 monthly averages that were computed from daily values. It can be seen that for two of the eight model simulations (Joe Bay and Long Sound), there was no statistically significant difference for any month between Base95 and IOP runs at the 80% level. However, for the other six simulations, there were at least 4 months where the difference between Base95 and IOP runs was significant at the 80% level. For Little Madeira Bay and Terrapin Bay the difference was significant at the 80% level for the all of the months of the dry season, November through May. At the North River station, the only station in this evaluation that receives direct flows from Shark River Slough, the difference was significant (80% level) for the months of October, November, December, and January. At Whipray Basin, the differences were significant (80% level) for all of the dry season months except January, and for the wet season months of July, September, and October. For Duck and Butternut Keys, the difference was significant (80% level) for the dry season months of January through May.

Therefore, these monthly comparisons indicate that there was a statistically significant effect of IOP operations on the salinity regime of the central near-shore embayments of Florida Bay, the downstream tidal reach of Shark River Slough (North River), and at the open-water stations in central and eastern Florida Bay at the monthly resolution. Although the effect was seen mostly during the dry season, an effect was also seen at two near-shore stations (Little Madeira Bay and Terrapin Bay) and one open-water station (Whipray Basin) during some of the months in the wet season.

Discussion

To be applied with confidence, MLR salinity models must be developed considering the physical phenomena that affect the salinity at a particular location and time in the estuary. The Everglades and the nearshore embayments of Florida Bay are a coastal aquifer system, with the freshwater body and the saltwater body competing to create an interface zone where there is a salinity gradient. In the Everglades, the surficial aquifer emerges above the ground most wet seasons and fresh water flows as

TABLE 4.5

Summary of Monthly Average Salinity Values Comparison for 31-Year Base95 and IOP Simulations at All Stations Using 80% Confidence Intervals

Joe Bay	Little Madeira Bay	Terrapin Bay	North River	Long Sound	Whipray Basin	Duck Key	Butternut Key
January	X	X	X			X	X
February	X	X			X	X	X
March	X	X			X	X	X
April	X	X			X	X	X
May	X	X			X	X	X
June		X					
July					X		
August							
September	X	X					
October		X	X		X		
November	X	X	X		X		
December	X	X	X		X		

Note: An "X" indicates that a significant difference was seen for that month.

sheet flow toward Florida Bay and the southwest coast of the Gulf of Mexico, creating the unique ecosystems that exist in Everglades National Park. Confined to a soil matrix, the interface zone is not affected by other factors that can affect a surface water body. In the absence of wind, direct rainfall, and evaporation, the "width" of the zone is relatively small. In an estuary, subject to these factors, the interface zone is relatively wide. Therefore, based on the coastal aquifer model presented above, it can be hypothesized that the salinity at a nearshore location is correlated in some manner to elevation of fresh water in the Everglades, sea level elevation, wind, and the watershed hydraulic gradient.

An ongoing research project for Everglades National Park had already shown that MLR salinity models prepared from observed data were capable of reasonably simulating the salinity regime in parts of Florida Bay (Marshall, 2003). For the IOP evaluation, these models were updated using an expanded data set of observed values, with new models prepared for Long Sound, Duck Key, and Butternut Key. The Base95 scenario was used as the existing conditions, and 31-year salinity simulations were prepared for both Base95 and IOP operational scenarios.

The results of monthly average and wet/dry season comparisons indicate that there was a potential dry season IOP effect shown by modeling at all locations except Joe Bay and Long Sound. As this was a modeling exercise, the effect can only be considered as potential, although knowing that there was a potential effect can help the understanding of the impact of the IOP operations.

In the case of Joe Bay, the effect of IOP operations (if there is an effect) may not have been discerned by this type of statistical analysis because of the large standard deviation of the observed and modeled salinity time series at this location. During some months in the Joe Bay 31-year simulation, the standard deviation was as large, if not larger, than the average monthly value. Time series and frequency distribution plots for Joe Bay (not presented) show that Joe Bay has a seasonal flashy variability, and salinity values drop rapidly to zero to signal the onset of the wet season most years. This may be caused by the relatively large inflow from Trout Creek into Joe Bay and the fact that the flow from this freshwater stream becomes small or ceases in the dry season many years. The monthly average values and standard deviations for Joe Bay salinity for each model run (Base95 and IOP) and the confidence intervals also were almost identical, which is not the case for any other stations except Long Sound. This may indicate that IOP operations were not affecting Joe Bay salinity. Additionally, aerial photos and site visits show that Joe Bay is relatively isolated from the rest of northeast Florida Bay, with flow in and out restricted to the narrow opening of Trout Cove into Florida Bay through the embayment-defining berms.

Long Sound was the only station that showed a wet season decrease in average value for IOP operations compared to Base95 model runs, although the difference was not significant at the 80% level. The

Long Sound model has a high R^2 value indicating that the model simulated the observed conditions well. Long Sound, like Joe Bay, also receives direct creek inflows during the wet season, creeks that also may cease flowing during the dry season. Also similar to Joe Bay, Long Sound drops to zero salinity and remains so during most wet seasons. By contrast, Little Madeira Bay and Terrapin Bay have a much smaller percentage of observed salinity values less than 10 psu. This may reflect the influence of more dispersed freshwater inflows to theses stations during the wet season from Taylor Slough and, perhaps, Shark River Slough, and wider openings in the mangrove-fringed berms surrounding these embayments that may allow the wind to push more salty water into these areas. Long Sound is also relatively isolated from the open waters of Florida Bay. Therefore, the fact that Base95 and IOP model runs for both Joe Bay and Long Sound showed no significant difference between Base95 and IOP may be explained by the physical conditions at these sites.

On the other hand, the comparisons of monthly average values over the 31-year period of the model runs indicated that there was a potential effect of IOP operations on salinity variation at the 80% level at Little Madeira Bay, Terrapin Bay, and North River for the near-shore locations. Dry season monthly average salinity values showed an increase at all of these stations, and some stations showed wet season monthly average value increases. The IOP operational scenario reduced the flow of fresh water to Shark River Slough, and it is supposed to increase the flow to Taylor Slough, compared to the Base95 conditions.

The level of influence of the Shark River Slough freshwater plume at Little Madeira Bay and Terrapin Bay has not been conclusively determined. However, it is thought that Taylor Slough flows are more of a determinant of the salinity at these stations although the volume of freshwater flow through Taylor Slough is much less than volume of flow through the Shark River Slough. While the North River is the only station directly downstream of Shark River Slough, the increase in dry season salinity for the 31-year IOP model run compared to the Base95 model run indicates that the reduced Shark River Slough flows are increasing downstream salinity at the monthly average level. Since IOP operations are intended to increase Taylor Slough flows, Little Madeira Bay and Terrapin Bay may be influenced more than previously thought by the Shark River Slough outflow. Another factor may be that more fresh water was discharged to the canals leading to the east coast through Biscayne Bay and was not diverted into Taylor Slough as planned.

A similar effect was seen at the open-water stations of Whipray Basin, Duck Key, and Butternut Key, with dry season monthly average values that increased for 31-year simulations of IOP operations compared to Base95 operations. This is not unexpected, as the models for these three open-water stations are a function of Little Madeira Bay and Terrapin Bay salinities. At the 0.999 significance level used for model development, Joe Bay and Long Sound salinities (unlagged and lagged) were not significant parameters in the open-water station models. When a model was developed as part of the previous Everglades National Park project for Whipray Basin using Everglades water levels instead of salinity along with historic wind and sea level, the only water level station that was significant at the 0.999 significance level was P33, a station in Shark River Slough. This may be additional evidence that Little Madeira Bay and Terrapin Bay salinity may be influenced by the outflow of Shark River Slough.

The R^2 value for the Whipray Basin model is higher than the R^2 values for the Duck Key and Butternut Key models, perhaps indicating that Whipray Basin may be more closely linked to the salinity variation in Little Madeira Bay and Terrapin Bay than the other two open-water stations, as would be expected by proximity. Whipray Basin salinity is considered an important indicator of hypersaline conditions in Florida Bay. Therefore, a statistically significant increase in Whipray Basin salinity for the IOP and ISOP runs at the monthly level, supported by similar increases at Duck and Butternut Keys, raises a "red flag" as a potential IOP effect. The statistical modeling effort to date supports the contention that changes to the salinity in the open-water areas of Florida Bay may be more influenced by changes in the volume of water flowing out of Shark River Slough than by changes in the volume of water flowing out of Taylor Slough. If that is the case, increases in Taylor Slough flow from IOP operations may not be enough to offset the change at the open-water stations that is caused by the decrease in Shark River Slough flows.

The consistent dry season effect that was seen at the 80% level at six of the eight stations for the monthly average comparisons supports the hypothesis that IOP operations in the Everglades may have an effect on the salinity regime in Florida Bay compared to the existing conditions. If the effect had been seen at only a few stations one might question whether the effect was real or whether this evaluation process was effective. This consistent result is an important finding because the evaluation procedure using 31-year simulations was completely model oriented, and the lack of consistent results may have complicated the analysis of the results.

A general evaluation of all comparisons indicates that the IOP operations have altered the salinity regime in Florida Bay in the central near-shore embayments and in the open-water areas compared to the existing conditions. In Joe Bay and Long Sound to the northeast, there appears to be no significant change. The results seem to indicate that the affected areas are more highly influenced by changes to Shark River Slough flows than by changes to the Taylor Slough flow. If this is thought to be representative of the effect of the IOP operations, the result at Whipray Basin may indicate a negative impact because higher monthly average conditions may mean longer and more frequent hypersaline periods at this important Florida Bay hypersalinity indicator station.

Conclusions

Updated MLR salinity models were prepared for eight locations in Florida Bay to evaluate changes to the operation of parts of the surface water management system for the Everglades. These models are physically defensible with terms in each model that are reasonable given the overall knowledge of the south Florida hydrologic system. By far, the most important Everglades water level station for modeling purposes was Craighead Pond (CP), which appeared in all of the near-shore models. Some combination of wind vectors also appeared in all models (including all open-water locations) except the North River model, which was as expected. Sea level (tide) appeared in most models, but not all. Because the significance level was set at a very high level for inclusion of a parameter in a model (0.999), it is expected that there are other parameters that would have been significant had the significance level been specified at a lower level more typically seen in other statistical evaluations (say, 0.95 or 0.90). However, the fact that the significance level is so high for parameter inclusion means that there is little doubt regarding the importance of the parameters in the models in explaining the variation in salinity when all of the other parameters are also included.

Comparisons of simulations for 31 years that do not represent actual operations must be made carefully. The results of the evaluations show that the MLR salinity models have done their job, simulating salinity and providing consistent results that are supported by the current level of knowledge in the hydrology and physiography of Florida Bay. The simulated salinity time series show that they can adequately estimate salinity in a manner that will allow the comparisons to be made.

The results of the wet/dry season and monthly average comparisons using the 31-year model runs showed that there is a potential effect of IOP operations at the 80% level for monthly comparisons and at a slightly lower significance level for wet/dry season comparisons in the central embayments and at the open-water stations. Exceptions to this are Joe Bay and Long Sound. The effect is somewhat more pronounced for monthly averages in the dry season, although some effect of IOP operations is also seen at some locations for monthly averages in the wet season. Although there is only one station downstream of the Shark River Slough discharge (North River), from a spatial point of view it appears that the effect at that location can be linked directly to a reduction of freshwater outflows from Shark River Slough. Because the IOP operations should have increased Taylor Slough flows, the monthly average increases at the central near-shore stations may indicate that these central stations are more influenced by Shark River Slough outflows than originally thought. The increase in monthly average salinity values in the near-shore central embayments is transferred to the open-water stations of Whipray Basin, Duck Key, and Butternut Key where the increase in salinity is also seen in 31-year simulations.

Notably absent from this effect are Joe Bay and Long Sound. Instead of indicating that there is a problem with the statistical analysis, the consistency with which the effect is expressed at Little Madeira Bay, Terrapin Bay, North River, Whipray Basin, Duck Key, and Butternut Key and the consistency with which the effect is not expressed at Joe Bay and Long Sound indicate that the lack of a response at these more isolated northeast Florida Bay stations is a plausible result.

The conclusions of this study can be summarized as follows:

1. Statistical models can be used in Florida Bay for the reasonable simulation of salinity using multivariate linear regression techniques.
2. The evaluation procedure using the MLR models with 2X2 model output for Everglades water levels and historical data for wind and tide to simulate long-term operations for Base95 and IOP water delivery scenarios shows an increase in salinity values at the following locations, primarily during the dry season, for monthly average values (80% significance level):
   - Little Madeira Bay
   - Terrapin Bay
   - North River
   - Whipray Basin
   - Duck Key
   - Butternut Key
3. No effect of IOP operations compared to Base95 31-year simulations was seen in the salinity regime of Joe Bay and Long Sound.
4. Time-series MLR models coupled with statistical analysis provide an effective tool for the simulation of salinity in Florida Bay, thereby extending the use of salinity as an estuarine indicator.

Acknowledgments

Funding for this project was provided by U.S. Department of the Interior, National Park Service, Everglades National Park Cooperative Agreements 5280-01-020. The author thanks DeWitt Smith and David M. Nickerson for their efforts leading to this chapter.

References

Brockwell, P. J. and R. A Davis. 1996. *Introduction to Time Series and Forecasting.* Springer-Verlag, New York.

Cosby, B. J. 1993. An Examination of the Relationships of Stage, Discharges and Meteorology in the Panhandle and Taylor Slough Areas of Everglades National Park to Salinity in Upper Florida Bay, Vol. 1–5. University of Virginia, Charlottesville.

CROGEE. 2002. Florida Bay Research Programs and Their Relation to the Comprehensive Everglades Restoration Plan. National Academies Press, Washington, D.C.

Everglades National Park. 1997a. Everglades National Park Marine Monitoring Network 1994 Data Summary. Everglades National Park, Homestead, FL.

Everglades National Park. 1997b. Everglades National Park Marine Monitoring Network 1995 Data Summary. Everglades National Park, Homestead, FL.

Kashigan, S. K. 1991. *Multivariate Statistical Analysis.* Radius Press, New York, pp. 117–192.

Marshall, F. E., III. 2000. Florida Bay Salinity Transfer Function Analysis, Vol. 1 of 2: Final Report. Cetacean Logic Foundation, New Smyrna Beach, FL.

Marshall, F. E., III. 2003. Salinity Simulation Models for North Florida Bay Everglades National Park. Cetacean Logic Foundation, New Smyrna Beach, FL.

Nuttle, W. K. 1997. Central and Southern Florida Project Restudy: Salinity Transfer Functions for Florida Bay and West Coast Estuaries, Vol. 1: Main Report. Southeast Environmental Research Program, Florida International University, Miami.

SAS Institute, Inc. 1989. SAS/STAT© User's Guide, Version 6, 4th ed., Vol. 2. SAS Institute, Inc., Cary, NC.

SAS Institute, Inc. 1993. SAS/ETS© User's Guide, Version 6, 2nd ed. SAS Institute, Inc., Cary, NC.

Scully, S. P. 1986. Florida Bay Salinity Concentration and Groundwater Stage Correlation and Regression. South Florida Water Management District, West Palm Beach, FL.

Smith, D. 1997. Everglades National Park Marine Monitoring Network 1996 Data Summary. Everglades National Park, Homestead, FL.

Smith, D. 1998. Everglades National Park Marine Monitoring Network 1997 Data Summary. Everglades National Park, Homestead, FL.

Smith, D. 1999. Everglades National Park Marine Monitoring Network 1998 Data Summary. Everglades National Park, Homestead, FL.

Smith, D. 2001. Everglades National Park Marine Monitoring Network 1999 Data Summary. Everglades National Park, Homestead, FL.

Tabb, D. 1967. Prediction of Estuarine Salinities in Everglades National Park, Florida by the Use of Ground Water Records. Ph.D. dissertation, University of Miami, Miami, FL.

U.S. Army Corps of Engineers and South Florida Water Management District, 2002. Project Management Plan–Florida Bay and Florida Keys Feasibility Study. USACOE, Jacksonville, FL.

5

Using Satellite Imagery and Environmental Monitoring to Interpret Oceanographic Influences on Estuarine and Coastal Waters

Brian D. Keller and Billy D. Causey

CONTENTS

Introduction

The Florida Keys, off the southeastern tip of the United States, extend southwest more than 350 km from Biscayne Bay to the Dry Tortugas and form the southeastern margin of Florida Bay (Figure 5.1). Extensive seagrass beds and mangroves surround the Florida Keys, and coral reefs of the Florida Reef Tract occur offshore of the Florida Keys along the Atlantic side. These environments support diverse and productive biological communities, making this area nationally significant because of its high conservation, recreational, commercial, ecological, historical, scientific, educational, and aesthetic values (Causey, 2002).

The greatest threat to the environment, natural resources, and economy of the Florida Keys has been degradation of water quality (Kruczynski and McManus, 2002), especially over the past two decades. Some of the reasons for degraded water quality include managed diversions of freshwater flows away from the Everglades, Florida Bay, and the southern coast of Florida; nutrients from domestic wastewater in the Florida Keys via shallow-well injection, cesspools, and septic tanks; storm water runoff containing heavy metals, fertilizers, insecticides, and other contaminants; marinas and live-aboard vessels; poor flushing of canals and embayments; accumulation of dead seagrasses and algae along the shoreline; sedimentation; infrequency of hurricanes in recent decades; and environmental changes associated with global climate change and rising sea level (Causey, 2002).

There also are important regional influences on the marine environment of the Florida Keys (Lee et al., 2002). Circulation patterns and exchange processes of South Florida coastal waters create strong physical linkages between the Keys and regions to the north. South Florida coastal waters are bounded by major currents, the Loop and Florida Currents, which connect South Florida to more remote regions of the Gulf of Mexico and Caribbean Sea (Lee et al., 2002). Weaker currents advect coastal waters of southwestern Florida across the Southwest Florida Continental Shelf to the Florida Keys.

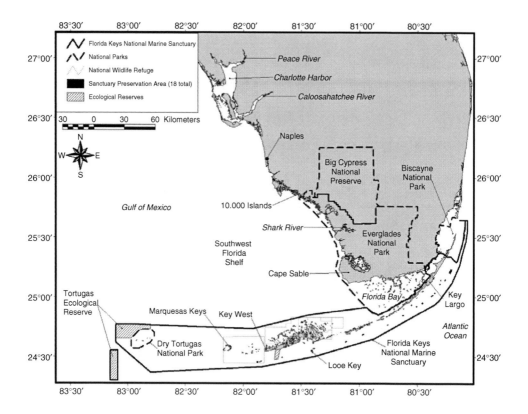

FIGURE 5.1 Map of South Florida. Surface currents generally flow in a southerly direction across the Southwest Florida Continental Shelf and through passes between the Florida Keys (Lee et al., 2002). The network of 24 fully protected marine zones within the sanctuary includes the Tortugas Ecological Reserve, the Western Sambo Ecological Reserve (hatched quadrangle east of Key West), and 18 small sanctuary preservation areas and four research-only areas, of which only Looe Key is labeled (see http://floridakeys.noaa.gov/research_monitoring/map.html for details).

A low-salinity plume from the Mississippi and Atchafalaya Rivers can extend hundreds of kilometers into the Gulf of Mexico (Paul et al., 2000; Wawrik et al., 2003). At an extreme, floodwaters from the Mississippi River during 1993 resulted in surface salinities in the Florida Keys that were substantially lower than normal (Ortner et al., 1995; Gilbert et al., 1996). In addition, satellite imagery shows that eastern Gulf of Mexico circulation patterns connect estuaries along the coast of southwestern Florida with the Florida Keys and the Florida Keys National Marine Sanctuary (http://coast-watch.noaa.gov/hab/bulletins_ms.htm).

Here we wish to demonstrate how estuarine and coastal conditions can be investigated and interpreted through the use of satellite imagery coupled with large-scale monitoring of water quality and other oceanographic observations. We worked with colleagues to apply this approach to interpret a blackwater event that affected the Florida Keys National Marine Sanctuary in 2002. This understanding resulted in better-informed communications with the public, colleagues, and news media. Retrospective investigations of this sort may lead to greater capacity to forecast coastal and estuarine conditions.

The Florida Keys National Marine Sanctuary

The National Marine Sanctuary Program of the U.S. National Oceanic and Atmospheric Administration (NOAA) has managed segments of the Florida Reef Tract since 1975. The Key Largo National Marine Sanctuary was established at that time to protect 353 km² of coral reef habitat offshore of the upper

Florida Keys. In 1981, the 18-km^2 Looe Key National Marine Sanctuary was established to protect the heavily used Looe Key Reef in the lower Florida Keys (Figure 5.1). These two National Marine Sanctuaries were, and continue to be, managed very intensively (Causey, 2002).

By the late 1980s it had become evident that a broader, more holistic approach to protecting and conserving the health of coral reef resources had to be implemented. Irrespective of the intense management of small areas of the reef tract, sanctuary managers were witnessing declines in water quality and the health of corals that apparently had a wide range of sources. The most obvious causes of decline were non-point-source discharges, habitat degradation because of development and overuse, and changes in reef fish and invertebrate populations because of overfishing.

The threat of oil drilling in the mid- to late 1980s off the Florida Keys, combined with reports of deteriorating water quality throughout the region (Kruczynski and McManus, 2002; Leichter et al., 2003), occurred at the same time scientists were assessing adverse affects of coral bleaching (Glynn, 1993), the 1983 die-off of the long-spined sea urchin (Lessios, 1988), loss of living coral cover on reefs (Dustan and Halas, 1987; Porter and Meier, 1992), a major seagrass die-off (Robblee et al., 1991), declines in reef fish populations (Ault et al., 1998), and the spread of coral diseases (Porter et al., 2001; Sutherland et al., 2004). These were topics of major scientific concern and the focus of several scientific workshops (e.g., Ginsburg, 1993).

In the fall of 1989, subsequent to the catastrophic *Exxon Valdez* oil spill in Alaska, three large ships became grounded on the Florida Reef Tract within a brief, 18-day period. These major physical impacts to the reef in conjunction with the cumulative effects of environmental degradation prompted the U.S. Congress to take action to protect the unique coral reef ecosystem of the Florida Keys. In November 1990, President George H.W. Bush signed into law the Florida Keys National Marine Sanctuary and Protection Act (FKNMS Act; DOC, 1996).

The FKNMS Act designated 9515 km^2 of coastal waters surrounding the Florida Keys as the Florida Keys National Marine Sanctuary and addressed two major concerns. There was an immediate prohibition on oil drilling, including mineral and hydrocarbon leasing, exploration, development, or production within the sanctuary. In addition, the legislation prohibited the operation of vessels longer than 50 m in an internationally recognized "Area to Be Avoided" within and near the boundary of the sanctuary.

The U.S. Congress recognized the critical role of water quality in maintaining sanctuary resources and directed the U.S. Environmental Protection Agency to develop a comprehensive Water Quality Protection Program for the sanctuary. The FKNMS Act also called for the development of a comprehensive management plan and implementation of regulations to achieve protection and preservation of the resources of the Florida Keys marine environment. The Final Management Plan (DOC, 1996), including a network of fully protected marine zones, was implemented in 1997 (Causey, 2002).

The Water Quality Protection Program includes three long-term monitoring projects, which are a key element of monitoring the marine environment and natural resources of the sanctuary. Additional monitoring projects along with the Water Quality Protection Program are providing extensive data on environmental conditions and the status and trends of natural resources in the sanctuary.

Sanctuary Monitoring Programs

Monitoring projects within and near the Florida Keys National Marine Sanctuary provide baseline data on the marine ecosystem. Three monitoring projects of the Water Quality Protection Program provide long-term, status-and-trends data on key components of the ecosystem. These are water quality (http://serc.fiu.edu/wqmnetwork/FKNMS-CD/index.htm), seagrasses (http://www.fiu.edu/~seagrass/), and coral reef and hard-bottom communities (http://www.floridamarine.org/features/category_sub .asp?id=2360).

Projects in the Marine Zone Monitoring Program compare ecological processes, populations, and communities inside and outside fully protected marine zones. Studies include coral recruitment; reef fish herbivory; benthic community structure; and reef fish, queen conch, and spiny lobster populations. There also are studies of human uses and perceptions of sanctuary resources. Summary findings of the Marine Zone Monitoring Program, Water Quality Protection Program, and other projects are posted at http://www.fknms.nos.noaa.gov/research_monitoring/, and socioeconomic reports are posted at: http://marineeconomics.noaa.gov/pubs/welcome.html. The Marine Zone Monitoring Program, like the Water Quality Protection Program, provides long-term, status-and-trends data on natural resources in the sanctuary.

Finally, there are four additional monitoring activities:

1. Bimonthly oceanographic research cruises conducted jointly by the NOAA Atlantic Oceano-graphic and Meteorological Laboratory (AOML) and the University of Miami's Rosenstiel School of Marine and Atmospheric Science (RSMAS) (http://www.aoml.noaa.gov/5fp/data.html)
2. Near-real-time meteorological and oceanographic monitoring at six data buoys within the sanctuary and a seventh in northwestern Florida Bay (http://www.coral.noaa.gov/ /seakeys/)
3. 32 fixed thermograph stations positioned throughout the sanctuary (http://coralreef.gov/pro-ceedings/Day%202%20PDF/1-Billy%20Causey.pdf, slide 23)
4. Near-real-time oceanographic monitoring at a recently activated data buoy near Looe Key Reef (Figure 5.1; http://www.looekeydata.net/)

Collectively, these programs and projects, along with monitoring within Dry Tortugas, Everglades, and Biscayne National Parks and elsewhere in coastal South Florida, provide a wealth of baseline data to evaluate changes in coastal environments and natural resources. For example, the Water Quality Monitoring Project (http://serc.fiu.edu/wqmnetwork/FKNMS-CD/index.htm) documented changes in water quality parameters between 1995 and 2000, when a trend analysis revealed statistically significant increases in the concentration of total phosphorus in several regions of the sanctuary and Southwest Florida Continental Shelf (http://floridakeys.noaa.gov/research_monitoring/monitoring_report_2000.pdf). Over this period, concentrations of total phosphorus tripled, from ~0.1 to ~0.3 μM. In contrast, no trends in total phosphorus were observed in the relatively isolated waters of Florida Bay. Large increases in nitrate concentrations also occurred in several regions of the sanctuary and the Southwest Florida Continental Shelf; in many areas values increased by two orders of magnitude, from <0.05 to >1 μM (http://floridakeys.noaa.gov/research_monitoring/monitoring_report_2000.pdf). In contrast to these two increases, concentrations of total organic nitrogen declined between 1995 and 2000. The authors (R.D. Jones and J.N. Boyer, Southeast Environmental Research Center, Florida International University) inferred that these trends, which ended in 2001 (http://floridakeys.noaa.gov/research_monitoring/ 2001_sci_rept.pdf), were caused by regional circulation patterns arising from the Loop and Florida Currents.

The 2002 Blackwater Event

Oceanic influences on the Florida Keys originating from the southwestern Florida coast were particularly evident during early 2002. In mid- to late January 2002, commercial fishers began observing a mass of dark water off the southwest coast of Florida. These observations were the first accounts of what became known as the "Blackwater Event of 2002." A reporter with the *Naples Daily News* brought this unusual event to the attention of Florida Keys National Marine Sanctuary staff. Initial reports came from commercial fishers who had observed a large area of dark discoloration offshore and south of Naples, Florida (http://web.naplesnews.com/02/03/naples/d599686a.htm). The "black water" was moving toward the sanctuary, and sanctuary staff called commercial fishers and colleagues to learn about the nature of the event and its likely consequences. News media enquiries began immediately, and it was critical for sanctuary staff to respond with the best-available science.

One concern was that this discolored water was the result of polluted runoff from South Florida agricultural lands. However, a similar blackwater event had been observed in 1878 coming from the South Florida mainland and moving past the Florida Keys between Key West and the Dry Tortugas (Mayer, 1903, cited in SWFDOG, 2002). This past account of a blackwater episode suggested that these might be infrequent events, with a long history, that are not necessarily influenced by recent human activities. Furthermore, major agricultural activities such as the Everglades Agricultural Area (EAA) south of Lake Okeechobee are quite distant from the Southwest Florida Continental Shelf. Surface runoff from the EAA likely does not reach coastal waters of southwestern Florida (Brand, 2002; Nelsen et al., 2002; see Lapointe et al., 2002 for contrasting views).

Working with colleagues, FKNMS staff developed the following scenario of the Blackwater Event of 2002 (SWFDOG, 2002). On 9 January 2002 Drs. Frank Müller-Karger and Chuanmin Hu of the University of South Florida (USF) collected a SeaWiFS true-color image that showed an area of blackness just west of Cape Sable (Figure 5.1) at the southwest tip of Florida (Figure 2a in SWFDOG, 2002; also available at http://imars.marine.usf.edu/~hu/black_water/imgs/swf/true-color/S010902.JPG). Total absorption coefficients (Figure 2c in SWFDOG 2002) were relatively high near the mouth of Shark River. Imagery prepared and analyzed by Dr. Richard Stumpf of NOAA/Center for Coastal Monitoring and Assessment (CCMA) indicated that the color of the black water was consistent with a source from wetlands of the 10,000 Islands and Everglades regions of South Florida, i.e., high levels of tannins and humic acid (R. Stumpf, pers. commun.). Additional imagery (NOAA/CCMA) showed a relatively high concentration of chlorophyll (a measure of phytoplankton density) in the same area (R. Stumpf, pers. commun.).

Managers compared these satellite data with routine water-quality samples collected by Drs. Ronald Jones and Joseph Boyer (Southeast Environmental Research Center, Florida International University) during 10–13 January 2002 along transects across part of the Southwest Florida Continental Shelf (http://serc.fiu.edu/wqmnetwork/CONTOUR%20MAPS/ContourMaps.htm). This is part of a long-term, water quality monitoring program for South Florida coastal waters, which includes the sanctuary's Water Quality Monitoring Project. Sampling showed that high concentrations of chlorophyll *a* matched the shape of the black water, indicating that a phytoplankton bloom was juxtaposed on the black water. It also showed a plume of low-salinity water emanating from the mainland and a high daytime concentration of dissolved oxygen. Research conducted by Dr. Gary Hitchcock's laboratory (RSMAS), including studies by Dr. Jennifer Jurado, has shown that plumes from Shark River transport silicate and nitrogen, which are critical nutrients for diatom blooms that occur year after year, generally October–December, and sometimes with a second peak in April (Jurado et al., 2003; see also http://nsgl.gso.uri.edu/flsgp/flsgpg01006.pdf for a Florida Bay Watch Report on this topic).

A true-color satellite image collected on 4 February 2002 (USF) showed a larger area of black water that had moved somewhat farther offshore and to the south since January (Figure 2b in SWFDOG, 2002; also available at http://imars.marine.usf.edu/~hu/black_water/imgs/swf/true-color/S020402.JPG). Enhanced imagery (NOAA/CCMA) on 4 February showed this large blackwater area superimposed on a high concentration of chlorophyll (R. Stumpf, pers. commun.). A 21 February to 1 March 2002 oceanographic research cruise conducted in the region by Dr. Peter Ortner (AOML) with additional sample analyses by RSMAS showed a high concentration of chlorophyll *a* centered on the area of black water and showed a continuing plume of low-salinity water coming from the mainland (P. Ortner, pers. commun.). A satellite-tracked surface drifter released by AOML/RSMAS near the mouth of Shark River (Figure 5.1) on 22 February 2002 moved slowly in a west-southwesterly arc, then slowly to the north during the last 2 weeks of March (http://mpo.rsmas.miami.edu/flabay/latest_29526.gif). This was consistent with the slow movement of the blackwater area evident in satellite imagery.

By mid-March 2002, the area of black water had moved south-southwesterly into the lower Florida Keys. There was little blackish coloration left and plankton samples showed signs of an aging diatom bloom (Florida Marine Research Institute, Dark Water Update: http://www.floridamarine.org/features/view_article.asp?id=21893). The bloom enveloped the lower Florida Keys, and an outflow as far west as the Marquesas Keys (Figure 5.1) was apparent in late March to early April (NOAA/CCMA; http://coastwatch.noaa.gov/hab/bulletins/hab20020320_200201_a.pdf and USF; http://imars.marine.usf.edu/~hu/black_water/imgs/swf/true-color/S040402.JPG).

The Blackwater Event of 2002 may have been unusually large and persistent because of a prolonged drought that was followed by heavy rains in late 2001. One of the worst droughts in Florida since the initiation of weather records occurred between 1998 and 2001; parts of the state reported the driest conditions in more than 100 years of record keeping (http://www.ncdc.noaa.gov/oa/climate/research/2001/preann2001/events.html#us). The drought conditions, coupled with the very slow rate of water flow across the Everglades (Holling et al., 1994), may have caused an accumulation of nutrients and organic material on the mainland. The 3-year drought began to end with the onset of heavy rains in mid-July 2001, followed by the passage of Tropical Storm Barry in late August, Tropical Storm Gabrielle in mid-September, and a glancing blow from Hurricane Michelle (South Florida) in early November 2001 (http://www.drought.unl.edu/dm/archive.html). The runoff resulting from this series of heavy rains probably flushed large quantities of nutrients and organic compounds off the mainland and into coastal waters. Inputs of silicate and nitrogen probably contributed to a massive diatom bloom (Jurado et al., 2003) that darkened coastal waters, and organic compounds such as tannins and humic acid apparently added a blackish cast to the bloom.

The 2002 Blackwater Event apparently caused some die-offs in benthic communities. On 27 March, an experienced diver (Ken Nedimyer) reported dead and dying sponges and corals in a channel northwest of Key West (Figure 5.1). Another experienced diver (Don DeMaria) reported dead and dying sponges in shallow water on the Gulf of Mexico side of the lower Florida Keys (Summerland Key). On 2 April, Dr. Niels Lindquist (University of North Carolina at Chapel Hill) dived at four sites near Key West. He reported a sponge die-off that appeared to be somewhat species specific, with the most severe effects on the sponges *Callyspongia* and *Niphates* and, to a lesser degree, *Amphimedon*. By contrast, the sponges *Aplysina* spp. and *Ircinia* spp. appeared healthy. The sponge die-off appeared to become less severe south and east of western Key West, away from the blackwater event. In addition, there was a report of coral die-offs at two long-term coral reef and hard-bottom community monitoring sites (Hu et al., 2003).

The 2003 Blackwater Event

Another blackwater event occurred in October 2003, with some important differences from the 2002 event. Extensive, dense plankton blooms occurred along the South Florida coast between Charlotte Harbor (Figure 5.1) and the Florida Keys during much of the month of October 2003 (http://modis.marine.usf.edu/). Red Tide Status reports (http://www.floridamarine.org/features/default .asp?id=1018) stated that both dinoflagellates, including the red tide species (*Karenia brevis*), and diatoms were present in nearshore and offshore waters during this month. Enhanced satellite imagery (NOAA/CCMA) showed an area of black water stained by organic compounds south of Charlotte Harbor on 18 October 2003, which dissipated over the next 2 weeks (R. Stumpf, pers. commun.). In the 2003 blackwater event, the source of tannins and humic acid appeared to be Charlotte Harbor, which is influenced by the Caloosahatchee and Peace Rivers (Figure 5.1), rather than the 10,000 Islands and Everglades regions of South Florida, as in 2002. Furthermore, the black water was not as extensive and persistent as it was during the 2002 event. Finally, the 2003 blackwater event may have had strong anthropogenic influences, unlike the 2002 event, because of releases of fresh water from Lake Okeechobee into the Caloosahatchee River (http://myfwc.com/fishing/pdf/toho-nov03.pdf) and additional nutrient inputs into the Caloosahatchee and Peace Rivers. Interestingly, blooms were reported south of Charlotte Harbor in January 2000 and October 2001 (Stumpf et al., 2003).

Summary and Conclusions

The Blackwater Event of 2002 probably was a large and persistent plankton bloom, with additional blackish staining by organic compounds, which slowly crossed the Southwest Florida Continental Shelf into the western end of the Florida Keys National Marine Sanctuary. Routine water quality monitoring in January and February 2002 reported very high concentrations of phytoplankton, and reports from boat-based observers were that the water appeared greenish brown at the surface. The blackness apparent in true-color satellite imagery may have been caused by a high degree of light absorption by the dense plankton bloom. Spectral analysis of satellite imagery also indicated actual blackish discoloration from decomposing vegetation while the bloom was adjacent to outflows from the Everglades.

Water samples to identify the types of phytoplankton responsible for the bloom were not collected until its late stages in March 2002, once the scientific community had been alerted to the event. Several samples had high concentrations of diatoms, and earlier research showed that diatom blooms have been common in this region of Florida over the past decade (Jurado et al., 2003; http://nsgl .gso.uri.edu/flsgp/flsgpg01006.pdf). Some samples contained harmful algal bloom (HAB) species as well; however, there were no reports of extensive fish kills associated with the 2002 bloom, which indicates that it was not a HAB during its later stages. The bloom had an associated die-off of certain sponge species and corals (Hu et al., 2003) near Key West.

A plankton bloom of this magnitude requires suitable environmental conditions (e.g., temperature, salinity, and light) and a substantial source of nutrients. In this region of South Florida, nitrogen rather than phosphorus may be a growth-limiting nutrient for phytoplankton (Boyer and Jones, 2002). Diatoms also require silicate, and both nutrients flow into coastal waters from Shark River (Jurado et al., 2003). Although this bloom was unusual, it appeared to result from a combination of natural events, including a slowly spinning gyre that apparently contributed to the cohesiveness and duration of the bloom.

By contrast, the smaller, more ephemeral blackwater event in 2003 was associated with Charlotte Harbor and inflows from the Peace and Caloosahatchee Rivers, and probably had strong anthropogenic influences. A unifying theme for both events was the utility of satellite imagery in monitoring and interpreting large-scale phenomena (http://coastwatch.noaa.gov/hab/bulletins_ms.htm), including plankton blooms, and in directing field sampling to locations of particular interest. Satellite imagery, long-term water quality monitoring, and collaboration of resource managers and scientists enabled retrospective analyses, which were central to interpreting and explaining the Blackwater Event of 2002.

Acknowledgments

We thank Drs. Frank Müller-Karger and Chuanmin Hu (University of South Florida), Dr. Richard Stumpf (NOAA/Center for Coastal Monitoring and Assessment), Drs. Ronald Jones and Joseph Boyer (Southeast Environmental Research Center, Florida International University), Dr. Peter Ortner (NOAA/Atlantic Oceanographic and Meteorological Laboratory), Ken Nedimyer, Don Demaria, Dr. Niels Lindquist (University of North Carolina at Chapel Hill), and John Hunt and Beverly Roberts (Florida Fish and Wildlife Conservation Commission, Fish and Wildlife Research Institute) for so generously sharing data and information. We also thank Dr. Steve Bortone (Sanibel-Captiva Conservation Foundation) for inviting us to participate in the Estuarine Indicators Workshop and for providing comments on the manuscript. Kevin Kirsch (Florida Keys National Marine Sanctuary) prepared the figure.

References

Ault, J. S., J. A. Bohnsack, and G. A. Meester. 1998. A retrospective (1979–1996) multispecies assessment of coral reef fish stocks in the Florida Keys. *Fishery Bulletin* 96:395–414.

Boyer, J. N. and R. D. Jones. 2002. A view from the bridge: external and internal forces affecting the ambient water quality of the Florida Keys National Marine Sanctuary (FKNMS). In *The Everglades, Florida Bay, and Coral Reefs of the Florida Keys: An Ecosystem Sourcebook*, J. W. Porter and K. G. Porter (eds.). CRC Press, Boca Raton, FL, pp. 609–628.

Brand, L. E. 2002. The transport of terrestrial nutrients to South Florida coastal waters. In *The Everglades, Florida Bay, and Coral Reefs of the Florida Keys: An Ecosystem Sourcebook*, J. W. Porter and K. G. Porter (eds.). CRC Press, Boca Raton, FL, pp. 361–413.

Causey, B. D. 2002. The role of the Florida Keys National Marine Sanctuary in the South Florida Ecosystem Restoration Initiative. In *The Everglades, Florida Bay, and Coral Reefs of the Florida Keys: An Ecosystem Sourcebook*, J. W. Porter and K. G. Porter (eds.). CRC Press, Boca Raton, FL, pp. 883-894.

Department of Commerce (DOC). 1996. Final environmental impact statement/final management plan for the Florida Keys National Marine Sanctuary. National Oceanic and Atmospheric Administration, Silver Spring, MD. Available online: http://floridakeys.noaa.gov/regs/.

Dustan, P. and J. C. Halas. 1987. Changes in the reef-coral community of Carysfort Reef, Key Largo, Florida: 1974 to 1982. *Coral Reefs* 16:91–106.

Gilbert, P. S., T. N. Lee, and G. P. Podesta. 1996. Transport of anomalous low-salinity waters from the Mississippi River flood of 1993 to the Straits of Florida. *Continental Shelf Research* 16:1065–1085.

Ginsburg, R. N. (compiler). 1993. Case studies for the colloquium and forum on global aspects of coral reefs: health, hazards and history. Rosenstiel School of Marine and Atmospheric Science, University of Miami, Miami, FL.

Glynn, P. W. 1993. Coral reef bleaching: ecological perspectives. *Coral Reefs* 12:1–17.

Holling, C. S., L. H. Gunderson, and C. J. Walters. 1994. The structure and dynamics of the Everglades system: guidelines for ecosystem restoration. In *Everglades: The Ecosystem and Its Restoration,* S. M. Davis and J. C. Ogden (eds.). St. Lucie Press, Delray Beach, FL, pp. 741–756.

Hu, C., K. E. Hackett, M. K. Callahan, S. Andréfouët, J. L. Wheaton, J. W. Porter, and F. E. Müller-Karger. 2003. The 2002 ocean color anomaly in the Florida Bight: a cause of local coral reef decline? *Geophysical Research Letters* 30(3):1151.

Jurado, J. L., G. L. Hitchcock, and P. B. Ortner. 2003. The roles of freshwater discharge, advective processes and silicon cycling in the development of diatom blooms in coastal waters of the southwestern Florida Shelf and northwestern Florida Bay. In *Florida Bay Program and Abstracts, Joint Conference on the Science and Restoration of the Greater Everglades and Florida Bay Ecosystem,* April 13–18, 2003, Palm Harbor, FL, pp. 119–121.

Kruczynski, W. L. and F. McManus. 2002. Water quality concerns in the Florida Keys: sources, effects, and solutions. In *The Everglades, Florida Bay, and Coral Reefs of the Florida Keys: An Ecosystem Sourcebook,* J. W. Porter and K. G. Porter (eds.). CRC Press, Boca Raton, FL, pp. 827–881.

Lapointe, B. E., W. R. Matzie, and P. J. Barile. 2002. Biotic phase-shifts in Florida Bay and fore reef communities of the Florida Keys: linkages with historical freshwater flows and nitrogen loading from Everglades runoff. In *The Everglades, Florida Bay, and Coral Reefs of the Florida Keys: An Ecosystem Sourcebook,* J. W. Porter and K. G. Porter (eds.). CRC Press, Boca Raton, FL, pp. 629–648.

Lee, T. N., E. Williams, E. Johns, D. Wilson, and N. P. Smith. 2002. Transport processes linking South Florida coastal ecosystems. In *The Everglades, Florida Bay, and Coral Reefs of the Florida Keys: An Ecosystem Sourcebook,* J. W. Porter and K. G. Porter (eds.). CRC Press, Boca Raton, FL, pp. 309–342.

Leichter, J. J., H. L. Stewart, and S. L. Miller. 2003. Episodic nutrient transport to Florida coral reefs. *Limnology and Oceanography* 48:1394–1407.

Lessios, H. A. 1988. Mass mortality of *Diadema antillarum* in the Caribbean: what have we learned? *Annual Review of Ecology and Systematics* 19:371–393.

Mayer, A. G. 1903. The Tortugas, Florida, as a station for research in biology. *Science* 17:190–192.

Nelsen, T. A., G. Garte, C. Feathersone, H. R. Wanless, J. H. Trefry, W.-J. Kang, S. Metz, C. Alvarez-Zarikian, T. Hood, P. Swart, G. Ellis, P. Blackwelder, L. Tedesco, C. Slouch, J. F. Pachut, and M. O'Neal. 2002. Linkages between the South Florida peninsula and coastal zone: a sediment-based history of natural and anthropogenic influences. In *The Everglades, Florida Bay, and Coral Reefs of the Florida Keys: An Ecosystem Sourcebook,* J. W. Porter and K. G. Porter (eds.). CRC Press, Boca Raton, FL, pp. 415–449.

Ortner, P. B., T. N. Lee, P. J. Milne, R. G. Zika, M. E. Clarke, G. P. Modesto, P. K. Swart, P. A. Tester, L. P. Atkinson, and W. R. Johnson. 1995. Mississippi River flood waters that reached the Gulf Stream. *Journal of Geophysical Research* 100(C7):13595–13601.

Paul, J. H., A. Alfreider, J. B. Kang, R. A. Stokes, D. Griffin, L. Campbell, and E. Ornolfsdottir. 2000. Form IA *rbcL* transcripts associated with a low salinity/high chlorophyll plume ("Green River") in the eastern Gulf of Mexico. *Marine Ecology Progress Series* 198:1–8.

Porter, J. W. and O. W. Meier. 1992. Quantification of loss and change in Floridian reef coral populations. *American Zoologist* 32:625–640.

Porter, J. W., P. Dustan, W. C. Jaap, K. L. Patterson, V. Kosmynin, O. W. Meier, M. E. Patterson, and M. Parsons. 2001. Patterns of spread of coral disease in the Florida Keys. *Hydrobiologia* 460:1–24.

Robblee, M. B., T. R. Barber, P. R. Carlson, M. J. Durako, J. W. Fourqurean, L. K. Muehlstein, D. Porter, L. A. Yarbro, R. T. Zieman, and J. C. Zieman. 1991. Mass mortality of the seagrass *Thalassia testudinum* in Florida Bay (USA). *Marine Ecology Progress Series* 71:297–299.

South-West Florida Dark-Water Observations Group (SWFDOG). 2002. Satellite images track "black water" event off Florida coast. *Eos, Transactions, American Geophysical Union* 83: 281, 285.

Stumpf, R. P., M. E. Culver, P. A. Tester, M. Tomlinson, G. J. Kirkpatrick, B. A. Pederson, E. Truby, V. Ransibrahmanakul, and M. Soracco. 2003. Monitoring *Karenia brevis* blooms in the Gulf of Mexico using satellite ocean color imagery and other data. *Harmful Algae* 2:147–160.

Sutherland, K. P., J. W. Porter, and C. Torres. 2004. Disease and immunity in Caribbean and Indo-Pacific zooxanthellate corals. *Marine Ecology Progress Series* 266:273–302.

Wawrik, B., J. H. Paul, L. Campbell, D. Griffin, L. Houchin, A. Fuentes-Ortega, and F. Muller-Karger. 2003. Vertical structure of the phytoplankton community associated with a coastal plume in the Gulf of Mexico. *Marine Ecology Progress Series* 251:87–101.

6

Development and Use of Assessment Techniques for Coastal Sediments

Edward R. Long and Gail M. Sloane

CONTENTS

Introduction

Potentially toxic chemicals enter waters dissolved in water or attached to suspended particulate matter. Most waterborne toxic substances are hydrophobic and bond to particulates. As particulates and associated toxicants become increasingly dense, they can sink to the bottom of lakes, rivers, estuaries, and bays in low-energy areas where they become incorporated into sediments. Therefore, sediments that have accumulated in depositional zones where they are not disturbed by physical processes or other factors can provide a relatively stable record of toxicant inputs (NRC, 1989; Power and Chapman, 1992). As a result, sediments are an important medium in which to estimate the degree and history of chemical contamination of our national waters.

In 1989, the U.S. National Research Council (NRC) Committee on Contaminated Marine Sediments examined the issue of chemical contamination and its effects in the nation's estuarine and marine waters (NRC, 1989, p. 1). The committee concluded, "sediment contamination is widespread throughout U.S. coastal waters and potentially far reaching in its environmental and public health significance." Furthermore, the NRC (1989, p. 1) determined, "The problem of contaminated marine sediments has emerged as an environmental issue of national importance." Relatively high chemical concentrations in sediments were reported for many sites near urban centers. However, the committee recognized the lack of sufficient data for assessing the severity and extent of sediment contamination in many areas and the need for better, more reliable assessment tools. The committee further recommended a more comprehensive national network to monitor and evaluate the condition of U.S. sediments.

Assessments of sediment quality are most comprehensive when conducted with a "sediment quality triad" approach, which consists of chemical analyses, toxicological tests, and metrics of benthic community structure (Long and Chapman, 1985; Chapman et al., 1987, 1998). Chemical analyses of sediments can provide information on the presence and concentrations of mixtures of potentially toxic substances in sediment samples. There are multiple sets of numerical guidelines with which to interpret the chemical data (U.S. EPA, 1992), including those developed for the state of Florida (MacDonald et al., 1996; MacDonald and Ingersoll, 2003). However, information gained from these analyses alone provides no direct measure of the toxicological significance of the chemicals.

Laboratory testing of the toxicity of sediments has become a widely used assessment tool commonly applied to a number of regulatory, monitoring, and scientific issues (Swartz, 1989; Hill et al., 1993; Long et al., 1996). The results of laboratory tests can stand alone as powerful indicators of degraded conditions and do not require either field validation or concordance with sediment chemistry (Chapman, 1995). However, measures of the structure and function of benthic populations and communities can provide important indicators of *in situ* adverse effects of toxicity among local resident biota (Canfield et al., 1994).

A considerable amount of research is conducted annually in the United States, Canada, and internationally to improve the tools used in sediment assessments. Research conducted to determine spatial status and temporal trends in coastal sediment quality using the triad of measures is becoming increasingly more common, and important to differing programs. In addition, considerable research is conducted to determine the factors that influence chemical contamination, the chemical concentrations that can cause toxicity and degraded benthos, and the degree of benthic resource losses associated with acute toxicity.

The purpose of this chapter is to provide an overview of current techniques used in sediment quality assessments and to summarize the relative quality of sediments in selected regions as examples of the uses of the various multimetric techniques. Some of the most commonly used methods to generate data for each of the sediment quality triad components are described along with summaries of results from selected large-scale studies and inventories, including studies completed in Florida.

Much of the data and information presented here were generated by two federal agencies in the United States. The National Oceanic and Atmospheric Administration (NOAA) in its National Status and Trends (NS&T) Program (Wolfe et al., 1993) conducted sediment quality assessments in numerous marine bays and estuaries from 1990 to the present time. The U.S. Environmental Protection Agency (U.S. EPA) managed both the National Sediment Quality Survey (U.S. EPA, 1997) and the Environmental Monitoring and Assessment Program (EMAP) estuary surveys (Paul et al., 1992). The EMAP surveys of estuaries were conducted over large portions of the U.S. coasts, whereas the NOAA surveys were conducted in specific bays and estuaries (Long, 2000). In addition, the U.S. Geological Survey (USGS) laboratory in Columbia (MO) has conducted many equivalent studies in fresh water, i.e., sediment quality assessments and derived sediment quality guidelines (SQGs) for fresh water.

Classification of Sediment Contamination and Interpretive Tools and Guidelines

Chemical analyses performed in sediment quality assessments typically involve quantification of the concentrations of numerous chemicals, except in cases where a small-scale site is investigated and where the list of chemicals of concern is known. Typically in status and trends monitoring, dredged material evaluations, and other enforcement actions, chemical data are generated for numerous trace metals and organic compounds. In addition, physical-chemical variables are measured that may aid in the interpretation of the data. Typically, chemical analyses are performed for the following materials:

Polynuclear aromatic hydrocarbons	Chlorinated pesticides
Phthalates	Polychlorinated biphenyls
Trace metals, metalloids	Mercury
Sediment texture	Total organic carbon
Ammonia	

In recent years, some studies of sediment quality have included analyses for other compounds, including:

Phenols, chlorinated phenols	Butyl tins
Pharmaceuticals	Personal hygiene products
Fire retardants	Detergents
Dioxins, dioxin-like compounds	Organic acids
Porewater chemistry	

Sediment studies evaluate these contaminants to help determine ecological and/or human health effects in the environment of concern. These may be a concern for myriad reasons ranging from permitting investigations to watershed/tribal/state/national characterizations. With a lack of regulatory "standards" for the majority of potentially toxic chemicals in sediments, there is a need to determine acceptable concentrations or risk.

Florida Sediment Quality Guidelines and Interpretive Tools

There are two basic approaches to the derivation of SQGs: (1) methods that involve comparisons to reference area conditions and (2) methods that involve associations with adverse biological effects. The State of Florida derived a method to identify anthropogenically enriched trace metal concentrations in companion with a set of effects-based SQGs.

Recognizing the need for a simple method to screen sediment data for evidence of metal contamination, the Florida Department of Environment Protection (FDEP) prepared an interpretative tool for sediment metals in Florida estuaries (Schropp and Windom, 1988) and fresh waters (Carvalho et al., 2002). The interpretative tool is based on the relatively constant relationships that exist between metals and a reference element (aluminum) in natural sediments. By normalizing metal concentrations to aluminum, the tool allows a simple determination of whether estuarine sediment metals are within or outside of the expected natural ranges (Figure 6.1). Simple to apply, the tool relies on metal concentration data from total digestion of bulk sediment samples. Others have developed similar tools based on uncontaminated reference sediments or on historical data from sediment cores from specific sites or regions (Loring, 1991; Daskalakis and O'Connor, 1995).

In a subsequent effort, FDEP developed a set of interpretative biological effects–based ("effects-based") guidelines based on a comprehensive review of biological responses to coastal sediment contamination (MacDonald, 1994) and for freshwater sediments (MacDonald et al., 2003).

Both tools adopted by FDEP provide a simple method to screen sediment quality data to determine whether the measured metal concentrations represent metal enrichment or potential harm to ecosystems. Evaluating the results using these indicators provides a better understanding of chemical, physical, and biological properties associated with the sediments and, thus, better assurance that a management decision will be founded on a weight of scientific evidence.

U.S. and International Sediment Quality Guidelines

Effects-based SQGs have been derived by several agencies in the United States and Canada with one of several empirical approaches or theoretical (or mechanistic) approaches. The U.S. EPA has promulgated national guidelines for five trace metals and has the lead responsibility to develop other effects-based guidelines for nationwide use. Effects-based SQGs were derived for NOAA (Long et al., 1995), FDEP (MacDonald, 1996, 2002), and the Great Lakes National Program Office of U.S. EPA (Ingersoll et al., 1996, 2000). The province of Ontario, Canada, has developed guidelines, using benthic community data (Persaud et al., 1992). Washington State has promulgated sediment quality standards for

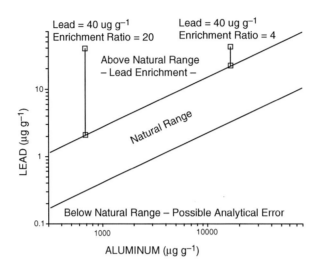

FIGURE 6.1 Example of use of geochemical normalization tool for metals. (From Schropp, S.J. and H.L. Windom. 1988. A Guide to the Interpretation of Metal Concentrations in Estuarine Sediments. Prepared for Florida Department of Environmental Regulation, Tallahassee, FL, 53 pp.)

use in enforcement and monitoring, and is the only state with enforceable sediment quality standards approved by its legislature (WDOE, 1995).

Other U.S. states and Canadian provinces have either adopted guidelines developed elsewhere through a formal process, or they use them occasionally either informally or in enforcement actions (including Alaska, California, Hawaii, Maine, Massachusetts, New Jersey, New York, Oregon, South Carolina, Texas, British Columbia, Quebec). The Netherlands has developed standards with both reference area- and effects-based approaches, primarily for dredged material evaluations (Stronkhorst et al., 2001). New Zealand and Australia, together, have adopted North American guidelines (ANZECC, 2002) and, more recently, Brazil has taken steps to do the same. SQGs have been used in monitoring and dredged material programs in Hong Kong, South Korea, and the United Kingdom. Effects-based SQGs are used to interpret chemical data collected routinely in many regional monitoring programs in the United States. Regions in which sediment chemical analyses are conducted annually or periodically include the Great Lakes, Puget Sound, Columbia River estuary, San Francisco Bay, Southern California Bight, Tampa Bay, South Carolina estuaries, Chesapeake Bay, and the New York/New Jersey Harbor. Estuarine sediment quality surveys conducted as a part of the EMAP–Estuaries studies of the U.S. EPA, the NS&T Program of NOAA, and various sediment quality assessments of the USGS in fresh water included use of SQGs to interpret the data.

Incidence of Chemical Contamination of Sediments

The U.S. EPA compiled chemical information from more than 21,000 locations sampled nationwide during the years 1980 through 1993 (U.S. EPA, 1997). Data were compiled from several federal and state databases for both fresh water and salt water. Most of the data were collected as requirements of compliance monitoring programs and represented conditions influenced by specific sources, and, therefore, were not a result of random site selections. However, some data from other programs, such as those from the NS&T Program (Daskalakis and O'Connor, 1994) and EMAP–Estuaries, in which samples were collected with random sampling designs, also were included in the inventory.

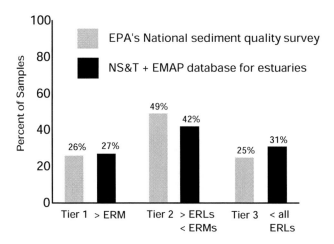

FIGURE 6.2 Classification of sediment contamination.

The U.S. EPA chose to characterize sediments using a tiered approach. Each sampling station entered into the U.S. EPA (1997) database was classified as Tier 1, Tier 2, or Tier 3, depending on the concentrations of chemicals in the sediments. Samples were classified as Tier 1 if one or more substances equaled or exceeded at least two "mid-range" numerical guidelines. These guidelines included concentrations derived with equilibrium-partitioning (EQP) modeling approaches (U.S. EPA, 1997), the effects range-median (ERM) values (Long et al., 1995), probable effects levels (PEL; MacDonald et al., 1996), and apparent effects thresholds (AET) derived by the State of Washington (WDOE, 1995). Some stations, in addition, were classified as Tier 1 based on results of at least two laboratory toxicity tests or the potential risks of adverse effects to fish and wildlife via bioaccumulation. Stations classified as Tier 2 were toxic in one toxicity test or had at least one chemical concentration that exceeded "low-range" values, including effects range-low (ERL) or threshold effects levels (TEL), but none of the mid-range concentrations. Stations in which none of the chemical concentrations equaled or exceeded any of these numerical guidelines or were nontoxic were classified as Tier 3.

U.S. EPA (1997) classified 26% of the samples as Tier 1, 49% as Tier 2, and the remaining 25% as Tier 3 (Table 6.1 and Figure 6.2). Samples classified as Tier 1 (contaminated) were scattered throughout the nation, most frequently in rivers, lakes, bays, and estuaries near large urban centers.

In 1998, chemical and toxicity data from the NS&T Program and EMAP analyses were compiled (Long et al., 1998) to determine the predictive ability of SQGs prepared by Long et al. (1995) and MacDonald et al. (1996). This database consisted of chemical and toxicity data from 1068 estuarine stations sampled along the Atlantic Coast from Boston (Massachusetts) to Charleston (South Carolina), the Gulf of Mexico coast from Tampa (Florida) to the U.S./Mexico border, and the Pacific Coast along portions of southern California. Among the 1068 samples, 27% had at least one chemical concentration that exceeded an ERM value, roughly equivalent to Tier 1 in the U.S. EPA (1997) study. Another 42% exceeded at least one ERL value but none of the ERMs (equivalent to Tier 2), and 31% did not have any chemical concentrations that equaled or exceeded the ERLs (equivalent to Tier 3) (Figure 6.2, Table 6.1).

The percentages of samples classified as chemically contaminated in these two national inventories were comparable to those in selected regional inventories (e.g., the Puget Sound SEDQUAL database, North Carolina estuaries, and Southern California Bight 1998 survey; Table 6.1). The percentages of samples classified as contaminated in the PSAMP/NOAA survey of Puget Sound, the NOAA survey of Biscayne Bay, in a database for Tampa Bay, and in the Regional Monitoring Program for San Francisco Bay (excluding data for nickel) were much lower. The incidence of contamination was highest in surveys conducted of toxic harbors and bays of California and in Pearl Harbor, Hawaii. When the data were

TABLE 6.1

Percentages of Sediment Samples in Which One or More SQGs Were Exceeded and the Spatial Area That They Represented in Different Databases and Estuarine Regions of the United States

Location, Database, and Criteria Used[a]	No. of Samples Exceeding at Least One SQG Value		As Percent of Study Area		Source of Data
	Ratio	Percent	km²	Percent	
National Inventories					
U.S. EPA 1996 National Sediment Quality Inventory					
Exceeded two or more SQGs or were toxic	5460::21,000	26.0			U.S. EPA, 1997
U.S. NOAA/EMAP data base for estuaries					Long et al., 1998
Exceeded at least one ERM value	291::1068	27.2			
Exceeded at least one PEL value	385::1068	36.0			
Regional Inventories: Estuaries					
PSAMP/NOAA survey of Puget Sound					
Exceeded at least one ERM value	39::300	13.0	30.7	1.3	Long et al., in prep.
Puget Sound SEDQUAL database					
Exceeded at least one sediment quality criterion (i.e., SQS)	2319::8523	27.2			Wash. Dept. of Ecology, SEDQUAL database
NOAA survey of Biscayne Bay, FL	33::226	14.6	3.5	0.7	Long et al. 2002
NOAA/EMAP database for North Carolina estuaries	44::175	25.1	1855.4	21±5	Hyland et al., 2000
EMAP–Louisiana estuaries				5±5	U.S. EPA/EMAP Web site
EMAP–Mississippi estuaries				0.0	U.S. EPA/EMAP Web site
EMAP–Alabama estuaries				29±30	U.S. EPA/EMAP Web site
EMAP–Florida Panhandle estuaries				4.0	U.S. EPA/EMAP Web site
Mid-Atlantic Integrated Assessment estuaries				6.0	U.S. EPA/EMAP Web site
Tampa Bay, FL estuary surveys	7::37	1.2			Steve Grabe, Hillsborough Co.
Southern California Bight shelf survey (1994)	51::261	19.5	3520.0	12.3	SCCWRP Web site
Southern California Bight shelf, bays, harbors survey (1998)	78::290	26.9		14.7	SCCWRP Web site
San Francisco Estuary Institute RMP data (1993–2000)					Bruce Thompson, SFEI
Exceeded at least one ERM value (all chemicals considered)	381::397	96.0			
Exceeded at least one ERM value (excluding nickel)	20::397	5.0			
Regional Inventories: Industrial harbors					
New York/New Jersey Harbor R-EMAP survey; 1993/94			250.5	50	Darvene Adams, U.S. EPA Region 2
New York/New Jersey Harbor R-EMAP survey; 1998			235.5	47	Darvene Adams, U.S. EPA Region 2
California BPTCP database for harbors and bays	406::568	71.4			Russell Fairey, CalState, Moss Landing
Pearl Harbor, U.S. Navy survey	176::219	80.4			Jeff Grovhoug, U.S. Navy, San Diego

[a] Unless indicated as otherwise, all data were calculated as incidence of samples in which one or more ERM values were exceeded.

expressed as percentages of the study area, the areas affected ranged from 0% (Mississippi estuaries) to 50% (New York/New Jersey harbor).

Classification of Toxicity of Sediments

A variety of laboratory tests have been used in recent years to classify sediments as either toxic or nontoxic (Chapman, 1988; Swartz, 1989; Lamberson et al., 1992; Long, 2002). These include tests of survival, reproductive success, growth, avoidance, burrowing ability, metabolic activity, morphological development, and bioaccumulation of toxicants in tissues. Tests are performed on whole (solid-phase) sediments, pore waters extracted from the sediments, sediment/water mixtures (i.e., elutriates), or organic solvent extracts. Tests of sediment toxicity are most frequently conducted with invertebrates, most often with the adult forms or in some tests with the gametes or embryos. Sublethal end points are entering the mainstream of toxicity testing, such as tests that involve measures of bacterial bioluminescence activity and induction of cytochrome P450 activity.

There are testing-methods manuals that have been developed by the American Society for Testing Materials (ASTM), the U.S. EPA, and the Army Corps of Engineers. No federal criteria or standards, per se, have been promulgated with which to declare samples as toxic. However, there are statistical methods developed to aid in classification of samples with the results of some tests (Thursby et al., 1997; Phillips et al., 2001). Tests of amphipod survival, echinoderm embryo development, and polychaete growth currently are used in Puget Sound to classify sediment quality along with numerical criteria developed as State of Washington standards (WDOE, 1995).

The test most frequently used in North America, whether in monitoring programs or in regulatory evaluations, is the amphipod survival test. The survival of amphipods exposed to whole sediments for 10 days is compared to that in a nontoxic control sediment or reference area sample. These tests are included in the dredged material assessment manuals for both fresh water and estuaries (U.S. EPA and the U.S. Army Corps of Engineers, 1991). They are commonly included in federal status and trends monitoring studies of estuaries, including those conducted as a part of the EMAP–Estuaries and NS&T programs (Long, 2000, 2002). They have been included in annual or periodic regional monitoring programs, including those in Puget Sound, the Great Lakes, San Francisco Bay, Columbia River estuary, Southern California Bight, Chesapeake Bay, Massachusetts Bay, New York/New Jersey Harbor, Tampa Bay, and South Carolina estuaries. Amphipod survival tests are often used in remedial investigation feasibility studies of hazardous waste sites. In fresh water they frequently have been included in waste site surveys, monitoring studies, and in the derivation and evaluation of SQGs (Ingersoll et al., 2001). These toxicity tests are key elements of dredged material assessments in Canada and the Netherlands and have been used occasionally in Australia, Belgium, France, Germany, Hong Kong, New Zealand, and the United Kingdom.

Spatial Extent of Sediment Toxicity

Estimates of the spatial (or surficial) extent of toxicity have been made in both the NS&T Program and individual EMAP estuarine studies. In both programs, sampling locations were chosen randomly to avoid biasing the data and results are expressed as both square kilometers and percentages of total study areas (Paul et al., 1992; Long et al., 1996).

During the period of 1991–1999, NOAA collected samples throughout 28 estuaries and marine bays in the United States. The survey areas extended from Boston Harbor to Biscayne Bay on the Atlantic Coast of the United States, from Tampa Bay to Galveston Bay (TX) on the Gulf of Mexico coast, and from the Tijuana River estuary (CA) to Puget Sound (WA) along the U.S. Pacific Coast (Long, 2000). Estimates of the spatial extent of toxicity were determined with three individual tests performed on subsamples of the sediments. The three bioassays included tests of reduced amphipod survival in exposures to solid-phase sediments (Swartz et al., 1985; ASTM 1993), diminished microbial bioluminescence in exposures to organic solvent extracts (Schiewe et al., 1985), and decreased fertilization success of sea urchin eggs following exposure of sperm to pore waters extracted from the sediments (Carr and Chapman, 1992). These tests provided nonduplicative, yet complementary, estimates of the severity and extent of toxicity based on tests of three different phases of sediments and three different toxicological end points.

Amphipod survival tests were performed in all surveys, but proved to be the least sensitive test of the three. Toxicity affected the largest areas in the Hudson–Raritan estuary (NY/NJ; 133.3 km²) and Delaware Bay (DE; 145.4 km²), but affected the largest percentages of survey areas (>50%) in Newark Bay (NJ), San Diego Bay (CA), California coastal lagoons, Tijuana River (CA) estuary, and Long Island Sound (NY/CT) bays (Table 6.2). The percentages of areas affected were intermediate in Boston Harbor (MA), Biscayne Bay (FL), and San Pedro Bay (CA). Toxicity was least widespread in most estuaries of the southeastern United States, including Charleston Harbor (SC), Savannah River and St. Simons Sound (GA), Tampa Bay (FL), Pensacola Bay (FL), other bays of the Florida Panhandle, and Sabine Lake (TX/LA). Only one of the 300 samples from Puget Sound was classified as toxic in these tests, and it represented less than 0.1% of that survey area. The low spatial extent of toxicity in southeastern estuaries was corroborated in EMAP studies with the same tests (Hyland et al., 1996). By combining results from tests performed in all NOAA surveys conducted through 1999, an overall average of about 4% of the combined survey area was classified as toxic in this test.

TABLE 6.2

Spatial Extent of Toxicity (expressed as km² and percentages of total survey area) in Amphipod Survival Tests[a] Performed with Solid-Phase Sediments from 28 U.S. Estuaries and Marine Bays

Survey Areas	Year Sampled	No. of Samples	Total Survey Area (km²)	Area Toxic — Toxic Area (km²)	Area Toxic — Estimated Toxic Area
Newark Bay (NJ)	93	57	13.0	10.8	85.0%
San Diego Bay (CA)	93	117	40.2	26.3	65.8%
California coastal lagoons	94	30	5.0	2.9	57.9%
Tijuana River (CA)	93	6	0.3	0.2	56.2%
Long Island Sound bays (NY/CT)	91	60	71.9	36.3	50.5%
Hudson–Raritan Estuary (NY/NJ)	91	117	350.0	133.3	38.1%
San Pedro Bay (CA)	92	105	53.8	7.8	14.5%
Biscayne Bay (FL)	95/96	226	484.2	62.3	12.9%
Boston Harbor (MA)	93	55	56.1	5.7	10.0%
Delaware Bay (DE)	97	73	2346.8	145.4	6.2%
Savannah River (GA)	94	60	13.1	0.2	1.2%
St. Simons Sound (GA)	94	20	24.6	0.1	0.4%
Tampa Bay (FL)	92/93	165	550.0	0.5	0.1%
Central Puget Sound (WA)	98	100	737.4	1.0	0.1%
Pensacola Bay (FL)	93	40	273.0	0.04	0.0%
Northern Chesapeake Bay (MD)	98	53	2265.0	0.0	0.0%
Southern Puget Sound (WA)	99	100	858	0	0.0%
Galveston Bay (TX)	96	75	1351.1	0.0	0.0%
Northern Puget Sound (WA)	97	100	773.9	0.0	0.0%
Choctawhatchee Bay (FL)	94	37	254.5	0.0	0.0%
Sabine Lake (TX/LA)	95	66	245.9	0.0	0.0%
Apalachicola Bay (FL)	94	9	187.6	0.0	0.0%
St. Andrew Bay (FL)	93	31	127.2	0.0	0.0%
Charleston Harbor (SC)	93	63	41.1	0.0	0.0%
Winyah Bay (SC)	93	9	7.3	0.0	0.0%
Mission Bay (CA)	93	11	6.1	0.0	0.0%
Leadenwah Creek (SC)	93	9	1.7	0.0	0.0%
San Diego River (CA)	93	2	0.5	0.0	0.0%
National estuarine average (1999)		1796	11139	432.8	3.9%

[a] Test animal was *Ampelisca abdita* except in California where *Rhepoxynius abronius* was used.

EMAP studies covered much broader estuarine regions than the NOAA surveys and, significantly, did not focus on urbanized bays as did the NOAA surveys. Data from amphipod tests performed with the same protocols were reported for Virginian, Louisianian, Carolinian, and offshore Californian province sediments in the EMAP studies (Summers and Macauley, 1993; Strobel et al., 1995; Hyland et al., 1996; Bay, 1996; respectively). The percentages of these study areas in which amphipod survival was significantly reduced were 10, 8.4, 2, and 0%, respectively, and with data from all four provinces combined represented about 7.3% of the combined areas.

Because they are more sensitive than the amphipod survival tests, results of sea urchin fertilization tests in undiluted pore waters and microbial bioluminescence tests in organic solvent extracts identified larger percentages of the study areas as inducing significant responses in the NOAA surveys (Long, 2000). With data combined from 24 survey areas ($n = 1468$, total area = 9840 km^2), responses were significant in samples tested for sea urchin fertilization success that represented about 25% of combined study areas. Data were generated in microbial bioluminescence tests in 20 surveys ($n = 1378$, total area = 10162 km^2) and the responses were significant in samples that represented about 30% of the combined survey areas.

Classification of Sediment Quality with Benthic Indices

The third component of the sediment quality triad consists of measures of the relative quality of the resident infaunal benthos. Because the benthic community represents an important component of the ecosystem that warrants protection, it can be argued that this is the most important element of the triad. However, because the composition of the benthos can be affected significantly by a complex variety of interacting natural factors, attribution of degradation to toxic chemicals in the sediments can be difficult and controversial.

To aid in the simplification and interpretation of benthic community data, benthic ecologists have developed a battery of numerical indices with which to characterize the abundance, biomass, and diversity of the assemblages of organisms found in a sample. Some of these indices are described and summarized by U.S. EPA (2000) for estuaries and coastal marine waters. They include total biomass of the organisms retained on a standard-size sieve and total numbers of identifiable organisms (i.e., total abundance). Often, indices are calculated of species diversity (e.g., the Shannon–Wiener diversity index, H′), species evenness (e.g., Pielou's evenness index, J), and dominance (e.g., Swartz's dominance index). Often benthic ecologists look for the presence of pollution-tolerant organisms such as capitellid worms and/or the absence of pollution-sensitive organisms such as infaunal amphipods (Long et al., 2002). A review of matching benthic community data and laboratory amphipod survival data showed that the amphipods and other sensitive crustaceans often were missing in the most toxic estuarine samples (Long et al., 2001).

In recent years, a variety of multiparameter (i.e., multivariate) benthic infaunal indices have been developed and evaluated for application in marine bays and estuaries of the United States and Canada (Engle et al., 1994; Weisberg et al., 1997; van Dolah et al., 1999; Llansó et al., 2002a,b; Janicki Environmental, 2003). These indices usually incorporate measures of species diversity and the presence/absence or relative abundance of selected species that are important local determinants of infaunal benthic composition. They often follow the principles and models of incremental changes in the composition of the benthos with increasing stress as summarized by Pearson and Rosenberg (1978). Some species are more tolerant than others of stresses, such as lower dissolved oxygen content and higher toxicant chemical concentrations. As the degree of stress increases, some species are able to proliferate in abundance while others decrease in abundance or are excluded. Ultimately, as the degree of stress increases to extreme levels, the sediments can become azoic.

The relationships among anthropogenic stresses, physical factors such as water depth, geochemical variables such as sediment texture, and benthic indices can be complex and interwoven (Bergen et al., 2001; Smith et al., 2001). The composition of the benthos may be a function of numerous biological and abiotic factors that are inseparable (Long et al., in prep.). However, there is a growing body of evidence that suggests that the composition of the benthos in estuaries and marine bays is a more sensitive biological indicator of degraded sediment quality than survival of test animals in laboratory bioassays

(Hyland et al., 1999, 2003). In Puget Sound, significant changes to the benthos were apparent, including azoic conditions, in samples that were not acutely toxic in amphipod survival tests (Long et al., in prep.).

Benthic community analyses have been conducted for many decades and there are numerous manuals that describe analytical methods (e.g., PSEP, 1987; U.S. EPA, 2000). However, there are no federal standards or criteria for the indices of benthic community composition with which to classify sediments as degraded. Benthic studies are not included in federal dredged material testing manuals and, therefore, are rarely conducted as a part of dredged material evaluations. The federal agencies that conduct estuarine and freshwater status and trends monitoring (U.S. EPA, USGS, and NOAA) often, but not always, include a benthic component in their programs. Benthic community analyses are important elements of many regional monitoring programs, such as those in Puget Sound, San Francisco Bay, Southern California Bight, Tampa Bay, South Carolina estuaries, New York/New Jersey Harbor, and the Great Lakes. Much of the pioneering work on benthic community composition was conducted in Scandinavia, the United Kingdom, Western Europe, and Canada.

Spatial Extent of Degraded Benthic Communities

The most geographically broad database with which to evaluate degraded estuarine benthic conditions is that developed in the EMAP–Estuaries studies (U.S. EPA, 2001). Based on analyses of samples collected along the lengths of the Atlantic and Gulf of Mexico coastlines in the 1990s, it was estimated that 22% of the combined area sampled supported benthos in a poor condition. That is, the communities were less diverse or abundant than expected, were dominated by pollution-tolerant species, and/or had relatively few pollution-sensitive species. Samples with the benthos in fair condition represented another 22% of the total area, and 56% of the area had benthos in good condition.

Benthic conditions were reported for the EMAP studies of the Virginian, Lousianian, and Carolinian provinces and the Regional EMAP study of the New York/New Jersey Harbor (Summers et al., 1993; Strobel et al., 1995; Hyland et al., 1996; Adams et al., 1996; respectively). Estimates were reported for low infaunal abundance, low species richness, and low benthic index scores. Among the three EMAP provinces, the spatial extent of degraded benthic communities ranged from 7 to 22% for measures of low infaunal abundance, 4 to 10% for low species richness, and 20 to 31% for low benthic index scores. In contrast, approximately 53% of the New York/New Jersey Harbor area had impacted benthos quality as indicated by the EMAP benthic index.

In the Carolinian province, the areas with impacted benthos (10% for low species richness and 22% for low abundance) corresponded fairly well with those that were contaminated (16% spatial coverage as defined by Hyland et al., 1996). Areas of overlap in which significant contamination occurred and either low species richness or low abundance also occurred represented 7 and 12% of the total area. Correspondence among measures of sediment quality also was relatively high in the New York/New Jersey Harbor REMAP study (Adams et al., 1996). Within the area (53% of total) in which benthic assemblages were affected, 74% had chemical concentrations that exceeded at least one ERM value and 89% had either high chemical concentrations or evidence of sediment toxicity.

Analyses of matching triad data from Tampa Bay often indicate similar and low percentages of samples (i.e., <10%) were acutely toxic and had impaired benthos when sediment contamination was lowest (MacDonald et al., 2004). In samples with intermediate levels of contamination, the incidence of benthic impairment often was higher than the incidence of acute toxicity, but this relationship was reversed in the most highly contaminated samples. In a study of Biscayne Bay and the adjoining lower Miami River, the degree of benthic impairment in contaminated samples was much higher than the degree of mortality in amphipod bioassays (Long et al., 2002). In Puget Sound, between-site differences in benthic indices spanned orders of magnitude as chemical contamination increased, but only 1 of 300 samples was classified as toxic to amphipods (Long et al., in prep.). A sediment quality triad index was applied to the Puget Sound data and indicated that 138 of the 300 samples were high quality. These samples represented about 68% of the bottom of Puget Sound (Table 6.3). There were 125 samples classified as intermediate in quality (representing 31% of the area) and the remaining 37 samples (representing 1% of the area) were degraded. As was observed in Biscayne Bay, differences in benthic indices

TABLE 6.3

Estimated Spatial Extent of Relative Sediment Quality in Puget Sound (Washington State, USA) Based on the Sediment Quality Triad of Data

Sediment Quality Category	No. of Stations	% of Stations	Area (km²)	% of Area
High	138	46.0	1616.1	68.4
Intermediate	125	41.7	724.1	30.7
Degraded	37	12.3	23.1	1.0

Note: Data are expressed as numbers and percentages of 300 total stations in each category and area (km²) and percentages of total area (2363 km²) that these stations represented.

Source: From Long, E. R. et al., Washington State Department of Ecology Pub. No. 03-03-048, Olympia, WA, 2003.

(i.e., dominance, diversity, and abundance of arthropods and echinoderms) between samples spanned orders of magnitude whereas amphipod survival was very similar among the same samples.

Discussion and Conclusions

A wide variety of assessment tools have been developed with which to describe the quality of sediments. These include the reference metals-to-aluminum ratios, sets of effects-based SQGs, batteries of acute and sublethal laboratory toxicity tests, and an increasing variety of indices of benthic condition. The level of interest and the amount of effort expended in sediment quality assessments continue to increase as more programs focus their attention on sediments. Some of these programs involve determinations of spatial status and temporal trends in estuarine sediment quality, using combinations of the elements of the sediment quality triad.

One of the major findings of the NRC (1989) review of contaminated sediments was that there were very little data available then that could be used to quantify the spatial extent of the problem. Data that are currently available from large-scale estuarine studies suggest that biologically significant chemical contamination and acute toxicity of sediments are scattered widely throughout U.S. coastal waters. Highest degrees of chemical contamination and toxicity generally occur in urbanized and industrialized regions of bays and estuaries (i.e., in definable waterways, bayous, and harbors).

Chemical concentrations that exceeded mid-range SQGs and, therefore, were sufficiently high to be of toxicological concern, occurred in about 26 to 27% of samples in the United States. The incidence of chemical contamination in some regions ranged from about 1 to 80%, indicative of the spatial heterogeneity in sediment quality. The area affected in acute amphipod survival tests of toxicity averaged about 4% of the total surveyed estuarine areas nationwide. As with the chemical data, the percentage of areas affected in these acute toxicity tests ranged from 0 to 85% among regions in the United States. Areas affected in two sublethal toxicity tests were higher on a national scale, about 25 to 30%. The benthos was in relatively poor condition in 22% of estuarine areas tested, corresponding with the estimates for sublethal toxicity tests.

Biologically diverse and abundant benthic communities are necessary to support coastal fisheries and other equally important resources. Protection of these resources is dependent in part on habitats, including sediments, of high quality. U.S. EPA (1997) concluded in its national survey of sediment quality, "sediment contamination is widespread and is an important national concern." If sufficiently severe, chemical contamination of sediments can have demonstrable effects beyond those observed either in laboratory tests of toxicity or in analyses of benthic community structure.

Measures of chemical exposures, physiological responses, and adverse histopathological effects have been reported in many species of demersal fish and bivalve mollusks in areas with high chemical concentrations in sediments (Becker et al., 1987; Malins et al., 1988; Spies et al., 1990; Johnson et al., 1992; Salazar and Salazar, 1995). Despite the high mobility of fishes, statistical relationships between the prevalence of lesions and other disorders in fish and the concentrations of contaminants in sediments can be highly significant. In data collected from many locations along the Pacific Coast, the concordance between concentrations of polynuclear aromatic hydrocarbons in sediments and the prevalence of liver lesions in English sole were sufficiently strong to warrant development of a predictive model (Horness et al., 1998). In some areas, the effects of chemical toxicants on reproductive success of fishes have been reported (Spies et al., 1990; Johnson et al., 1992; Long, 1996). Despite the difficulties in establishing the specific causes of decreases in fishery populations where multiple stressors occur (Hoss and Engel, 1996), there is a considerable body of evidence from numerous studies with which to convincingly link the effects of chemical pollution to losses of important marine and estuarine resources (Sindermann, 1997).

In conclusion, estuarine sediments represent a relatively stable and ecologically important medium in which to characterize the concentrations and biological effects of chemical contamination. Using multiple lines of evidence and a tiered approach provides a solid, defensible approach to evaluating sediment quality. Information gathered from chemical analyses, or toxicity tests, or benthic community composition analyses, or combinations of the components of the sediment quality triad can be used to effectively describe and compare relatively quality of estuaries.

References

Adams, D. A., J. S. O'Connor, and S. B. Weisberg. 1996. Sediment Quality of the NY/NJ Harbor System. Draft final report. U.S. EPA Region 2, Edison, NJ.

ASTM. 1993. Standard guide for conducting solid phase, 10-day, static sediment toxicity tests with marine and estuarine infaunal amphipods. ASTM E 1367-92. American Society for Testing and Materials, Philadelphia, PA.

Batley, G. E. and W. A. Maher. 2001. The development and application of ANZECC and ARMCANZ sediment quality guidelines. *Australasian Journal of Ecotoxicology* 7:81–92.

Bay, S. M. 1996. Sediment toxicity on the mainland shelf of the Southern California Bight in 1994. In SCCWRP Annual Report 1994–1995. Southern California Coastal Water Research Project, Westminster, CA, pp. 128–137.

Becker, D. S., T. C. Ginn, M. L. Landolt, and D. B. Powell. 1987. Hepatic lesions in English sole (*Parophrys vetulus*) from Commencement Bay, Washington (USA). *Marine Environmental Research* 23:153–173.

Bergen, M., S. B. Weisberg, R. W. Smith, D. B. Cadien, A. Dalkey, D. E. Montagen, J. K. Stull, R. G. Velarde, and J. A. Ranasinghe. 2001. Relationship between depth, sediment, latitude, and the structure of benthic infaunal assemblages on the mainland shelf of southern California. *Marine Biology* 138:637–647.

Canfield, T. J., N. E. Kemble, W. G. Grumbaugh, F. J. Dwyer, C. G. Ingersoll, and J. F. Fairchild. 1994. Use of benthic invertebrate community structure and the sediment quality triad of evaluate metal-contaminated sediment in the Upper Clark Fork River, Montana. *Environmental Toxicology and Chemistry* 13:1999–2012.

Carr, R. S. and D. C. Chapman. 1992. Comparison of solid-phase and pore water approaches for assessing the quality of marine and estuarine sediments. *Chemistry and Ecology* 7:19–30.

Carvalho, A., S. J. Schropp, G. M. Sloane, T. P. Biernacki, and T. L. Seal. 2002. Development of an Interpretive Tool for Assessment of Metal Enrichment in Florida Freshwater Sediment. Prepared for the Florida Department of Environmental Protection, Tallahassee, FL. 61 pp.

Chapman, P.M. 1988. Marine sediment toxicity tests. In *Chemical and Biological Characterization of Sludges, Sediments, Dredge Spoils, and Drilling Muds,* STP 976, J. J. Lichtenberg, F. A. Winter, C. I. Weber, and L. Fradkin (eds.). American Society for Testing and Materials, Philadelphia, PA, pp. 391–402.

Chapman, P. M. 1995. Do sediment toxicity tests require field validation? *Environmental Toxicology and Chemistry* 14:1451–1453.

Chapman, P. M., R. N. Dexter, and E. R. Long. 1987. Synoptic measures of sediment contamination, toxicity and infaunal community composition (the Sediment Quality Triad) in San Francisco Bay. *Marine Ecology Progress Series* 37:75–96.

Chapman, P. M. et al. 1998. General guidelines for using the Sediment Quality Triad. *Marine Pollution Bulletin* 34(6):368–372.

Daskalakis, K. D. and T. P. O'Connor. 1994. Inventory of chemical concentrations in coastal and estuarine sediments. NOAA Technical Memorandum NOS ORCA 76. National Oceanic and Atmospheric Administration, Silver Spring, MD.

Daskalakis, K. D. and T. P. O'Connor. 1995. Normalization and Elemental Sediment Contamination in the Coastal United States. *Environmental Science and Technology* 29(2):470–477.

Di Toro, D. M., C. S. Zarba, D. J. Hansen, W. J. Berry, R. C. Swartz, C. E. Cowan, S. P. Pavlou, H. E. Allen, N. A. Thomas, and P. R. Paquin. 1991. Technical basis for establishing sediment quality criteria for nonionic organic chemicals using equilibrium partitioning. *Environmental Toxicology and Chemistry* 10:1–43.

Engle, V. D., J. K. Summers, and G. R. Gaston. 1994. A benthic index of environmental condition of Gulf of Mexico estuaries. *Estuaries* 17:372–384.

Hill, I. R., P. Matthiessen, and F. Heimbach. 1993. Guidance document on sediment toxicity tests and bioassays for freshwater and marine environments, presented at Workshop on Sediment Toxicity Assessment, Slot Moermond Congrescentrum, Renesse, the Netherlands. Society of Environmental Toxicology and Chemistry–Europe.

Horness, B. H., D. P. Lomax, L. L. Johnson, M. S. Myers, S. M. Pierce, and T. K. Collier. 1998. Sediment quality thresholds: estimates from hockey stick regression of liver lesion prevalence in English sole (*Pleuronectes vetulus*). *Environmental Toxicology and Chemistry* 17:872–882.

Hoss, D. E. and D. W. Engel. 1996. Sustainable development in the Southeastern Coastal Zone: environmental impacts on fisheries. In *Sustainable Development in the Southeastern Coastal Zone*, F. J. Vernberg, W. B. Vernberg, and T. Siewicki (eds.). Belle W. Baruch Library in Marine Science 20. University of South Carolina Press, Columbia, SC, pp. 171–186.

Hyland, J., T. Herrlinger, T. Snoots, A. Ringwood, B. VanDolah, C. Hackney, G. Nelson, J. Rosen, and S. Kokkinakis. 1996. Environmental quality of estuaries of the Carolinian Province: 1994. NOAA Technical Memorandum 97. National Oceanic and Atmospheric Administration, Charleston, SC.

Hyland, J. L., R. F. Van Dolah, and T. R. Snoots. 1999. Predicting stress in benthic communities of southeastern U.S. estuaries in relation to chemical contamination of sediments. *Environmental Toxicology and Chemistry* 18(11):2557–2564.

Hyland, J. L., W. L. Balthis, C. T. Hackney, and M. Posey. 2000. Sediment quality of North Carolina estuaries: an integrative assessment of sediment contamination, toxicity, and condition of benthic fauna. *Journal of Aquatic Ecosystem Stress and Recovery* 8:107–124.

Hyland, J. L., W. L. Balthis, V. D. Engle, E. R. Long, J. F. Paul, J. K. Summers, and R. F. Van Dolah. 2003. Incidence of stress in benthic communities along the U.S. Atlantic and Gulf of Mexico coasts within different ranges of sediment contamination from chemical mixtures. *Environmental Monitoring and Assessment* 81(1–3), 149–161.

Ingersoll, C. G., P. S. Haverland, E. L. Brunson, T. J. Canfield, F. J. Dwyer, C. E. Henke, N. E. Kemble, D. R. Mount, and R. G. Fox. 1996. Calculation and evaluation of sediment effect concentrations for the amphipod *Hyalella azteca* and the midge *Chironomus riparius*. *Journal of Great Lakes Research* 22:602–623.

Ingersoll, C. G., D. D. MacDonald, N. Wang, J. L. Crane, L. J. Field, P. S. Haverland, N. E. Kemble, R. A. Lindskoog, C. Severn, and D. E. Smorong. 2001. Predictions of sediment toxicity using consensus-based freshwater sediment quality guidelines. *Archives of Environmental Contamination and Toxicology* 41:8–21.

Janicki Environmental. 2003. Development of a benthic quality index for use as a management tool in establishing sediment quality targets for the Tampa Bay Estuary. Prepared for Tampa Bay Estuary Program, St. Petersburg, FL.

Johnson, L. L., J. E. Stein, T. K. Collier, E. Casillas, B. McCain, and U. Varanasi. 1992. Bioindicators of contaminant exposure, liver pathology, and reproductive development in pre-spawning female winter flounder (*Pleuronectes americanus*) from urban and nonurban estuaries on the Northeast Atlantic Coast. NOAA Technical Memorandum NMFS-NWFSC-1. National Marine Fisheries Service, Seattle, WA.

Lamberson, J. O., T. H. DeWitt, and R. C. Swartz. 1992. Assessment of sediment toxicity to marine benthos. In *Sediment Toxicity Assessment*, G. A. Burton, Jr. (ed.). Lewis Publishers, Boca Raton, FL, pp. 183–240.

Llansó, R. J., L. C. Scott, J. L. Hyland, D. M. Dauer, D. E. Russell, and F. W. Kutz. 2002a. An estuarine benthic index of biotic integrity for the mid-Atlantic region of the United States. I. Classification of assemblages and habitat definition. *Estuaries* 25:1219–1230.

Llansó, R. J., L. C. Scott, J. L. Hyland, D. M. Dauer, D. E. Russell, and F. W. Kutz. 2002b. An estuarine benthic index of biotic integrity for the mid-Atlantic region of the United States. II. Index development. *Estuaries* 25:1231–1242.

Long, E. R. 1996. The use of biological measures in assessments of toxicants in the coastal zone. In *Sustainable Development in the Southeastern Coastal Zone,* F. J. Vernberg, W. B. Vernberg, and T. Siewicki (eds.). Belle W. Baruch Library in Marine Science 20. University of South Carolina Press, Columbia, SC, pp. 187–219.

Long, E. R. 2000. Spatial extent of sediment toxicity in U.S. estuaries and marine bays. *Environmental Monitoring and Assessment* 64:391–407.

Long, E. R. 2002. Toxicity tests of marine and estuarine sediment quality: applications in regional assessments and uses of the data. In *Handbook on Sediment Quality,* R. C. Whittemore (ed.). Water Environment Federation. Alexandria, VA, pp. 259–316.

Long, E. R. and P. M. Chapman. 1985. A sediment quality triad: measures of sediment contamination, toxicity and infaunal community composition in Puget Sound. *Marine Pollution Bulletin* 16:405–415.

Long, E. R., D. D. MacDonald, S. L. Smith, and F. D. Calder. 1995. Incidence of adverse biological effects within ranges of chemical concentrations in marine and estuarine sediments. *Environmental Management* 19:81–97.

Long, E. R., A. Robertson, D. A. Wolfe, J. Hameedi, and G. M. Sloane. 1996. Estimates of the spatial extent of sediment toxicity in major U.S. estuaries. *Environmental Science and Technology* 30:3585–3592.

Long, E. R., L. J. Field, and D. D. MacDonald. 1998. Predicting toxicity in marine sediments with numerical sediment quality guidelines. *Environmental Toxicology and Chemistry* 17:714–727.

Long, E. R., C. B. Hong, and C. G. Severn. 2001. Relationships between acute sediment toxicity in laboratory tests and the abundance and diversity of benthic infauna in marine sediments: a review. *Environmental Toxicology and Chemistry* 20(1):46–60.

Long, E. R., M. J. Hameedi, G. M. Sloane, and L. Read. 2002. Chemical contamination, toxicity, and benthic community indices in sediments of the lower Miami River and adjoining portions of Biscayne Bay, Florida. *Estuaries* 25:622–637.

Long, E. R., M. Dutch, S. Aasen, and K. Welch. 2003. Chemical contamination, acute toxicity in laboratory tests, and benthic impacts in sediments of Puget Sound. A summary of results of the joint 1997–1999 Ecology/NOAA survery. Washington State Department of Ecology Pub. No. 03-03-048. Department to Ecology, Olympia, WA.

Long, E. R., M. Dutch, S. Aasen, K. Welch, M. J. Hameedi, T. P. Cardoso, and L. B. Read. In prep. Relationships among variables of the sediment quality triad in Puget Sound (Washington, USA).

Loring, D. H. 1991. Normalization of heavy-metal data from estuarine and coastal sediments. *ICES Journal of Marine Science* 48:101–115.

MacDonald, D. D. 1994. Approach to the Assessment of Sediment Quality in Florida Coastal Waters. Vol. 1: Development and Evaluation of Sediment Quality Guidelines. Prepared for the Florida Department of Environmental Protection, Tallahassee, FL, 126 pp.

MacDonald, D. D., R. S. Carr, F. D. Calder, E. R. Long, and C. G. Ingersoll. 1996. Development and evaluation of sediment quality guidelines for Florida coastal waters. *Ecotoxicology* 5:253–278.

MacDonald, D. D., C. G. Ingersoll, D. E. Smorong, R. A. Lindskoog, G. Sloane, and T. Biernacki. 2003. Development and Evaluation of Numerical Sediment Quality Assessment Guidelines for Florida Inland Waters. Technical Report prepared for Florida Department of Environmental Protection, Tallahassee, FL, 150 pp.

MacDonald, D. D., R. S. Carr, D. Eckenrod, H. Greening, S. Grabe, C. G. Ingersoll, S. Janicki, T. Jankick, R. A. Lindskoog, E. R. Long, R. Pribble, G. Sloane, and D. E. Smorong. 2004. Development, evaluation, and application of sediment quality targets for assessing and managing contaminated sediments in Tampa Bay, Florida. *Archives of Environmental Contamination and Toxicology* 46:147–161.

Malins, D. C., B. B. McCain, J. T. Landahl, M. S. Myers, M. M. Krahn, D. W. Brown, S.-L. Chan, and W. T. Roubal. 1988. Neoplastic and other diseases in fish in relation to toxic chemicals: an overview. *Aquatic Toxicology* 11:43–67.

NRC (National Research Council). 1989. Contaminated Marine Sediments — Assessment and Remediation. National Academy Press. Washington, D.C.

Paul, J. F., K. J. Scott, A. F. Holland, S. B. Weisberg, J. K. Summers, and A. Robertson. 1992. The estuarine component of the US EPA's environmental monitoring and assessment program. *Chemistry and Ecology* 7:93–116.

Pearson, T. H. and R. Rosenberg. 1978. Macrobenthic succession in relation to organic enrichment and pollution of the marine environment. *Oceanography and Marine Biology Annual Review* 16:229–311.

Persuad, D., R. Jaagaumagi, and A. Hayton. 1992. Guidelines for the Protection and Management of Aquatic Sediment Quality in Ontario. Ontario Ministry of the Environment, Water Resources Branch, Toronto, ON, Canada.

Phillips, B. M., J. W. Hunt, B. S. Anderson, H. M. Puckett, R. Fairey, C. J. Wilson, and R. Tjeerdema. 2001. Statistical significance of sediment toxicity test results: threshold values derived by the detectable significance approach. *Environmental Toxicology and Chemistry* 20(2):371–373.

Power, E. A. and P. M. Chapman. 1992. Assessing sediment quality. In *Sediment Toxicity Assessment,* G. A. Burton, Jr. (ed.). Lewis Publishers, Boca Raton, FL, pp. 1–18.

Puget Sound Estuary Program (PSEP). 1987. Recommended Protocols for Sampling and Analyzing Subtidal Benthic Macroinvertebrate Assemblages in Puget Sound. Prepared for U.S. EPA Region 10, Office of Puget Sound, Seattle, WA and Puget Sound Water Quality Authority, Olympia, WA by Tetra Tech, Inc., Bellevue, WA, 32 pp.

Salazar, M. H. and S. M. Salazar. 1995. *In situ* bioassays using transplanted mussels: I. Estimating chemical exposure and bioeffects with bioaccumulation and growth. In *Third Symposium on Environmental Toxicology and Risk Assessment,* J. S. Hughes, G. R. Biddinger, and E. Mones (eds.). ASTM STP 218. American Society for Testing and Materials, Philadelphia, PA, pp. 216–241.

Schiewe, M. H., E. G. Hawk, D. I. Actor, and M. M. Krahn. 1985. Use of a bacterial bioluminescence assay to assess toxicity of contaminated marine sediments. *Canadian Journal of Fisheries and Aquatic Science* 42:1244–1248.

Schropp, S. J. and H. L. Windom. 1988. A Guide to the Interpretation of Metal Concentrations in Estuarine Sediments, Tallahassee, FL. Prepared for Florida Department of Environmental Regulation, 53 pp.

Sindermann, C. J. 1997. The search for cause and effect relationships in marine pollution studies. *Marine Pollution Bulletin* 34:218–221.

Smith, R. W., M. Bergen, S. B. Weisberg, D. Cadien, A. Dalkey, D. Montagne, J. K. Stull, and R. G. Velarde. 2001. Benthic response index for assessing infaunal communities on the Southern California mainland shelf. *Ecological Applications* 11(4):1073–1087.

Spies, R. B., J. J. Stegeman, D. W. Rice, Jr., B. Woodin, P. Thomas, J. E. Hose, J. N. Cross, and M. Prieto. 1990. Sublethal responses of *Platichthys stellatus* to organic contamination in San Francisco Bay with emphasis on reproduction. In *Biomarkers of Environmental Contamination,* J.F. McCarthy and L.R. Shugart (eds.). Lewis Publishers, Boca Raton, FL, pp. 87–122.

Strobel, C. J., H. W. Buffum, S. J. Benyi, E. A. Petrocelli, D. R. Reifsteck, and D. J. Keith. 1995. Statistical Summary. EMAP–Estuaries. Virginian Province — 1990 to 1993. U.S. EPA/620/R-94/-26. U.S. Environmental Protection Agency, Narragansett, RI.

Stronkhorst, J., C. A. Schipper, J. Honkoop, and K. van Essen. 2001. Disposal of dredged material in Dutch coastal waters. A new effect-oriented assessment framework. RIKZ/2001.030. Directorate-General of Public Works and Water Management, National Institute for Coastal and Marine Management, The Hague, the Netherlands, 41 pp.

Summers, J. K. and J. M. Macauley. 1993. Statistical summary: EMAP–Estuaries Louisianian Province — 1991. EPA/600/R-93-001. U.S. EPA, Office of Research and Development, Washington, D.C.

Swartz, R. 1989. Marine sediment toxicity tests. In Contaminated Marine Sediments — Assessment and Remediation. National Research Council, National Academy Press, Washington, D.C., pp. 115–129.

Swartz, R. C., W. A. DeBen, J. K. P. Jones, J. O. Lamberson, and F. A. Cole. 1985. Phoxocephalid amphipod bioassay for marine sediment toxicity. In *Aquatic Toxicology and Hazard Assessment: Seventh Symposium,* R. D. Cardwell, R. Purdy, and R. C. Bahner (eds.). ASTM STP 854. American Society for Testing and Materials, Philadelphia, PA, pp. 284–307.

U.S. EPA. 1992. Sediment Classification Methods Compendium. U.S. Environmental Protection Agency, Washington, D.C.

U.S. EPA. 1997. The Incidence and Severity of Sediment Contamination in Surface Waters of the United States. Vol. 1. National Sediment Quality Survey. EPA 823-R-97-006. U.S. Environmental Protection Agency, Washington, D.C.

U.S. EPA. 2000. Estuarine and Coastal Marine Waters: Bioassessment and Biocriteria Technical Guidance. EPA-822-B-00-024. U.S. Environmental Protection Agency, Office of Water, Washington, D.C.

U.S. EPA. 2001. National Coastal Condition Report. EPA-620/R-01/005. U.S. Environmental Protection Agency, Office of Research and Development, Washington, D.C.

U.S. EPA and U.S. ACOE. 1991. Evaluation of Dredged Material Proposed for Ocean Disposal. Testing Manual. EPA-503/8/91/001. U.S. Environmental Protection Agency and U.S. Army Corps of Engineers, Washington, D.C.

Van Dolah, R. F, J. L Hyland, A. F Holland, J. S Rosen, and T. R Snoots. 1999. A benthic index of biological integrity for assessing habitat quality in estuaries of the southeastern United States. *Marine Environmental Research* 48:1–15.

WDOE. 1995. Sediment management standards. Publ. .96-252. Washington Department of Ecology, Olympia, WA.

Weisberg, S. B., J. A. Ranasinghe, D. M. Dauer, L. C. Schaffner, R. J. Diaz, and J. B. Frithsen. 1997. An estuarine benthic index of biotic integrity (B-IBI) for Chesapeake Bay. *Estuaries* 29(1):149–158.

Wolfe, D. A., E. R. Long, and A. Robertson. 1993. The NS&T Bioeffects Surveys: design strategies and preliminary results. In *Coastal Zone '93. Proceedings of the 8th Symposium on Coastal and Ocean Management,* Vol. 1, O. T. Magoon, W. S. Wilson, H. Converse, and L. T. Tobin (eds.). American Society of Civil Engineers, New York, pp. 298–312.

7

Sediment Habitat Assessment for Targeted Near-Coastal Areas

Michael A. Lewis

CONTENTS

Introduction

The Gulf of Mexico is an economic and natural resource that receives anthropogenic contamination from a variety of sources (Truax and Daniel, 1991). Approximately 3700 wastewaters, the most for any U.S. coastal area, are discharged into near-coastal waters. In addition, four of the five states that led the nation in surface water discharges of toxic chemicals are located in the Gulf region (U.S. EPA, 1994a) and approximately 4.5×10^6 kg of registered pesticides were applied to estuarine drainage areas in 1987, the most for any coastal region (Pait et al., 1992). As a result of these anthropogenic inputs, approximately

62% of Gulf of Mexico estuaries are negatively affected by contaminants (U.S. EPA, 2001a), and 233 waters are listed as impaired (U.S. EPA, 2001a). More specifically, about 2850 km² of Florida's estuaries are partially impaired and only 16 of 10460 km² attain all designated uses (FDEP, 2002). Despite the environmental degradation, the relative contributions of specific contaminant sources to the impairments are usually unknown.

The fate of most anthropogenic contaminants is in the sediments. Sediment contamination is considered a major problem in U.S. coastal areas and has emerged as an important environmental issue (U.S. EPA, 1996, 1997a). It is understood that management of contaminated sediments is necessary to maintain sustainable ecosystems. The determination of the magnitude, extent, sources, and causes of sediment contamination are important to the development of national and regional sediment quality criteria and to the successful implementation of Section 303(d) of the Clean Water Act (U.S. EPA, 2000) as related to the determination of total maximum daily loads in impaired waters. Surveys to determine the condition of sediments in the Gulf of Mexico region have occurred at different geographical scales using a variety of assessment techniques (FDEP, 1994a; Macauley et al., 1995, 1999; Carr et al., 1996; Long et al., 1997; Johnson and Long, 1998, U.S. EPA, 2001a). The primary focus of these surveys was the characterization of the extent and magnitude of contamination and not the identification of sources and causes of impacts.

To provide additional insight on the quality of sediments in the Gulf of Mexico region, a series of sediment surveys were conducted from 1993 to 2000 for near-coastal areas in Florida affected by point- and non-point-source contamination, areas designated for special environmental protection, and near-coastal habitats at risk, seagrass beds. These surveys were conducted using targeted sampling sites with the intent that some would represent worst-case examples of contamination. The primary objectives of this summary are to provide an overview of the chemical and biological results of these surveys and rank several stressor sources for severity of impact. The information will be useful as a reference database for future stressor source comparisons that will be needed for effective regulatory management of this important habitat and will also add to the continuing effort to validate the relevance of current diagnostic procedures used in the sediment hazard assessment process.

Methods

Study Areas

Sediments were collected during 1993–2000 from 23 locations (Table 7.1) in coastal rivers, bays, and estuaries located along the Florida coastline (Figure 7.1). Multiple nonrandom sampling sites were selected at each location based, in part, on historical use. Many of the 97 sites were targeted to areas of known or suspected contamination resulting from 11 treated wastewater outfalls (municipal, industrial, power generating, pulp mill) and storm water runoff from agriculture (the Everglades–Florida Bay transitional zone), urban development (three bayous), and a coastal golf complex. The sampling sites were located in wastewater outfall mixing zones or immediately below the coastal entry of culverts and drainage streams of runoff. In addition, sediments were collected from 13 seagrass beds dominated by *Thallassia testudinum* Banks ex König (turtle grass) and from estuarine areas of the Suwannee and Withlacoochee Rivers, which are designated Florida Outstanding Waters (FDEP, 2002). Only chemical quality was determined for seagrass-vegetated sediments.

Sediment Collection and Preparation

Sediments were collected to a depth of 13 cm with a Ponar® grab (volume = 2.1 L) or by hand (seagrass beds). Sediments were collected seasonally from the same 20 sites receiving urban storm water runoff (four to six collections — for a 2-year period) and from four sites in estuaries of the Suwannee and Withlacoochee Rivers (three collections — for a 1-year period) to provide a perspective on temporal variability in results. It was assumed that the identical site was sampled and small-scale spatial differences did not occur.

TABLE 7.1

Sampling Locations for Near-Coastal Sediments in Florida

Area	Area
1. Eleven Mile Creek (ww)	4. Suwannee River Estuary[b]
Perdido Bay (ww)	5. Withlacoochee River Estuary[b]
Escambia Bay (ww)	6. Florida Bay (ag)
Escambia River (ww)	Trout Creek (ag)
Pensacola Bay (ww)	Taylor River (ag)
Bayou Texar (us)	Shell Creek (ag)
Bayou Chico (us)	Canal C-111 (ag)
Bayou Grande (us)	7. Ohio Key Channel[a]
Santa Rosa Sound (gc)	8. Little Duck Key Channel[a]
Little Sabine Bay[a]	
2. Choctawhatchee Bay[a]	
3. Bonito Bay[a]	
St. Andrew Bay (ww)	
St. Joseph Bay (ww)	

Note: See Figure 7.1 for locations of the eight general areas. Stressor categories also shown. ww = wastewater outfall, us = urban storm water. gc = golf complex, ag = agricultural runoff.

[a] Seagrass bed.

[b] Florida Outstanding Water (FDEP, 2002).

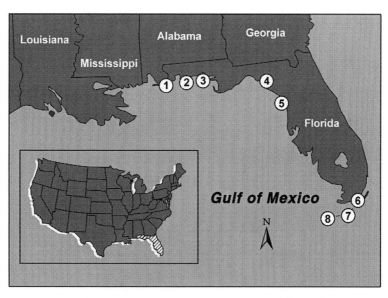

FIGURE 7.1 Sampling locations along Florida's coastline. See Table 7.1 for additional details.

Replicate samples collected at each site were combined, mixed, and fractionated into subsamples for chemical analysis and use in sediment toxicity tests. Two additional samples were collected for benthic community analysis. Pore waters were obtained for toxicity tests by centrifugation at 6000 rpm (5858 *g*) for 20 min. The extracted pore water was either used immediately or stored at 4°C until use within 48 h. Storage and preparation of the whole sediments used in the toxicity tests followed procedures described by the American Society for Testing and Materials (1993a). Prior to storage at 4°C, the samples were passed through a 1-mm stainless steel sieve to remove predators and vegetation. Most samples were stored for less than 2 weeks before use in the bioassays.

Particle Size Distribution and Total Organic Carbon

Particle size distribution and total organic carbon were determined for most sediments using standard methods (APHA et al., 1998). The sediments were classified as muds (>80% silt/clay), sands (<20% silt/clay), or muddy sand (20 to 80% silt/clay) based on criteria reported in U.S. EPA (2001a).

Sediment Chemical Quality

Concentrations of 10 trace metals, 25 chlorinated pesticides, 17 PCBs (polychlorinated biphenyls), and 23 PAHs (polycyclic aromatic hydrocarbons) compounds were determined for most sediment samples. The results are expressed in terms of dry weight. Sediments were analyzed for trace metals following standard U.S. Environmental Protection Agency (U.S. EPA) techniques (1997b) for concentrated nitric acid extraction, cleanup, and analysis. Metal concentrations were analyzed using a Jarrell-Ash Atomcomp Series 800 ICP (Fisher Scientific, Franklin, MA). The method detection limits (MDL) were between 0.2 and 2.1 µg/g dry wt. Mercury concentrations were determined with a Leeman PS200 Automated Mercury Analyzer (Leeman Labs, Hudson, NH) using mercury cold vapor atomic absorption analysis with tin (IV) as the reductant. The MDL was 0.2 ng/g dry wt.

Sediment samples were solvent extracted (acetone/acetonitrile) for 30 min. prior to analysis for chlorinated pesticides, PCBs, and PAHs using U.S. EPA techniques (U.S. EPA, 1997b). The elutriates were analyzed using an HP-5890 Series II Gas Chromatograph (Hewlett Packard, Palo Alto, CA) equipped with an HP-5 fused silica analytical column and mass spectrum temperature detector. The MDL values were 1.0 ng/g dry wt (chlorinated pesticides, PCBs) and 400.0 ng/g dry wt (PAHs).

The inorganic and non-nutrient organic analyses included multilevel calibration and internal standards for peak identification and quantitation. All analytical data met objectives for blanks, spiked samples, and duplicates. A standard reference material (SRM 2704: Buffalo River sediment) was used for calibration, and percent recoveries were between 65 and 80%.

Concentrations of arsenic, cadmium, chromium, copper, lead, nickel, and zinc were compared to expected background concentrations using a geochemical approach of normalization to aluminum (Windom et al., 1989). Sediment contaminant concentrations were compared also to effects-based sediment quality assessment guidelines proposed for Florida coastal areas (FDEP, 1994b,c; MacDonald et al., 1996). Concentrations exceeding the threshold effects level (TEL) but less than the probable effects level (PEL) represent a potential risk to aquatic organisms. Concentrations exceeding PEL guidelines are usually associated with adverse biological effects.

Sediment Biological Quality

Biological quality was determined using acute toxicity tests conducted with whole sediments (four species) and pore waters (one species), whole sediment chronic toxicity tests (two species), and tests for pore water genotoxicity (microbial) and whole sediment phytotoxicity (two vascular rooted plants). Structural measures of abundance and diversity were also determined for the benthic macroinvertebrate community.

Whole Sediment and Pore Water Toxicity: Animal Species

Whole sediment acute toxicity tests were conducted with *Americamysis bahia* (shrimp), *Leptocheirus plumulosus* (camphipod), and *Ampelisca abdita* (amphipod) using standard test methodologies (ASTM, 1993b, 1995; U.S. EPA, 1994b, 1996, 1998). Pore water toxicity was analyzed using embryos of *Palaemonetes pugio* (grass shrimp) following guidelines of Lewis and Foss (2000). The whole sediment and pore water acute toxicity tests ranged in duration from 4 to 10 days, depending on the test species. All tests were conducted in a temperature-controlled environmental chamber under a 16 h light/8 h dark photoperiod. A reference sediment (Perdido Bay, FL) and seawater control (Santa Rosa Sound, FL) were included in toxicity tests conducted with the benthic invertebrates and *P. pugio*, respectively. Salinity, pH, temperature, and dissolved oxygen of the test waters during the toxicity tests were determined using portable instrumentation. Toxicity was considered significant when mortality exceeded 20% (control corrected).

Chronic toxicity tests were conducted with the burrowing-amphipod *L. plumulosus* and the epibenthic mysid *Americamysis bahia*, and 27 whole sediments collected from coastal areas receiving treated wastewater and runoff from agriculture and a golf complex. Effects on reproduction and growth were determined after 7 to 28 days of exposure following the methodologies of ASTM (1993b) and U.S. EPA (2001b).

Whole Sediment Phytotoxicity: Early Seedling Growth

Whole sediments collected from estuarine areas affected by wastewaters and storm water runoff were evaluated for toxicity usually to early seedlings of two vascular plants. Procedures for sediment preparation, seed germination, seedling culture, and exposure followed those of Walsh et al. (1991) and Weber et al. (1995). Seedlings of *Spartina alterniflora* Loisel (cordgrass), and *Scirpus robustus* Pursh (saltmarsh bulrush) were more commonly used and grown either from seeds (*Spartina alterniflora*) obtained from a commercial source (Environmental Concern, St. Michaels, MD) or field-collected from mature plants (*Scirpus robustus*). The seedlings were exposed for 14 to 28 days to whole sediment, after which effects on root and shoot dry weight biomass were determined following standard methods (APHA et al., 1998).

Microbial Genotoxicity

The *in vitro* Mutatox™ assay (Microbics Corporation, Carlsbad, CA) was used to detect mutagenicity of 68 pore waters obtained from sediments affected by the golf complex, wastewater, and urban storm water runoff. The methodology uses a dark mutant strain (M169) of *Vibrio fisheri* that exhibits light production when grown in the presence of sublethal concentrations of genotoxins. Ten 1:2 serial dilutions of each pore water (range = 2.6 to 100%) were analyzed in the presence and absence of an inductive metabolic activation system (S-9 enzymes–rat liver microsome mix). The direct assays of the pore waters were conducted by coincubation of the bacteria, media, and pore water at 27°C for a maximum of 24 h. Exogenous activation of promutagens was initiated at 35°C and was followed by incubation at 27°C. Mutatox media, cofactors, and S-9 media were prepared in accordance with the manufacturer's recommendations (Microbics Corporation, 1993).

Benthic Community Composition

The macrobenthos were removed from replicate sediment samples collected from each site using a 500-μm-mesh sieve. Organisms were preserved in 60% isopropanol and identified to species when possible using regional taxonomic keys and video and high-power light microscopy. Number of taxa, organism abundance, and the Shannon–Wiener diversity index (Shannon and Weaver, 1949) were calculated for each replicate. The Shannon–Wiener diversity index is the most widely used measure of benthic community diversity (Clarke and Warwick, 1994) and it is recommended for evaluations of this type (U.S. EPA, 2002). Sediment quality was considered poor if the diversity index value was 2.0 or less and good if values were 3.3 or greater. These values are based on a frequency distribution of Ponar diversity values reported by Friedman and Hand (1989) for Florida estuaries. Abundances of 200 orgs/m^2 or less in a sediment sample were considered indicative of poor conditions, and values between 200 and 500 orgs/m^2 were considered representative of marginal conditions based on criteria reported by Macauley et al. (1995).

Results

The results discussed below represent a summary of unpublished and published data and are reported in general terms. More detailed information is available for study areas affected by urban storm water (Lewis et al., 2001a; Butts and Lewis, 2002), treated wastewater (Lewis et al., 2000), golf complex runoff (Lewis et al., 2001b), and agriculture runoff (Goodman et al., 1999; Lewis et al., 1999), and for the value of genotoxicity (Lewis et al., 2002) and phytotoxicity (Lewis et al., 2001c) as indicators of sediment contamination.

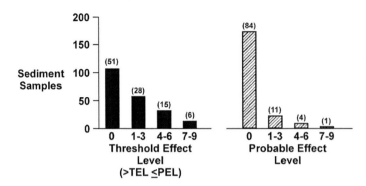

FIGURE 7.2 Exceedances of numerical, effects-based, sediment quality guidelines proposed by MacDonald et al. (1996). Numbers in parentheses are percentages.

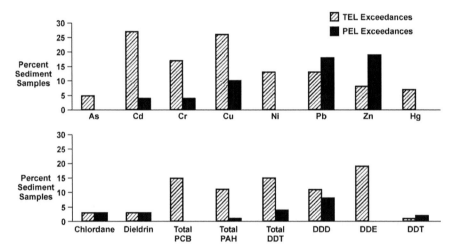

FIGURE 7.3 Exceedances of sediment quality guidelines proposed for Florida near-coastal areas. Values represent percent exceedances based on 319 total analyses (trace metals, above) and 236 total analyses (organic compounds, below).

Particle Size Distribution and Total Organic Carbon (TOC)

The percent particle sizes (mean values) categorized as muds (>80% silt clay), sand (<20% silt and clays), and mud–sand (20 to 80% silt clays) were as follows: 36, 25, and 39 (urban runoff), 30, 40, and 30 (wastewaters), 56, 0, and 44 (agriculture runoff), 0, 58, and 42 (Suwannee and Withalacoochee River estuaries), and 0, 100, and 0 (seagrass-vegetated sediments). TOC values ranged from 0.09 to 14.3% (mean = 0.78, standard deviation = 1.3). Approximately 85% of the TOC concentrations were 1.0% or less; concentrations exceeded 2.0% for 7% of the sediments.

Chemical Quality

Effects-Based Exceedances

About 51% of the sediments did not contain contaminant concentrations that exceeded individual TEL guidelines and concentrations in 84% of the sediments were less than individual PEL guidelines (Figure 7.2). Concentrations of eight trace metals (range of exceedances = 5 to 27% total analyses) and eight organic analytes (range of exceedances = 3 to 19% of total analyses) were between TEL and PEL guideline values (Figure 7.3). Exceedances occurred more frequently for cadmium (27%), copper (26%), and *p,p*-DDE (19%). Concentrations of five trace metals and six organic analytes exceeded PEL guidelines, more frequently for zinc (19%), lead (18%), and copper (10%). Sediments collected from bayous affected by urban storm water were more contaminated based on total (Table 7.2) and analyte-specific (Table 7.3) exceedances.

Concentrations of contaminants historically known for their bioaccumulative properties and usually of more concern than others (total PAHs, total PCBs, total DDT, total mercury) appear in Figure 7.4. TEL guidelines were exceeded for 7 to 15% of the analyses for these contaminants; PEL guidelines were exceeded only for total PAHs (1%) and total DDT (4%). Exceedances, as for trace metals, were more common for sediments collected from urbanized bayous.

Geochemically Based Enrichment

Cadmium, copper, and zinc exceeded background concentrations (based on normalization to aluminum) more frequently than other trace metals (Figure 7.5). Sediments receiving urban storm water runoff and treated wastewater discharges contained the most enriched trace metals (five); and the frequency of enrichment ranged from 24 to 86% (urban storm water runoff, 118 sediments) and 2 to 27% (treated wastewater, 60 sediments). Sediments collected from estuarine areas of two Florida Outstanding Waters were less enriched than most others. In contrast, 5 to 42% of seagrass-vegetated sediments contained cadmium, copper, zinc, and lead above background concentrations. These concentrations, although elevated, were usually below sediment quality guideline concentrations.

Acute and Chronic Toxicities

Acute toxicity (mortality > 20%) to at least one test species was observed for approximately 9% of the sediments. These toxic sediments were collected from areas affected by runoff from the golf complex, urban storm water, and agriculture (Figure 7.6). Acute toxicity of whole sediments occurred in approximately 8 and 12% of the tests conducted with *Americamysis bahia* and *Ampelisca abdita,* respectively, and 9% of those conducted with *Palaemonetes pugio* and pore waters.

Chronic toxicity, based on statistically significant changes in growth and reproduction of either *Americamysis bahia* or *Leptocheirus plumulosus,* occurred for 9 of 26 whole sediments (for example, Figure 7.7). Toxicity occurred more frequently for sediments collected from areas associated with a golf complex (4 of 8 sediments) and those receiving to treated wastewater (5 of 12 sediments). No chronic effects were observed for *A. bahia* after exposure to six sediments collected from coastal areas receiving agriculture runoff.

Microbial Genotoxicity

Approximately, 41% of 68 pore waters evaluated in the various surveys were genotoxic. Four of 68 pore waters exhibited only direct genotoxic activity, 18 were genotoxic after enzyme activation, and 9 pore waters exhibited both direct and activated genotoxicity. Genotoxicity was detected in 24% (4 of 17), 38% (14 of 37), and 71% (10 of 14) of the pore waters associated with sediments collected near the golf complex, wastewater outfalls, and an urbanized bayou, respectively. The lowest mean pore water concentrations (%) first causing a direct genotoxic response were 1.8 (one value), 14.5 ($\pm$ 9.3) and 32.1 ($\pm$ 6.0), respectively, for pore waters collected from areas affected by the same stressor sources as noted above. Mean first effect pore water concentrations (%) after enzyme activation were 20.8 ($\pm$ 7.2), 27.2 ($\pm$ 27.5), and 10.3($\pm$ 13.2) for the golf complex, wastewater, and urban storm water areas, respectively.

Phytotoxicity: Early Seedling Growth

Statistically significant effects were observed in 22 of 45 phytotoxicity tests conducted with seedlings of either *Spartina alterniflora* Loisel or *Scirpus robustus* Pursh and whole sediments collected from areas affected by wastewater and urban storm water (for example, Figure 7.8). Phytostimulation occurred for 13 sediments; plant biomass increased an average of 141%. Decreases in dry weight biomass were observed for nine sediments and averaged 68%. Phytoinhibition and phytostimulation were more

TABLE 7.2

Comparison of Frequency Distribution of Proposed Sediment Quality Guideline Exceedances at Sampling Locations

Guideline	Total No. Exceeded	Non-Point-Source Runoff			Treated Wastewater Outfalls	Suwannee River Estuary	Withlacoochee River Estuary	Seagrass Beds
		Urban Storm Water	Golf Complex	Agricultural				
TEL	0	16	60	79	57	50	75	75
	1–3	14	33	21	36	42	17	25
	4–6	46	3	0	7	8	8	0
	7–9	24	4	0	0	0	0	0
PEL	0	30	80	100	99	100	100	88
	1–3	41	20	0	1	0	0	12
	4–6	24	0	0	0	0	0	0
	7–9	5	0	0	0	0	0	0
N		37	30	19	76	12	12	12

Note: Values represent percent of the total number of sites (N) for which result was observed. TEL = threshold effects level; PEL = probable effects level.

TABLE 7.3

Comparison of Exceedances (%) of Threshold and Probable Effects Level Guidelines for Trace Metals and Selected Organic Contaminants at Sediment Collection Areas

Analyte	Non-Point Runoff			Treated Wastewater Outfalls	Suwannee River Estuary	Withlacoochee River Estuary	Seagrass Beds
	Urban Storm Water	Golf Complex	Agriculture				
As	7 (0)[a]	2	5	5	0	0	8
Cd	59 (11)	7	0	14	25	0	0
Cr	30 (10)	0	19	8	8	8	0
Cu	46 (26)	7 (2)	25	21	0	0	8
Ni	24 (0)	0	0	11	0	0	8
Pb	34 (47)	0	0	0	0	0	0
Zn	19 (49)	2	0	2	0	0	0
Hg	—	5	0	8	8	8	0
Chlordane	5 (15)[a]	6 (6)	20	0	0	0	0 (12)
Dieldrin	13 (21)	6	5	0	0	0	0
TPCB	51 (3)	0	0	19	8	8	0
TPAH	56 (5)	0	0	1	0	8	0
TDDT	51 (18)	27 (6)	0	6	8	0	25
p,p-DDD	23 (36)	15 (9)	20	9 (1)	8	0	12
p,p-DDE	67 (0)	30	0	7	16	0	12
p,p-DDT	5 (5)	3 (6)	0	0	0	0	0

Note: Number of analyses for the 16 analytes ranged from 8 to 118 (TEL) and 8 to 69 (PEL) for the various locations.

[a] Exceedance for TEL (> TEL ≤ PEL) and PEL in parentheses.

Source: Sediment quality guidelines from MacDonald et al. (1996).

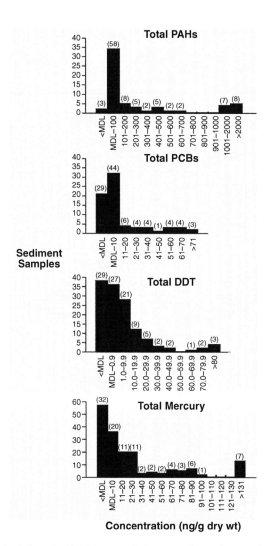

FIGURE 7.4 Concentrations (ng/g dry wt) of four bioaccumulative contaminants. Percent of total sediments in parentheses.

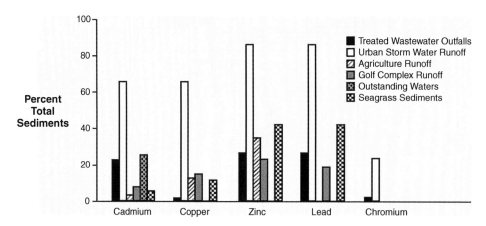

FIGURE 7.5 Trace metal enrichment based on normalization to aluminum (Windom et al., 1989). Values represent percent of 60 (wastewater), 118 (urban storm water), 23 (agriculture runoff), 26 (golf complex), 12 (Florida Outstanding Waters), and 19 (seagrass-vegetated) sediments.

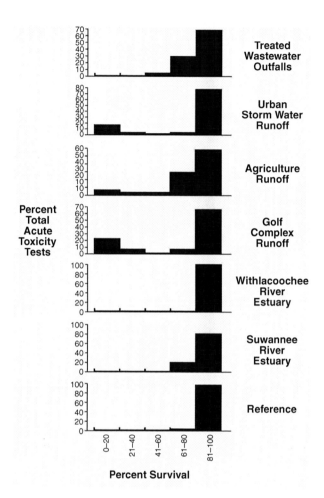

Percent Survival

FIGURE 7.6 Percent survival of benthic invertebrate test species exposed to whole sediments and pore waters collected from reference and nonreference near-coastal areas. Based on results for 188 acute toxicity tests. Toxicity was not determined for seagrass-vegetated sediments. Scale varies with each graph.

common for whole sediments collected from urban storm water areas (17 of 30 sediments) than those receiving treated wastewaters (5 of 15 sediments).

Benthic Macroinvertebrate Community

More than half of the sediment grab samples contained fewer than 20 taxa and 100 organisms (Figure 7.9). The percent of sediments for which the number of taxa was 20 or less was 100% (wastewater), 94% (urban runoff), 83% (Suwannee River estuary), 65% (golf complex runoff), 50% (agriculture runoff), and 16% (Withlacoochee River estuary). Approximately 13% of the sediments had poor benthic abundance (<200 orgs/m^2), and the abundance of benthos in an additional 25% of the sediments was characterized as marginal (200 to 500 orgs/m^2). These sediments were collected only from areas affected by wastewater and golf complex runoff.

About 61% of the Shannon–Wiener diversity index values were 2.0 or less (Figure 7.10) and indicative of poor diversity. Only 2% of the index values indicated good diversity (value ≥ 3.3). Benthic diversity was the least for benthos collected from wastewater-impacted sediments and was usually greater for benthos collected from areas receiving golf complex runoff and from the Withlacoochee River estuary (Figure 7.11).

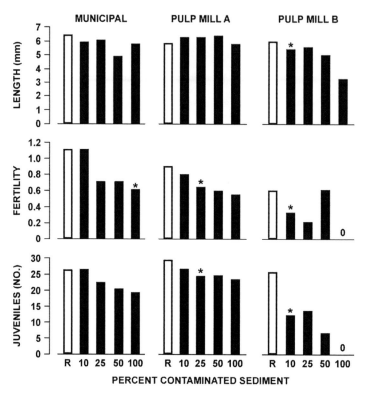

FIGURE 7.7 Results of 28-day chronic toxicity tests conducted with *Leptocheirus plumulosus* and four dilutions of sediment collected below three wastewater outfalls. * = Lowest observed effect concentration (*P* < 0.05). *R* = reference sediment. (Results adapted from Lewis et al., 2000.)

Temporal Variability

Temporal variability, expressed as coefficients of variation (CV), was greater for contaminant concentrations than for biological effects. Mean CV values (%) based on the combined results for eight trace metals and four organic analytes and sediments collected from the same sites in areas receiving urban storm water and from the Suwanee and Withlacoochee River estuaries were 77 (± 49) and 65 (± 27), respectively. The mean CV values based on acute toxicity test results and Shannon–Wiener diversity index values were 16 (± 7) and 9 (± 14) and 19 (± 5) and 21 (± 12), respectively, for sediments collected from the same sites and locations noted above.

Discussion

Stressor Ranking

One of the primary objectives of this chapter was to provide a perspective on the severity of sediment contamination relative to stressor source. Coastal sediments receiving urban storm water runoff and treated wastewater were the more impaired based on several measures of chemical and biological quality (Table 7.4). This ranking is obviously specific to the study areas and metrics used in the comparisons. Furthermore, it assumes that the observed effects are attributable to the contaminant sources alone, which, although likely due to the location of the sampling sites, is not certain. Therefore, the rankings should be considered preliminary and in need of validation, although the conclusion that urban storm water is the major source of sediment contamination in Florida coastal areas has been reported elsewhere (FDEP, 1994a).

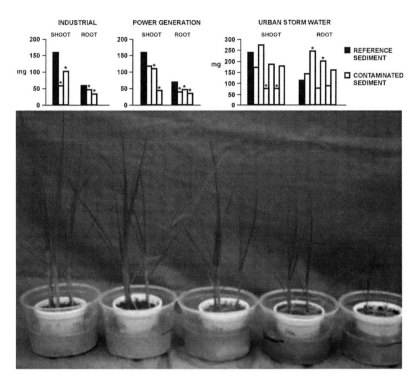

FIGURE 7.8 Phytotoxicities of whole sediments collected below two point-source discharges and from a bayou receiving urban storm water. Values represent dry weight (mg) of shoots and roots of *Echinochloa crusgalli* (L.) Beauv. (wastewaters) and *Spartina alterniflora* Loisel (urban storm water). The range of the early-seedling response of *S. alterniflora* after 28 d exposure to several of the above sediments is shown. * = Significant difference relative to artificial or reference sediment (*P* < 0.05). (Data adapted from Lewis et al., 2000 and Lewis et al., 2001c.)

Metric Concordance

Sediment quality guidelines are usually used to identify contaminants and areas of concern before detailed biological investigations are conducted. Their ability to predict biological impairment has been confirmed for some coastal areas although not without some uncertainty (Carr et al., 1996; Long et al., 1998; Long and MacDonald, 1998). Based on the results of this summary, concordance or agreement between sediment quality guidelines proposed for Florida coastal areas and results of toxicity and benthic community evaluations occurred for 60% (benthic invertebrate acute toxicity), 58% (pore water microbial genotoxicity), 65% (early seedling phytotoxicity), and 56% (benthic macroinvertebrate diversity) of the comparisons (Table 7.5). In other words, at least one individual TEL guideline was exceeded and biological effects occurred or no individual guidelines were exceeded and no biological effects occurred. Biological effects were observed in 1 to 31% of sediments for which no sediment quality guidelines were exceeded (Type II error, false negatives) and no biological effects were observed when sediment quality guidelines were exceeded in 13 to 39% of the comparisons (Type I error, false positives).

In addition to the ability of sediment quality guidelines to predict biological impairment, it is important to recognize the potential of temporal variability in sediment chemical quality on which SQG exceedances are based. The reliance on one analytical measurement to predict biological impairment of sediments is not likely a realistic scenario. Future investigation is needed to provide insight on the importance of this issue.

Results for whole-sediment acute toxicity tests were not consistently reliable in predicting genotoxicity, phytotoxicity, and low benthic diversity (Table 7.5). In addition, sublethal toxicity occurred in about 33% of the sediments and their associated pore waters for which mortality to several invertebrates was

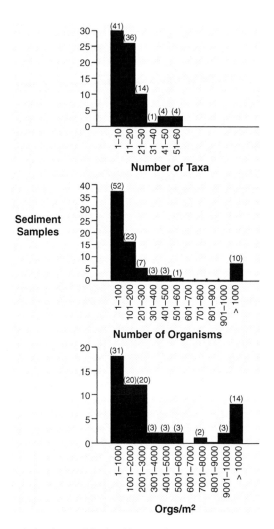

FIGURE 7.9 Number of taxa and abundance of the benthic macroinvertebrate community for all sediments. Percent of total sediments in parentheses.

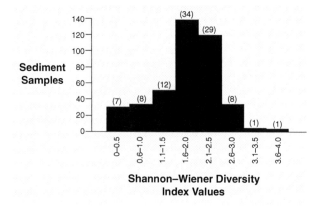

FIGURE 7.10 Shannon–Wiener diversity index values (Shannon and Weaver, 1949) for all sediments. Percent of total sediments in parentheses.

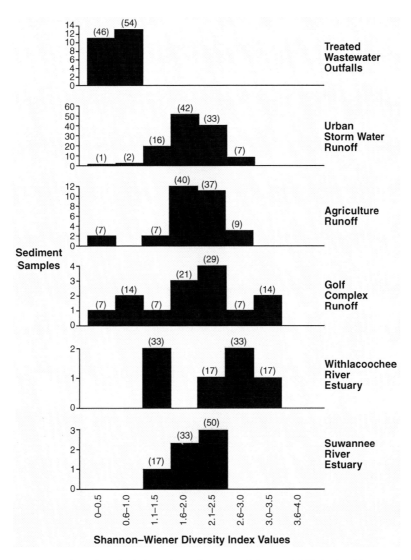

FIGURE 7.11 Shannon–Wiener diversity index values (Shannon and Weaver, 1949) for benthic macroinvertebrates in sediments collected from anthropogenically affected areas and two Florida Outstanding Waters. Values not available for seagrass-vegetated sediments.

20% or less (data not shown). Consequently, use of results for whole-sediment acute toxicity alone in sediment hazard assessments provides a limited perspective on biological impairment.

Data Comparisons

The surveys summarized in this chapter utilized a targeted sampling design with the intent to provide, in some cases, a worst-case perspective of the magnitude of sediment contamination to be expected in Florida near-coastal areas. The results of this summary are compared primarily to those of two surveys conducted in the Gulf of Mexico region during 1991–1994. These surveys utilized a probabilistic sampling design and were conducted for chemical quality, acute toxicity, and benthic community composition at 154 sites located between Anclote Anchorage, FL, and Rio Grande, TX (U.S. EPA Environmental Monitoring and Assessment Program, EMAP Louisianian Province; Macauley et al., 1995) and for chemical quality and acute and chronic toxicity for 123 sediments collected from four

TABLE 7.4

Relative Ranking of Stressor Sources in Decreasing Order of Severity

Metric	Ranking
Sediment Quality Guidelines[a]	
TEL exceedances (>TEL ≤ PEL)	US > WW > GC > AG
PEL exceedances	US > GC > WW > AG
Trace metal enrichment[b]	US = WW > GC > AG
Acute toxicity[c]	GC > US > AR > WW
Chronic toxicity[d]	GC > WW > AG
Genotoxicity (pore water)[e]	US > WW > GC
Benthic diversity index[f]	WW > US > AG > GC
Macroinvertebrate abundance[g]	WW > GC > US = AG
Number of taxa[h]	WW > US > GC > AG

Notes: US = urban storm water runoff in three bayous. WW = 11 treated wastewaters,
      GC = runoff from two golf courses, AG = agriculture runoff in Everglades–Florida
      Bay transitional zone.

[a] From MacDonald et al. (1996).
[b] Based on total number of trace metals after normalization to aluminum (Windom et al., 1989).
[c] Based on total number of toxicity tests for *Americamysis bahia, Ampelisca abdita,* and *Palaemonetes pugio* where mortality was 20% or more (after control correction).
[d] Not determined for sediments affected by urban storm water, based on results for *L. plumulosus* and *Americamysis bahia.*
[e] Not determined for agriculture runoff pore waters.
[f] Based on Shannon–Wiener diversity index values of 2.0 or less.
[g] Based on 500 orgs/m^2 or less.
[h] Based on 20 taxa or less.

bays located in northwest Florida (NOAA Status and Trends; Long et al., 1997). The comparison among surveys is limited, however, to only a few parameters due to differences in experimental techniques.

Common sediment contaminants in this summary based on exceedances of proposed sediment quality guidelines were in decreasing order for trace metals: copper > cadmium = lead > zinc > chromium > nickel > mercury > arsenic and for organic contaminants: total DDT > total PCBs > total PAHs > chlordane = dieldrin. Many of these same contaminants were found to exceed the sediment quality guidelines of Long et al. (1995) in the two previous surveys. Contaminants exceeding numerical guidelines included, among others, copper, lead, zinc, total DDT, total PAHs, and dieldrin (Long et al., 1997) and chromium, nickel, mercury, arsenic, zinc, and dieldrin (Macauley et al., 1995). In earlier surveys (1982–1991) summarized by FDEP (1994a), common trace metals in Florida estuarine areas were reported in decreasing order: lead > zinc > mercury > cadmium > copper > chromium > nickel > arsenic.

Several biological measures of sediment quality were similar among surveys, such as the evaluation of acute toxicity of whole sediments to *Ampelisca abdita*. This is the most widely used test species in whole-sediment toxicity tests, although it is considered relatively insensitive. In this summary, about 12% of the sediments were acutely toxic to this species compared to approximately 1% (NOAA survey; Long et al., 1997) and 1 (± 3)% (EMAP survey; Macauley et al., 1995). About 10 (± 7)% of the sediments were acutely toxic to *Americamysis bahia* in the EMAP survey relative to 8% in this summary. Genotoxicity, as measured by the Mutatox assay, occurred in 22 of 52 (42%) solvent extracts of sediments (Long et al., 1997) compared to 31 of 68 (46%) centrifuged pore waters in this summary. Macroinvertebrate abundance was characterized as poor and marginal for 13 and 25% of the sediments surveyed in this study, respectively, relative to 5 (± 5)% and 15 (±8)% of sediments in the EMAP Louisianian Province survey.

In conclusion, contaminants in about 50 and 16% of the sediments exceeded individual TEL and PEL guidelines, respectively. Acute toxicity was observed in approximately 9% of the sediments and the benthos in 13 and 61% of the sediments were in poor condition based on abundance and diversity, respectively. Adverse effects were usually more common and pronounced for areas receiving urban

TABLE 7.5

Comparison of Exceedances of Individual Sediment Quality Guidelines (SQGs) and Occurrence of Biological Effects Observed for the Same Sediment Sample

	Benthic Invertebrate Acute Toxicity[a]	Genotoxicity[b]	Phytotoxicity[c]	Benthic Macroinvertebrate Diversity[d]
SQG exceedances and biological effects	9	20	53	31
SQG exceedances and no biological effects	39	20	32	13
No SQG exceedances and biological effects	1	22	3	31
No SQG exceedances and no biological effects	51	38	12	25
Number comparisons	350	40	38	62
Benthic invertebrate acute toxicity and biological effects	—	5	11	4
Benthic invertebrate acute toxicity and no biological effects	—	7	5	3
No benthic invertebrate acute toxicity and biological effects	—	44	45	45
No benthic invertebrate acute toxicity and no biological effects	—	44	39	48
Number comparisons	—	40	38	78

Note: Values represent percent of comparisons where outcome was observed.

[a] Survival 20% or less (control-corrected) for *Americamysis bahia, Ampelisca abdita, Palaemonetes pugio*.

[b] Direct or activated response observed.

[c] Statistically significant phytoinhibition or phytostimulation.

[d] Shannon–Wiener diversity index value < 2.0.

runoff where "hot spots" were evident based on frequent exceedances of PEL guidelines and the common occurrence of adverse biological effects. It was evident that current sediment diagnostic techniques have several limitations. The sometimes lack of agreement between proposed sediment quality guidelines and biological effects coupled with possible temporal variability in sediment contaminant concentrations (or the ability to measure them consistently) are factors to consider when using this screening tool. Use of acute toxicity tests alone, as would be expected, is a limited approach if a realistic toxicity profile is desired. Benthic community diversity appears to be a more sensitive and consistent indicator of sediment quality, which is not surprising as the benthos is relatively immobile and integrates long-term responses to contaminated sediments. However, this conclusion is based on the results for one metric (Shannon–Wiener diversity) and its value relative to other structural characteristics of this biota (see Washington, 1994; Clarke and Warwick, 1994) is unknown.

Recommendations

Sediment quality guidelines proposed for Florida coastal areas must be updated to include new information, particularly that available for different levels of biological organization. The use of chronic toxicity tests with rooted vascular plants and animal species must be increased in sediment hazard assessments to better determine interspecific sensitivities and to augment a relatively sparse sublethal toxicity database. Reference conditions for chemical and biological parameters need identification to better judge the significance of results for regional sediment surveys. Finally, there must be an increased focus on determining the cause or causes of sediment contamination and biological effects, which usually have been assumed to be the measured anthropogenic contaminants. Causal identification should go beyond statistical procedures and focus on biological methods such as the use of toxicity or pollutant identification evaluations (Ho et al., 2002). These techniques have not been utilized for toxicity tests and contaminated sediment and pore waters in previous surveys conducted in the Gulf of Mexico region. Determination of the cause of impairments to benthic flora and fauna will be difficult but is necessary for prevention of sediment contamination and for effective management of current sediment quality.

Acknowledgments

Field sampling and toxicity tests were conducted by the following U.S. EPA personnel: Carol Daniels, David Weber, Matthew MacGregor, Jim Patrick, Bob Quarles, Larry Goodman, Roman Stanley, Peggy Harris, and Darrin Dantin. Peggy Rogers (NCBA) prepared the manuscript. Steve Embry (CSC Corporation, Gulf Breeze, FL) provided the graphics. Chemistry support was provided by Avanti Corporation (Gulf Breeze, FL) and TDI Brooks International (College Station, TX). The U.S. Environmental Protection Agency through its Office of Research and Development funded and managed the research described here. It has been subjected to the agency's peer and administrative review and has been approved for publication as an EPA document.

References

APHA (American Public Health Association), American Water Works Association, Water Environment Federation. 1998. *Standard Methods for the Examination of Water and Wastewater,* 20th ed. American Public Health Association, Washington, D.C.

ASTM (American Society for Testing and Materials). 1993a. Guide for collection, storage, characterization and manipulation of sediments for toxicological testing. In *ASTM Standards on Aquatic Toxicology and Hazard Evaluation.* E 1391-90. American Society for Testing and Materials, Philadelphia, PA, pp. 321–335.

ASTM (American Society for Testing and Materials). 1993b. Guide for Conducting Life-Cycle Tests with Saltwater Mysids. E 1191-90. ASTM, Philadelphia, PA, pp. 122–137.

ASTM (American Society for Testing and Materials). 1995. Standard test methods for measuring the toxicity of sediment-associated contaminants with freshwater invertebrates. E 1706-95b. In *Annual Book of ASTM Standards,* Vol. 11.05. ASTM, Philadelphia, PA, pp. 294–320.

Butts, G. and M. A. Lewis. 2002. A survey for chemical quality, sediment toxicity and marcoinvertebrate community composition in three Florida urbanized bayous. *Gulf of Mexico Science* 1:1–11.

Carr, R. S., E. R. Long, H. L Windom, D. C. Chapman, G. Thursby, G. M. Sloane, and D. A. Wolfe. 1996. Sediment quality assessment studies of Tampa Bay, Florida. *Environmental Toxicology and Chemistry* 15:1218–1231.

Clarke, K. R. and R. M. Warwick. 1994. Change in Marine Communities: An Approach to Statistical Analysis and Interpretation. Natural Environment Research Council, U.K., 144 pp.

FDEP (Florida Department of Environmental Protection). 1994a. Florida Coastal Sediment Contaminants Atlas. FDEP, Tallahassee, FL.

FDEP (Florida Department of Environmental Protection). 1994b. Approach to the Assessment of Sediment Quality in Florida Coastal Waters, Vol. 2, Application of the Sediment Quality Assessment Guidelines. Office of Water Policy. Tallahassee, FL.

FDEP (Florida Department of Environmental Protection). 1994c. Approach to the Assessment of Sediment Quality in Florida Coastal Waters, Vol. 1, Development and Evaluation of Sediment Quality Assessment Guidelines. Office of Water Policy, Tallahassee, FL.

FDEP (Florida Department of Environmental Protection). 2002. Florida's Water Quality Assessment 2002 305 (b) Report. FDEP, Tallahassee, FL.

Friedman, M. and J. Hand. 1989. Typical Water Quality Values for Florida Lakes, Streams and Estuaries. Florida Department of Environmental Protection, Tallahassee, FL.

Goodman, L., M. Lewis, J. Macauley, R. Smith, and J. Moore. 1999. Preliminary survey of chemical contaminants in water, sediment and aquatic biota at selected sites in northeastern Florida Bay. *Gulf of Mexico Science* 1:1–16.

Ho, K. T., R. M. Burgess, M. C. Pelletier, J. R. Serbst, S. A. Ryba, M. G. Cantwell, A. Kuhn, and P. Raczelowski. 2002. An overview of toxicant identification in sediments and dredged materials. *Marine Pollution Bulletin* 44:286–293.

Johnson, B. T. and E. R. Long. 1998. Rapid toxicity assessment of sediments from estuarine ecosystems: a new tandem *in vitro* testing approach. *Environmental Toxicology and Chemistry* 17:1099–1106.

Lewis, M. A. and S. S. Foss. 2000. A caridean grass shrimp (*Palaemonetes pugio* Holthius) as an indicator of sediment quality in Florida coastal areas affected by point and non-point source contamination. *Environmental Toxicology* 15:234–242.

Lewis, M. A., L. Goodman, and D. Weber. 1999. Periphyton and sediment bioassessment as indicators of site-specific environmental condition in the Florida Bay area. *Environmental Monitoring and Assessment* 65:535–540.

Lewis, M. A., D. E. Weber, and B. Albrecht. 2000. Sediment chemical quality and toxicity in Gulf of Mexico coastal areas below ten wastewater discharges. *Environmental Toxicology and Chemistry* 19:192–203.

Lewis, M. A., L. R. Goodman, J. M. Patrick, J. C. Moore, and D. E. Weber. 2001a. The effects of urbanization on the spatial and temporal quality of three tidal estuaries in the Gulf of Mexico. *Water, Air, and Soil Pollution* 127:65–91.

Lewis, M. A., S. S. Foss, S. Harris, R. S. Stanley, and J. C. Moore. 2001b. Sediment chemical contamination and toxicity associated with a coastal golf course. *Environmental Toxicology and Chemistry* 20:1390–1398.

Lewis, M. A., D. E. Weber, B. Albrecht, and R. S. Stanley. 2001c. Plants as indicators of near-coastal sediment quality: interspecific variation. *Archives of Environmental Contamination and Toxicology* 40:25–34.

Lewis, M. A., C. Daniels, J. C. Moore, and T. Chen. 2002. Potential genotoxicity pore water with comparison to sediment toxicity and macrobenthic community composition. *Environmental Toxicology* 17:63–73.

Long, E. R. and D. D. MacDonald. 1998. Perspective: recommended uses of empirically derived, sediment quality guidelines for marine and estuarine ecosystems. *Human and Ecological Risk Assessment* 4:1019–1039.

Long, E. R., D. D. MacDonald, S. L. Smith, and F. D. Calder. 1995. Incidence of adverse biological effects within ranges of chemical concentrations in marine and estuarine sediments. *Environmental Management* 19:81–97.

Long, E. R., G. M. Sloane, R. S. Carr, T. Johnson, J. Biedenbach, K. J. Scott., G. B. Thursby, E. Grecelius, C. Oenen, H. L. Windom., R. D. Smith, and B. Loganathon. 1997. Magnitude and extent of sediment toxicity in four bays of the Florida Panhandle: Pensacola, Choctawhatchee, St. Andrew and Apalachicola. NOAA Technical Memorandum NOS ORCA 117, NOAA, Silver Spring, MD.

Long, E. R., L. J. Field, and D. D. MacDonald. 1998. Predicting toxicity in marine sediments with numerical sediment quality guidelines. *Environmental Toxicology and Chemistry* 17:714–727.

Macauley, J. M., J. K. Summers, V. D. Engle, D. T. Heitmuller, and A. M. Adams. 1995. Statistical summary: EMAP–Estuaries Louisianian Province–1993. EPA/620/R-96/003, Office of Research and Development, Washington, D.C.

Macauley, J. M., J. K. Summers, and V. D. Engle. 1999. Estimating the ecological condition of the Gulf of Mexico. *Environmental Monitoring and Assessment* 57:59–83.

MacDonald, D. D., R. S. Carr, F. D. Calder, E. R. Long, and G. C. Ingersoll. 1996. Development and evaluation of sediment quality guidelines for Florida coastal water. *Ecotoxicology* 5:253–278.

Microbics Corporation. 1993. Mutatox manual. 1986. *Mutation Research* 172:89–96.

Pait, A. S., A. E. DeSouza, and D. R. G. Farrow. 1992. Agricultural Pesticide Use in Coastal Areas: A National Summary. National Oceanic and Atmospheric Administration, Rockville, MD.

Shannon, C. E. and W. Weaver. 1949. *The Mathematical Theory of Communication.* University of Illinois Press, Urbana, IL.

Truax, D. and J. Daniel. 1991. *Proceedings, Workshop on Contaminated Sediments and the Gulf of Mexico. U.S. Gulf of Mexico Program,* August 27–29, Stennis, MS.

U.S. EPA (U.S. Environmental Protection Agency). 1994a. Toxic substances and pesticides: Action agenda for the Gulf of Mexico first generation management committee report. EPA-800-B-94-005. Technical report. U.S. Gulf of Mexico Program, Stennis, MS.

U.S. EPA (U.S. Environmental Protection Agency). 1994b. Methods for assessing the toxicity of sediment associated contaminants with estuarine and marine amphipods. EPA 600/R-94/025. Technical report. Narragansett, RI.

U.S. EPA (U.S. Environmental Protection Agency). 1996. The national sediment quality survey: a report to Congress on the extent and severity of sediment contamination on surface waters of the United States. EPA-823-D-96-002. Office of Science and Technology, Washington, D.C.

U.S. EPA (U.S. Environmental Protection Agency). 1997a. The incidence and severity of sediment contamination in surface waters of the United States. Vol. 3. National sediment contaminant point source inventory. EPA 823-R-97-008. Technical report. Washington, D.C.

U.S. EPA (U.S. Environmental Protection Agency). 1997b. Methods for the determination of chemical substances in marine and estuarine environmental matrices, 2nd ed. EPA/600/R-97/072. Washington, D.C.

U.S. EPA (U.S. Environmental Protection Agency). 1998. Method for assessing the chronic toxicity of sediment-associated contaminants with *Leptocheirus plumulosus*. Office of Research and Development, Duluth, MN.

U.S. EPA (U.S. Environmental Protection Agency). 2000. Final TMDL rule: fulfilling the goals of the Clean Water Act. EPA 841-F-00-008. Office of Water, Washington, D.C.

U.S. EPA (U.S. Environmental Protection Agency). 2001a. National Coastal Condition Report. EPA-620/R-01/005. Office of Research and Development/Office of Water, Washington, D.C.

U.S. EPA (U.S. Environmental Protection Agency). 2001b. Methods for assessing the chronic toxicity of marine and estuarine sediment-associated contaminants with the amphipod, *Leptocheirus plumulosus*. EPA/1600/R-01/020. Office of Water, Washington, D.C.

U.S. EPA (U.S. Environmental Protection Agency). 2002. Estuarine and Coastal Marine Waters: Bioassessment and Biocriteria Technical Guidance. EPA-822-B-00-024. Office of Water, Washington, D.C.

Walsh, G. E., D. E. Weber, T. L. Simon, L. K. Brashers, and J. C. Moore. 1991. In *Use of Marsh Plants for Toxicity Testing of Water and Sediments Plants for Toxicity Assessment,* Vol. 2, J. W. Gosuch, W. R. Lower, W. Wang, and M. A. Lewis (eds.). ASTM STP 1115. ASTM, Philadelphia, PA, pp. 341–354.

Washington, H. G. 1994. Diversity, biotic and similarity indices — a review with special relevance to aquatic ecosystems. *Water Research* 18:653–694.

Weber, D. E., G. E. Walsh, and M. A. MacGregor. 1995. In *Use of Vascular Aquatic Plants in Phytotoxicity Studies with Sediments. Environmental Toxicology and Risk Assessment,* Vol. 3, J. S. Hughes, G. R. Biddinger, and E. Mones (eds.) ASTM STP 1218, ASTM, Philadelphia, PA, pp. 187–200.

Windom, H. L., S. J. Schropp, F. D. Caldu, J. D. Ryan, R. G. Smith, L. C. Burney, F. G. Lewis, and C. H. Rawlinson. 1989. Natural trace metal concentrations in estuarine and coastal marine sediments of the southeastern United States. *Environmental Science and Technology* 23:314–320.

8

Bacterial Communities as Indicators of Estuarine and Sediment Conditions

Eric C. Milbrandt

CONTENTS

Introduction

Restoration of tidal wetlands has emerged as a viable mechanism to recover lost habitat function, biological diversity, and ecological value (Frenkel and Morlan, 1991; Costanza et al., 1997). Restoration activities include hydrologic improvement, exotic removal, and ecological engineering and can return predisturbance ecosystem function to a severely degraded habitat (National Research Council, 1992). Past failures to recover predisturbance structure and function (Lewis, 1982) have increased the demand for universal restoration guidelines and predictable models for postrestoration recovery.

Colonization, secondary succession, and competition are key ecological principles at work in the days, weeks, and months following a hydrologic restoration (Zedler, 2000). These principles can be used to direct the collection of data to guide an adaptive management strategy or to build a predictive model of ecological recovery. The ecological principles mentioned above require data about community structure and function. These data, when collected, can be used to better understand tideland communities, to improve the probability of successful restoration, and to provide recommendations for future mitigation and restoration projects (Zedler and Callaway, 1999). The outcome of a restoration project should be based on structural and functional attributes of the predisturbance condition; however, there are often few or, worse, no data on the predisturbance condition. A common source for data about target community structure and function comes from nearby pristine, undisturbed habitats. Presumably the adjacent site used to set restoration goals has reached a dynamic equilibrium with the environment. The difficulty in using this approach is that locality can have a significant impact on the conditions affecting community composition and function and therefore may not be representative of the target parameters. Given this limitation, it has become routine to choose several reference sites to establish the ecological goals of the project (Portnoy and Giblin, 1997a). A second approach for determining realistic restoration guidelines is through experimentation. Mesocosms provided hydrologic replication to test the predictions of multiple ecological treatments (Rumrill and Cornu, 1995; Callaway et al., 1997; Boyer and Zedler, 1999).

One neglected biotic component of tideland restoration projects is the community of bacteria that live in the sulfidic muds. Bacteria use potential energy from oxygen, sulfate, and ferrous iron to decompose and recycle large pools of proteins, amino acids, and lipids in shallow coastal ecosystems (Fenchel et al., 1998). By using an abundant supply of reductants, communities of bacteria metabolize carbon molecules and grow. Their biomass provides food for bactivorous invertebrates, and the products of their metabolism dictate the condition of intertidal sediments. The condition of intertidal sediments following hydrologic restoration may limit colonization by emergent vegetation (Boyer et al., 2000; Lindig-Cisneros and Zedler, 2002). An important recycling function has been hypothesized between sediment microbes and emergent vegetation; however, few data have been collected to demonstrate the relationship among plant biomass, nutrient additions, and microbial diversity (Piceno and Lovell, 2000a,b).

Microbial communities have two characteristics that may be useful for applications in restoration ecology. Many recent descriptions of bacteria community composition show that bacteria are extremely diverse and are found in a range of habitat types. Bacteria have been discovered deep in the lithosphere (Ghiorse, 1997), in hot springs (Pace, 1997), and around hydrothermal vents (Reysenbach and Cady, 2001). Therefore, it is not unreasonable to suspect that every proposed restoration project in tidal wetlands will include a bacterial community component. Second, bacteria are metabolically versatile and have short generation times. The colonization and eventual recovery of plant or infaunal invertebrate communities likely depend on the condition of the sediment. Any cause of degradation, whether it is fossil fuel contamination, impoundment, or urbanization, results in changes in the available pool of carbon, nitrogen, and phosphorus. Community composition dictates the efficiency and mode of carbon metabolism in both the water column and sediments (Labrenz et al., 2000; Rappe et al., 2002). Shifts in the microbial community have the potential to indicate the need for planting, nitrogen additions, or additional hydrologic manipulations.

During the past two decades, the temporal and spatial scales of microbial community diversity from several habitats have been described (Torsvic et al., 1990; Wawer and Muyzer, 1995; Cifuentes et al., 2000; Nadeau et al., 2001). There are also descriptions of distinct microbial communities along environmental gradients (Bouvier and del Giorgio, 2002) and associated with attached particles in estuaries (Crump et al., 1999). Bacteria are ubiquitous in the biosphere, but their organization is neither haphazard nor random. Progress in describing microbial communities is the result of several novel techniques that employ sensitive DNA detection. These tools can be used to follow the colonization, modification, and secondary succession in muds, a difficult natural *in situ* substrate to apply traditional cultivation and microscopic methods.

In one of the first applications of molecular techniques to microbial ecology, a survey of genetic diversity from the Sargasso Sea was reported (Giovannoni et al., 1990). The small subunit ribosomal gene, common to all members of the Domain Bacteria, was extracted and polymerase chain reaction (PCR) amplified from environmental samples. PCR amplification targets regions of the genome and provides a high number of copies of a gene or gene fragment for subsequent analysis. The amplified genes were then cloned into a plasmid vector and transformed into *Escherichia coli* cells. This technique was widely used in the 1970s to isolate physiologically important genes and to screen coding genes from noncoding DNA. The ability to amplify specific regions of the genome made cloning an ideal tool for dissecting a genetically diverse community into its component species. Because bacteria are primarily asexual, the biological species concept has elicited considerable debate and has led to suggestion of alternative classification strategies. Traditionally, microbiologists used a hierarchical series of physiological and diagnostic tests on cultivated strains to determine the phylogenetic classification of the unknown specimen. This innovative DNA-based approach circumvented the cultivation requirement and allowed direct identification based on the sequence order of the highly conserved 16S ribosome (Woese, 1987). Estimates of the diversity of cultivated bacteria to the actual diversity of bacteria in the environment suggested that 99% of the known bacterial diversity had never been brought into culture (Amann et al., 1995). This approach was widely published and serves as the foundation for the next generation of molecular techniques. A limitation of cloning and sequencing is that time and expense prohibited ecological replication. Many of the gene sequences were new to science and thus required complete characterization. One replicate was used to classify free-living, estuarine bacteria communities (Crump et al., 1999). Ecological replication is critical to determine the statistical probability and predictability of ecological patterns.

Denaturing gradient gel electrophoresis (DGGE) allows as many as 36 communities (replicates, samples) to be run side by side, so that community composition can be compared both spatially and temporally. This apparatus has enormous versatility, and it can be applied to laboratory experiments and field samples. The technique follows similar procedures of direct extraction, purification, and amplification as cloning. The difference lies in separating like-sized fragments with a denaturing acrylamide gel, rather than using plasmids, *E. coli*, and screening. The principle behind separation lies in the melting characteristics of double-stranded nucleic acids. The composition and order of bases in a strand of DNA determine the denaturation characteristics. Denaturation, or melting, can be accomplished either by heating up the DNA or by increasing the salt concentration in the solution. High temperature or salt ions interfere with the hydrogen bonds that hold the two single strands together. As the double strand separates into single strands, the migration rate increases. Therefore, each lane is a community fingerprint, where a band represents a population of bacteria with a unique genetic sequence. The gel can be analyzed at the community level by comparing the shared presence of similarly migrating bands. Alternatively, the genetic sequence of each band can be obtained by excising the band, extracting the DNA, and sequencing the molecule. A review of DGGE and its application to microbial ecology can be found in Shäfer and Muyzer (2000).

The following case study describes an application of the aforementioned molecular techniques to a restoration project. The project addresses the scale and magnitude of variability in the composition of bacterial communities. The geographic localities sampled, a restoration site and several pristine reference sites, were revisited to determine the temporal stability of the community. This chapter familiarizes the reader with the types of data available through molecular analysis of bacteria communities and shows the potential for application of these tools to tideland restoration.

Case Study

South Slough National Estuarine Research Reserve (SSNERR) is a sub-basin of the Coos Estuary system in Oregon, along the northwest Pacific Coast, USA. South Slough is a relatively undisturbed representative of the Lower Columbian River coastal ecosystem (Rumrill, 2002). Estuarine wetlands in Pacific Northwest estuaries, including the Coos Estuary system, characteristically have fringing or pocket salt marshes with extensive mudflats. One pristine reference locality was chosen to compare the bacterial communities within the estuary and to provide a snapshot of what the restoration site should mimic. A hydrologic restoration site was also chosen whose earthen levees were removed in 1997.

Hidden Creek, the reference locality, is broadly typical of a mature wetland based on two attributes, wetland age and marsh surface elevation relative to the tide (Callaway, 2001). Hidden Creek tidal wetland formed in a protected embayment where a lower-order perennial freshwater stream flowed into a higher-order tidal channel. The delta at the confluence of Hidden Creek consists of a relatively large and persistent mudflat and sinuous tidal creeks. Each tidal cycle delivered and deposited sediment and organic material. A combination of several dynamic feedback processes affecting the relative elevation of the marsh surface resulted in the presumed steady state for most of the intertidal mudflats in South Slough. A steady state had been previously defined in intertidal marshes by the elevation of the marsh surface (Redfield, 1972). Sediments were stratified, with a black band that extended 2 cm below the sediment surface.

Kunz Marsh, the restored locality, was once a mature tidal wetland that was restricted from tidal flooding at the turn of the 20th century. Channelized drainage channels and tide gates were installed by early settlers to drain the estuarine wetland for crop production and livestock grazing. The hydrologic modifications decoupled marsh primary production from the estuarine food web and eliminated critical habitat for waterfowl, fish, and invertebrates. Linear drainage channels and marsh subsidence pooled fresh water on the landward side of the levee. Waterlogged soils provided ideal conditions for colonization by the cattail *Typha latifolia* and other freshwater marsh plants. The impounded freshwater marsh (*sensu* Montague et al., 1987) produced a thick layer of peat similar to that described in a New England impoundment (Portnoy and Giblin, 1997b). Restoration of tidal flooding to Kunz Marsh and recovery of estuarine wetlands was initiated in 1996. The restoration project removed an earthen levee and filled linear drainage

FIGURE 8.1 Aerial photos of the restoration (Kunz Marsh, KM) site before (left) and after (right) dike removal. (Photographs courtesy of South Slough National Estuarine Research Reserve.)

channels (Figure 8.1). Natural drainage channels were allowed to develop after reintroduction of tidal flooding. Following restoration, the site was devoid of marsh plants and the sediments were watery and black.

Materials and Methods

Four replicate core samples were collected haphazardly from unvegetated intertidal mudflats at each locality and depth-fractionated into 1-cm increments. The sampling occurred in September 2000, and was repeated in September 2001 and 2002. An elevation of 1.2 m above MLLW (mean lower low water) was targeted for sampling in both sites. Core samples were collected with a modified 140-ml syringe with a diameter of 4 cm (Walters and Moriarty, 1993). Upon returning to the laboratory, DNA was extracted from the sediments immediately.

Replicate profiles of relative redox potential were collected with an ORP® Redox Combination Electrode (Lazar Laboratories, Inc., Los Angeles, CA). Redox profiles were measured within the Kunz Marsh and Hidden Creek Marsh in September 2000. Duplicate profiles were run at two randomly chosen locations within the site. A platinum-banded metallic sensor was used in combination with a double-junction reference electrode connected to a digital mV meter. Depth of the electrode tip was measured using a ruler on a micromanipulator to the nearest 0.10 mm. Relative redox potential (Δ mV) quantifies the change in electron flow at a specified depth relative to surface sediments (0 mm). Absolute redox potentials were not reported because electron flow of surface sediment against a standard hydrogen electrode was not measured (Skoog and Leary, 1992).

DNA extraction was conducted by a direct lysis method modified from Zhou et al. (1996): 100 µl of 5 M NaCl was added to 1 g of sediment; then 750 µl of cetyltrimethylammonium bromide lysis buffer was added (pH 8.0, 1% cetyltrimethylammonium bromide 0.05 M Tris, 0.05 M ethylenediaminetetraacetic acid, 0.05 M NaH$_2$PO$_4$, 1.5 M NaCl). The sediment slurry was incubated at 37°C for 30 min after proteinase K (5 µl, 20 mg/ml) was added. After 100 µl of 20% sodium dodecyl sulfate was added, the mixture was incubated at 65°C for 1 h, –80°C for 1 h, and 65°C for 1 h. The boil/freeze/boil step was followed by a 25:24:1 phenol/chloroform/isoamyl extraction. The aqueous extract was removed and precipitated with 0.6 volumes isopropanol, and incubated overnight at –20°C. Precipitated nucleic acids were pelleted for 30 min at 15,000g, washed with 70% ethanol, and air-dried. Pellets were resuspended in 100 µl Sigma molecular-grade water and incubated overnight at 4°C. Crude extracts were not amendable to enzymatic reactions or restriction digestion, so they were agarose-gel-purified (Li and Ownby, 1993). We attempted to minimize PCR inhibitors by gel-purifying all crude extractions prior to PCR analysis.

Then 10 µl (~40 ng) of gel-purified template was used in a 50 µl PCR reaction to amplify small subunit 16S rDNA. Primers gc338F and 519R amplified a hypervariable region of the 16S rRNA gene (Muyzer et al., 1993). These PCR products were used in DGGE to observe profiles of bacteria communities. After differences in DGGE profiles were observed between sites, two clone libraries were constructed from the same template used for DGGE. A PCR reaction with primers 8F and 1392R amplified the 16S rDNA gene. Each reaction contained 5 μl 10× buffer, 5 μl 25 m*M* MgCl$_2$ (Applied Biosystems) 1.2 μl 2.5 m*M* dNTP, 10 U Amplitaq LD, 1.5 μl 10 pmol forward primer, and 1.5 μl 10 pmol reverse primer. The reaction was brought to 95°C for 3 min then cycled 35 times — 95°C, 30 s; 48°C, 30 s; 72°C 30 s — and held at 72°C for 10 min. A potential source of PCR bias is the presence of contaminating DNA. Given the universal nature of 16S rRNA (it is a gene found in all members of the Domain Bacteria), the Domain-level primers (8F, gc338F, 519R, 1392R), and the ubiquity of bacteria in the environment, contamination of samples was of great concern. The apparatus and supplies were frequently checked for contamination by running negative controls; that is, a PCR reaction was run with no template DNA and contained all of the necessary reagents and thermostable enzyme. Amplification of DNA in the negative control was not observed, and therefore contamination in this data set was not a factor.

A Biorad® D Code gel rig was used to separate similar-sized amplified rDNAs by melting domain (Biorad, Hercules, CA). All reagents were prepared as described in the D Code Instruction Manual and Applications Guide. The denaturing gradient was 20 to 60% by weight urea and formamide; 1 mm, 7% gels were loaded with 15 μl of the PCR reaction plus 5 μl of loading dye (0.01% Bromphenol Blue, 40% glycerol), and run for 3 h at 200 V. Gels were stained with SYBR® gold nucleic acid gel stain (Molecular Probes, Eugene, OR) and photographed on Polaroid® 770 film. Polaroid gel images were scanned into digital format and inverted for analysis.

DGGE provided a profile of a mixed community amplified by PCR to compare replicates from environmental samples visually. Gels were analyzed following van Hannen et al. (1998, 1999). Individual bands that composed the community profile were identified using Gel Pro 4.0, 1-D gel routine. All bands, including heteroduplexes, were included in the analysis since their appearance was characteristic of the samples and informative. A binary matrix was derived by identifying presence (1) or absence (0) of a band. Binary matrices from Kunz Marsh and Hidden Creek Marsh were used to calculate a Bray-Curtis similarity coefficient with PRIMER® v.5.2 (Clarke and Gorley, 2002). A non-metric multidimensional scaling ordination was plotted from the similarity matrix (Clarke and Warwick, 2001). Further statistical testing of similarity among sites was performed using ANOSIM, a simple, nonparametric permutation procedure applied to the similarity matrix underlying the multidimensional scaling (MDS) ordination (Clarke and Warwick, 2001).

The DGGE gels provided a fingerprint of the bacterial community in that sample and were used to calculate a similarity value from the presence/absence of each numbered band. There were no sequence data derived from the DGGE gels, only a similarity value to communities in the other samples. It was necessary, based on the results from the DGGE gels, to completely characterize two "representative" communities. A community from the 1- to 2-cm anaerobic fraction from the reference site was characterized along with a sample from the 1- to 2-cm anaerobic fraction of the restoration site. To completely characterize these representative communities and to confirm the findings of DGGE, a sample from each of the two localities was cloned and all of the phylotypes were sequenced. Gene sequences were generated for 27 unique phylotypes in the restored site and 7 phylotypes in the restoration site. The following methodology describes this characterization. Amplified 16S rDNAs (8F, 1392R) were cloned with an Invitrogen TOPO TA cloning kit. Inserts were amplified with M13F, M13R primers and screened using double-digest restriction fragment length polymorphism (RFLP). Abnormally sized inserts were removed from analysis and 10 μl of PCR product was digested. Then 1 μl 10× buffer, 1 μl BSA, 0.2 μl Alu I, 0.4 μl Msp I, 7.4 μl water were incubated with amplified inserts for 1 h at 37°C; 10 μl of the digest was run on a 2% Agarose gel at 100 V for 1.5 h.

Results

Redox potential (Δ mV) was measured in three reference sites, including Hidden Creek. The redox potential from the restoration site is also shown (Figure 8.2). The restored site showed a steep negative gradient in mV as the probe was carefully lowered into the sediments. The sediments in this site were

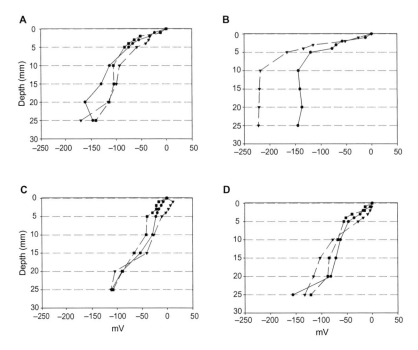

FIGURE 8.2 Vertical microelectrode redox potential profiles from restored site (A) plus three mature intertidal mudflats (B, C, D). Δ mV is the change in redox potential from the sediment water interface to the depth of the probe tip. (A) Kunz Marsh, (B) Hidden Creek. Gray bars denote the depth range where restored sediments showed steeper gradients than any of the three mature sediments. A = the restored site.

highly reducing relative to the other sites where redox potential was measured. The reference sites, including Hidden Creek, showed a gradual redox gradient and were broadly typical of other estuarine intertidal sediments (Fenchel and Riedl, 1970).

Molecular DGGE "fingerprint" data were collected at the restoration site and one reference site in 2000 (subsequent years data (2001, 2002) sampled in these sites plus two additional reference sites, which were not included in this analysis for clarity). To serve as a point of reference, microbial diversity data collected from an outer coast (outside the estuarine embayment) and from a molecular standard were included. The gel shown in Figure 8.5 (below) is a digitized image of the gel generated from the restoration site. The bands in each lane were detected by Gel Pro software and used to generate a binary matrix. A total of 28 bands were identified in all 3 years of data collected from these locations. Bray-Curtis similarity values were used to generate the non-metric MDS plots shown in Figure 8.3. The larger MDS shown contains all 3 years of data from restored, reference, coastal sites plus the molecular marker. This plot showed that, over the 3-year period sampled, the restoration and reference sites showed partially overlapping communities. A one-way ANOSIM tested the dissimilarity among sites when all 3 years were considered and showed that the Global R was 0.50. Since the statistic is a value between 0 and 1, the test indicated statistically significant differences ($p < 0.01$) among the reference, restored, and coastal, sites.

Pairwise comparisons showed that the R-statistic between reference and restoration sites over the 3-year period was 0.46 ($p < 0.01$). Further, there was no statistically significant difference among years when site groups were averaged (Global $R = 0.02$, $p < 0.32$). To show the differences between bacterial communities in the restoration and reference sites, a nonmetric MDS was presented for each year (Figure 8.3). The MDS for 2000 data revealed strong separation by site and additional clusters within both the restoration and reference sites. The genetic diversity in the restoration site was considerably higher than in the restored site. The restored site had eight or more prominent bands per lane, whereas the mature site had two prominent bands and several faint bands. The average number of bands per lane, averaged across replicates, was higher in the restored site (mean = 8.9, n = 12, sd = 2.7) than in the reference

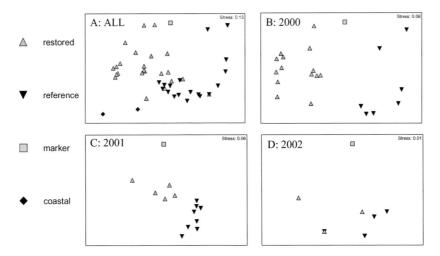

FIGURE 8.3 Non-metric MDS ordination of restored and pristine sites for (A) all years, (B) 2000, (C) 2001, (D) 2002.

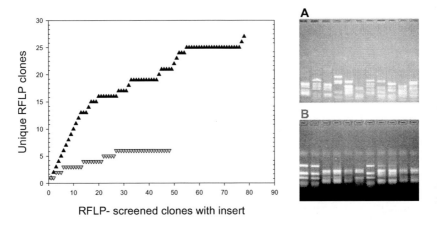

FIGURE 8.4 Rarefaction curve for 16S rDNA clones with unique RFLP profiles amplified with universal primers (8F, 1392R). A clone library was constructed for the restoration site, Kunz Marsh (A), and the mature estuarine wetland, Hidden Creek (B).

site (mean = 3.75, *n* = 12, sd = 2.6). The dissimilarity in the community composition shown in the MDS was also a reflection of community composition.

Molecular cloning was used as an alternative molecular approach to confirm the observation that genetic diversity was higher in the restoration site. In all, 80 clones containing the ~1400 basepair insert were screened from the restoration site, yielding 27 unique clones. Only six unique RFLP profiles were found in the reference site of the 50 clones that were screened (Figure 8.4). These tests confirmed the results of the DGGE analysis from 2002 and reflected the greater genetic diversity in the restored site compared to the restoration site (Figure 8.5). One explanation of this observation can be found by further examining the MDS from 2000. There are two clusters within the restoration site; one cluster from samples collected at 0 to 2 cm depth, and one cluster from 2 to 4 cm depth. A similar but not as obvious pattern was shown in the reference site. Given the steep change in mV observed in the restored site, it is likely that the clusters reflect a difference in the communities above and below the redox discontinuity layer (RPD). The clusters (and differences among communities) do not appear to be as great within the reference site. In subsequent years (2000, 2001), there does not appear to be strong evidence of within-

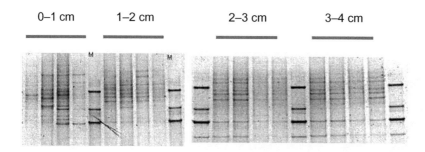

FIGURE 8.5 DGGE gel image from restored site; depth fractions are labeled; M indicates marker lanes used to standardize gels.

site clustering among 0 to 2 cm and 2 to 4 cm. The similarity between restoration and reference appears to be higher in 2001 and 2002, as the sampled communities are, in some cases, overlapping.

Discussion

Estuarine and marine sediments have several notable characteristics that should be considered when evaluating the condition of estuarine and marine sediments. There is a gradient of oxygen and reduction potential as one moves deeper into the sediments (Zobell, 1946). The strength of the redox potential gradient and the position of the redox potential discontinuity layer were shown to be related to sediment size (Fenchel and Riedl, 1970), water velocity, and the presence of bioturbators (Hines, 1991). Organic loading, a sign of hypereutrophication and restricted water flow, is also important. The nature of the redox gradient at a given site and across a gradient of organic loading is a useful tool to quantify the degree of loading and deterioration and degradation of sedimentary conditions. Estuarine sediments have a characteristic 3 to 4 cm aerobic zone overlying the anaerobic zone (Chester, 1990). From the data presented above, the redox potential and soil condition are not only the key to understanding the biological data (bacterial community), but also appear to be a central theme to understanding the recovery of formerly impounded sites.

The higher genetic diversity observed in the restoration site can be attributed to the presence of two distinct communities of bacteria found in the 0 to 2 cm fraction and the 2 to 4 cm fraction. The steep gradient in mV was between 0 and 1 cm, when measured with the platinum electrode. Because the redox potential was not measured in the sediments where the core sample was collected, it is not possible to clearly coordinate the steep mV to a shift in community composition. However, the MDS plot clearly shows two clusters within the reference site between the shallow fractions and deeper fractions. A similar pattern was observed in the reference site, but the differences between the shallow and deep bacterial communities and the differences in mV were not as great. In subsequent years, the differences between shallow and deep bacterial communities were not significant. An effort is under way to increase the sensitivity of molecular characterization by identifying only those bacteria that are active (Choi et al., 1999).

The strong reduction gradient observed in the restoration (Kunz Marsh) site is likely due to the impounded history of the locality. Core samples collected from Kunz Marsh contained a thick layer of peat 2 cm below the sediment–air interface. The peat layer contained roots, branches, leaves, and other large pieces of organic material. Stratigraphic analysis of impounded and mature wetlands showed 30 cm of freshwater wetland peat overlying dead *Spartina* rhizomes (Portnoy and Giblin, 1997b). The residual peat in the restoration (Kunz Marsh) site likely influenced the amount and type of carbon source available to the bacterial community causing a distinct community to develop. Carbon sources in a pristine estuarine wetland include complex structural polysaccharides from wetland plants and algae as well as more labile phytoplankton exudates (Haddad and Martens, 1987). Greater available carbon resources also increase the electrons needed for oxidation (Chester, 1990), which may have been partly responsible for the development of two unique communities within the restoration site. Through time, the distinctness of the two depth-related communities within the restoration site decreased. In 2001 and 2002, the difference between shallow and deeper communities was less obvious and reflected the organization of the communities from the reference site.

Several authors in this volume acknowledge the value and limitations of using community structure to evaluate ecosystem and ecological performance. It has a high level of ecosystem relevance, but a low level of precision in identifying the ecological response of chronic stress. To be an effective indicator, community shifts in response to perturbation should be understood relative to natural scales of habitat complexity, such as redox potential. This approach will lead to generalizations about the response of biological communities to restoration-specific environmental conditions and recovery.

Acknowledgments

The author acknowledges the advice, patience, and financial support of his major professor Dr. Lynda Shapiro. The work was conducted at the Oregon Institute of Marine Biology in partial fulfillment of a Ph.D. in Biology. Dr. Steve Rumrill, Craig Cornu, and the South Slough National Estuarine Research Reserve provided enlightened instruction and discussion about the field of restoration ecology in coastal areas. An anonymous reviewer greatly improved the manuscript in its early stages. The research was funded by a NERR Graduate Research Fellowship to E.C.M. and a grant from the University of Oregon to Dr. Shapiro.

References

Amann, R. I., W. Ludwig, and K.-H. Schleifer. 1995. Phylogenetic identification and *in situ* detection of individual microbial cells without cultivation. *Microbiological Reviews* 59:143–169.

Bouvier, T. C. and P. del Giorgio. 2002. Compositional changes in free-living bacterial communities along a salinity gradient in two temperate estuaries. *Limnology and Oceanography* 47:453–470.

Boyer, K. E. and J. B. Zedler. 1999. Nitrogen addition could shift plant community composition in a restored California salt marsh. *Restoration Ecology* 7:74–85.

Boyer, K. E., J. C. Callaway, and J. B. Zedler. 2000. Evaluating the progress of restored cordgrass (*Spartina foliosa*) marshes: belowground biomass and tissue nitrogen. *Estuaries* 23:711–721.

Callaway, J. C. 2001. Hydrology and substrate. In *Handbook for Restoring Coastal Wetlands,* J. Zedler (ed.). CRC Press, Boca Raton, FL, pp. 89–115.

Callaway, J. C., J. B. Zedler, and D. L. Ross. 1997. Using tidal salt marsh mesocosms to aid wetland restoration. *Restoration Ecology* 5:135–146.

Chester, R. 1990. *Marine Geochemistry.* Unwin Hyman, London.

Choi, J., B. Sherr, and E. Sherr. 1999. A large fraction of ETS-inactive marine bacterioplankton cells, as assessed by reduction of CTC, can become ETS-active with incubation and substrate addition. *Aquatic Microbial Ecology* 18:105–115.

Cifuentes, A., J. Anton, S. Benlloch, A. Donnelly, R. A. Herbert, and F. Rodriguez-Valera. 2000. Prokaryotic diversity in *Zostera noltii*-colonized marine sediments. *Applied and Environmental Microbiology* 66:1715–1719.

Clarke, K. R. and R. M. Warwick. 2001. *Change in Marine Communities: An Approach to Statistical Analysis and Interpretation,* 2nd ed. PRIMER-E, Plymouth, U.K.

Clarke, K. R. and R. N. Gorley. 2002. PRIMER v5 (Plymouth Routines in Multivariate Ecological Research). PRIMER-E, Plymouth, U.K.

Costanza, R., R. dArge, R. deGroot, S. Farber, M. Grasso, B. Hannon, K. Limburg, S. Naeem, R. V. Oneill, J. Paruelo, R. G. Raskin, P. Sutton, and M. vandenBelt. 1997. The value of the world's ecosystem services and natural capital. *Nature* 387:253–260.

Crump, B. C., E. V. Armbrust, and J. A. Baross. 1999. Phylogenetic analysis of particle-attached and free-living bacterial communities in the Columbia River, its estuary, and the adjacent coastal ocean. *Applied and Environmental Microbiology* 65:3192–3204.

Fenchel, T. and R. Riedl. 1970. The sulfide system: a new biotic community underneath the oxidized layer of marine sand bottoms. *Marine Biology* 7:255–268.

Fenchel, T., G. M. King, and T. H. Blackburn. 1998. *Bacterial Biogeochemistry*. Academic Press, San Diego.

Frenkel, R. and J. Morlan. 1991. Can we restore our salt marshes? Lessons from the Salmon River, OR. *Northwest Environmental Journal* 7:119–135.

Ghiorse, W. C. 1997. Subterranean life. *Science* 275:789–790.

Giovannoni, S. J., T. B. Britschgi, C. L. Moyer, and K. Field. 1990. Genetic diversity in Sargasso Sea bacterioplankton. *Nature* 345:60–63.

Haddad, R. I. and C. S. Martens. 1987. Biogeochemical cycling in an organic-rich coastal marine basin. *Geochimica et Cosmochimica Acta* 51:2991–3002.

Hines, M. 1991. The role of certain infauna and vascular plants in the mediation of redox reactions in marine sediments. Diversity of environmental biogeochemistry. *Developments in Geochemistry* 6(6):275–286.

Labrenz, M., G. Druschel, T. Thomsen-Ebert, and B. Gilbert. 2000. Formation of sphalerite (ZnS) deposits in natural biofilms of sulfate-reducing bacteria. *Science* 290:1744–1747.

Lewis, R. R. 1982. *Creation and Restoration of Coastal Plant Communities*. CRC Press, Boca Raton, FL.

Li, Q. and C. L. Ownby. 1993. A rapid method for extraction of DNA from agarose gels. *Biotechniques* 15:976–978.

Lindig-Cisneros, R. and J. B. Zedler. 2002. Halophyte recruitment in a salt marsh restoration site. *Estuaries* 25:1174–1183.

Montague, C. L., A. V. Zale, and H. F. Percival. 1987. Ecological effects of coastal marsh impoundments: a review. *Environmental Management* 11:743–756.

Muyzer, G., E. C. D. Wall, and A. G. Uitterlinden. 1993. Profiling of complex microbial populations by denaturing gradient gel electrophoresis analysis of polymerase chain reaction-amplified genes coding for 16S rRNA. *Applied and Environmental Microbiology* 59:695–700.

Nadeau, T. L., E. C. Milbrandt, and R. W. Castenholz. 2001. Evolutionary relationships of cultivated Antarctic oscillatorians (Cyanobacteria). *Journal of Phycology* 37:650–654.

National Research Council. 1992. Restoration of Aquatic Ecosystems: Science, Technology, and Public Policy. National Academy Press, Washington, D.C.

Pace, N. R. 1997. A molecular view of microbial diversity and the biosphere. *Science* 276:734–740.

Piceno, Y. M. and C. R. Lovell. 2000a. Stability in natural bacterial communities: I. Nutrient addition effects on rhizosphere diazotroph assemblage composition. *Microbial Ecology* 39:32–40.

Piceno, Y. M. and C. R. Lovell. 2000b. Stability in natural bacterial communities: II. Plant resource allocation effects on rhizosphere diazotroph assemblage composition. *Microbial Ecology* 39:41–48.

Portnoy, J. and A. Giblin. 1997a. Biogeochemical effects of seawater restoration to diked salt marshes. *Ecological Applications* 7:1054–1063.

Portnoy, J. and A. Giblin. 1997b. Effects of historic tidal restrictions on salt marsh sediment chemistry. *Biogeochemistry* 36:275–303.

Rappe, M. S., S. A. Connon, K. L. Vergin, and S. J. Giovannoni. 2002. Cultivation of the ubiquitous SAR11 marine bacterioplankton clade. *Nature* 418:630–633.

Redfield, A. C. 1972. Development of a New England saltmarsh. *Ecological Applications* 42:201–237.

Reysenbach, A. L. and S. L. Cady. 2001. Microbiology of ancient and modern hydrothermal systems. *Trends in Microbiology* 9:79–86.

Rumrill, S. 2002. The Ecology of South Slough Estuary: Site Profile of the South Slough National Estuarine Research Reserve. NOAA, Charleston, OR.

Rumrill, S. and C. Cornu. 1995. South Slough coastal watershed restoration: a case study in integrated ecosystem restoration. *Pacific Northwest Reports* 13:53–57.

Shäfer, H. and G. Muyzer. 2000. Denaturing gradient gel electrophoresis in marine microbial ecology. In *Methods in Microbiology, Marine Microbiology*, J. Paul (ed.). Academic Press, San Diego, CA, pp. 425–468.

Skoog, D. A. and J. J. Leary. 1992. *Principles of Instrumental Analysis,* 4th ed. Saunders College Publishing, Ft. Worth, TX.

Torsvic, V., J. Goksoyr, and L. Daae. 1990. High diversity in DNA of soil bacteria. *Applied and Environmental Microbiology* 56:782–787.

Van Hannen, E. J., M. P. Van Agterveld, H. J. Gons, and H. J. Laanbroek. 1998. Revealing genetic diversity of eukaryotic microorganisms in aquatic environments by denaturing gradient gel electrophoresis. *Journal of Phycology* 34:206–213.

Van Hannen, E. J., W. Mooij, M. P. Van Agterveld, H. J. Gons, and H. J. Laanbroek. 1999. Detritus-dependent development of the microbial community in an experimental system: qualitative analysis by denaturing gradient gel electrophoresis. *Applied and Environmental Microbiology* 65:2478–2484.

Walters, K. and D. J. W. Moriarty. 1993. The effects of complex trophic interactions on a marine microbenthic community. *Ecology* 74:1475–1489.

Wawer, C. and G. Muyzer. 1995. Genetic diversity of *Desulfovibrio* spp. in environmental samples analyzed by denaturing gradient gel electrophoresis of [NiFe] hydrogenase gene fragments. *Applied and Environmental Microbiology* 61:2203–2210.

Woese, C. R. 1987. Bacterial evolution. *Microbiological Reviews* 51:221–271.

Zedler, J. B. 2000. Progress in wetland restoration ecology. *Trends in Ecology and Evolution* 15:402–407.

Zedler, J. B. and J. C. Callaway. 1999. Tracking wetland restoration: do mitigation sites follow desired trajectories? *Restoration Ecology* 7:69–73.

Zhou, J., M. Bruns, and J. Tiedje. 1996. DNA recovery from soils of diverse composition. *Applied and Environmental Microbiology* 62:316–322.

Zobell, C. 1946. Studies on redox potential of marine sediments. *Bulletin of the American Association of Petroleum Geologists* 30:477–513.

9

Microbial Biofilms as Integrative Sensors of Environmental Quality

Richard A. Snyder, Michael A. Lewis, Andreas Nocker, and Joe E. Lepo

CONTENTS

Introduction

Recognition of correlations between the occurrence of microbial organisms and various environmental conditions is as old as microscopy (Dobell, 1958). Formalization of the concept as a biological indicator also has a long history. The Saprobic (Saprobien) system represents an early European effort to standardize this type of approach (Kolkwitz and Marsson, 1908). Ruth Patrick in the United States established periphyton analysis as a standard for freshwater work (Patrick et al., 1954; Patrick, 1967; Weitzel, 1979), and the dominance of diatoms within open-water biofilms has led to the description of these films as "periphyton." The autotrophic component is, however, only part of the entire community structure. These surface films are composed of bacteria, microalgae, protists, and small metazoans in a polymer matrix (Figure 9.1), making the term *aufwuchs* technically more correct, and inclusive of biofilms occurring in light-limited conditions. Efforts to tap these communities as indicators of water quality have focused on both the algal and protist components (Stewart et al., 1985; Foissner et al., 1992; Eaton et al., 1995; Foissner and Berger, 1996; U.S. EPA, 1997; FDEP, 1998; Kanhere and Gunale, 1999; Madoni and Bassanini, 1999; Wu, 1999). Analysis of the prokaryotic portion of environmental microbial biofilms to define environmental conditions has been a more recent approach (Guckert et al., 1992; Mohamed et al., 1998; Manz et al., 1999; Piceno and Lovell, 2000).

This chapter presents a review of biofilm research related to the factors controlling community structure and function, the conceptual basis for using biofilms as biological indicators, and case studies of several applications the authors have undertaken to answer specific questions.

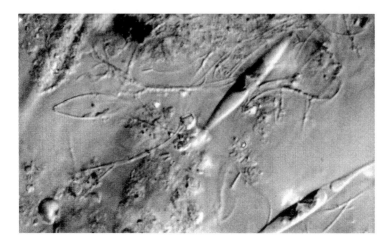

FIGURE 9.1 Aufwuchs removed from an artificial substrate incubated in the Pensacola Bay estuary, FL. These communities are a mixture of autotrophic and heterotrophic organisms, both prokaryotes and eukaryotes. Nikon DIC image at 1750× original magnification.

Biofilm Ecology

Immersion of clean surfaces in aquatic environments results in a variety of physical and biological events (Wahl, 1989; Cooksey and Wigglesworth-Cooksey, 1995; Wimpenny and Colasanti, 1997). Initially, the development of biofilms is dominated by allochthonous processes, with autochthonous factors becoming more important with time (Wahl, 1989). The relative abundance of colonizing particles often determines the speed and apparent sequence of biofilm colonization, although some colonizers clearly respond to chemical signals from previously colonized biofilm organisms. Adsorption of dissolved organic material (DOM) proceeds rapidly, changes surface charge and hydrophobicity, and provides substrates for bacterial growth. Suspended bacterial cells adsorb by diffusive movement in still or laminar flow conditions, and by advective processes in turbulent conditions (Escher and Characklis, 1988). Bacteria vary in their colonizing ability (Dalton et al., 1996) suggesting the process is not entirely passive. Bacteria colonizing surfaces in a tidal marsh study were predominantly of the Roseobacter group, which were also found in nearby, existing biofilms (Dang and Lovell, 2000). Other studies show greater diversity in colonizers and shifts of community structure over time that may be habitat specific (Sonak and Bhosle, 1995).

Individual cells may become irreversibly adsorbed via extracellular polysaccharides (EPS). Growth of these colonists results in microcolonies (Lawrence and Caldwell, 1987) and further production of EPS, which develops into a biofilm matrix. Biofilms developing under oligotrophic conditions develop "mushroom" pillars of EPS and internally dispersed cells with a low biomass-to-EPS ratio (Møller et al., 1997). Pillars may fuse at the tops to form a highly complex system of channels and conduits through the biofilm. At the other extreme, biofilms developing in high-nutrient (DOM and inorganic nutrients) conditions tend to have less structural variability and have high ratios of cell biomass to EPS (Wimpenny and Colasanti, 1997). At a chemical/structural diversity level, EPS also changes with (1) nutrient availability (as a determinant of bacterial physiological state), (2) composition (quality) of organic substrates available for growth, (3) ionic and physical conditions (Decho, 1990), and (4) predation by protists and metazoans (Hunter, 1983; Lawrence et al., 2002).

Anaerobic microzones develop in the deeper spaces harboring sulfate-reducing prokaryotes, nitrogen cycling prokaryotes, and methylotrophic prokaryotes (Kuhl and Jorgensen, 1992; Amann et al., 1992; Raskin et al., 1995; Santegoeds et al., 1998; Hunter et al., 1998; Kolari et al., 1998; Okabe et al., 1999; Schramm et al., 1999; Tolker-Nielsen and Molin, 2000). The speed of development and extent of the anaerobic component appear to be tied to the availability of DOM and inorganic nutrients.

Once established, the microbial communities within biofilms may display considerable resiliency. In a laboratory experiment with a biofilm of nitrifying bacteria developed under continuous flow of ammonium, ammonium starvation for 43 days reduced cell number but did not cause a lag in response to reintroduction of ammonium (Batchelor et al., 1997). It has been suggested that the polymer matrix may function as a buffer and concentrator for growth substrates and nutrients that dampen environmental stochasticity of these parameters (Sinsabaugh et al., 1991; Freeman and Lock, 1995).

Biofilms concentrate and integrate organic and inorganic constituents of the milieu in which they develop, making them sites of intensified biogeochemical process relative to their environs (Zobell, 1943; Jeffrey and Paul, 1986). This property of microbial aufwuchs makes them more sensitive indicators of ambient conditions than water column parameters, especially in the initial phase of development where allochthonous factors have a more pronounced influence on biofilm growth and activity. As the community matures, bacterivorous and algivorous protists colonize biofilms as both sessile and mobile components, and provide a steady flux of internally recycled nutrients to the biofilm (Pederson, 1982; Jackson and Jones, 1991; Rao et al., 1997). Biofilm nutrient content (nitrogen and phosphorus) has been documented to be related to protist presence, and is usually higher than ambient water concentrations, even in high light levels that provide a demand for nutrients into autotrophic (diatom) biomass (Rao et al., 1997).

Extensive laboratory work has analyzed microbial film bacterial communities for physical structure, taxonomic diversity, and activity. The advent of confocal laser microscopy has permitted nondestructive analysis and has dramatically added to our understanding of the physical structure and distribution of species within these complex communities (Surman et al., 1996; Neu and Lawrence, 1997; Norton et al., 1998). Fluorescent *in situ* hybridizations (FISH) with taxon-specific probes and the use of fluorescent activity probes have permitted visualization of the distribution patterns of species and their activities (Raskin et al., 1995; Santegoeds et al., 1998; Kolari et al., 1998; Okabe et al., 1999; Manz et al., 1999; Bartosch et al., 1999; Schramm et al., 1999; Araya et al., 2003). In at least one case, nucleic acid probes for microbial diversity provided results that were contrary to culture data (Roske et al., 1998).

Intertidal biofilms represent a special situation with periodic air exposure and interactions with the surface microlayer of estuaries. These biofilms develop in similar fashion to submerged biofilms when flooded by tides, but they also receive significant contributions of DOM, particulate organic matter (POM), and inorganic material deposited from the surface microlayer as the tide falls past them. The surface microlayer is a site of concentrated organics (low-molecular-weight compounds, proteins, lipids, EPS), cells, detritus from both allochthonous and autochthonous sources (Hale and Mitchell, 1997; Gasparovi et al., 1998; Falkowska, 1999a,b), and toxic materials, especially metals (Liu and Dickhut, 1997).

Submerged microbial biofilms also adsorb toxic materials from the environment. Metals and organic toxins are sequestered within the matrix by adsorption and complexation reactions with polymers (Wolfaardt et al., 1998) and interactions with iron and manganese oxides. These reactions are facilitated by uronic acid and ketal-linked pyruvate residues, and although the concentration of toxins from the environment is enhanced, binding reactions may also limit the bioavailability of the toxic compounds (Decho, 1990; Said and Lewis, 1991).

In addition to being a direct food source for many estuarine organisms, biofilms may also control the development and structure of macrobenthic communities through the stimulation or inhibition of settlement and metamorphosis reactions in invertebrate larvae (Phelps and Mihursky, 1986; Chang et al., 1996; Wieczorek and Todd, 1997; Unabia and Hadfield, 1999; Rodriguez and Epifanio, 2000; Huang and Hadfield, 2003). Thus, monitoring the structure of microbial biofilm communities in estuaries has the potential to predict impacts on the distribution and abundance of higher trophic levels. In freshwater environments, the metazoan component of surface communities is not as much of a factor as in the estuarine and marine environment, where extensive colonization by sessile invertebrates may occur, many of which have calcareous shells. Allowing development of biofilms in estuarine and marine systems to proceed to this extent may therefore pose problems for collection of biomass and bulk chemical analyses.

Biofilms as Estuarine Indicators

The health of estuarine ecosystems is reflected in biogeochemical cycles of sulfur, nitrogen, phosphorus, and carbon in the water column and benthos. These processes determine nutrient availability and are catalyzed by, or closely linked to, microorganisms. Measurement of key processes controlling biogeochemical fluxes and storage of sulfur, nitrogen, and phosphorus in the ecosystem would provide an index of ecosystem status. However, the inherent *in situ* variability of measurements of such processes (Meyer-Reil, 1994) may complicate their use as effective indicators. Biofilms developed on introduced standardized substrata may provide a mechanism to control microhabitat variability resulting from differences in natural substrata, and yet provide macrohabitat-specific responses, quantifiable by a suite of parameters varying in sophistication and information content.

Current ecosystem indicators do not meet all criteria of the U.S. Environmental Protection Agency (EPA) as reflections of biogeochemical ecosystem function, resiliency, and integrity (Jackson et al., 2000). Microbial biofilms fulfill EPA criteria for ecosystem status indicators. They are based on high numbers of physiologically and taxonomically diverse populations that mediate most biogeochemical processes critical to ecosystem function and are reactive to changes in nutrients, toxics, and other stressors. We hypothesize that the species diversity of biogeochemical functional groups, for example, autotrophs, denitrifying bacteria, sulfate-reducing prokaryotes (SRBs), nitrogen-fixing prokaryotes, will be responsive to nutrient loading and other stressors. Diversity would be reduced through stressor impact, and ecosystem robustness (resiliency) would decrease.

Biofilms as a physical habitat provide buffering of minor fluctuations in ambient water quality parameters, resulting in integration of short-term variability. The technologies for population diversity assessment and biogeochemical process measurement are routine and quantitative. Microbial sentinels are hypothesized to be predictive of incipient impacts on higher trophic levels (fish, submerged aquatic vegetation [SAV]), and on general aesthetics (water clarity, odor). Microbial biofilm analytical end points can be categorized into a tiered system based on information content, technical sophistication, and cost of implementation.

Sampler Design

The Catherwood diatometer using glass slides as artificial substrates is the most widely used sampler for periphyton analysis. However, many different types of artificial substrate samplers have been employed from glass cover slips and scanning electronic microscope (SEM) stubs to acrylic plates and machined rock surfaces (Austin et al., 1981; Bamforth, 1982; Aloi, 1990). We have found that the size and type of substrate used needs to be matched to the intended analysis. For our work, we have designed and constructed inexpensive frames of PVC (polyvinylchloride) pipe to hold acrylic plates, and we have used acrylic racks to hold glass slides on these same sampling platforms (Figure 9.2). Recognition of the importance of benthic processes in estuaries led to the development of a benthic frame sampler with a concrete ring base (Figure 9.2).

FIGURE 9.2 Inexpensive frames with acrylic plates for biofilm sampling. Floating frame (left) and benthic sampling frames (right).

A variety of standard methods are available to analyze biofilm community structure and function, used directly or with minor modification from such sources as Eaton et al. (1995), Gerhardt et al. (1994), and Kemp et al. (1993). Typically, biofilm material is removed from plates by scraping or "squeegee." This material can then be collected on glass fiber or other filter types (depending on the intended application) and processed immediately or stored frozen until analysis. Typical analyses include relatively simple bulk chemical characterizations (dry weight; carbon, nitrogen, and phosphorus content) and optical properties (optical density, spectral reflectance; Gitelson et al., 1999). More sophisticated chemical analyses, such as stable isotope signatures, or analysis of specific organic fractions (pigments, lipids, carbohydrates) can provide useful information. Biofilms also concentrate waterborne contaminants and are amenable to standard analyses for metals and organic toxic compounds. Process-oriented analyses (e.g., total heterotrophic activity, phosphatase activity, and nitrogen fixation/denitrification) can be conducted with material removed from substrates by standard techniques and for intact biofilms with minor modifications to accommodate substrates in the incubations. In addition to the traditional microscopic analysis of diatom community structure, phospholipid fatty acid profiles and molecular methods for determining community structure and shifts in specific functional groups of microorganisms can be accomplished using methods that are becoming routine.

Estuarine Periphyton as an Environmental Diagnostic Indicator

The use of the autotrophic and heterotrophic components of aufwuchs as an indicator of environmental condition has been more common in freshwater environments (for example, McCormick and Stevenson, 1998; Hill et al., 2000) than in near-coastal areas. Consequently, the database describing the development of basic characteristics as a response to different water quality conditions is limited.

In a series of surveys conducted in coastal areas receiving agricultural runoff (Lewis et al., 2000, 2004a), treated wastewater discharges (Lewis et al., 2001a, 2002a), dredging activity (Lewis et al., 2001b), golf course runoff (Lewis et al., 2002b, 2004b) and urban storm water runoff, we evaluated biofilm characteristics. Acrylic substrates (0.1 m² each) were immersed in floating racks (see Figure 9.1) approximately 0.3 m below the surface for up to 6 weeks in these studies. Analyses included species community composition and nontaxonomic structural characteristics such as biomass and pigment analysis (chlorphylls *a*, *b*, and *c*, carotenes, phaeophytin). Biomass was determined as dry and ash-free weights and expressed as an autotrophic index by comparison to chlorophyll *a*. In addition, trace metal, polycyclic aromatic hydrocarbons (PAH), polychlorinated biphenyls (PCB), and pesticide residues were determined for the biofilm matrix. These concentrations were compared to the corresponding contaminants in surface water to determine bioconcentration factors.

The algal component of the biofilms was dominated by species of *Chrysophyta* and *Chlorophyta* at all locations after the 3- to 6-week colonization periods. For example, 106 species representing 56 genera colonized the substrates in the Everglades–Florida Bay transitional zone. Of these, diatoms represented 33 genera and 76 species and 15 species of the *Chlorophyta* were identified representing 11 genera. The results of the nontaxonomic structural analyses were variable; spatial and temporal variation in biomass and pigment concentrations were usually fourfold or less for spatial variation in three urbanized bayous. Ash-free dry weight was significantly greater for biofilms colonized in areas receiving urban storm water and golf course runoff (Figure 9.3A).

Residues for organic contaminants in the periphyton were usually below detection. However, trace metals were commonly detected and were significantly greater for the periphyton colonized in urbanized coastal areas (Figure 9.3B). Metals concentrations were greater than in surface water samples, but in similar concentrations as in local sediments (Figure 9.4). Spatial and temporal differences for the same trace metal were usually fourfold or less for the biofilms colonized in the various surveys. The bioconcentration factors (BCF; biofilm:water column concentration ratio) ranged from 150 to 28000 for the biofilm matrix relative to water column samples. The BCF was greatest in most cases for mercury, and was similar to those calculated from transplanted caged oysters (Goodman et al., 1999).

The results of the various surveys suggested that use of biofilms, and more specifically periphytic algae, may be useful indicators of environmental condition in shallow near-coastal areas characteristic

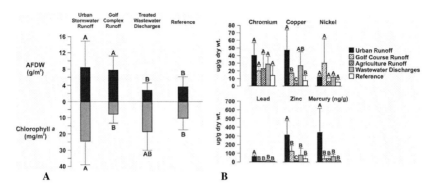

FIGURE 9.3 Analysis of biofilms from floating samplers. (A) Comparison of ash-free dry weight (AFDW) and chlorophyll *a* concentrations for biofilms colonized in reference and nonreference near-coastal areas for 21 days. Letters designate significant differences among means (ANOVA, Duncan's Multiple Range Test, *P* = 0.05). (B) Comparison of trace metal bioaccumulation in biofilms colonized in reference and nonreference coastal areas. Values represent means (standard deviation) in μg/g dry weight (ng/g dry weight for mercury). Letters represent significant differences among means (*P* = 0.05).

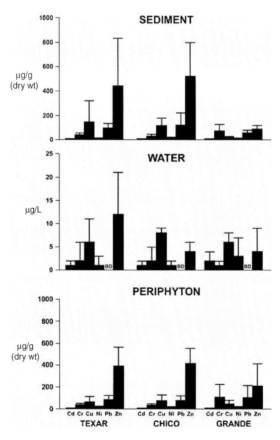

FIGURE 9.4 Trace metal residues in 21-day-old biofilms, water column samples, and sediments from three urbanized bayous, Pensacola Bay system. Values represent mean (standard deviation) in μg/g dry weight.

of the Gulf of Mexico region. However, the technique is not simple. The collection and analysis of the taxonomic and chemical residue data, although routine, is labor intensive. The frequency of colonization and number of colonization sites need consideration since they may affect the relevancy of results and the ability to extrapolate data among and within coastal areas if not considered in the experimental design. Furthermore, the effect of settled substrate solids on contaminant residue results will need determination to ensure the relevancy of these results.

Analysis of Spatial Patterns of Biofilm Growth

Problems encountered in using biofilm growth as an absolute indicator of condition by comparison to a "library" or database of cause-and-effect responses may be avoided by using the samplers for relative comparison purposes. As in the standard periphyton method, a comparison between an impacted and a reference site can be the focus of the sampling design. However, at a larger ecosystem scale, biofilms may be used to examine patterns of nutrient loading, or the extent of loadings impacts, as was done in the Baltic by Mattila and Räisänen (1998). One must also be careful to match the resolution of an indicator to the environmental stochasticity of the system. A common complaint of the standard periphyton method is the variability between slides in a sampler. Often this is high enough to preclude meaningful statistical comparison between sites. Some of this variability may be due to sample size, especially in more oligotrophic settings, where the founder effect of colonizing microalgae may dominate a portion of a slide and uncolonized areas may persist. The variance between slides may be higher in oligotrophic or relatively "clean sites" while, at sites with excess nutrients, such spatial heterogeneity would not persist with growth.

Thus, variation that is often seen as a problem in the periphyton method may indeed be as valuable a signal as the species composition and biomass accumulation estimates that are the usual focus. Biofilms are directly amenable to addressing spatial scaling and spatial variance patterns within estuaries. We have used a simple estimate of biomass by optical density for this type of ecosystem characterization, but chlorophyll content, ash-free dry weight, or other measures of biomass accumulation would also be appropriate. By using relatively simplistic and inexpensive analysis techniques, sample size and number of stations can be increased. We have focused on an incubation period of 7 days to maximize allochthonous influences on biofilm growth, maximize biomass for analysis, and limit colonization of calcareous invertebrates.

We scanned dried biofilms at resolution of 200 dpi on a flat-bed scanner in transparency mode. Digitized plate images were then analyzed using a standard image analysis software package (NIH Image; available free from http://rsb.info.nih.gov/nih-image/). Mean and standard deviations of whole plate images were used for comparisons between plates for total density and spatial variance.

A continuous temporal relationship is definable for these parameters in the benthic biofilms from a single site, with the standard deviation increasing as the mean pixel density increases through spring, summer, and fall highs (Figure 9.5A). The June 2003 sampling, however, provides an interesting contrast. Here we found similar biomass estimates as the preceding and subsequent months (May, July), but the variance estimate, as a measure of biofilm spatial pattern, quantifies the uniformity of the sample apparent in Figure 9.5A. This sampling event was coincident with a transition period in the water column (Figure 9.5B). The spatial structure, as variability of pixel densities, of the biofilms indicates similar patterns for the time periods before and after the transition period, despite the overall change in suspended particulate matter (Figure 9.5B). This type of approach indicates how deviations from a "normal" pattern may indicate ecosystem change. In this case, an increased trophic condition of the water column is reflected in the physical structure of the biofilm, whereas the system at the prior or subsequent trophic state yields a similar biofilm structure. More sophisticated analysis of the biofilm biological information content (see Molecular Methods section below) can provide resolution to establish differences in community structure, if any, between the two end points of similar physical structure.

At the ecosystem level, a discrepancy has been noted between phytoplankton biomass response and nutrient availability as determined by periphyton growth (Welch et al., 1972). Estuaries are by nature hydrodynamically complex, with mixing and residence times within a system changing daily with tidal

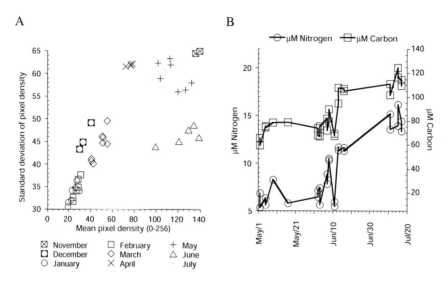

FIGURE 9.5 Temporal analysis of biofilms. (A) Standard deviation of digitized biofilm pixel densities plotted as a function of mean pixel density. Note that the June samples have similar mean density, but lower standard deviations than the preceding May samples or the subsequent July samples, and deviate from a consistent temporal trend defined by the remainder of the samples. (B) Water column particulate analysis for the 3-month period May, June, July 2003. Note the trophic transition as indicated by an overall increase in water column particles and an increase in the nitrogen content of the water column particles.

flow, periodically and seasonally with river flow and temperature changes, and spatially with changes in physical dimensions along the estuarine salinity gradient. These same characteristics define differences between estuaries. These patterns of natural variation within estuarine systems require either encompassing the estuarine gradients in the ecosystem assessment or carefully controlling how comparative sites are chosen within the gradients. The change in residence time affects the ability of phytoplankton (Welch et al., 1972) and bacteria (Crump et al., 1999) to maintain populations within specific areas of the estuary and utilize available nutrients. Analyzing biofilm growth at fixed locations provides the means to examine integrated nutrient loading patterns, independent of the hydrodynamic constraints on the accumulation of phytoplankton biomass.

Estuaries are benthic-dependent systems (Day et al., 1989), yet most of our routine monitoring involves point samples or profiles of the water column at a limited number of sites. Using the benthic frames spaced on a 1 nautical mile grid to obtain biofilms from the estuarine bottom, we were able to estimate the magnitude and spatial variability of benthic production. Clearly, nutrient flux out of the sediments will be intercepted to some degree by benthic microalgae, so this is not a true measure of benthic production. Still, as a relative measure, the results are instructive and can provide information on the status of the system. A strong depth dependence on benthic biofilm development was expected and should be a function of water column turbidity. Depth limitation of biofilm growth was not observed. Indeed, the amount of biofilm accumulation on benthic samplers was equal to floating samplers overall, and for some stations, benthic production levels were much greater than surface production estimates, although the variance between benthic samplers was greater than between floating samplers. This pattern was found for both Escambia and East Bay estuaries, despite a difference in overall magnitude of biofilm growth between the two systems (Table 9.1).

Some of the spatial variance in benthic production can be attributed to the spatial distribution of ecological foci of high primary or secondary production: oyster reefs, shallow areas supporting intense benthic microalgal growth, seagrasses, patches of invertebrate populations, etc. At the other extreme, deeper areas of the estuary that experience reduced light availability and hypoxic or anoxic conditions will have reduced biofilm growth.

TABLE 9.1

Comparison of Benthic and Floating Biofilm Densities after a 7-Day Incubation

Bay	Mean Pixel Density	SD	CV
Floating Biofilms			
Escambia	33.099	11.849	0.358
East	18.543	8.370	0.451
Benthic Biofilms			
Escambia	41.280	32.555	0.789
East	16.378	14.156	0.864

Note: Pixel density values are determined on a gray scale from 0–256.

Molecular Approaches

In the spirit of Ruth Patrick's taxonomic analysis of the diatom component of biofilms as an indicator, we are using molecular analysis to apply this concept to the prokaryotic component of these communities. Standard molecular techniques are becoming faster and cheaper to use. The existing technology allows us to access the tremendous information content of the microbial community structure of biofilms, and to identify shifts in particular functional groups, target genes for specific enzymes, and quantify specific targets. The most commonly applied, culture-independent molecular techniques in microbial ecology are denaturing gradient gel electrophoresis (DGGE), amplified ribosomal DNA restriction analysis (ARDRA), single-strand conformation polymorphism (SSCP), and terminal restriction fragment length polymorphism (T-RFLP). With increasing knowledge about biofilm community composition and specific targets, emerging technology may also permit the monitoring of gene expression patterns and quantification of functional groups using "DNA Chips" and real-time quantitative polymerase chain reaction (QPCR) approaches (e.g., Cho and Tiedje, 2002).

Microbial biofilm communities in the Pensacola Bay estuary were characterized using ARDRA with subsequent sequencing of unique clones and T-RFLP analysis, which has turned out to be a highly reproducible and robust technique, yielding high-quality fingerprints of bacterial assemblages (Clement et al., 1998; Osborn et al., 2000; Nocker et al., in press). The corresponding marker genes (mostly 16S rDNA) were PCR-amplified using genomic DNA as a template and with at least one of the primers carrying a fluorescent label. The amplicon mixture was subsequently digested with a suitable restriction endonuclease, generating fragments of different sizes. The digestion products were then size-sorted on an automated sequencer, which detects the labeled terminal fragments generating a community profile based on fragment lengths. A size standard, run with the sample, allowed the software to calculate the size of the fragments. The digitized image was then displayed as an electropherogram. Ideally, fragment lengths thus generated are characteristic of certain microorganisms or at least of certain phylogenetic groups. The taxonomic resolution of this approach can be increased by combining the information generated with different restriction enzymes. Despite well-known biases inherent to all PCR-based techniques, T-RFLP analyses have been shown to provide an accurate reflection of the ratios of 16S rRNA templates in model communities with defined amounts of 16S rRNA gene copies from selected organisms (Lueders and Friedrich, 2003) and, hence, are reflective of relative abundance.

We have defined significant habitat specificity of benthic microbial biofilms for a healthy oyster reef and an adjacent open muddy-sand bottom at equal depth (Nocker et al., in press). Benthic biofilms were generated over a 7-day period. Total biomass and optical densities of dried biofilms showed dramatic differences for oyster reef vs. non-oyster reef biofilms. The molecular analysis approach should assess whether the observed spatial variation was reflected in the prokaryotic species composition. Indeed, the differences between these adjacent habitats turned out to be equally dramatic also in terms of species

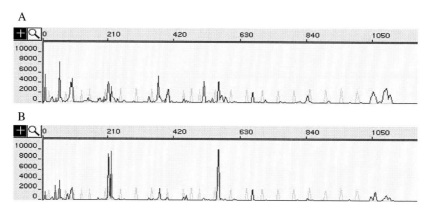

FIGURE 9.6 T-RFLP fingerprints from biofilms grown in an oyster reef (A) and a muddy-sand bottom habitat (B). *CfoI* (*HhaI*) was used as restriction enzyme. Fragment length (bp) is shown on the *x* axis. Signal intensity (fluorescence units) is shown on the *y* axis.

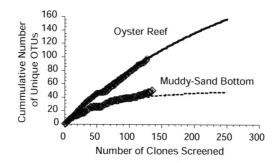

FIGURE 9.7 Estimation of microbial diversities in biofilms grown over 7 days in an oyster reef setting and a muddy-sand bottom setting. The numbers of the cumulative different/unique operational taxonomic units (OTUs) were plotted against the number of clones screened. The bold curves were calculated based on the experimental data. The thin lines represent curve fits based on the equation $y = x/(ax + b)$. Saturation values for these sampling curves estimate the total species richness for the oyster reef biofilm to be 417 OTUs and for the muddy-sand bottom biofilm to be 60 OTUs.

richness, community evenness, and in the shifts of different groups of prokaryotes. The fingerprint from the oyster-reef biofilm showed a greater number of peaks and a substantially more even community profile (Figure 9.6).

Moreover, cloned 16S rRNA PCR products were subjected to restriction fragment length analysis using three independent restriction enzymes. Based on the obtained restriction patterns, clones were grouped into distinct operational taxonomic units (OTUs). Fitting the sampling curve $y = x/(ax + b)$ (Sekiguchi et al., 1998) allowed the estimation of the total species richness for the two habitats (Figure 9.7). Total species richness was estimated to be 417 for the oyster reef and 60 for the muddy-sand bottom with only 10.5% of the total unique OTUs identified being shared between habitats.

The sequences from unique OTUs were matched against the GenBank® database (maintained by the National Center for Biotechnology Information) to determine the nearest phylogenetic neighbors. Major taxonomic grouping for the two biofilm communities indicated the vast majority of the bacteria in the oyster reef biofilm were related to members of the Phyla (Bergey's) δ- and γ-subdivisions of Proteobacteria, the Cytophaga-Flexibacter-Bacteroides cluster, the Phyla Planctomyces and Holophaga-Acidobacterium. The same groups were also present in the biofilm harvested at the muddy-sand bottom, with the difference that nearly half of the community consisted of representatives of the Planctomyces phylum. The results suggest that submerged artificial surfaces allow the formation of uniquely adapted communities that integrate and reflect habitat specificity, even within short spatial scales.

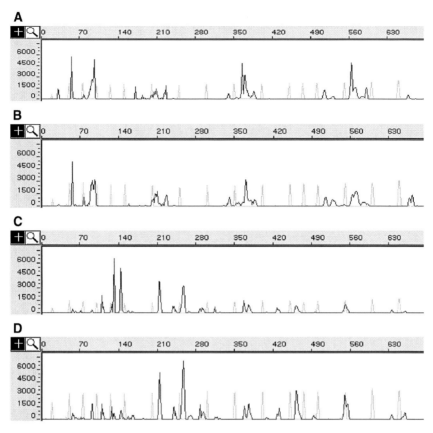

FIGURE 9.8 T-RFLP fingerprints from biofilms grown at a sewage outfall site and a nearby reference site. 16S patterns were generated using *CfoI* (*HhaI*) for the outfall (A) and reference site (B). *Dsr* patterns were generated using *NdeI* for the outfall (C) and the reference site (D). Fragment length (bp) is shown on the *x* axis. Signal intensity (fluorescence units) is shown on the *y* axis.

T-RFLP fingerprints also appear suitable for detecting small differences between similar microbial communities. We examined the prokaryotic community response to a sewage outfall compared to a nearby reference site at similar depth. Analysis of the extracted DNA was performed using PCR primers targeting either eubacterial 16S rRNA genes or *dsr* genes (Figure 9.8). The *dsr* gene codes for dissimilatory sulfite reductase, which catalyzes the central energy-conserving step of sulfate respiration (Odom and Peck, 1984). It is present in all known sulfate-reducing prokaryotes allowing the analysis of shifts within this specific functional group. A substantial amount of the organic carbon in estuaries is processed by these organisms (Day et al., 1989). The analysis of sulfate-reducing microorganisms is of particular interest as some of these organisms utilize environmental contaminants such as petroleum hydrocarbon constituents (e.g., benzene, toluene, polycyclic hydrocarbons, xylenes) or halogenated compounds directly as a source of carbon and energy (Ensley and Suflita, 1995; Zhang and Young, 1997). Their ability to utilize low-molecular-mass organic acids such as acetate, propionate, and butyrate as carbon sources also explains their abundance in anaerobic habitats (Sorenson et al., 1981; Balba and Nedwell, 1982; Parkes et al., 1989).

A comparison of T-RFLP fingerprints both of 16S rRNA as well as *dsr* genes at the two locations reveals a high degree of similarity in the community profiles (Figure 9.8). However, distinct differences in peak occurrences and peak heights can be observed. The relative sensitivity of these organisms to differences in environmental conditions needs to be established to fully utilize their potential to serve as indicator organisms.

Summary

Microbial biofilms, like other indicators, have limitations that must be understood to employ them effectively as environmental sensors. It is unlikely that they can be developed as an absolute indicator of condition, i.e., if we see "x" result from analysis of a biofilm, it means "y" for that location. Rather, they are perhaps best employed as a relative indicator of differences across spatial and temporal scales. As they comprise predominantly single-celled organisms, they are physiologically proximate to the conditions in which they develop, yet they are also integrative of conditions over the incubation period chosen. The former characteristic makes them reactive to environmental conditions and change in advance of other components of the ecosystem. The latter characteristic is perhaps most important in estuaries, where the dynamics of the system often preclude adequate characterization from single spatial and temporal point grab samples. Possible applications span a scale from simple bulk growth response to nutrients to sophisticated analysis of microbial community structure. They are also applicable to analysis of spatial patterns of condition or impacts, from determining site-specific conditions to whole ecosystem or watershed scales. We have only just begun to tap into the information content of these systems and realize their full potential as biosensors.

Acknowledgments

This research was supported by a grant from the U.S. Environmental Protection Agency's Science to Achieve Results (STAR) Estuarine and Great Lakes (EaGLe) Coastal Initiative through funding to the CEER-GOM Project, U.S. EPA Agreement EPA/R-82945801. We gratefully acknowledge assistance from Laura Pennington, Joseph Moss, and Anadiuska Rondon for biofilm processing.

References

Aloi, J.E. 1990. A critical review of recent freshwater periphyton field methods. *Canadian Journal of Fisheries and Aquatic Science* 47:656–670.

Amann, R.I., J. Stromley, R. Devereux, R. Key, and D.A. Stahl. 1992. Molecular and microscopic identification of sulfate-reducing bacteria in multispecies biofilms. *Applied and Environmental Microbiology* 58:614–623.

Araya, R, K. Tani, T. Takagi, N. Yamaguchi, and M. Nasu. 2003. Bacterial activity and community composition in stream water and biofilm from an urban river determined by fluorescent *in situ* hybridization and DGGE analysis. *FEMS Microbiology Ecology* 43:111–119.

Austin, A., S. Lang, and M. Pomeroy. 1981. Simple methods for sampling periphyton with observations on sampler design criteria. *Hydrobiologia* 85:33–47.

Balba, M. T. and D. B. Nedwell. 1982. Microbial metabolism of acetate, propionate and butyrate in anoxic sediment from the Colne Point Saltmarsh, Essex, U.K. *Journal of General Microbiology* 128:1415–1422.

Bamforth, S. S. 1982. The variety of artificial substrates useful for microfauna. In *Artificial Substrates,* J. Cairns (ed.). Ann Arbor Science, Ann Arbor, MI, pp. 115–130.

Bartosch, S., I. Wolgast, E. Spieck, and E. Bock. 1999. Identification of nitrite-oxidizing bacteria with monoclonal antibodies recognizing the nitrite oxidoreductase. *Applied and Environmental Microbiology* 65:4126–4133.

Batchelor, S. E., M. Cooper, S. R. Chhabra, L. A. Glover, G. S. A. B. Stewart, P. Williams, and J. I. Prosser. 1997. Cell density-related recovery of starved biofilm populations of ammonia-oxidizing bacteria. *Applied and Environmental Microbiology* 63:2281–2286.

Chang, E. Y., S. L. Coon, M. Walch, and R. Weiner. 1996. Effects of Hyphomonas PM-1 biofilms on the toxicity of copper and zinc to *Crassostrea gigas* and *Crassostrea virginica* larval set. *Journal of Shellfish Research* 15:589–595.

Cho, J.-C. and J. M. Tiedje. 2002. Quantitative detection of microbial genes by using DNA microarrays. *Applied and Environmental Microbiology* 68:1425–1430.

Clement, B. G., L. E. Kehl, K. L. Debord, and C. L. Kitts. 1998. Terminal restriction fragment patterns (TRFPs), a rapid, PCR-based method for the comparison of complex bacterial communities. *Journal of Microbiological Methods* 31:135–142.

Cooksey, K. E. and B. Wigglesworth-Cooksey. 1995. Adhesion of bacteria and diatoms to surfaces in the sea: a review. *Aquatic Microbial Ecology* 9:87–96.

Crump, B. C., E. V. Armbrust, and J. A. Baross. 1999. Phylogenetic analysis of particle-attached and free-living bacterial communities in the Columbia River, its estuary, and the adjacent coastal ocean. *Applied and Environmental Microbiology* 65:3192–3204.

Dalton, H. M., A. E. Goodman, and K. C. Marshall. 1996. Diversity in surface colonization behavior in marine bacteria. *Journal of Industrial Microbiology and Biotechnology* 17:228–234.

Dang, H. and C. R. Lovell. 2000. Bacterial primary colonization and early succession on surfaces in marine waters as determined by amplified rRNA gene restriction analysis an sequence analysis of 16S rRNA genes. *Applied and Environmental Microbiology* 66:467–475.

Day, J. W., C. A. S. Hall, W. M. Kemp, and A. Yanez-Arancibia. 1989. *Estuarine Ecology*. John Wiley & Sons, New York, pp. 90–96.

Decho, A. W. 1990. Microbial exopolymer secretions in ocean environments: their role(s) in food webs and marine processes. *Oceanography and Marine Biology Annual Review* 28:73–153.

Dobell, C. 1958. *Antony Van Leeuwenhoek and His "Little Animals."* Russell & Russell, New York.

Eaton, A. D., L. S. Clesceri, and A. E. Greenberg. 1995. *Standard Methods for the Examination of Water and Wastewater,* 19th ed. American Public Health Association; American Water Works Association, Water Environment Federation, Washington, D.C.

Ensley, B. D., and J. M. Suflita. 1995. Metabolism of environmental contaminants by mixed and pure cultures of sulfate-reducing bacteria. In *Sulfate-Reducing Bacteria*, L. L. Barton (ed.). Plenum Press, New York, pp. 293–332.

Escher, A. R. and W. G. Characklis. 1988. Microbial colonization of a smooth substratum: a kinetic analysis using image analysis. *Water Science and Technology* 20:277–283.

Falkowska, L. 1999a. Sea surface microlayer: a field evaluation of Teflon plate, glass plate and screen sampling techniques. Part. 1. Thickness of microlayer samples and relation to wind speed. *Oceanologia* 41:211–222.

Falkowska, L. 1999b. Sea surface microlayer: a field evaluation of Teflon plate, glass plate and screen sampling techniques. Part 2. Dissolved and suspended matter. *Oceanologia* 41:223–240.

FDEP. 1998. Impact Bioassessment Investigations, Florida Department of Environmental Protection, available at http://www.dep.state.fl.us/labs/biol/BAintro.html.

Foissner, W. and H. Berger. 1996. A user-friendly guide to the ciliates (Protozoa, Ciliophora) commonly used by hydrobiologists as bioindicators in rivers, lakes, and waste waters, with notes on their ecology. *Freshwater Biology* 35:375–482.

Foissner, W., A. Unterweger, and T. Henschel. 1992. Comparison of direct stream bed and artificial substrate sampling of ciliates (Protozoa, Ciliophora) in a mesosaprobic river. *Limnologica* 22:97–104.

Freeman, C. and M. A. Lock. 1995. The biofilm polysaccharide matrix: a buffer against changing organic substrate supply? *Limnology and Oceanography* 40:273–278.

Gasparovi, B., Z. Kozarac, A. Saliot, B. Osovi, and D. Moebius. 1998. Physicochemical characterization of natural and ex-situ reconstructed sea-surface microlayers. *Journal of Colloid and Interface Science* 208:191–202.

Gerhardt, P., R. G. E. Murray, W. A. Wood, and N. R. Krieg (eds.). 1994. *Methods for General and Molecular Bacteriology*. American Society for Microbiology, Washington, D.C., 791 pp.

Gitelson, A. A., J. F. Schalles, D. C. Rundquist, F. R. Schiebe, and Y. Z. Yacobi. 1999. Comparative reflectance properties of algal cultures with manipulated densities. *Journal of Applied Phycology* 11:345–354.

Goodman, L., M. Lewis, J. Macauley, R. Smith, and J. Moore. 1999. Preliminary survey of chemical contaminants in water sediment and aquatic biota at selected sites in Northeastern Florida Bay. *Gulf of Mexico Science* 1:1–161.

Guckert, J. B., S. C. Nold, H. L. Boston, and D. C. White. 1992. Periphyton response in an industrial receiving stream: lipid-based physiological stress analysis and pattern recognition of microbial community structure. *Canadian Journal of Fisheries and Aquatic Science* 49:2579–2587.

Hale, M. S. and J. G. Mitchell. 1997. Sea surface microlayer and bacterioneuston spreading dynamics. *Marine Ecology Progress Series* 147:269–276.

Hill, B. H., P. V. McCormick, P. R. Herlihy, P. R. Kauffman, F. H. McCormick, and C. B. Johnson, 2000. Use of periphyton assemblage data as an index of biotic integrity. *Journal of the American Benthological Society* 19:50–67.

Huang, S. and M. G. Hadfield. 2003. Composition and density of bacterial biofilms determine larval settlement of the polychaete *Hydroides elegans*. *Marine Ecology Progress Series* 260:161–172.

Hunter, C. H., E. Senior, J. R. Howard, and I. W. Bailey. 1998. The establishment and characterisation of a nitrifier population in a continuous-flow multi-stage model used to study microbial growth and interactions inherent in aquatic ecosystems. *Water SA* 24:85–91.

Hunter, R. D. 1983. Bioenergetic and community changes in intertidal aufwuchs grazed by *Littorina littorea*. *Ecology* 64:761–769.

Jackson, L. E., J. C. Kurtz, and W. S. Fisher (eds.). 2000. Evaluation Guidelines for Ecological Indicators. EPA.620/R-99/005. U.S. Environmental Protection Agency, Office of Research and Development, Research Triangle Park, NC, 107 pp.

Jackson, S. M. and E. B. G. Jones. 1991. Interactions within biofilms: the disruption of biofilm structure by protozoa. *Kieler Meeresforschungen Sonderheft* 8:264–268.

Jeffrey, W. H. and J. H. Paul. 1986. Activity measurements of planktonic microbial and microfouling communities in a eutrophic estuary. *Applied and Environmental Microbiology* 51:157–162.

Kanhere, Z. D. and V. R. Gunale. 1999. Evaluation of the saprobic system for tropical water. *Journal of Environmental Biology* 20:259–262.

Kemp, P. F., B. F. Sherr, E. B. Sherr, and J. J. Cole (eds.). 1993. *A Handbook of Methods in Aquatic Microbial Ecology*. Lewis Publishers, Boca Raton, FL, 777 pp.

Kolari, M., K. Mattila, R. Mikkola, and M. S. Salkinoja-Salonen. 1998. Community structure of biofilms on ennobled stainless steel in Baltic Sea water. *Journal of Industrial Microbiology and Biotechnology* 21:261–274.

Kollwitz, R. and M. Marsson. 1908. Okologie der pflanzlichen saprobien. *Berichte der Deutschen Botanischen Gesellshaft A* 26:505–519.

Kuhl, M. and B. B. Jorgensen. 1992. Microsensor measurements of sulfate reduction and sulfide oxidation in compact microbial communities. *Applied and Environmental Microbiology* 58:1167–1174.

Lawrence, J. R. and D. E. Caldwell. 1987. Behavior of bacterial stream populations within the hydrodynamic boundary layers of surface microenvironments. *Microbial Ecology* 14:15–27.

Lawrence, J. R., B. Scharf, G. Packroff, and T. R. Neu. 2002. Microscale evaluation of the effects of grazing by invertebrates with contrasting feeding modes on river biofilm architecture and composition. *Microbial Ecology* 44:199–207.

Lewis, M. A., D. E. Weber, L. R. Goodman, R. S. Stanley, W. G. Craven, J. M. Patrick, R. L. Quarles, T. H. Roush, and J. M. Macauley. 2000. Periphyton and sediment bioassessment in north Florida Bay. *Environmental Monitoring and Assessment* 65:503–522.

Lewis, M. A., D. E. Weber, and J. C. Moore. 2001a. Trace metal bioavailability below near-coastal wastewater discharges using periphyton. *Dimensions of Pollution* (India) 1:77–94.

Lewis, M. A., D. E Weber, R. S. Stanley, and J. C. Moore. 2001b. Dredging impact on an urbanized Florida estuary: benthos and algal-periphyton. *Environmental Pollution* 115:161–171.

Lewis, M. A., D. E. Weber, T. H. Roush, R. L. Quarles, D. Dantin, and R. S. Stanley. 2002a. In-situ phytoassessment as an indicator of nutrient enrichment from ten wastewaters discharged to the Gulf of Mexico near-coastal areas. *Archives of Environmental Contamination and Toxicology* 43:11–18.

Lewis, M., R. Boustany, D. Dantin, R. Quarles, J. C. Moore, and R. S. Stanley. 2002b. Effect of a coastal golf course complex on water quality, periphyton and seagrass. *Ecotoxicology and Environmental Safety* 53:154–162.

Lewis, M., L. Goodman, and J. Macauley. 2004a. Sediment toxicity and macrobenthos and periphytic-algal community composition in the Everglades–Florida Bay transitional zone. *Ecotoxicology* 13:231–244.

Lewis, M. A., R. L. Quarles, D. D. Dantin, and J. C. Moore. 2004b. Evaluation of a Florida golf complex as a local and watershed source of bioavailable contaminants. *Marine Pollution Bulletin* 48:254–262

Liu, K. and R. M. Dickhut. 1997. Surface microlayer enrichment of polycyclic aromatic hydrocarbons in southern Chesapeake Bay. *Environmental Science and Technology* 31:2777–2781.

Lueders, T. and M. W. Friedrich. 2003. Evaluation of PCR amplification bias by T-RFLP analysis of SSU rRNA and *mcr*A genes using defined template mixtures of methanogenic pure cultures and soil DNA extracts. *Applied and Environmental Microbiology* 69:320–326.

Madoni, P. and N. Bassanini. 1999. Longitudinal changes in the ciliated protozoa communities along a fluvial system polluted by organic matter. *European Journal of Protistology* 35:391–402.

Manz, W., K. Wendt-Potthoff, T. R. Neu, U. Szewzyk, and J. R. Lawrence. 1999. Phylogenetic composition, spatial structure, and dynamics of lotic bacterial biofilms investigated by *in situ* hybridization and confocal laser scanning microscopy. *Microbial Ecology* 37:225–237.

Mattila, J. and R. Räisänen. 1998. Periphyton growth as an indicator of eutrophication; an experimental approach. *Hydrobiologia* 377:15–23.

McCormick, P. V. and R. J. Stevenson. 1998. Periphyton as a tool for ecological assessment and management in the Florida Everglades. *Journal of Phycology* 34:726–733.

Meyer-Reil, L.-A. 1994. Microbial life in sedimentary biofilms — the challenge to microbial ecologists. *Marine Ecology Progress Series* 112:303–311.

Mohamed, M. N., J. R. Lawrence, and R. D. Robarts. 1998. Phosphorous limitation of heterotrophic biofilms from the Fraser River, British Columbia, and the effect of pulp mill effluent. *Microbial Ecology* 36:121–130.

Møller, S., D. R. Korber, G. M. Wolfaardt, S. Molin, and D. E. Caldwell. 1997. Impact of nutrient composition on a degradative biofilm community. *Applied and Environmental Microbiology* 63:2432–2438.

Neu, T. R. and J. R. Lawrence. 1997. Development and structure of microbial biofilms in river water studied by confocal laser scanning microscopy. *FEMS Microbiology Ecology* 24:11–25.

Nocker, A., J. E. Lepo, and R. A. Snyder. In press. Influence of an oyster reef on the development of the microbial heterotrophic community of an estuarine biofilm. *Applied and Environmental Microbiology.*

Norton, T. A., R. C. Thompson, J. Pope, C. J. Veltkamp, B. Banks, C. V. Howard, and S. J. Hawkins. 1998. Using confocal laser scanning microscopy, scanning electron microscopy and phase contrast light microscopy to examine marine biofilms. *Aquatic Microbial Ecology* 16:199–204.

Odom, J. M. and H. D. Peck, Jr. 1984. Hydrogenase, electron transfer proteins, and energy coupling in the sulfate-reducing bacteria *Desulfovibrio. Annual Review of Microbiology* 38:551–592.

Okabe, S., H. Satoh, and Y. Watanabe. 1999. *In situ* analysis of nitrifying biofilms as determined by *in situ* hybridization and the use of microelectrodes. *Applied and Environmental Microbiology* 65:3182–3191.

Osborn, A. M., E. R. B. Moore, and K. N. Timmis. 2000. An evaluation of terminal-restriction fragment length polymorphism (T-RFLP) analysis for the study of microbial community structure and dynamics. *Environmental Microbiology* 2:39–50.

Parkes, R. J., G. R. Gibson, I. Mueller-Harvey, W. J. Buckingham, and R. A. Herbert. 1989. Determination of the substrates for sulphate-reducing bacteria within marine and estuarine sediments with different rates of sulphate reduction. *Journal of General Microbiology* 135:175–187.

Patrick, R. 1967. The effect of invasion rate, species pool, and size of area on the structure of the diatom community. *Proceedings of the National Academy of Sciences of the United States of America* 58:1335–1342.

Patrick, R., M. H. Hohn, and J. H. Wallace. 1954. A new method for determining the pattern of the diatom flora. *Notulae Naturae* 259:1–12.

Pederson, K. 1982. Factors regulating microbial biofilm development in a system with slowly flowing seawater. *Applied and Environmental Microbiology* 44:1196–1204.

Phelps, H. L. and J. A. Mihursky. 1986. Oyster (*Crassostrea virginica* Gmelin) spat settlement and copper in aufwuchs. *Estuaries* 9:127–132.

Piceno, Y. M. and C. R. Lovell. 2000. Stability in natural bacterial communities: I. Nutrient addition effects on rhizosphere diazotroph assemblage composition. *Microbial Ecology* 39:32–40.

Rao, T. S., P. G. Rani, V. P. Venugopalan, and K. V. K. Nair. 1997. Biofilm formation in a freshwater environment under photic and aphotic conditions. *Biofouling* 11:265–282.

Raskin, L., R. I. Amann, L. K. Poulsen, B. E. Rittmann, and D. A. Stahl. 1995. Use of ribosomal RNA-based molecular probes for characterization of complex microbial communities in anaerobic biofilms. *Water Science and Technology* 31:261–272.

Rodriguez, R. A. and C. E. Epifanio. 2000. Multiple cues for induction of metamorphosis in larvae of the common mud crab *Panopeus herbstii. Marine Ecology Progress Series* 195:221–229.

Roske, I., K. Roske, and D. Uhlmann. 1998. Gradients in the taxonomic composition of different microbial systems: comparison between biofilms for advanced waste treatment and lake sediments. *Water Science and Technol*ogy 37:159–166.

Said, W. A. and D. L. Lewis. 1991. Quantitative assessment of the effects of metals on microbial degradation of organic chemicals. *Applied and Environmental Microbiology* 57:1498–1503.

Santegoeds, C. M., T. G. Ferdelman, G. Muyzer, and D. deBeer. 1998. Structural and functional dynamics of sulfate-reducing populations in bacterial biofilms. *Applied and Environmental Microbiology* 64:3731–3739.

Schramm, A., D. de Beer, J. C. van den Heuvel, S. Ottengraf, and R. Amann. 1999. Microscale distribution of populations and activities of *Nitrosospira* and *Nitrospira* spp. along a macroscale gradient in a nitrifying bioreactor: quantification by in situ hybridization and the use of microsensors. *Applied and Environmental Microbiology* 65:3690–3696.

Sekiguchi, Y., Y. Kamagata, K. Syutsubo, A. Ohashi, H. Harada, and K. Nakamura. 1998. Phylogenetic diversity of mesophilic and thermophilic granular sludges determined by 16S rRNA gene analysis. *Microbiology* 144:2655–65.

Sinsabaugh, R. L., D. Repert, T. Weiland, S. W. Golladay, and A. E. Linkins. 1991. Exoenzyme accumulation in epilithic biofilms. *Hydrobiologia* 222:29–37.

Sonak, S. and N. Bhosle. 1995. Observations on biofilm bacteria isolated from aluminium panels immersed in estuarine waters. *Biofouling* 8:243–254.

Sorensen, J., D. Christensen, and B. B. Jorgensen 1981. Volatile fatty acids and hydrogen as substrates for sulfate-reducing bacteria in anaerobic marine sediment. *Applied and Environmental Microbiology* 42:5–11.

Stewart, P. M., J. R. Pratt, J. Cairns, Jr., and R. L. Lowe. 1985. Diatom and protozoan species accrual on artificial substrates in lentic habitats. *Transactions of the American Microscopical Society* 104:369–377.

Surman, S. B., J. T. Walker, D. T. Goddard, L. H. G. Morton, C. W. Keevil, W. Weaver, A. Skinner, K. Hanson, D. Caldwell, and J. Kurtz. 1996. Comparison of microscope techniques for the examination of biofilms. *Journal of Microbiological Methods* 25:57–70.

Tolker-Nielsen, T. and S. Molin. 2000. Spatial organization of microbial biofilm communities. *Microbial Ecology* 40:75–84.

Unabia, C. R. C. and M. G. Hadfield. 1999. Role of bacteria in larval settlement and metamorphosis of the polychaete *Hydroides elegans*. *Marine Biology* 133:55–64.

U.S. Environmental Protection Agency. 1997. Revision to Rapid Bioassessment Protocols for Use in Streams and Rivers: Periphyton, Benthic Macroinvertebrates, and Fish. EPA 841-D-97-002. U.S. Environmental Protection Agency, Washington, D.C.

Wahl, M. 1989. Marine epibiosis. I. Fouling and antifouling: some basic aspects. *Marine Ecology Progress Series* 58:175–189,

Weitzel, R. L. 1979. *Methods and Measurements of Periphyton Communities: A Review.* ASTM STP 690, American Society for Testing and Materials, Philadelphia.

Welch, E. B., R. M. Emery, R. I. Matsuda, and W. A. Dawson. 1972. The relation of periphytic and planktonic algal growth in an estuary to hydrographic factors. *Limnology and Oceanography* 17:731–737.

Wieczorek, S. K. and C. D. Todd. 1997. Inhibition and facilitation of bryozoan and ascidian settlement by natural multi-species biofilms: effects of film age and the roles of active and passive larval attachment. *Marine Biology* 128:463–473.

Wimpenny, J. W. T. and R. Colasanti. 1997. A unifying hypothesis for the structure of microbial biofilms based on cellular automaton models. *FEMS Microbiology Ecology* 22:1–16.

Wolfaardt, G. M., J. R. Lawrence, R. D. Robarts, and D. E. Caldwell. 1998. *In situ* characterization of biofilm exopolymers involved in the accumulation of chlorinated organics. *Microbial Ecology* 35:213–223.

Wu, J. 1999. A generic index of diatom assemblages as bioindicator of pollution in the Keelung River of Taiwan. *Hydrobiologia* 397:79–87.

Zhang, X. and L. Y. Young. 1997. Carboxylation as an initial reaction in the anaerobic metabolism of naphthalene and phenanthrene by sulfidogenic consortia. *Applied and Environmental Microbiology* 63:4759–4764.

Zhou, X. and K. Mopper. 1997. Photochemical production of low-molecular-weight carbonyl compounds in seawater and surface microlayer and their air-sea exchange. *Marine Chemistry* 56:201–213.

Zobell, C. E. 1943. The effect of solid surfaces upon bacterial activity. *Journal of Bacteriology* 46:39–56.

10

Diatom Indicators of Ecosystem Change in Subtropical Coastal Wetlands

Evelyn Gaiser, Anna Wachnicka, Pablo Ruiz, Franco Tobias, and Michael Ross

CONTENTS

Introduction

Coastal ecosystems often support a diverse benthic microalgal community that, together with associated bacteria, fungi, and macroalgae, forms prolific periphyton growths on sediments and the grasses and/or wet forest vegetation that inhabit the coastline. Particularly in the subtropics and tropics, coastal periphyton communities form the base of a productive and diverse food web both in the marsh and the adjacent offshore marine environment as tides transport both periphyton products and consumers across the marine–freshwater interface (Admiraal, 1984; Day et al., 1989). Coastal wetlands at this interface present a diversity of environmental conditions because of the strong gradients in salinity, water availability, and nutrient supply inherent in this transitional environment. A variety of habitat types result (depending on latitude), including interior freshwater forested marshes, supertidal graminoid marshes, intertidal estuarine lagoons, hypersaline pools, mangrove swamps, and grassy salt marshes. Consequently, coastal periphyton communities contain some of the most compositionally diverse algal floras in the world (de Wolf, 1982). Because algae are strongly influenced by their surrounding chemical and structural environment, they provide a useful tool for environmental monitoring in complex coastal systems (Vos and de Wolf, 1993; Sullivan, 1999; Cooper et al., 1999).

Several anthropogenic influences threaten the existence and viability of coastal systems worldwide, including nutrient enrichment, overharvesting of consumable resources, landscape modification, and saltwater encroachment (National Research Council, 1993). Documentation of detrimental ecological effects of the last has, in recent decades, been increasing in frequency and extent around the globe (Park et al., 1989), as the rate of saltwater encroachment into coastal ecosystems increases due to sea-level rise exacerbated by diversion and depletion of coastward overland freshwater flow. The history of coastal ecosystems in South Florida provides an unfortunate example of the magnitude and complexity of effects that decades of canalization and sea-level rise can have on intertidal communities. Rates of saltwater encroachment in coastal South Florida exceed 400 m per decade in some areas (Ross et al., 2000), resulting in the disappearance of vast areas of freshwater marsh and interior migration of mangrove swamps.

Because salinity has an overriding influence on microbial community composition, algae (particularly diatoms) have been used to track rates of saltwater encroachment in both modern monitoring and paleoecological studies (Gasse et al., 1983; Juggins, 1992; Ross et al., 2001). Algal populations respond on timescales of weeks to months to changes in environmental conditions, integrating much of the small-scale temporal variation that is often the source of unwanted "noise" in continuous salinity recording data (Snoeijs, 1999). Transfer functions have been created from the modern distribution of diatoms along salinity gradients (in coastal areas and closed-basin "saline" lakes, e.g., Campeau et al., 1995; Fritz et al., 1999, respectively) that allow salinity to be predicted from diatom community composition with a very high degree of accuracy. However, while many coastal diatom taxa are thought to be widely distributed, application of salinity preferences for diatoms collected in regions (e.g., Baltic Sea, Snoeijs, 1999; Thames River, England, Juggins, 1992; Chesapeake Bay, Cooper, 1995; Mississippi salt marsh, Sullivan, 1982) other than South Florida would be problematic because there would likely be a low degree of taxonomic overlap with these data sets. Subtropical wetlands in general and specifically the Everglades have been poorly explored taxonomically, resulting in incompletely defined ecological and range size distributions. Further, coastal environments of the subtropics are dominated by mangrove swamps, and other than studies by Siqueiros-Beltrones and Castrejón (1999, Balandra Lagoon, Baja CA), Navarro and Torres (1987, Indian River, FL), Sullivan (1981, Mississippi salt marsh), Reimer (1996, Bahamas), and Podzorski (1985, Jamaica), there have been few explorations of coastal mangrove diatoms. The composition and range size distribution of mangrove diatoms and associated microflora, and their response to environmental variation, are practically unknown.

The objectives of the present study were to survey the algal flora of periphyton communities in coastal wetlands in the Everglades of southeast Florida. Periphyton mats are a dominant feature in both fresh-water and saline Everglades wetlands (Browder et al., 1982; Ross et al., 2001). The specific purposes of this work were to (1) document the taxonomic composition of algal assemblages, particularly diatoms, in periphyton of the coastal Everglades and (2) determine environmental drivers of assemblage composition, in order to (3) create algae-based inference models that could be used to track trajectories of environmental change. Our goal was to produce a taxonomic guide to aid in identifying subtropical coastal diatoms and to create algae-based environmental inference models that can be employed in long-term monitoring and/or paleoecological studies to document ecological response to habitat alteration along the South Florida coastline.

Methods

Study Site

The southeastern edge of Florida was historically characterized by expansive coastal mangrove wetlands that were dissected by tidal creeks flowing from the freshwater Everglades to the coast. Egler (1952) was able to distinguish distinct vegetation zones lying in bands parallel to the coast, driven by gradients of salinity, water availability, nutrients, and susceptibility to drought and fire, including a coastward sequence of graminoid freshwater wetlands (to the interior), followed by dwarf mangrove scrub swamps in intertidal areas, bounded by fringing mangrove forest on the coast. Throughout the last several decades, an extensive network of drainage canals has been constructed in South Florida, effectively draining much of the interior and coastal Everglades for urban and agricultural development. By the turn of the 21st century, the wetland bands had been diminished to the periphery of the coastline: freshwater graminoid

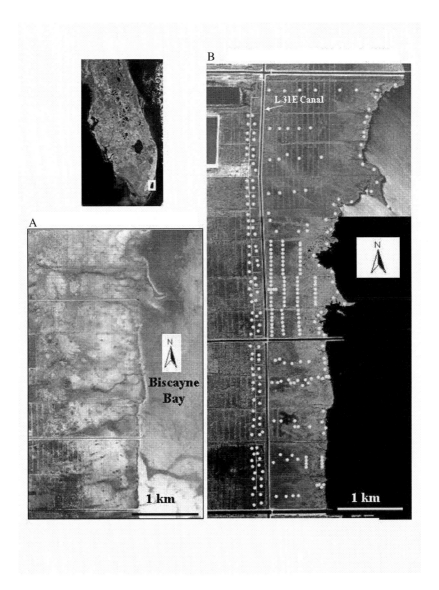

FIGURE 10.1 Location of study area in southeast Florida. (A) Aerial photograph from 1940 showing east–west canals, north–south drainage ditches, and remnant tidal creeks. (B) Aerial photograph from 1990 showing additional canals built since 1940, including the L-31E canal, the disappearance of tidal creeks, and the distribution of collecting sites among the 13 wetland sub-basins.

marshes had been largely displaced by an enroaching mangrove scrub community and most tidal creeks had disappeared (Ross et al., 2000; Figure 10.1B).

The present study focuses on an area of remnant coastal wetlands, parts of which are protected in Biscayne National Park (Figure 10.1). The ~7 km long study area is bounded to the north and south by major east–west drainage canals (Princeton and Mowry, respectively) and bisected north–south by a secondary canal (L-31E). The region is dissected by many smaller east–west ditches, which compartmentalize the area longitudinally into 13 hydrologically distinct wetland basins that range in width from about 0.5 to 2 km. To the west of the L-31E canal, freshwater marshes are now hydrologically isolated from the coast and bounded to the west by agricultural lands, the periphery of which is heavily invaded by exotic trees including *Schinus terebinthifolius* (Brazilian peppertree) and *Casuarina equisetifolia*

(Australian pine). To the east of the L-31E canal, mangrove communities predominate, with strands of upland forest now occupying the remnant tidal creek beds.

We used a stratified–random design to select study sites within each of the 13 sub-basins. Using aerial photos of the area, each sub-basin was divided into four to six units, including, to the west of the L-31E canal, a freshwater swamp forest dominated by exotics that have invaded abandoned agricultural land and remnant freshwater graminoid marsh and, to the east of the L-31E canal, mangrove forests that can be characterized by canopy height and cover as dwarf, transitional, and fringing (along the coastline). Within each unit a north–south transect was randomly located, and one to five sampling stations were evenly distributed along its length. A total of 226 stations were sampled within the 12-km^2 area (Figure 10.1B).

Data Collection and Processing

At each station, we assessed the vegetation community structure, roughly described the sediments, and sampled periphyton and several chemical parameters in surface and/or pore water. Vegetation was assessed using methods of Ross et al. (2001), where species cover and canopy height were estimated separately for upper (~2 m height) and lower (<2 m) strata in repeated quadrats. Depth of sediments to the limestone bedrock was measured at five stations with a probe-rod, and using a soil auger, sediments were extracted to measure depths of readily apparent compositional and textural transitions. Using a polyvinylchloride (PVC) pipe, five small (3.8 cm^2, 1 to 2 cm thick) sections of surface soil, commonly occupied by periphyton, were extracted from each location and composited. A portable meter was used to measure pH and conductivity in surface water, if present, or in pore water that filled the auger hole. Conductivity (µS cm^{-1}) was converted to salinity (ppt) using a model provided from a previous study in a nearby basin where both variables were directly measured (Ross et al., 2001).

In the laboratory, periphyton was picked free of large plant fragments, homogenized, diluted, and subsampled for analysis of dry weight (DM, 2 days at 100°C), ash-free dry weight (AFDM, 1 h at 500°C), total phosphorus (TP, by automated colorimetry), and soft-algae and diatom composition. Diatoms were cleaned of calcite and organic matter by chemical oxidation and permanently fixed to a glass microslide using Naphrax® mounting medium. At least 500 diatom valves were counted on random, measured transects on a compound light microscope at 1000×. Nondiatom algae ("soft algae") were analyzed from one station within each unit in sub-basins 1 to 8 by preparing semipermanent water-mounted slides. At least 500 units (cells, colonies, or filaments) were counted and identified on random transects on the slide at 400 to 1000× magnification. Abundance estimates were converted to biovolume using critical dimensions (length, width, breadth) of 20 representatives of each morphologically distinct unit and applying volumetric formulas for the closest geometric shape. Diatom and soft algal samples, permanent slides, photographs of all taxa, database links, and all references used in taxonomic determination can be accessed through our Web site at http://serc.fiu.edu/periphyton/index.htm and are archived in a curated collection in the microscopy laboratory at Florida International University.

Data Analysis

Stations were sorted into five vegetation type categories based on survey data and aerial photographs, including a freshwater swamp forest, freshwater graminoid marsh, and dwarf, transitional, and fringing mangrove forest. The distinctiveness of the categories based on relative cover of species present in more than 5% of the sites was confirmed using analysis of similarity (among community types) employing the Bray-Curtis similarity metric in PRIMER-E/ANOSIM® software. Plant species significantly influencing the five community types were identified using Dufrene and Legendre's (1997) "Indicator Species Analysis," where taxa having an indicator value (based on relative abundance and frequency among sites) above 40% of perfect indication ($P < 0.05$) were considered reliable indicators.

Using the spatial modeling and analysis (V2.0) module in Arcview GIS 3.2®, we mapped the distribution of the vegetation community types and other environmental variables (soil depth, canopy height, salinity, and periphyton AFDM and TP content). To interpolate between points, we used the IDW method, which weights the value of each point by the distance that point is from the cell being analyzed and then averages the values. The output grid cell size was 10 m and the number of neighbors was 3 points. Means of each parameter were calculated within each vegetation type and compared using a Student's *t*-test, and correlations among parameters were determined using the Pearson correlation coefficient on log-transformed data, with $P < 0.001$.

Patterns in relative abundances and biovolumes of diatom and nondiatom taxa, respectively, were determined using nonmetric multidimensional scaling ordination (NMDS), analysis of similarity, and weighted-averaging regression. Species by station data matrices were established and species present in fewer than 1% of samples and having a mean relative abundance (when present) of <0.05% were removed prior to analysis. Assortment of sites in the NMDS ordinations based on the Bray-Curtis similarity metric were related to environmental variables using vector fitting. The significance of algal community patterns relative to vegetation type (a categorical variable) was determined using analysis of similarity on the same similarity matrix as used for the NMDS.

We used weighted-averaging regression and calibration to determine the strength of the relationship of species composition to salinity and vegetation type. This approach assumes that species abundance responses can be characterized by an optimum or mode where abundances are greatest and a tolerance that defines the breadth of appearance along a gradient. The value of an environmental variable can then be calculated for a sample from an unknown environment, using the average of the optima of the species present, weighted by their abundances and possibly tolerances. Using the weighted-averaging program C2 (Juggins, 2003), we estimated the salinity and vegetation optimum and tolerance for each species as the average among sites in which the taxon occurred and then tested the prediction power by estimating the salinity and vegetation type from a random set of sites (bootstrapping with replacement) and plotted predictions against observed values. Predicted values for salinity and vegetation type from diatom and soft-algae calibration models were mapped using the same approach as for the environmental variables (described above).

Results

Vegetation

The five major vegetation community types distinguished through interpretation of aerial photographs and used to determine selection of sampling sites were confirmed to be compositionally distinct based on relative cover of 45 of the most abundant of the 84 plant species found in the study area (ANOSIM, all combinations, global $R = 0.48$, $P < 0.01$). Compositional differences within freshwater units (upland forest and freshwater marsh) and interior mangrove units (dwarf and transitional) were less ($R = 0.2$ and 0.3, $P < 0.05$, respectively) than differences between freshwater and mangrove units (mean $R = 0.4$, $P < 0.01$), and the fringing mangrove forest was highly distinguishable from all other units (mean $R = 0.8$, $P < 0.001$). Although the coastward sequence of vegetation zones was consistent among sub-basins, there was variation in the breadth of each zone along the 7-km study area (Figure 10.2A), and we acknowledge that additional distinct community types occur within these units, most notably including a densely vegetated, heavily canopied mangrove forest growing in historic drainages that meander through adjoining units and forests occupying tree islands that punctuate all units of the landscape. Vegetation canopy height was significantly higher in the upland forest and transitional and fringing mangroves than in the freshwater marsh and dwarf mangrove community (see Figure 10.4A below).

Environmental Variation

Compositional differences among units were associated with variation in several environmental parameters. In pore water, while no significant pattern was observed in pH (mean = 7.2), a strong west–east increase in salinity was observed in many of the sub-basins, with the L-31E clearly separating freshwater (salinity < 5 ppt) from marine (5 to 20 ppt) conditions (Figure 10.3A). Soils were significantly deeper in the fringing mangrove forest than other units (126 cm vs. mean 104 cm, respectively), and although

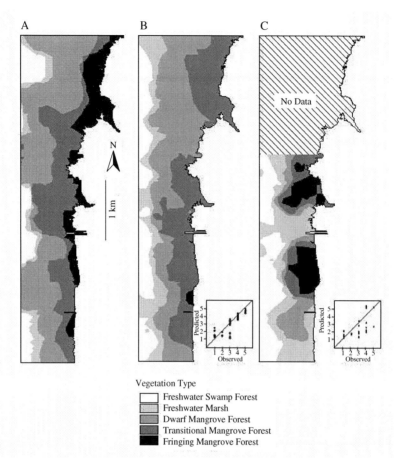

Vegetation Type

☐ Freshwater Swamp Forest
▨ Freshwater Marsh
▨ Dwarf Mangrove Forest
▨ Transitional Mangrove Forest
■ Fringing Mangrove Forest

FIGURE 10.2 Observed distribution of the five major vegetation types within the study area (A) and distribution of vegetation types predicted from diatom community composition (B) and nondiatom algal community composition (C) using weighted-averaging regression. Insets are plots of observed vs. inferred vegetation type based on diatom and soft-algae optima and tolerances (R^2 = 0.69, 0.42 and RMSE = 0.77 and 1.2, for diatoms and algae, respectively). Plant species significantly associated with each community type were (1) Freshwater swamp forest: *Casuarina equisetifolia* (Australian pine), *Conocarpus erectus* (buttonwood), *Schinus terebinthifolius* (Brazilian pepper); (2) Freshwater marsh: *Cladium jamaicense* (sawgrass), *Juncus rhomerianus* (black rush), *Typha domingensis* (cattail); (3) Dwarf mangrove forest: *Laguncularia racemosa* (white mangrove), *Rhizophora mangle* (red mangrove); (4) Transitional mangrove forest: *Avicennia germinans* (black mangrove); and (5) Fringing mangrove forest: *R. mangle, L. racemosa, A. germinans.*

nearly all cores were characterized by an upper heavily rooted peat, this layer was deepest in the fringing mangroves and gradually became shallower to the interior freshwater marsh (66 cm vs. 12 cm, respectively; Figure 10.4B). With implications for linking biotic patterns to environmental variation, several variables were significantly correlated with each other, including soil, peat depth, salinity, and pH.

Periphyton Biomass and TP Content

Algae were organized into periphyton communities of considerable mass throughout the wetland units (Figure 10.4C). Periphyton DM was highest in the dwarf mangrove and freshwater marsh units (903 and 575 g m^{-2}, respectively) and lower in the forested units (mean = 266 g m^{-2}). A considerable portion of this mass in all units was composed of calcite, particularly in the dwarf mangrove and freshwater units, such that when this portion that is not combustible is subtracted from the dry mass (in the AFDM calculation), some of the pattern in periphyton distribution disappears, although AFDM biomass remains significantly higher in the dwarf mangrove forest than other units (Figure 10.4C). Likewise, the portion

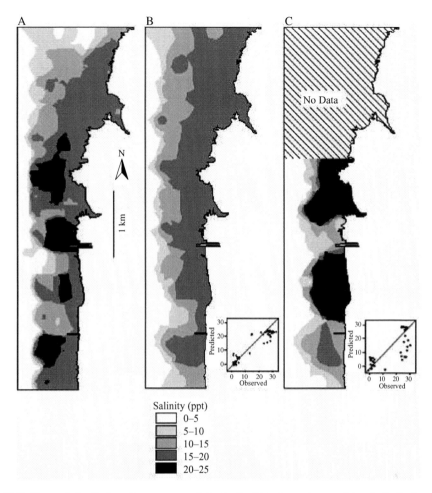

FIGURE 10.3 Observed distribution of pore water salinity (ppt) within the study area (A) and distribution of salinity predicted from diatom community composition (B) and nondiatom algal community composition (C) using weighted-averaging regression. Insets are plots of observed vs. inferred salinity based on diatom and soft-algae optima and tolerances (R^2 = 0.91, 0.58 and RMSE = 0.14, 0.34 for diatoms and soft-algae, respectively).

of the periphyton composed of organic (rather than calcitic) mass was significantly higher in the forested units than in the dwarf mangroves and freshwater marsh. The DM, AFDM, and organic carbon content of the periphyton mats were, by nature of their analysis, correlated and also strongly negatively related to canopy height, and less so to peat depth in the sediments.

Very strong trends in the TP content of periphyton mats were evident in the system, with periphyton in the freshwater marsh having significantly lower P than all other units and mats in the transitional and fringing mangrove forest having more than an order of magnitude higher P content than other units (Figure 10.4D). Patterns of variation in periphyton TP content were positively correlated with peat depth, canopy height, and salinity.

Algal Community Composition

A total of 405 diatom taxa representing 64 genera were collected from periphyton in the study area. Genera represented by the most taxa (number given in parentheses) were *Amphora* (59), *Navicula* (55), *Mastogloia* (51), *Nitzschia* (39), *Fragilaria* (21), *Achnanthes* (16), and *Diploneis* (15). The NMDS ordination (two dimensions, stress = 0.12) of relative abundance of 133 of the most abundant taxa found clear separation of diatom communities occupying the freshwater units (forest and marsh) from the

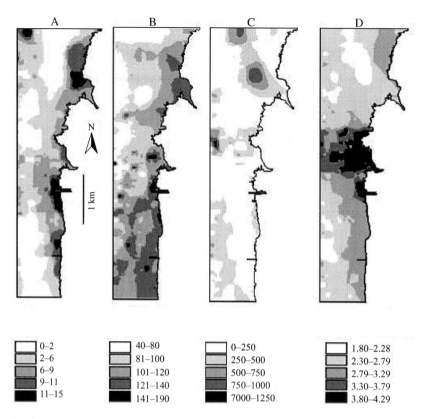

FIGURE 10.4 Distribution of (A) mean vegetation canopy height (m), (B) soil depth (cm), (C) periphyton AFDM (g m⁻²), and (D) periphyton tissue total phosphorus concentration (log μg g⁻¹) within the study area.

marine mangrove units. This pattern was verified by the analysis of similarity, which showed that significant separation between freshwater units 1 and 2 vs. marine units 3, 4, and 5 (global $R > 0.6$ for all comparisons, $P < 0.001$), but little distinction in comparisons within freshwater and marine units (global $R < 0.2$ for all comparisons, $P > 0.1$). While the ANOSIM analysis suggested two groups (freshwater vs. marine), the weighted-averaging regression model revealed a more linear gradient from interior to coastal communities (Figure 10.2B). A total of 35 indicator taxa were identified, 6 for the upland forest, 5 for the freshwater marsh, 2 for the dwarf mangroves, 7 for the transitional mangrove forest, and 15 for the fringing mangrove forest (pictured in Figure 10.5 and Figure 10.6). When mapped spatially, diatom-based vegetation type predictions appear similar to measured values (Figure 10.2B).

The NMDS ordination also revealed significant patterns in diatom composition among sites relative to salinity, canopy height, organic content, peat depth, and TP (maximum vector $R^2 = 0.34$, 0.30, 0.29, 0.24, and 0.23, respectively). Effects of canopy height and TP on diatom composition were positively correlated and together negatively correlated with the influence of organic content of the periphyton mats. The effect of salinity, the strongest variable influencing composition, was correlated with that of peat depth. Because salinity had an overriding effect on composition and was only correlated with one other variable, we examined this relationship further using weighted-averaging regression.

Because the frequency distribution of salinity values among sites was bimodal, with sites in the freshwater units confined to the west of the L-31 E canal having much lower values than mangrove sites to the east, the linear model used in the weighted-averaging regression may not provide the best fit to these data. Even so, the model has strong predictive power because most of the taxa incorporated in the model have well-defined salinity optima and narrow tolerances (provided in the appendix to this chapter). When mapped spatially, diatom-based salinity predictions appear similar to measured values (Figure 10.3B).

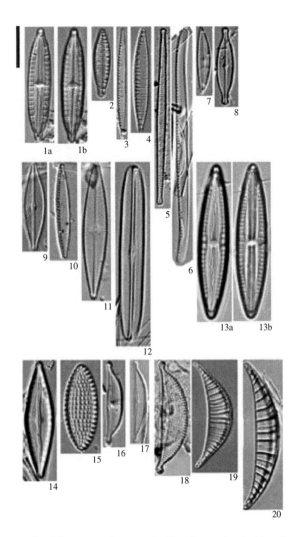

FIGURE 10.5 Digital photographs of diatom taxa that were significantly associated with each vegetation community type. From the freshwater forest: (1) *Mastogloia smithii* (a = midvalve focus showing internal partectae and b = surface of valve), (2) *Nitzschia semirobusta*, (3) *N. amphibia* f. *frauenfeldii*, (4) *N. amphibia*, (5) *Fragilaria synegrotesca*, and (6) *N. nana*; from the freshwater marsh: (7) *Encyonema evergladianum*, (8) *Brachysira neoexilis* (Typ 3), (9) *B. neoexilis* (Typ 2), (10) *N. palea* var. *debilis*, and (11) *Navicula podzorski*; from the dwarf mangrove forest: (12) *N. palestinae* and (13) *M. reimeri* (a = surface of valve and b = midvalve focus showing internal partectae); and from the transitional mangrove forest: (14) *M. angusta*, (15) *Tryblionella granulata*, (16) *Amphora* cf. *fontinalis*, (17) *A. coffeaeformis* var. *aponina*, (18) *A. costata*, (19) *Rhopalodia acuminata*, and (20) *R. gibberula*. Scale bar = 10 μm; original magnification: figures 1 to 15, 17, 19, and 20, ×1008; figure 16, ×1600; figure 18, ×1250.

In the study, 57 additional nondiatom algal taxa were found and identified co-occurring with diatoms in the periphyton communities at the reduced set of sites. The soft-algae flora was taxonomically dominated by coccoid and filamentous cyanophytes (39 and nine taxa, respectively), but also included two coccoid, two desmid, and three filamentous chlorophyte taxa, one dinoflagellate taxon and one purple-sulfur bacterium (non-algal, but included in counts). Taxa comprising more than 1% of the total biovolume of soft algae included, in decreasing order of abundance, the three filamentous chlorophytes (undetermined branching filaments resembling *Rhizoclonium*; 42%), followed by the blue-green filament *Scytonema* cf. *hofmannii* C. Agardh *ex* Bornet (35%) and three other unidentifiable blue-green filaments (resembling *Schizothrix* spp., 6.5%), seven *Chroococcus* spp. (5.8%), five *Gloeothece* spp. (3.4%), six *Aphanothece* spp. (2.5%), and the purple-sulfur bacterium (1.3%).

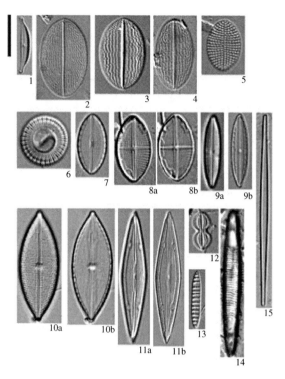

FIGURE 10.6 Digital photographs of diatoms taxa that were significantly associated with the fringing mangrove forest: (1) *Amphora subacutiuscula*, (2) *Cocconeis placentula*, (3) *C. placentula* var. *lineata*, (4) *C. placentula* var. *euglipta*, (5) *C. scutellum*, (6) *Cyclotella distinguenda*, (7) *Mastogloia ovalis*, (8) *M. crucicula* (a = surface of valve and b = midvalve focus showing internal partectae), (9) *M. pusilla* (a = surface of valve and b = midvalve focus showing internal partectae), (10) *M. nabulosa* (a = surface of valve and b = midvalve focus showing internal partectae), (11) *M. erythraea* (a = surface of valve and b = midvalve focus showing internal partectae), (12) *Diploneis caffra*, (13) *Denticula subtilis*, (14) *Rabdonema adriaticum*, and (15) *Hyalosynedra leavigata*. Scale bar = 10 μm; original magnification ×1008.

Both nondiatom and diatom algae responded similarly to measured environmental variables. The NMDS ordination (two-dimensional stress = 0.11) of relative biovolume of 35 of the most abundant taxa separated freshwater forest and marsh sites from marine mangrove units, and this distinction was shown to be significant in the analysis of similarity ($P < 0.001$). Several sites were distinctly grouped apart from other sites because they were uniquely dominated by a filamentous chlorophyte-resembling *Rhizoclonium*. These included most of the coastal sites in sub-basins 4 and 7. The ANOSIM analysis showed clear separation of algal communities occupying the freshwater units (forest and marsh) from the marine mangrove units. The weighted-averaging regression model for habitat types was strong but less predictive than the diatom-based model (Figure 10.2C). Five species were significantly indicative of three of the vegetation units, including, for the freshwater forest, two blue-green filaments resembling *Schizothrix calcicola* (Agardh) Gomont and the coccoid blue-green *Gomphosphaeria semenvitis*; for the dwarf mangrove scrub, an unidentified *Gloeothece* sp.; and for the fringing mangroves, an unidentified chlorophyte resembling *Rhizoclonium*. When mapped spatially, algae-based vegetation type predictions appear similar to measured values (Figure 10.2C).

The NMDS ordination showed the same variables to be important in explaining soft-algal distribution as the diatoms, including salinity, peat depth, canopy height, TP, and organic content (maximum vector R^2 = 0.34, 0.33, 0.25, 0.21, and 0.20, respectively). Effects of canopy height, TP, and peat depth on soft-algal species composition were positively correlated and together negatively correlated with the influence of organic content of the periphyton mats. The effect of salinity, the strongest variable influencing composition, was independent of other variables, and we examined this relationship further using weighted averaging regression. The model to predict salinity from algal species composition is strong. When mapped spatially, the algal-based predictions are less consistent with actual measured values (in comparison to diatoms; Figure 10.3C), although sites to the east and west of the L-31E canal can be clearly distinguished.

Discussion

The freshwater–saltwater ecotone lining the coast of southeast Florida coast is migrating rapidly westward. In our 7-km study area, four canals constructed over the last several decades now discharge the majority of overland freshwater flow directly into Biscayne Bay. Comparing the current locations of the five vegetation zones to those observed in the 1940 aerial photograph, several changes are obvious: (1) tidal creeks linking the interior freshwater marsh to the coast have disappeared, so that most fresh water is now delivered in large volumes to point locations where canals terminate, (2) the freshwater marsh has been drained and native vegetation displaced to the west by invasive exotic trees and to the east by the expanding dwarf mangrove forest, and (3) all coastal vegetation bands are now restricted to the east of canals that, by running parallel to the coastline, prohibit natural mixing of fresh water and salt water during the tidal cycle. Together with sea-level rise (the regional rate is estimated to be 3 to 4 mm yr^{-1}; Wanless et al., 1994), massive freshwater drainage has caused a rapid rate of saltwater encroachment that forces mangrove communities to shift landward to the canal boundary, which disrupts natural exchange across the coastal ecotone.

Evidence of shifts in the width and location of the vegetation zones can be interpreted from the soil profiles. All of our soil cores contained a substantial marl layer below a surficial peat. In the Everglades, marl soils are generally associated with freshwater marsh communities, particularly wet-prairie meadows (defined as graminoid marshes that are inundated 6 to 9 months per year). Peat soils are indicative of deeper water, more prolonged flooding, and mangrove communities along coastlines. We interpret the peat layer upper soils across the wetland basins to be indicative of (1) invasion by forest elements into the freshwater marsh due to water diversion from areas west of the L-31E canal, and (2) invasion of mangroves into areas previously occupied by freshwater marsh in areas east of the L-31E. Sedimentation rates from studies in similar communities nearby (1 to 3 mm/yr^{-1}; Scholl et al., 1969) suggest the contact is coincident with the establishment of the drainage canal network.

The microbial community in the Biscayne Bay coastal wetlands was, in most areas, organized into a cohesive periphyton mat. Organic biomass, measured by AFDM, was high (mean = 190 g m^{-2}) throughout the study area exceeding values found in a nearby mangrove marsh (5 to 20 g m^{-2}; Ross et al., 2001), where saltwater encroachment has caused a rapid expansion of a broad band of low productivity (referred to as the "white zone"). The highest values in the shallow freshwater marsh units (mean = 317 g m^{-2}) were comparable to marshes in the interior Everglades, where periphyton biomass can exceed that of emergent plants (Browder et al., 1982). Notably, the percent organic carbon in periphyton mats was highest toward the coast where marl deposition is minimized.

Of the measured environmental variables, only canopy height was correlated with periphyton biomass, with lowest values in heavily canopied fringing forests and highest in open areas of freshwater marsh. It is not unexpected that light availability would control periphyton biomass, although Beanland and Woelkerling (1983) found no effect of canopy on periphyton algal biomass in an *Avicennia* forest in South Australia, and, in our study, biomass was still relatively high in the heavily canopied fringe. Ambient daytime irradiance to the surface of Everglades periphyton mats can exceed 1000 μmol m^{-2} s^{-1}, an intensity that has been shown to photo-inhibit photosynthesis (Underwood, 2002). Typically, periphyton mats have distinct vertical structure, with green productive layers underlying a calcitic, inactive (possibly light-inhibited) surface. In this study, mats in open areas in this study were thick and structured, whereas periphyton in shaded areas usually comprised a thin, green film growing attached to roots and leaf litter. This community may be encouraged by the higher TP availability in coastal areas, and contribute to the inverse relationship between periphyton organic content and P availability. It was somewhat remarkable that although P availability (as measured by the TP content of the periphyton mats) varied an order of magnitude in the study area, there was no measurable correlation with periphyton DM or AFDM biomass, although it has been shown to control periphyton biomass in other areas of the Everglades (Gaiser et al., 2004). However, although biomass may be similar to or lower than interior areas, turnover of periphyton in the fringing mangroves may be higher as a result of increased nutrient availability. Other studies have found particularly high algal productivity in fringing mangrove ecosystems and in neighboring seagrass beds (Koch and Madden, 2001). Total algal biomass in one *Rhizophora mangle* forest in Puerto Rico was actually found to exceed the total annual leaf litterfall (Rodriguez and Stoner, 1989), pointing to the impact of benthic periphyton to the food web of mangroves and neighboring lagoons and estuaries.

Periphyton mat biomass was high across the broad range of salinity experienced by this system. This shows that the complicated, intricately connected communities forming highly structured periphyton mats can be created in both freshwater and marine conditions, even though the composition (at all levels, macro- and microalgal, bacterial, fungal, etc.) differs due to the strong osmotic gradient represented from freshwater to marine environments.

Indeed, we did find that salinity had an overriding control on algal composition throughout the coastal wetlands. While mats were dominated throughout the system by green algal filaments, *Scytonema* spp., and small coccoid blue-green algae, their morphologies (and, thus, our taxonomic designations) differed along the salinity gradient. The difficulty of assigning names to most of the taxa stems from the fact that coastal mangrove microalgal communities are poorly explored taxonomically. That we could not find many of the taxa collected here listed, described, or pictured in studies from similar system suggests either that these studies misdiagnosed taxa because of the paucity of appropriate taxonomic literature or that there is more regionality to the coastal microalgal flora than previously thought.

However, at higher levels of taxonomic organization, this flora did resemble that of other microbial mangrove mats collected elsewhere and, at those levels, responded similarly to salinity variation in the system. Phillips et al. (1994) found that horizontal zonation of algae on pneumatophores of *Avicennia marina* in South Africa was controlled primarily by salinity and wetting frequency, with the green alga *Rhizoclonium* dominating wet areas and providing support for numerous filamentous cyanophyte taxa (notably, *Lygbya confervoides* and *Microcoleus chthonoplastes*, belonging to genera also found in this study). *Rhizoclonium* and other green-algal filaments are often abundant in mangrove periphyton communities, often forming a tertiary layer over the macroalgae that are directly attached to the mangrove roots (Phillips et al., 1994). The macroalgae have been shown to be an important component of mangrove marshes, both in terms of their own productivity and diversity and also through their support of a diverse epiphytic community: it is not uncommon to find upward of 20 species of macroalgae inhabiting mangrove benthos, providing support for hundreds of microalgal taxa (Collado-Vides, 2000). Although we excluded macroalgae from our detailed analyses, we did note in field collections that *Bostrichia* was abundant on prop-roots and often coated with a thin green-algal mat (likely *Rhizoclonium* spp.). These *Rhizoclonium*-based communities were particularly important in coastal sites in sub-basins 4 and 7, which was what influenced the separation of these sites in the NMDS ordination.

The filamentous chlorophytes and macroalgae were joined by cyanobacterial filaments, particularly *Scytonema* and *Schizothrix* species, which often form the backbone of microbial mats across the full salinity range, from shallow, freshwater calcareous wetlands in the Everglades and Belize (Rejmánková and Komárková, 2000) to intertidal mangroves (Collado-Vides, 2000) to subtidal marine stromatolites (Rasmussen et al., 1993)) and hypersaline lagoons (Hussain and Khoja, 1993). These genera both contain species representing the full salinity spectrum, and indeed some of the species (*Scytonema hofmannii*) appear capable themselves of thriving in vastly different salinity regimes. In this study, the *Scytonema* and *Schizothrix* were most abundant in the freshwater marsh where they appeared, upon microscopic examination, to be coated with calcium carbonate crystals, which has been noted elsewhere (Browder et al., 1982). These were displaced by noncalcite precipitating blue-green algae in communities closer to the coast. Similar *Lyngbya*- and *Microcoleous*-dominated blue-green algae have been collected from mangrove pneumatophores elsewhere (Hussain and Khoja, 1993). While the periphyton matrix appears throughout the system to be macroscopically strung together by filamentous green or blue-green algae, the interstices of this web are often "glued" together by mucilaginous polysaccharide produced by abundant and diverse coccoid blue-green algae, which may increase desiccation resistance, provide a barrier to fluctuations in salinity, and concentrate nutrients and enzymes that control nutrient cycling.

The most diverse algal component in the periphyton mats studied here was the diatoms. It is common to find a large number of diatom genera in estuaries and near-coast environments because typically genera are confined to either fresh or salt water, and rarely mix except in brackish situations (Snoeijs, 1999). The dominance of *Amphora* and *Mastogloia* in the coastal flora is similar to findings in other parts of Florida and the Caribbean (Montgomery, 1978; Sullivan, 1981; Navarro, 1982; Foged, 1984; Podzorski, 1985; Reimer, 1996). These genera, together with *Navicula, Nitzschia, Cocconeis, Fragilaria,* and *Achnanthes,* are probably important in coastal floras circumglobally, at least in the Northern Hemisphere. At lower taxonomic levels we found several taxa with consistent morphologies that have not appeared elsewhere (or only in the regional literature — Montgomery, 1978; Navarro, 1982; Foged, 1984; Podzorski, 1985) that may be unique to the subtropical/tropical Atlantic Coast.

Diatoms organized into distinct freshwater and marine assemblages on either side of the L-31 E canal that effectively deterred mixing of tidal and overland freshwater flow. To the east of the canal, the freshwater marsh flora was dominated by *Encyonema evergladianum, Brachysira neoexilis,* and *Nitzschia palea* var. *debilis,* which are all common in un-enriched periphyton mats throughout the freshwater Everglades (Slate, 1998; Cooper et al., 1999). The freshwater swamp forest contained many of the same taxa as the freshwater marsh, but was the preferred habitat for *Mastogloia smithii, Fragilaria synegrotesca,* and four species of *Nitzschia* (*N. semirobusta, N. amphibia* f. *frauenfeldii, N. amphibia,* and *N. nana*). While these are all common elsewhere in the Everglades (Slate, 1998; Cooper et al., 1999), the predominance of *Nitzschia* taxa in the forest relative to the marsh is notable, and may reflect a higher stress tolerance for members of this genus (i.e., disturbance and low light intensities).

To the east of the L-31 E canal, the mangrove system was dominated by pennate benthic taxa. Mangrove-inhabiting taxa appear to be capable of withstanding a broad range of salinity and frequent desiccation. Taxa in the genus *Amphora, Achnanthes,* and *Tryblionella* became gradually more dominant toward the coast, indicating an affinity for higher salinities. These taxa appeared to assort better along the salinity gradient than by the vegetation type categories, likely because of the effect of tidal transport from the coastline to the canal levee. Transport was also probably responsible for the presence of notably marine planktonic taxa, such as *Cyclotella striata, Catacombas gaillonii, Biddulphia* spp., and *Terpsinoë musica,* in benthic samples.

Applications

The algal flora of coastal South Florida is not only prolific in terms of biomass and richness, but is highly correlated with salinity and vegetation type — two factors that will be influenced most by continued saltwater encroachment. Models provided here allow salinity to be predicted from diatom composition with an error of <10% of the actual value. Considering the high degree of variation in continuous salinity recordings, diatoms not only offer a means of monitoring salinity more accurately in the modern environment, but also provide a tool for reconstructing past salinity from fossil assemblages. Further, diatom composition offers a tool for "hindcasting" an ecological variable (vegetation type) from past communities. The predictive power of these models can be strengthened by those of Ross et al. (2001) who created a diatom-based transfer function that predicts distance from the coast in a neighboring South Florida wetland with 100 m resolution. The use of diatoms in coastal environments should receive increased attention in coming years as the realization of their tight linkages to the strong zonation typical of coastal environments is recognized in different regions. While long-term preservation of diatoms in sediments of coastal mangrove systems is sometimes poor (Ross et al., 2001), locations slightly displaced offshore appear to offer better preserved records that have been useful in salinity reconstructions (Huvane and Cooper, 2001). This work strongly advocates the use of diatoms in tracking habitat shifts in response to restoration at the freshwater–marine coastal interface.

Appendix

Number of occurrences, maximum relative abundances, and weighted-averaging (WA) salinity optima and tolerances (ppt) of the 132 most common diatom taxa in the southeast Florida coastal wetland study area. Taxa are listed in order of estimated WA salinity optima.

Taxon	No. Occ.	Max. Abund.	Salinity Optimum	Salinity Tolerance
Achnanthidium minutissimum (Kütz.) Czar.	8	0.29	1.80	0.20
Nitzschia nana Grun. in V. H.	11	0.14	1.83	0.72
Navicula subrostellata Hust.	5	0.07	1.84	0.21
Encyonopsis microcephala (Grun.) Kr.	18	0.43	1.84	0.85
Nitzschia amphibia f. *frauenfeldii* (Grun.) L-Bert.	6	0.13	1.84	0.28
Pinnularia maior (Kütz.) Rab.	9	0.02	1.95	0.35
Encyonema carina L-Bert. & Kr.	5	0.04	1.96	0.23
Brachysira neoexilis L-Bert. (Typ2)	16	0.09	2.04	0.66
Nitzschia palea var. *debilis* (Kütz.) Grun. in Cl. & Grun.	23	0.18	2.06	1.24
Nitzschia amphibia Grun.	13	0.06	2.08	1.08
Nitzschia semirobusta L-Bert.	18	0.25	2.19	0.86
Navicula cryptotenella L-Bert.	17	0.32	2.25	1.84
Encyonema evergladianum Kr.	33	0.57	2.25	1.51
Encyonema neomesianum Kr.	14	0.20	2.26	1.09
Encyonema silesiacum (Bl.) Mann	13	0.07	2.32	1.47
Diploneis ovalis (Hilse in Rab.) Cl.	28	0.10	2.39	1.44
Gomphonema intricatum var. *vibrio* (Ehr.) Cl.	5	0.01	2.44	1.44
Mastogloia smithii Thw. *ex* Sm.	38	0.65	2.44	1.53
Diploneis vacilans (Schm.) Cl.	5	0.05	2.62	1.78
Brachysira neoexilis L-Bert. (Typ1)	9	0.04	2.69	1.77
Navicula veneta Kütz.	18	0.09	2.89	2.36
Brachysira neoexilis L-Bert. (Typ3)	7	0.06	2.92	1.52
Navicella pusilla Kr.	29	0.39	3.00	1.88
Nitzschia palea (Kütz.) Sm.	14	0.04	3.11	2.31
Caponea caribbea Podz.	14	0.02	3.40	2.47
Fragilaria synegrotesca L-Bert.	29	0.16	3.41	2.68
Rhopalodia operculata (Ag.) Håkansson	6	0.21	3.66	2.24
Rhopalodia gibba (Ehr.) Müller	9	0.04	3.68	2.89
Nitzschia bergii Cl.-Eul.	4	0.11	3.72	2.23
Amphora sulcata A. Schm.	20	0.57	3.87	2.61
Nitzschia intermedia Hantzsch *ex* Cl. & Grun.	4	0.02	3.89	2.30
Nitzshia dissipata (Kütz.) Grun.	8	0.05	3.94	3.00
Selaphora stroemii (Hust.) Mann	5	0.01	4.15	3.17
Mastogloia smithii var. *lacustris* Grun.	13	0.02	4.20	2.75
Diploneis litoralis (Donkin) Cl.	9	0.03	4.36	3.06
Diploneis oblongella (Nae.g.) Cl.-Eul.	39	0.35	4.58	2.90
Diploneis parma Cl.	14	0.02	5.22	3.53
Nitzschia sigmoidea (Nitzsch) Sm.	5	0.01	5.34	3.23
Nitzschia serpentiraphe L-Bert.	4	0.07	5.89	2.83
Navicula erifuga L-Bert.	16	0.02	5.98	3.85
Kolbesia amoena (Hust.) Kingston	10	0.04	6.14	3.61
Gomphonema vibrioides Reich. & L-Bert.	8	0.01	6.31	3.68
Navicula podzorski L-Bert.	11	0.06	6.46	3.49
Rhopalodia musculus (Kütz.) Müller	5	0.03	7.07	2.89
Nitzschia gracilis Hantzsch	14	0.03	7.20	3.66
Caloneis sp. 02L31E	4	0.02	7.31	3.04
Navicula sp. 03L31E	23	0.34	8.17	3.03
Fragilaria fasciculata (Ag.) L-Bert.	18	0.27	8.85	3.98

(continued)

Taxon	No. Occ.	Max. Abund.	Salinity Optimum	Salinity Tolerance
Nitzschia graciliformis L-Bert. & Simonsen	5	0.01	9.15	4.41
Seminavis strigosa (Hust.) Danielidis & Economou-Amilli	14	0.17	9.45	4.15
Fragilaria capensis Grun.	4	0.02	10.37	2.25
Mastogloia braunii Grun.	15	0.05	10.55	3.60
Fragilaria femelica (Kütz.) L-Bert.	31	0.36	11.60	2.51
Amphora sp. 22L31E	8	0.10	11.79	1.57
Nitzschia microcephala Grun. in Cl. & Möller	27	0.05	11.97	3.20
Navicula tenelloides Hust.	12	0.07	12.41	3.27
Nitzschia sigma (Kütz.) Sm.	4	0.09	12.74	2.85
Navicula cryptocephala Kütz.	5	0.05	13.07	3.92
Rhopalodia gibberula (Ehr.) Müller	25	0.69	13.09	3.88
Nitzschia scallpeliformis (Grun.) Grun. in Cl. & Grun.	4	0.09	13.13	3.79
Amphora subacutiuscula Sch.	23	0.30	13.26	2.44
Brachysira aponina Kütz.	12	0.11	13.54	3.78
Amphora cymbifera Greg	7	0.03	13.73	2.00
Mastogloia halophila John	20	0.27	13.89	2.80
Cyclotella meneghiniana Kütz.	12	0.10	14.38	3.88
Proszkinia bulnheimii Grun. Karayeva	4	0.08	14.43	1.47
Nitzschia frustulum (Kütz.) Grun.	12	0.16	14.44	3.82
Amphora eulensteinii Grun.	9	0.05	14.60	2.16
Amphora coffeaeformis Kütz.	15	0.18	14.61	2.02
Entomoneis sp.02L31E	11	0.01	14.69	1.92
Denticula subtilis Grun.	20	0.21	15.18	3.03
Grammatophora oceanica (Ehr.) Grun.	6	0.01	15.30	1.61
Amphora acutiuscula Kütz.	10	0.04	15.31	1.31
Amphora coffeaeformis var. *borealis* (Kütz.) Cl.	4	0.01	15.51	1.40
Amphora coffeaeformis var. *aponina* (Kütz.) Arch. & Sch.	27	0.36	15.55	1.98
Mastogloia reimeri John	11	0.06	15.69	2.96
Cocconeis placentula var. *lineata* (Ehr.) V. H.	9	0.16	15.75	1.77
Bacillaria paradoxa Gmelin	8	0.06	16.02	0.70
Mastogloia erythraea var. *grunowii* Foged	7	0.02	16.06	0.88
Planothidium rostratum (Østrup) Round & Bukhtiyarova	4	0.05	16.34	3.07
Cocconeis scutellum var. *ornata* Grun.	6	0.04	16.48	0.39
Amphora normani Hust.	9	0.09	16.54	0.94
Rhopalodia constricta (Sm.) Kr.	8	0.03	16.59	1.24
Rhopalodia acuminata Kr.	22	0.49	16.60	2.11
Cocconeis placentula Ehr.	10	0.24	16.84	0.96
Mastogloia recta var. *pumila* Hust.	4	0.04	16.85	0.54
Melosira sp. 01L31E	10	0.02	17.02	2.69
Navicula recens L-Bert.	8	0.05	17.02	2.50
Mastogloia angusta Hust.	17	0.17	17.06	2.38
Cocconeis scutellum Ehr.	4	0.13	17.12	0.55
Mastogloia ovalis Schm.	6	0.06	17.24	0.26
Achnanthes nitidiformis L-Bert.	5	0.02	17.28	0.51
Amphora sp. 02L31E	8	0.04	17.30	0.57
Hyalosynedra leavigata (Grun.) Will. & Round	16	0.41	17.31	1.48
Cyclotella distinguenda Hust.	22	0.21	17.38	1.12
Mastogloia braunii f. *minuta* Voigt	10	0.02	17.45	2.41
Licmophora normaniana (Grev.) Wahrer	11	0.03	17.50	0.80
Mastogloia crucicula (Grun.) Cl.	5	0.05	17.57	0.54
Rhabdonema adriaticum Kütz.	5	0.21	17.57	0.68
Navicula palestinae (Gerloff)	17	0.37	17.61	1.70
Seminavis sp. 02L31E	4	0.00	17.62	1.22
Mastogloia pumila (Grun.) Cl.	4	0.01	17.67	0.99

(continued)

Taxon	No. Occ.	Max. Abund.	Salinity Optimum	Salinity Tolerance
Amphora sp. 39L31E	9	0.03	17.74	0.80
Cocconeis placentula var. *euglipta* (Ehr.) Grun.	5	0.08	17.90	0.91
Mastogloia pusilla Grun.	6	0.03	18.23	0.81
Mastogloia sp. 04L31E	5	0.08	18.29	0.88
Diploneis gruendleri (Schm.) Cl.	6	0.02	18.33	0.73
Tryblionella granulata (Grun. in Cl. & Möller) Mann	6	0.05	18.38	0.92
Mastogloia biocellata Navarino & Muftah	7	0.01	18.40	1.27
Mastogloia erythraea Grun.	12	0.06	18.43	2.22
Amphora costata W. Sm.	6	0.13	18.45	0.84
Navicula sp. 01L31E	4	0.11	18.59	0.76
Tryblionella coarctata (Grun. in Cl. & Grun.) Mann	4	0.00	18.66	0.90
Seminavis gracilenta (Grun. *ex* Schm.) Mann	12	0.02	18.72	1.12
Grammatophora macilenta Sm.	4	0.00	18.74	1.16
Navicula cincta (Ehr.) Ralfs in Pritchard	5	0.07	18.86	0.75
Mastogloia cyclops Voigt	9	0.03	18.93	0.86
Diploneis caffra Giffen	14	0.21	19.07	0.98
Amphora ostrearia var. *lineata* (Bréb. *ex* Kütz.) Cl.	8	0.14	19.11	0.76
Navicula tripunctata (Müller) Bory	5	0.01	19.12	1.21
Amphora veneta Kütz.	4	0.13	19.16	1.04
Planothidium lanceolatum (Bréb.) Round & Bukhtiyarova	4	0.07	19.28	1.08
Melosira nummuloides (Dillwyn) Ag.	4	0.09	19.32	1.29
Nitzschia vitrea Norman	9	0.03	19.38	1.85
Seminavis robusta Danielidis & Mann	12	0.01	19.67	1.01
Navicula pseudocrassirostris (Hust.)	8	0.03	19.73	1.33
Mastogloia nabulosa Voigt	10	0.09	20.00	0.92
Fragilaria tenera (Sm.) L-Bert.	8	0.04	20.25	1.39
Caloneis sp. 01L31E	4	0.00	20.52	1.06
Tryblionella debilis Arnott	7	0.07	20.86	1.33
Mastogloia elegans Levis	13	0.09	20.99	1.03
Amphora sp. 24L31E	6	0.02	21.38	0.66

References

Admiraal, W. 1984. The ecology of estuarine sediment-inhabiting diatoms. *Progress in Phycological Research* 3:269–322.

Beanland, W. R. and W. Woelkerling. 1983. *Avicennia* canopy effects on mangrove algal communities in Spencer Gulf, South Australia. *Aquatic Botany* 17:309–313.

Browder, J. A. et al. 1982. Biomass and primary production of microphytes and macrophytes in periphyton habitats of the southern Everglades. Report T-662. South Florida Research Center, Homestead.

Campeau, S., A. Hequette, and R. Pienitz. 1995. The distribution of modern diatom assemblages in coastal sedimentary environments of the Canadian Beufort Sea: an accurate tool for monitoring coastal changes. In *Proceedings of the 1995 Canadian Coastal Conference*, Vol. 1. Canadian Coastal Science and Engineering Association, Dartmouth, Nova Scotia, pp. 105–116.

Collado-Vides, L. 2000. A review of algae associated with Mexican mangrove forests. In *Aquatic Ecosystems of Mexico: Status and Scope,* M. Munawar et al. (eds.). Ecovision World Monograph Series. Backhuys Publishers, Leiden, the Netherlands, pp. 353–365.

Cooper, S. R. 1995. Chesapeake Bay watershed historical land use: Impact on water quality and diatom communities. *Ecological Applications* 5:703–723.

Cooper, S. R. 1999. Estuarine paleoenvironmental reconstruction using diatoms. In *The Diatoms: Applications for the Environmental and Earth Sciences,* E. F. Stoermer and J. P. Smol (eds.). Cambridge University Press, New York, pp. 352–373.

Cooper, S. R. et al. 1999. Calibration of diatoms along a nutrient gradient in Florida Everglades Water Conservation Area-2A, USA. *Journal of Paleolimnology* 22:413–437.

Day, J. W. et al. 1989. *Estuarine Ecology.* John Wiley, New York.

De Wolf, H. 1982. Method of coding ecological data from diatoms for computer utilization. *Mededelingen Rijks Geologische Dienst* 36:95–98.

Dufrene, M. and P. Legendre. 1997. Species assemblages and indicator species: the need for a flexible asymmetrical approach. *Ecological Monographs* 67:345–366.

Egler, F. E. 1952. Southeast saline Everglades vegetation, Florida, and its management. *Vegetation Acta Geobotanica* 3: 213–265.

Foged, N. 1984. Freshwater and littoral diatoms from Cuba. *Bibliotheca Diatomologica* 5:1–243.

Fritz, S. C. et al. 1999. Diatoms as indicators of hydrologic and climatic change in saline lakes. In *The Diatoms: Applications for the Environmental and Earth Sciences,* E. F. Stoermer and J. P. Smol (eds.). Cambridge University Press, New York, pp. 41–72.

Gaiser, E. et al. 2004. Phosphorus in periphyton mats provides the best metric for detecting low-level P enrichment in an oligotrophic wetland. *Water Research* 38:507–516.

Gasse, F., J. F. Talling, and P. Kilham. 1983. Diatom assemblages of East Africa: classification, distribution and ecology. *Revue d'Hydrobiologie Tropicale* 16:3–34.

Hussain, M. and T. Khoja. 1993. Intertidal and subtidal blue-green algal mats of open and mangrove areas in the Farasan Archipelago (Saudi-Arabia), Red Sea. *Botanica Marina* 36:377–388.

Huvane, J. K. and S. R. Cooper. 2001. Diatoms as indicators of environmental change in sediment cores from northeastern Florida Bay. In Paleoecological Studies of South Florida. *Bulletins of American Paleontology* 361:145–158.

Juggins, S. 1992. Diatoms in the Thames estuary, England. Ecology, paleoecology, and salinity transfer function. *Bibliotheca Diatomologica* 25:1–216.

Juggins, S. 2003. *C2 User Guide. Software for Ecological and Palaeoecological Data Analysis and Visualisation.* University of Newcastle, Newcastle-upon-Tyne, U.K., 69 pp.

Koch, M. S. and C. J. Madden. 2001. Patterns of primary production and nutrient availability in a Bahamas lagoon with fringing mangroves. *Marine Ecology Progress Series* 219:109–119.

Montgomery, R. T. 1978. Environmental and Ecological Studies of Diatom Communities Associated with the Coral Reefs of the Florida Keys. Ph.D. dissertation. Florida State University, Tallahassee.

National Research Council. 1993. Managing Wastewater in Coastal Urban Areas. National Academy Press, Washington, D.C.

Navarro, J. N. 1982. Marine diatoms associated with Mangrove Prop Roots in the Indian River, Florida, U.S.A. *Bibliotheca Phycologica* 61:1–151.

Navarro, N. and R. Torres. 1987. Distribution and community structure of marine diatoms associated with mangrove prop roots in the Indian River, Florida, U.S.A. *Nova Hedwigia* 45:101–112.

Park, R. A. et al. 1989. Coastal wetlands in the twenty-first century: profound alterations due to rising sea level. In *Wetlands: Concerns and Successes. Proceedings of the American Water Resources Association,* Tampa, FL, pp. 71–80.

Phillips, A. et al. 1994. Horizontal zonation of epiphytic algae associated with *Avicennia marina* (Forssk) Vierh pneumatophores at Beachwood Mangroves Nature Reserve, Durban, South Africa. *Botanica Marina* 37:567–576.

Podzorski, A. C. 1985. An illustrated and annotated check-list of diatoms from the Black River waterways, St. Elisabeth, Jamaica. *Biblioteca Diatomologica* 7:1–177.

Rasmussen, K. A., I. F. MacIntyre, and L. Prufert. 1993. Modern stromatolite reefs fringing a brackish coastline, Chetumal Bay, Belize. *Geology* 21:199–202.

Reimer, C. W. 1996. Diatoms from some surface waters on Great Abaco Island in the Bahamas (Little Bahama Bank). *Beiheft zu Nova Hedwigia* 112:343–354.

Rejmánková, E. and J. Komárková. 2000. A function of cyanobacterial mats in phosphorus-limited tropical wetlands. *Hydrobiologia* 431:135–153.

Rodriguez, C. and A. W. Stoner. 1989. The epiphyte community of mangrove roots in a tropical estuary: distribution and biomass. *Aquatic Botany* 36:117–126.

Ross, M. S. et al. 2000. The Southeast Saline Everglades revisited: a half-century of coastal vegetation change. *Journal of Vegetation Science* 11:101–112.

Ross, M. S. et al. 2001. Multi-taxon analysis of the "white zone," a common ecotonal feature of South Florida coastal wetlands. In *The Everglades, Florida Bay and Coral Reefs of the Florida Keys: An Ecosystem Sourcebook,* J. Porter and K. Porter (eds.). CRC Press, Boca Raton, FL, pp. 205–238.

Scholl, D. W., F. C. Craighead, and M. Stuiver. 1969. Florida submergence curve revisited: its relation to coastal sedimentation rates. *Science* 163:562–564.

Siqueiros-Beltrones, D. A. and E. S. Castrejón. 1999. Structure of benthic diatom assemblages from a mangrove environment in a Mexican subtropical lagoon. *Biotropica* 31:48–70.

Slate, J. 1998. Inference of present and historical environmental conditions in the Everglades with diatoms and other siliceous microfossils. Ph.D. dissertation. University of Louisville, Louisville, KY.

Snoeijs, P. 1999. Diatoms and environmental change in brackish waters. In *The Diatoms: Applications for the Environmental and Earth Sciences,* E. F. Stoermer and J. P. Smol (eds.). Cambridge University Press, New York, pp. 298–333.

Sullivan, M. J. 1981. Effects of canopy removal and nitrogen enrichment on *Distichlis spicata*–edaphic diatom complex. *Estuarine and Coastal Shelf Science* 13:119–129.

Sullivan, M. J. 1982. Distribution of edaphic diatoms in a Mississippi salt marsh: a canonical correlation analysis. *Journal of Phycology* 18:130–133.

Sullivan, M. J. 1999. Applied diatom studies in estuaries and shallow coastal environments. In *The Diatoms: Applications for the Environmental and Earth Sciences,* E. F. Stoermer and J. P. Smol (eds.). Cambridge University Press, New York, pp. 334–351.

Underwood, G. J. C. 2002. Adaptations of tropical marine microphytobenthic assemblages along a gradient of light and nutrient availability in Suva Lagoon, Fiji. *European Journal of Phycology* 37: 449–462.

Vos, P. and H. de Wolf. 1993. Diatoms as a tool for reconstructing sedimentary environments in coastal wetlands: methodological aspects. *Hydrobiologica* 269/270:297–296.

Wanless, H. R., R. W. Parkinson, and L. P. Tedesco. 1994. Sea level control on stability of Everglades wetlands. In *Everglades: The Ecosystem and Its Restoration,* S. M. Davis and J. C. Ogden (eds.). St. Lucie Press, Delray Beach, FL, pp. 199–223.

11

Using Microalgal Indicators to Assess Human- and Climate-Induced Ecological Change in Estuaries

Hans W. Paerl, Julianne Dyble, James L. Pinckney, Lexia M. Valdes, David F. Millie,
Pia H. Moisander, James T. Morris, Brian Bendis, and Michael F. Piehler

CONTENTS

Introduction

Estuaries represent a formidable challenge when it comes to determining status and trends in water quality, habitat, and ecological condition. These systems are dynamic and complex from hydrologic, nutrient cycling, and biotic resource perspectives. Hydrologically, freshwater runoff interacts with tidal saltwater exchange and upwelling, leading to complex circulation and mixing patterns. These patterns vary from minutes to weeks and meters to many kilometers, strongly shaping the chemical and biological characteristics of these ecosystems. In addition, there are strong seasonal and interannual shifts in climatic forcing (i.e., temperature, irradiance, rainfall, wind) that can vary substantially. Last, but not least, human activity is an additional and often dominant source of stress and change. At least half the world's population resides in estuarine watersheds (Vitousek et al., 1997; Culliton, 1998), and this percentage continues to grow. Human development in coastal river basins has greatly increased nutrient and sediment loads to downstream estuarine and coastal waters (Peierls et al., 1991; Nixon, 1995; Paerl, 1997), resulting in deterioration of water quality, loss of fisheries habitat and resources, and an overall decline in ecological and economic condition of the coastal zone (Costanza et al., 1997; National Research Council, 2000; Boesch et al., 2001). Given the overall importance of estuarine ecosystems, there is an urgent need to develop sensitive, definitive, and broadly applicable indicators of water quality, habitat condition, biodiversity, and overall ecological change. The Committee on Environmental and Natural Resources (1997) summarized the need as follows: "To link stressors to biotic responses across diverse estuarine ecosystems, specific, yet broadly-applicable and integrative indicators that can couple biotic community structure to function in the context of ecological condition and change are needed."

Anthropogenic and natural stressors frequently interact. For example, nutrient, sediment, and toxin inputs may be affected by climatic, geological, and other forms of natural change. Certain manifestations of climate change, including tropical storm and hurricane frequency, may also be increasing (Goldenberg et al., 2001). It is therefore useful to develop indicators that can help distinguish human from natural perturbations. This goal is compounded by the fact that these perturbations may be identical, overlap, or act synergistically, potentially blurring this distinction.

In addressing the need to assess estuarine ecological change in response to diverse stressors, environmental and resource management agencies, for example, U.S. Environmental Protection Agency (U.S. EPA), National Oceanic and Atmospheric Administration (NOAA), U.S. Geological Survey (USGS), have developed regional networks over which estuarine and coastal condition can be determined and compared based on a suite of water quality and habitat indicators. These indicators have been used to develop water quality criteria, including designations of nutrient-sensitive waters and total maximum daily (nutrient) loads (TMDLs) (U.S. EPA, 1993, 1998a,b; National Research Council, 1994, 2001). In 1990, the U.S. EPA Environmental Monitoring and Assessment Program (EMAP) launched a survey aimed at developing a comparative analysis of water and habitat quality in diverse U.S. estuarine and coastal waters (National Research Council, 1994; U.S. EPA, 2001). Coastal EMAP has generated regional databases for identifying undesirable conditions and for gauging trends in biological structure and function (National Research Council, 1994). The NOAA National Estuarine Eutrophication Assessment (Bricker et al., 1999) further documented deficiencies in water and habitat quality for many U.S. estuaries.

Collectively, these studies have identified nutrient-enhanced primary production, or eutrophication, and its unwanted consequences (algal blooms, hypoxia, finfish and shellfish disease, and mortality) as a primary cause of estuarine water quality degradation, food web alterations, habitat loss, and overall ecosystem impairment. Microalgae, including prokaryotic cyanobacteria and eukaroytic algal groups (e.g., chlorophytes, chrysophytes, cryptophytes, diatoms, dinoflagellates), account for a major amount of primary production and play a central role in carbon, nutrient (i.e., nitrogen and phosphorus), and oxygen cycling in estuaries. Microalgae have fast growth rates (i.e., doubling times of a day or less) and rapidly respond to diverse chemical (nutrients, toxicants) and physical (light, temperature, turbulence) stresses over a wide range of concentrations and intensities. Changes in microalgal community structure and activity often precede larger-scale, longer-term changes in ecosystem function, including shifts in material flux, oxygen balances, food webs and fisheries, and habitat.

Using recently developed microalgal indicators, we examine and evaluate ecological and biogeochemical impacts of a range of physical, chemical, and biotic perturbations in geographically distinct estuaries varying in water residence time, climate, and trophic state. The focus is on the causes and effects of nutrient overenrichment, a common and expanding stressor (Smetacek et al., 1991; Nixon, 1995).

Methods

Looking into the "Green Box": Diagnostic Pigment Indicators of Phytoplankton Community Composition and Activity

Planktonic microalgal, or phytoplankton, communities are dynamic multispecies assemblages that exhibit spatial patchiness (microns to meters) over timescales ranging from minutes to days (Dustan and Pinckney, 1989). Detailed characterizations of the community-level processes that structure phytoplankton species composition are essential for developing accurate conceptual and mathematical ecological models. A critical need for characterizing these processes is the ability to determine the species composition of natural phytoplankton communities with reliability and accuracy. A reliable technique for enumerating single species in mixed phytoplankton samples is microscopic counts, which are tedious

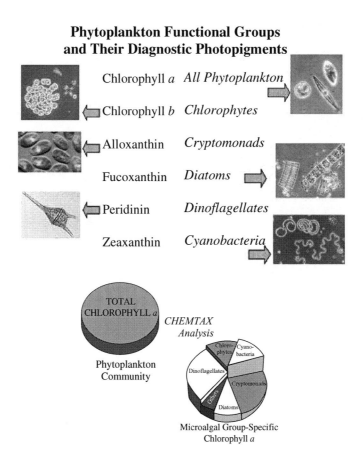

FIGURE 11.1 (Color figure follows p. 266.) (Top) Chlorophyll and carotenoid photopigments commonly used for identifying and quantifying phytoplankton functional groups, representative photomicrographs of which are shown. (Bottom) Diagram, showing how the ChemTax matrix factorization program is used to determine the proportions of total phytoplankton biomass (as chl *a*) attributable to phytoplankton taxonomic groups, based on HPLC separation and quantification of the diagnostic photopigments shown in the upper frame.

and require a high level of expertise. However, chemosystematic pigments encoding specific phytoplankton taxonomic groups (PTGs) (i.e., diatoms, chlorophytes, dinoflagellates, cyanobacteria, cryptomonads, etc.) can also be used (Jeffrey et al., 1997). In particular, PTG-specific carotenoids provide diagnostic biomarkers for determining the relative abundance of PTGs in mixed species assemblages. Photopigment mixture extracts from natural samples are separated and quantified by high-performance liquid chromatography (HPLC) coupled to diode array spectrophotometry (PDAS) (Wright et al., 1991; Millie et al., 1993, 1995b). Photopigment composition is usually significantly (linearly) correlated with species cell counts or biovolume estimates (Tester et al., 1995; Descy and Métens, 1996; Roy et al., 1996; Woitke et al., 1996; Wright et al., 1996; Millie et al., 2002).

ChemTax® is a matrix factorization program used to calculate the absolute and relative abundances of major algal groups from concentrations of chemosystematic photopigment biomarkers (Mackey et al., 1996, 1998; Wright et al., 1996) (Figure 11.1). This program uses a steepest descent algorithm to determine the best fit based on an initial estimate of pigment ratios for algal classes. Input for the program consists of a raw data matrix of photopigment concentrations obtained by HPLC analyses and an initial pigment ratio file. Relatively large errors in the initial estimates of pigment ratios have little influence on the final determination of algal class abundances (Mackey et al., 1996; Schlüter et al., 2000). The data matrix is subjected to a factor minimization algorithm that calculates a best-fit pigment ratio matrix and a final phytoplankton class composition matrix. The class composition matrix can be expressed as relative or absolute values for specified photopigments. The absolute chlorophyll (chl) *a* contribution of each class is particularly useful because it partitions the total chl *a* into major PTGs (Figure 11.1).

A limitation of the HPLC-ChemTax approach is that it cannot determine individual phytoplankton species, but instead classifies a mixed sample in terms of major algal groups. In practice, this is not as large a limitation as it seems. Frequently, in estuaries PTGs are dominated by only one or at most a few species during significant productivity and bloom events; in which case the technique captures the key "players" dominating productivity. Numerous studies have shown that examining phytoplankton community dynamics and successional changes at the PTG level often provides excellent insight into the environmental controls of shifts in productivity, biogeochemical fluxes, and food web dynamics (Cottingham and Carpenter, 1998; Pinckney et al., 2001). Studies comparing ChemTax and microscopy have demonstrated the reliability of ChemTax to accurately assess such changes with taxonomic reliability (Jeffrey et al., 1999; Schluter et al., 2000; Wright and van den Enden, 2000).

HPLC-ChemTax can detect significant changes in community composition over a broad range of timescales (<24 h to decades) and thus is well suited for monitoring programs designed to assess short- and long-term trends and interannual variability in PTG composition and biomass. Applications include: examining phytoplankton community changes in response to large-scale hydrographic (circulation, upwelling) forcing features (Gieskes and Kraay, 1986; Tester et al., 1995), nutrient enrichment (Pinckney et al., 1997, 1999, 2001; Wear et al., 1999), and climatic and hydrologic perturbations (floods, droughts) (Harding et al., 1999; Paerl et al., 2001, 2003). In addition, "top-down" effects of grazing (Burkill et al., 1987; Strom and Welschmeyer, 1991; Head and Harris, 1994; Meyer-Harms and von Bodungen, 1997) have been examined using this technique. Routine monitoring of HPLC-derived photopigments has proved useful as a method of ground-truthing remotely sensed estimates of phytoplankton biomass and bloom events (Millie et al., 1992, 1995c; Harding et al., 1999). This application has provided significant improvements in "scaling up," i.e., mapping the spatial distributions of phytoplankton groups over large geographical areas not amenable to routine field sampling (Millie et al., 1993, 2002), evaluating the effectiveness of nutrient management strategies (Luettich et al., 2000; Paerl et al., 2002), acting as an early warning system for blooms of nuisance or toxic species (harmful algae blooms, or HABs; Millie et al., 1995a, 1997), and developing sensitive bioindicators of ecosystem-scale water quality conditions (Pinckney et al., 1999, 2001). HPLC can be used to assess ecophysiological properties of phytoplankton, including growth rates (Redalje, 1993; Pinckney et al., 1996), and palatability (e.g., production of grazing deterrents, food value) (Kleppel et al., 1988; Buffan-Dubau et al., 1996; Irgoien et al., 2000; Guisande et al., 2002). Last, HPLC-ChemTax can be used to distinguish abiotic from biotic (phytoplankton) turbidity in waters.

Long-term monitoring and assessment programs can benefit from this technology by establishing production and community compositional baselines against which ecological change can be assessed. In addition, invasive phytoplankton species may be detected in the early stages of colonization. Toxin-producing algal species, such as the red tide dinoflagellae *Karenia brevis* and some cyanobacteria, may result in the widespread mortality of estuarine biota, the closure of fisheries, and have negative impacts on tourism and human health. When incorporated into estuarine water quality monitoring programs, these indicators can help provide a determination of potential causal factors for these detrimental conditions, an evaluation of the extent of the problem, and a mechanism for evaluating the effectiveness of management efforts. Diagnostic pigment measurements may be used for evaluating total (allowable) maximum daily loads (TMDL) of nutrients, and whether water bodies meet state and federal standards (many are already based on chl *a*) for various uses (drinking, swimming, fishing).

Use of Neural Networks

Artificial neural networks (ANN) have recently found numerous computational applications in ecology and environmental science (e.g., Barciela et al., 1999; Lek and Guegan, 1999; Karul et al., 2000) and show promise for classifying remote imagery (Foody and Cutler, 2003, Tapiador and Casanova, 2003) and modeling phytoplankton dynamics, including those of HAB species (Recknagel et al., 1997; Maier et al., 1998; Richardson et al., 2002; Lee et al., 2003). ANNs use historical data sets to approximate the relationship between input and corresponding output variables (Maier et al., 1998). Through iterative presentation of the data and intrinsic mapping characteristics of neural topologies, ANNs identify correlated patterns between input data sets (e.g., environmental conditions) and corresponding output values (e.g., phytoplankton biomass). Neural network modeling has some advantages over ChemTax in that it does not require the use of a proprietary software package (MATLAB®), is quicker than the trial-and-error type of algorithm of ChemTax, and does not require a fixed data matrix of unchanging ratios of secondary pigments to chl *a* in each of the major taxonomic groups likely to be present in a sample (Table 11.1). A feedforward, backpropagating ANN is "trained" on existing data sets with known output values from which "learned" models are developed to predict output values for a new, independent input data set (Lek and Guegan, 1999).

To train the ANN, a computer program was written to randomly choose the chl *a* distributions of the algal taxa in each of hundreds of simulated water samples, and then to use a pigment ratio matrix (Table 11.1) to construct the pigment vector (Table 11.1, column z) associated with each sample. Five groups of data were constructed, each representing a different degree of variability in the individual ratios ranging from 0 to ±20% deviation from the ideal ratios. This was done to simulate the actual variation in the pigment ratios that arise due to changes in environmental condition and algal physiology, and changes in the relative abundance of algal species within a taxon. The ANN was trained on one combined data set consisting of all five groups. Inputs to the ANN were the pigment vectors, and the taxon-specific chl *a* distributions were the outputs. Independent sets of randomly constructed samples were used for validation and testing. These sets were analyzed by group and sorted by the degree of variability (0 to 20%) allowed in the pigment ratios. These test data were also analyzed using ChemTax. Figure 11.2 shows that both ChemTax and ANN are good to excellent predictors of the concentrations of algal taxa when the pigment ratios of the algae in the sample do not differ significantly from the ideal ratios in the matrix. However, in samples in which the actual pigment ratios differ from the ideal ratios in the matrix, ANN outperforms ChemTax for some taxa.

Molecular Approaches for Taxa-Specific Identification and Characterization

Molecular analyses can be useful in identifying and detecting species-specific responses to environmental change. Characterizing microbial populations based on DNA analysis has been used to identify microbial diversity in a wide variety of environments, including hot springs (Reysenback et al., 1994), microbial mats (Zehr et al., 1995; Steppe et al., 2001), oceanic phytoplankton (Giovannoni et al., 1990; Rappe et al., 1997; Zehr et al., 1998), and sediments (Widmer et al., 1999; Gordon et al., 2000), among many others. While much of the early analysis of microbial communities was based on structural genes like 16S rRNA, more recent work has focused on using functional genes to look at specific groups of microbes and the expression of those genes to investigate how those microbes are responding to environmental change. An example of such a functional gene is *nifH*, which encodes the dinitrogenase reductase enzyme involved in nitrogen fixation, an ecologically important process that can be a significant source of "new" nitrogen in nitrogen-limited water bodies. Many populations of N_2 fixers have been characterized by *nifH* sequence analysis, and this part of the *nif* operon has been very useful in differentiating genera of both cyanobacterial and heterotrophic (i.e., some sulfate reducers, methanogens) diazotrophs (Ben-Porath et al., 1993; Zehr et al., 1995, 1997, 1998; Dyble et al., 2002). The *nifH* sequences isolated from environmental samples are compared to sequences in the GenBank database for identification of novel sequences and for determining the degree of genetic similarity between populations. This similarity is visualized in the construction of phylogenetic trees that group together the most similar sequences in clusters.

Once a sufficient number of sequences have been identified for a specific functional group, it is possible to target that group in a mixed environmental population. Polymerase chain reaction (PCR) primers can be designed to amplify a specific group of organisms to the exclusion of others and, when applied to a bulk DNA extract from an environmental sample, will identify the presence of those sets of sequences. For example, PCR primers were designed to specifically amplify the *nifH* gene from the cyanobacterial diazotroph *Cylindrospermopsis raciborskii* (Dyble et al., 2002; see below). Identification of this toxic, bloom-forming cyanobacterial diazotroph is important in many lakes and reservoirs used for drinking

TABLE 11.1

Example of a Pigment Ratio Matrix Used in Chemtax

	Pigment Ratio Matrix A													Chl a	Total Pigments
	Cyano	Prochlo	Eugleno	Chloro	Prasino	Dino	Hapto	Crypto	Diatom	Chryso	Pelago	Karenia		
Chc	0	0	0	0	0.047	0.219	0.199	0.221	0.327	0	0.397	0.2	5	0.6
Perid	0	0	0	0	0	0.533	0	0	0	0	0	0	1	0
But	0	0	0	0	0	0	0.023	0	0	0	0.51	0.1		0.3
Fuco	0	0	0	0	0	0	0.304	0	0.779	0.23	0.732	0.291		1.333
Hex	0	0	0	0	0	0	0.27	0	0	0	0	0.109	2	0.327
Neo	0	0	0.069	0.043	0.096	0	0	0	0.001	0.001	0	0		0.088
Viola	0	0	0.007	0.032	0.138	0	0	0	0	0.058	0	0		0.18
Ddx	0	0	0.189	0	0	0.234	0.113	0	0.317	0.058	0.14	0.186	0	0.674
Diat	0	0	0.015	0	0	0.042	0.042	0	0.074	0.005	0.088	0.01		0.04
Allox	0	0	0	0	0	0	0	0.405	0	0	0	0		0
Lut	0	0	0	0.16	0.034	0	0	0	0	0	0	0		0.32
Zeax	1.25	0.114	0.056	0.041	0.061	0	0	0	0	0.029	0	0	2	6.504
Ch_b	0	0.286	0.219	0.277	0.606	0	0	0	0	0	0	0		0.84
Bcar	0.376	0.029	0.03	0.053	0.101	0.106	0.028	0.045	0.066	0.046	0.021	0.01	3	2.137
Gyro	0	0	0	0	0	0	0	0	0	0	0	0.1		0.3

($\times$ between Karenia and Chl a columns; $=$ between Chl a and Total Pigments columns)

Note: The column on the right (call it y) shows a hypothetical distribution of secondary pigment concentrations in a water sample. The ratio matrix (A) contains the ratios of the secondary pigments to chl a, and when multiplied by the concentrations of chl a associated with each taxon (middle column, z) yields the pigment concentrations (z). Or, $Az=y$.

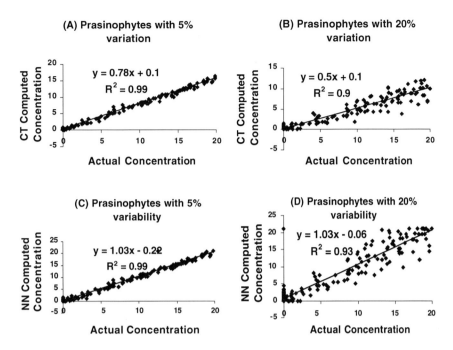

FIGURE 11.2 Typical results for a representative taxonomic group (Prasinophytes) when ChemTax (A and B) and a trained neural network (C and D) are applied to the same set of synthetic data. Synthetic data were generated assuming that the variability in the ratios of secondary pigments to chl *a* was ±5% (A and C) or as great as ±20% (B and D).

water and recreation and PCR-based methods are often a quicker and more reliable means than labor-intensive microscopy; especially when this cyanobacterium lacks characteristic cells like heterocysts used for identification.

Results and Discussion

Case Studies

Photopigment, molecular, and other recently developed indicators have been used to examine phytoplankton community responses to anthropogenic nutrient enrichment and hydrologic perturbations (droughts and floods). Examples are provided for the Neuse River Estuary, North Carolina; Galveston Bay, Texas; and the St. Johns River Estuary, Florida.

Impacts of Human and Climatic Perturbations in the Neuse River Estuary, North Carolina

The Neuse River Estuary (NRE) is one of three major tributaries of North Carolina's Pamlico Sound System, the nation's second largest estuarine complex and a key fisheries nursery for the Mid-Atlantic and Southeast Atlantic regions (Figure 11.3). The NRE is downstream of rapidly expanding agricultural (hog, poultry, and rowcrop operations), urban (Raleigh–Durham Research Triangle), and industrial activities in NC coastal watersheds. Excessive nutrient (largely non-point nitrogen) discharge associated with this expansion has promoted eutrophication, nuisance cyanobacterial and dinoflagellate blooms, hypoxia, altered nutrient cycling, toxicity, and food web modifications over the past three decades (Copeland and Gray, 1991; Paerl et al., 1995, 1998; Figure 11.3). In response to scientific evidence of nitrogen-driven eutrophication, the North Carolina General Assembly mandated a 30% reduction in external nitrogen loading to the NRE to be in place in 2004. This large-scale manipulation provides opportunities to examine the effects of nutrient perturbations on estuarine phytoplankton dynamics.

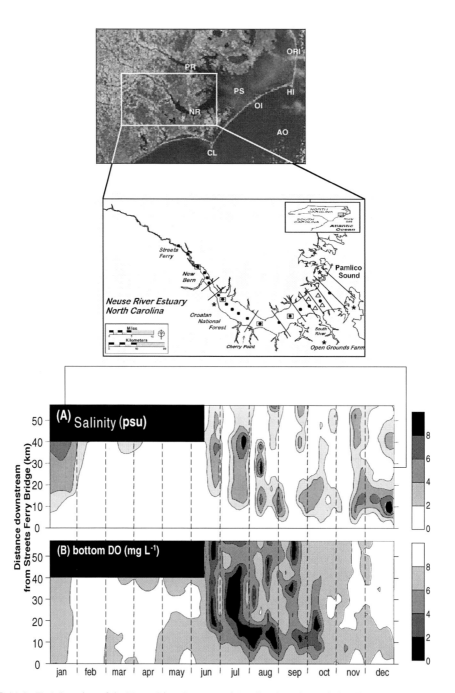

FIGURE 11.3 (Top) Location of the Neuse River Estuary and Pamlico Sound, North Carolina. Shown are the Atlantic Ocean (AO), Oregon, Hatteras, and Ocracoke Inlets (ORI, HI, OI, respectively), Cape Lookout (CL), Pamlico Sound (PS), and the Pamlico and Neuse Rivers (PR and NR). The NRE sampling sites for mid-river water quality (19 filled circles) and continuous in-stream monitoring (4 open boxes) are shown. Triangles indicate sites for diel and other periodic studies during which additional samples are collected. (Bottom) Spatiotemporal contour plots of salinity and dissolved oxygen characteristics of the Neuse River Estuary during an annual cycle. The sampling locations from which the data were derived are shown. Near-surface and near-bottom samples were collected biweekly as part of the Neuse River Modeling and Monitoring Program (www.marine.unc.edu/neuse/modmon). Samples were collected at the surface and bottom. The data are plotted along a transect spanning upstream freshwater (Streets Ferry Bridge (SFB, designated 0 km), to a downstream mesohaline location (50 km downstream of Streets Ferry Bridge), approximately midway between Minnesott Beach and the entrance to Pamlico Sound (50 km location). Data were plotted using Surfer Plot software.

FIGURE 11.4 (Color figure follows p. 266.) (Top) NASA SeaWiFS ocean color satellite images of the landfalls of Hurricanes Dennis and Floyd in Eastern North Carolina, during September 1999. (Bottom left) SeaWiFS image of the Pamlico Sound and surrounding watershed and coastal region. This image was recorded on 23 September 1999, approximately 1 week after landfall of Hurricane Floyd. The sediment-laden floodwaters from the cumulative rainfall of Hurricanes Dennis and Floyd (up to 1 m in some parts of the sound's watershed) can be seen moving across Pamlico Sound. Also note the turbid overflow from Pamlico Sound entering the Atlantic Ocean. The sediment plume was advecting northward into the coastal ocean by the Gulf Stream, which was located approximately 40 km offshore at the time this image was recorded. (Bottom right) Edge of the sediment-laden floodwaters moving across the Pamlico Sound. The floodwater caused strong vertical salinity stratification, hypoxic bottom waters, and stimulated phytoplankton production throughout the sound (see Paerl et al., 2001).

This system has also been under the influence of natural perturbations such as droughts, hurricanes, and flooding. During the fall of 1999, Hurricanes Dennis, Floyd, and Irene inundated coastal North Carolina with as much as 1 m of rainfall, causing a 100-year flood in the watershed of the Pamlico Sound (Figure 11.4). Sediment and nutrient-laden floodwaters displaced more than 80% of the sound's volume, depressed salinity by more than 70%, and accounted for half the annual nitrogen load to this nitrogen-sensitive system (Paerl et al., 2001). Biogeochemical and ecological effects included hypoxic (<4 mg O_2 l^{-1}) bottom waters, major changes in nutrient cycling, a threefold increase in algal biomass, altered fish distributions and catches, and an increase in fish disease (Paerl et al., 2001).

HPLC diagnostic photopigment determinations coupled to ChemTax analyses have been applied in the NRE to characterize spatial and temporal trends in phytoplankton community structure since 1994. The absolute concentrations of five common PTGs (chlorophytes, cryptophytes, cyanobacteria, diatoms, and dinoflagellates) were determined from biweekly water samples collected from fixed sampling stations in the NRE (Figure 11.3, Figure 11.5, and Figure 11.6). During this time, the NRE was influenced by increases in anthropogenic nutrient loading and variable hydrologic conditions including droughts and, since 1996, an increase in the frequency of tropical storms and hurricanes. These anthropogenic and natural disturbances have provided an opportunity to examine how specific PTGs respond to changes in nutrient loading and hydrology through time.

The effect of hydrologic variability on the abundance of each of the PTGs and on chl *a* is illustrated for a mid-river long-term monitoring station located at the bend in the Neuse River where flow changes from a southeast to a northeast direction (Figure 11.3 and Figure 11.5). Seasonal and hurricane-induced pulses in river discharge, and the resulting changes in estuarine flushing and water residence times, have differentially affected PTGs as a function of their contrasting growth characteristics. Chlorophyte and

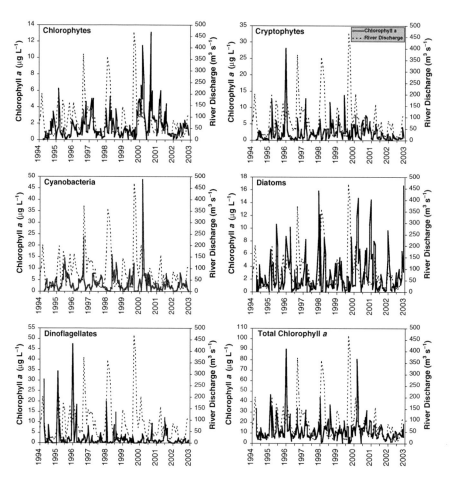

FIGURE 11.5 Absolute concentrations of common phytoplankton taxonomic groups (chlorophytes, cryptophytes, cyano-bacteria, diatoms, and dinoflagellates) and total chl *a* (all phytoplankton groups) at a midriver long-term monitoring station in the Neuse River Estuary (solid line). Values were determined from ChemTax analysis of phytoplankton pigments that were measured from biweekly surface water samples collected during 1994 through 2002. River discharge values (dashed line) represent monthly means of mean daily river discharge rates measured at the USGS stream gauge station 2089500, located upstream of the Neuse River Estuary in Kinston, NC.

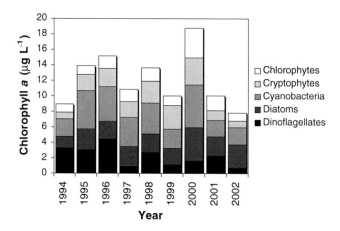

FIGURE 11.6 Mean annual contribution of chlorophytes, cryptophytes, cyanobacteria, diatoms, and dinoflagellates to total surface chl *a* at a midriver station in the Neuse River Estuary.

diatom abundance coincided with periods of elevated river flow, cyanobacteria and dinoflagellate growth was reduced during these events, while cryptophyte abundance was variable in response to hydrologic change. It is hypothesized that the efficient growth rates and enhanced nutrient uptake rates of chlorophytes and diatoms allow for the rapid utilization of pulsed nutrient supplies accompanying high flushing rates (short residence time). On the contrary, cyanobacteria were more abundant when river discharge was minimal. Their growth seems to be optimal during periods of long residence time and water column stratification, which typically occur during the summer.

The heavy rainfall, elevated river discharge rates, and increased nutrient loading from the three sequential hurricanes that struck during the fall of 1999 had profound effects on the relative abundances of these PTGs (Figure 11.5). During the 6 weeks of flooding that followed these storms, the abundance of all PFGs was reduced as they were flushed out of the NRE into the Pamlico Sound. However, once the floodwaters receded, high chlorophyte abundance was observed from late fall 1999 through early summer 2000, while diatoms increased in abundance during the spring of 2000. In addition, cyanobacteria, which normally do not show high concentrations during the spring, demonstrated a large spring peak. The overall effect of these peaks was a significant increase in total chl *a* in 2000 when compared to the other years studied (Figure 11.6). These results highlight the strong influence that increased flow rates can have on phytoplankton community structure in this system.

Further evidence that changes in hydrologic conditions may have altered phytoplankton community structure is provided by trends in dinoflagellate abundance (Figure 11.5 and Figure 11.6). Since the increase in hurricane frequency in 1996, the typical late winter–early spring blooms of the dinoflagellate *Heterocapsa triquetra* that regularly frequented the NRE from the 1970s through the mid-1990s have essentially disappeared (Figure 11.5). The relatively slow growth rates of dinoflagellates may have led to their reduced abundance during these high river discharge events. It appears as though the winter–spring blooms of *H. triquetra* were returning in January 1998, following an approximate 1-year lapse in hurricane activity; however, the magnitude of this bloom was reduced compared to the previous winter–spring blooms of 1994–1996. Following the high spring runoff that occurred in 1998 and the hurricanes of 1999, these winter–spring dinoflagellate blooms were once again absent at this mid-river location. Interestingly, following a 3-year hiatus in hurricane activity since 1999, dinoflagellate blooms were once again starting to return to this mid-estuarine region (fall 2000, summer 2001), however, at comparatively lower concentrations. Interestingly, the most recent hurricane (Hurricane Isabel) that passed directly over the North Carolina coastal estuaries in the fall of 2003 had little rainfall and hence minimal associated flushing. As expected, *H. triquetra* is continuing to increase in abundance following this particular storm.

Hydrologically induced changes in PFGs may have potentially altered the trophodynamics and nutrient cycling processes in the NRE throughout these years. We are examining potential links between these changes and altered numbers and diversity of estuarine-dependent finfish and shellfish species (L. Crowder, pers. commun.), as well as size spectra of planktivorous and carnivorous fish (E. Houde et al., pers. commun.).

Galveston Bay, Texas: The Case of the "Pink Oysters"

Galveston Bay (GB) supports a large, commercial fishery for the eastern oyster (*Crassostrea virginica*), with annual harvests of near 400 metric tons. Phytoplankton is a primary food source for oysters and individual algal species vary in nutritional quality. Texas oystermen have recently expressed concern over a peculiar red/pink coloration of oysters ("pink oyster") from some commercial reefs in GB. Although the conspicuous color has no apparent effect on oyster condition and is not known to pose a human health hazard, the coloration adversely affects consumer acceptance of GB "pink oysters." In addition, these oysters reportedly have an "off-taste" that further detracts from their marketability. Pink-oyster events appear to be increasing in GB, suggesting that this is a growing problem.

The coloration is caused by the phytoplankton upon which the oysters feed. Accessory photosynthetic pigments (carotenoids, phycobilins) from the algae accumulate in the oysters, leading to the red-pink coloration. During the December 2000 pink-oyster event, the gut contents of both oyster types were

FIGURE 11.7 Galveston Bay, Texas. The dashed line down the center of the bay indicates the sampling transect (from 0.0 to 60.0 km) used to construct the spatiotemporal contour plots in Figure 11.8 and Figure 11.9. The reference point shows the location of several large commercial oyster reefs in the bay where pink oysters have been collected.

analyzed by HPLC to determine phytoplankton groups present in the oyster guts. A comparison between the two oyster types revealed a high concentration of the red pigment, peridinin, in the guts of pink oysters.

Microscopic examinations of water samples during this period suggested that the color could be due to the abundant dinoflagellate *Prorocentrum minimum*. However, gut pigment analysis can be misleading because pigment degradation rates differ depending on the type of pigment and chemical conditions within the gut during digestion (McLeroy-Etheridge and McManus, 1999; Goericke et al., 2000; Bustillos-Guzman et al., 2002). Cryptophytes, which also have red accessory pigments (water-soluble phycoerythrin), were also present in high abundance during December 2000 and small amounts of alloxanthin (the carotenoid pigment indicative of cryptophytes) were detected in gut contents. Although the HPLC method used for these analyses cannot detect phycoerythrin, the presence of alloxanthin in the oyster guts does suggest that the oysters were grazing on the phycoerythrin-containing cryptophytes.

An examination of the water quality conditions and phytoplankton community composition from 1999 to 2001 offered insights into potential causal mechanisms for the occurrence and magnitude of pink oyster events (Figure 11.7 through Figure 11.9). Salinity was relatively high in GB during the fall and winter 1999. A tropical storm in May 2000, high rainfall, and subsequent freshwater inputs in late September 2000 resulted in lower salinities in the bay. Similarly, high rainfall in September–October 2001 lowered salinities. Riverine freshwater inputs resulted in elevated concentrations of dissolved inorganic nitrogen (DIN), in excess of 25 μM nitrogen and fostered phytoplankton blooms within the bay. The location of these blooms overlapped with the commercial oyster reefs.

Phytoplankton community composition determined with HPLC indicates that cryptophytes and peridinin-containing dinoflagellates were the most abundant phytoplankton groups present when pink oysters were harvested (Figure 11.8). A comparison of the spatiotemporal distributions of cryptophytes and dinoflagellates suggests that cryptophytes were the primary contributor to the pink coloration of oysters. The timing of cryptophyte blooms and the occurrence of pink oysters seem to be more closely linked than for dinoflagellates and pink oysters (Figure 11.9). The dinoflagellate blooms may be linked to the cryptophyte blooms because the cryptophytes provide an abundant food source for the mixotrophic

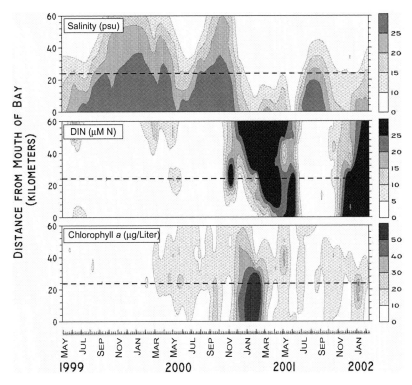

FIGURE 11.8 Spatiotemporal contour plots of salinity, total DIN, and total chl *a* (phytoplankton biomass) along the transect shown in Figure 11.7. The horizontal dashed line indicates the location of the reference point shown in Figure 11.7.

dinoflagellate *P. minimum* (the major dinoflagellate species in these blooms). These observations illustrate the applicability and underscore the importance of routine phytoplankton monitoring, supplemented by diagnostic tools, for understanding the linkages between system-level "driving" features (i.e., nutrient enrichment, phytoplankton blooms) and the "condition" of commercial oysters in GB and other U.S. estuaries.

Cyanobacterial Bloom Dynamics in the St. Johns River System, Florida

The St. Johns River system (SJRS), a 300-mile-long estuarine system located in northeastern Florida, has undergone eutrophication as a result of accelerating point and non-point nutrient loading. Although the upper and lower reaches of the system differ greatly with respect to salinity regimes, both are composed of a series of lakes, tributaries, riverine segments, and springsheds (Figure 11.10). Cyanobacterial blooms are indicative of eutrophication and often occur at freshwater and oligohaline sites during the "wet" summer months, a period typically characterized by increased storm-water inflows accompanied by nutrient enrichment from the watershed. These blooms have been associated with fish kills, loss of submerged vegetation (from reduced water clarity), wildlife mortalities, and human health issues.

Freshwater inflows throughout the SJRS affect phytoplankton dynamics by providing biologically available nutrients (Pigg et al., 2004) and may "fuel" and/or sustain blooms by infusion of transient phytoplankton throughout the SJRS. We have been developing diagnostic photopigment, molecular, and microbiological indicators of overall water quality and cyanobacterial expansion throughout the entire SJRS. Collective use of these techniques has allowed us to better understand phytoplankton, specifically, cyanobacterial, dynamics in the SJRS.

Total chl *a* concentrations throughout the lower, oligo-/meso-haline portions of the SJRS are variable throughout the year, with the greatest concentrations occurring in late spring and early summer. Phytoplankton composition is diverse and comprises diatoms, chlorophytes, cryptophytes, and cyanobacteria. Although diatoms typically dominate most assemblages, with relative abundances ranging from 25 to

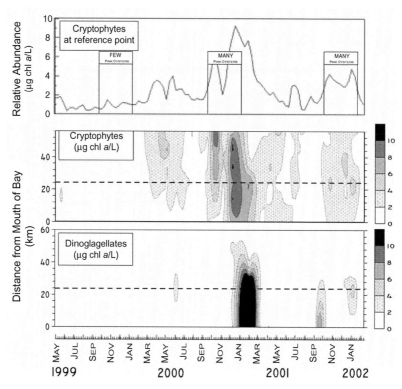

FIGURE 11.9 Spatiotemporal contour plots of the relative abundance of cryptophytes and dinoflagellates along the transect shown in Figure 11.7. The horizontal dashed line in the lower two panels indicates the location of the reference point shown in Figure 11.7. The graph in the top panel illustrates the relative abundance of cryptophytes at the reference point in Figure 11.7, the time period when pink oysters occur, and the prevalence of pink oysters during the 3 years.

75% of total chl *a*, cyanobacterial blooms are common during summer months within the "feeder" tributaries (e.g., Cedar/Ortega Rivers), connecting lakes (e.g., Doctor's Lake, Crescent Lake, Lake George), and lake-like portions of the river's main stem (e.g., Palatka to Jacksonville). (Figure 11.10). Cyanobacterial populations in the lower SJRS are dominated by *Microcystis* spp. and *Anabaena* spp., with relative abundances ranging up to about 65% of the total chl *a*.

Nitrogen fixation can be a significant source of nitrogen in aquatic systems (Horne, 1977), and many N_2-fixing cyanobacteria are toxic (Chorus and Bartram, 1999) and/or less palatable than other phytoplankton. *In situ* nutrient manipulation bioassays have been conducted in the upper SJRS over seasonal cycles to gain a better understanding of the interactions of the native phytoplankton community and changing nutrient regimes. In particular, the response of bloom-forming cyanobacteria (both N_2 fixing and non-N_2 fixing) to varying nutrient regimes has been examined. Bioassays were initiated by obtaining water from upstream and downstream sites in the SJRS, followed by incubation under ambient light and temperatures in 10-l polyethylene Cubitainers® that transmit 85% of photosynthetically active radiation (PAR). Nutrient treatments included dissolved inorganic nitrogen as nitrate, dissolved inorganic phosphorus as phosphate, the combination of nitrogen and phosphorus, and an unamended control. Following 4-day incubations, N_2 fixation (acetylene reduction assay), phytoplankton community structure (HPLC photopigment analysis), and microscopic counts of potential N_2 fixing cyanobacteria were assessed.

Data are presented from two occasions when N_2-fixing cyanobacteria were prevalent in Lake George in the upper, freshwater portion of the SJRS (Figure 11.11 and Figure 11.12). In July 2000, rates of N_2 fixation were stimulated by the addition of phosphorus and the addition of nitrogen and phosphorus together (Figure 11.11). The proportion of the phytoplankton community that was cyanobacterial increased in the phosphorus-addition treatment and decreased in the nitrogen-addition treatment (Figure 11.11). Microscopic examination of the potential N_2-fixing cyanobacteria revealed that specific N_2-fixing

FIGURE 11.10 (Color figure follows p. 266.) (Top left) Map of the St. Johns River. This northward flowing river is the largest river in Florida with the tidally driven outlet near Jacksonville. Upstream lakes often have high densities of cyanobacterial species. In particular, Lake George (L. George), a large feeder lake for this system, constitutes one of the largest innoculum of cyanobacteria to the system. (Courtesy of St. Johns Water Management District, Palatka, FL.) Other frames: Photographs of cyanobacterial (*Microcystis* sp., *Cylindrospermopsis*, *Anabaena* spp., and *Aphanizomenon flos aquae*) blooms along the St. Johns River system. (Lower lefthand photograph courtesy J. Burns.)

cyanobacteria also increased in abundance in some nutrient additions. However, phosphorus did not appear to be the major stimulant of N_2-fixing cyanobacteria. The filamentous, heterocystous (heterocysts are morphologically differentiated N_2-fixing cells) species *Cylindrospermopsis raciborskii* and the non-heterocystous *Planktolyngbya undulata* and *P. contorta* were equally stimulated by the addition of nitrogen alone and nitrogen and phosphorus in combination, but far less stimulated by phosphorus alone (Figure 11.11). The complex response of the phytoplankton community, particularly among the N_2-fixing cyanobacteria, required analysis at the species level in order to understand phytoplankton community responses to changing nutrient regimes.

In October 2000, cyanobacteria were again a large proportion of the Lake George phytoplankton community (Figure 11.12B). Once more, N_2 fixation was stimulated by the addition of phosphorus alone and nitrogen and phosphorus in combination, but was unchanged by the addition of nitrogen (Figure 11.12A). However, in this case some of the N_2-fixing cyanobacteria were stimulated by the addition of phosphorus. *Cylindrospermopsis raciborskii* was most abundant following phosphorus addition, unlike the July experiment in which phosphorus alone had no effect on its abundance. Other non-heterocystous cyanobacteria (*P. undulata* and *Pseudoanabaena* sp.) showed very different responses (Figure 11.12C). These data underscore the importance of sufficient resolution (e.g., species and strain level) to understand phytoplankton community responses to changing nutrient regimes.

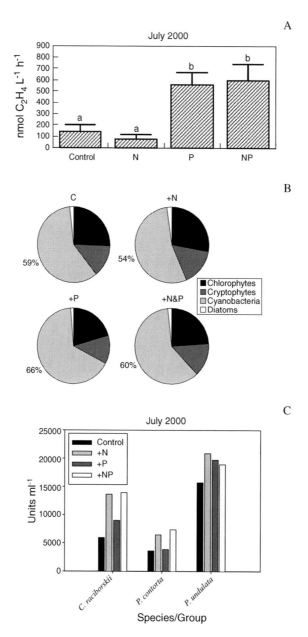

FIGURE 11.11 Data from a nutrient bioassay conducted in July 2000 in Lake George, FL. Nutrient additions were C (control, no addition), N (nitrate 20 μM), P (phosphate 5 μM), and NP (nitrate and phosphate). (A) Response of nitrogenase activity to nutrient additions, (B) changes in PTG group distribution, (C) changes in numbers of N_2-fixing cyanobacteria from microscopic counts.

Molecular detection, PCR amplification, and sequencing of the N_2-fixing gene *nifH* were used to identify *C. raciborskii* strains in the St. Johns River and in 16 lakes in central and northern Florida to determine the genetic similarity among these populations (Dyble et al., submitted). The high degree similarity between *C. raciborskii nifH* genes sequenced within lakes (97.7 to 100%) and between lakes (97.18 to 100%) suggests that this invasive cyanobacterium originated from a common source (Dyble et al., submitted).

In addition to differentiating populations based upon genetic similarity, the expression of functional genes can be used to measure the activity of a population in response to environmental changes (Pichard

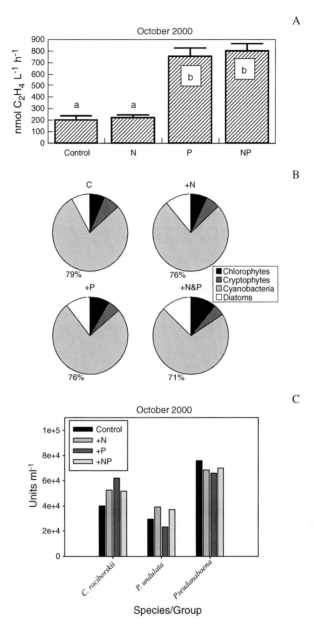

FIGURE 11.12 Data from a nutrient bioassay conducted in October 2000 in Lake George, FL. Nutrient additions were C (control, no addition), N (nitrate 20 μM), P (phosphate 5 μM), and NP (nitrate and phosphate). (A) Response of nitrogenase activity to nutrient additions, (B) changes in PTG distribution, (C) changes in numbers of N_2-fixing cyanobacteria from microscopic counts.

et al., 1997; Zehr et al., 2003). Analysis of the mRNA expression of the *nifH* gene was used to identify temporal patterns of N_2 fixation in *C. raciborskii* on scales ranging from time of day to time of year and spatial patterns throughout the water column using reverse transcription (RT) PCR (Dyble et al., submitted). In this method, mRNA is extracted from an environmental sample, reverse-transcribed into cDNA, and PCR-amplified using *C. raciborskii*-specific PCR primers. The assumption made is that if the *nifH* gene in *C. raciborskii* is "turned on" and making mRNA transcripts, then it is likely that this species is actively fixing N_2, thus allowing the identification of N_2 fixation activity of *C. raciborskii* in a mixed diazotrophic population. This method was applied in Lake George (in the headwaters of the SJR) to

A

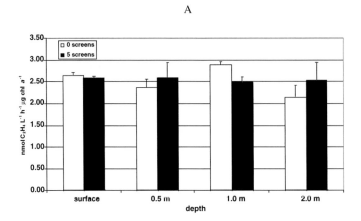

B

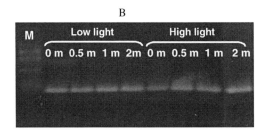

FIGURE 11.13 (A) Nitrogenase activity (nmol C_2H_4 l^{-1} h^{-1} µg chl a) at four depths (0, 0.5, 1, 2 m) in Lake George in June 2002. Samples from each of these depths were incubated under either 0 layers (unfilled bars) or 5 layers (hatched bars) of neutral density screening. Error bars are 1 standard deviation. (B) For each of the above samples, the expression of the *nifH* gene in *Cylindrospermopsis raciborskii* was identified using RT-PCR with primers specific to this species. Low light refers to the samples incubated under 5 screens (3% of surface irradiance), high light refers to those incubated under 0 screens (surface irradiance), and "M" is a X174/*Hae*III molecular weight marker. The PCR product is 225 base pairs in length.

identify the depth at which *C. raciborskii* fixes N_2. Blooms of *C. raciborskii* are often observed to be concentrated at depth below the surface, perhaps due to an ability of this species to utilize low light levels, and thus N_2 fixation rates would be expected to also occur at depth (Dokulil and Mayer, 1996; Fabbro and Duivenvoorden, 1996; Padisak, 1997). Nitrogenase activity (NA) rate measurements for the entire diazotrophic phytoplankton population showed a lack of significant differences in NA in phytoplankton communities originating from different depths (0, 0.5, 1, 2 m) in Lake George or incubated under different light levels (0 to 94% light attenuation) (Figure 11.13). RT-PCR data demonstrated that *C. raciborskii* in particular was exhibiting *nifH* expression at all four depths and under both high and low light levels and thus was likely responsible for at least a portion of the nitrogen fixed throughout the water column (Figure 11.13). Despite the early stage of development and application, PCR approaches for mRNA studies should gain wide use since they generally require small sample size and are amenable to routine field monitoring.

Indicator Deployment and Data Acquisition

A rapidly expanding array of new techniques and approaches is now available to detect and characterize phytoplankton responses to environmental stressors and perturbations ranging from cellular to ecosystem and regional scales. Taxa-specific identification and quantification techniques such as those described above complement production, growth, and nutrient cycling rate measurements, biomass determinations, and estimates of microbially mediated material flux in estuarine ecosystems. This will facilitate quantifying the causes and effects of biogeochemical and trophic changes in these ecosystems. These detection methodologies, used in combination and integrated into a multiplatform sampling network, may allow for continuous assessment of phytoplankton assemblages and relevant environmental forcing features throughout estuarine systems. Such data acquired from both invasive sampling and autonomous instrument platforms can be used to support modeling/forecasting efforts for determining synoptic- and meso-scale environmental perturbations and phytoplankton blooms.

"Real-time" acquisition of physical, chemical, and biological data by automated sensors can be transmitted to local data portals, fed into current, regional-scale modeling/forecasting efforts, and, if appropriate, advisories for potential (public) health risks can be generated. An array of monitoring programs and platforms are available to collect space- and time-intensive data needed to assess ecological conditions over relevant scales and to ground-truth and calibrate remote sensing efforts aimed at scaling up. These include boat-based surveys, instrumenting channel markers, bridges, piers/docks, and buoys with unattended monitoring equipment, deploying moorings, and outfitting ferries and other vessels using regular routes as "ships of opportunity."

An example of a portable instrument platform for autonomous, *in situ* acquisition of near-real-time water quality data is MARVIN (Merhab Autonomous Research Vessel for IN-situ sampling). This platform was deployed and successfully tested within the Trout River tributary of the St. Johns River. It is based on a pontoon boat deck, permitting safe and convenient maintenance and portability among monitoring sites. The platform houses an extensive sensor array recording various biological, physical, chemical, and meteorological data (Table 11.2) and is programmed to record (hourly) measurements from both surface and near-bottom waters. Water samples can be remotely acquired via an in-line water sampler using a pre-programmed "trigger" and/or a timed collection scenario.

Such high-resolution sampling provides opportunities to detect conditions not resolved within a typical invasive monitoring program over extremely short timescales (minutes to hours). For example, a bloom of the nuisance cyanobacteria *Microcystis* spp. developed following a rain in July 2002, and was immediately detected by MARVIN via *in situ* fluorescence of chlorophyll *a*. Short-term, tidal, and diel influences on algal biomass and phosphate (with greater concentrations during low tides, especially low tides during high PAR values in the afternoon) were subsequently identified. Nitrate concentrations varied approximately sevenfold, with the lowest concentrations occurring during the night and at high

TABLE 11.2

Specifications of Instruments Deployed on the Autonomous Sampling Platform, MARVIN

Instrument	Parameter	Resolution	Accuracy
YSI 6600	DO% saturation	0.1%	±2%
	DO mg/L	0.01 mg/L	±0.2 mg/L
	Conductivity	0.001–0.1 mS/cm	±0.5% of reading
	Salinity	0.01 ppt	±1% or 0.1 ppt
	Temperature	0.01°C	±0.15°C
	pH	0.01 unit	±0.2 unit
	Turbidity	0.1 NTU	±5% of reading (2 NTU)
	Chlorophyll/relative fluorescence	0.1 µg/L	—
MET one windset	Wind speed	0.9 mph	±0.25 mph at <22.7 mph
	Wind direction	0.5°	±4°
CS barometric pressure sensor	Barometric pressure		±2 mb @ 0–40°C
RM Young rain gauge	Rain fall	±0.05 in.	±0.1 in.
HMP45C	Air temperature		~0.2°C
	Relative humidity (RH)		±2% RH @ 20°C (0–90%)
Licor quantum sensor	PAR (in air and underwater)	5 µA/1000 µmol/s/m^2	—
Sontek Argonaut-SL	Water velocity	0.1 cm/s	±0.5 cm/s
	Water temperature	0.01°C	±0.1°C
	Water level (pressure)	—	0.1% of full scale
EnviroTech ANA	Nitrate + nitrite	0.05 µM	2%
	Phosphate	0.06 µM	3%
WETstar Fluorometer	Relative fluorescence/chlorophyll	0.03 µg/L	—

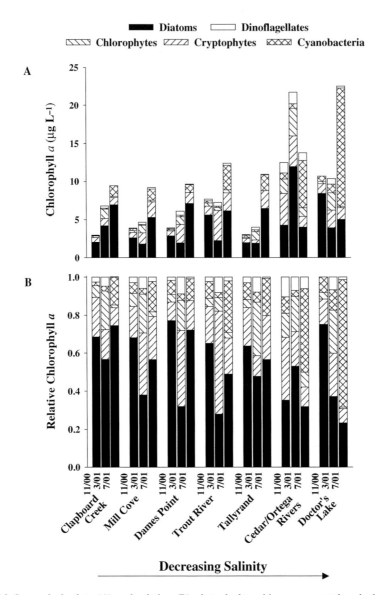

FIGURE 11.14 Seasonal absolute (A) and relative (B) phytoplankton biomass, as total and phylogenetic-group chlorophyll *a* (chl *a*), within the lower St. Johns River system for November 2000 through July 2001; data are means, *n* = 3 to 6. Phylogenetic-group chlorophyll concentrations were determined using ChemTax matrix factorization incorporating suites of photopigments derived using HPLC (see Table 11.2). Note that total chl *a* concentrations were variable among sampling dates, but typically increased with decreasing salinity. Phytoplankton assemblages mostly comprised diatoms, cryptophytes, and cyanobacteria.

tides and increasing concentrations increasing during ebb and flood tides. In addition, acquisition of water-column dissolved oxygen concentrations for 24 h during the aforementioned period provided for high-resolution characterization of system-level gross/net production and respiration (Figure 11.14). Within this tributary, respiration typically exceeded gross production. Determination of net ecosystem metabolism values, a proxy for the system's trophic condition (Caffrey, 2003), subsequently indicated that the production dynamics of this segment of the SJRS are highly variable and that both autochthonous and allochthonous sources of organic matter can dominate over short timescales (Figure 11.15).

Autonomous sampling from ferries can collect near-real-time physical-chemical-biological data (including chl *a* and diagnostic photopigments) for assessing phytoplankton community structural and

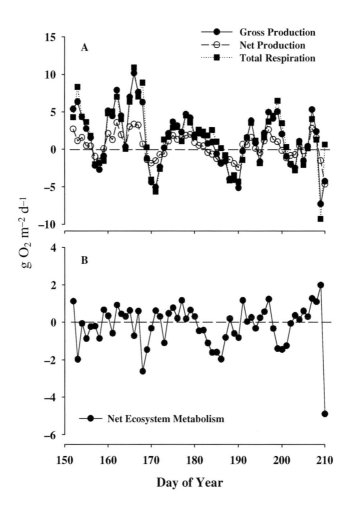

FIGURE 11.15 Production, respiration, and net ecosystem metabolism (NEM) for the Trout River tributary within the lower St. Johns River system during June and July 2001 (after Caffrey, 2003). Dashed lines within each panel indicate the positive/negative threshold. (A) Gross/net production and total respiration were derived from oxygen flux calculations incorporating diel dissolved oxygen concentrations, air–water oxygen exchange (diffusion), and water depth. (B) NEM was derived from estimates of gross production and total respiration. If NEM values are positive, autochthonous sources of organic matter likely dominate the system (autotrophism), whereas if values are negative, allochthonous sources of organic matter likely dominate (heterotrophism). Note that respiration and production processes are highly dynamic over short-time intervals (hours to days) and the system is heterotrophic over daily intervals.

functional responses to environmental change in large estuarine and coastal ecosystems not amenable to routine monitoring. One example is the use of the North Carolina Department of Transportation (DOT) ferries for water quality monitoring. This program, "FerryMon" (www.ferrymon.org), has been designed to (1) assess and predict the relationships between human nutrient and other pollutant inputs, algal blooms, and associated water quality changes, and ecosystem response; (2) provide critical information to long-term water quality and fishery management; and (3) develop FerryMon as a national model for real-time assessment of coastal water quality.

Three ferries crossing the Pamlico Sound and the Neuse River Estuary have been equipped with automated water quality monitoring systems (Figure 11.16). A continuous-flow, automated system monitors surface waters along the ferry route for temperature, salinity, pH, dissolved oxygen, turbidity,

FerryMon Routes

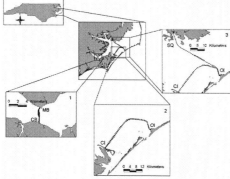

FerryMon Water Quality Monitoring Parameters

- **Temperature**   - **Turbidity**

- **Salinity**   - **Chlorophyll**

- **Dissolved oxygen**   - **Diagnostic algal**
  pigments

- **Nutrients (inorganic N and P, Si, organic N, DOC)**

- **Microbiological/molecular analyses**

FIGURE 11.16 Routes traveled and water quality parameters collected by the North Carolina ferry-based water quality monitoring program, FerryMon (www.ferrymon.org). Shown is one of 3 North Carolina Department of Transportation ferries equipped to collect water quality data. The ferry crossings include (1) the Neuse River, between Cherry Branch and Minnesot Beach (operates from 5 A.M. until midnight daily), (2) the Cedar Island to Ocracoke crossing (transecting Southwestern Pamlico Sound) (6 A.M. until midnight), and (3) the Swan Quarter to Ocracoke crossing (transecting western-central Pamlico Sound) (6 A.M. until midnight). The following water quality parameters are measured in real time: pH, dissolved oxygen, temperature, turbidity, salinity, and chl *a* by fluorescence. Water samples for nutrients, diagnostic photopigments, and other microbial analyses are collected by an in-line refrigerated sampler at prescribed intervals. These samples are brought to the laboratory for analysis.

chlorophyll biomass, and is accompanied by geographic position (GPS) referencing of the data (Buzzelli et al., 2003). An automated and refrigerated discrete sampler collects samples for measurement of nutrients and diagnostic algal pigments, colored dissolved organic matter (CDOM), and total suspended solids (TSS) (Figure 11.16). High-frequency data collection (minutes to hours) ensures that all-important spatial and temporal scales are represented (Figure 11.17). The data are archived in digital form and made available for scientific analysis, modeling efforts, and management needs.

FerryMon also provides water samples for nucleic acid (16S rRNA) analysis of microbial community composition and function.

Concluding Remarks

As remote sensing, modeling, and statistical (i.e., neural networks) approaches for "scaling up" microalgal-based estimates of activity, biomass, and composition improve in resolution, sensitivity, and specificity, we will be able to more easily transcend the range of scales relevant for assessing ecological change, and establishing nutrient, other pollutant, and stressor (i.e., turbidity, hydrology) thresholds needed to quantify ecosystem tolerance, resilience, and recovery in response to such stressors.

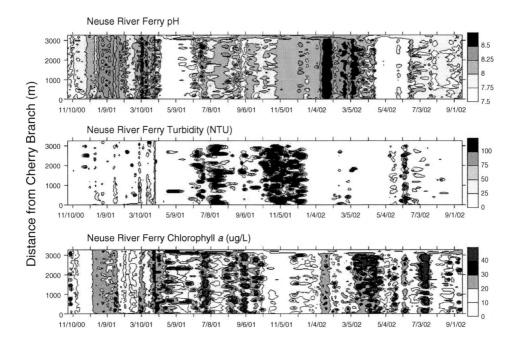

FIGURE 11.17 Example of spatially and temporally intensive data collected by FerryMon on its Neuse River Estuary crossings. Shown are pH, turbidity, and chl *a* over a period from November 2000 to January 2002. Note the fine-scale detail of physical-chemical changes and phytoplankton bloom events (chl *a* levels in excess of 30 μg chl *a* l⁻¹). Some of these short-lived but significant bloom events are not captured in routine biweekly water quality monitoring conducted by state and university researchers.

Interdisciplinary, cross-media efforts will be highly effective and in many cases essential for identifying and formulating the causes, consequences, and controls of microalgal-mediated ecological change. We have illustrated this by combining ecophysiological, chemical, and molecular approaches with statistical, process-based modeling, spatially and temporally intensive unattended and real-time monitoring and remote sensing to enable assessments of ecological change across habitat, ecosystem, and regional boundaries that are often crossed by physical processes (e.g., atmospheric transport and deposition, advection, and estuarine-coastal circulation). Last, it is important to employ sensing and characterization techniques on seasonal and longer terms (interannual, decadal, and longer) to establish baseline and reference conditions against which to gauge ecological change.

Acknowledgments

We appreciate the technical assistance and input of C. Buzzelli, J. Fleming, N. Hall, M. Harrington, A. Joyner, K. McFarlin, B. Peierls, J. Ramus, V. Winkelmann, A. Chapman, and P. Wyrick. This work was supported by the National Science Foundation (DEB 9815495 and OCE 9905723, Graduate Fellowship to J. Dyble), U.S. Department of Agriculture NRI Project 00-35101-9981, U.S. EPA STAR Projects R82-5243-010 and R82867701, NOAA/North Carolina Sea Grant Program R/MER-43, and the North Carolina Department of Natural Resources and Community Development/University of North Carolina Water Resources Research Institute (Neuse River Monitoring and Modeling Project, ModMon).

References

Barciela, R. M., E. Garcia, and E. Fernandez. 1999. Modelling primary production in a coastal embayment affected by upwelling using ecosystem models and artificial neural networks. *Ecological Modelling* 120: 199–211.

Ben-Porath, J., E. J. Carpenter, and J. P. Zehr. 1993. Genotypic relationships in *Trichodesmium* (Cyanophyceae) based on *nifH* sequence comparisons. *Journal of Phycology* 29:806–810.

Boesch, D. F., E. Burreson, W. Dennison, E. Houde, M. Kemp, V. Kennedy, R. Newell, K. Paynter, R. Orth, and W. Ulanowicz. 2001. Factors in the decline of coastal ecosystems. *Science* 293:629–638.

Bricker, S. B., C. G. Clement, D. E. Pirhalla, S. P. Orlando, and D. R. G. Farrow. 1999. National Estuarine Eutrophication Assessment: Effects of Nutrient Enrichment in the Nation's Estuaries. NOAA, National Ocean Service, Special Projects Office, and the National Centers for Coastal Ocean Science, Silver Spring, MD.

Buffan-Dubau, E., R. de Wit, and J. Castel. 1996. Feeding selectivity of the harpacticoid copepod *Canuella perplexa* in benthic muddy environments demonstrated by HPLC analyses of chlorin and carotenoid pigments. *Marine Ecology Progress Series* 137:71–82.

Burkill, P., R. Mantoura, C. Llewellyn, and N. Owens. 1987. Microzooplankton grazing and selectivity of phytoplankton in coastal waters. *Marine Biology* 93:581–590.

Bustillos-Guzman, J., D. Lopez-Cortes, M. E. Mathus, and E. Harnandez. 2002. Dynamics of pigment degradation by the copepodite stage of *Pseudodiaptomus euryhalinus* feeding on *Tetraselmis suecica*. *Marine Biology* 140:143–149.

Buzzelli, C. P., J. R Ramus, and H. W. Paerl. 2003. Ferry-based monitoring of surface water quality in North Carolina estuaries. *Estuaries* 26:975–984.

Caffrey, J. M. 2003. Production, respiration, and net ecosystem metabolism in U. S. estuaries. *Environmental Monitoring and Assessment* 81:207–219.

Committee on Environment and Natural Resources (CENR). 1997. Integrating the Nation's Environmental Monitoring and Research Networks and Programs: A Proposed Framework. Office of Science and Technology Policy, Washington, D.C.

Copeland, B. J. and J. Gray. 1991. Status and Trends Report for the Albemarle-Pamlico Estuary. In Albemarle-Pamlico Estuary Study Report 90-01, J. Steel, (ed.), North Carolina Department of Environment, Health and Natural Resources, Raleigh, NC.

Costanza, R., R. D. Arge, R. De Groot, S. Farber, M. Grasso, B. Hannon, K. Limburg, S. Naeem, R. V. O'Neill, J. Pareuello, R. G. Raskin, P. Sutton, and M. van den Belt. 1997. The value of the world's ecosystems services and natural capital. *Nature* 387:253–260.

Cottingham, K. L. and S. R. Carpenter. 1998. Population, community, and ecosystem variates as ecological indicators: phytoplankton responses to whole-lake enrichment. *Ecological Applications* 8:508–530.

Culliton, T. J. 1998. Population: distribution, density and growth. NOAA State of the Coast Report. Silver Spring, MD. Available: http//state-of-coast.noaa.gov/topics/html/pressure.html.

Descy, J.-P. and A. Metens. 1996. Biomass-pigment relationships in potamoplankton. *Journal of Plankton Research* 18:1557–1566.

Dokulil, M. T. and J. Mayer. 1996. Population dynamics and photosynthetic rates of a *Cylindrospermopsis–Limnothrix* association in a highly eutrophic urban lake, Alte Donau, Vienna, Austria. *Algological Studies* 83:179–195.

Dustan, P. and J. L. Pinckney. 1989. Tidally induced phytoplankton patchiness. *Limnology and Oceanography* 34:408–417.

Dyble, J., H. W. Paerl, and B. A. Neilan. 2002. Genetic characterization of *Cylindrospermopsis raciborskii* (cyanobacteria) isolates from diverse geographic origins based on *nifH* and *cpcBA*-IGS nucleotide sequence analysis. *Applied and Environmental Microbiology* 68:2567–2571.

Dyble, J., K. Havens, P. H. Moisander, T. F. Steppe, A. Chapman, and H. W. Paerl. Submitted. Genetic diversity among *Cylindrospermopsis raciborskii* populations in Florida lakes based on *nifH* sequence analysis. *Journal of Phycology.*

Fabbro, L. D. and L. J. Duivenvoorden. 1996. Profile of a bloom of the cyanobacterium *Cylindrospermopsis raciborskii* (Woloszynska) Seenaya and Subba Raju in the Fitzroy River in tropical central Queensland. *Marine and Freshwater Research* 47:685–694.

Fisher, T. R., L. W. Harding, Jr., D. W. Stanley, and L. G. Ward. 1988. Phytoplankton, nutrients, and turbidity in the Chesapeake, Delaware, and Hudson estuaries. *Estuarine and Coastal Shelf Science* 27:61–93.

Foody, G. M. and M. E. J. Cutler. 2003. Tree biodiversity in protected and logged Bornean tropical rain forests and its measurement by satellite remote sensing. *Journal of Biogeography* 30:1053–1066.

Gieskes, W. and G. Kraay. 1986. Floristic and physiological differences between the shallow and deep nanophytoplankton community in the eutrophic zone of the open tropical Atlantic revealed by HPLC analysis of pigments. *Marine Biology* 91:567–576.

Giovannoni, S. J., T. B. Britschgi, C. L Moyer, and K. G. Field. 1990. Genetic diversity in Sargasso Sea bacterioplankton. *Nature* (London) 345:148–149.

Goericke. R., S. L. Strom, and M. A. Bell. 2000. Distribution and sources of cyclic pheophorbides in the marine environment. *Limnology and Oceanography* 45:200–211.

Goldenberg, S. B., C. W. Landsea, A. M. Mestas-Nuzes, and W. M. Gray. 2001. The recent increase in Atlantic Hurricane Activity: causes and implications. *Science* 293:474–479.

Gordon, D. A., J. Priscu, and S. Giovannoni. 2000. Origin and phylogeny of microbes living in permanent Antarctic lake ice. *Microbial Ecology* 39:197–202.

Guisande, C., I. Maneiro, I. Riveiro, A. Barreiro, and Y. Pazos Y. 2002. Estimation of copepod trophic niche in the field using amino acids and marker pigments. *Marine Ecology Progress Series* 239:147–156.

Harding, L. W., D. Degobbis, and R. Precali. 1999. Production and fate of phytoplankton: annual cycles and interannual variability. In *Ecosystems at the Land-Sea Margin: Drainage Basin to Coastal Sea,* T. C. Malone et al. (eds.). American Geophysical Union, Coastal and Estuarine Studies, Vol. 55. Washington, D.C., pp. 131–172.

Head, E. and L. Harris. 1994. Feeding selectivity by copepods grazing on natural mixtures of phytoplankton determined by HPLC analysis of pigments. *Marine Ecology Progress Series* 110:75–83.

Irigoien, X., R. N. Head, R. P. Harris, D. Cummings, and D. Harbour. 2000. Feeding selectivity and egg production of *Calanus helgolandicus* in the English Channel. *Limnology and Oceanography* 45:44–54.

Jeffrey, S., R. Mantoura, and S. Wright (eds.). 1997. *Phytoplankton Pigments in Oceanography: Guidelines to Modern Methods.* UNESCO, Paris.

Jeffrey, S. W., S. W. Wright, and M. Zapata. 1999. Recent advances in HPLC pigment analysis of phytoplankton. *Marine and Freshwater Research* 50:879–896.

Karul, C., S. Soyupak, A. Cilesiz, N. Akbay, and E. Germen. 2000. Case studies on the use of neural networks in eutrophication modeling. *Ecological Modelling* 134:145–152.

Kleppel, G., D. Frazel, R. Pieper, and D. Holliday. 1988. Natural diets of zooplankton off southern California. *Marine Ecology Progress Series* 49:231–241.

Lee, J. H. W., Y. Huang, M. Dickman, and A. W. Jayawardena. 2003. Neural network modeling of coastal algal blooms. *Ecological Modelling* 159:179–201.

Lek, S. and J.F. Guegan. 1999. Artificial neural networks as a tool in ecological modeling, an introduction. *Ecological Modelling* 120:65–73.

Luettich, R. A., Jr., J. E. McNinch, H. W. Paerl, C. H. Peterson, J. T. Wells, M. Alperin, C. S. Martens, and J. L. Pinckney. 2000. Neuse River Estuary modeling and monitoring project stage. 1: Hydrography and circulation, water column nutrients and productivity, sedimentary processes and benthic-pelagic coupling. Report UNC-WRRI-2000-325B, Water Resources Research Institute of the University of North Carolina, Raleigh, 172 pp.

Mackey, D. J., H. W. Higgins, M. D. Mackey, and D. Holdsworth. 1998. Algal class abundances in the western equatorial Pacific: estimation from HPLC measurements of chloroplast pigments using CHEMTAX. *Deep-Sea Research* I 45:1441–1468.

Mackey, M., D. Mackey, H. Higgins, and S. Wright. 1996. CHEMTAX — a program for estimating class abundances from chemical markers: application to HPLC measurements of phytoplankton. *Marine Ecology Progress Series* 144:265–283.

Maier, H. R., G. C. Dandy, and M. D. Burch. 1998. Use of artificial neural networks for modeling cyanobacteria *Anabaena* spp. in the River Murray, South Australia. *Ecological Modelling* 105:257–272.

McLeroy-Etheridge, S. L. and G. B. McManus. 1999. Food type and concentration affect chlorophyll and carotenoid destruction during copepod feeding. *Limnology and Oceanography* 44:2005–2011.

Meyer-Harms, B. and B. von Bodungen. 1997. Taxon-specific ingestion rates of natural phytoplankton by calanoid copepods in an estuarine environment (Pomeranian Bight, Baltic Sea) determined by cell counts and HPLC analysis of marker pigments. *Marine Ecology Progress Series* 153:181–190.

Millie, D. F., M. C. Baker, C. S. Tucker, B. T. Vinyard, and C. P. Dionigi. 1992. High-resolution airborne remote sensing of bloom-forming phytoplankton. *Journal of Phycology* 28: 281–290.

Millie, D. F., H. W. Paerl, and J. Hurley. 1993. Microalgal pigment assessments using high performance liquid chromatography: a synopsis of organismal and ecological applications. *Canadian Journal of Fisheries and Aquatic Science* 50:2513–2527.

Millie, D. F., G. J. Kirkpatrick, and B. T. Vinyard. 1995a. Relating photosynthetic pigments and *in vivo* optical density spectra to irradiance for the Florida red-tide dinoflagellate *Gymnodinium breve*. *Marine Ecology Progress Series*, 120:65–75.

Millie, D. F., O. M. Schofield, C. P. Dionigi, and P. B. Johnsen. 1995b. Assessing noxious phytoplankton in aquaculture systems using bio-optical methodologies: a review. *Journal of the World Aquaculture Society* 12:329–345.

Millie, D. F., B. T. Vinyard, M. C. Baker, and C. S. Tucker. 1995c. Testing the temporal and spatial validity of site-specific models derived from airborne remote sensing of phytoplankton. *Canadian Journal of Fisheries and Aquatic Science* 52:1094–1107.

Millie, D. F., O. M. Schofield, G. B. Kirkpatrick, G. Johnsen, P. A. Tester, and B. T. Vinyard. 1997. Detection of harmful algal blooms using photopigment and absorption signatures: a case study of the Florida red-tide dinoflagellate, *Gymnodinium breve*. *Limnology and Oceanography* 42:1240–1251.

Millie, D. F., G. L. Fahnenstiel, H. J. Carrick, S. E. Lohrenz, and O. Schofield. 2002. Phytoplankton pigments in coastal Lake Michigan: distributions during the spring isothermal period and relation with episodic sediment resuspension. *Journal of Phycology* 38:639–648.

National Research Council (NRC). 1994. Review of EPA's Environmental Monitoring and Assessment Program: Forest and Estuaries Components. National Academy Press, Washington, D.C.

National Research Council (NRC). 2000. Clean Coastal Waters: Understanding and Reducing the Effects of Nutrient Pollution. National Academy Press, Washington, D.C.

National Research Council (NRC). 2001. Assessing the TMDL Approach to Water Quality Management. National Academy Press, Washington, D.C.

Nixon, S. W. 1995. Coastal marine eutrophication: a definition, social causes, and future concerns. *Ophelia* 41:199–219.

Padisak, J. 1997. *Cylindrospermopsis raciborskii* (Woloszynska) Seenayya et Subba Raju, an expanding, highly adaptive cyanobacterium: worldwide distribution and review of its ecology. *Archiv fur Hydrobiologie* Supplement 107:563–593.

Paerl, H. W. 1997. Coastal eutrophication and harmful algal blooms: Importance of atmospheric deposition and groundwater as "new" nitrogen and other nutrient sources. *Limnology and Oceanography* 42:1154–1165.

Paerl, H. W., M. A. Mallin, C. A. Donahue, M. Go, and B. L. Peierls. 1995. Nitrogen loading sources and eutrophication of the Neuse River Estuary, NC: direct and indirect roles of atmospheric deposition. Report 291, UNC Water Resources Research Institute, Raleigh, NC, 119 pp.

Paerl, H. W., J. L. Pinckney, J. M. Fear, and B. L. Peierls. 1998. Ecosystem responses to internal and watershed organic matter loading: consequences for hypoxia in the eutrophying Neuse River Estuary, North Carolina, USA. *Marine Ecology Progress Series* 166:17–25.

Paerl, H. W., J. D. Bales, L. W. Ausley, C. P. Buzzelli, L. B. Crowder, L. A. Eby, J. M. Fear, M. Go, B. L. Peierls, T. L. Richardson, and J. S. Ramus. 2001. Ecosystem impacts of 3 sequential hurricanes (Dennis, Floyd and Irene) on the US's largest lagoonal estuary, Pamlico Sound, NC. *Proceedings of the National Academy of Sciences of the United States of America* 98:5655–5660.

Paerl, H. W., J. Dyble, P. H. Moisander, R. T. Noble, M. F. Piehler, J. L. Pinckney, L. Twomey, and L. M. Valdes. 2003. Microbial indicators of aquatic ecosystem change: current applications to eutrophication studies. *FEMS Microbial Ecology* 46(3): 233–246.

Peierls, B. L., N. F. Caraco, M. L. Pace, and J. J. Cole. 1991. Human influence on river nitrogen. *Nature* 350:386–387.

Pigg, R. J., D. F. Millie, B. J. Bendis, and K. A. Steidinger. 2004. Relating cyanobacterial abundance to environmental parameters in the lower St. Johns River Estuary. In *Proceedings of the 10th International Conference on Harmful Algae*, Steidinger, K. A., Lansberg, J. H., Tomas, C. R., and Vargo, G. A. (eds.), Harmful Algae 2002. Florida Fish and Wildlife Conservation Commission and Intergovernmental Oceanographic Commission of UNESCO. Precision Litho Service, St. Petersburg, Fl.

Pinckney, J. L., D. F. Millie, K. E. Howe, H. W. Paerl, and J. Hurley. 1996. Flow scintillation counting of 14C-labeled microalgal photosynthetic pigments. *Journal of Plankton Research* 18:1867–1880.

Pinckney, J. L., D. F. Millie, B. Vinyard, and H. W. Paerl. 1997. Environmental controls of phytoplankton bloom dynamics in the Neuse River Estuary (North Carolina, USA). *Canadian Journal of Fisheries and Aquatic Science* 54:2491–2501.

Pinckney, J. L., H. W. Paerl, and M. B. Harrington. 1999. Responses of the phytoplankton community growth rate to nutrient pulses in variable estuarine environments. *Journal of Phycology* 35:1455–1463.

Pinckney, L. J., T. L. Richardson, D. F. Millie, and H. W. Paerl. 2001. Application of photopigment biomarkers for quantifying microalgal community composition and in situ growth rates. *Organic Geochemistry* 32:585–595.

Rappe, M. S., P. F. Kemp, and S. J. Giovannoni. 1997. Phylogenetic diversity of marine coastal picoplankton 16S rRNA genes cloned from the continental shelf off Cape Hatteras, North Carolina. *Limnology and Oceanography* 42:811–826.

Recknagel, F., M. French, P. Harkonen, and K.-I. Yabunaka. 1997. Artificial neural network approach for modeling and prediction of algal blooms. *Ecological Modelling* 96:11–28.

Redalje, D. 1993. The labeled chlorophyll *a* technique for determining photoautotrophic carbon specific growth rates and carbon biomass. In *Handbook of Methods in Aquatic Microbial Ecology,* P. Kemp, B. Sherr, E. Sherr, and J. Cole (eds.). Lewis Publishers, Boca Raton, FL, pp. 563–572.

Reysenbach, A., G. S. Wickham, and N. R. Pace. 1994. Phylogenetic analysis of the hyperthermophilic pink filament community in Octopus Spring, Yellowstone National Park. *Applied and Environmental Microbiology* 60:2113–2119.

Richardson, A. J., M. C. Pfaff, J. G. Field, N. F. Silulwane, and F. A. Shillington. 2002. Identifying characteristic chlorophyll *a* profiles in the coastal domain using an artificial neural network. *Journal of Plankton Research* 24:1289–1303.

Roy, S., J. P. Chanut, M. Gosselin, and T. Sime-Ngando. 1996. Characterization of phytoplankton communities in the lower St. Lawrence Estuary using HPLC-detected pigments and cell microscopy. *Marine Ecology Progress Series* 142:55–73.

Schluter, L., F. Mohlenberg, H. Havskum, and S. Larsen. 2000. The use of phytoplankton pigments for identifying and quantifying phytoplankton groups in coastal areas: testing the influence of light and nutrients on pigment/chlorophyll *a* ratios. *Marine Ecology Progress Series* 192:49–63.

Smetacek, V., U. Bathman, E. M. Nothig, and R. Scharek. 1991. Coastal eutrophication: causes and consequences. In *Ocean Margin Processes in Global Change*, R. C. F. Mantoura, J. M. Martine, and R. Wollast (eds.). John Wiley & Sons, New York, pp. 251–279.

Steppe, T. F., H. W. Paerl, and R. P. Reid. 2001. Diazotrophy in modern Bahamian stromatolites. *Microbial Ecology* 41:36–44.

Strom, S. and N. Welschmeyer. 1991. Pigment-specific rates of phytoplankton growth and microzooplankton grazing in the open subarctic Pacific Ocean. *Limnology and Oceanography* 36:50–63.

Tapiador, F. J. and J. L. Casanova. 2003. Land use mapping methodology using remote sensing for the regional planning directives in Segovia, Spain. *Landscape and Urban Planning* 62: 103–115.

Tester, P. A., M. R. Geesey, C. Guo, H. W. Paerl, and D. F. Millie. 1995. Evaluating phytoplankton dynamics in the Newport River Estuary (North Carolina, USA.) by HPLC-derived pigment profiles. *Marine Ecology Progress Series* 124:237–245.

U.S. EPA. 1998a. National Strategy for the Development of Regional Criteria. Office of Water, Washington, D.C. (On line). Available at http://www.epa.gov/OST/standards/nutrient.htmal (1999, June 30).

U.S. EPA. 1998b. Condition of the Mid-Atlantic Estuaries. Office of Research and Development, Washington, D.C.

U.S. EPA. 1999. Total Maximum Daily Load (TMDL) Program. Office of Water, Washington, D.C. (On line). Available at http://www.epa.gov/OWOW/tmdl (1999, December 20).

U.S. EPA. 2001. Clean Water Action Plan. National Coastal Condition Report. EPA-620-R-00-004. Office of Research and Development. Available at www.cleanwater.gov.

Vitousek, P. M., H. A. Mooney, J. Lubchenko, and J. M. Mellilo. 1997. Human domination of Earth's ecosystem. *Science* 277:494–499.

Wear, D. J., M. J. Sullivan, A. D. Moore, and D. F. Millie. 1999. Phytoplankton primary production: a neural network case study. *Ecological Modelling* 120:213–223.

Widmer, F., B. Shaffer, L. A. Porteous, and R. J. Seidler. 1999. Analysis of *nifH* gene pool complexity in soil and litter at a Douglas fir forest site in the Oregon Cascade range. *Applied and Environmental Microbiology* 65:374–380.

Woitke, P., P. Schiwietz, K. Teubner, and J.-G. Kohl.. 1996. Annual profiles of photosynthetic lipophilic pigments in four freshwater lakes in relation to phytoplankton cultures as well as nutrient data. *Archiv für Hydrobiologie* 137:363–384.

Wright, S., S. Jeffrey, R. Mantoura, C. Llewellyn, T. Bjørnland, D. Repeta, and N. Welschmeyer. 1991. An improved HPLC method for the analysis of chlorophylls and carotenoids from marine phytoplankton. *Marine Ecology Progress Series* 77:183–196.

Wright, S., D. Thomas, H. Marchant, H. Higgins, M. Mackey, and D. Mackey. 1996. Analysis of phytoplankton of the Australian sector of the Southern Ocean: comparisons of microscopy and size frequency data with interpretations of pigment HPLC data using the "Chemtax" matrix factorization. *Marine Ecology Progress Series* 144:285–298.

Wright, S. W. and R. L. van den Enden. 2000. Phytoplankton community structure and stocks in the East Antarctic marginal ice zone (BROKE survey, January–March 1996) determined by CHEMTAX analysis of HPLC pigment signatures. *Deep-Sea Research II* 47:2363–2400.

Zehr, J. P., M. Mellon, S. Braun, W. Litaker, T. Steppe, and H. W. Paerl. 1995. Diversity of heterotrophic nitrogen fixation genes in a marine cyanobacterial mat. *Applied and Environmental Microbiology* 61:2527–2532.

Zehr, J. P., M. T. Mellon, and W. D. Hiorns. 1997. Phylogeny of cyanobacterial *nifH* genes: evolutionary implications and potential applications to natural assemblages. *Microbiology* 143:1443–1450.

Zehr, J. P., M. T. Mellon, and S. Zani. 1998. New nitrogen-fixing microorganisms detected in oligotrophic oceans by amplification of nitrogenase (*nifH*) genes. *Applied and Environmental Microbiology* 64: 3444–3450.

Further Reading

Affourtit, J., J. P. Zehr, and H. W. Paerl. 2001. Distribution of nitrogen-fixing microorganisms along the Neuse River Estuary, North Carolina. *Microbial Ecology* 41:114–123.

Buzzelli, C. P., J. R Ramus, and H. W. Paerl. 2002. Ferry-based monitoring of surface water quality in North Carolina estuaries. *Estuaries* 26:975–984.

Chorus, I. and J. Bartram (eds.). 1999. *Toxic Cyanobacteria in Water.* E&F Spon, London.

Cloern, J. E. 2001. Our evolving conceptual model of the coastal eutrophication problem. *Marine Ecology Progress Series* 210:223–253.

Collos, Y. 1989. A linear model of external interactions during uptake of different forms of inorganic nitrogen by microalgae. *Journal of Plankton Research* 11:521–533.

D'Elia, C.F., J. G. Sanders, and W. R. Boynton. 1986. Nutrient enrichment studies in a coastal plain estuary: phytoplankton growth in large scale, continuous cultures. *Canadian Journal of Fisheries and Aquatic Science* 43:397–406.

Diaz, R. J. and R. Rosenberg. 1995. Marine benthic hypoxia: a review of its ecological effects and the behavioral responses of benthic macrofauna. *Oceanography and Marine Biology Annual Review* 33:245–303.

Fisher, T. R., L. W. Harding, Jr., D. W. Stanley, and L. G. Ward. 1988. Phytoplankton, nutrients, and turbidity in the Chesapeake, Delaware, and Hudson estuaries. *Estuarine and Coastal Shelf Science* 27:61–93.

Goericke, R. and N. Welschmeyer. 1993a. The carotenoid-labeling method: measuring specific rates of carotenoid synthesis in natural phytoplankton communities. *Marine Ecology Progress Series* 98:157–171.

Goericke, R. and N. Welschmeyer. 1993b. The chlorophyll-labeling method: measuring specific rates of chlorophyll a synthesis in cultures and in the open ocean. *Limnology and Oceanography* 38:80–95.

Harding, L. W., Jr. 1994. Long-term trends in the distribution of phytoplankton in Chesapeake Bay: roles of light, nutrients, and streamflow. *Marine Ecology Progress Series* 104:267–291.

Harding, L. W. and E. S. Perry. 1997. Long-term increase of phytoplankton biomass in Chesapeake Bay, 1950–1994. *Marine Ecology Progress Series* 157:39–52.

Harrington, M. B. 1999. Responses of Natural Phytoplankton Communities from the Neuse River Estuary, NC to Changes in Nitrogen Supply and Incident Irradiance. M.Sc. thesis, University of North Carolina, Chapel Hill, 89 pp.

Harrison, P., and D. Turpin. 1982. The manipulation of physical, chemical, and biological factors to select species from natural phytoplankton populations. In *Marine Mesocosms: Biological and Chemical Research in Experimental Ecosystems,* G. Grice and M. Reeve (eds.). Springer-Verlag, New York, pp. 275–287.

Harrison, W., T. Platt, and M. Lewis. 1987. F-ratio and its relationship to ambient nitrate concentration in coastal waters. *Journal of Plankton Research* 9:235–245.

Horne, A. J. 1977. Nitrogen fixation — a review of this phenomenon as a polluting process. *Progress in Water Technology* 8:359–372.

Jackson, J. B. C., M. X. Kirby, W. H. Berger et al. 2001. Historical overfishing and recent collapse of coastal ecosystems. *Science* 293(July):629–638.

Jørgensen, B. B. and K. Richardson. 1996. *Eutrophication of Coastal Marine Systems.* American Geophysical Union, Washington, D.C.

Kirshtein, J. D., J. P. Zehr, and H. W. Paerl. 1993. Determination of N_2 fixation potential in the marine environment: application of the polymerase chain reaction. *Marine Ecology Progress Series* 95:305–309.

MacGregor, B. J., B. Van Mooy, B. J. Baker, M. Mellon, P. H. Moisander, H. W. Paerl, J. P. Zehr, D. Hollander, and D. A. Stahl. 2001. Microbiological, molecular biological, and stable isotopic evidence for nitrogen fixation in the open waters of Lake Michigan. *Environmental Microbiology* 3:205–219.

Malone, T. C., D. J. Conley, T. R. Fisher, P. M. Gilbert, L. W. Harding, and K. G. Sellner. 1996. Scales of nutrient-limited phytoplankton productivity in Chesapeake Bay. *Estuaries* 19(2B):371–385.

Molloy, C. and P. Syrett. 1988. Interrelationships between uptake of urea and uptake of ammonium by microalgae. *Journal of Experimental Marine Biology* 118:85–95.

Murray, A. E., J. T. Hollibaugh, and C. Orrego. 1996. Phylogenetic compositions of bacterioplankton from two California estuaries compared by denaturing gradient gel electrophoresis of 16S rDNA fragments. *Applied and Environmental Microbiology* 62:2676–2680.

Olson, J. B., R. W. Litaker, and H. W. Paerl. 1999. Ubiquity of heterotrophic diazotrophs in marine microbial mats. *Aquatic Microbial Ecology* 19:29–36.

Oviatt, C, P. Doering, B. Nowicki, L. Reed, J. Cole, and J. Frithsen. 1995. An ecosystem level experiment on nutrient limitation in temperate coastal marine environments. *Marine Ecology Progress Series* 116:171–179.

Paerl, H. W. 1987. Dynamics of blue-green algal (*Microcystis aeruginosa*) blooms in the lower Neuse River, NC: causative factors and potential controls. Report 229. UNC Water Resources Research Institute, Raleigh, NC, 164 pp.

Paerl, H. W. 1988. Nuisance phytoplankton blooms in coastal, estuarine, and inland waters. *Limnology and Oceanography* 33:823–847.

Paerl, H. W. 1995. Coastal eutrophication in relation to atmospheric nitrogen deposition: current perspectives. *Ophelia* 41:237–259.

Paerl, H. W. 1996. A comparison of cyanobacterial bloom dynamics in freshwater, estuarine and marine environments. *Phycologia* 35:25–35.

Paerl, H. W. and N. D. Bowles. 1987. Dilution bioassays: their application to assessments of nutrient limitation in hypereutrophic waters. *Hydrobiologia* 146:265–273.

Paerl, H. W. and J. Kuparinen. 2002. Microbial aggregates and consortia. In *Encyclopedia of Environmental Microbiology,* Vol. 1, G. Bitton (ed.). John Wiley & Sons, New York, pp. 160–181.

Paerl, H. W. and D. F. Millie. 1996. Physiological ecology of toxic cyanobacteria. *Phycologia* 35(6):160–167.

Paerl, H. W. and J. P. Zehr. 2000. Nitrogen fixation. In *Microbial Ecology of the Oceans,* D. Kirchman (ed.). Academic Press, New York, pp. 387–426.

Paerl, H. W., J. Rudek, and M. A. Mallin. 1990. Stimulation of phytoplankton production in coastal waters by natural rainfall inputs: nutritional and trophic implications. *Marine Biology* 107:247–254.

Paerl, H. W., J. Dyble, P. H. Moisander, R. T. Noble, M. F. Piehler, J. L. Pinckney, L. Twomey, and L. M. Valdes. 2003. Microbial indicators of aquatic ecosystem change: current applications to eutrophication studies. *FEMS Microbial Ecology* 46(3): 233–246.

Pichard, S. L., L. Campbell, and J. H. Paul. 1997. Diversity of the ribulose bisphosphate carboxylase/oxygenase form I gene (rbcL) in natural phytoplankton communities. *Applied and Environmental Microbiology* 63:3600–3606.

Richardson, K. 1997. Harmful or exceptional phytoplankton blooms in the marine ecosystem. *Advances in Marine Biology* 31:302–385.

Riegman, R. 1995. Nutrient-related selection mechanisms in marine phytoplankton communities and the impact of eutrophication on the planktonic food web. *IAWQ SIL Conference on Selection Mechanisms Controlling Biomass Distribution.* Noordwykerhout, the Netherlands 32:4.

Rowan, K. 1989. *Photosynthetic Pigments of the Algae.* Cambridge University Press, New York, 334 pp.

Rudek, J., H. W. Paerl, M. A. Mallin, and P. W. Bates. 1991. Seasonal and hydrological control of phytoplankton nutrient limitation in the lower Neuse River Estuary. NC. *Marine Ecology Progress Series* 75:133–142.

Ryther, J. H., and C. B. Officer. 1981. Impact of nutrient enrichment on water uses. In *Estuaries and Nutrients*, B. J. Neilson and L. E. Cronin (eds.). Humana Press, Clifton, NJ, pp. 247–261.

Scardi, M., and L. W. Harding, Jr. 1999. Developing an empirical model of phytoplankton primary production: a neural network case study. *Ecological Modelling* 120:213–223.

Smith, V. H. 1990. Nitrogen, phosphorus, and nitrogen fixation in lacustrine and estuarine ecosystems. *Limnology and Oceanography* 35:1852–1859

Steppe, T. F., J. B. Olson, H. W. Paerl, and J. Belnap. 1996. Consortial N_2 fixation: a strategy for meeting nitrogen requirements of marine and terrestrial cyanobacterial mats. *FEMS Microbiology Ecology* 21:149–156.

Stolte, W., T. McCollin, A. Noordeloos, and R. Riegman. 1994. Effect of nitrogen source on the size distribution within marine phytoplankton populations. *Journal of Experimental Marine Biology and Ecology* 184:83–97.

Syrett, P. 1981. Nitrogen metabolism in microalgae. Physiological bases of phytoplankton ecology. *Canadian Bulletin of Fisheries and Aquatic Science* 210:182–210.

U.S. EPA. 1993. Guidance Specifying Management Measures for Sources of Nonpoint Pollution in Coastal Waters. EPA-840-B-93-001c. Office of Water, Washington, D.C.

Van Heukelem, L., A. Lewitus, T. Kana, and N. Craft. 1994. Improved separations of phytoplankton pigments using temperature-controlled high performance liquid chromatography. *Marine Ecology Progress Series* 114:303–313.

Vollenweider, R. A., R. Marchetti, and R. Viviani (eds.). 1992. *Marine Coastal Eutrophication.* Elsevier Science, New York.

Wyman, M., J. P. Zehr, and D. G. Capone. 1996. Temporal variability in nitrogenase gene expression in natural populations of the marine cyanobacterium *Trichodesmium thiebautii. Applied and Environmental Microbiology* 62:1073–1075.

Zehr, J. P., B. D. Jenkins, S. M. Short, and G. F. Steward. 2003. Nitrogenase gene diversity and microbial community structure: a cross-system comparison. *Environmental Microbiology* 5:539–554.

12

A Hierarchical Approach to the Evaluation of Variability in Ecoindicators of the Seagrass Thalassia testudinum

John W. Hackney and Michael J. Durako

CONTENTS

Introduction

Seagrasses are the dominant communities in many coastal environments of tropical and subtropical zones (den Hartog, 1970). As the climax communities in these systems (den Hartog, 1977), seagrass beds fulfill several functions: they stabilize bottom sediments, form structural substrate for epiphytic growth, provide nursery habitat and shelter to many organisms, and, most important, fix large amounts of carbon by photosynthesis, which becomes available via direct herbivory and the detrital food web, both within the system and by export to other systems (Zieman, 1982; Thayer et al., 1984; Kenworthy et al., 1988; Duarte, 1989). The shallow distribution and close proximity to the land/sea interface of seagrass beds cause them to be sensitive to changes in estuarine and nearshore marine environments. Therefore, the distribution, abundance, and condition of seagrasses may be indicative of the health of coastal ecosystems.

Seagrasses have recently been affected by catastrophic mortality events (Orth and Moore, 1983; Larkum and West, 1990; Robblee et al., 1991), and a worldwide decline in the areal extent of seagrass beds has been reported (Short and Wyllie-Echeverria, 1996; Duarte, 2002). Because seagrass habitats are crucial for the productivity of fisheries and wildlife (Zieman et al., 1989), the distribution and abundance of many faunal species are closely linked to the condition of seagrass meadows (Thayer and Chester, 1989; Thayer et al., 1999). Declines in seagrass-associated faunal communities have been shown to be associated with losses in their habitat (Matheson et al., 1999). One such change is the rapid and widespread mortality of *Thalassia testudinum* (turtle grass) that has been documented in Florida Bay since 1987 (Fourqurean and Robblee, 1999). A number of stress-inducing agents have been proposed as the cause of this mass mortality, including unusually high salinities and temperatures, reduced freshwater input from the Everglades watershed, reduced frequency of tropical storms, sulfide toxicity, self-shading and chronic hypoxia of *T. testudinum* roots and rhizomes caused by biomass accumulation, an epidemic of a pathogenic marine slime mold (*Labyrinthula* sp.), or a combination of one or more of these factors (Fourqurean and Robblee, 1999). The resultant die-off of *T. testudinum* has caused eutrophication and increased turbidity in the Florida Bay system, leading to a systemwide disturbance that threatens the stability of this valuable ecosystem (Butler et al., 1995).

Realization of the importance of seagrass habitats and the key role they play in coastal ecosystems has led to increased efforts to quantify their condition using various structural and dynamic characters to gain insight into the health of entire systems (Duarte et al., 1994; Durako, 1995). Structural and dynamic characters of *T. testudinum* include leaf width, leaf length, number of leaves per short-shoot, leaf area, number of leaf scars, leaf productivity, and leaf turnover (which are shoot-specific characters, along with plastochrone interval); area-specific characters include short-shoot and rhizome densities, leaf area index, leaf productivity, and biomass (Durako, 1995). Phillips and Lewis (1983) specifically correlated leaf width with environmental stress and observed that seagrasses occur over ecological, spatial, and temporal gradients where plant characters (such as leaf dimensions) are significantly associated with environmental factors. Durako (1995) also observed that changes in structural characters, such as leaf length, width, and shoot-specific leaf area, may indicate response to environmental conditions at intermediate timescales between acute and chronic stress.

This chapter describes the trends and patterns of the morphometric characters of *T. testudinum* in ten basins within Florida Bay during two sampling seasons. *Thalassia testudinum* is one of the most studied seagrasses; however, there are few studies with data sets as large as the one presented here. *Thalassia testudinum* was collected during spring 1998 and 1999; for each sample year, shoot density, shoot morphometrics, shoot age, and standing crop were measured at more than 300 stations in ten basins of Florida Bay. Our hypotheses were that the morphometric characters of *T. testudinum* vary significantly across the bay and between years. These hypotheses were tested using a hierarchical statistical approach. The goal was to determine if the variability in *T. testudinum* morphometrics exhibited spatial coincidence with previously defined ecozones in Florida Bay and to determine which morphometric characters were most responsive to environmental variability.

Materials and Methods

Study Area

Plant material for this study was collected in Florida Bay (approximately 25°05′N, 81°45′W; UTM 424370E, 2774385N), a shallow, seagrass-dominated estuary that separates the Florida Keys from the southern tip of the Florida peninsula (Figure 12.1). The bay is traversed by a series of reticulating mud banks that subdivide the bay into 49 basins; these mud banks also restrict circulation and dampen tidal influence (Robblee et al., 1991; see Fourqurean and Robblee, 1999 for a detailed description of Florida Bay). The average water depth in the 2000 km^2 area of the bay within Everglades National Park (ENP) is ~1 m but varies from less than 1 m to about 3 m (Schomer and Drew, 1982). Although seagrass communities cover most of the bay, community development increases in a strong gradient from the enclosed northeastern sections of the bay to the more open western sections (Zieman et al., 1989). *Thalassia testudinum* occurs in both monospecific and mixed-species seagrass beds throughout the bay; in the northeast, communities are dominated by mostly sparse beds of *T. testudinum* with localized denser areas, but increasingly dense beds of *T. testudinum* are often intermixed with the seagrasses *Halodule wrightii* Aschers. (shoal grass), *Syringodium filiforme* Kützing (manatee grass), and *Halophila englemanii* Aschers. (star grass) toward the west. Salinity and water clarity are highly variable throughout the bay (Zieman, 1982). As a negative estuary, the bay is normally subject to periods of hypersalinity (Boyer et al., 1999). After a period of extremely high salinities in 1989–1990, Zieman et al. (1999) found mean salinity in the bay from 1991–1995 to oscillate between 29 and 32 psu in winter and between

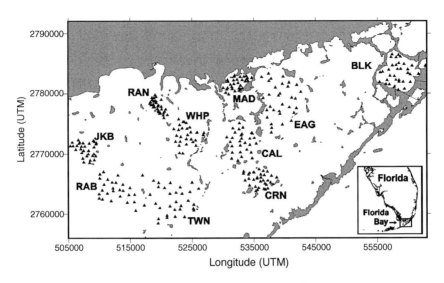

FIGURE 12.1 Florida Bay showing the locations of sites sampled in 1998 and 1999.

35 and 39 psu in summer. The waters of Florida Bay historically have been clear, but water clarity in the bay is subject to broad regional and temporal differences. Increased light attenuation, due to microalgal blooms and resuspended sediments, has become apparent in recent years in many parts of the bay (Stumpf et al., 1999). Turbidity has increased in the years since the *T. testudinum* die-off; Fourqurean and Zieman (1991) found diffuse light attenuation (k_d) to be generally low in the bay (mean $k_d = 0.5$ m^{-1}) prior to 1990, but in 1993–1994 k_d ranged from 0.7 to 2.8 m^{-1} (Phlips et al., 1995). Light attenuation is generally greatest in the eastern and south-central regions of the bay and lowest in the north-central and western regions (Phlips et al., 1995). Water temperature is more constant among basins, but it shows much seasonal variation due to the shallow nature of the bay (Zieman, 1982). Fine-grained to muddy-sand carbonate sediments increase in depth in a gradient from northeast to southwest (Zieman et al., 1989).

Sampling Methodology

Seagrass samples were collected during the spring sampling of the Florida Bay Fish Habitat Assessment Program (FHAP) in 1998 and 1999. This program was established to assess status and trends in the benthic fish habitat of Florida Bay as part of a multiagency coordinated monitoring program implemented in this area to detect and avert regional-scale seagrass loss (Durako et al., 2002; Fourqurean et al., 2002). Spatial assessment of the bay was achieved by examining macrophyte species distribution, structure, and relative abundance at a baywide scale. Sampling was conducted in 10 of the 49 basins (Table 12.1), which represent a gradient of conditions across the bay. Each basin was subdivided into 28 to 33 fixed, tessellated hexagonal grids from within which station locations were randomly chosen to yield a total of about 320 stations per sample period (Figure 12.1). Stations were located using a handheld GPS (global positioning system) receiver and one basin was sampled per day. Water depth and temperature, salinity, and Secchi depth were recorded at each station. This sampling design resulted in semisynoptic systematic random sampling, with sampling effort scaled to the size of the basin (see Durako et al., 2002, for a detailed description of the FHAP sampling design).

For this study *T. testudinum* was collected as part of the quantitative assessment of the benthic macrophyte communities during the May 1998 and May 1999 FHAP sampling; the 1998 sites were revisited in 1999. A single PVC (polyvinylchloride) core sample (177 cm^2) was taken at each station. Plant material from the cores was washed free of sediment in the field, stored in plastic bags, and frozen for subsequent analysis. After thawing, seagrasses were sorted by species, short-shoot density (number m^{-2}) was determined from the material in each core, the plant material was rinsed in 10% HCl to remove

TABLE 12.1

List of Basins Sampled in Florida Bay
and Their Abbreviations

Abbreviation	Basin
BLK	Blackwater Sound
CAL	Calusa Key
CRN	Crane Key
EAG	Eagle Key
JKB	Johnson Key
MAD	Madeira Bay
RAB	Rabbit Key
RAN	Rankin Lake
TWN	Twin Key
WHP	Whipray Bay

carbonates, and leaves were scraped carefully with a razor blade to remove epiphytes. Epiphyte loads were not quantified. Only cores containing live *T. testudinum* short-shoots were analyzed. In 1998, a total of 318 cores were collected; of these, 211 (66%) contained live *T. testudinum* short-shoots. Of the 314 cores taken in 1999, 232 (74%) contained at least one live short-shoot. For each live short-shoot, the number, length (cm) from point of attachment to the short-shoot leaf tip, and width (cm) just above the leaf sheath of all green blades were recorded, and the shoot age in plastochrone intervals (i.e., the number of leaf scars plus the number of green and white blades) was calculated.

After morphometric measurements were obtained, green leaves (aboveground biomass) were dried to constant weight at 60°C and weighed to obtain standing crop (g m^{-2}). Live short-shoot and live rhizomes and roots (white or brown and crispy) were also dried to constant weight (60°C) and weighed to obtain belowground biomass (g m^{-2}). These data were used to generate two types of characters of *T. testudinum*, shoot-specific characters and area-specific characters. Shoot-specific characters are leaves shoot^{-1}, maximum leaf length, mean leaf length, maximum leaf width, shoot-specific leaf area (the sum of the leaf area of a short-shoot, cm^2), and shoot age (leaf scars shoot^{-1}). Area-specific characters are short-shoot density, leaf area index (LAI, mean shoot-specific leaf area per short-shoot density, m^2 m^{-2}), standing crop (g m^{-2}), and the ratio of above- to belowground biomass.

Statistical Analyses

Samples from both years were used in hierarchical statistical analyses to assess year-to-year changes in shoot-specific and area-specific morphometric and biomass characters at two spatial scales: bay level and basin level. Thus, within-year and between-year variability of the bay were described, as was the variability between years of each basin. Graphical and statistical analyses were performed on the data both to visualize and quantify trends in each measured or derived parameter. Morphometric characters were compared with a multitiered approach. First, paired *t*-tests were used to assess bay-scale differences in the six shoot-specific characters and four area-specific characters between 1998 and 1999. Data were logarithm (log$_{10}$)-transformed if necessary to approach normality and homogeneity of variance; however, Mann–Whitney Rank Sum analyses were used in lieu of *t*-tests when data could not be transformed to meet these assumptions. The criterion for significant differences was $p < 0.05$ throughout. Because of the possible relatedness of the individual ramets within a core, data from the short-shoots within each core were averaged prior to subsequent analysis. As a result, each core was considered an independent sample unit and each short-shoot within a core was considered a subsample.

Second, box-and-whisker diagrams were created for each morphometric character in each basin to depict the distribution of data around the mean and median. Paired *t*-tests and Mann–Whitney Rank Sum tests were used to assess differences in the morphometric characters between 1998 and 1999 within each basin. In addition, differences among basins for each year were tested with a one-way analysis of variance (ANOVA), after which basin means were grouped into similar subsets using Duncan's Multiple Range tests for each variable to assess regional trends within the bay.

After these tests were performed, some patterns became evident. Certain basins consistently grouped together into similar Duncan's subsets. To assess the statistical significance of these patterns, principal components analysis (PCA) was used to extract principal components from the means of the data and the ten basins were grouped into four zones or regions based on PCA similarities. Basin data from each of the four zones were then pooled and differences among the four zones were assessed with a second series of one-way ANOVAs of the zone-pooled data. Zones were then grouped into similar subsets using Duncan's Multiple Range tests for each variable to confirm the independent assortment of the zones suggested by the PCA results. All statistical tests were performed using the SAS® statistical program (SAS, Cary, North Carolina) or SigmaStat® (Jandel Scientific, San Rafael, California).

Results

Biological Parameters

Morphometric characters of *T. testudinum* in Florida Bay in 1998 and 1999 are summarized in Table 12.2. In this table, data from all ten basins were pooled for each year. Large ranges were measured and standard deviations generally were high due to the wide spatial sampling plan. Leaf width and leaf number exhibited the lowest variation in the pooled data, each with a coefficient of variation of about 30%. The

TABLE 12.2

Summary of Structural Characteristics of *T. testudinum* in Florida Bay in 1998 and 1999, All Samples Combined

Year	Characteristic	n	Mean (Median)	STD	Range
Shoot-Specific Characteristics					
1998	Leaves shoot^{-1}	1494	3.29	1.082	1–7
1999	Leaves shoot^{-1}	1622	3.15	1.045	1–9
1998	Max leaf length (cm)	1494	11.27	7.143	0.1–45.2
1999	Max leaf length (cm)	1622	12.01	6.880	0.1–58.5
1998	Mean leaf length (cm)	1494	8.60	5.398	0.1–34.5
1999	Mean leaf length (cm)	1622	8.93	4.980	0.1–37.6
1998	Max leaf width (cm)	1494	0.48	0.168	0.1–1.4
1999	Max leaf width (cm)	1622	0.52	0.174	0.1–1.1
1998	Leaf area shoot^{-1} (cm^2)	1494	17.39	21.371	0.02–250.3
1999	Leaf area shoot^{-1} (cm^2)	1622	18.03	19.228	0.04–179.1
1998	Leaf scars shoot^{-1}	1494	33.89 (30.0)	19.429	4–159
1999	Leaf scars shoot^{-1}	1622	36.41 (32.0)	19.817	3–138
Area-Specific Characteristics					
1998	Short-shoots m^{-2}	211	400.76	333.930	56.6–1754.6
1999	Short-shoots m^{-2}	232	394.01	351.840	56.6–2037.6
1998	LAI (m^2 m^{-2})	211	0.697	0.893	0.008–5.304
1999	LAI (m^2 m^{-2})	232	0.710	0.969	0.007–6.682
1998	Standing crop (g m^{-2})	211	27.56	33.521	0.38–183.95
1999	Standing crop (g m^{-2})	232	27.30	35.242	0.28–205.80
1998	Aboveground:belowground biomass	211	0.14	0.129	0.006–0.921
1999	Aboveground:belowground biomass	232	0.12	0.096	0.003–0.642

Note: The median value is given in parentheses for leaf scars shoot^{-1} (shoot age) because of the skewed nature of most age data.

other characters had coefficients of variation from 60 to >100% of the mean. The means of all shoot-specific characters increased from 1998 to 1999, except leaf number, which decreased. If the short-shoots within each core were treated as individuals (resulting in the high sample sizes reported in Table 12.2), each of the shoot-specific characters displayed significant interannual differences (Mann–Whitney Rank Sum tests). However, if each core was treated as a replicate, only leaf number exhibited a significant difference from 1998 to 1999; shoots had 4% fewer leaves in 1999 ($p < 0.005$). Area-specific characters were not significantly different between 1998 and 1999 ($p > 0.05$).

Box-and-whisker plots of core means of *T. testudinum* shoot-specific and area-specific characters in the ten sampled basins in Florida Bay in 1998 and 1999 are shown in Figure 12.2 through Figure 12.5. The similarity in spatial patterns among five of the six shoot-specific characters, leaves per shoot (Figure 12.2), leaf width (Figure 12.3A), maximum leaf length (Figure 12.3B), mean leaf length (Figure 12.3C), and shoot-specific leaf area (Figure 12.3D), is striking. With few exceptions, the means and medians of these five parameters tended to increase in a gradient from EAG in the northeast to RAB and JKB in the southwest. Blackwater Sound (in the northeast) and TWN (in the south-central bay) were notable exceptions. Three of the area-specific characters (excluding density) also followed this general trend; density (Figure 12.4A) increased in a gradient from the northeast to a high in the central bay, where it decreased toward the southwest. Shoot age exhibited no east–west gradient (Figure 12.4B). Although leaf area index (Figure 12.4C) and standing crop (Figure 12.4D) are greatly influenced by density, these characters are also affected by shoot-specific characters and reflected those trends, as did the ratio of above- to belowground biomass (Figure 12.5).

Short-shoots in JKB had significantly more leaves than in all other basins in 1998 and 1999 (Figure 12.2). Otherwise, no other basin was significantly distinct from any other, and, excluding JKB, leaf number was fairly uniform across the bay in both years. In 1998, CAL in the east had the fewest leaves per shoot (3.0) and JKB had the most (4.4), nearly 50% more leaves per shoot than in CAL. In 1999, JKB again had the most leaves per shoot (5.3) while the shoots of CRN had the fewest (2.8), a difference of almost 100%. Shoots in BLK, CRN, and MAD had significantly fewer leaves in 1999 than in 1998.

Leaf width was more specific to regions within the bay (Figure 12.3A). In both sampling seasons, leaves in the western bay (JKB, 0.8 cm; RAB, 0.7 cm) on average were as much or more than twice as wide as those in the eastern bay (EAG, CAL, CRN, MAD, 0.3 to 0.4 cm). Leaves of shoots in the central bay (and in BLK) were of intermediate width. Only leaves in WHP displayed a significant interannual difference, and for the bay as a whole, width was almost identical between 1998 and 1999. Maximum leaf length (Figure 12.3B), or canopy height, was much greater in the western JKB and RAB basins in 1998 (more than 20 cm) than in basins to the east (less than 15 cm). The mean of the longest leaves in JKB was 24 cm, four times the mean of the longest leaves in EAG (5.9 cm). Again, the four eastern basins had the shortest canopy height, and leaf length increased toward the southwest. In 1999, a decline in maximum length in JKB and RAB (significant in RAB), combined with increases in all other basins (significant in CAL and WHP), made this character slightly less variable across the bay than in 1998. As in the previous year, shoots in JKB had the longest leaves and EAG the shortest.

Mean leaf length (Figure 12.3C) showed a similar pattern to maximum length. Short-shoots in JKB had the longest leaves on average (18 cm). The four eastern basins (excluding BLK) had the shortest leaves (less than 7 cm) in both years, about 35% as long as the leaves in JKB. As with maximum length, leaves in CAL and in MAD were significantly longer in 1999 than in 1998, and those in RAB were significantly shorter. Differences in leaf length were reflected in the plots of shoot-specific leaf area (Figure 12.3D). Shoots in JKB had much greater total area (79.6 cm^2 in 1998, 74.3 cm^2 in 1999), almost twice that of the shoots in RAB and more than an order of magnitude greater than the shoots in EAG. Shoots in WHP and CAL had significantly greater specific leaf areas in 1999, while leaf area decreased in RAB.

Short-shoots in EAG had the highest number of leaf scars (a mean of more than 40 scars per shoot) compared to those in other basins in both 1998 and 1999 (Figure 12.4B). Eagle Key shoots had 26 to 38% more leaf scars than those of WHP and RAN, the basins with the youngest shoots in 1998 and 1999, respectively. Seven of the ten basins showed an increase in the number of leaf scars per short-shoot between 1998 and 1999, but the increases were only significant in MAD and WHP.

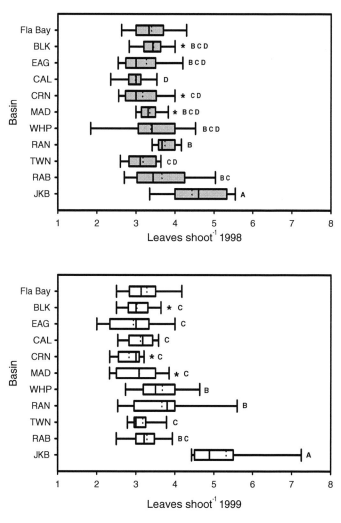

FIGURE 12.2 Number of leaves per individual of *T. testudinum* short-shoots in ten basins in Florida Bay in 1998 (shaded) and 1999 (unshaded). Basins with the same letter designation did not have significantly different means within a year based on Duncan's Multiple Range test. Basins marked with an asterisk (*) exhibited significant differences between years based on Mann–Whitney Rank Sum tests or Student's *t*-tests. Box-and-whisker diagrams: boxes enclose interquartile range, vertical line within box represents median, dashed vertical line represents mean, whisker caps represent the 5th and 95th percentiles. Values for Florida Bay shown for comparison.

Although *T. testudinum* mean short-shoot density decreased in the bay in 1999 compared to 1998 (Table 12.2), this decrease was statistically insignificant, and *T. testudinum* was more widespread in 1999 (present in 232 of 314 cores in 1999 vs. 211 or 318 cores in 1998). Twin Key Basin had the highest density in both years, about 650 shoots m^{-2} (Figure 12.4A). In 1998, JKB had the lowest density (170 shoots m^{-2}), and in 1999 EAG had the lowest density (178 shoots m^{-2}), both of which were about 25% that of TWN. Six of the ten basins displayed insignificant decreases in density while the other four had insignificant increases, between 1998 and 1999.

Leaf area index (LAI) increased in a gradient from the northeast (not including BLK) to the southwest; LAI was greatest in the western basins of JKB and RAB in both sampling years, more than ten times that of the basin with the lowest value, EAG (Figure 12.4C). In 1998 leaf area index was 1.69 m^2 m^{-2} in RAB and 0.12 m^2 m^{-2} in EAG, and in 1999 LAI was 1.75 m^2 m^{-2} in JKB and 0.11 m^2 m^{-2} in EAG. A statistically insignificant increase in LAI was seen in the bay as a whole from 1998 to 1999, the only basin that showed a significant change (an increase) between years was CAL.

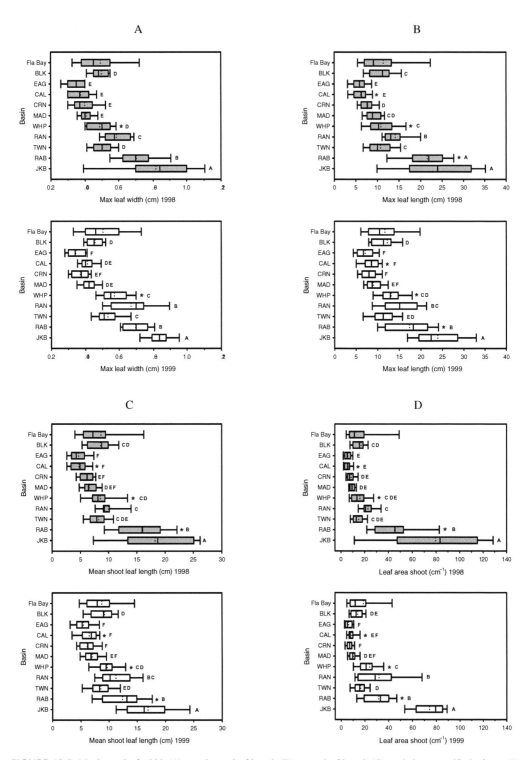

FIGURE 12.3 Maximum leaf width (A), maximum leaf length (B), mean leaf length (C), and shoot-specific leaf area (D) of short-shoots of *T. testudinum* in ten basins in Florida Bay in 1998 and 1999. See Figure 12.2 for description of box-and-whisker plots, shading, letters, and asterisk (*).

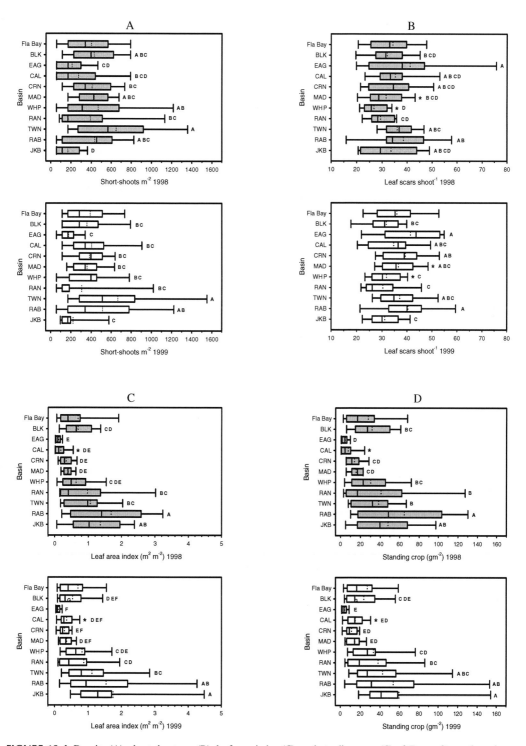

FIGURE 12.4 Density (A), short-shoot age (B), leaf area index (C), and standing crop (C) of *T. testudinum* short-shoots in ten basins in Florida Bay in 1998 and 1999. See Figure 12.2 for description of box-and-whisker plots, shading, letters, and asterisk (*).

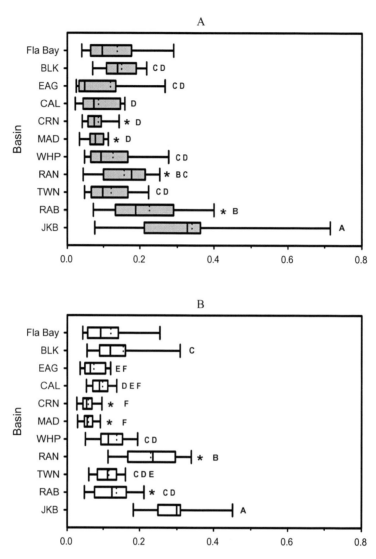

FIGURE 12.5 Ratio of aboveground-to-belowground biomass of *T. testudinum* in ten basins in Florida Bay in 1998 (A) and 1999 (B). See Figure 12.2 for description of box-and-whisker plots, shading, letters, and asterisk (*).

The plots of *T. testudinum* standing crop closely mirror those of leaf area index (Figure 12.4D). Again, the lowest values were seen in the northeast (EAG, 4 to 5 g m^{-2} in both years) and the highest in the west (~65 g m^{-2} in RAB in 1998 and ~60 g m^{-2} in JKB in 1999). There was no significant change in standing crop in the bay between 1998 and 1999, and again, the only significant difference at the basin level was an increase in CAL. Above- and belowground biomass ratio showed similar trends as well with a maximum in the southwest and a minimum in the northeast (Figure 12.5). Shoots in JKB had the highest ratio in both years (~30% aboveground biomass), and shoots in MAD and CRN had the lowest ratios (less than 10% aboveground biomass). The bay showed an insignificant decrease in this ratio from 1998 to 1999, a trend probably driven by significant decreases in CRN, MAD, and RAB; RAN was the only basin which showed a significant increase between years.

Principal Components Analysis

PCA confirmed that certain basins grouped together, based on morphometric and biomass characters, as was frequently indicated in the Duncan's tests (Figure 12.6). In 1998, the first principal component axis (PC1) explained 53% of the variance and the second principal component axis (PC2) explained 21% of the variance; these composite variables captured 74% of the variation in the original data. In 1999, the first principal component axis (PC1) explained 52% of the variance and the second principal component axis (PC2) explained 23% of the variance; these composite variables captured 75% of the variation in the original data. The basin groupings were used to divide Florida Bay into four zones, which exhibited similar trends for the variables examined. The Eastern Bay community consisted of

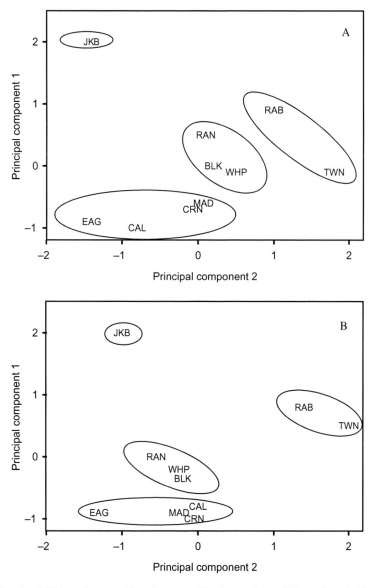

FIGURE 12.6 Results of PCA on shoot-specific and area-specific characteristics of *T. testudinum* in Florida Bay in 1998 (A) and 1999 (B) with groupings into communities of biological similarity. Eastern Bay community = Eagle Key, Calusa Key, Crane Key, and Madeira Bay Basins; Central Bay community = Rankin Lake, Whipray Basin, and Blackwater Sound; Western Bay community = Rabbit Key and Twin Key Basins; Johnson Key community = Johnson Key Basin.

TABLE 12.3

Results of One-Way ANOVA to Test for Spatial Differences in Shoot-Specific and Area-Specific Characteristics of *T. testudinum* between Four Communities in Florida Bay in 1998 and 1999, Followed by Duncan's Multiple Range Tests for Comparisons among Communities

Year	Characteristic	One-Way ANOVA			Duncan's Homogeneous Subsets			
		n	F	P				
Shoot-Specific Characteristics								
1998	Leaves shoot^{-1}	211	12.4	<0.001	Jkb	<u>Cent</u>	<u>West</u>	<u>East</u>
1999	Leaves shoot^{-1}	232	39.3	<0.001	Jkb	<u>Cent</u>	<u>West</u>	<u>East</u>
1998	Max leaf width	211	81.4	<0.001	Jkb	West	Cent	East
1999	Max leaf width	232	113.6	<0.001	Jkb	West	Cent	East
1998	Max leaf length	211	64.9	<0.001	Jkb	West	Cent	East
1999	Max leaf length	232	63.7	<0.001	Jkb	West	Cent	East
1998	Mean leaf length	211	63.3	<0.001	Jkb	West	Cent	East
1999	Mean leaf length	232	61.5	<0.001	Jkb	West	Cent	East
1998	Leaf area shoot^{-1}	211	82.5	<0.001	Jkb	West	Cent	East
1999	Leaf area shoot^{-1}	232	128.0	<0.001	Jkb	West	Cent	East
1998	Leaf scars shoot^{-1}	211	3.6	0.015	<u>West</u>	<u>East</u>	<u>Jkb</u>	<u>Cent</u>
1999	Leaf scars shoot^{-1}	232	6.4	<0.001	<u>West</u>	<u>East</u>	<u>Cent</u>	<u>Jkb</u>
Area-Specific Characteristics								
1998	Short-shoots m^{-2}	211	6.8	<0.001	<u>West</u>	<u>Cent</u>	<u>East</u>	Jkb
1999	Short-shoots m^{-2}	232	9.5	<0.001	West	<u>Cent</u>	<u>East</u>	<u>Jkb</u>
1998	LAI	211	25.2	<0.001	<u>West</u>	<u>Jkb</u>	Cent	East
1999	LAI	232	23.5	<0.001	Jkb	West	<u>Cent</u>	<u>East</u>
1998	Standing crop	211	23.8	<0.001	<u>West</u>	<u>Jkb</u>	Cent	East
1999	Standing crop	232	21.0	<0.001	<u>Jkb</u>	<u>West</u>	Cent	East
1998	Aboveground: belowground biomass	211	22.7	<0.001	Jkb	<u>West</u>	<u>Cent</u>	East
1999	Aboveground: belowground biomass	232	45.6	<0.001	Jkb	Cent	West	East

Abbreviations: Jkb = Johnson Key Basin, West = Western Bay, Cent = Central Bay, East = Eastern Bay.

EAG, MAD, CAL, and CRN. The Central Bay community consisted of BLK, RAN, and WHP; although geographically BLK is in northeastern Florida Bay, statistically it grouped most often with RAN and WHP and not with the other eastern basins examined. The Western Bay community consisted of RAB and TWN. Finally, JKB in the northwest was consistently significantly different from the other groupings and, thus, was placed as its own community.

Results of one-way ANOVA for shoot-specific and area-specific characters for 1998 and 1999 among the pooled data for the communities identified by PCA and Duncan groupings of similar subsets are shown in Table 12.3. All shoot-specific and area-specific characters displayed significant zone-level differences. The JKB community was significantly different in both years from the other three communities in the number of leaves per shoot, leaf width, leaf length, average shoot leaf length, and shoot leaf area. In 1998, *T. testudinum* did not show significant differences among the Central Bay, Western Bay, and Eastern Bay communities in the number of leaves per shoot. In 1999, *T. testudinum* in the Central Bay and Western Bay did not show significant differences in leaf number, and leaf number in the Western Bay and Eastern Bay was not significantly different. All four communities were significantly different from each other in both sample years in maximum leaf width, maximum leaf length, mean shoot leaf length, and shoot-specific leaf area. In 1998, the Western Bay, Eastern Bay, and JKB communities were similar to each other in leaf scars per shoot, and in 1999 the Western and Eastern Bay communities were in similar subsets of leaf scars, and the Central Bay and JKB communities were in similar subsets.

Significant trends in area-specific characters among communities were less obvious, but still present. Short-shoot density was highest in the Western Bay and lowest in JKB. In 1998, the JKB community had significantly different short-shoot density from all other communities; the Western Bay and Central Bay communities were similar, and the Central Bay was also similar to the Eastern Bay. In 1999, the Western Bay community was significantly different from the other communities in shoot density, and the Central Bay, Eastern Bay, and JKB communities were not significantly different from each other. Leaf area index was highest in the Western Bay and JKB communities. In 1998, leaf area index was similar in the Western Bay and JKB communities, while the Central and Eastern Bay communities were distinct. In 1999, leaf area index was distinct in the Western Bay and JKB communities while the Central and Eastern Bay communities were in a homogeneous subset. As with LAI, standing crop was highest in the Western Bay and JKB, intermediate in the Central Bay, and lowest in the Eastern Bay. In both sampling years *T. testudinum* did not show significant differences in standing crop between the Western Bay and JKB communities while the Central Bay community and the Eastern Bay community were distinct subsets. The ratio of aboveground to belowground biomass was highest in both sampling seasons in the JKB community and lowest in the Eastern Bay community; in 1998 the Western and Central Bay communities were a similar subset while in 1999 all four communities were significantly different from each other.

Discussion

Morphometric and biomass characters of *T. testudinum* varied at the two spatial scales of Florida Bay examined in this study. Basin-level variation indicated that Florida Bay is not a uniform environment (Phlips et al., 1995; Boyer et al., 1997, 1999). Because seagrass morphometrics can also vary seasonally (Duarte, 1989; Alcoverro et al., 1995; Durako, 1995; van Tussenbroek, 1998; Irlandi et al., 2002), the effect of seasonal variation was minimized in this study with year-to-year comparisons of samples taken at approximately the same time of year. Of the measured morphometric characters of *T. testudinum*, only mean leaf number showed a significant (though minimal) interannual difference at the bay level. Mean area-specific (density-dependent) characters showed no bay-level interannual differences in this study.

Shoot-specific and area-specific characters of *T. testudinum* had high degrees of variability within Florida Bay, at the level of individual basins. These characters were not constant across Florida Bay (thus, we accepted our original hypotheses). The results of our analyses confirm those of earlier studies demonstrating an increase in *T. testudinum* abundance from northeastern to southwestern Florida Bay (Zieman et al., 1989; Hall et al., 1999). In addition, short-shoots were generally larger and had more leaves in the western basins. Leaf number and size decreased in a gradient from west to east, following trends in phosphorus availability (Fourqurean et al., 1992a,b), sediment depth (Zieman et al., 1989), and iron concentration (Chambers et al., 2001). Three of the four area-specific characters (standing crop, leaf area index, and aboveground/belowground biomass) measured showed essentially the same trend; short-shoot density was slightly more uniform across the bay.

The ranges of measured morphological and biomass characters were relatively large but were within reported values for this species. The station-specific differences in *T. testudinum* characters observed are most likely regulated by differences in environmental parameters. Seagrass growth is regulated by light availability, temperature, and nutrient supply. Local environmental conditions probably form microhabitats in Florida Bay that account for differences in structural patterns among sites. Clonal plants often display morphological plasticity, which may be adaptive by allowing them to persist in a wide range of environmental conditions (Hutchings, 1988). *Thalassia testudinum* exhibits phenotypic plasticity in its leaf dimensions (van Tussenbroek, 1996). In addition to responding to seasonal effects, leaf characters, standing crop, and density of *T. testudinum* and other seagrasses have been correlated with several environmental factors, including salinity (Dawes et al., 1985), low light availability because of depth (McMillan and Phillips, 1979) and turbidity (Phillips and Lewis, 1983; Lee and Dunton, 1997), latitude (McMillan, 1978), nutrient availability (Short, 1983; Lee and Dunton, 2000), freshwater input (Irlandi et al., 2002), and possibly intra- and interspecific competition (Rose and Dawes, 1999; Davis and Fourqurean, 2001).

The relatively low proportion of aboveground to belowground biomass (~15%) of *T. testudinum* compared to other seagrass species in Florida Bay accounts for its high light requirements (Fourqurean and Zieman, 1991), and it also may be indicative of phosphorus limitation (Pérez et al., 1995). The values measured in this study (14 and 12% in 1998 and 1999, respectively) are similar to values reported by Iverson and Bittaker (1986) for southern Gulf of Mexico beds in and around Florida Bay, but slightly lower than those reported by Fourqurean and Zieman (1991). However, our measurements were taken before maximum leaf biomass was reached in midsummer (Dawes et al., 1985). The aboveground-to-belowground biomass ratio is a function of density and leaf area, and it can be influenced by either factor. Some of the highest ratios of aboveground to belowground biomass were seen in Johnson Key Basin and Rankin Lake. Although short-shoot densities were relatively low in Rankin Lake and Johnson Key Basin (<200 shoots m^{-2}) compared to other basins in the bay, the short-shoots in these areas were leafier. Short-shoots here had more leaves on average than those in other basins, their leaves were among the widest and longest, and they had high shoot-specific leaf areas. These two basins have been severely affected by die-off and chronic turbidity (Durako, 1995; Phlips and Badylak, 1996; Hall et al., 1999); the increased leafiness may indicate a morphological response to light limitation (Dawes and Tomasko, 1988).

To facilitate characterization and study of Florida Bay, several investigators have divided the bay into distinct subenvironments or ecological zones (Figure 12.7A,B,C), including those based on bank morphology and dynamics (Wanless and Taggett, 1989), benthic plant communities (Zieman et al., 1989), water quality (Boyer et al., 1999), and light availability for planktonic and benthic primary production (Phlips et al., 1995). Zieman et al. (1989) used macrophyte distribution, standing crop, and productivity, along with sediment type and depth and water depth, to divide the bay into six ecological regions with similar biological and physical characters (Figure 12.7A). They included two other seagrass species and four types of macroalgae, in addition to *T. testudinum*, which was the most abundant macrophyte. Phlips et al. (1995) used regional differences in light attenuation to define four ecological zones in the bay (Figure 12.7B), and Boyer et al. (1999) divided the bay into three "zones of similar influence" based on water quality data (Figure 12.7C). These three schemes are similar in that they all have groupings of western, central, and eastern basins in some fashion.

The consistent basin-level statistical groupings of morphometric parameters we observed led us to define four ecological zones, based on shoot-specific and area-specific characters of *T. testudinum*. These zones were statistically robust and support the concept of ecologically distinct regions in Florida Bay. In addition, our results demonstrate that the significant plasticity of *T. testudinum* morphology is under the control of the physical and chemical environment of the bay. The four zones we defined (Johnson Key, Western Bay, Central Bay, and Eastern Bay communities, Figure 12.7D) are most similar to the four zones of Phlips et al. (1995). Our Johnson Key community corresponds to the West region of Phlips et al. (1995), our Western Bay community to their South-Central region, our Central Bay community to their North-Central region, and our Eastern Bay community to their East region.

The most obvious disparity between our classification scheme compared to previous classification schemes is the inclusion of Blackwater Sound with the Central Bay community (Figure 12.7D). This basin is very diverse. Blackwater Sound is subject to the influence of freshwater flow from the South Florida Water Management District's C-111 canal (which drains the eastern Everglades) and to oceanic influences due to its proximity to several cuts through the Florida Keys that lead to the Atlantic. The sound has the deepest areas of any of the studied basins (Table 12.2), areas of extremely clear water as well as very turbid areas near the Intracoastal Waterway, localized areas of shallow coarse sediments, deeper fine sediments and bedrock outcrops, and four of the five species of seagrass that occur in Florida Bay are present in the sound (personal observations). Biologically and physically, Blackwater Sound represents a microcosm of Florida Bay. Geographically, Blackwater Sound is in the East, but in terms of morphometric characters of *T. testudinum*, it is most similar to Rankin Lake and Whipray Basin in the central interior of Florida Bay, which also share many of the same environmental characters. The box-and-whisker diagrams of shoot-specific and area-specific characters of *T. testudinum* and the PCA of these traits demonstrate how short-shoots in Blackwater Sound were similar to those found in central Florida Bay. The case of Blackwater Sound succinctly illustrates how an understanding of morphology,

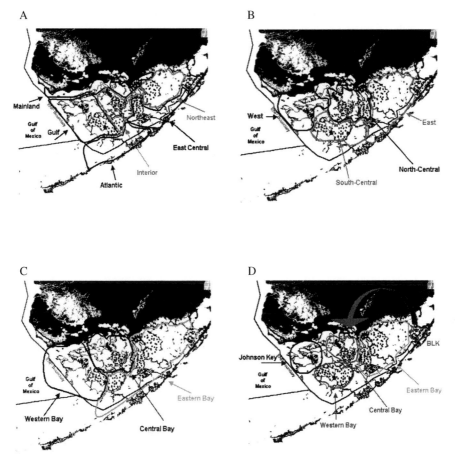

FIGURE 12.7 Ecological zones in Florida Bay as defined by (A) Zieman et al. (1989), (B) Phlips et al. (1995), (C) Boyer et al. (1999), and (D) this study.

distribution, and abundance over larger spatial scales is necessary to indicate the ecological status of *T. testudinum* within this ecosystem.

Conclusions

Small-scale sampling or large-scale pooling of morphometric and biomass characters of *T. testudinum* in Florida Bay would mask the level and spatial patterns of variability of these characters among basins. In this study, variability among basins of shoot-specific and area-specific characters of *T. testudinum* was much greater than the interannual variation. Phenotypic variability within this single seagrass species illustrated the environmental heterogeneity of the bay. Local sources of variation could be missed if samples were taken at fewer sites across the bay, or pooled. For effective management and restoration of Florida Bay, there is a need to better define regional differences in nutrient inputs and cycling, freshwater and tidal mixing effects, sediment resuspension, and algal blooms and how these various factors affect the regional trends in the morphology of *T. testudinum*. Because of the fundamental roles that seagrasses play in support of fish and wildlife resources and in maintenance of water quality, it is essential to document trends and ecological condition in seagrass populations. Morphometric characters of seagrasses are relatively easy to collect and can provide evidence of changing environments within coastal habitats. Spatially extensive, long-term monitoring of sensitive parameters may reveal fundamental changes in the ecological condition of seagrass communities to future management changes.

Acknowledgments

Support for this research was provided by USGS Biological Resources Division (No. 98HQAG2186), Everglades National Park, and the National Undersea Research Center. We thank the staff of the Center for Marine Science, Everglades National Park, and the NOAA/National Undersea Research Center for logistical support. We also thank Jill Paxson, Ron Dean, Leanne Rutten, Jen Kunzelman, and Paula Whitfield from UNCW/CMS and Donna Burns, Dr. Penny Hall, and Manuel Merello from the Florida Marine Research Institute for their help and support in the field and in the laboratory.

References

Alcoverro, T., C. M. Duarte, and J. Romero. 1995. Annual growth dynamics of *Posidonia oceanica*: contribution of large-scale versus local factors to seasonality. *Marine Ecology Progress Series* 120:230–210.

Boyer, J. N., J. W. Fourqurean, and R. D. Jones.1997. Spatial characterization of water quality in Florida Bay and Whitewater Bay by multivariate analyses: zones of similar influence. *Estuaries* 20:743–758.

Boyer, J. N., J. W. Fourqurean, and R. D. Jones. 1999. Seasonal and long-term trends in the water quality of Florida Bay (1989–1997). *Estuaries* 22:417–430.

Butler, M. J. et al. 1995. Cascading disturbances in Florida Bay, USA: cyanobacterial blooms, sponge mortality, and implications for juvenile spiny lobsters *Panulinus argus*. *Marine Ecology Progress Series* 129:119–125.

Chambers, R. M., J. W. Fourqurean, S. A. Macko, and R. Hoppenot. 2001. Biogeochemical effects of iron availability on primary producers in a shallow marine carbonate environment. *Limnology and Oceanography* 46:1278–1286.

Davis, B. C. and J. W. Fourqurean. 2001. Competition between the tropical alga, *Halimeda incrassata*, and the seagrass, *Thalassia testudinum*. *Aquatic Botany* 71:217–232.

Dawes, C. J., M. O. Hall, and R. K. Riechert. 1985. Seasonal biomass and energy content in seagrass communities on the west coast of Florida. *Journal of Coastal Research.* 1:255–262.

den Hartog, C. 1970. *The Seagrasses of the World.* North-Holland, Amsterdam.

den Hartog, C. 1977. Structure, function and classification in seagrass communities. In *Seagrass Ecosystems: A Scientific Perspective,* C. P. McRoy and C. Helferrich (eds.). Marcel Dekker, New York, pp. 89–121.

Duarte, C. M. 1989. Temporal biomass variability and production/biomass relationships of seagrass communities. *Marine Ecology Progress Series* 51:269–276.

Duarte, C. M. 2002. The future of seagrass meadows. *Environmental Conservation* 29:192–206.

Duarte, C. M. et al. 1994. Reconstruction of seagrass dynamics: age determinations and associated tools for the seagrass ecologist. *Marine Ecology Progress Series* 107:195–209.

Durako, M. J. 1995. Indicators of seagrass ecological condition: an assessment based on spatial and temporal changes associated with the mass mortality of the tropical seagrass *Thalassia testudinum*. In *Changes in Fluxes in Estuaries: Implications for Science to Management,* K. R. Dyer and R. J. Orth (eds.). Olsen & Olsen, Fredensborg, Denmark, pp. 261–266.

Durako, M. J., M. O. Hall, and M. Merello. 2002. Patterns of change in the seagrass dominated Florida Bay hydroscape. In *The Everglades, Florida Bay, and Coral Reefs of the Florida Keys: An Ecosystem Sourcebook,* J. W. Porter and K. G. Porter (eds.). CRC Press, Boca Raton, FL, pp. 523–537.

Fourqurean, J. W. and M. B. Robblee. 1999. Florida Bay: A history of recent ecological changes. *Estuaries* 22:345–357.

Fourqurean, J. W. and J. C. Zieman. 1991. Photosynthesis, respiration and whole plant carbon budget of the seagrass *Thalassia testudinum*. *Marine Ecology Progress Series* 69:161–170.

Fourqurean, J. W., J. C. Zieman, and G. V. N. Powell. 1992a. Relationships between porewater nutrients and seagrasses in a subtropical carbonate environment. *Marine Biology* 114:57–65.

Fourqurean, J. W., J. C. Zieman, and G. V. N. Powell. 1992b. Phosphorus limitation of primary production in Florida Bay: evidence from C:N:P ratios of the dominant seagrass *Thalassia testudinum*. *Limnology and Oceanography* 37:162–171.

Fourqurean, J. W., M. J. Durako, M. O. Hall, and L. N. Hefty. 2002. Seagrass distribution in South Florida: a multi-agency coordinated monitoring program. In *The Everglades, Florida Bay, and Coral Reefs of the Florida Keys: An Ecosystem Sourcebook*, J. W. Porter and K. G. Porter (eds.). CRC Press, Boca Raton, FL, pp. 497–522.

Hall, M. O., M. J. Durako, J. W. Fourqurean, and J. C. Zieman. 1999. Decadal changes in seagrass distribution and abundance in Florida Bay. *Estuaries* 22:445–459.

Hutchings, M. J. 1988. Differential foraging for resources and structural plasticity in plants. *Trends in Ecology and Evolution* 3:200–204.

Irlandi, E. et al. 2002. The influence of freshwater runoff on biomass, morphometrics, and production of *Thalassia testudinum*. *Aquatic Botany* 72:67–78.

Iverson, R. L. and H. F. Bittaker. 1986. Seagrass distribution and abundance in eastern Gulf of Mexico coastal waters. *Estuarine and Coastal Shelf Science* 22:577–602.

Kenworthy, W. J., G. W. Thayer, and M. S. Fonseca. 1988. The utilization of seagrass meadows by fishery organisms. In *The Ecology and Management of Wetlands*, Vol. 1, D. D. Hook et al. (eds.). Timber Press, Portland, OR, pp. 548–560.

Larkum, A. W. D. and R. J. West. 1990. Long-term changes of seagrass meadows in Botany Bay, Australia. *Aquatic Botany* 37:55–70.

Lee, K. S. and K. H. Dunton. 1997. Effects of *in situ* light reduction on the maintenance, growth and partitioning of carbon resources in *Thalassia testudinum* Banks ex König. *Journal of Experimental Marine Biology and Ecology* 210:53–73.

Lee, K. S. and K. H. Dunton. 2000. Effects of nitrogen enrichment on biomass allocation, growth, and leaf morphology of the seagrass *Thalassia testudinum*. *Marine Ecology Progress Series* 196:39–48.

Matheson, R. E., Jr., S. M. Sogard, and K. A. Bjorgo. 1999. Changes in seagrass-associated fish and crustacean communities on Florida Bay mud banks: the effects of recent ecosystem changes? *Estuaries* 22:534–551.

McMillan, C. 1978. Morphogeographic variation under controlled conditions in five seagrasses, *Thalassia testudinum, Halodule wrightii, Syringodium filiforme, Halophila englemanii*, and *Zostera marina*. *Aquatic Botany* 4:169–189.

McMillan, C. and R. C. Phillips. 1979. Differentiation in habitat response among populations of new world seagrasses. *Aquatic Botany* 7:185–196.

Orth, R. J. and K. A. Moore. 1983. Chesapeake Bay: an unprecedented decline in submerged aquatic vegetation. *Science* 222:51–53.

Peréz, M., C. M. Duarte, J. Romero, K. Sand-Jensen, and T. Alcoverro. 1995. Growth plasticity in *Cymodocea nodosa* stands: the importance of nutrient supply. *Aquatic Botany* 47:249–264.

Phillips, R. C. and R. L. Lewis. 1983. Influence of environmental gradients on variations in leaf widths and transplant success in North American seagrasses. *Marine Technology Society Journal* 17:59–68.

Phlips, E. J. and S. Badylak. 1996. Spatial variability in phytoplankton standing crop and composition in a shallow inner-shelf lagoon, Florida Bay, Florida. *Bulletin of Marine Science* 58:203–216.

Phlips, E. J., T. C. Lynch, and S. Badylak. 1995. Chlorophyll *a*, tripton, color, and light availability in a shallow tropical inner-shelf lagoon, Florida Bay, USA. *Marine Ecology Progress Series* 127:223–234.

Robblee, M. B. et al. 1991. Mass mortality of the tropical seagrass *Thalassia testudinum* in Florida Bay (USA). *Marine Ecology Progress Series* 71:297–289.

Rose, C. D. and C. J. Dawes. 1999. Effects of community structure on the seagrass *Thalassia testudinum*. *Marine Ecology Progress Series* 184:83–95.

Schomer, N. S. and R. D. Drew. 1982. An Ecological Characterization of the Lower Everglades, Florida Bay, and the Florida Keys. U.S. FWS/OBS-82/58.1. U.S. Fish and Wildlife Service, Washington, D.C., 246 pp.

Short, F. T. 1983. The seagrass, *Zostera marina* L.: plant morphology and bed structure in relation to sediment ammonium in Izembek Lagoon, Alaska. *Aquatic Botany* 16:149–161.

Short, F. T. and S. Wyllie-Echeverria. 1996. Natural and human-induced disturbance of seagrasses. *Environmental Conservation* 23:17–27.

Stumpf, R. P., M. L. Frayer, M. J. Durako, and J. C. Brock. 1999. Variations in water clarity and bottom albedo in Florida Bay from 1985 to 1997. *Estuaries* 22:431–444.

Thayer, G. W. and A. J. Chester. 1989. Distribution and abundance of fishes among basin and channel habitats in Florida Bay. *Bulletin of Marine Science* 44:200–219.

Thayer, G. W., W. J. Kenworthy, and M. S. Fonseca. 1984. The Ecology of Eelgrass Meadows of the Atlantic Coast: a Community Profile. FWS/OBS-84/02, U.S. Fish and Wildlife Service, Washington, D.C., 147 pp.

van Tussenbroek, B. I. 1995. *Thalassia testudinum* leaf dynamics in a Mexican Caribbean coral reef lagoon. *Marine Biology* 122:33–40.

van Tussenbroek, B. I. 1996. Leaf dimensions of transplants of *Thalassia testudinum* in a Mexican Caribbean reef lagoon. *Aquatic Botany* 55:133–138.

van Tussenbroek, B. I. 1998. Above- and below-ground biomass and production by *Thalassia testudinum* in a tropical reef lagoon. *Aquatic Botany* 61:69–82.

Wanless, H. R. and M. G. Taggett. 1989. Origin, growth and evolution of carbonate mudbanks in Florida Bay. *Bulletin of Marine Science* 44:454–489.

Zieman, J. C. 1982. The Ecology of the Seagrasses of South Florida: A Community Profile. FWS/OBS-82/25, U.S. Fish and Wildlife Service, Washington, D.C., 158 pp.

Zieman, J. C., J. W. Fourqurean, and R. L. Iverson. 1989. Distribution, abundance and productivity of seagrasses and macroalgae in Florida Bay. *Bulletin of Marine Science* 44:292–311.

Zieman, J. C., J. W. Fourqurean, and T. A. Frankovich. 1999. Seagrass die-off in Florida Bay: long-term trends in abundance and growth of turtle grass, *Thalassia testudinum*. *Estuaries* 22:460–470.

13

Evaluating Indicators of Seagrass Stress to Light

Patrick D. Biber, Hans W. Paerl, Charles L. Gallegos, and W. Judson Kenworthy

CONTENTS

Introduction

Estuaries and coastal waters are highly productive, ecologically and societally valuable ecosystems. They are under increasing stress from both anthropogenic factors, such as nutrient enrichment and sedimentation, and altered frequencies and intensities of natural disturbances arising across many scales, from inputs to a watershed to those wrought by global climate change. Seagrasses are often dominant primary producers that can play a central role in the stability, nursery function, biogeochemical cycling, and trophodynamics of coastal ecosystems and, as such, are important for sustaining a broad spectrum of organisms (Thayer et al., 1984; Hemminga and Duarte, 2000). For example, they stabilize sediments, which are easily resuspended if the plants are lost, resulting in increased and prolonged turbidity that reduces available light reaching the seafloor. For these reasons, seagrasses are widely recognized as "barometers" of estuarine water quality, being perhaps the most parsimonious integrator of estuarine water quality throughout the range of their current and historic distribution (Dennison et al., 1993). Thriving seagrass communities signal a productive, diverse, and biogeochemically and trophically well-coupled coastal ecosystem (Harlin and Thorne-Miller, 1981; Thayer et al., 1984; Fonseca et al., 1998; Hauxwell et al., 2001). Accordingly, the presence or absence of seagrass is a useful measure of estuarine condition, but reliance on presence/absence as an indicator implicitly requires significant degradation of estuarine water quality (Zimmerman et al., 1991; Short and Wyllie-Echeverria, 1996). By focusing on seagrass decline, we are restricted to detecting conditions when water quality is already so degraded that there is virtually no time for corrective actions. Therefore, early detection of sublethal stress thresholds in seagrass plants is crucial for effective conservation of this resource.

The role of seagrasses as indicators of estuarine condition, particularly decreased water clarity, was proposed in the early 1990s (Kenworthy and Haunert, 1991; Neckles, 1994). Dennison et al. (1993) concluded that seagrasses were potentially sensitive indicators of declining water quality because of their high light requirements (15 to 25% surface irradiance) compared to those of other aquatic primary producers, such as macroalgae and benthic microalgae, with much lower light requirements (Markager and Sand-Jensen, 1992, 1996; Agusti et al., 1994). To develop predictive indicators of estuarine suitability for seagrasses, both water-quality-driven stressors and the whole-plant integrated responses should be assessed. Potential predictive indicators need to respond clearly and reliably to abiotic factors that cause suboptimal seagrass growth (e.g., light limitation) and should come from a suite of approaches over a range of hierarchical levels.

Bio-Optical Modeling

Provided that all other environmental parameters are suitable, the light environment during the growing season is the most important abiotic factor determining survival of seagrasses (Moore et al., 1997; Batiuk et al., 2000; Dixon, 2000a). Light attenuation by the water column is a major variable related to seagrass decline (Dennison and Alberte, 1982, 1985; Bulthuis, 1983). Another is epiphytic light attenuation, which Dixon (2000b) found to be more important in the subtropical waters of Florida. Low light levels, below some minimum physiological requirement (typically 15 to 25% of incident surface light = I_o) may result in a loss of seagrasses. Light is attenuated down the water column resulting in less light available at the bottom (I_z) than at the surface (I_o) by factors including:

1. Turbidity, expressed as total suspended particulate matter (SPM)
2. Phytoplankton, which both absorb and scatter light, expressed in chlorophyll concentration (chl *a*)
3. Colored dissolved organic matter (CDOM) leaching from decaying vegetation
4. Macroalgae and epiphytic microalgae that grow on the seagrass, which are usually more problematic when eutrophication is taking place (Harlin and Thorne-Miller, 1981; Hauxwell et al., 2001)

Light attenuation (Figure 13.1A) can be expressed as PLW (percent light through water) or the combined effects of contributions due to turbidity, chlorophyll, and color, resulting in significantly reduced light with increasing depth. Additional light attenuation can occur at the leaf surface due to epiphyte fouling, PLL (percent light at leaf), which occurs primarily under heavily eutrophic conditions (Batiuk et al., 2000). However, this may not always be the case; for example, Dixon (2000b) found light attenuation due to epiphytes averaging 34% of all attenuation of light in lower Tampa Bay where chl *a* has an annual average that is less than 5 mg m^{-3} (not heavily eutrophic). One of the goals of our research has been to refine a bio-optical water quality model (Gallegos, 1994, 2001) by determining the importance of SPM, chl *a*, and CDOM on light attenuation to seagrasses in North Carolina during different seasons.

Graphically, the results of the bio-optical model are shown as in Figure 13.1B, where two components of light attenuation (chl *a*, TSS) are presented on the *x* and *y* axes. The third constituent, CDOM, can be plotted on the *z* axis on a three-dimensional (3D) plot to show all three components of attenuation. Median concentrations for one water quality sample are plotted on this graph and compared to a minimum light water quality requirement for a given depth (for simplicity, depicted as a line of constant attenuation), which is calculated using a radiative-transfer model and knowledge of seagrass species light requirements. Target minimum water clarity requirements for seagrass survival are found at the intersection of vectors perpendicular to the axes or the origin from the median sample concentration. The target concentrations in this figure suggest that both TSS and chl *a* need to be reduced to meet the minimum light requirements of this seagrass species.

Determining the minimum light requirements for a given species requires either monitoring and identifying the deep edge of existing seagrass beds within an estuary, or an experimental approach where the minimum light requirements for survival are determined over a period of time. An understanding of

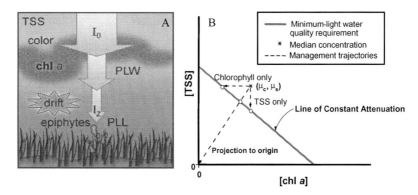

FIGURE 13.1 (A) Conceptual diagram of light attenuation down the water column (PLW), and the relative contributions of turbidity, chlorophyll *a*, and color to attenuation. Additional attenuation can occur at the leaf surface due to epiphyte fouling (PLL), which occurs primarily under heavily eutrophic conditions. (B) Graphical representation of the bio-optical model. Components of light attenuation in the water column are presented along the axes. Median concentrations for one sample are plotted on this graph and compared to a minimum-light water quality requirement for a given depth (line of constant attenuation). Target minimum water clarity requirements for seagrass survival are found at the intersection of vectors perpendicular to the axes or the origin from the median sample concentration. The target concentrations in this figure suggest that both TSS and chl *a* need to be reduced to meet the minimum light requirements of this seagrass species.

the physiological processes occurring within plants subject to chronic light limitation is, therefore, important when applying the bio-optical model.

Plant Physiology

For a seagrass-dominated ecosystem to be judged sustainable, the seagrasses must be vital and capable of long-term persistence by growth and reproduction. We are focusing on a common set of photosynthetic and physiological metrics that measure seagrass condition and may provide *early warning* of plant stress or demise, before extensive loss of seagrass occurs.

A promising approach needing further investigation is chlorophyll fluorescence of photosystem II, also known as P680, where a molecule of water is split to form O_2 and a reductant in the light reaction of photosynthesis (Stryer, 1981). To measure this, we utilized a photosynthetic efficiency analyzer (PEA) that was developed to measure the health of crops (e.g., Critchley and Smilie, 1981; Havaux and Lannoye, 1983; Bowyer et al., 1991; Filiault and Stier, 1999), and because all plants utilize the same fundamental processes in photosynthesis, this instrument can also be used on marine macrophytes.

The PEA consists of a computer and a photo-emitter/sensor unit. Clips are supplied with the apparatus, which are attached to a plant leaf and serve to occlude all but a small area of the leaf needed for photosynthetic testing. A small area of the plant leaf is dark-adapted with the clip and then a shutter built into the clip is opened, exposing the leaf area under the clip to high-intensity light provided by nearly monochromatic light emitting diodes (LEDs) located in the photo-emitter/sensor unit. The chlorophyll in the dark-adapted area of the leaf fluoresces and the PEA measures this. The initial fluorescence (F_o) and the maximal fluorescence (F_m) are recorded and the difference between the maximal and initial fluorescence levels ($F_m - F_o$) is called the variable fluorescence (F_v). The PEA computer calculates the ratio F_v/F_m, or photosynthetic efficiency; the greater the fluorescence per unit incoming light, the higher the efficiency of the photosystem, which equates with a plant under low physiological stress. Conversely, low F_v/F_m indicates a plant under stress (Schulze and Caldwell, 1990; Krause and Weis, 1991; Rohacek and Bartak, 1999).

This approach has been used to measure acute stress in seagrasses, such as desiccation (Adams and Bates, 1994; Bjork et al., 1999), temperature, or salinity shifts (Ralph et al., 1998; Ralph, 1999), and even changes in ambient light over short time durations (Beer and Bjork, 2000; Major and Dunton, 2002). However, to our knowledge this technique has not been evaluated in seagrass plants subject to chronic stress, such as light limitation arising from reduced water clarity.

There are two components to our research that are discussed here:

1. Monitoring water quality to calibrate a bio-optical model and using this modeling approach to understand water quality changes that result in low-light stress
2. Determining seagrass photophysiological indicators as a measure of individual plant responses to light stress

Methods

Calibrating a Bio-Optical Model

To refine the bio-optical model as an indicator for seagrass habitat suitability in the mid-Atlantic region of the eastern coastal United States, we have initiated a monthly water-sampling program in North River, North Carolina, an estuary that contains sizable seagrass beds along a gradient in water clarity. We collect water samples at nine stations in North River, North Carolina (Figure 13.2). Additionally, at each station we profile water quality with YSI® 6600 multiparameter probes (temperature, salinity, dissolved oxygen, or DO, pH, turbidity, and chlorophyll fluorescence), as well as collecting light attenuation data using a LICOR® 4π sensor. The light data are used to calculate attenuation coefficients to compare with Secchi disk readings and laboratory-measured optical properties of the water samples. Water samples are analyzed for total absorption (referenced to pure water) and scattering coefficients using the methods described in Gallegos (1994, 2001).

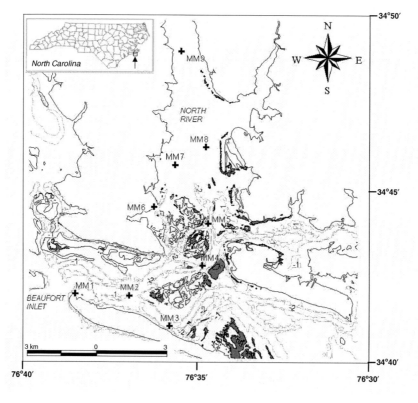

FIGURE 13.2 Map of nine sampling locations in North River, a small estuary in the southern Pamlico Sound region of North Carolina. Seagrass beds are indicated by shaded polygons; 1 and 2 m depths are indicated by the dashed contour lines.

Light-Stress Experiments

A light gradient (Table 13.1) was created in a large indoor tank under controlled environmental (light, air temperature) conditions to determine minimum integrated light requirements for seagrass survival. A bank of overhead Halogen® lights of increasing wattage provided the different light treatments. Lights were placed on a timer to create a 12:12 light:dark (L:D) environment. Water was provided as a flow-through system and temperature was monitored at 30 min intervals at both ends of the light gradient. Water quality data were collected with a YSI 6000 multiparameter water quality probe. Two seagrass species were tested in 2003: *Zostera marina* (eelgrass) seedlings were grown for 20 weeks (27 January to 6 June 2003), and *Halodule wrightii* (shoalgrass) plants were grown for 16 weeks (8 July to 27 October 2003) in a range of light intensities (Table 13.1).

Two tubs (Rubbermaid® 2951) were planted with nine *Z. marina* seedlings or nine *H. wrightii* plants (3 × 3 arrangement) and placed into each of the seven light gradient treatments after allowing 10 days recovery from transplant stress. Initial morphological measurements were taken on the day the plants were placed into the light gradient tank and weekly thereafter until the experiment was terminated. Biweekly, mature leaves from a random subset of plants were collected to measure width and destructive analysis of tissue constituents, including chlorophyll extraction and tissue nutrient content (C:N ratio) using standard analytical methods.

For the weekly chlorophyll fluorescence measurements, the second leaf on the terminal apical shoot was clipped about half way along the adaxial side as per the recommendations of Durako and Kunzelman (2002). Dark adaptation of the clipped portion was 10 min, and the PEA recorded chlorophyll fluorescence from the leaf for 1 s. Typically, the maximum fluorescence in green plants occurs within the first second after exposure to the high-intensity LED lights (Srivastava et al., 1999).

TABLE 13.1

Seven Irradiance Treatments Used for Indoor Light Experiment[a]

Instantaneous Flux (μmol m^{-2} s^{-1})	Integrated Daily Flux (mol m^{-2} d^{-1})	Hsat (h)	Hcomp (h)
Zostera marina			
277.03	11.97	5.49	11.54
161.49	6.98	0.50	6.54
69.80	3.02	−3.46	2.58
33.29	1.44	−5.04	1.01
10.93	0.47	−6.01	0.04
4.21	0.18	−6.30	−0.25
0.00	0.00	−6.48	−0.43
Halodule wrightii			
248.20	10.72	4.24	10.29
95.91	4.14	−2.34	3.71
59.96	2.59	−3.89	2.16
31.11	1.34	−5.14	0.91
6.90	0.60	−5.88	0.16
2.96	0.26	−6.22	−0.18
0.00	0.00	−6.48	−0.43

[a] Expressed as instantaneous irradiance, integrated daily irradiance for a 12-h day, number of hours (Hsat) that irradiance exceeded a saturation intensity (150 μmol m^{-2} s^{-1}), and number of hours (Hcomp) that irradiance exceeded a compensation intensity (10 μmol m^{-2} s^{-1}) where photosynthesis balances respiration. Negative numbers indicate insufficient light to meet physiological requirements.

Results

Water Quality Monitoring; North River, North Carolina

Temperature followed an approximate sinusoidal pattern during the year with minimum temperatures in February and maximum temperatures in July and August (Figure 13.3). Shallower upriver stations had more rapid temperature changes than the tidally influenced downstream stations, which reflected oceanic water temperatures. Salinity was higher and more stable in the downstream, ocean-influenced section of North River than the upstream portion, which was more influenced by terrestrial runoff from surrounding salt-marsh drainage streams. Salinity dropped after heavy rainfall events, which was especially evident during the wet spring and summer of 2003 (Figure 13.3); this year was the wettest on record with 2337 mm (92 in.) recorded, 50% greater than average (National Weather Service data).

Chlorophyll concentrations (mg m^{-3}) were lowest during the cold winter months and increased during the spring and summer when phytoplankton were more abundant (Figure 13.3). Highest values were observed at the farthest upstream station corresponding with increased nitrogen concentrations measured at this station. High chlorophyll concentrations were also recorded during storm events and co-occurred with elevated turbidity, likely due to resuspension of benthic microalgae (Figure 13.3). Turbidity was similar at all stations during a given sampling date. In general, there was a trend of increasing turbidity from the downstream to the upstream stations (Figure 13.3). Turbidity was lower in winter than other seasons. The highest turbidity values were recorded at shallow upstream sites during storms (e.g., September 2002).

The effect of chlorophyll and turbidity on measured light penetration to the bottom at each of the nine stations was expressed as an attenuation coefficient (K_d), derived from the negative exponential

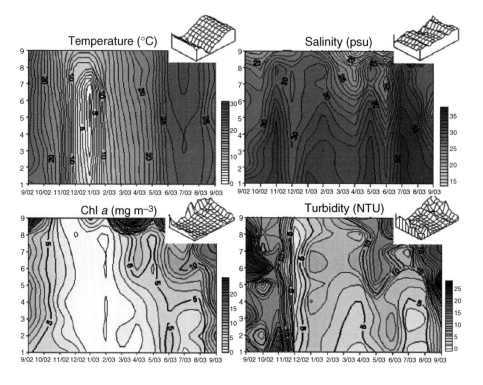

FIGURE 13.3 Selected results from our monthly water quality monitoring program in North River. Time–space contour plots of temperature, salinity, chl *a* concentration, and turbidity (NTU) measured monthly from September 2002 to September 2003. Sites are numbered from downstream (1) to most upstream (9). Insets are same data plotted as 3D wireframe plots.

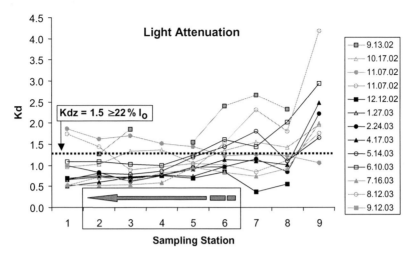

FIGURE 13.4 The effects of chlorophyll and turbidity on measured light attenuation with depth (K_d) at each of the nine stations. Filled symbols are samples collected during floodtide, open symbols are ebb tide samples. For seagrass survival K_d should not exceed 1.5, i.e., 22% of surface light reaches the bottom (dotted bar). Seagrass beds are found in North River between stations 2 and 6. The gray arrow indicates relative abundance of seagrasses.

Beer-Lambert function. Higher values of K_d indicate more light attenuation (i.e., more turbid water). K_d values increased with distance from the Beaufort Inlet at station 1 to the highest values found at station 9, where chl *a* and turbidity were also typically highest. During 2 months (September 2002 and March 2003), samples were collected during storm conditions (tropical storm and northeaster, respectively) with some of the highest attenuation values seen, indicating the importance of natural events in driving extreme values in this estuary. For seagrass survival to 1 m, K_d should not exceed 1.5 m^{-1}, i.e., 22% of surface light (I_o) reaching the bottom, as indicated by the dotted line (Figure 13.4). This condition was typically met at the stations near where seagrass are found in North River (stations 2 through 6). K_d values sufficient for seagrass survival were less frequently seen in the upstream stations.

Using our initial calibration values from the bio-optical model, monthly water quality conditions were plotted for North River sampling stations (Figure 13.5). Most samples fall in the region of acceptable light quantities (22% I_o) reaching 1 m depth, except for a few samples that were collected during storm conditions (September 2002 and March 2003). This indicates that water clarity in the North River is sufficient to support seagrasses to 1 m depth, and this agrees with observations made on the multidecadal persistence of seagrass beds in this estuary (since the early 1970s). Interestingly, the line of constant attenuation for the North River (solid line) allows much higher turbidity concentrations than the same light-at-depth requirement for Chesapeake Bay seagrasses (dashed line). The reason for this (based on initial results) is that per unit mass suspended sediments in the North River absorb and scatter much less light than those in the mesohaline Chesapeake Bay, due to differences in sediment properties related primarily to particle size (and perhaps also phytoplankton composition) between these regions.

Additionally, the North Carolina State estuarine water quality criteria of 40 mg m^{-3} chl *a* and 25 NTU turbidity (NC-DENR, 2003) are shown on the plot (Figure 13.5, black box). Almost all water quality samples from North River fall within the acceptable state criteria, and within the submerged aquatic vegetation (SAV) survival criteria based on the bio-optical model calibrated to North River conditions. There are combinations of turbidity and chlorophyll concentration permitted by state criteria that would *not* support seagrasses according to the bio-optical model (Figure 13.5, shaded triangle); interestingly, no water quality observations fell in this region.

The results from the bio-optical model can be corroborated by results from the permanent monitoring stations. Three locations in North River are designated as permanent stations, with continuous unattended water quality and light monitoring using YSI and LICOR sensors and data loggers. These locations are near stations 2, 6, and 9 in North River (Figure 13.2). Station 6 has been continuously monitored since September 2002. Integrated daily light (mol m^{-2} d^{-1}) received at two sensor depths, 1 and 1.5 m, was

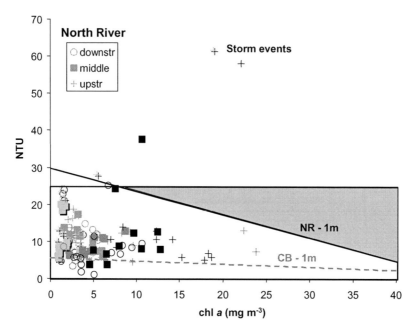

FIGURE 13.5 Median monthly water quality conditions plotted for North River, based on the bio-optical model. Symbol tints indicate season (gray shadow = winter, light gray = spring and fall, black = summer), symbol shape indicates location within North River. Additionally, the North Carolina State estuarine water-quality criteria of 40 mg m^{-3} chl *a* and 25 NTU are shown (rectangle outlined with thick lines). The bio-optical model was used to predict water quality criteria for survival of seagrass to 1 m depth in Chesapeake Bay (dashed line) and North River (solid line). The different slopes reflect regional differences in sediments and phytoplankton.

averaged for each month and the number of hours that irradiance exceeded the photosynthesis saturation intensity (150 µmol m^{-2} s^{-1}) for *Z. marina* (Dennison and Alberte, 1982) calculated (Table 13.2). Light available for photosynthesis was high in December to February (clear water) and June to July (strong solar irradiance) even though summertime water clarity was much reduced compared to the winter condition; i.e., stronger solar irradiance may compensate for higher chl *a* concentrations in the summer.

Light Stress Experiment

Comparing and contrasting the results of the light gradient experiments using *Z. marina* seedlings and *H. wrightii* plants showed similar responses by both species to light limitation stress. Both instantaneous and integrated (12 h) irradiance fluxes during the two experiments are given in Table 13.1, as well as integrated irradiance greater than the saturation irradiance (~150 µmol) and compensation irradiance (~10 µmol) for *Z. marina* (Dennison and Alberte, 1982) and *H. wrightii*.

For both species, all but the two greatest light treatments resulted in a deficit of light to saturate photosynthesis, ultimately resulting in mortality of many of the plants. This result indicates that irradiances consistently lower than the saturation threshold will ultimately result in plant death. Observations made during weekly measurements suggested that mortality in response to light limitation occurred from the base of the plant upward, with the meristem becoming nonviable about 1 to 2 weeks before the leaves were apparently dead (*Z. marina*) or broke off at the sheath (*H. wrightii*). This observed lag-time of 1 to 2 weeks may be critical in monitoring seagrass mortality under natural situations where light limitation is the cause of mortality.

During the *Z. marina* experiment, water temperatures increased from 10 to 28°C. The upper favorable temperature for this species is around 25°C (Thayer et al., 1984), which was attained in May (Figure 13.6A). This high-temperature stress resulted in reduced growth, as seen by a reduction in leaf area for the remainder of the experiment. In contrast, temperatures started to decline during the *H. wrightii*

TABLE 13.2

Light Data from 1- and 1.5-m-Deep Sensors Collected at
Station 6 from 2000 to 2003[a]

| 2002–2003 | mol m^{-2} d^{-1} | | Hsat | |
Month	1 m	1.5 m	1 m	1.5 m
Sept	6.22	1.57	5.00	0.14
Oct	5.01	1.61	3.42	0.45
Nov	7.70	4.94	4.87	4.13
Dec	12.02	6.02	6.29	4.35
Jan	12.29	5.16	7.00	4.15
Feb	12.95	5.99	6.72	4.39
Mar	8.06	3.03	4.72	1.66
Apr	8.81	3.55	5.24	2.36
May	7.66	2.36	4.33	1.15
Jun	12.90	4.74	7.17	3.41
Jul	9.36	5.34	6.27	3.85
Aug	6.99	3.61	4.65	2.31
Sept	2.98	1.49	1.86	0.36
Mean	8.35	3.79	5.04	2.52

[a] Expressed as mean daily integrated irradiance (mol m^{-2} d^{-1}) each month, and the number of hours (Hsat) that irradiance exceeded a saturation intensity (150 µmol m^{-2} s^{-1}). The growth period of *Z. marina* is from December to June, and the growth period for *H. wrightii* is from May to November. Both integrated and Hsat values are about twice as great at 1 m vs. 1.5 m depths.

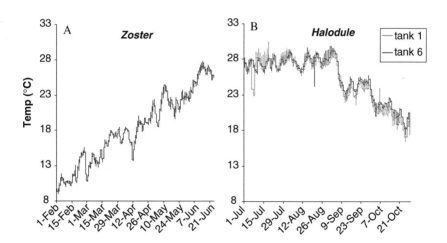

FIGURE 13.6 Water temperatures recorded in the indoor tank during the two light gradient experiments. Water temperatures were recorded at 30 min time intervals in the highest light (tank 1) and lowest light (tank 6) treatments.

experiment (Figure 13.6B). They remained at 28°C until early September; then temperatures declined to 18°C by the end of the experiment.

Plant growth and photosynthesis responses for *Z. marina* and *H. wrightii* in the light gradient experiment showed some similarities as well as differences with respect to the minimum integrated light requirements for survival. Both single shoot *Z. marina* seedlings and two shoot *H. wrightii* plants were grown for the duration of their growth season. For *Z. marina*, no branching or spatial expansion was observed in any of the seven irradiance treatments. Of the 117 seedlings, 5 flowered (4.3%) all at irradiances of 33 µmol or higher (Figure 13.7A). In *H. wrightii* plants, branching or spatial expansion was observed in the highest irradiance treatment only (Figure 13.7B). Shoot loss and mortality were observed at irradiances of 60 µmol or lower in this species. These integrated light thresholds indicate that tropical *H. wrightii* plants have somewhat higher minimum light requirements for survival than temperate *Z. marina* seedlings.

The length of the longest leaf on each surviving seedling was measured as a proxy for canopy height in both species. In *Z. marina*, leaf growth (linear extension) occurred only in the two highest light levels

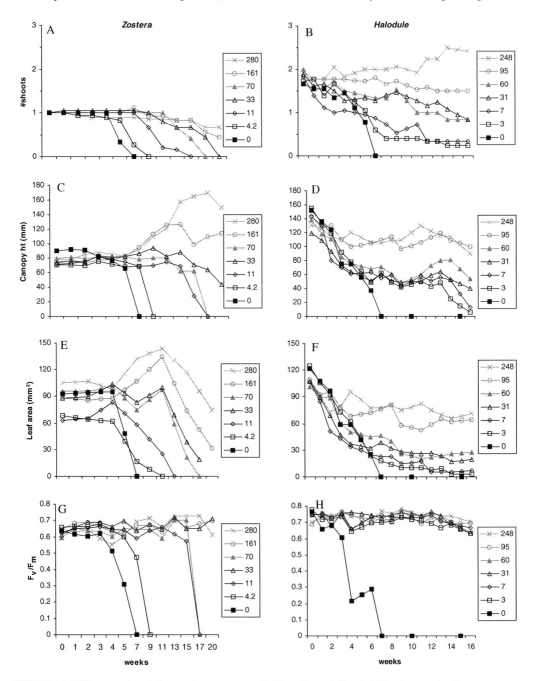

FIGURE 13.7 Plant growth and photosynthesis responses for *Z. marina* and *H. wrightii* in the seven irradiance treatments (units of µmol m^{-2} s^{-1}) in the light gradient experiment. Selected variables shown are mean number of shoots per plant ($n = 18$), length of longest leaf measured as a proxy for canopy height, leaf area (single surface), and photosynthesis yield ratio (F_v/F_m).

(irradiance > saturation). Early mortality (leaf length = 0) was observed in the two lowest light levels with evidence of plant stress by weeks 3 and 5, respectively, about half the time to mortality (Figure 13.7C). In *H. wrightii*, leaf length was maintained only at the two highest light levels (irradiance ≥ 95 µmol). Early mortality was observed in the lowest light level with evidence of plant stress by weeks 3 to 4, about half the time to mortality of all the plants in this treatment (Figure 13.7D).

Leaf area (single surface) was calculated for the strap-like leaves by multiplying maximum length by average width in both species. Width was measured on alternative sampling dates on a random subset of three to five plants with linear interpolation for nonsampled dates. Area was primarily influenced by leaf length, as width remained approximately constant within a given light treatment (Figure 13.7E and F).

Zostera marina seedling photosynthesis was measured with an OptiSciences® OS-30 PEA (Plant Efficiency Analyzer), and *H. wrightii* photosynthesis was measured with a Hansatech PEA that had increased sensitivity, an important consideration given the narrow leaves of this species. Photosynthetic yield, expressed as the ratio F_v/F_m, ranged from 0.6 to 0.75 in healthy *Z. marina* seedlings (Figure 13.7G) and from 0.65 to 0.78 in healthy *H. wrightii* plants (Figure 13.7H). In both species F_v/F_m fell below 0.5 in plants that were dead or dying. No significant difference in F_v/F_m was observed in those plants remaining alive across all seven light treatments, indicating acclimation of the photosynthetic apparatus to the ambient light field and temperature conditions. In summary, the yield ratio (F_v/F_m) is not a sensitive measure of chronic stress conditions, either light or temperature, to which tropical *H. wrightii* and temperate *Z. marina* can acclimate.

Discussion

Seagrass is an important estuarine habitat that is declining globally (Green and Short, 2003). Much of this decline can be attributed to decreased light availability to the plants because of reductions in water clarity due to degrading water quality (Dennison et al., 1993; Hauxwell et al., 2001). Seagrasses have relatively high light requirements compared to other marine primary producers and so are more susceptible to low light stress (Kenworthy and Haunert, 1991; Kenworthy and Fonseca, 1996; Dixon, 2000a). For this reason, they have been proposed as indicators of estuarine change (Dennison et al., 1993). However, it is desirable to have an indicator that provides an early warning of potential seagrass demise, rather than waiting until after the fact.

Water quality criteria, particularly those related to the optical water quality (i.e., water clarity) needed for the survival and growth of seagrasses, have been the subject of considerable research (Kenworthy and Haunert, 1991; Neckles, 1994; Gallegos and Kenworthy, 1996; Kenworthy and Fonseca, 1996; Longstaff et al., 1999; Zimmerman, 2003). A general conclusion of those workshops and research programs was that water column clarity needs to be greatly increased to provide light conditions suitable for the survival of most seagrasses (Kenworthy and Fonseca, 1996; Moore et al., 1996; Batiuk et al., 2000).

Water Quality Stress Indicators

We are developing a water quality stress indicator using a bio-optical model of water clarity that can identify harmful trends in water quality and make explicit what is causing the increased light attenuation. This will enable early preventative management actions to be taken to remediate water quality before it becomes critically limiting to seagrass survival, resulting in loss of this important habitat.

The bio-optical model is useful because it permits us to determine the relative contributions of the different water quality parameters to light attenuation at different positions in the estuary. This is a very important step in managing water quality as it makes explicit the relative contribution of suspended particles, phytoplankton, and color to light attenuation. From this knowledge, target reductions or target loading rates for one or more of these components of water quality can be set to achieve a desired water clarity. Comparing light attenuation calculated at the deep survival limit of seagrasses with water quality concentrations measured there allows the determination of ranges of water clarity that permit expansion or cause contraction of the seagrass bed. Basing the model on inherent optical properties has the advantage that extrapolation beyond the range of water quality concentrations encountered during model development is possible, because the absorption and scattering coefficients are linearly related to the relevant water quality concentrations (Figure 13.1B). Such an exercise can be used to determine the availability of light at the edge of the seagrass bed in response to hypothetical scenarios, such as accelerated eutrophication resulting from increased nutrient loading in the watershed. This bio-optical model has already been calibrated to conditions typical for Chesapeake Bay, Maryland (Gallegos, 2001) and Indian River Lagoon, Florida (Gallegos and Kenworthy, 1996), and recently North River, North Carolina (Biber and Gallegos, unpubl. data), all estuaries with significant seagrass habitats.

Based on the initial results of the bio-optical model for North River seagrass habitats, we can conclude that, even though water quality appears to be adequate to sustain stable seagrass beds in this estuary, the long-term survival of these mostly shallow seagrass communities in North River is still uncertain, especially when considering the likelihood of sea-level rise projected to occur within the next century. More stringent water clarity criteria with respect to seagrass habitats should be adopted by the state to ensure seagrasses receive adequate light. Similar results have been found by researchers in Chesapeake Bay, where large declines in submerged aquatic vegetation (SAV) have been ongoing and are almost impossible to reverse (Orth and Moore, 1983; Dennison et al., 1993; Batiuk et al., 2000; U.S. EPA, 2003).

Seagrass communities in North Carolina present a unique situation at the overlap of temperate and tropical biogeographic regions resulting in the coexistence of *Z. marina* (temperate, winter dominant) and *H. wrightii* (tropical, summer dominant), allowing research on the competitive interactions between these two widely distributed seagrasses (Green and Short, 2003). Results are likely to be conservative as plants may already be experiencing some stress related to their presence at extremes of distribution. Further differences exist in that *Z. marina* reproduces sexually with annual seedling recruitment being a significant contribution to the population, whereas *H. wrightii* has never been observed to reproduce sexually in North Carolina, dispersing instead by clonal growth and asexual fragmentation (Thayer et al., 1984; Ferguson et al., 1993). These different reproductive strategies affect the dispersal ability and population recovery strategies after severe stress events, with potentially important repercussions in the face of possible global climate change scenarios. Because of the extreme importance of annual seedling recruitment to eelgrass population dynamics, we suggest that future research be focused on understanding the stress tolerance of this life stage, so that relevant differences in survival ability can be incorporated into management plans.

We suggest that the higher irradiance penetrating through the water column in the winter months, when water is clearer, may be a critical window of opportunity for seedling growth for *Z. marina*. Similarly for *H. wrightii*, high light levels, due to strong solar irradiances, in early summer may be critical for this species to grow rapidly early in the season, even though summer water clarity is much reduced compared to the winter condition. Similar critical windows of increased irradiance, due to higher water clarity, have been identified in the polyhaline Chesapeake Bay, from 1 March to 31 May, the early part of growing season for *Z. marina* (U.S. EPA, 2003). Management implications of this are that better water quality needs to be present during these critical periods.

Plant Physiology Indicators

Most current approaches do not address the integrated light requirements of seagrass, focusing instead on "instantaneous" measures of irradiance flux and seagrass photosynthetic rates, e.g., photosynthesis–irradiance curves (Dennison and Alberte, 1982, 1985; Goodman et al., 1995; Zimmerman et al., 1995; Bintz and Nixon, 2001). This approach implicitly ignores cumulative stress effects, so that important questions regarding how the duration of exposure or the frequency of exposure to a given level of environmental degradation might influence survival of the seagrasses are overlooked. Only recently has the frequency and duration of stressful conditions started to be investigated for the survival of seagrass (Dunton and Tomasko, 1994; Onuf, 1996; Moore et al., 1997). It is this whole-plant integrated response that is biologically and ecologically relevant to seagrass management. This is also the scale at which we can link water quality to seagrass photophysiology.

Recent technological advances and research into plant photosynthesis have resulted in the commercial production of instruments that can measure chlorophyll fluorescence to detect sublethal stress thresholds *in situ*, even underwater (e.g., Diving PAM, Walz, Germany). This technique has opened a new vista for seagrass physiologists, allowing rapid, nondestructive measurements of plant photosynthetic performance (e.g., Ralph and Burchett, 1995; Beer et al., 1998; Ralph et al., 1998; Bjork et al., 1999; Beer and Bjork, 2000; Durako and Kunzelman, 2002).

In our light-limitation experiments, seagrass chlorophyll induction kinetics were measured using a PEA. Contrary to expectation, the photosynthetic efficiency (F_v/F_m) indicated that both *Z. marina* and *H. wrightii* adapt their photosystem to the ambient light and temperature conditions within less than a week, as indicated by a near constant F_v/F_m ratio. Because of this adaptation, declines in the F_v/F_m ratio in low light treatments failed to precede losses in shoot number (see Figure 13.7A and G). F_v/F_m ratio alone is, therefore, unlikely to be a suitable tool to measure sublethal chronic stress responses in seagrass because of its invariance in the face of photosystem change. However, other ratios and indices derived from chlorophyll fluorescence induction kinetics that can be collected by PEA show more promise.

Chlorophyll fluorescence induction kinetics are extremely rich in terms of quality and quantity of different information (Strasser et al., 1995) and can only be measured by a PEA with a 1-μs sampling rate or higher, but not a pulse amplitude modulated (PAM) fluorometer, which has only a 30-μs sampling rate, as the lower sampling frequency misses the very early induction dynamics (Rohacek and Bartak, 1999). Because the induction kinetics are determined by both the physiological state of a plant and the ambient and past physical and chemical environmental conditions, they can be used to detect and predict sublethal stress thresholds not apparent in F_v/F_m. Strasser et al. (1995, 1996) have developed a method (JIP-test analysis) to further analyze the data on chlorophyll fluorescence induction kinetics gathered by PEA into ratios and indices of photosynthetic performance of the photosystem.

Upon illumination chlorophyll exhibits a fast fluorescence rise from an initial fluorescence intensity, F_o, to a maximal intensity, F_P ($= F_m$). Between these two extremes, the fluorescence intensity usually shows two intermediate steps: F_J at about 2 ms, and F_I at about 30 ms (Strasser and Govindjee, 1992a,b; Strasser et al., 1995) followed by F_P at about 300 ms. The labeling of these steps follows an alphabetic order, from the initial (F_o) to the final (F_P) part of the transient. Based on the information inherent in the O-J-I-P fluorescence transient, a test has been developed (called "JIP-test" after the steps of the transient), which can be used as a tool for rapid screening of many samples providing additional information about the structure, conformation, and function of their photosynthetic apparatus (Strasser and Strasser, 1995; Strasser et al., 1996).

As seagrass leaves age, many of the JIP-test ratios and indices change (Figure 13.8). Standardized methods for collecting PEA data for JIP-test analysis from seagrasses have not yet been adopted, in part because this is a new approach, and in part because of lack of knowledge on the variability inherent in natural, healthy seagrass populations (Figure 13.9). To better understand seagrass photophysiology, PEA and JIP-test will need to be further investigated to determine the inherent variability of these ratios and indices in seagrass species under stress (e.g., light limitation). We will continue to test the responses of *Z. marina* and *H. wrightii* to chronic light stress using PEA and JIP-test analyses to determine whether this may provide potentially powerful early warning of imminent demise of seagrass plants.

The ability to combine both water quality stressors, i.e., the light environment linked explicitly to constituents of attenuation, with plant physiological responses to light stress forms the basis of an integrative indicator of water quality that will permit evaluation and determination of the suitability of water quality in an estuary for continued seagrass sustainability.

Conclusions

Once the optical properties of the bio-optical model are calibrated to regional conditions, the next step is to utilize the seagrass ecophysiological information to set light requirement thresholds for seagrass plant survival, growth, and reproduction. Each of these three thresholds will be successively greater than the previous one because of the increased cumulative light resource requirement by these plants to achieve the successive stages of development; i.e., resources (light) required for growth, a net increase

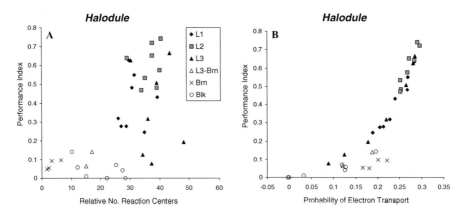

FIGURE 13.8 Results of PEA-derived JIP-test analyses on *H. wrightii* leaves of different ages, L1 = youngest, L3 = oldest, Brn and Blk denote dead leaves in increasing stages of decomposition. (A) Relationship between the relative number of active PSII reaction centers and the performance index. (B) Relationship between the probability that an electron will be transported beyond quinone A and the performance index.

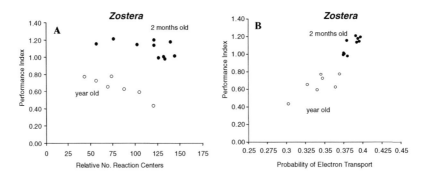

FIGURE 13.9 Results of JIP-test on *Z. marina* plants at 2 months and 1 year of age. (A) Relationship between the relative number of active PSII reaction centers per unit area and the performance index. (B) Relationship between the probability that an electron will be transported beyond quinone A and the performance index.

in biomass, will be greater than those required for survival (maintenance metabolism or no net increase in mass) and similarly for growth to reproductive maturity. The final stage of this process will be to use the bio-optical model as a tool to forecast the results of possible altered water quality scenarios and determine the impacts of these changes on existing seagrass habitats.

Acknowledgments

Grateful thanks to the cadre of persons who were instrumental in assisting with data collection especially in the field, including Christine Addison, Craig Bonn, John Burke, John Brewer, Priscilla Delano, Teri Denault, Don Field, Nikki Fogarty, Tom Gallo, Kamille Hammerstrom, Kar Howe, Loretta Leist, Amit Malhotra, Ryan McQuinn, Joe Purifoy, Karen Rossignol, Stopher Slade, Amy Uhrin, Rich Weaver, Paula Whitfield, and Pam Wyrick. Additional support was provided by the Center for Coastal Fisheries and Habitat Research, National Centers for Coastal Ocean Science (NCCOS), National Ocean Service (NOS), National Oceanic and Atmospheric Administration (NOAA), Beaufort, North Carolina. Thanks to Glen Thursby for providing the tools to do the JIP-test analyses. This research was funded by EPA-STAR Grants (R82867701) to the ACE-INC EaGLes Center (www.aceinc.org) and (R82868401) to the Atlantic Slope Consortium EaGLes Center (www.asc.psu.edu).

References

Adams, J. B. and G. C. Bates. 1994. The tolerance to desiccation of the submerged macrophytes *Ruppia cirrhosa* (Petagna) Grande and *Zostera capensis* Setchell. *Journal of Experimental Marine Biology and Ecology* 183:53–62.

Agusti, E. S., S. Enriquez, H. Frost-Christensen, K. Sand-Jensen, and C. M. Duarte. 1994. Light-harvesting among photosynthetic organisms. *Functional Ecology* 8:273–279.

Batiuk, R. A., P. Bergstrom, M. Kemp, E. Koch, L. Murray, J. C. Stevenson, R. Bartleson, V. Carter, N. B. Rybicki, J. M. Landwehr, C. Gallegos, L. Karrh, M. Naylor, D. Wilcox, K. A. Moore, S. Ailstock, and M. Teichberg. 2000. Chesapeake Bay Submerged Aquatic Vegetation Water Quality and Habitat-based Requirements and Restoration Targets: A Second Technical Synthesis. Chesapeake Bay Program, U.S. EPA, Washington, D.C., 231 pp.

Beer, S. and M. Bjork. 2000. Measuring rates of photosynthesis of two tropical seagrass by pulse amplitude modulated (PAM) fluorometry. *Aquatic Botany* 66:69–76.

Beer, S., B. Vilenkin, A. Weil, M. Veste, L. Susel, and A. Eshel. 1998. Measuring photosynthetic rates in seagrasses by pulse amplitude modulated (PAM) fluorometry. *Marine Ecology Progress Series* 174:293–300.

Bintz, J. C. and S. W. Nixon. 2001. Responses of eelgrass, *Zostera marina*, seedlings to reduced light. *Marine Ecology Progress Series* 223:133–141.

Bjork, M., J. Uku, A. Weil, and S. Beer. 1999. Photosynthetic tolerances of desiccation of tropical intertidal seagrasses. *Marine Ecology Progress Series* 191:121–126.

Bowyer, J. R., P. Camilleri, and W. F. J. Vermaas. 1991. Photosystem II and its interaction with herbicides. In *Herbicides. Topics in Photosynthesis*, Vol. 10, N. R. Baker and M. P. Percival (eds.). Elsevier, Amsterdam, pp. 27–85.

Bulthuis, D. A. 1983. Effects of in situ light reduction on density and growth of the seagrass *Heterozostera tamanica* in Western Port, Victoria, Australia. *Journal of Experimental Marine Biology and Ecology* 67:91–103.

Critchley, C. and R. M. Smilie. 1981. Leaf chlorophyll fluorescence as an indicator of photoinhibition in *Cucumis sativus* L. *Australian Journal of Plant Physiology* 8:133–141.

Dennison, W. C. and R. S. Alberte. 1982. Photosynthetic responses of *Zostera marina* to *in situ* manipulations of light intensity. *Oecologia* 55:137–144.

Dennison, W. C. and R. S. Alberte. 1985. Role of daily light period in the depth distribution of *Zostera marina*. *Marine Ecology Progress Series* 25:51–61.

Dennison, W. C., R. J. Orth, K. A. Moore, J. C. Stevenson, V. Carter, S. Kollar, P. W. Bergstrom, and R. A. Batiuk. 1993. Assessing water quality with submersed aquatic vegetation. *Bioscience* 43:86–94.

Dixon, L. K. 2000a. Establishing light requirements for the seagrass *Thalassia testudinum*: an example from Tampa Bay, Florida. In *Seagrasses: Monitoring, Ecology, Physiology and Management*, S. A. Bortone (ed.). CRC Press, Boca Raton, FL, pp. 9–32.

Dixon, L. K. 2000b. Light requirements of Tampa Bay Seagrasses: nutrient related issues still pending. In *Seagrass Management: It's Not Just Nutrients! Proceedings of a Symposium*, H. Greening (ed.). August 22–24, 2000. St. Petersburg, FL.

Dunton, K. H. and D. A. Tomasko. 1994. In situ photosynthetic performance in the seagrass *Halodule wrightii* in a hypersaline subtropical lagoon. *Marine Ecology Progress Series* 107:281–293.

Durako, M. J. and J. I. Kunzelman. 2002. Photosynthetic characteristics of *Thalassia testudinum* measured *in situ* by pulse-amplitude modulated (PAM) fluorometry: methodological and scale-based considerations. *Aquatic Botany* 73:173–185.

Ferguson, R. L., B. T. Pawlak and L. L. Wood. 1993. Flowering of the seagrass *Halodule wrightii* in North Carolina. *Aquatic Botany* 46:91–98.

Filiault, D. L. and J. C. Stier. 1999. The use of chlorophyll fluorescence in assessing the cold tolerance of three turfgrass species. *Wisconsin Turfgrass Research Report* 16:109–110.

Fonseca, M. S., W. J. Kenworthy, and G. W. Thayer. 1998. Guidelines for the conservation and restoration of seagrasses in the United States and adjacent waters. NOAA Coastal Ocean Program Decision Analysis Series, No. 12. NOAA Coastal Ocean Office, Silver Spring, MD, 222 pp.

Gallegos, C. L. 1994. Refining habitat requirements of submersed aquatic vegetation: role of optical models. *Estuaries* 17:187–199.

Gallegos, C. L. 2001. Calculating optical water quality targets to restore and protect submersed aquatic vegetation: overcoming problems in partitioning the diffuse attenuation coefficient for photosynthetically active radiation. *Estuaries* 24:381–397.

Gallegos, C. L. and J. W. Kenworthy. 1996. Seagrass depth limits in the Indian River Lagoon (Florida): application of an optical water quality model. *Estuarine and Coastal Shelf Science*, 42:267–288.

Goodman, J. L., K. A. Moore, and W. C. Dennison. 1995. Photosynthetic responses of eelgrass, *Zostera marina*, to light and sediment sulfide in a shallow barrier island lagoon. *Aquatic Botany* 50:37–47.

Green, E. P. and F. T. Short. 2003. *World Atlas of Seagrasses*. University of California Press, Berkeley, 320 pp.

Harlin, M. M. and B. Thorne-Miller. 1981. Nutrient enrichment of seagrass beds in a Rhode Island coastal lagoon. *Marine Biology* 65:221–229.

Hauxwell, J., J. Cebrian, C. Furlong, and I. Valiela. 2001. Macroalgal canopies contribute to eelgrass (*Zostera marina*) decline in temperate estuarine ecosystems. *Ecology* 82:1007–1022.

Havaux, M. and R. Lannoye. 1983. Chlorophyll fluorescence induction: A sensitive indicator of water stress in maize plants. *Irrigation Science* 4:147–151.

Hemminga, M. A. and C. M. Duarte. 2000. *Seagrass Ecology*. Cambridge University Press, Cambridge, U.K., 298 pp.

Kenworthy, J. W. and M. S. Fonseca. 1996. Light requirements of seagrasses *Halodule wrightii* and *Syringodium filiforme* derived from the relationship between diffuse light attenuation and maximum depth distribution. *Estuaries* 19:740–750.

Kenworthy, W. J. and D. E. Haunert. 1991. The Light Requirements of Seagrasses: Proceedings of a Workshop to Examine the Capability of Water Quality Criteria, Standards and Monitoring Programs to Protect Seagrasses. NOAA Technical Memorandum NMFS-SEFC-287, NOAA.

Krause, G. H. and E. Weis. 1991. Chlorophyll fluorescence and photosynthesis: the basics. *Annual Review of Plant Physiology and Plant Molecular Biology* 42:313–349.

Longstaff, B. J., N. R. Loneragan, M. J. O'Donohue, and W. C. Dennison. 1999. Effects of light deprivation on the survival and recovery of the seagrass *Halophila ovalis*. *Journal of Experimental Marine Biology and Ecology* 234:1–27.

Major, K. M. and K. H. Dunton. 2002. Variations in light-harvesting characteristics of the seagrass *Thalassia testudinum*: evidence for photoacclimation. *Journal of Experimental Marine Biology and Ecology* 275:173–189.

Markager, S. and K. Sand-Jensen. 1992. Light requirements and depth zonation of marine macroalgae. *Marine Ecology Progress Series* 88:83–92.

Markager, S. and K. Sand-Jensen. 1996. Implications of thallus thickness for growth irradiance relationships of marine macroalgae. *European Journal of Phycology* 31:79–87.

Moore, K. A., H. A. Neckles, and R. J. Orth. 1996. *Zostera marina* (eelgrass) growth and survival along a gradient of nutrients and turbidity in the lower Chesapeake Bay. *Marine Ecology Progress Series* 142:247–259.

Moore, K. A., R. L. Wetzel, and R. J. Orth. 1997. Seasonal pulses of turbidity and their relations to eelgrass (*Zostera marina*) survival in an estuary. *Journal of Experimental Marine Biology and Ecology* 215:115–134.

NC Department of Environment and Natural Resources. 2003. Classifications and Water Quality Standards. North Carolina Administrative Code: Section 15A NCAC 2B.0200.

Neckles, H. A. 1994. Indicator Development: Seagrass Monitoring and Research in the Gulf of Mexico. EPA/620/R-94/029. U.S. EPA, ORD-ERL, Gulf Breeze, FL, 64 pp.

Onuf, C. P. 1996. Seagrass responses to long-term light reduction by brown tide in upper Laguna Madre, Texas: distribution and biomass patterns. *Marine Ecology Progress Series* 138:219–231.

Orth, R. J. and K. A. Moore. 1983. Chesapeake Bay: an unprecedented decline in submerged aquatic vegetation. *Science* 222:51–53.

Ralph, P. J. 1999. Photosynthetic response of *Halophila ovalis* to combined environmental stress. *Aquatic Botany* 65:83–96.

Ralph, P. J. and M. D. Burchett. 1995. Photosynthetic responses of the seagrass *Halophila ovalis* to high irradiance stress using chlorophyll *a* fluorescence. *Aquatic Botany* 51:55–66.

Ralph, P. J., R. Gademann, and W. C. Dennison. 1998. *In situ* seagrass photosynthesis measured using a submersible, pulse-amplitude modulated fluorometer. *Marine Biology* 132:367–373.

Rohacek, K. and M. Bartak. 1999. Techniques of modulated chlorophyll fluorescence: basic concepts, useful parameters, and some applications. *Photosynthetica* 37:339–363.

Schulze, E. D. and M. M. Caldwell (eds.). 1990. *Ecophysiology of Photosynthesis.* Springer, Berlin, 432 pp.

Short, F. T. and S. Wyllie-Echeverria. 1996. Natural and human-induced disturbance of seagrasses. *Environmental Conservation* 23:17–27.

Srivastava, A., R. J. Strasser, and Govindjee. 1999. Greening of peas: parallel measurements of 77K emission spectra, OJIP chlorophyll *a* fluorescence transient, period four oscillation of the initial fluorescence level, delayed light emission and P700. *Photosynthetica* 37:365–392.

Strasser, B. J. and R. J. Strasser. 1995. Measuring fast fluorescence transients to address environmental questions: the JIP-test. In *Photosynthesis: From Light to Biosphere,* Vol. 5, P. Mathis (ed.). Kluwer Academic, Dordrecht, pp. 977–980.

Strasser, R. J. and Govindjee. 1992a. The Fo and the O-J-I-P fluorescence rise in higher plants and algae. In *Regulation of Chloroplast Biogenesis*, J. H. Argyroudi-Akoyunoglou (ed.). Plenum Press, New York, pp. 423–426.

Strasser, R. J. and Govindjee. 1992b. On the O-J-I-P fluorescence transient in leaves and D1 mutants of *Chlamydomonas reinhardtii*. In *Research in Photosynthesis,* Vol. 4, N. Murata (ed.). Kluwer Academic, Dordrecht, pp. 29–32.

Strasser, R. J., A. Srivastava, and Govindjee. 1995. Polyphasic chlorophyll *a* fluorescence transient in plants and cyanobacteria. *Photochemistry and Photobiology* 61:32–42.

Strasser, R. J., P. Eggenberg, and B. J. Strasser. 1996. How to work without stress but with fluorescence. *Bulletin de la Société Royal des Sciences de Liège* 65:330–349.

Stryer, L. 1981. *Biochemistry,* 2nd ed. W.H. Freeman, New York, 950 pp.

Thayer, G. W., W. J. Kenworthy, and M. S. Fonseca. 1984. The Ecology of Eelgrass Meadows of the Atlantic Coast: A Community Profile. FWS/OBS-84/02. U.S. Fish and Wildlife Service, Washington, D.C., 147 pp.

U.S. EPA. 2003. Ambient Water Quality Criteria for Dissolved Oxygen, Water Clarity and chlorophyll *a* for the Chesapeake Bay and Its Tidal Tributaries. EPA903-R-03-002. U.S. Environmental Protection Agency, National Service Center for Environmental Publications, Washington, D.C.

Zimmerman, R. C. 2003. A bio-optical model of irradiance distribution and photosynthesis in seagrass canopies. *Limnology and Oceanography* 48:568–585.

Zimmerman, R. C., J. L. Reguzzoni, S. Wyllie-Echeverria, M. Josselyn, and R. S. Alberte. 1991. Assessment of environmental suitability for growth of *Zostera marina* L. (eelgrass) in San Francisco Bay. *Aquatic Botany* 39:353–366.

Zimmerman, R. C., J. L. Reguzzoni, and R. S. Alberte. 1995. Eelgrass, *Zostera marina*, transplants in San Francisco Bay: role of light availability on metabolism, growth and survival. *Aquatic Botany* 51:67–86.

14

Significance of Considering Multiple Environmental Variables When Using Habitat as an Indicator of Estuarine Condition

Melody J. Hunt and Peter H. Doering

CONTENTS

Introduction

Many organisms depend on estuaries during part of their life cycle (Gunter, 1961; Day et al., 1989). One of the more salient ecological or resource functions attributed to estuaries is their role as nursery areas for larval and juvenile stages of many species including commercially important fish and shellfish (Gunter, 1961; Rozas and Hackney, 1983, 1984). As much as 90% of the annual fisheries value for the Gulf of Mexico fisheries can be attributed to estuarine-dependent species (Seaman, 1988). Within estuaries, three nursery habitats have been recognized in the literature: wetlands (i.e., salt marshes, mangroves, and mudflats), the low-salinity region at the head of the estuary, and submerged aquatic grass beds (Day et al., 1989).

The ecological role of submerged aquatic vegetation in estuarine and coastal systems is well documented (Thayer et al., 1984; Day et al., 1989). Because of the many ecological functions attributed to submerged aquatic vegetation (SAV) and their potential economic value (e.g., as fish habitat), SAV is often a focal point for restoration efforts (SJRWMD, 1994; SFWMD 2003). SAV is also thought to be a sensitive barometer or indicator of anthropogenic impacts and is a prime candidate for inclusion in monitoring programs (Tomasko et al., 1996).

Extensive anthropogenic modifications of watersheds, ultimately affecting estuarine coastal areas, have been implemented in Florida to ensure both water supply and flood protection to a rapidly growing population. These modifications have altered the timing and magnitude of freshwater delivery to estuaries. In turn, altered water delivery has resulted in wide fluctuations in salinity within unnaturally short time periods (Doering et al., 2002).

In the Caloosahatchee Estuary, located on the southwest coast of Florida, the South Florida Water Management District (SFWMD) has been using *Vallisneria americana* Michx. to help meet management challenges associated with alterations in the quantity and timing of freshwater delivery. *Vallisneria americana* (tape grass, wild celery) is a salt-tolerant freshwater angiosperm that often occurs in the oligohaline reaches of estuaries in the northeastern and southeastern United States (Bourn, 1932; Lowden, 1982). The salinity tolerance of *V. americana* has been used to identify freshwater inflows that will maintain important grass bed habitat and ensure the persistence of a low-salinity region in the upper estuary (SFWMD, 2003).

While *V. americana* responds to salinity fluctuation in the field and in the laboratory (Twilley and Barko, 1990; Doering et al., 2001, 2002), it also responds to changes in other environmental variables such as light availability (Harley and Findlay, 1994) and temperature (Zamuda, 1976). Like salinity, light availability also varies with freshwater discharge in the Caloosahatchee due to changes in color, suspended solids, and chlorophyll *a* concentration (Doering and Chamberlain, 1999). Low temperature and high salinity, both of which inhibit growth, are most likely to coincide during the winter dry season. Such confounding of environmental variables and their effects makes changes in plant performance difficult to interpret and hinders use of *V. americana* as an indicator of the success or failure of attempts to manage the quantity of fresh water delivered to the estuary. Numerical modeling of SAV can be a useful tool to help elucidate impacts of multiple environmental variables (Hunt and Madden, 2001; Madden et al., 2001).

A numerical model was constructed as a means to integrate both field and laboratory data and to predict the effect of environmental variables on the growth, survival, and reestablishment of *V. americana*. The model includes the effects of three environmental variables (salinity, light attenuation, and temperature) and predicts several measures of growth including biomass, shoot density, and blade density.

Methods

Study Site

The Caloosahatchee River and Estuary are located on the southwest coast of Florida (Figure 14.1). The Caloosahatchee River runs from Lake Okeechobee to the Franklin Lock and Dam (S-79), where it empties into the estuary. The estuary extends west and south from the dam for approximately 40 km and terminates at Shell Point. The Caloosahatchee River is the major source of fresh water to the estuary. Enough fresh water enters the estuary at S-79 to fill its volume more than eight times per year (Doering and Chamberlain, 1999).

The hydrology of the watershed has been altered over time. The Caloosahatchee River was modified to bring about a permanent connection to Lake Okeechobee and about 20% of the freshwater entering the estuary now comes from the lake. Fresh water is released primarily to maintain the lake at a prescribed water level. The river has also been straightened and deepened, and three water control structures have been added (Figure 14.1A). The last structure, S-79, was completed in 1966 to act, in part, as a salinity barrier (Flaig and Capece, 1998).

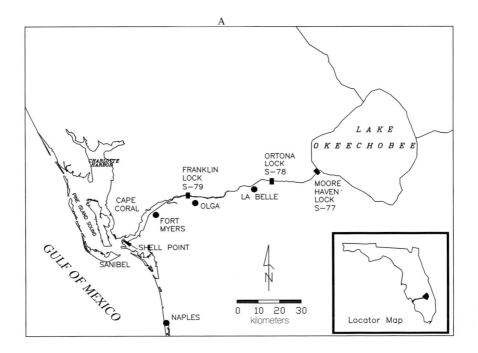

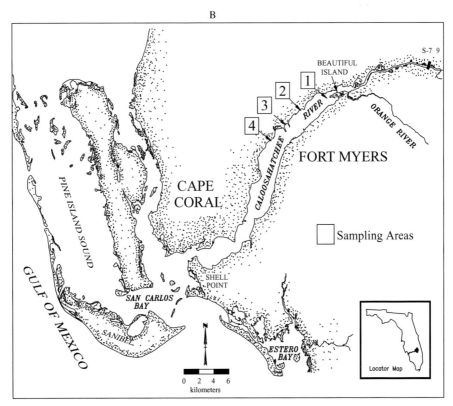

FIGURE 14.1 Site map. (A) Map of Caloosahatchee River and Estuary showing the connection to Lake Okeechobee and the locations of the water control structures. (B) Map of Caloosahatchee River and Estuary showing the locations of the *V. americana* sampling stations.

Beds of *V. americana* can occur up to 30 km downstream of S-79 but grow most luxuriantly upstream of Station 3 (Figure 14.1B), especially around Beautiful Island (Hoffacker, 1994). The beds may be important habitat for fish and shellfish, as described previously and also may comprise a food source for a resident population of West Indian manatees (*Trichechus manatus*). *Vallisneria americana* can be a major dietary component of the manatee and is sometimes preferred over other SAV (Packard, 1981). When conditions are favorable, *V. americana* exhibits a seasonal pattern of growth, with highest biomass achieved in the late summer, flowering in the late summer–early fall, and a winter decline in biomass (Bortone and Turpin, 2000). *Vallisneria americana* does not completely die back in winter (Dawes and Lawrence, 1989) and typically overwinters as a small rosette. Consistent with the southern ecotype of *V. americana* reported by Smart and Dorman (1993), no overwintering buds or tubers have been reported for *V. americana* in the Caloosahatchee Estuary, and it is assumed reestablishment occurs via seedbank.

In the upper estuary, fluctuations in salinity can be quite large and may cause *V. americana* densities to fluctuate as well. Salinity fluctuations are driven by seasonal changes in rainfall and by modifications to the Caloosahatchee River and its watershed. During the wet season (May–October) the upper estuary can be entirely fresh, while during the dry season (November–April) salinity can rise to over 20 ppt. The density of *V. americana* begins to decline as salinity rises above 10 ppt (Doering et al., 2002). Anecdotal, qualitative observations and quantitative data from monitoring indicate that shoot density and canopy height of *V. americana* beds vary interannually. In some years growth may be extensive and quite lush in both wet and dry seasons, while in other years only sparse populations consisting of small plants may be found.

Field Monitoring

A field-monitoring program was initiated in 1998 as described by Bortone and Turpin (2000). This program was designed to obtain information on the life history of *V. americana* in the upper estuary and to make associations of growth response with environmental variables. Ongoing monthly measurements of water quality parameters (i.e., Secchi-disk depth, light, color, total suspended solids) and *V. americana* metrics (i.e., biomass, shoot density, blade density, blade length and width) have been made at four stations since 1998.

Model Description

A numerical model was developed to estimate growth of *V. americana* under varying environmental conditions in the upper Caloosahatchee Estuary. The model consists of a system of three simultaneous finite-difference equations, one for each of three variables — total mass, number of shoots, and number of blades — solved by Euler numerical integration with a time step of 1 day. Primary inputs are water temperature, incident photosynthetically active radiation (PAR), Secchi-disk depth, water depth, and salinity (Table 14.1). Information based on laboratory and field efforts is integrated in the model. It is assumed that neither nutrients nor epiphyte growth on leaves limits growth. Physical processes including waves, currents, and burial were not included in the model.

TABLE 14.1

Summary of Input Data

Input Data	Source (Frequency)
Salinity (ppt)	Regression model developed from field data (daily avg.) (SFWMD, 2003)
Secchi-disk depth (water transparency [m])	Field measurement at each station (monthly)
Incident PAR ($\mu E/m^2 s$)	Estero Bay Station with continuous recording (daily avg.)
Water depth (m)	Field measurement at each station (monthly)
Water temperature (°C)	Field measurement at each station (monthly)

Model Formulation

Separate equations for mass (g dry weight carbon/m^2), blade density (number of blades/m^2), and shoot density (number of shoots/m^2) were formulated and parameterized independently. The discussion here presents the equations for the variable blade density. The basic equations are the same for each remaining variable. Blade density is represented by

$$\text{Blade Density } (t) = \text{Blade Density } (t - dt) + \text{Productivity} - \text{Loss} \tag{14.1}$$

where

$$\text{Productivity} = f \text{ (Blade Density, Salinity, Temperature, Light)} \tag{14.2}$$

and

$$\text{Loss} = f \text{ (Senescence, Acute Stress Mortality, Respiration)} \tag{14.3}$$

Loss

Senescence is considered to be triggered by day of the year and temperature cues. These cues are based on observations in the upper estuary and calibration. Losses from respiration and acute stress mortality are temperature dependent and have the form:

$$\text{(Acute Stress Mortality * Respiration Coefficient) * (Blade Density}^2\text{) *}$$
$$[0.63 * \exp^{(0.092*\text{water temperature})}] \tag{14.4}$$

The acute stress mortality term includes separate multiplicative coefficients for light, salinity, and temperature. These coefficients are only utilized when conditions fall below specified acute stress levels, which cause rapid mortality. If conditions are not outside specified tolerance levels, only the respiration coefficient is utilized in the calculation. The acute stress mortality levels included in the model are light < 17 µE/m^{2}s, salinity > 18 ppt, and temperature < 16°C. These values are within reported ranges (Wilkinson, 1963; Harley and Findlay, 1994; Doering et al., 1999).

Productivity

Maximum productivity is multiplied by a series of reduction factors that range from 0 to 1, with 1 representing productivity at optimal environmental conditions and 0 representing no growth. The reduction factors include the effects of salinity, light, and temperature. Maximum productivity is a density-dependent, self-limiting term determined by calibration that represents the carrying capacity of the environment. Relative growth effect relationships for salinity, light, and temperature were developed based on field data, experimental studies using *V. americana* obtained from the upper estuary, and information reported in the literature (Table 14.2).

TABLE 14.2

Summary of Productivity Variables

Variable	Input Data	Relationship	Parameters Required	Source
Salinity	Salinity	Graphical	Growth rate at different salinities corresponding to small and large plants	Doering et al., 1999
Light	Incident PAR, Secchi-disk water depth	*P/I* curve	Small rosette: $I_k = 200$ µE/m^{2}s Large plant: $I_k = 150$ µE/m^{2}s	Hunt et al., 2004
Temperature	Water temperature	Empirical equation	$Q_{10} = 2$ Optimum Temp. = 34°C Maximum Temp. = 50°C	Wilkinson, 1963

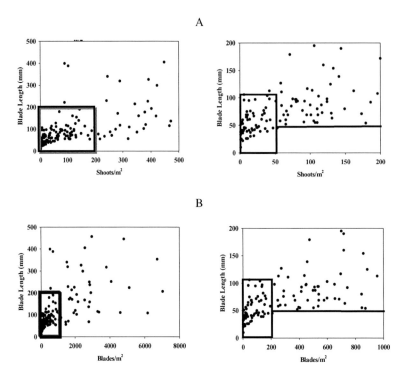

FIGURE 14.2 Relationship between *V. americana* densities and blade length in the Caloosahatchee. Data consists of monthly field measurements from 1 January 1998 through 30 September 2003 at four stations in the upper Caloosahatchee Estuary. (A) When shoot density < 50/m², then blade length < 100 mm, and when shoot density > 50/m², then blade length > 50 mm. (B) When blade density < 200/m², then blade length < 100 mm, and when blade density > 200/m², then blade length > 50 mm.

Salinity Effect

Relationships were developed between relative growth rates and salinity in the range from 0 to 15 ppt based on mesocosm studies using *V. americana* obtained from the Caloosahatchee Estuary (Doering et al., 1999). These researchers report two different rates based on wet season (large plants) and dry season (small rosettes) experiments. Salinity effect relationships were developed for shoots and blades based on these experimental data and differentiated in the model according to plant density. Based on field data, larger plants (>50 mm average blade length) occur when the number of blades/m² are >200 and/or number of shoots/m² are >50 (Figure 14.2).

Light Effect

The light available for photosynthesis is modeled based on a relationship between photosynthesis and irradiance (*P/I*) (Blackman, 1905). The amount of light reaching the bottom at any given location is assumed to be the amount of light available for photosynthesis. It is recognized that this is a conservative formulation most appropriate for small rosettes and likely underestimates the amount of light available to larger plants with blades extending into the water column. The amount of light reaching the bottom is determined by the following computation:

$$\text{Bottom PAR} = [\text{PAR} * (1 - \text{Surface Reflectance})] * \exp^{(-K * \text{Bottom Depth})} \tag{14.5}$$

where

$$\text{Surface Reflectance} = 0.10 \tag{14.6}$$

and

$$K = 1.65/\text{Secchi-disk depth} \tag{14.7}$$

The relationship for vertical light attenuation (K) and Secchi-disk depth is a conversion based on measurements made by eight independent researchers (Giesen et al., 1990) valid in the range from 0.5 to 2.0 m (U.S. EPA, 1992). The calculated bottom PAR is then used to calculate the effect of light changes relative to growth by the following relationship:

$$\text{Light Effect} = \text{Bottom PAR}/I_k \tag{14.8}$$

Plant size was differentiated using the same relationship developed from field data (Figure 14.2) as described for salinity effects. Light saturation (I_k) was established at one of two fixed values according to plant size: 200 μE/m^{2}s for the small rosettes and 150 μE/m^{2}s for the larger plants. The I_k values were determined from laboratory mesocosm experiments (Hunt et al., 2003) and are consistent with other reported values for *V. americana* corresponding to varying seasonal growth in the Hudson River (Harley and Findlay, 1994). When bottom PAR is greater than I_k, then light effect is assumed to be 1 (optimal available light). The likelihood of photoinhibition is very small in the upper estuary and is not considered in this formulation.

Temperature Effect

The effect of temperature on relative growth is modeled using the following equation (O'Neill et al., 1972):

$$k_t = k_{max}U^x e^{(XV)} \tag{14.9}$$

where

$$U = (T_{max} - T)/(T_{max} - T_{opt}) \tag{14.10}$$

$$V = (T - T_{opt})/(T_{max} - T_{opt}) \tag{14.11}$$

$$X = W^2 (1 + (\text{SQT}(1 + 40/W))^2)/400 \tag{14.12}$$

$$W = (Q_{10} - 1)(T_{max} - T_{opt}) \tag{14.13}$$

In this formulation, k_t is the rate of process at temperature T, and k_{max} is the rate of process at the optimum growth temperature (T_{opt}). In the model k_{max} is 1, Q_{10} is 2, optimum growth temperature (T_{opt}) is 34°C, and the upper lethal temperature (T_{max}) is 50°C (Wilkinson, 1963). Figure 14.3 illustrates the temperature growth curve with different Q_{10} values and the experimentally reported values of T_{opt} and T_{max}. The Q_{10} parameter dictates the shape of the growth curve, which is undetermined for *V. americana*. Thus, Q_{10} is a calibration parameter in the model.

Results

Model Calibration

Vallisneria americana densities were calibrated to monthly field measurements at two stations within the upper estuary during the period 1998–2001. Mass was calibrated to the 1 year of available above-ground mass (1998) and was observed for the subsequent years to be consistent with blade and shoot model results. Calibrations are shown for shoot and blade density at Station 2 (Figure 14.4A and B). Although salinity is generally lower and thus more favorable to growth at the upstream location

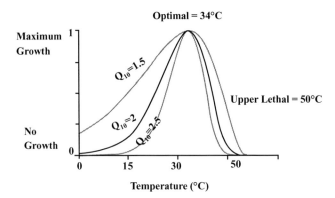

FIGURE 14.3 Temperature growth curve. Temperature growth curve formulation showing different Q_{10} values and reported optimal and upper lethal values. The precise shape of this curve is not known for *V. americana* and Q_{10} is a calibration parameter.

(Station 1), light limits *V. americana* densities at Station 1 such that that decreased densities result relative to Station 2.

The 4-year input data set includes a range of environmental conditions measured at the sites used for model calibration. During the calibration period, salinity ranged from fresh water (0 ppt) to over 22 ppt. Secchi-disk measurements ranged from 0.25 to 1.8 m, and water temperature ranged from 14°C to 31°C.

The calibration period illustrates the importance of all three environmental variables of interest (salinity, light, and temperature) on *V. americana* growth in the estuary. Field measurements and model results reveal that during the first year of calibration (1998) a large standing crop of *V. americana* was produced at Station 2, despite low initial densities (initial shoot density = 70/m² and blade density = 300/m². Throughout the year, salinity was relatively low and water transparency was relatively high, representing ideal conditions for growth. To varying degrees, diminished growth of *V. americana* occurred in the years subsequent to 1998, associated with a combination of elevated salinity, reduced water transparency, and low temperatures (Figure 14.4C and D). In 1999, salinity stress (>10 psu persisting more than 1 month) at the beginning of the growing season resulted in slightly decreased density. However, growth resumed when salinity returned to a more favorable level. In 2000, the salinity remained fairly low; however, algal blooms in the estuary created periodic, light-limiting conditions below the tolerance level for growth. These blooms were noted by SFWMD field personal and manifested as reduced transparencies in the water column. Densities dropped and stayed very low throughout this year, even though the salinity and temperature were favorable for growth following the severely light limiting conditions. By the end of 2000, *V. americana* was reduced significantly. The water temperature early in 2001 was 14°C, which is below the reported growth range for *V. americana* and likely inhibitory to growth. Due to drought conditions in South Florida, high salinity persisted during the remainder of 2001 in the upper estuary, and although light and temperature returned to near optimal levels for growth later in the year, *V. americana* beds did not survive. The calibration period thus shows that growth and survival of *V. americana* in the upper estuary is influenced, at times, by all three environmental variables: salinity, light, and temperature.

Validation Period: Modeling Recovery

Model validation was performed for the period 1 January 2002 through 1 October 2003. This represents a recovery period following the complete loss of *V. americana* beds at Stations 1 to 4 in the previous year (Figure 14.4). The model reasonably predicts field measurements for this period (Figure 14.5). Extremely limited reestablishment occurred in the estuary in the spring of 2002 (<10 shoots/m²) and there were no 2002–2003 overwintering plants detected at the sampling stations. Flowering was not observed in the upper estuary during 2001–2003.

Despite a return to salinities appropriate for growth during most of the year, the watershed discharge was highly colored (exceeding 150 platinum cobalt units) throughout the summer and fall, which resulted

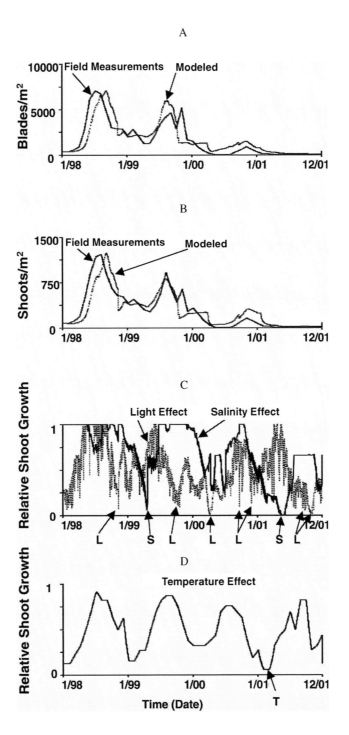

FIGURE 14.4 Station 2 calibration and variable effect. (A) Modeled and field measurements for *V. americana* blade density from 1 January 1998 through 31 December 2001 at Station 2. (B) Modeled and field measurements for *V. americana* shoot density from 1 January 1998 through 31 December 2001 at Station 2. (C and D) Relative shoot growth for each variable during the calibration period is illustrated. On the *y* axis, 1 represents optimal growth conditions and 0 represents no growth. There were seven instances of acute light stress (L), two instances of acute salinity stress (S), and one instance of acute temperature stress (T) during the calibration period.

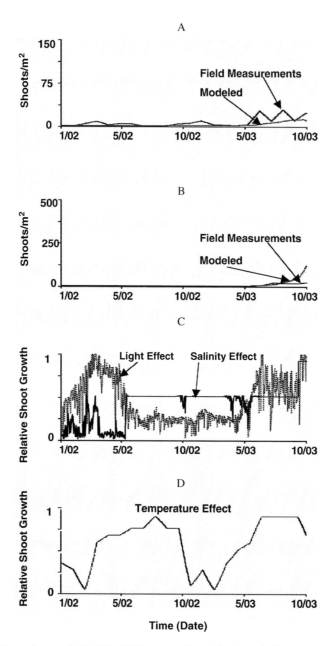

FIGURE 14.5 Validation and regrowth. Model validation was performed for the period 1 January 2002 through 1 October 2003. This period represents a recovery period following the complete loss of *V. americana* beds. Note that a considerably smaller scale is shown than for the calibration period. (A) Shoot density recovery is predicted reasonably well at Station 1. (B) Shoot density recovery is predicted reasonably well at Station 2. (C) Light levels remained consistently low throughout the growing season of 2002. In 2003 there were numerous instances of short-term acute stress levels. (D) There were acutely low water temperatures in the winter of 2002 and 2003.

in low Secchi measurements. Some reestablishment occurred again during the summer of 2003; however, shoot densities did not exceed 20/m² throughout the summer, and large plants did not develop. This observation is consistent with the density-dependent growth pattern found in the field data (Figure 14.2). These field measurements and our observed reestablishment in the Caloosahatchee Estuary are also consistent with observations of *V. americana* reestablishment in the lower St. Johns River, Florida, following salinity exposure as a result of a regional drought in 2001 (Steinmetz and Dobberfuhl, 2003).

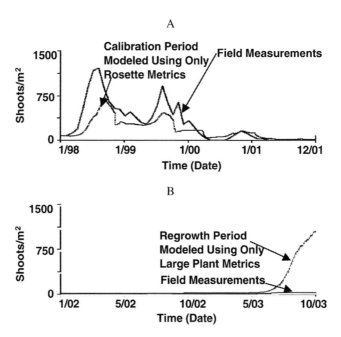

FIGURE 14.6 Static representation of salinity growth metrics. (A) Comparison of modeled and field measurements over the 4-year calibration period using only the rosette plant metrics at Station 2. The shoot density was underpredicted during the periods when large plants were present. (B) Comparison of modeled and field measurements over the regrowth period using only the large plant metrics at Station 2. The shoot density was overpredicted during this period when only rosette plants were present. Consistent results were obtained using static representation of light.

Vallisneria americana cover exhibited a faster recovery rate than blade length in the lower St. Johns River, and it is proposed that reserved resources and energy are allocated preferentially to the colonizing of new plants prior to blade elongation when recovering from long-term exposure of high salinities (Steinmetz and Dobberfuhl, 2003).

Representation of Environmental Variables

The formulation of the relationship between growth and the environmental variables included in the model greatly influences the predictive capability. Plant metrics are an important consideration when predicting plant growth. Based on experimental data, the model includes two separate growth relationships for both salinity and light — one for large plants and one for the small rosettes. To illustrate the importance of including both growth relationships, model runs were performed using the individual relationships separately and compared to the field measurements (growth prediction using static representation of salinity shown in Figure 14.6). As would be expected, the modeled predictions over the 4-year calibration period using growth relationships for small rosette plants only underestimate field densities during periods when large plants are prevalent (1998–1999). During periods when densities were low (2000–2001), reasonable predictions were obtained using the relationship for small rosettes (Figure 14.6A). Additionally, the modeled results based only on the relationships for large plants, overestimated regrowth during the validation period when small rosettes were present (Figure 14.6B).

In addition to salinity and light, model predictions were also sensitive to the temperature relationship, and results varied depending on the temperature curve used. The temperature relationship using $Q_{10} = 2$ provided the best fit to measured field densities. Formulating the growth curve using $Q_{10} = 1.5$ overestimated density while using $Q_{10} = 2.5$ underpredicted density during the 4-year calibration period (Figure 14.7).

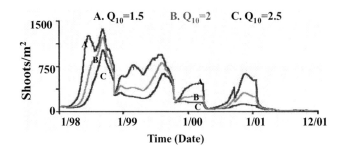

FIGURE 14.7 Comparison of shoot density using different temperature curves. The model prediction for shoot density varied over the 4-year calibration period using different Q_{10} values. $Q_{10} = 2$ was used in the model formulation and most closely corresponds to field measurements.

Effects of Acute Stress

At one time or another during the 4-year calibration period, each of the environmental variables reached values below the acute stress levels (Figure 14.4). Salinity levels above the survival threshold (18 ppt) persisted in the upper estuary for a prolonged period during the drought period of 2001 and severely affected *V. americana* survival. Water temperature reached critical levels (<16°C) in the early winter months of 2001. Acute stress levels for light (<17 µE/m²s) were reached numerous times throughout the 4-year period. When the acute stress mortality coefficients were removed from the model, the resulting predictions overestimated growth for each of the three environmental variables (not shown). This result was most apparent when the acute stress mortality coefficient for light was removed. Although the acute light limitations were of short duration (<1 week), they had a clear impact on the predicted *V. americana* density in the estuary. Although occurring less frequently, the inclusion of acute stress coefficients for salinity and temperature was also critical for successful model application. Without all the acute stress mortality coefficients included, the model predictions resulted in low densities remaining in December 2001. The significance of representing these acute stress conditions in the model that resulted in the demise of *V. americana* beds in 2001 is shown in Figure 14.8. Even a small overestimation of shoot densities (5/m²) at the end of 2001 resulted in significant overestimation both in timing and quantity of predicted reestablishment in 2002–2003. The identification of acute stress or tolerance levels for environmental variables is a necessary component in the model to accurately predict *V. americana* reestablishment after being severely impacted in the estuary.

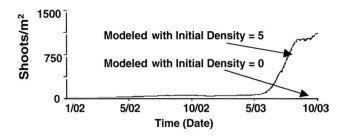

FIGURE 14.8 Importance of initial density on regrowth prediction. Simulation for Station 2 is shown. A very small overestimation of initial density (5 shoots/m² depicted) following the drought period of 2001 in which bed loss occurred results in significant overestimation in both timing and quantity of shoot reestablishment in 2002–2003. When initial density was set to 0, results closely match field measurements of 2002–2003 (see Figures 14.5B and 14.6B). Comparable results were found for blade density and mass at Stations 1 and 2.

Discussion

Salinity

Salinity is an important environmental variable regulating the growth and distribution of *V. americana* in the upper Caloosahatchee Estuary (Doering et al., 1999). Salinity also plays a role in the development of management criteria for the estuary. The relationship between freshwater inflow and the spatial distribution of salinity in the estuary serves as a central basis for establishing minimum freshwater flow requirements (SFWMD, 2003). The minimum flow requirement is intended to establish salinity conditions that will protect submerged *V. americana* beds in the upper estuary from significant harm. A major assumption of this approach is that salinity conditions that protect *V. americana* will also protect other key organisms in the environment (SFWMD, 2003). Thus, the salinity requirement of *V. americana* is employed as an indicator of upper estuarine condition.

While several studies have assessed salinity tolerances of *V. americana* (Bourn, 1932, 1934; Haller et al., 1974; Twilley and Barko, 1990), estimates do not agree, and there is little information concerning factors that might modify salinity tolerance (see discussion in Doering et al., 1999). Laboratory results suggest that for *V. americana* from the Caloosahatchee growth is low or nil in the 10 to 15 ppt salinity range and mortality occurs at salinities greater than 15 ppt (Doering et al., 1999, 2001). This agrees well with transplant experiments conducted in the Caloosahatchee, which indicated mortality at salinities > 15 ppt (Kraemer et al., 1999). Seasonal differences may also be an important factor in the growth response relative to salinity changes. Depending on the site location and climatic patterns, seasonal differences measured may include prior exposure and plant size. In laboratory experiments using *V. americana* obtained from the Caloosahatchee Estuary, Doering et al. (1999) found seasonal differences in growth at salinities in the range of 0 to 9 ppt. Experimental results show that growth in response to salinity variation may differ according to plant size, and successful predictions using the model require this consideration. Because of the variation in reported results and differences in growth patterns for *V. americana*, site-specific information should be utilized when possible.

Light

The central role of light availability for SAV has been demonstrated in numerous field, laboratory, and modeling studies. Changes in water clarity can affect density, depth distribution, and species in a given area. *Vallisneria americana* is generally considered light adaptable as it acclimates rapidly to increasing light and efficiently uses low light (Meyers et al., 1943; Titus and Adams, 1979; Harley and Findlay, 1994). However, its limited elongation potential may be a disadvantage in deep turbid water (Barko et al., 1984, 1991), and water clarity may be an important factor regulating growth and survival, especially for seedlings or immature rosettes (Kimber et al., 1995).

The life cycles of northern and southern populations of *V. americana* differ (Dawes and Lawrence, 1989), and light requirements for seedlings may be of special relevance in a southern estuarine environment such as the Caloosahatchee. Northern populations overwinter as a dormant winter bud buried in the sediment while the aboveground biomass disintegrates (Titus and Hoover, 1991). The production of winter buds at the end of the growing season and the sprouting of leaves from these buds in the spring determine the vegetative persistence of *V. americana* (Korschgen et al., 1997). While we have observed *V. americana* plants to flower both in the field and laboratory, we have not observed winter bud formation in either place. The vegetative persistence of *V. americana* in the Caloosahatchee Estuary may depend largely on the survival of rosettes during the winter when both saltwater intrusions and colder water temperatures, nearing acute growth tolerances levels for both parameters, are possible. Thus, in the event that *V. americana* beds do not survive the winter months, such as occurred during 2001, reestablishment via seedbank may serve as the only means of bed recovery. When freshwater conditions return after a drought period, this water is typically highly colored. The color may limit light availability and pose a particular disadvantage for seedling growth and subsequent *V. americana* bed recovery in the estuary.

The light-limiting conditions in the upper estuary during 2002–2003 were likely inhibitory to seedling development and resulted in the slow reestablishment of *V. americana* beds after the 2001 drought.

Development of *P/I* relationships is standard practice to represent plant productivity over a range of light levels. There are notable limitations to using this approach because many factors may influence photosynthesis at a particular light level. These relationships are a static representation of a plant's ability to utilize light and do not incorporate change across environmental conditions. Several *P/I* relationships for marine SAV have been shown to vary with depth and season (Drew, 1978; Dawes and Tomasko, 1988). Harley and Findlay (1994) report differing *P/I* relationships in the Hudson River for *V. americana* across the growing season. Recent mescosm experiments (Hunt et al., 2004) indicate that the *P/I* relationship for *V. americana* in the Caloosahatchee Estuary changes with salinity conditions as well as plant size. The information for metrics is included in the model by using separate *P/I* curves for rosettes and large plants. Other factors that may influence photosynthesis at a particular light level include the age of the leaves, the orientation of the leaves with respect to the light field, and the physiological health of the leaves (Fourqurean and Zieman, 1991).

Temperature

Temperature is the least known of the three parameters in the model, and no site-specific information was found relating temperature to *V. americana* growth. In general, temperature changes primarily influence growth of SAV over predictable seasonal cycles. Similar to salinity tolerances, various temperature growth ranges (minimum to maximum) have been reported for *V. americana* (Meyers et al., 1943; Hunt, 1963; Wilkinson et al., 1963; Barko et al., 1982, 1984). This is not surprising considering that these values are determined in populations growing in different climates and under different environmental conditions. Titus and Adams (1979) report a temperature optimum for *V. americana* obtained from University Bay, Madison, Wisconsin, of 32.6°C. In laboratory tests conducted by Wilkinson et al. (1963), *V. americana* grew best within a water temperature range of 33°C to 36°C. Arrested growth occurred below 19°C, and plants became limp and disintegrated above 50°C. The determination of the acute lower tolerance level may be particularly relevant for the southern ecotype found in the South Florida. Given the span in water temperatures in the Caloosahatchee Estuary, ranging from optimal conditions to below tolerance levels in any given year, and the importance of overwintering survival, water temperature may be an important variable that influences *V. americana* survival. The minimum, optimal, and maximum values in the model, as well as the structure of the curve (as determined by the Q_{10} parameter), were chosen based on observation and model calibration and thus involve obvious limitations.

Similar to the other environmental parameters discussed, relationships developed for temperature are also limited by the conditions under which determinations are made (typically optimal growth conditions). There is reason to be cautious of the assumption that they are valid over a broad range of environmental conditions. Bulthuis (1987) reports that under nonsaturating and low-light conditions, temperature optimums may not remain constant for marine SAVs. He reports lower temperature optimum values during periods of low light relative to higher or saturating light conditions. Further, given the differences in growth according to plant size already noted for salinity and light, it is possible that such a relationship also exists for temperature. Owing to the model sensitivity to changes in the formulation of this relationship (Figure 14.7), uncertainties in curve shape (Figure 14.3), the range of water temperatures in the estuary, and the potential for changes in the relationship with other environmental conditions, further experimental work relating temperature to growth under different conditions is warranted.

Conclusions

The utilization of SAV as an indicator of estuarine condition calls for predictive tools that facilitate the assimilation of available information from a variety of sources such as field studies, site-specific experimental data, mesocosm results, and the literature. The ability to integrate the effects of a number of environmental variables through the development and application of the numerical model presented here provides a valuable means to examine the factors regulating the growth, reestablishment, and survival of *V. americana* in the Caloosahatchee Estuary. This work demonstrates that all three environmental variables examined (salinity, light, and temperature) can play an important role in both maintaining and establishing *V. americana* habitat in the upper Caloosahatchee Estuary. To construct a numerical model, quantifiable information relating plant growth and survival to the variables of interest is needed. Owing to the highly variable and rapidly shifting physiochemical parameters in most estuaries, information included in model formulation should span a range of expected environmental conditions. Our work indicates that the acute stress levels for environmental variables should be well represented. Model analysis also suggests that consideration should be given to developing growth response relationships in SAV models with the ability to relate growth and the effect of environmental variables to different stages in plant development. This is especially relevant when different plant developmental stages are expected and when predicting reestablishment. In general, carefully constructed SAV models can provide a valuable framework to determine the relevance of different environmental variables of interest and elucidate the importance of timing, frequency of stress, and interactions with other variables. The model presented is a useful interpretive tool that helps identify the factors to which the indicator *V. americana*, responds.

Acknowledgments

We thank Robert Chamberlain, Kathy Haunert, Robin Bennett, Dan Crean, Julian DiGialleonardo, Keith Donohue, Matt Giles, and Steve Bortone for field and laboratory efforts. We thank Chenxia Qiu for providing salinity input via hydrodynamic modeling.

References

Barko, J. W., D. G. Hardin, and M. S. Matthews. 1982. Growth and morphology of submerged freshwater macrophytes in relation to light and temperature. *Canadian Journal of Botany* 60:877–887

Barko, J. W., D. G. Hardin, and M. S. Matthews. 1984. Interactive Influences of Light and Temperature on the Growth and Morphology of Submerged Freshwater Macrophytes. Technical Report A-84-3, U.S. Army Corps of Engineers, Waterways Experimental Station, Vicksburg, MS, 24 pp.

Barko, J. W., R. M. Smart, and D. G. McFarland. 1991. Interactive effects of environmental conditions on the growth of submerged aquatic macrophytes. *Journal of Freshwater Ecology* 6(2):199–207

Blackman, F. F. 1905. Optima and limiting factors. *Annals of Botany* 19:281–295.

Bortone, S. A. and R. K. Turpin. 2000. Tapegrass life history metrics associated with environmental variables in a controlled estuary. In *Seagrass Monitoring, Ecology, Physiology and Management,* S. A. Bortone (ed.). CRC Press, Boca Raton, FL, pp. 65–79.

Bourn, W. S. 1932. Ecological and physiological studies on certain aquatic angiosperms. *Contributions from Boyce Thompson Institute* 4:425–496.

Bourn, W. S. 1934. Sea-water tolerance of *Vallisneria spiralis* L. and *Potomogeton foliosus* Raf. *Contributions from Boyce Thompson Institute* 6:303–308.

Bulthuis, D. A. 1987. Effects of temperature on photosynthesis and growth of seagrasses. *Aquatic Botany* 27:27–40.

Dawes, C. J. and J. M. Lawrence. 1989. Allocation of energy resources in the freshwater angiosperms *Vallisneria americana* Michx. and *Potomogeton pectinatus* L. in Florida. *Florida Scientist* 52: 59 – 63.

Dawes, C. J. and D. A. Tomasko. 1988. Depth distribution of *Thalassia testudinum* in two meadows on the west coast of Florida: a difference in effect of light availability. *Marine Ecology* 9:123–130.

Day, J. W., Jr., C. A. S. Hall, W. M. Kemp, and A. Yanez-Arancibia. 1989. *Estuarine Ecology.* John Wiley & Sons, New York.

Doering, P. H. and R. H. Chamberlain. 1999. Water quality and the source of freshwater discharge to the Caloosahatchee Estuary, FL. *Water Resources Bulletin* 35:793–806.

Doering, P. H., R. H. Chamberlain, K. M. Donohue, and A. D. Steinman. 1999. Effect of salinity on the growth of *V. americana* Michx. from the Caloosahatche Estuary, FL. *Florida Scientist* 62(2):89–105.

Doering, P. H., R. H. Chamberlain, and J. M. McMunigal. 2001. Effects of simulated saltwater intrusions on the growth and survival of wild celery, *Vallisneria americana*, from the Caloosahatchee Estuary (South Florida). *Estuaries* 24 (6A):894–903.

Doering, P. H., R. H. Chamberlain, and D. E. Haunert. 2002. Using submerged aquatic vegetation to establish minimum and maximum freshwater inflows to the Caloosahatchee Estuary, Florida. *Estuaries* 25(6B):1343–1354.

Drew, E. A. 1978. Factors affecting photosynthesis and its seasonal variation in the seagrasses *Cymodocea nodosa* (ueria) Aschers, and *Posidonia oceanica* (L.) Delile in the Mediterranean. *Journal of Experimental Marine Biology and Ecology* 31:173–194.

Flaig, E.G., and J. Capece 1998. Water use and runoff in the Caloosahatchee watershed. In Proceedings of the Charlotte Harbor Public Conference and Technical Symposium, March 15–16, 1997, Punta Gorda, FL, S. F. Treat (ed.). Charlotte Harbor National Estuary Program Technical Report 98-02, pp. 73–80.

Fourqurean, J. W. and. J. C. Zieman. 1991. Photosynthesis, respiration and whole plant carbon budget of the seagrass *Thalassia testudinum*. *Marine Ecology Progress Series* 69:161–170.

Giesen, W. B. J. T., M. M. van Katijk, and C. Den Hartog. 1990. Eelgrass condition and turbidity in the Dutch Wadden Sea. *Aquatic Botany* 37:71–85.

Gunter, G. 1961. Some relations of estuarine organisms to salinity. *Limnology and Oceanography* 6:182–190.

Haller, W. T., D. L. Sutton, and W. C. Barlowe. 1974. Effects of salinity on growth of several aquatic macrophytes. *Ecology* 55:891–894.

Harley, M. T. and S. Findlay.1994. Photosynthesis–irradiance relationships for three species of submerged macrophytes in the tidal freshwater Hudson River. *Estuaries* 17(1B):200–205.

Hoffacker, V. A. 1994. 1993 Caloosahatchee River Submerged Grass Observations. W. Dexter Bender and Associates, Inc. Report and Map. South Florida Water Management District, Ft. Myers Service Center, Ft. Myers, FL.

Hunt, G. S. 1963. Wild celery in the lower Detroit River. *Ecology* 44:360–370.

Hunt, M. J. and C. J. Madden. 2001. Use of simulation models to determine effects of multiple stressors on seagrasses in Florida Bay. In *Conference Proceedings of the 16th Biennial Conference of Estuarine Research Federation*, St. Petersburg, FL, November.

Hunt, M. J., P. H. Doering, R. H. Chamberlain, and K. M. Haunert. 2004. Grass bed growth and estuarine condition: is plant size a factor worth considering? In *Conference Proceedings of the Southeastern Estuarine Research Society*, April 15–17, 2004, Harbor Branch Oceanographic Institution, Fort Pierce, FL.

Kimber, A., C. E. Korschgen, and A. G. Van der Valk. 1995. The distribution of *Vallisneria americana* seeds and seedling light requirements in the Upper Mississippi River. *Canadian Journal of Botany* 73(12):1966–1973.

Kraemer, G. P., R. H. Chamberlain, P. H. Doering, A. P. Steinman, and M. D. Hanisak. 1999. Physiological responses of transplants of the freshwater angiosperm *Vallisneria americana* along a salinity gradient in the Caloosahatchee Estuary (Southwestern Florida). *Estuaries* 22(1):138–148.

Korschgen, C. E., W. L. Green, and K. P. Kenow. 1997. Effects of irradiance on growth and winter bud production by *Vallisneria americana* and consequences to its abundance and distribution. *Aquatic Botany* 58:1–9.

Lowden, R. M. 1982. An approach to the taxonomy of *Vallisneria* L. (hydrocharitaceae). *Aquatic Botany* 13:269–298.

Madden, C. J., M. J. Hunt, W. M. Kemp, and D. F. Gruber. 2001. Seagrass habitat recovery and everglades restoration: use of an ecological model to assess management strategies in Florida Bay. In *Proceedings Florida Bay Science Conference*, Key Largo, FL, April 23–26.

Meyers, B. S., F. H. Bell, L. C. Thompson, and E. I. Clay. 1943. Effect of depth of immersion on apparent photosynthesis in submerged vascular aquatics. *Ecology* 24:393–399.

O'Neill, R. V., R. A. Goldstein, H. H. Shugart, and J. B. Manki. 1972. Terrestrial Ecosystem Energy Model. U.S. IBP Eastern Deciduous Forest Biome Memo Report 72–19. Oak Ridge National Laboratory, Oak Ridge, TN.

Packard, J. M. 1981. Abundance, Distribution, and Feeding Habits of Manatees (*Trichechus manatus*) Wintering between St. Lucie and Palm Beach Inlets, Florida. Report to U.S. Fish and Wildlife Service Contract 14-16-0004-80-105, 142 pp.

Rozas, L. P. and C. T. Hackney. 1983. The importance of oligohaline estuarine wetland habitats to fisheries resources. *Wetlands* 3:77–89.

Rozas, L. P. and C. T. Hackney. 1984. Use of oligohaline marshes by fishes and macrofaunal crustaceans in North Carolina. *Estuaries* 7(3):213–224.

Seaman, W., Jr. 1988. *Florida Aquatic Habitat and Fisheries Resources.* Florida Chapter, American Fisheries Society, Eustis, FL.

Smart, R. M. and J. D. Dorman. 1993. Latitudinal differences in the growth strategy of a submerged aquatic plant: ecotype differences in *Vallisneria americana? Bulletin of the Ecological Society of America* 74(Suppl.):439.

South Florida Water Management District. 2003. Technical Documentation to Support Development of Minimum Flows and Levels for the Caloosahatchee River and Estuary, Status Update Report. May 2003.

St. Johns River Water Management District and South Florida Water Management District. 1994. Indian River Lagoon Surface Water Improvement and Management (SWIM) Plan. Palatka and West Palm Beach, FL, 120 pp.

Steinmetz, A. and D. Dobberfuhl. 2003. Drought related effects on *Vallisneria americana* in the lower St. Johns River, FL. Poster presentation at the 17th Biennial Conference of the Estuarine Research Federation, Seattle, WA. September 14–18.

Thayer, G. W., W. J. Kenworthy, and M. S. Fonseca. 1984. The Ecology of Eelgrass Meadows of the Atlantic Coast: A Community Profile. U.S. Fish and Wildlife Service Report FWS/OBS-84/02, 147 pp.

Titus, J. E. and M. S. Adams. 1979. Coexistence and the comparative light relations of the submerged macrophytes *Myriophylulum spicatum* L. and *Vallisneria americana* Michx. *Oecologia* 40:273–286.

Titus, J. E. and D. T. Hoover. 1991. Toward predicting reproductive success in submerged freshwater angiosperms. *Aquatic Botany* 41:111–136.

Tomasko, D. A., C. J. Dawes, and M. O. Hall. 1996. The effects of anthropogenic enrichment on turtle grass (*Thalassia testudinum*) in Sarasota Bay, Florida. *Estuaries* 19(2B):448–456.

Twilley, R. R. and J. W. Barko. 1990. The growth of submerged macrophytes under experimental salinity and light conditions. *Estuaries* 13:311–321.

U.S. EPA. 1992. Chesapeake Bay Submerged Aquatic Vegetation Habitat Requirements and Restoration Targets: A Technical Synthesis, Annapolis, MD, December 1992. U.S. Environmental Protection Agency for the Chesapeake Bay Program, Contract 68-WO-00043, pp. 15–19.

Wilkinson, R. E. 1963. Effects of light intensity and temperature on the growth of water stargrass, coontail, and duckweed. *Weeds* 11:287–289.

Zamuda, C. D. 1976. Seasonal Growth and Decomposition of *Vallisneria americana* in the Pamlico River Estuary. M.S. thesis, East Carolina University, Greenville, NC, 86 pp.

15

Using Seagrass Coverage as an Indicator of Ecosystem Condition

Catherine A. Corbett, Peter H. Doering, Kevin A. Madley, Judith A. Ott, and David A. Tomasko

CONTENTS

Introduction

Concern about the condition of natural resources faced with threats from human impacts have led to efforts to monitor and assess the effects of these impacts and the efficacy of the management regimes that attempt to mitigate them. Increasingly, the use of indicators that reflect biological, chemical, or physical attributes of the "health" of an ecosystem has been encouraged, and since the 1993 U.S. Government Performance and Results Act (GPRA), government agencies have been required to develop performance reports that measure management success using indicators and goals (U.S. EPA, 2000). The U.S. EPA has developed 15 evaluation guidelines for developing environmental indicators that includes, among others, the following:

- Relevance to the assessment — The proposed indicator should be responsive to an identified question and provide information useful for management decisions.
- Temporal variability across years — While indicator responses may show interannual variability even as environmental conditions remain stable, the indicator should reflect true trends in ecological condition for the assessment question. To determine variability across years, monitoring must proceed for several years at relatively ecologically stable sites.
- Discriminatory ability — The indicator should reflect differences among sites along a known condition gradient.

- Linkage to management action — An indicator is useful only if it can provide adequate information to support management decisions or quantify the success of past decisions (U.S. EPA, 2000).

Examples of indicators of ecological condition include direct measurements (e.g., total nitrogen concentration), indices (e.g., macroinvertebrate condition index), and multimetrics (e.g., fish assemblage) (U.S. EPA, 2000).

Evaluating seagrass coverage and condition has become one method of monitoring the condition of coastal regions worldwide. Monitoring and conserving seagrass is a global issue illustrated by the incidences of seagrass declines attributed to eutrophication that have been documented in Denmark, Australia, Bermuda, and the United States (Kemp, 2000). Within the United States, several federal agencies use aerial photography to map and monitor coastal wetlands (Dobson et al., 1995; Kiraly et al., 1990; Coyne et al., 2001; Kendall et al., 2001). In addition, federal, state, and local government partnerships have resulted in seagrass monitoring programs in Chesapeake Bay (Chesapeake Executive Council, 1993), Florida (Virnstein and Morris, 1996; Kurz et al., 2000), North Carolina (Kiraly et al., 1990), Padilla Bay (Shull and Bulthuis, 2002), and Texas (Moulton et al., 1997).

In Southwest Florida, including Tampa and Sarasota Bays, many research and restoration efforts have focused on seagrass meadows as indicators of estuarine condition. In Tampa Bay, where historical losses have been linked to both direct and indirect impacts, resource managers have set goals for restoring seagrass coverage to approximately 95% of the coverage present in 1950. The Nitrogen Management Consortium, a partnership among private and public entities, was formed to develop methods of reducing anthropogenic nitrogen loads to the bay to reduce algae blooms within the water column and epiphytes on the surface of seagrass blades, both of which reduce the supply of light available to the plants. Recent increases (1980s to 1996) in seagrass coverage in Tampa Bay (and also Sarasota Bay) are thought resultant from decreasing anthropogenic nitrogen loads to these estuaries (Kurz et al., 2000). Nonetheless, seagrass coverage in some areas of Tampa Bay may not be increasing as expected even though water clarity appears to be adequate to support seagrass growth in these areas (TBEP, 2000), and it appears that many factors other than nutrient loads may play an important role in driving seagrass coverage in some areas of the bay.

In contrast to Tampa and Sarasota Bays, resource managers in Charlotte Harbor, located on the west coast of Florida south of Tampa and Sarasota Bays and contiguous with these estuaries, have not used seagrass coverage as an environmental indicator of the health of the estuary to date. Even with its close proximity to these other estuaries, Charlotte Harbor experiences disparate issues than its northern neighbors. Charlotte Harbor itself is approximately 700 km^2 of open water and has a watershed of well over 11,300 km^2, a ratio of watershed to open water of over 16:1 (SWFWMD, 2000). The estuarine system is very dynamic due to the influence of the freshwater inflows from its major tributaries, and natural interannual variability in seagrass coverage mars the ability to ascribe changes in seagrass coverage to specific anthropogenic impacts within the watershed.

This chapter reviews historical and current seagrass coverage information as well as 1999 to 2001 fixed transect monitoring data for the Charlotte Harbor region to examine some of the problems associated with using seagrass as an indicator of ecosystem condition. Although current seagrass monitoring efforts in Charlotte Harbor are extensive, insufficient data, natural variability, and confounding indirect and direct impacts within the region coalesce to create perplexing problems with the use of seagrass as an environmental indicator of ecosystem condition.

Charlotte Harbor, Florida

Charlotte Harbor is the second largest open water estuary in the state, and it is also generally considered one of Florida's most productive and a relatively healthy estuary (Figure 15.1). The estuarine complex includes numerous interconnected estuaries from Lemon Bay south to Estero Bay. The majority of the harbor is surrounded by an extensive conservation buffer system of well over 21,610 ha. Much of the shoreline in this buffer system is unaltered mangrove and salt marsh habitats, thereby providing abundant

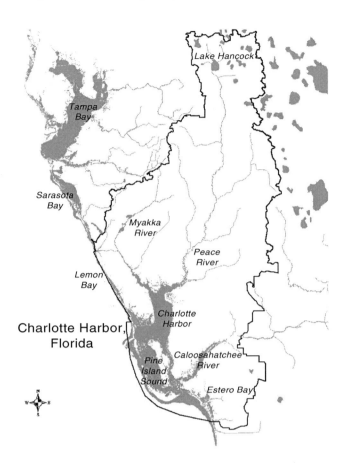

FIGURE 15.1 Map of Charlotte Harbor watershed. (Data provided by CHNEP, NOAA, and SWFWMD.)

food and shelter for juveniles of many of the harbor's estuarine species. The Charlotte Harbor watershed extends approximately 210 km (130 mi) from the northern-most headwaters of the Peace River to southern Estero Bay. Three large rivers, the Peace (a 6090 km² basin), the Myakka (a 1560 km² basin), and the Caloosahatchee River (3570 km² basin extending to Moore Haven), serve as the major sources of fresh water to the Charlotte Harbor estuary (Hammett, 1990). Six species of seagrass are found within the Charlotte Harbor region: *Halodule wrightii, Thalassia testudinum, Syringodium filiforme, Halophila englemanni, H. decipiens,* and *Ruppia maritime. Halodule* sp. is the most common species within the harbor, found in all the estuary segments, whereas *Thalassia* sp. and *Syringodium* sp. are more abundant in those estuary segments where they are found (Staugler and Ott, 2001).

Agriculture encompasses the major land use in the Charlotte Harbor watershed and is second only to tourism in economic impact. In 1995, a total of 114,520 ha within the watershed was dedicated to citrus crops, one third of all Florida citrus extent (CHNEP, 2000), while in 1990 more than 404,680 ha was devoted to rangeland or pasture for cattle (CHNEP, 1999). Simultaneously, Florida leads the United States in conversion of farmland to urban lands, and along the coast especially, residential and urban development is rapidly expanding. In 2020, the region is projected to have a population of almost 2 million residents, a 424% increase from the 1960s population of 363,200 (cited in CHNEP, 2000). Finally, there is an extensive phosphate mining industry within the middle and upper reaches of the northern watershed. The "Bone Valley" phosphate deposit of more than 202,342 ha (500,000 acres) lies primarily within the Peace River sub-basin. This phosphate deposit provides almost 75% of the nation's phosphate supply and 25% of the world's (CHNEP, 2000). Future mining is expected to move southward toward the harbor and last an additional 30 years.

Within this rapidly changing region, the Charlotte Harbor ecosystem currently faces a number of considerable hydrologic and water quality challenges. In the southern region of the Charlotte Harbor system, the Caloosahatchee River, which contributes an annual average inflow to the lower harbor of approximately 57 m³/s (Hammett, 1990), was channelized and artificially connected to Lake Okeechobee in the late 1800s. A series of locks and dams were constructed along the river, one of which, the W. P. Franklin Locks and Dam, artificially truncates the river's estuarine system by blocking the natural gradient of fresh to saltwater that historically extended upstream during the dry season. Currently, it is common that water immediately downstream of the locks has nearly one third the salinity of the Gulf of Mexico while water immediately upstream of the structure is fresh (SFWMD, 2000). Also, freshwater releases from Lake Okeechobee to the estuary for several weeks during late spring each year to lower lake levels before the summer rainy season exceed 71 to 85 m³/s (2500 to 3000 ft³/s), estimated as maximum high flows for estuarine resources (see Chamberlain and Doering, 1997). There are projects under way to restore a more natural variation in flows to the Caloosahatchee estuary stemming from the Comprehensive Everglades Restoration Plan (South Florida Ecosystem Task Force, 2002); nonetheless, the past hydrological alterations of the watershed have dramatically changed the natural quantity, quality, timing, and distribution of flows to the estuary with limited regard to maintaining the biological integrity of the ecosystem (SFWMD, 2000). In addition, the Caloosahatchee River is the major source of nutrients to the Caloosahatchee estuary (Environmental Research and Design, 2003), and more than a decade ago, the Florida Department of Environmental Regulation reported that the estuary had reached its nutrient loading limits as indicated by elevated chlorophyll *a* and depressed dissolved oxygen levels (DeGrove, 1981; DeGrove and Nearhoof, 1987; Baker, 1990).

In the northern region of the watershed, the Upper Peace River has gone from a gaining stream with continuous flows throughout the year via groundwater contribution to a losing stream (i.e., predominantly groundwater recharge area) since the 1950s (SWFWMD, 2002). In contrast to historical conditions, flow measurements at the northern Bartow, Florida, USGS (U.S. Geological Survey) flow gauge now often exceed the flow measurements of the downstream Zolfo Springs gauge. The potentiometric surface of the Upper Floridan aquifer within the Upper Peace River basin has been lowered 6 to 12 m (20 to 40 ft) in the dry season (April–May), and areas of past artesian flow, such as Kissingen Spring, have ceased discharge. While the aquifer apparently is rebounding, the Upper Peace River now experiences periods of no or little flow for several weeks during the dry season (SWFWMD, 2002).

The Peace River is also the major source of nutrient loads to northern Charlotte Harbor, and the entire stretch of the river is considered "fair" to "poor" water quality by Florida standards (FDEP, 2003). Three water bodies within the upper Peace River watershed, Lake Parker, Banana Lake, and Lake Hancock (along with their tributaries), consistently exhibit some of the poorest water quality within the state (FDEP, 2003). Lake Hancock, a 1840-ha lake within the headwaters of the Peace River, is a hypereutrophic lake that has accumulated a layer of flocculent organic sediments approximately 1.7 m thick or an estimated 13.8 million m³ (Camp Dresser & McKee, Inc., 2001). It is considered one of the most degraded lakes within the State of Florida, and from 1985 to 1999 it had a mean chlorophyll value of 170 mg/m³ (cited in Camp Dresser & McKee, Inc., 2001). The estimated yield of nitrogen from this lake is approximately 5700 kg/mi²/year (Tomasko, 2001) and represents roughly 19% of the total nitrogen loadings to the Peace River (cited in Camp Dresser & McKee, Inc., 2001). Also, the river has naturally high levels of phosphorus, and average total phosphorus concentrations from 1993 to 1995 exceeded 0.70 mg/L to place it within the poorest 20% of Florida streams (FDEP, 2003). Water quality data collected in the region indicate that algae blooms and chlorophyll *a* levels exceeding 60 to 80 µg/L in the tidal Peace River region have occurred seasonally since monitoring began in 1976 (FDEP, 2003). Seasonal chlorophyll *a* levels of this magnitude are considered indicative of hypereutrophic conditions in some estuarine water quality classification systems (e.g., NOAA, 1996), and chlorophyll *a* concentrations exceeding 20 µg/L, considered "high" by classification systems (e.g., NOAA, 1996), were consistently observed in both the tidal Peace and Myakka Rivers during July through September from data collected in 1993–1996 (Morrison et al., 1997).

Methods

Historical Seagrass Coverage Estimates

To represent historical conditions, black-and-white photographs from 1946 to 1951 (referred to in the original document as the 1945 set) were acquired, and 1:24,000 scale positive false-color, infrared transparencies were produced from flights in April 1982 for current conditions. In 1983, the Florida Department of Natural Resources (FDNR) and the Florida Department of Transportation (FDOT) produced a document with associated maps that examined the historic and current land uses of Charlotte Harbor and Lake Worth Lagoon, Florida (Harris et al., 1983). The 1982 photographs were analyzed with stereoscopic visual equipment; the 1945 photographs did not have the requirements needed to perform stereoscopic analyses. For the 1945 and 1982 maps, seagrass beds were delineated onto Mylar® overlays, and then the data were digitized into the FDOT proprietary point-vector database. Final maps were produced at 1:24,000 scale.

Seagrass meadows, as determined from the 1945 aerial photography, were classified to only one category (submerged aquatic vegetation) because the quality of the photographs did not allow multiple category classification. The 1982 seagrass coverage was classified to three categories, but for the purposes of this report, coverages for all data sets, including subsequent mapping efforts, are reported as combinations of all seagrass classification categories employed (i.e., sparse, dense, patchy).

Subsequently, Florida Marine Research Institute (FMRI) staff redigitized the seagrass polygons derived from the 1982 photographs to create a digital Arc/Info® file. Because these 1982 data exist in digital format, calculations of seagrass coverage based on the harbor sub-basins were possible for this report and are used as historical data for comparison with more recent mapping results. However, lack of digital data for the 1945 effort has limited the examination of the 1945 seagrass coverage by sub-basin for this effort. Seagrass coverage for the 1945 maps was evaluated by USGS quadrangle areas rather than the 14 sub-basins that have been used to define coverage for subsequent mapping efforts. Also, the geographic boundaries for the 1945 study area do not match the boundaries of the 1982 and subsequent year analyses. Therefore, a comparison of total seagrass coverage from the 1945 to recent efforts is not possible. The 1983 study did not include Lemon Bay or southern Estero Bay. To fill a void in the aerial photographic coverage for 1982, FMRI interpreted Estero Bay coverage from photographs taken in 1990. Because of these issues, Estero and Lemon Bays are not included in the discussion of changes in seagrass coverages within this chapter. Finally, the black-and-white photographs used in the 1945 effort were of low quality for the purpose of delineating seagrass coverage, and the absence of ground verification during the year the photographs were produced is reason for caution when examining these data and resulting maps.

Current Seagrass Mapping Efforts

The Charlotte Harbor watershed falls within the jurisdiction of two water management districts, the South Florida Water Management District (SFWMD) and Southwest Florida Water Management District (SWFWMD), and these two agencies divide the responsibilities of mapping seagrass coverage within the Charlotte Harbor region. In northern Charlotte Harbor, the SWFWMD has conducted seagrass mapping efforts on a roughly biennial basis since 1988, while the SFWMD recently initiated the undertaking of biennial seagrass mapping efforts within their jurisdiction of southern Charlotte Harbor in 1999. In 1999, the most recent comprehensive mapping effort for the harbor, the two districts used somewhat dissimilar methodologies in their mapping efforts in that the minimum mapping units, map accuracy standards, and classification of "patchy" vs. "continuous" seagrass coverage (see below for explanation) varied slightly.

Seagrass maps are produced through a multistep process. First, aerial photographs are obtained during times of good water clarity and moderately high seagrass biomass — usually November or December (*Note:* The 2002 photographs were taken in January 2002) — after the summer rains have ceased. True color photographs at a scale of 1:24,000 are used. Appropriate environmental conditions, such as 2-m depth for water clarity, wave height less than 0.6 m, and wind speed less than 5 m/s (10 knots), are required on the day the photographs are obtained.

Next, investigators examine bottom cover *in situ* at various locations to allow identification of distinct photographic signatures and to investigate unusual signatures. In the office, the field classifications are matched to signatures on the photographs. Seagrass signatures are divided into two classes: continuous coverage (approximately <25% unvegetated bottom visible within a polygon) and patchy (approximately >25% unvegetated bottom visible within a polygon). However, mapping accuracy of these two polygon classifications could not be ground-truthed by site visits. For this effort, coverages for both polygon classes were combined. The minimum mapping unit is approximately 0.2 ha. It should be noted that in 1999 the SFWMD required a minimum mapping unit of 0.1 ha and defined "continuous" seagrass beds as polygons with approximately >85% cover.

For earlier mapping efforts (1988, 1990, 1992, 1994, and 1996) the individual polygons were delineated on transparent Mylar® sheets placed on top of the aerial photographs. A zoom transfer scope was used to transfer the delineated polygons to USGS quadrangles. Next, the polygons were digitally transferred to an ARC/Info database. The resulting seagrass maps meet USGS National Map Accuracy Standards for 1:24,000 scale maps. For the 1999 and 2002 Southwest Florida Water Management District seagrass maps, tighter ground control and more sophisticated mapping techniques were utilized to meet 1:12,000 National Map Accuracy Standards while still using 1:24,000 scale photographs. Analytical stereo-plotters were used for photointerpretation in lieu of stereoscopes. This method allowed for the production of a georeferenced digital file of the photo-interpreted images without the need for additional photographs to be transferred to maps. Instead of drawing complete polygons each year, effort and errors have been reduced by using the previous effort's digital coverage as the baseline and delineating any changes to seagrass extent for the current effort. This method has provided a change analysis as well as current seagrass coverage.

Hard-copy plots were produced and checked for errors. Finally, a number of randomly chosen points were identified and plotted for a post-map production classification accuracy assessment. The points were randomly selected using Arc/Info processes by first defining the coordinates of the study area, generally the BND coordinates of the Arc/Info coverage. The point selection then involved the random generation of numbers based on the minimum and maximum values of the X and Y coordinates of the study area. The numbers that were generated were stored as variables and a selection was made from the Arc/Info coverage to see if they fit the criteria specified (i.e., seagrass codes = "patchy" or "continuous"). A variable was also set up to be used as a counter, and set to a value of zero (0). If the area did not fit the selection criteria, the "counter" variable was not calculated and the loop ran itself again. If the area fit the selection criteria, a point was placed at the position, the coordinates were stored in the variable, and the "counter" variable was calculated with the next value. This process was repeated until approximately 10 to 20 points per estuary region were selected. Field staff then utilized the coordinates for the randomly chosen sites, a site map and a global positioning system (GPS) to visit the locations in the field and classify the bottom cover. These *in situ* inspections were compared to the map classifications to develop an unbiased determination of the map's classification accuracy.

Transects

As a supplement to the water management districts' aerial photography mapping efforts to estimate seagrass extent, the Florida Department of Environmental Protection–Charlotte Harbor Aquatic Preserves Office established a series of 50 transects, distributed over the various sub-basins of the harbor, to quantify seagrass species composition, distribution, and abundance. Beginning in 1999, these 50 fixed transects (Figure 15.2) have been visited annually during September through November to determine changes in seagrass conditions by detecting differences in seagrass depth distributions, abundance, epiphyte coverage, short-shoot densities, and species composition. Each transect consists of a fixed line, determined by a compass heading and marked with PVC (polyvinylchloride) stakes, extending from the shoreward seagrass edge out to the deep edge of the meadow. Program researchers collect depth measurements, seagrass species abundance (Braun-Blanquet Cover Scale), blade length, sediment type, and epiphyte coverage and type at 50-m intervals along each transect (or 10-m intervals for transects shorter than 50 m) from shore to edge of bed (Staugler and Ott, 2001). Depth measurements were adjusted to mean water depth by adjusting the tide level observed in the field to mean water based on the 12 National Oceanographic and Atmospheric Administration (NOAA) tide stations located throughout the study area.

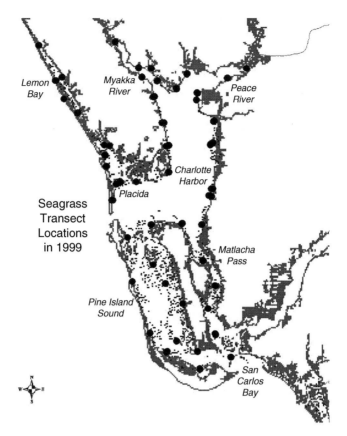

FIGURE 15.2 Graphic of transect locations. (Data provided by CHNEP, FDEP, SFWMD, and SWFWMD.)

Results

Seagrass Coverage

The first comprehensive evaluation of seagrass coverage in the Charlotte Harbor region was completed in the 1983 FDNR and FDOT effort (Harris et al., 1983). The study compared seagrass coverage of Charlotte Harbor derived from black-and-white photographs from 1946 to 1951 (referred to in this report as 1945) to data derived from positive false-color infrared transparencies produced from flights in April 1982. The study reported the coverage in acres by USGS quadrangles (Figure 15.3); for this report, the coverages have been converted to hectares (Table 15.1).

The report documented a 29% decrease in seagrass, from 33,572 ha (82,959 acres) to 23,672 ha (58,495 acres) between 1945 and 1982 for the harbor, excluding Lemon and southern Estero Bays. All quadrangles within Charlotte Harbor demonstrated losses, ranging from 6 to 87%; however, the study determined that 40% of the total region-wide loss was located solely within the Captiva quadrangle region. When combined with the loss of Pine Island Center, Wulfert, and Sanibel quadrangles, this loss equaled 57% of the total Charlotte Harbor region-wide loss of 29%.

The 1945 data exist neither in digital format nor are they reported with comparable geographic boundaries as the 1982 and subsequent coverage data. Also, the quality of the photographs used to obtain the 1945 seagrass coverage was questionable for the purpose of delineating seagrass. The absence of ground-verification during the years the photographs were produced is reason for caution when examining these data. Thus, the 1945 data will be discussed only briefly in the following analyses.

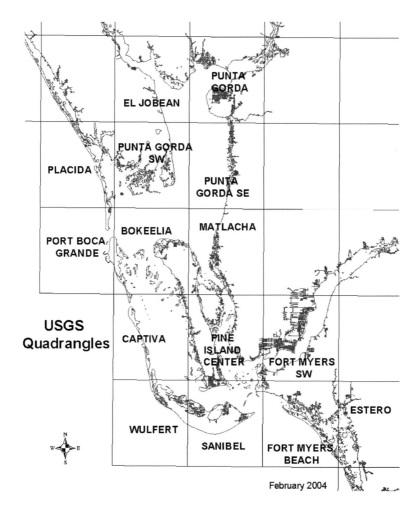

FIGURE 15.3 Graphic of USGS quads used for Harris et al. (1983). (Data provided by CHNEP, USGS, and NOAA.)

   Results of current mapping efforts in Charlotte Harbor from 1982 forward are reported according to water management district geopolitical regions (northern and southern Charlotte Harbor) and also further segmented into 14 sub-basins (Figure 15.4; Table 15.2).

   The first reliable seagrass interpretation work for Charlotte Harbor was created with the 1982 mapping effort. Table 15.3 represents the seagrass coverage data collected to date with comparable methodologies for the Charlotte Harbor region. Apparent from this table is that there exists more data within the northern Charlotte Harbor region than in the southern region, and there exists only two harbor-wide data sets, 1982 and 1999. A change analysis by sub-basin of these two data sets is presented in Table 15.4. However, the 1982 data set does not include seagrass coverage for the Lemon Bay sub-basin, and the data for the Estero Bay sub-basin incorporate data derived from photographs taken in 1990. Thus, for the following comparisons of seagrass extent in Charlotte Harbor, area calculations for the Lemon and Estero Bays sub-basins are not included.

   Interpretation of the 1982 photographs resulted in harbor-wide total of 23,127 ha of seagrass, excluding Lemon and Estero Bays. Combining both northern and southern Charlotte Harbor mapping efforts in 1999 produced seagrass estimates of 21,802 ha for the entire region, a 6% (1325 ha) decrease in coverage.

   Combined estimates for the seven sub-basins in Charlotte Harbor under the SWFWMD jurisdiction (Myakka River, Peace River, East Wall, West Wall, Middle Harbor, Placida Region, and South Harbor) demonstrate no trends in seagrass coverage between 1982 and 1999; however, the region does demonstrate interannual variability. Coverage for the seven sub-basins has fluctuated within a variance of less

TABLE 15.1

Historical Seagrass Extent[a] (in ha)

USGS Quad Name	1945	1982	Change	% Change
El Jobean	660	362	–299	–45
Punta Gorda SW	2,785	2,331	–454	–16
Placida	1,056	634	–422	–40
Bokeelia	4,919	4,600	–318	–6
Port Boca Grande	155	27	–128	–83
Captiva	8,056	4,112	–3,944	–49
Wulfert	1,112	677	–435	–39
Sanibel	2,143	1,594	–549	–26
Punta Gorda	361	312	–49	–13
Punta Gorda SE	1,718	1,441	–277	–16
Matlacha	2,339	1,999	–340	–15
Pine Island Center	4,639	3,919	–720	–16
Fort Myers Beach	1,451	1,063	–388	–27
Fort Myers SW	593	76	–516	–87
Estero	1,585	523	–1,062	–67
Total	33,572	23,672	–9,900	–29

[a] Data converted from acreages reported in Harris et al. (1983). Sub-basins represent the reporting units for the northern and southern Charlotte Harbor region.

than 1000 ha since 1982. The 1999 extent is 10 ha less than the 1982 value, while the 2002 coverage is 16 ha greater (Figure 15.5).

The five Charlotte Harbor sub-basins within SFWMD jurisdiction (Pine Island Sound, Matlacha Pass, San Carlos Bay, Lower Caloosahatchee River, and Upper Caloosahatchee River) encompass the majority of the seagrass coverage in Charlotte Harbor — almost double that of the northern region in 1999. These five subsegments experienced an 8% (1315 ha) decrease in seagrass from 1982 to 1999. The Matlacha Pass, San Carlos Bay, and lower Caloosahatchee sub-basins alone account for approximately 77% of the overall 6% seagrass acreage decline in the entire Charlotte Harbor region from 1982 to 1999. It is interesting to note that between 1945 and 1982, there was a reported decrease of over 3944 ha in the Pine Island Sound region (approximately the Captiva quadrangle in the 1983 FDNR and FDOT effort), while between 1982 and 1999, the area (Pine Island Sound sub-basin) gained 631 ha in extent.

Deep Edge of Seagrass Beds

The seagrass transect data are analyzed using roughly the same sub-basin boundaries as the seagrass coverage data, except that there were no fixed transects within the Middle Harbor, Estero Bay, and Caloosahatchee sub-basins from 1999–2001. Also, data for the South Harbor sub-basin are incorporated into the Pine Island Sound and Placida sub-basins. Figure 15.6 shows the average maximum depths to which seagrass beds grew for each sub-basin in 1999, 2000, and 2001.

As a result of the relatively short period of record, these transect data have not been analyzed for statistical significance. Also, observed changes may be within the realm of error due to sampling and/or conversion to tide-normalized values. Accordingly, caution must be used in drawing conclusions from the following analyses. Between 1999 and 2000, the average maximum depth for seagrass beds increased in every sub-basin, except the East Wall of Charlotte Harbor, in which the depth remained constant at 114 cm, and in Pine Island Sound, which experienced a decrease of 8 cm in average maximum depth. From 2000 to 2001 seagrass beds in every sub-basin receded to shallower depths, except in Lemon Bay, which experienced an increase of 3 cm in maximum depth. For the entire 3-year period of 1999 to 2001, seagrass receded in average maximum depth in the majority of sub-basins and increased in depth only in the West Wall, Matlacha Pass, and Lemon Bay sub-basins. These preliminary results indicate that the maximum depths to which seagrass grow within each sub-basin may demonstrate interannual variability. However, further information is needed to determine if these changes are significant or due to sampling and/or conversion errors.

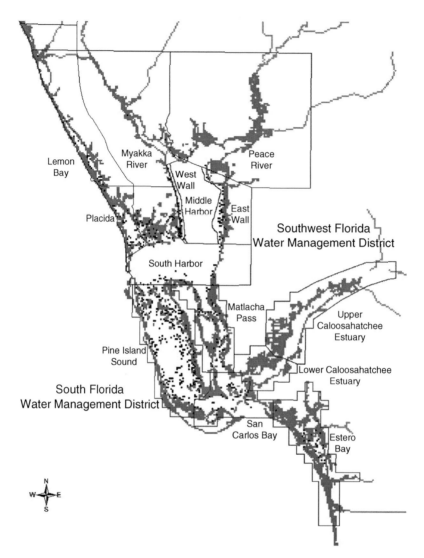

FIGURE 15.4 Map of 14 seagrass sub-basins. (Data provided by CHNEP, SFWMD, SWFWMD, and NOAA.)

TABLE 15.2

The 14 Sub-Basins Created for Analyses of Seagrass Coverage in the Greater Charlotte Harbor Region

SWFWMD Region (northern Charlotte Harbor)	SFWMD Region (southern Charlotte Harbor)
Lemon Bay	Charlotte Harbor:
Peace River	Pine Island Sound
Myakka River	Matlacha Pass
Charlotte Harbor:	San Carlos Bay
Middle Harbor	Lower Caloosahatchee River
West Wall	Upper Caloosahatchee River
East Wall	Estero Bay
Placida Region	
South Harbor	

TABLE 15.3

Seagrass Coverage (by ha) by Year in the 14 Sub-Basins of the Charlotte Harbor Region

Subsegment	Year						
	1982*	1988	1992	1994	1996	1999	2002
Lemon Bay		1,055		1,073	1,054	1,044	1,049
Myakka River	238	202	130	189	209	191	185
Peace River	378	158	166	196	225	109	137
Charlotte Harbor							
East Wall	1,548	1372	1,361	1,416	1,371	1,452	1,454
West Wall	672	585	495	675	794	699	699
Middle Harbor	70	50	50	60	76	63	64
Placida Region	948	1,408	1,376	1,337	1,450	1,503	1,531
South Harbor	3,513	3,684	3,636	3,633	3,626	3,340	3,313
Subtotal (excluding Lemon Bay)	7,367	7,458	7,214	7,505	7,751	7,357	7,383
Charlotte Harbor							
Pine Island Sound	9,853					10,484	
Matlacha Pass	3,245					2,456	
San Carlos Bay	2,420					1,504	
Upper Caloosahatchee River	0					0	
Lower Caloosahatchee River	242					1	
Estero Bay	2,504*					1,008	
Subtotal (excluding Estero Bay)	15,760					14,445	
Total	25,631	8,513	7,214	8,578	8,805	23,854	8,432

* FMRI interpreted Estero Bay from 1990 photographs to fill a void in the 1982 maps.

TABLE 15.4

Comparison of 1982 to 1999 Seagrass Coverage (in ha) by Sub-Basin

Subsegment	Year		Change	% Change
	1982	1999		
Myakka River	238	191	−47	−20
Peace River	378	109	−269	−71
Charlotte Harbor				
East Wall	1,548	1,452	−96	−6
West Wall	672	699	27	4
Middle Harbor	70	63	−7	−10
Placida Region	948	1,503	555	59
South Harbor	3,513	3,340	−173	−5
Subtotal	7,367	7,357	−10	0
Charlotte Harbor				
Pine Island Sound	9,853	10,484	631	6
Matlacha Pass	3,245	2,456	−789	−24
San Carlos Bay	2,420	1,504	−916	−38
Upper Caloosahatchee River	0	0	0	0
Lower Caloosahatchee River	242	1	−241	−100
Subtotal	15,760	14,445	−1315	−8
Total	23,127	21,802	−1325	−6

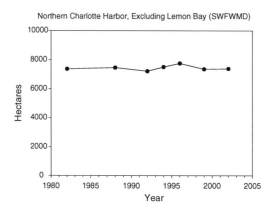

FIGURE 15.5 Seagrass extent in the northern Charlotte Harbor region since 1982.

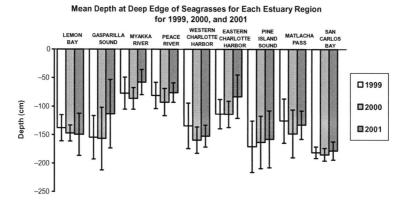

FIGURE 15.6 Comparison of average maximum depths (in centimeters) of seagrass beds by year by sub-basin.

Discussion

From the seagrass data compiled to date, it is difficult to ascribe changes in coverage to specific anthropogenic impacts and, thus, to utilize seagrass coverage as an environmental indicator of the condition of the Charlotte Harbor ecosystem. There are some data that suggest overall losses of seagrass coverage in the harbor. For example, the authors of the 1983 FDNR and FDOT effort (Harris et al., 1983) concluded there was a decrease in seagrasses in the deep edges of seagrass beds between the 1940s and 1982 and speculated that the loss was the result of decreasing water clarity with increasing pollutants and changing drainage patterns. But while this study represents the best available information of historical conditions, the methods used to obtain the 1945 data set are reason for caution in formulating strong conclusions. A comparison of 1982 to 1999 coverage data also shows an overall 6% harbor-wide decrease in seagrass extent; however, it is difficult to draw substantial conclusions from only two data sets.

Direct impacts to seagrass coverage may play a significant role in the potential loss of coverage since the 1940s. Harris et al. (1983) determined that 40% of the total region-wide loss was located within the lower Pine Island Sound area. When combined with the loss of southern Matlacha Pass and San Carlos Bay areas, this loss equaled 57% of the total Charlotte Harbor region-wide loss of 29%. The authors attributed this substantial loss to the dredging of the Intracoastal Waterway (ICW) and construction of a bridge and multiple causeway islands to Sanibel Island in the 1940s and 1960s, among other potential factors. Using nautical maps, the authors noted that a shallow bar less than 1.5 m in depth with deeper channels (2.4 to 4.6 m) on either side, had served as a tidal node that extended entirely across Pine Island

Sound. During an ebbing tide, flow occurred to the north and south of this bar. This shallow bar was apparently the location of one of the first channel dredging operations in the 1940s. In the 1960s, the ICW was dredged through Pine Island Sound and up the Caloosahatchee River, and the Sanibel Causeway, including its spoil islands, was constructed across San Carlos Bay. The authors reasoned that the diversion of water from the Caloosahatchee River into Pine Island Sound and the changes of circulation patterns as a result of these two projects have lowered the salinities in the lower Caloosahatchee River estuary, San Carlos Bay, Matlacha Pass, and Pine Island Sound areas. Also, during the dredging of the ICW and construction of the Sanibel Causeway, there was a direct loss of seagrasses due to the excavation and spoil deposition on the nearby seagrass beds (cited in Harris et al., 1983, and James Beever, FWC, written commun.). Following both projects, loss of seagrasses ensued due to turbidity and spoil spread (James Beever, FWC, written commun.).

Indirect impacts to seagrass extent in Charlotte Harbor from pollutant loads have not been documented to date. Relative to Tampa, Sarasota, and Lemon Bays to the northwest, Charlotte Harbor is highly influenced by the freshwater inflows of its large watershed. The surface area of the harbor is approximately 700 km^2 while the watershed is almost 11,300 km^2, a ratio of watershed to open water of more than 16:1 (SWFWMD, 2000). The result of this large ratio of watershed to open water is that the water clarity of the harbor is reduced due to the influx of tannins and suspended matter from the watershed. McPherson and Miller (1987) found that non-chlorophyll suspended matter (including detritus, cellular material, and minerals) accounts for an average of 72% of light attenuation in the water column, color (dissolved matter) accounts for 21%, phytoplankton chlorophyll for 4%, and water itself the remaining 3%, while Dixon and Kirkpatrick (1999) found that color, turbidity, and chlorophyll accounted for 66, 31, and 4% of light attenuation. Water clarity in the harbor increases with increasing salinity (McPherson and Miller, 1987, 1994; Dixon and Kirkpatrick, 1999; Tomasko and Hall, 1999). Thus, the light reaching the tops of seagrass beds is largely a factor of basin runoff and flows from the three major tributaries — the Peace, Myakka, and Caloosahatchee Rivers (McPherson and Miller, 1987; Doering and Chamberlain, 1999; Tomasko and Hall, 1999). In turn, seagrass coverage changes in the harbor and the tidal reaches of the rivers are thought to be a function of changes in these freshwater inflows. For example, between the 1996 and 1999 mapping events, the harbor experienced higher levels of rainfall due to an *El Niño* event. During that time period, there was a decrease in seagrass coverage. Conversely, between the 1999 and 2002 mapping events, the harbor experienced a drought period and also an increase in seagrass coverage in northern Charlotte Harbor. Generally, the maximum depths of seagrass beds increase with increasing distance from the mouths of the Peace and Myakka Rivers and increasing salinities (Dixon and Kirkpatrick, 1999).

Natural interannual variability of seagrass coverage helps obscure the reasons behind changes of extent. No significant trends in seagrass extent are evident within the northern Charlotte Harbor region between 1982 and 1999, with a mean coverage value of 7,432 ha. However, water quality data collected within the harbor at long-term sampling locations reveal that water clarity in the harbor may be on the decline, as data demonstrate increasing trends in suspended matter — one of the major light attenuators within the water column. Janicki Environmental, Inc. (2003) found significant increasing trends through 2000 of total suspended solids at numerous stations throughout the harbor, along with increasing turbidity within numerous watersheds in southern Charlotte Harbor. Analysis of seagrass extent from 1982 to 1999 exhibits an overall harbor-wide 6% decrease in seagrass extent. The majority, 77%, of this loss was located within the five sub-basins of the southern Charlotte Harbor region. However, considering the variability of the northern region of Charlotte Harbor, this loss may represent a loss due to the *El Niño* rainfall event and concomitant changes in salinity of the time period before the 1999 mapping effort rather than long-term changes in water clarity.

Conclusions

Despite changes in water quality and hydrology, long-term losses of seagrass have not been well documented in Charlotte Harbor. Current seagrass monitoring efforts in Charlotte Harbor are extensive; however, natural variability and confounding direct and indirect impacts as well as insufficient data, especially in the southern region, inhibit the ability to attribute changes in seagrass coverage to specific causes. Thus, seagrass coverage does not meet some of the EPA evaluation criteria for indicators of environmental condition (see U.S. EPA, 2000). For example, a management action to alleviate increases in turbidity and total suspended solids in southern Charlotte Harbor, such as increased treatment of storm water runoff, may not lead to increases in seagrass extent in the area. Seagrass coverage could be more strongly influenced by changes in salinity and the quantity and timing of freshwater flows from the Caloosahatchee River. Thus, based on the current available data for the Charlotte Harbor region, seagrass coverage fails to meet the "linkage to specific management action" guideline recommended by the EPA (see U.S. EPA, 2000). Additionally, the environmental conditions of the Charlotte Harbor watershed have not remained "relatively ecologically stable" (see U.S. EPA, 2000) since regular monitoring of seagrass began. Rapid urbanization and other land-use changes have caused hydrologic and water quality impacts to the watershed (Janicki Environmental, Inc., 2003), but seagrass coverage has remained relatively stable in northern Charlotte Harbor. Thus, seagrass fails to meet the "temporal variability across years" guideline recommended by the EPA as well (U.S. EPA, 2000).

In the northern Charlotte Harbor, the seagrass coverage has shown considerable year-to-year variation, but has remained stable since 1982. In the southern portion, some loss may have occurred between 1945 and 1982 and between 1982 and 1999. However, given only a few points in time (1945, 1982, and 1999) and the interannual variation to the north, it is difficult to determine if this loss is real, and if so, the specific causes of any such loss. What does seem clear is that interannual variation in northern Charlotte Harbor reflects interannual variation in freshwater input from major tributaries. In years with high discharge, seagrass coverage tends to be lower, and in years with little discharge, seagrass coverage tends to be higher. Reductions in coverage during high flow years may be due to low salinity and/or lower light availability associated with the increase of dissolved and suspended matter from the watershed. Thus, while seagrass coverage may not be an adequate indicator of long-term trends in the health of Charlotte Harbor, seagrass coverage may be a good indicator of interannual changes of freshwater input.

Overarching questions for the use of seagrass coverage as an indicator of ecosystem condition are the adequacy of coverage data and the period of record for these data. Many of the considerable water quality and hydrologic impacts to the watershed, such as the degradation of water quality within the Peace River basin and the artificial connection of the Caloosahatchee River with Lake Okeechobee, occurred before reliable seagrass monitoring began in 1982. It could be that seagrass coverage in Charlotte Harbor receded before baseline data were collected, and the current long-term (1982 to 1999) mean extent of 7432 ha in northern Charlotte Harbor already reflects impacts due to human influences.

Or possibly, seagrass extent in Charlotte Harbor has suffered little from changes to the watershed over time. Further monitoring and analysis, such as spatial and temporal changes of species composition and depth distributions, are needed to resolve these questions.

Acknowledgments

This chapter was made possible only by the tremendous effort of numerous individuals. Betty Staugler and Katie Fuhr performed fieldwork and analysis of the seagrass transect data. Tomma Barnes was the project manager for the southern Charlotte Harbor seagrass mapping efforts in 1999 (and currently). Tim Walker, Bob Diogo, and Tim Lieberman provided enormous support for the GIS-based graphics and analysis. Lesley Ward provided detailed information on post-map production processes for seagrass maps. Tom Ries helped establish original site locations for seagrass transects. The analysis of results was influenced immensely by discussions with James Beever, Paul Carlson, Aaron Adams, Brad Robbins, Bob Chamberlain, and Lisa Beever.

References

AGRA Baymont, 2001. Southwest Florida Seagrass Mapping Project: Final Report for Florida Fish and Wildlife Conservation Commission, Florida Marine Research Institute, South Florida Water Management District, Fort Myers, FL.

Baker, B. 1990. Draft Caloosahatchee water quality based effluent limitations documentation (Lee County). Florida Department of Environmental Regulation, Water Quality Technical Series 2, 121 pp.

Beever, J. 2002. E-mail correspondence to Catherine Corbett, February 25, 2002.

Camp Dresser & McKee, Inc. 1998. The Study of Seasonal and Spatial Patterns of Hypoxia in Upper Charlotte Harbor. Report to Southwest Florida Water Management District, Southwest Florida Water Management District, Venice, FL.

Camp Dresser & McKee, Inc. 2001. Lake Hancock Restoration Management Plan, Prepared for Polk County Board of County Commissioners and Florida Department of Environmental Protection, October 2001.

Chamberlain, R. H and P. H. Doering. 1997. Preliminary estimate of optimum freshwater inflow to the Caloosahatchee estuary: a resource-based approach. In *Proceedings of the Charlotte Harbor Public Conference and Technical Symposium*, Charlotte Harbor National Estuary Program, North Fort Myers, FL.

Chesapeake Executive Council. 1993. Directive No. 93-3 Submerged Aquatic Vegetation Restoration Goals. Annapolis, MD.

CHNEP. 1999. Synthesis of Existing Information, Volume 1: A Characterization of Water Quality, Hydrologic Alterations and Fish and Wildlife Habitat Loss in the Greater Charlotte Harbor Watershed, Charlotte Harbor National Estuary Program, North Fort Myers, FL.

CHNEP. 2000. Committing to our Future, Charlotte Harbor National Estuary Program, North Fort Myers, FL.

Coyne, M. S., M. E. Monaco, M. Anderson, W. Smith, and P. Jokiel. 2001. Classification Scheme for Benthic Habitats: Main Eight Hawaiian Islands. Available at http://biogeo.nos.noaa.gov/projects/mapping/pacific/main8/classification/. National Oceanic and Atmospheric Administration, Silver Spring, MD.

DeGrove, B. D. 1981. Caloosahatchee River wasteload allocation documentation. Florida Department of Environmental Regulation, Water Quality Technical Series 2, 52 pp.

DeGrove, B. and F. Nearhoof. 1987. Water quality assessment for the Caloosahatchee River. Florida Department of Environmental Regulation, Water Quality Technical Series 3, 19 pp.

Dixon, L. K. and G. J. Kirkpatrick. 1999. Causes of Light Attenuation with Respect to Seagrasses in upper and Lower Charlotte Harbor, Report to Southwest Florida Water Management District, SWIM Program and Charlotte Harbor National Estuary Program. Report available from Southwest Florida Water Management District, SWIM Program, Tampa, FL and Charlotte Harbor National Estuary Program, North Fort Myers, FL.

Dobson, J. E., E. A. Bright, R. L. Ferguson, D. W. Field, L. L. Wood, K. D. Haddad, H. Iredale III, J. R. Jensen, V. V. Klemas, R. J. Orth, and J. P. Thomas. 1995. NOAA Coastal Change Analysis Program (C-CAP): Guidance for Regional Implementation. NOAA Technical Report NMFS 123. U.S. Department of Commerce, Seattle, WA.

Doering, P. H. and R. H. Chamberlain. 1999. Water quality and the source of freshwater discharge to the Caloosahatchee Estuary, FL. *Water Resources Bulletin* 35:793-806.

Environmental Research and Design. 2003. Caloosahatchee water Quality Data Collection Program. Final Interpretive Report for Years 1–3. Report to the South Florida Water Management District, 3301 Gun Club Road, West Palm Beach, FL 33406.

FDEP. 2003. Draft Water Quality Status Report Sarasota Bay and Peace and Myakka Rivers, September 2003, Florida Department of Environmental Protection, Division of Water Resource Management, Tallahassee, FL.

Hammett, K. M. 1990. Land Use, Water Use, Streamflow and Water Quality Characteristics of the Charlotte Harbor Inflow Area, Florida. Water Supply Paper 2359-A. Prepared in cooperation with the Florida Department of Environmental Regulation. U.S. Geological Survey, Tallahassee, FL.

Harris, B. A., K. D. Haddad, K. A. Steidinger, and J. A. Huff. 1983. Assessment of Fisheries Habitat: Charlotte Harbor and Lake Worth, Florida, Final Report, Florida Department of Natural Resources. Available from Florida Fish and Wildlife Conservation Commission-Florida Marine Research Institute, St. Petersburg, FL.

Janicki Environmental, Inc. 2003. Water Quality Data Analysis and Report for the Charlotte Harbor National Estuary Program, August 27, 2003. Available from the Charlotte Harbor National Estuary Program, North Fort Myers, FL.

Kemp, W. M. 2000. Seagrass Ecology and Management: An Introduction. In *Seagrasses: Monitoring, Ecology, Physiology, and Management*, S. Bortone (ed.), CRC Press, Boca Raton, FL, pp. 1–6.

Kendall, M. S., C. R. Kruer, K. R. Buja, J. D. Christensen, M. Finkbeiner, and M. E. Monaco. 2001. Methods Used to Map the Benthic Habitats of Puerto Rico and the U.S. Virgin Islands, http://biogeo.nos.noaa.gov/projects/mapping/caribbean/startup.htm. NOAA National Ocean Service, Silver Spring, MD.

Kiraly, S. J., F. A. Cross, and J. D. Buffington. 1990. Overview and recommendations. In Federal Coastal Wetland Mapping Programs. Fish and Wildlife Service, Department of the Interior. Biological Report 90, 18 pp.

Kurz, R. C., D. A. Tomasko, D. Burdick, T. F. Ries, K. Patterson, and R. Finck. 2000. Recent trends in seagrass distributions in southwest Florida coastal waters. In *Seagrasses: Monitoring, Ecology, Physiology, and Management*, S. A. Bortone (ed.). CRC Press, Boca Raton, FL, pp. 157–166.

McPherson, B. F. and R. L. Miller. 1987. The vertical attenuation of light in Charlotte Harbor, a shallow, subtropical estuary, south-western Florida, *Estuarine, Coastal, and Shelf Science* 25:721–737.

McPherson, B. F. and R. L. Miller. 1994. Causes of light attenuation in Tampa Bay and Charlotte Harbor, southwestern Florida, *Water Resource Bulletin* 30(1):43–53.

McPherson, B. F., R. L. Miller, and Y. E. Stoker. 1996. Physical, Chemical, and Biological Characteristics of the Charlotte Harbor Basin and Estuarine System in Southwestern Florida — A Summary of the 1982–89 U.S. Geological Survey Charlotte Harbor Assessment and Other Studies, U.S. Geological Survey Water Supply Paper 2486. Prepared in cooperation with the Florida Department of Environmental Protection. Available from U.S. Geological Survey, Denver, CO.

Morrison, G., R. Montgomery, A. Squires, R. Starks, E. DeHaven, and J. Ott. 1997. Nutrient, chlorophyll and dissolved oxygen concentrations in charlotte harbor: existing conditions and long-term trends, in *Proceedings of the Charlotte Harbor Public Conference and Technical Symposium*, Charlotte Harbor National Estuary Program, North Fort Myers, FL.

Moulton, D. W., T. E. Dahl, and D. M. Dall. 1997. Texas Coastal Wetlands: Status and Trends, Mid-1950s to Early 1990s. U.S. Department of the Interior, Fish and Wildlife Service, Southwestern Region, Albuquerque, NM.

NOAA. 1996. NOAA's Estuarine Eutrophication Survey. Vol. 1: South Atlantic Region. National Oceanic and Atmospheric Administration, Office of Ocean Resources Conservation Assessment, Silver Spring, MD.

SFWMD. 2000. Technical Documentation to Support Development of Minimum Flows and Levels for the Caloosahatchee River and Estuary. South Florida Water Management District, West Palm Beach, FL.

South Florida Ecosystem Task Force. 2002. Coordinating Success: Strategy for Restoration of the South Florida Ecosystem and Tracking Success: Biennial Report for FY 2001–2002 of the South Florida Ecosystem Task Force to the U.S. Congress, Florida Legislature, Seminole Tribe of Florida and Miccosukee Tribe of Indians of Florida, August 2002.

Shull, S. and D. A. Bulthuis. 2002. A Methodology for Mapping Current and Historical Coverage of Estuarine Vegetation with Aerial Photography and ArcView. Washington State Department of Ecology (Publication 03-06-020), Padilla Bay National Estuarine Research Reserve, Mount Vernon, WA. Padilla Bay National Estuarine Research Reserve Technical Report 26, 52 pp.

Staugler, E. and J. Ott. 2001. Establishing Baseline Seagrass Health Using Fixed Transects in Charlotte Harbor, Florida: 2 Year Seagrass Monitoring Summary 1999–2000, Technical Report 1. Florida Department of Environmental Protection, Charlotte Harbor Aquatic Preserves, Punta Gorda, FL.

SWFWMD. 2000. Charlotte Harbor Surface Water Improvement and Management (SWIM) Plan Update — Spring 2000. December, 2000. SWIM Section of Southwest Florida Water Management District, Tampa, FL.

SWFWMD. 2002. Upper Peace River: An Analysis of Minimum Flows and Level, Draft Document, August 25, 2002. Southwest Florida Water Management District, Brooksville, FL.

TBEP. 2000. *Seagrass Management: It's Not Just Nutrients!* Proceedings from a Symposium held August 22–24, 2000, St. Petersburg, FL. Available from Tampa Bay Estuary Program, St. Petersburg, FL.

Texas Parks and Wildlife. 1999. Seagrass Conservation Plan for Texas. Available from Texas Parks and Wildlife, Resource Protection Division, Austin.

Tomasko, D. A. 2001. Assessing the need (if any) for a pollutant load reduction goal (PLRG) in Charlotte Harbor, Florida. Presentation to the Policy Committee of the Charlotte Harbor National Estuary Program, December 2001. Southwest Florida Water Management District, Tampa, FL.

Tomasko, D. A. and M. O. Hall. 1999. Productivity and biomass of the seagrass *Thalassia testudinum* along a gradient of freshwater influence in Charlotte Harbor, Florida, *Estuaries* 22(3A):592–602.

Tomasko, D. A., D. L. Bristol, and J. A. Ott. 2001. Assessment of present and future nitrogen loads, water quality, and seagrass (*Thalassia testudinum*) depth distribution in Lemon Bay, Florida. *Estuaries* 24(6A):926–938.

U.S. Environmental Protection Agency. 2000. Evaluation Guidelines for Ecological Indicators, L. E. Jackson, J. C. Kurtz, and W. S. Fisher. May 2000, EPA/620/R-99/005, U.S. Environmental Protection Agency, Office of Research and Development, Research Triangle Park, NC.

Virnstein, R. W. and L. J. Morris. 1996. Seagrass Preservation and Restoration: A Diagnostic Plan for the Indian River Lagoon, Technical Memorandum 14. St. Johns River Water Management District, Palatka, FL.

16

Mangroves as an Indicator of Estuarine Conditions in Restoration Areas

Kathy Worley

CONTENTS

Introduction

More than 50% of the world's mangrove forests have been destroyed (Lewis, 1999). The majority of these losses are attributable to human activities (Parks and Bonifaz, 1994). More specifically these activities include development along coastal areas and concomitant altered hydrology due to dredging, filling, diking and impounding wetlands (Turner and Lewis, 1997). Mangroves dominate many coastal areas in subtropical and tropical biomes, yet the remaining forests are being continually degraded by development, pollution, overexploitation, and use as disposal sites for garbage and toxic chemicals (Ball, 1988).

What Is a Mangrove?

Mangroves are a group of coastal plants with similar adaptations (but are not necessarily genetically related) that exceed a half meter in height and grow primarily in the intertidal zone (Duke, 1992). Mangroves are found in warm humid climates usually between 25N and 25S latitude. They exist as low shrubs in harsh conditions and can attain over 40 m in height under favorable conditions. They are viviparous and possess a variety of adaptations that allow them to survive in "nasty" habitats, salty water, and reduced soils. Mangroves have a competitive advantage over other plants in saltwater areas. They can survive in freshwater environments, but they are poor competitors there. Mangroves remove excessive salt by either excreting salt (in or near their leaves) or by excluding salt (at their roots). Mangroves also have special aeration structures that include above ground prop roots or pneumatophores to accommodate the lack of gaseous exchange in the stagnant muck, typical of their habitat.

Worldwide there are about 65 recognized species of mangrove plants in approximately 20 families (Tomlinson, 1986; Field, 1995). Florida mangrove species include *Rhizophora mangle* L. (red mangroves), *Laguncularia racemosa* L. (white mangroves), and *Avicennia germinans* L. (black mangroves). Red mangroves are easily identified by their distinctive aerial prop roots that extend down from their branches. Red mangroves are generally found along the fringe of estuarine waterways along the coast. They are salt excluders and have membranes in their roots that exclude more than 90% of the salt found in seawater. Reds are viviparous and are one of the few plant species that have "live births." Each propagule that drops from the tree is already a living seedling. White mangroves are identified by their fleshy ovate leaves, each with two distinct salt-secreting nectaries or glands at their base. White mangroves often grow peg roots out of the substrate to aid gaseous exchange in stagnant, waterlogged soils. White mangroves are the smallest species that often pioneers disturbed coastal areas in Florida and are generally located inland of red mangroves. Black mangroves have a rough, dark bark, and can be easily identified by fingerlike roots (pneumatophores) that poke out from the substrate. The pneumatophores allow the plant to survive in waterlogged or oxygen-poor soil. From a single tree, there are often thousands of pneumatophores that vary in height dependent on local hydrologic conditions. Black mangrove leaves have a whitish underside and are capable of excreting excess salt that is externally visible. Black mangroves are usually found inland of red mangroves at a slightly higher elevation and, therefore, are often located inland of either red or white mangroves. The buttonwood, *Conocarpus erectus* L., is often included as a component of mangrove forests in the south Florida coastal ecosystem. They are commonly found in coastal brackish-water environments, usually at higher elevations in association with white mangroves. Because buttonwood lacks many of the morphological specializations typical of mangroves (aerial roots, vivipary), it is often not considered a "true mangrove" and was not included in experimental designs.

Why Are Mangroves Important Indicators?

Mangroves are a pivotal component of Florida's estuaries. Mangroves function as land stabilizers, prevent coastal erosion, and provide habitat for a variety of organisms (Odum and McIver, 1990). The canopy is rich in epiphytes and provides nesting sites for many species of birds. Mangrove roots both above and below the water provide a safe haven for organisms such as mangrove crabs, shrimp, barnacles, sea squirts, insect larvae, oysters, and mussels, while their branches house a variety of insects. In Florida, 191 species of birds, along with numerous species of reptiles, amphibians, mammals, and insects utilize the mangrove habitat (USFWS, 1999). Mangroves provide nursery areas for many species of fish. At some point during their lives, approximately 75% of commercially caught local fish and prawns depend directly on mangroves to help juveniles avoid predation and to provide shelter and feeding sites for fish species such as the opossum pipefish, Atlantic sturgeon, least killifish, mangrove rivulus, snapper, snook, and many others. Additionally, mangroves import inorganic matter from terrestrial systems and export organic matter to estuaries and oceans (Odum and Heald, 1975). Mangroves produce approximately 1 kg of litter/m^2/year, and the density of bacteria in mangrove mud is one of the highest in the world. This indicates high productivity (i.e., an estimated 80% of the total estuarine productivity; Robertson and Blaber, 1992), along with the concomitant nutrients, supports a huge variety of sea life through intricate estuarine food webs (Teas, 1979). Thus, mangroves provide one of the key building blocks for a healthy Florida estuary. The mass of mangrove detritus found in an area is a good predictor of the local epibenthos stock, which in turn affects local fish density (Robertson and Blaber, 1992). Mangrove detritus in association with fungi, bacteria, and protozoa form a large part of the foundation of detritus-based food webs that support nearshore secondary production and primary and secondary estuarine consumers. Thus, the condition and viability of mangroves can be useful in indicating whether or not an estuary is healthy, in decline, or recovering.

Die-Offs and Restoration

Mangrove die-offs can set into motion a chain of events, which result in either changing or collapsing the ecosystem, affecting both physical and biological components. To understand underlying causes of mangrove die-offs, one must first understand a variety of factors that influence the health of these systems (Twilley, 1998). Nutrient limitation, soil waterlogging, and soil salinity have been hypothesized as the principal factors controlling mangrove growth (Davis, 1940). Factors including hydrology, tidal regimes, geology, soil, and nutrient chemistry are important in understanding what drives mangrove systems. One hypothesis is that extended hydroperiod, deeper surface water depths, higher groundwater levels, and lower topography negatively affect mangrove health by further reducing existing soil conditions and blocking aeration pathways to black mangrove pneumatophores. Extended flooding, in turn, facilitates substrate erosion, lowering elevation that subsequently results in expanding the die-off. Soil and its chemical state are affected by topography, tidal range and sedimentation patterns, climate, and long-term sea level changes. When soils are inundated with water, anaerobic conditions usually result, reducing the rate oxygen can diffuse through the soil. If frequent flooding and drying out occurs, hypersaline conditions can develop that negatively affect mangrove distribution and growth (Boto, 1984). However, in mangroves located near developments, freshwater inundation often occurs due to storm water runoff and flood control mechanisms directed at removing excess fresh water from nearby developments, negating hypersalinity problems. It is suspected that successful restoration is dependent on decreasing the retention time of floodwaters following storm events to prevent "drowning" of emerging vegetation because prolonged flooding in black mangrove systems can cause extended stress on aerial roots, resulting in mass mortality (Odum and McIver, 1990).

Hypothetical Chronology of a Die-Off

In a typical die-off scenario (Figure 16.1), commercial or residential development occurs next to the mangroves, resulting in soil compaction, which subsequently reduces natural interstitial water flow — a situation of altered natural hydrology. This situation is often accompanied by a change in tidal flow and increased freshwater runoff into the mangroves that can result in an altered hydroperiod. Soil compaction during building most likely prevents above- and belowground sheetflow of water to the ocean, which in turn contributes to higher floodwaters, longer surface water retention, and increased surface water levels. This results in pooling (particularly in areas with lower elevation), and (if tidal circulation is cut off) the water becomes stagnant. Constant waterlogging results in anoxic soil conditions that, in turn, can lead to a buildup of toxins that are the products of soil reduction reactions. As the surface water levels rise dramatically and do not drain or evaporate quickly, the black mangrove pneumatophores become submerged, blocking gaseous exchange to the roots. If the pneumatophores are submerged for a prolonged period of time, the black mangroves figuratively "drown" and the result is their mass mortality. Die-offs initiate belowground decay, which can lead to ground-level subsidence causing die-off expansion. To break this cycle, the retention time of floodwaters following storm events must be decreased to prevent "drowning" of emerging vegetation. Importantly, one can occasionally observe something similar when hurricanes affect a mangrove system, cutting off water flow and subsequently causing mangrove die-offs from altered hydrology. Over time this land could sink, becoming an isolated pond, or regenerate to its earlier condition.

Restoring estuaries that have been altered, disturbed, or destroyed by anthropogenic factors often has varying degrees of success. Mangroves can serve as useful indicators of estuary recovery, particularly in tropical or subtropical areas where seagrasses are sparse or nonexistent. The research investigation

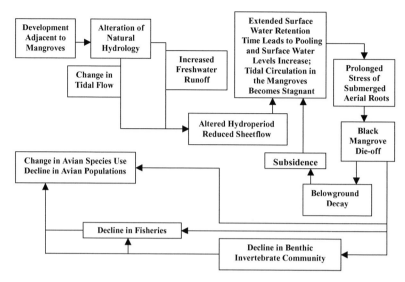

FIGURE 16.1 Chronology of a mangrove die-off.

described below was designed to assess the recovery of an estuarine system, post-restoration in a subtropical estuary located in Clam Bay, Collier County, Florida (USA).

Site Description

The Clam Bay estuary (Figure 16.2) consists of approximately 242.82 ha (600 acres) of bay and mangrove reserve and is one of the few dynamic estuarine systems remaining in the Cocohatchee-Gordon River Drainage System (Burch, 1990). The estuary includes a small, elongated system of shallow bays and mangrove swamps. Historically, Clam Bay was tidally connected to Wiggins Pass to the north and Doctor's Pass to the south. The north end of Clam Bay became isolated in the 1950s when road beds were constructed. Today, two deadend bays connect to the Gulf of Mexico at Clam Pass. Recent large-scale die-offs of black mangroves indicate that the impacts of intense development over the past three decades have influenced the demise of portions, if not all, of formerly pristine mangrove forests (Worley and Gore, 1995). In 1991, 5.67 ha of black mangroves died within 6 months of completion of a new hard-surfaced road atop new fill material. In 1995, a massive die-off of black mangroves began adjacent to the original die-back and began to extend southward along the western shore of upper Clam Bay (Figure 16.3). An unusually high rainfall season (the second highest precipitation year on record in the last 20 years) resulted in inundations of black mangrove pneumatophores for periods of 2 to 6 weeks, and soils remained saturated for more than 4 months. Altered soil chemistry, lack of tidal exchange, and high water surface water retention possibly contributed to the decline in productivity, growth, and eventual death of these mangroves. The black mangroves, were, in fact, probably slowly dying for many years and the rainfall event simply accelerated the process. Development surrounding Clam Bay (particularly to the north, which is almost completely enclosed by roads, walls, and houses) has caused major changes in the hydrology (Worley and Gore, 1995). The impact on soil during building has most likely prevented above- and belowground sheetflow of water to the Gulf of Mexico. Cutting off water flow to the Gulf has contributed to higher floodwaters and longer water retention times within the mangrove system. In 1999, the local government initiated a 10-year restoration project that included dredging Clam Pass and the main tributaries, clearing existing smaller tributaries, located near the main die-off, with dynamite, and establishing new hand-dug channels to try to drain out the excess surface water and encourage tidal flushing (Figure 16.4).

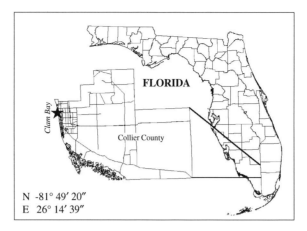

FIGURE 16.2 Clam Bay site location Naples, Florida.

FIGURE 16.3 Clam Bay estuary.

The objectives of this project are to (1) evaluate the general health of the Clam Bay estuary over time; (2) gauge mangrove recovery in areas that have died out; (3) compare pre- and post-restoration project recovery throughout Clam Bay.

Methods

A gradsect sampling regime, a variant of stratified random sampling regimes (Gillison and Brewer, 1985), was used to determine placement of monitoring plots to evaluate mangrove species occurrence and abundance. This environmental stratification approach uses key environmental variables such as substrate, hydrology, species, topography, mangrove condition (i.e., dead, stressed, or alive) and proximity to development to characterize the mangrove forest into three discreet "gradsects" or themes (i.e., die-off areas, stressed areas, or live "healthy" areas). These variables were based on prior knowledge of

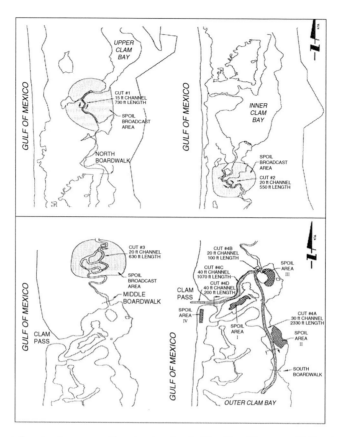

FIGURE 16.4 Excerpts from local government's restoration plan for Clam Bay. (Courtesy of Pelican Bay Service Division.)

the Clam Bay mangrove forest and were believed to be influential factors in forest deterioration, while providing an adequate representation of the key environmental gradients present in Clam Bay. Once the mangrove forest was classified into the three themes it resembled a mosaic of polygons from which monitoring plots were selected. A total of 12 plots were established pre-restoration and monitored semiannually, 4 plots were randomly established within die-off areas, 4 plots were randomly established within stressed areas, and 4 plots were randomly established in mangrove areas that were alive. These 12 plots were designated according to condition, species mix, topography, substrate, and hydrology and were classified pre-restoration as alive, or stressed, or in a die-off area, to facilitate an understanding of the condition of each plot and possible contributors to its existing condition (Figure 16.5).

Each plot was circular in shape with a radius of 6 m (Smith, 2000). The center of each plot was mapped using geographic coordinates. Distance and bearing of each mangrove (tree or seedling) were measured in relation to the center of the plot to determine exact location of each tree and seedling within each plot. All trees >1.4 m in height within each plot were identified semiannually to species, tagged, measured for diameter at breast height (DBH), and visually classified for condition (alive, dead, or stressed). For purposes of this study, a seedling was defined as a propagule at least 32 cm tall and identified to species, tagged, and measured (height). Canopy cover was also estimated using a densiometer.

Semiannual estimates of floristic composition were used to characterize and assess Clam Bay's recovery. Criteria used to assess plot condition over time included an evaluation of each individual tree and seedling for growth, health (branch and leaf loss, wilting, yellowing, galls, insect infestation, boring beetles in particular, and disease), percent canopy, and overall tree and seedling recruitment and mortality rates.

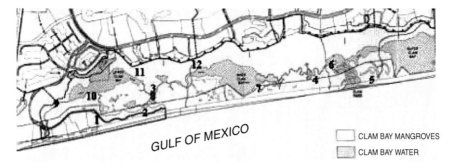

FIGURE 16.5 Clam Bay mangrove monitoring plots.

Results

Die-Off Areas

Of the original four plots established in mangrove die-off areas in Clam Bay, two plots have been upgraded to alive; one plot upgraded to stressed; and one plot has remained classified as a die-off area after 5 years of monitoring. Prior to restoration, these four die-off plots presented with the following characteristics: (1) more than 90% percent of the plot consisted of dead mangrove trees; (2) plots were subject to extended periods of inundation and high surface water levels (often for months at a time) due to freshwater runoff or tidal storm surges; (3) soils were waterlogged and redox levels were lower in comparison to stressed or alive "healthy" areas within Clam Bay; and (4) canopy cover was negligible or non-existent. Five years after restoration, three of the four plots exhibited signs of improvement. These areas became revegetated with seedlings, water levels and retention periods decreased, and canopy coverage increased by 14 to 45%.

Successful mangrove restoration can occur if the altered hydrology is restored, particularly with regard to the abatement of floodwater levels and retention periods. An indication of restoration success can be observed in a notable increase in mangrove seedling recruitment, followed by increased growth and viability over the years. This is what occurred in Plot 3. This plot had the most significant improvement throughout the 5-year monitoring period. Plot 3 is situated in the north, on the eastern side of the main tributary, which was dredged between the inner and upper bays in the summer of 1999 and had a hand-dug channel cut directly through it. Plot 3 was basically dead in the spring of 1998, with only a few stressed white and red mangrove trees present on the plot. Within 3 years, this plot became flooded with over 2500 mangrove seedlings, some of which became established as trees. White mangrove seedling recruitment peaked in 2001, and as seedling recruitment receded, white mangrove tree recruitment increased. White seedling mortality increased in the later years most likely due to natural interspecies competition for resources, and many of the white mangrove seedlings reached sufficient height and were reclassified as trees. Interestingly, although white mangrove seedling recruitment decreased, red mangrove seedlings continue to be recruited to the area (Figure 16.6). Future inspections will determine if red mangroves will eventually displace white mangroves in this die-off area.

Plot 3 exemplifies the process of mangrove recovery, with basal area and canopy cover steadily increasing over time. First, the area became naturally seeded with white mangroves. White mangroves act as a pioneer species invading disturbed areas (this plot had a canal cut right through it). Because white mangroves grow quickly, putting most of their energy into shooting upward (attaining tree height sometimes within 6 months), numerous white mangrove seedlings inhabited a small area appearing as a multitude of thin tall stems with a few leaves. The abundance of white mangroves within a small area apparently caused the individual trees and seedlings to show signs of stress (approximately 60% as of 2003) due to competition for space and resources. The taller and healthier white mangroves quickly outcompeted their smaller siblings. Smaller siblings had lower recruitment and higher mortality rates. In 2003, Plot 3 began to show the early signs of "succession" as red mangrove seedlings were becoming

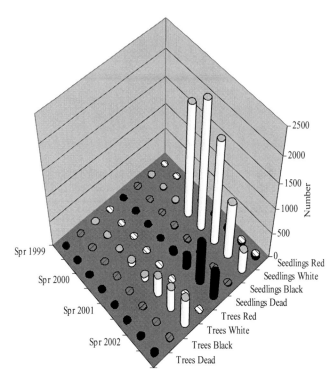

FIGURE 16.6 Plot 3 floristic composition 1999–2003.

established in the area and might ultimately replace the white mangroves over time. Red mangroves grow much slower, as they put the majority of their energy into their roots systems, and then develop thick stems and a more "bush-like" appearance, slowly growing to "treehood." The current status of an abundance of white trees and seedlings is perhaps facilitating subsequent changes to a red mangrove-dominated area in the future. It is possible that the area could experience a reemergence of the original black mangrove forest that was present prior to the die-off, if the substrate continues to build up and sufficient time elapses.

Stressed Areas

All of the original four stressed plots established in Clam Bay remain stressed after 5 years of monitoring. Prior to restoration, these four plots had the following characteristics: (1) more than 30% of the plot consisted of dead mangrove trees; (2) plots were subject occasional periods of extended inundation and high surface water levels due to freshwater runoff or tidal storm surges, but water levels receded to acceptable levels within a few weeks; (3) greater than 50% of the mangrove vegetation in the plot showed visible signs of stress in the form of galls, wilting, leaf and branch loss, and/or disease. Restoration initiatives did not appear to improve the condition of any of these plots, which continue to be subjected to developmental pressures.

For example, historically Plot 12 consisted entirely of mangroves, but recently this plot has been invaded by freshwater plant species such as palmettos and fern; even cattails have become proximate to this plot. These freshwater plants have been steadily increasing in this part of Clam Bay and are fast replacing the mangroves. This is most likely due to the increase in freshwater outflows from the surrounding development, which have created a freshwater wetland from a historically saline environment. There is evidence that this area once supported a large stand of black mangrove trees, but, by spring of 1999, only one black mangrove remained; subsequently dying in 2000. Most recently, red mangrove trees were more abundant, but white mangrove trees were larger and had greater girths. Tree and seedling recruitment was practically nonexistent, and mortality has outstripped recruitment

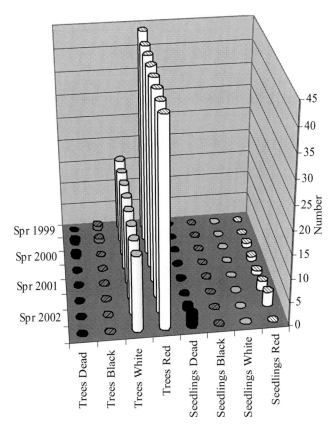

FIGURE 16.7 Plot 12 floristic composition 1999–2003.

(Figure 16.7). A few red mangrove seedlings periodically became established throughout the monitoring period, but none was successful due to the thick underbrush of palmetto and fern that became the dominant understory vegetation type in this area. Mangrove tree basal area steadily decreased and trees showed visible signs of stress. Trees were littered with galls, had lost a great deal of their leaves and branches, and showed evidence of moderate to severe insect infestation. Plot 12, therefore, remains classified as stressed due to the presence of larger mangrove trees that, although very stressed, remained as a remnant population, having so far survived the impact of better-adapted freshwater immigrants.

Alive "Healthy" Areas

After 5 years of monitoring, of the original four plots that were established in mangrove areas that were classified as "alive," two plots were downgraded to "stressed" and two plots remained classified as "alive." Prior to restoration, these four plots presented with the following characteristics: (1) less than 10% of the plot consisted of dead mangrove trees that presumably died due to natural occurrences such as lightning strikes or other forms of natural mortality; (2) plots were subjected to normal tidal cycles and rainfall events typical of mangrove forests; (3) less than 25% of the mangrove vegetation in the plot showed visible signs of stress in the form of galls, wilting, leaf and branch loss, and/or disease; and (4) the overstory was more developed than the understory, typical of mature mangrove forests.

An example of unexpected declines that could possibly be an unforeseen result of the restoration project in Clam Bay is illustrated in Plot 4. Plot 4 is located toward the southern end of Clam Bay, where significant dredging occurred during restoration for the purpose of increasing tidal flushing in the northern extremes of the estuary. Historically, Plot 4 was dominated by red mangrove trees and seedlings. Prior to restoration, there were 113 red mangrove trees present in this plot. The basal area within the plot was fairly consistent until 2001, when 58 trees died and the basal area subsequently decreased.

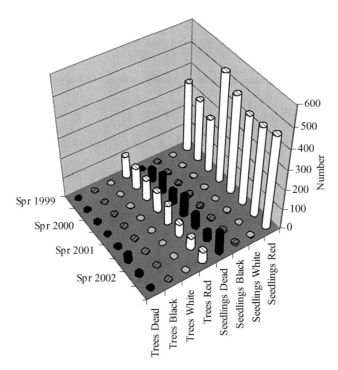

FIGURE 16.8 Plot 4 floristic composition 1999–2003.

Between the spring of 1999 and fall of 2001, red mangrove trees became stressed to the point that a severe infestation of boring beetles developed, resulting in heavy tree mortality. Most recently, the surviving red mangrove trees were stressed and vulnerable to insect infestation. Seedling recruitment, primarily red mangroves, showed wide swings in variability and peaked in the fall of 2000, coincidentally when the trees in this plot displayed signs of severe stress. As of 2003, all of the trees on this plot were stressed along with approximately 31% of the seedlings (Figure 16.8). Following dredging, tidal water was retained onsite for a longer period of time and at higher levels. This could account for increased stress to mangroves observed in this plot. However, the decline of red mangrove trees could also be attributed to a natural, but as yet unidentified occurring problem that frequents mangroves periodically in this area of Florida, but was never previously documented in this estuary. The outcome in this area of Clam Bay remains uncertain, as the natural hydrology of this tributary could return and/or the abatement of the boring beetle population could alleviate the mangrove decline in this area of the estuary.

Discussion

Over time, after restoration, a variety of changes occurred throughout the existing mangrove communities in Clam Bay. In summary, of the original 12 plots (4 in die-off areas, 4 in stressed areas, and 4 in alive "healthy" areas), a total of 5 plots have been reclassified. Three of the plots that were originally classified as die-off areas have shown signs of recovery and two plots that were originally classified as alive are now in a state of decline.

Ideally, one would like to see restoration efforts that result in the return of the estuary to its former pristine state. Realistically, restoration in estuarine areas that have been and will continue to be affected by surrounding development should not be expected to return to their former status. Development has permanently altered the landscape surrounding the estuary and has most likely altered the natural hydrology by changing the volume and timing of freshwater flows into the estuary and in some cases even altering the existing tidal regime. Typically, an increase in freshwater runoff into the estuary occurs.

Storm water protection actions typically divert excess freshwater from the developed areas into the estuary to prevent flooding in the developed areas adjacent to the estuary. This, in turn, elevates nutrient and urban pollution loads entering the estuary, particularly during storm events. Freshwater delivery often occurs in large pulses that enter the estuary directly, rather than naturally percolating through groundwater or slowly moving over the surface as sheetflow. Thus, in the case of Clam Bay, it is unrealistic to expect restoration initiatives to return the impacted area to a pristine old growth black mangrove forest.

More realistic restoration expectations in estuarine areas that have been affected by development should be geared at preventing further decline and restoring areas that have suffered major habitat loss to a state that the altered hydrology and topography can now support, with the understanding that we cannot totally negate developmental impacts, as long as development exists adjacent to the estuary. In the case of Clam Bay, restoration efforts should not be expected to return the die-off areas to the status of the past, but rather to a status in accord with what the current hydrology and topography can support now and in the future. Alteration of the Clam Bay estuary as a result of development has significantly changed the topography and hydrologic flows to the area, rendering it less suitable to revegetation by black mangroves, as the area is no longer as elevated or as dry as it was in the past. Restoration efforts successfully removed a substantial amount of standing fresh water from the die-off area by digging a rather extensive mosaic of channels throughout the die-off. As a result of this "channelization," the die-off areas became more suited to revegetation by white and red mangroves, as these types of mangroves generally do well in disturbed areas or areas that are subjected to frequent flooding.

Restoration success of mangrove areas that have been affected by the accumulation of excess water and prolonged flooding is often in direct proportion to the reduction of surface water levels and surface water retention periods. If restoration projects employ the use of channels to drain off excess surface water, the recovery in the die-off area is often directly proportional to the proximity of the channel and localized topography. For example, in the northern black mangrove die-off area in Clam Bay, following channelization, water levels and retention periods were reduced and the barren areas directly adjacent to channels became populated with white mangrove seedlings. Within a 3-year period following channel construction, white mangrove seedling emergence and establishment was the most successful within a 2-m swathe contiguous with the channel. Success waned proportionate to the distance of the plots from the channels. Saltwort and glasswort plant species also emerged throughout the die-off area, aiding in the reduction of standing water, as these plants absorb large quantities of water and are often present as a precursor to mangrove establishment. Within 5 years, the northern black mangrove die-off area in Clam Bay became "green" again, with a plethora of glass and saltwort species and primarily white and a few red mangrove seedlings. Dead spots remained in the die-off area where the topography slopes downward into these spots, subjecting the area to deeper standing water here during heavy rains. Whether or not this particular area within the Clam Bay die-off will recover and remain viable in the future will depend upon (1) how saturated the soil becomes, (2) how heavy the rainfall and runoff are, and (3) how long it takes for surface waters to recede following rain events.

Is the Clam Bay Restoration a Success?

After 5 years of monitoring, this coastal estuarine ecosystem can be categorized into areas that show promise of recovery and "trouble spots" can be pinpointed where mangroves are showing signs of decline. In the northern area, where the major 1995 die-off of black mangroves occurred, prior to restoration there was no visible sign of recovery. In fact, the die-off area was slowly becoming a stagnant area that formed small ponds during the wet season and the die-off was in fact expanding. Fiddler crabs, once abundant in the area, became absent, the soil became further reduced, and algal mats formed on the surface of the substrate. Wildlife usage shifted from terrestrial to aquatic as fish and wading birds immigrated and reptiles, mammals, and arboreal birds exited the area. Within 5 years of restoration, not only are white mangroves becoming established in this die-off area, but also because the surface water levels have declined fiddler crabs have returned and the area is not as stagnant. Whether or not the black mangrove forest of old can return to this area is unknown, but the presence of white mangrove trees and seedlings is encouraging for the health of this estuary. In the southern areas, however, within 5 years of restoration, some stressed areas show no sign of improvement and once "healthy" areas are in a state of decline. Whether or not this is due to the restoration project or a natural occurrence is uncertain. We hope that the next 5 years of monitoring this area will provide more definitive answers.

Using Mangroves as an Indicator of Successful Estuarine Restoration: Pros and Cons

Mangroves can prove a useful indicator in evaluating the success of tropical and subtropical coastal restoration projects. As the dominant habitat, particularly in estuaries devoid of seagrass (which have long been used as an indicator of estuary condition), mangroves provide a large portion of the foundation essential to primary and secondary consumers that inhabit estuaries. Thus, the condition of the mangroves is often a reflection of the state of the estuary.

Advantages of using mangroves as indicators are that equipment costs are negligible; the fieldwork, although repetitive and time-consuming, does not require in-depth training or special skills; and results are reproducible. Oppositely, mangrove monitoring can be labor intensive (if every individual tree and seedling is monitored) and fieldwork can be logistically challenging at times. Also, patience and persistence are required when using mangroves as an indicator of estuarine trends. Mangrove forests usually change slowly over time, and thus, to determine the outcome of a estuarine restoration project definitively, years and even decades of monitoring are needed to accurately predict the recovery and more importantly the relative "stability" of the forest in the future. Future direction includes determining which floristic factors, if any, are more statistically significant in order to streamline monitoring techniques and also to evaluate recruitment, mortality, and growth rates in different scenarios by species type, canopy cover, condition, inter- and intraspecific competition, and forest structure.

Conclusions

Development and roadway construction adjacent to Clam Bay reduced or even halted the natural drainage patterns within the mangroves, causing accumulation of surface water, which drained off slowly. Black mangrove pneumatophores were completely submerged by high surface water levels that were present for months at a time, effectively suffocating the mangroves and resulting in large-scale mangrove die-offs. A restoration project, geared at increasing tidal flushing in the northern ends of the system and reducing floodwaters in the die-off areas, has achieved its goals from an engineering perspective. Tidal flow has increased and the placement of channels throughout the system is effective at draining a considerable amount of surface water from the die-off areas and reducing high surface water level retention in areas near the channels. The results from a biological standpoint are guardedly optimistic for the die-off areas located in the northern and upper midsections of the Clam Bay estuary. These areas show evidence of revegetation of primarily white seedlings, growing alongside the channel cuts. However, the prognosis for the lower middle and southern parts of the system, which had not suffered large-scale die-offs prior to restoration, is uncertain, as some areas have become subjected to increased stress and other areas have even experienced small-scale mangrove die-backs.

The priority in any restoration project should be "First do no harm." It is prudent to advise caution in thinking that channelization is the remedy for stressed coastal mangrove estuary systems. We can never return to the past when development did not exist adjacent to estuaries. Engineering attempts to solve hydrologic problems can lead to difficulties in other areas of the estuary. One question that needs to be answered is whether or not the practice of channeling and subsequent lowering of water levels, faster and in greater volumes than would occur naturally, has any associative, long-term impacts. The occurrence of higher surface water levels and longer retention periods within the mangroves was the result of storm water runoff from the surrounding developments. Generally, this runoff is higher in nutrients and pesticides (among other features) due to landscaping and golf course maintenance practices. Channeling runoff into estuary tributaries could cause problems downstream and eventually into the ocean. Even if the pollutant amounts being discharged from the channels are minimal, in comparison to agricultural and industrial runoff, the cumulative effects from all coastal discharges into our estuaries and oceans could prove disastrous over time.

Acknowledgments

I thank the following people who spent innumerable hours slugging around in the mosquito-infested swamps with me over the years and without whose help this project would have never gotten off the ground: Markus Hennig, Jessica Servens, Ian Bartoszek, Melinda Schuman, Glen Buckner, Joshua Gates, Sarah Muffelman, and Kristen Kuehl. In addition, I thank The Conservancy of Southwest Florida, the Mangrove Action Group in Pelican Bay, Pelican Bay Foundation, and The National Estuarine Reserve, NOAA for their monetary assistance, and thank Dr. Daniel Childers, Dr. Stephen Bortone, and David Addison for their guidance and support. Special thanks to Captain and Mrs. C.R. Worley (USN RET) for their encouragement over the years.

References

Ball, M. C. 1988. Ecophysiology of mangroves. *Trees* 2:129–142.

Boto, K. G. 1984. Waterlogged saline soils. In *The Mangrove Ecosystem: Research Methods,* S. C. Snedaker and J. G. Snedaker (eds.). UNESCO, Bungay, U.K., pp. 114–259.

Burch, J. N. 1990. Coastal Barrier Management Plan. Department of Natural Resources, Collier County, Naples, FL.

Davis, J. H. 1940. *The Ecology and Geologic Role of Mangroves in Florida.* Publication 517. Carnegie Institute, Washington, D.C.

Duke, N. C. 1992. Mangrove floristics and biogeography. In *Tropical Mangrove Ecosystems,* A. I. Robertson and D. M. Alongi (eds.). American Geophysical Union, Washington, D.C., pp. 66–100.

Field, C. D. 1995. *Journey amongst the Mangroves.* International Society of Mangrove Ecosystem, Okinawa, Japan, p. 140.

Gillison, A. N. and K. R. W. Brewer 1985. The use of gradient directed transects or gradsects in natural resource survey. *Journal of Environmental Management* 20:103–127.

Lewis, R. R., III. 1999. Key concepts in successful ecological restoration of mangrove forests. In *TCE-Project Workshop II: Coastal Environmental Improvement in Mangrove/Wetland Ecosystems,* 18–23 August 1998, Danish-SE Asian Collaboration in Tropical Coastal Ecosystems (TCE), pp. 19–32.

Odum, W. E. and E. J. Heald. 1975. Mangrove forests and aquatic productivity. In *Coupling Land and Water,* A. Hasler (ed.). Springer-Verlag, New York, pp. 129–136.

Odum, W. E. and C. C. McIver. 1990. Mangroves. In *Ecosystems of Florida,* R. L. Myers and J. J. Ewel (eds.). Central Florida Press, Orlando, pp. 517–548.

Parks, P. J. and M. Bonifaz. 1994. Nonsuitable use of renewable resources: mangrove deforestation and mariculture in Ecuador. *Marine Resources Economics* 91:1–18.

Robertson, A. I. and S. J. M. Blaber. 1992. Plankton, epibenthos and fish communities. In *Tropical Mangrove Ecosystems,* A. I. Robertson and D. M. Alongi (eds.). American Geophysical Union, Washington, D.C., pp. 173–224.

Smith, T. J., III. 2000. It's Been Five Years Since Hurricane Andrew: Long-Term Growth and Recovery in Mangrove Forests Following Catastrophic Disturbance. U.S. Geological Survey, BRD, Florida International University, SERP, OE-148, Miami, FL.

Teas, H. 1979. Silviculture with saline water. In *The Biosaline Concept*, A. Hollaender (ed.). Plenum Press, New York, pp. 117–161.

Tomilinson, P. B. 1986. *The Botany of Mangroves.* Cambridge University Press, New York.

Turner, R. E and R. R. Lewis III. 1997. Hydrologic restoration of coastal wetlands. *Wetlands Ecology and Management* 4(2):65–72.

Twilley, R. R. 1998. Mangrove wetlands. In *Southern Forested Wetlands Ecology and Management.* M. G. Messina and W. H. Conner (eds.). CRC Press, Boca Raton, FL, pp. 445–473.

U.S. Fish and Wildlife Service. 1999. South Florida Multi-Species Recovery Plan. U.S. Fish and Wildlife Service, Atlanta, GA, pp. 525–531.

Worley, K. and J. A. Gore. 1995. Status of the Clam Bay Ecosystem — Phase I. Technical Report. The Conservancy, Inc., Naples, FL.

17

Molecular and Organismal Indicators of Chronic and Intermittent Hypoxia in Marine Crustacea

Marius Brouwer, Nancy J. Brown-Peterson, Patrick Larkin, Steve Manning, Nancy Denslow, and Kenneth Rose

CONTENTS

Introduction and Background

Human population growth in coastal regions and their watersheds, accompanied by agricultural, industrial, and urban development, has led to an unprecedented acceleration of contaminant and nutrient inputs into estuaries. Both the nutrients that fuel primary productivity and the near-coastal hydrodynamics that generate water column stratification contribute to the formation of hypoxic zones. Of the total estuarine area in the Gulf of Mexico that was surveyed in 1994–1995, oxygen depletion (anoxia or hypoxia) events occurred in 32 of 38 estuaries (U.S. EPA, 1999), whereas an expansive area of seasonal hypoxia/anoxia develops yearly on the Louisiana continental shelf (Turner and Rabalais, 1994). Eutrophication and ensuing bottom water hypoxia and anoxia are regarded as major factors responsible for declines in habitat quality and harvestable resources in estuarine ecosystems (Justic et al., 1993; Turner and Rabalais, 1994; Paerl et al., 1998). In addition, increased nutrient loading amplifies cyclic dissolved oxygen (DO) patterns that often develop in shallow waters during the summer months, leading to conditions of intermittent hypoxia. Estuarine organisms are therefore not only at risk of being subjected to chronic hypoxic conditions, but also face increases in duration and frequency of hypoxic–normoxic cycles (Ringwood and Keppler, 2002). However, because we lack fundamental information regarding sublethal effects of chronic or intermittent hypoxia on estuarine organisms, indicators of adaptive responses to these conditions are largely unknown.

Currently, the effects of hypoxia on biota are often inferred from measurements of low oxygen levels, which coincide with a reduction of demersal fish and death of benthic fauna (May, 1973; Seliger et al., 1985; Winn and Knott, 1992). The consequences are immediate as well as long term due to the elimination of sensitive species and reduction in overall abundances of food in the benthic communities. Laboratory experiments have shown that, when possible, fish and crustaceans will avoid or move out of hypoxic conditions (Wannamaker and Rice, 2000; Wu et al., 2002). Prolonged exposure to hypoxia results in behavioral or physiological changes such as increased ventilation frequency and cardiac output in both fish and crustaceans (McMahon, 2001; Wu et al., 2002; Robb and Abrahams, 2003). However, while these responses are valuable indices of low oxygen conditions, they cannot serve as effective biomarkers of hypoxia since they involve *in situ* measurements of the organisms, which is not practical for monitoring or management.

Indicators of hypoxia at the organismal/cellular level are needed that can be used to assess the onset, duration, and severity of chronic as well as intermittent hypoxia and its effect on biota. The challenge of scaling these molecular indicator responses to population level responses is addressed later in this chapter. Genes and their products (mRNA and proteins) that respond to hypoxia have great potential to serve as indicators of hypoxic stress, including enzymes of the glycolytic pathway, which increase anaerobic ATP production, glucose transporters, enzymes involved in amino acid metabolism and gluconeogenesis, which maintain blood glucose levels (Semenza et al., 1994; Hochachka et al., 1996; Gracey et al., 2001), and heat shock proteins involved in protein repair and refolding (Hightower, 1993). In general, genes that encode proteins involved in energy production, protein synthesis and degradation, lipid and carbohydrate metabolism, locomotion and contraction, and antioxidant defense are also potential biomarkers of hypoxic stress (Hochachka et al., 1996). While much of the work on gene and protein expression as related to hypoxia has focused on vertebrates, there is growing evidence that these molecular indicators, including ribosomal proteins, antioxidant defense enzymes, and oxygen-carrying proteins, can also effectively indicate hypoxia exposure in invertebrates (DeFur et al., 1996; Mangum, 1997; Choi et al., 2000; McMahon, 2001; Brouwer et al., 2004).

We briefly review the importance of the use of genomics and proteomics in developing indicators for detection of environmental stressors to provide the reader, especially coastal resource managers who are potential end users of these novel indicators, with a background on the importance of this emerging technology. Examples of the use of these molecular indicators for evaluating effects of chronic and intermittent hypoxia in estuarine crustacea, and the link between these indicators and whole animals/populations, are discussed.

Brief Review of Molecular Biomarkers

The utility of molecular biomarkers is that they allow scientists to identify a potential environmental problem prior to any phenotypic or toxicological expression in the organism, thus providing an "early warning system" of a developing problem. Changes in behavior, physiological processes, and metabolic pathways can help determine specific mechanisms of action for a variety of chemical and physical environmental stressors (Denslow et al., in press). Transcriptions of genes that encode specific proteins that can deal with perceived stressors are usually the first measurable biomarkers that can be assessed. However, examining gene regulation by itself does not give a complete picture as it is also necessary to quantify the protein activity to ascertain that increased gene transcription also results in increased protein levels. Changes in both specific gene and protein expression are excellent indicators that the organism has mobilized metabolic pathways in response to a specific stimulus.

The classic method for measuring expression levels of mRNA has been Northern blotting, in which the relative abundance of genes is determined by probing electrophoretically separated RNA bound to a nylon membrane with radiolabeled cDNA that is complementary to the sequence of interest. However,

Northern blotting can only be used for measuring a small number of genes at a time, is less sensitive and quantitative, and has a lower throughput than more recently developed methods for measuring gene expression (Larkin et al., 2003b). Nevertheless, Northern blots have been successfully used to illustrate the downregulation of a variety of genes in aquatic species (Buhler et al., 2000; Funkenstein, 2001; Larkin et al., 2003a).

The recent availability of complete gene sequences for a variety of organisms has resulted in the rapid development of the field of genomics. Emerging technologies now make it possible to screen for changes in up- or downregulation of a large number of different genes at one time. One such technology, quantitative real-time polymerase chain reaction (Q-PCR), emerged in the 1990s (Heid et al., 1996) and is a fairly quick and efficient method to accurately quantify mRNA, allowing the determination of expression levels of the gene or genes of interest. The advantages of this procedure include the sensitivity of the assay, the small amount of total RNA needed, the elimination of radio-isotope labeling, the ability to process a large number of samples, and the potential to measure several genes at once (Larkin et al., 2003b). However, researchers must have already identified the genes they think will be up- or downregulated to effectively utilize this technology. Furthermore, Q-PCR can only be used for genes for which at least a partial sequence has been identified. Q-PCR is a powerful tool that has been used successfully to examine gene expression and induction in a number of aquatic organisms (Trant et al., 2001; Kumar et al., 2001; Dixon et al., 2002).

In situations where scientists want to determine which genes respond to chemical or environmental stimuli, a more global, open-ended technology than Q-PCR is required. Two techniques, subtractive hybridization and gene arrays, can be used to initially identify a large number of genes of interest. Subtractive hybridization enables the researcher to compare mRNA from two different treatments or populations and obtain cDNA clones that are differentially expressed in one group compared to the other (Lisitsyn and Wigler, 1993; Ermolaeva and Sverdlov, 1996). Subtractive hybridization has the potential to identify genetic biomarkers for specific types of environmental contaminants (Larkin et al., 2003b), although this use in aquatic species has only been recently documented (Blum et al., 2004).

Gene arrays, also called micro- or macroarrays, can be used to measure simultaneously the expression of hundreds or thousands of genes affected by a particular compound or exposure condition. These arrays are like reverse Northern blots, in that the RNA is reverse-transcribed with fluorescent or radiolabeled markers and then hybridized with DNA sequences (specific genes) attached to a solid support matrix such as nylon membranes or glass slides (Schena et al., 1998). Up- or downregulation of gene expression can be visualized by the intensity of the spots on the membrane or slide. Arrays are limited in that they may not be sensitive enough to detect rare genes, and the function of many genes spotted onto an array may not be known (Larkin et al., 2003b). However, microarrays are invaluable for identifying a suite of genes that are affected by chemical or environmental stimulation, and several studies suggest that specific contaminants may have their own specific profiles of gene expression (Bartosiewicz et al., 2001; Larkin et al., 2003b). The use of gene arrays to study environmental contaminants in aquatic systems has been applied to fish (Hogstrand et al., 2002; Larkin et al., 2003a; Williams et al., 2003). Furthermore, gene arrays have been successfully used to examine the effects of hypoxia in fish and crabs (Gracey et al., 2001; Ton et al., 2002; Brouwer et al., 2004). This emerging technology will no doubt become routine for assessing a variety of bioindicators in the next decade.

The classic method for quantifying protein levels is Western blotting, the protein analog to Northern blots. Proteins from homogenized tissue samples are electrophoretically separated in denaturing poly-acrylamide gels, transferred to a membrane, and then incubated with primary antibodies to the proteins. Secondary antibodies with a covalently linked enzyme, which catalyzes formation of a colored or chemiluminescent product from an appropriate substrate, are used to visualize and quantify the protein–primary antibody complex. Western blots are an excellent choice for confirming protein expression from a limited number of up- or downregulated genes, and they have been used to confirm changes in gene expression in blue crab (Brouwer et al., 2004).

Proteomics refers to the large-scale study of proteins, including understanding the complex interactions among proteins that occur within cells. Expression proteomics is a subfield that focuses on understanding the factors that affect protein expression, and thus is complementary to Q-PCR, subtractive hybridizations, and gene arrays (Denslow et al., in press). Proteomics analysis requires separation of proteins by two-dimensional (2D) gel electrophoresis or ion-exchange chromatography followed by mass spectrometry of tryptic digests of the separated proteins. A number of proteins can be identified in this manner, but the procedure is costly and labor intensive. Changes in protein patterns and expression in response to environmental and chemical factors have recently been identified in fish (Kultz and Somero, 1997; Bradley et al., 2002) and invertebrates (Shepard and Bradley, 2000; Meiller and Bradley, 2002).

Indicators of Hypoxia in Crustacea

While the ultimate goal of bioindicators is effectiveness and usefulness in the field, the development of new bioindicators requires testing, calibration, and validation under controlled laboratory conditions. As a first step toward attaining this goal, the effects of chronic and intermittent hypoxia on blue crab (*Callinectes sapidus*), brown shrimp (*Farfantopenaeus aztecus*), and grass shrimp (*Palaemonetes pugio*), were examined in the laboratory. The identification of potential molecular bioindicators of hypoxia in blue crab will enable examination of the efficacy of these indicators to identify hypoxic exposure in wild-caught blue crabs.

Materials and Methods

Exposure Methods

Wild-caught, adult male blue crabs (85 to 152 mm carapace width [CW] 42.5 to 191.3 g wet weight [ww]) were maintained in the laboratory at 15‰ and 27 ± 1°C for 6 to 28 days prior to experimentation. During acclimation and experimentation periods, crabs were fed commercial shrimp pellets once daily. For the chronic hypoxia study, six crabs were placed in each of six hypoxic (2 to 3 ppm DO) and two normoxic (6 to 8 ppm DO) 35-L tanks under flowthrough conditions (Manning et al., 1999). Oxygen in the flowthrough dilution water was maintained at supersaturation (14 ppm, hypoxic tanks, and 18 to 20 ppm, normoxic tanks). The high DO concentrations were necessary because of the high rates of oxygen consumption by the study crabs. Ten control (normoxic) crabs were sacrificed at days 0 and 15. Ten hypoxic crabs were sacrificed at days 5, 10, and 15. To examine the effects of cyclic, intermittent hypoxia, five crabs were placed in each of four cyclic DO (2 to 3 ppm DO to 8 to 9 ppm DO cycle every 24 h) and four normoxic (8 ppm DO) 35-L tanks under flowthrough conditions for 10 days. This exposure was conducted by limiting the oxygen addition to the dilution water reservoir during the evening hours, which resulted in a drop in DO in the treatment aquaria to 2 to 3 mg/L by early morning. In the morning the frequency of oxygen aeration was increased in the reservoir, which increased the aquaria to normoxic levels by late morning or early afternoon. Ten control (normoxic) crabs were sacrificed at days 0 and 10 and ten intermittent hypoxic crabs were sacrificed at days 5 and 10. The hepatopancreas of the crabs was removed and frozen at −70°C for protein analysis or stored in RNAlater (Ambion®, Austin, Texas) at −20°C for nucleic acid extraction.

Wild-caught, adult brown shrimp (64 to 138 mm total length, TL, 1.85 to 17.55 g) were maintained in flowthrough tanks in the laboratory at 15‰ and 27 ± 1°C for 16 to 30 days prior to experimentation. Shrimp were fed commercial shrimp pellets once daily. To determine survival of brown shrimp under chronic hypoxic conditions, 12 shrimp were placed in each of 6 hypoxic (2 to 3 ppm DO) and 12 normoxic (6 to 8 ppm DO) 35-L tanks under flowthrough conditions for 12 days. Survival of brown shrimp subjected to cyclic DO variations, as described above, was assessed by placing 12 shrimp in each of 6 cyclic DO (2 to 3 ppm DO to 8 to 10 ppm DO cycle every 24 h) 35-L tanks and 6 normoxic (6 to 8 ppm DO) 35-L tanks under flowthrough conditions for 13 days.

Wild-caught, adult grass shrimp (25 to 38 mm TL, 0.1711 to 0.4925 g) were maintained in the laboratory at 15‰ and 27 ± 1°C for 6 to 182 days prior to experimentation. During acclimation and experimentation periods, grass shrimp were fed twice daily, once with commercial maintenance flake and once with *Artemia* sp. *nauplii* (24 to 48 h post-hatch). Individual grass shrimp were isolated in mesh containers to facilitate observations of molting and egg production. The 25 containers were placed into each of eight hypoxic (2 to 3 ppm DO) and four normoxic (6 to 8 ppm DO) 35-L tanks under flowthrough conditions for 14 days. The 20 control (normoxic) female shrimp were sampled at day 0, and the 16 hypoxic and 16 normoxic female shrimp were sacrificed at days 3, 7, and 14. After 14 days of exposure, male and female pairs were placed into the mesh containers for an additional 4 weeks; 16 pairs were maintained under hypoxia, 32 pairs consisted of hypoxia-exposed females maintained under normoxia for the reproductive portion of the experiment, and 32 pairs were maintained under normoxia. Female shrimp were sacrificed 2 to 5 days after the appearance of a clutch of eggs. A second experiment, to determine the effects of extreme hypoxia (1 to 2 ppm DO) on grass shrimp, was conducted following the same protocol, with the exception that the reproductive pairs of shrimp were maintained for an additional 6 weeks in hypoxic or normoxic conditions following the initial 14-day exposure. Upon sacrifice, grass shrimp were weighed (0.0001 g) and measured (TL, mm), and eggs were removed and counted. The thorax of the shrimp samples was frozen at $-70°C$ for protein analysis or stored in RNAlater at $-20°C$ for nucleic acid extraction.

Cloning and Macroarrays of Blue Crab Genes

Ten genes, including heat shock protein 70 (Hsp70), mitochondrial and cytosolic manganese superoxide dismutase (MnSOD) (Brouwer et al., 2003), hemocyanin (Brouwer et al., 2002), Cu and Cd-metallothionein (Syring et al., 2000), beta-actin, and ribosomal proteins S15, L23, and S20, were cloned and sequenced from hepatopancreas tissue of blue crab. The blue crab genes were PCR-amplified and then robotically spotted in duplicate onto neutral nylon membranes. Total hepatopancreatic mRNA was extracted from six to nine blue crabs per treatment group using Stat-60® (TEL-TEST, Friendswood, Texas). mRNA was transcribed into radiolabeled cDNA, and hybridized to the membranes. For each cDNA clone, the general background of each membrane was subtracted from the average value of the duplicate spots on the membrane. The values were then normalized to a beta-actin cDNA clone; the resulting values were used to determine which gene transcripts were up- or downregulated by hypoxia (Larkin et al., 2003a).

Subtractive Hybridization for Identification of Hypoxia-Responsive Genes in Blue Crabs and Grass Shrimp

Subtractive hybridizations were performed by EcoArray® LLC (Alachua, Florida). For the blue crab subtractive hybridizations (Clontech® PCR-Select cDNA Subtraction Kit, Palo Alto, California), mRNA samples from seven day 0 (control) and seven day 5 (hypoxic) crabs were reverse-transcribed into cDNA and heat-denatured, and the cDNA pools were then hybridized together. Two subtractive hybridizations were performed to obtain grass shrimp hypoxia-responsive genes. mRNA samples from day 0 normoxic (control) and day 3 and 5 moderate hypoxic (2 to 3 ppm DO) and day 3 severe hypoxic (1.5 ppm DO) shrimp were reverse transcribed into cDNA. The cDNAs that remained unhybridized were PCR-amplified, cloned, and subsequently sequenced for identification. Subtraction libraries were run in both the forward and reverse directions to obtain both up- and downregulated genes. The blue crab clones obtained through subtractive hybridization were spotted in duplicate onto expanded macroarrays as described above.

Western Blots

Homogenized hepatopancreas tissue samples from seven to ten blue crabs from each treatment group were subjected to SDS-PAGE on 10 and 12.5% polyacrylamide gels for hemocyanin and MnSOD, respectively. Separated proteins were transferred electrophoretically to PVDF membranes and visualized using the Pierce SuperSignal West Dura Extend Substrate system for chemiluminescent detection of proteins. The *Callinectes sapidus* antihemocyanin antibody was prepared by DLAR Vivarium (Duke University, Durham, North Carolina). The MnSOD antibody was commercially available from Stressgen Biotechnologies Corporation (Victoria, British Columbia, Canada).

Statistical Analyses

Significant differences in survival between normoxic and hypoxic crabs and shrimp were determined by chi-square analysis of 2×2 contingency tables. Differences in gene expression between control and normoxic crabs were determined with a Student's t-test. Percentage data were arcsine square root transformed prior to analysis (Sokal and Rohlf, 1995). Results were considered significant if $p < 0.05$.

Results

Laboratory Survival under Hypoxic Conditions

Blue crab survival decreased during the 15 days of the chronic hypoxia experiment, but there was no significant difference between normoxic and hypoxic crabs (Table 17.1). Survival of crabs in the cyclic DO experiment decreased in both the normoxic and cyclic DO treatments by day 10, with crabs exposed to cyclic DO showing higher, but statistically not significant mortality than those in normoxic conditions (Table 17.1).

TABLE 17.1

Percent Survival of Blue Crabs Exposed to Chronic or Intermittent Hypoxia

| | Days of Exposure | | | |
Treatment	Day 3	Day 5	Day 10	Day 15
Hypoxic	88	78	55	50
Normoxic	83	77	67	53
Intermittent Hypoxic	96.7	90	47.8	—
Normoxic	80	80	71.4	—

Survival of brown shrimp consistently decreased during the 12-day exposure to chronic hypoxia (Figure 17.1A), and there was a significant difference in survival between normoxic and hypoxic brown shrimp at days 10 and 12. Intermittent hypoxia appeared to affect brown shrimp survival more severely than chronic hypoxia, and there was a significant difference in survival between brown shrimp exposed to normoxic or cyclic DO within 3 days exposure (Figure 17.1B).

Survival of grass shrimp exposed to normoxia or 2 to 3 ppm DO hypoxia over a 6-week period was nearly 100% (data not shown). However, grass shrimp exposed to severe hypoxia (1.5 ppm DO) began to show increased mortality by week 5, which became significant by week 8, when there was only 56% survival of the hypoxic shrimp (Figure 17.2).

Changes in Gene and Protein Expression in Blue Crabs in Response to Hypoxia

Gene macroarrays appear to be an effective method to detect changes in gene expression in blue crabs following exposure to hypoxia. Macroarrays spotted with ten blue crab genes plus several internal controls show that five genes appear to be downregulated (less intense spotting) after 5 days exposure to hypoxia: HSP70, mitochondrial MnSOD, hemocyanin, and ribosomal proteins S15 and L23 (Figure 17.3). Data from all membranes combined show subtle, but significant, differences in gene expression between control and 5 day hypoxia-exposed crabs for hemocyanin, mitochondrial MnSOD, and ribosomal protein S15 (Figure 17.4A). In general, all genes with the exception of Cu-metallothionein showed modest downregulation after 5 days exposure to hypoxia. This generic downregulation of gene expression after 5 days appears to be a transient phenomenon since gene transcription was restored to normal levels after 15 days exposure to hypoxia (Figure 17.4B).

Hemocyanin protein concentrations in the hepatopancreas were significantly elevated after 5 days of hypoxia exposure (Figure 17.5A), but this difference was no longer observed after 15 days of hypoxia exposure (Figure 17.5B). Western blots showed unexpected changes in molecular weight of MnSOD

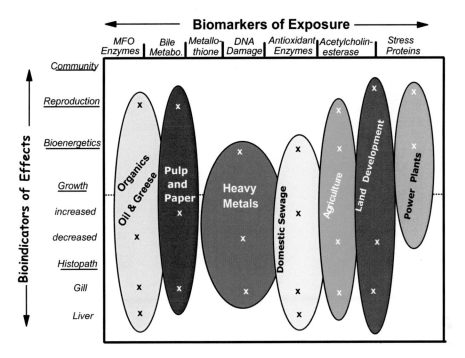

COLOR FIGURE 2.4 Biomarker–bioindicator response profiles characteristic of several major types of anthropogenic activities. Use of such characteristic profiles can help diagnose and identify sources of stress in estuarine systems affected by multiple stressors.

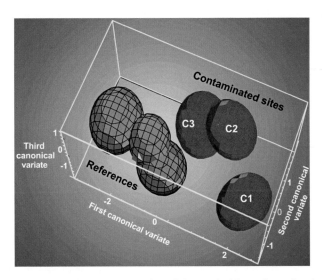

COLOR FIGURE 2.5 Integrated health responses of sunfish sampled from three sites in a contaminated river and three reference sites. Boundaries of each ellipse are based on the 95% confidence radii of the integrated site means.

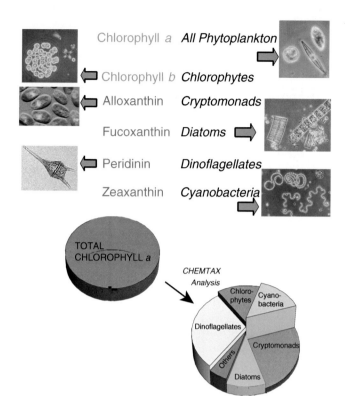

Chlorophyll *a* *All Phytoplankton*

Chlorophyll *b* *Chlorophytes*

Alloxanthin *Cryptomonads*

Fucoxanthin *Diatoms*

Peridinin *Dinoflagellates*

Zeaxanthin *Cyanobacteria*

TOTAL CHLOROPHYLL *a*

CHEMTAX Analysis

Chlorophytes

Cyanobacteria

Dinoflagellates

Cryptomonads

Others

Diatoms

COLOR FIGURE 11.1 (Top) Chlorophyll and carotenoid photopigments commonly used for identifying and quantifying phytoplankton taxonomic groups, representative photomicrographs of which are shown. (Bottom) Diagram, showing how the ChemTax matrix factorization program is used to determine the proportions of total phytoplankton biomass (as chl *a*) attributable to phytoplankton taxonomic groups, based on HPLC separation and quantification of the diagnostic photopigments shown in the upper frame.

Dennis, 30 Aug., 1999

Floyd, 15 Sept., 1999

COLOR FIGURE 11.4 (Top) NASA SeaWiFS ocean color satellite images of the landfalls of Hurricanes Dennis and Floyd in Eastern North Carolina, during September 1999. (Bottom left) SeaWiFS image of the Pamlico Sound and surrounding watershed and coastal region. This image was recorded on 23 September 1999, approximately 1 week after landfall of Hurricane Floyd. The sediment-laden floodwaters from the cumulative rainfall of Hurricanes Dennis and Floyd (up to 1 m in some parts of the sound's watershed) can be seen moving across Pamlico Sound. Also note the turbid overflow from Pamlico Sound entering the Atlantic Ocean. The sediment plume advected northward into the coastal ocean by the Gulf Stream, which was located approximately 40 km offshore at the time this image was recorded. (Bottom right) Edge of the sediment-laden floodwaters moving across the Pamlico Sound. The floodwater caused strong vertical salinity stratification, hypoxic bottom waters, and stimulated phytoplankton production throughout the sound (see Paerl et al., 2001).

COLOR FIGURE 11.10 (Top left) Map of the St. Johns River. This northward flowing river is the largest river in Florida with the tidally driven outlet near Jacksonville. Upstream lakes often have high densities of cyanobacterial species. In particular, Lake George (L. George), a large feeder lake for this system, constitutes one of the largest innoculum of cyanobacteria to the system. (Courtesy of St. Johns Water Management District, Palatka, FL.) Other frames: Photographs of cyanobacterial (*Microcystis* sp., *Cylindrospermopsis*, *Anabaena* spp., and *Aphanizomenon flos aquae*) blooms along the St. Johns River system. (Lower left photograph courtesy J. Burns.)

COLOR FIGURE 25.1 Six natural, tidally influenced rivers and three canals in the Ten Thousand Islands of southwest Florida. Canals on either side of U.S. Highway 92 are labeled 92 Canal West and 92 Canal East. The dots on the river and canal transects designate locations where fish traps and crab traps were placed to catch juvenile goliath grouper, *Epinephelus itajara*, from 1999–2000.

COLOR FIGURE 25.2 Photographs of typical eroded shorelines in the Ten Thousand Islands of southwest Florida. The erosion along the mangrove shorelines provides for underwater habitat underneath the mangrove overhangs.

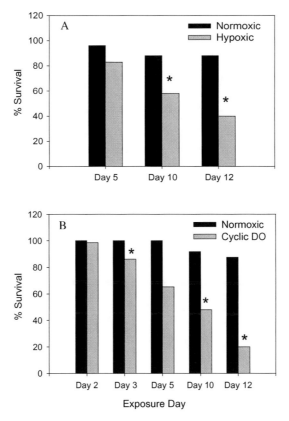

FIGURE 17.1 Survival of brown shrimp, *Farfantopenaeus aztecus*, exposed to hypoxic conditions. (A) Chronic hypoxia (2 to 3 ppm DO). (B) Diurnal fluctuations in hypoxia (2 to 3 ppm DO to 6 to 8 ppm DO over a 24-h period). * = Significantly reduced survival relative to the controls ($p < 0.05$).

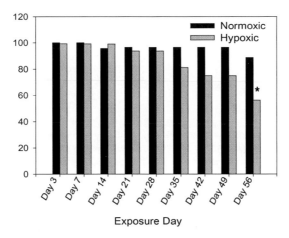

FIGURE 17.2 Survival of grass shrimp, *Palaemonetes pugio*, exposed to severe chronic hypoxia (1.5 ppm DO). * = Significantly reduced survival relative to the controls ($p < 0.05$).

protein in the hepatopancreas of hypoxia-treated crabs. Control, normoxic crabs showed 22- and 25-kDa bands corresponding to the mitochondrial and cytosolic forms of this protein (Brouwer et al., 2003). After 5, 10, and 15 days of exposure to hypoxia, significant cross-linking of proteins occurred, resulting in high-molecular-weight (~70 to 90 kDa) aggregates (Figure 17.6).

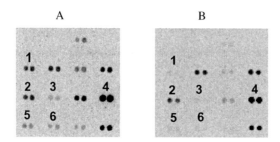

FIGURE 17.3 Gene macroarrays of blue crabs exposed to hypoxia in the laboratory. (A) Normoxia (6 to 8 ppm DO). (B) Chronic hypoxia (2 to 3 ppm DO) exposure for 5 days. 1: HSP 70; 2: actin; 3: mitochondrial MnSOD; 4: hemocyanin; 5: ribosomal protein S15; 6: ribosomal protein L23.

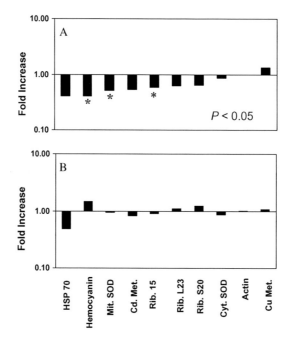

FIGURE 17.4 Fold differences in blue crab gene expression, as determined from macroarray analysis, following chronic hypoxia (2 to 3 ppm DO) exposure. (A) Difference between day 0 normoxia and 5 day chronic hypoxia. (B) Difference between day 15 normoxia and day 15 chronic hypoxia. HSP 70: heat shock protein 70; Mit. SOD: mitochondrial-MnSOD; Cd. Met.: Cd-metallothionein; Rib. 15: ribosomal protein S15; Rib. L23: ribosomal protein L23; Rib. S20: ribosomal protein S20; Cyt. SOD: cytosolic-MnSOD; Cu Met.: Cu-metallothionein. * = Significant difference.

   Diurnal fluctuations in DO did not significantly affect gene or protein expression in blue crab, although the overall patterns were similar to those observed for crabs exposed to chronic hypoxia. For example, hepatopancreas hemocyanin protein concentration was higher at day 5 and 10 diurnal hypoxia than in the day 0 and day 10 normoxic controls (data not shown). All genes appear to be slightly downregulated after 5 days of exposure to diurnal DO cycles (Figure 17.7A). Similar to results from the chronic hypoxia study, this generic pattern changes after 10 days of exposure to diurnal DO fluctuations (Figure 17.7B).

   Due to only a modest response in gene regulation following exposure to chronic or diurnal hypoxia, subtractive hybridizations were performed to identify more potentially hypoxia-responsive genes in blue

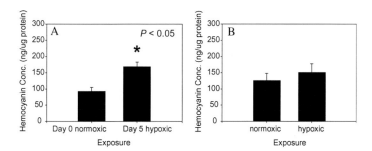

FIGURE 17.5 Hemocyanin protein concentration in hepatopancreas of normoxia- and hypoxia-exposed blue crabs. (A) Short-term (5 day) chronic hypoxia (2 to 3 ppm DO) exposure. (B) Long-term (15 day) chronic hypoxia (2 to 3 ppm DO) exposure. * = Significant difference.

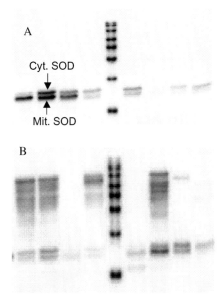

FIGURE 17.6 SDS-PAGE gels of MnSOD protein from hepatopancreas of blue crab, *Callinectes sapidus*. (A) Normoxic crabs showing the cytosolic and mitochondrial forms of blue crab MnSOD. (B) Day 5 and 10 hypoxic crabs showing cross-linked MnSOD. The ladder represents molecular weight markers of 20, 30, 40, 50, 60, 80, 100, and 120 kDa.

crab. An additional 14 genes were identified, and their expression levels determined in blue crabs exposed for 10 days to either normoxic or diurnal DO fluctuations (Figure 17.7B, right half of graph). The new set of genes did not show greater up- or downregulation compared to the first set (Figure 17.7B).

Effects of Hypoxia on Grass Shrimp

Subtractive hybridization using mRNA from grass shrimp exposed to chronic hypoxia for 3 days resulted in a suite of 74 potentially differentially regulated genes. A large number of these genes are related to specific biological functions that are influenced by hypoxic conditions, such as ATP metabolism, oxygen transport, protein synthesis, and gluconeogenesis (Table 17.2). There is a high likelihood that future testing of this suite of genes on macroarrays with shrimp exposed to chronic hypoxia will result in biologically relevant molecular biomarkers of hypoxia.

The effects of chronic hypoxia on grass shrimp fecundity were examined. Grass shrimp exposed to both chronic moderate (2 to 3 ppm DO) and severe (1.5 ppm DO) hypoxia showed significant differences in relative fecundity (number of eggs/g shrimp; Figure 17.8). In both experiments, grass shrimp

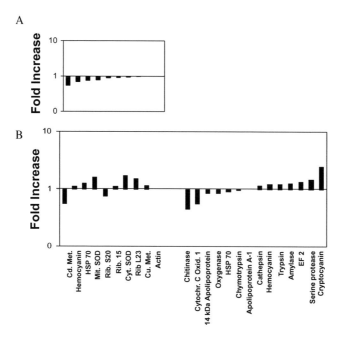

FIGURE 17.7 Fold differences in blue crab gene expression, as determined from analysis of macroarrays, following diurnal hypoxia (2 to 3 ppm DO to 8 to 10 ppm DO) exposure. (A) Difference between day 0 normoxia and 5 day diurnal hypoxia ($n = 8$). (B) Difference between day 10 normoxia and day 10 diurnal hypoxia ($n = 8$). Note additional genes identified through subtractive hybridization on right side of graph. Cd. Met.: Cd-metallothionein; HSP 70: heat shock protein 70; Mit. SOD: mitochondrial-MnSOD; Rib. S20: ribosomal protein S20; Rib. S15: ribosomal protein S15; Cyt. SOD: cytosolic-MnSOD; Rib. L23: ribosomal protein L23; Cu. Met.: Cu-metallothionein; Cytochr. C Oxid. 1: cytochrome *c* oxidase subunit 1; EF2: elongation factor 2.

TABLE 17.2

Potential Hypoxia Responsive Genes Identified in Grass Shrimp, *Palaemonetes pugio*, through Subtractive Hybridization

Biological Function	No. of Genes
Energy production—ATP metabolism	7
Energy production—Oxygen transport	6
Protein synthesis	9
Protein degradation	6
Protein folding	3
Lipid metabolism	7
Carbohydrate metabolism	4
Gluconeogenesis	3
Locomotion/contraction	4
Other functions	25

Note: Genes are grouped by general biological function.

maintained in hypoxic conditions for 6 (Figure 17.8A) or 8 (Figure 17.8B) weeks had a significantly higher relative fecundity compared to shrimp maintained in normoxia for the same time, and also compared to shrimp exposed to hypoxia for 2 weeks and then transferred to normoxia for the remaining 4 or 6 weeks (Figure 17.8). However, there was no difference in relative fecundity between grass shrimp exposed to hypoxia for only 2 weeks and those maintained in normoxia for the entire experiment (Figure 17.8).

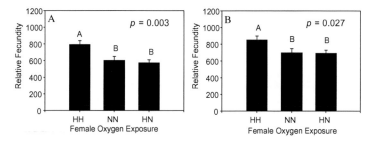

FIGURE 17.8 Relative fecundity ($\bar{x} \pm 1$ SE) of grass shrimp, *Palaemonetes pugio*, exposed to chronic hypoxia: (A) 6-week exposure to moderate (2 to 3 ppm DO) hypoxia; (B) 8-week exposure to severe (1.5 ppm DO) hypoxia. Letters indicate groups that are significantly different. HH, hypoxic females; NN, normoxic females; HN, females maintained in hypoxic conditions during the first 2 weeks of the experiment and then transferred to normoxic conditions.

Discussion

Laboratory data provide convincing evidence of the effects of hypoxia on crustaceans, from the whole-animal to the molecular level. Of the three species examined here, tolerance to chronic and intermittent hypoxia, as measured by survival, decreases from grass shrimp to blue crabs to brown shrimp. For both crabs and brown shrimp intermittent hypoxia appeared to have a greater effect than chronic hypoxia. The dramatic decrease in survival of brown shrimp during both chronic and intermittent hypoxia indicates this species cannot tolerate low DO. Indeed, independent studies have shown that brown shrimp strongly avoided oxygen concentrations of 2 mg/L (Renaud, 1986) and 1 mg/L (Wannamaker and Rice, 2000), whereas blue crabs do not show any significant avoidance behavior (Das and Stickle, 1994). Recent studies have shown that localized shrimp catch is negatively related to the amount of local coverage by hypoxia and that shrimp migration to offshore habitat is blocked by hypoxia (Zimmerman and Nance, 2001). It appears therefore that chronic and intermittent hypoxia may have serious consequences for shrimp fisheries by limiting availability of suitable habitat.

Fecundity of grass shrimp appears to be affected by hypoxia, with shrimp held under chronic hypoxia having a higher fecundity than normoxic animals (Figure 17.8). While this seems counterintuitive, it may represent a strategy of maximizing reproductive output prior to death. The reproductive studies reported here measured fecundity based on the first brood observed. Experiments are planned to determine if second and third broods (if any) decrease in size under hypoxic conditions, and if time between production of broods increases. Combination of whole-animal information such as this with data from molecular biomarkers is crucial for developing models of the effects of hypoxia at the ecosystem level (see below).

Molecular biomarkers could be potentially useful for assessing exposure and effects of hypoxia on estuarine organisms that are hypoxia tolerant, such as blue crab and grass shrimp. Significant differences in both gene and protein expression in crabs were observed after 5 days exposure to chronic hypoxia, indicating measurable changes are occurring on the molecular level that are not manifested at the whole-organism level. One of the first responses to hypoxia is downregulation of protein synthesis to save energy (Hochachka and Lutz, 2001). This was evident in blue crab by the downregulation of gene expression of ribosomal proteins, heat shock protein, hemocyanin, and mitochondrial MnSOD after 5-day exposure to both chronic and intermittent hypoxia (Figure 17.4A and Figure 17.7A). Downregulation of MnSOD is a typical cellular response to hypoxia in both invertebrates and vertebrates (Russell et al., 1995; Choi et al., 2000). After 15 days of hypoxia exposure, the expression level of most of the genes had returned to normoxic levels. This suggests that the initial response to hypoxia is reversible and that the crab has additional regulatory mechanisms to stabilize energy consumption at a new, hypometabolic steady state, as found for other hypoxia-tolerant organisms (Hochachka and Lutz, 2001).

Despite the fact that hemocyanin mRNA levels had decreased after 5 days of hypoxia, hemocyanin protein levels had increased significantly. This observation stresses the importance of validating the results from gene arrays with protein-level determinations. The apparent discrepancy may be explained as follows. Subtractive hybridization experiments suggest that transcription of the lysosomal proteolytic enzyme cathepsin is downregulated in response to hypoxia. Thus, the existing pool of intracellular proteins may be preserved through deactivation of proteolytic pathways, resulting in discrepancies between mRNA and protein levels.

Intermittent hypoxia and chronic hypoxia affect gene regulation differently. For example, genes for enzymes that detoxify reactive oxygen species produced during reoxygenation, such as MnSOD, are upregulated in response to cyclic DO (Hochachka and Lutz, 2001), but downregulated in response to chronic hypoxia (Russel et al., 1995; Choi et al., 2000). Blue crabs respond in a similar manner, as mitochondrial MnSOD appears to be upregulated after 10 days of intermittent hypoxia (Figure 17.7A), but downregulated in response to chronic hypoxia (Figure 17.4A). The effect of chronic and intermittent hypoxia on MnSOD protein is different as well. Under hypoxic conditions, a cross-linked, high-molecular-weight species is the predominant form of MnSOD, whereas under normoxic and intermittent hypoxic conditions the monomeric form of the protein predominates. It is of interest to point out that cytosolic MnSOD, which occurs in a monomer–dimer equilibrium, is found only in crustacea that use hemocyanin for oxygen transport (Brouwer et al., 2003). The monomer–dimer equilibrium may lend itself to further polymerization, by an as yet not understood mechanism. While the differences we have demonstrated in gene expression appear promising, they are rather subtle and often not statistically significant. Verification of the most promising gene expression results using Q-PCR is therefore necessary to obtain a complete and accurate picture of the molecular responses of blue crab to hypoxia, as well as increase the reliability of the gene arrays as a tool for detection of DO stress in field-collected animals.

The results of the subtractive hybridizations for discovery of hypoxia-responsive genes in grass shrimp are very promising. Most of the genes identified are known to play a role in adaptation to hypoxia or in antioxidant defense. In all, 13 genes are involved in energy (ATP) production, the use of which must be limited to survive prolonged hypoxia/anoxia (Hochachka et al., 1996). Other downregulated genes are those involved in protein synthesis and turnover, which will result in turning protein synthesis rates down to pilot-light levels (Hochachka and Lutz, 2001). Still other genes are involved in lipid and carbohydrate metabolism, including enzymes that are catalyzing the conversion of amino acids into glucose (gluconeogenesis), which maintains blood glucose levels under hypoxic conditions (Gracey et al., 2001). Finally, several proteins involved in locomotion are downregulated, similar to what is observed in hypoxic fish (Gracey et al., 2001). Macroarrays constructed with the identified grass shrimp genes should provide unequivocal biomarkers of hypoxia for this important estuarine species, which is amenable to reproductive studies, which, in turn, will allow us to determine if the hypoxia-responsive gene arrays can be used as predictive indicators of reproductive impairment.

Scaling Molecular Biomarker Responses to Population Responses

The greatest limitation of many indicators of coastal condition is the lack of linkage with the cause or causes for change (Suter et al., 2002). Molecular indicators of stress in individuals have considerable potential to provide information on cause-and-effect relationships, and to serve as indicators of early stages of ecological change. However, for this potential to be realized, it will be necessary to develop modeling tools that will make it possible to scale molecular responses in individuals to higher, ecologically more relevant, levels of biological organization. This scaling up of the indicators can be accomplished by using carefully designed laboratory experiments that provide relationships between indicator values in controlled conditions and laboratory end points such as fecundity and embryo survival.

Predicting the population-level consequences of environmental stressor exposure is difficult because many of the potentially important effects are sublethal. Incorporating sublethal effects into population dynamics models is not currently possible because they lack sufficient detail for such effects to be represented as changes in model parameters (Rose, 2000). However, a suite of appropriately scaled models — physiological/statistical, individual-based (IBM), and matrix projection models — for linking sublethal physiological or behavioral effects to long-term population dynamics is an effective method for addressing this problem (Figure 17.9). This approach has been applied to predict population effects of contaminants (PCB) on fish (Rose et al., 2003), and can be adapted to predict the population-level effects of hypoxia on grass shrimp.

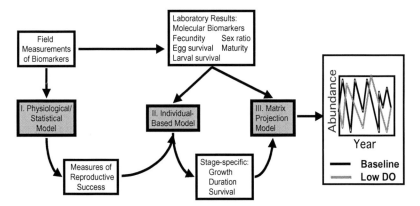

FIGURE 17.9 Conceptual model for scaling individual-level effects (biomarkers) of environmental stressors to the population level.

As shown in Figure 17.9, when correlations between biomarkers and reproductive success are known from laboratory experiments, field-measured biomarkers may be directly converted to relevant inputs to the IBM. When these correlations are not known, it is necessary to model physiological or behavioral responses of individuals to convert biomarkers to measures of reproductive success for inputs to the IBM. The IBMs will link the various measures of reproductive output (from experiments and from the output of the physiological/statistical model) to changes in the inputs of the matrix projection model, which will then simulate the long-term dynamics of the shrimp population. Since both approaches use molecular biomarkers as inputs, they will enable the prediction of ecologically relevant population effects in situations when only biomarkers are measured in shrimp collected in the field.

Conclusions

The data presented here demonstrate that current technology, in the form of gene arrays and subtractive hybridizations, shows considerable promise to provide diagnostic tools for the identification of hypoxic indicators in estuarine crustacea. Additionally, whole-animal indicators of hypoxia, such as mortality and changes in reproductive parameters, are important in combination with the molecular indicators. The true test of these indicators, identified under controlled laboratory conditions, will be their ability to distinguish crustacea exposed to hypoxic conditions in the field. The ability to incorporate information obtained from these molecular and whole-animal indicators into models to predict population-level effects will provide resource managers with sensitive early-warning indicators of incipient ecological change due to chronic and intermittent hypoxia. Such indicators are vitally important for effective management of ecological systems, because they will allow pro-active rather than re-active strategies for restoring ecosystem health. In addition, the indicators will be useful for evaluating the success of remediation efforts in a timely manner.

Acknowledgments

We appreciate the technical expertise of C. King, T. Brouwer, and W. Grater and their help with Western blotting, cDNA cloning, and sequencing. This research was supported by grants from the National Science Foundation (MCB-0080075) and the U.S. Environmental Protection Agency's Science to Achieve Results (STAR) Estuarine and Great Lakes (EaGLe) program through funding to the Consortium for Estuarine Ecoindicator Research for the Gulf of Mexico (CEER-GOM; U.S. EPA Agreement R82945801).

References

Bartosiewicz, M., S. Penn., and A. Buckpit. 2001. Applications of gene arrays in environmental toxicology: fingerprints of gene regulation associated with cadmium chloride, benzo(*a*)pyrene, and trichloroethylene. *Environmental Health Perspectives* 109:71–74.

Blum, J. L., I. Knoebl, P. Larkin, K. J. Kroll, and N. D. Denslow. 2004. Use of suppressive subtractive hybridization and cDNA arrays to discover patterns of altered gene expression in the liver of dihydrotestosterone and 11-ketotestosterone exposed adult male largemouth bass (*Micropterus salmoides*). *Marine Environmental Research* 58:565–569.

Bradley, B. P., E. A. Shrader, D. G. Kimmel, and J. C. Meiller. 2002. Protein expression signatures: an application of proteomics. *Marine Environmental Research* 54:373–377.

Brouwer, M., R. Syring, and T. H. Brouwer. 2002. Role of a copper-specific metallothionein of the blue crab, *Callinectes sapidus,* in copper metabolism associated with degradation and synthesis of hemocyanin. *Journal of Inorganic Biochemistry* 88:228–239.

Brouwer, M., T. Hoexum Brouwer, W. Grater, and N. Brown-Peterson. 2003. A novel cytosolic Mn-superoxide dismutase (MnSOD) has replaced cytosolic Cu,ZnSOD in crustacea that use copper (hemocyanin) for oxygen transport. *Biochemical Journal* 374:219–228.

Brouwer, M., P. Larkin, N. Brown-Peterson, C. King, S. Manning, and N. Denslow. 2004. Effects of hypoxia on gene and protein expression in the blue crab, *Callinectes sapidus. Marine Environmental Research* 58:787–792.

Buhler, D. R., C. L. Miranda, M. C. Henderson, Y. H. Yang, S. J. Lee, and J. L. Wang-Buhler. 2000. Effects of 17 beta-estradiol and testosterone on hepatic mRNA/protein levels and catalytic activities of CYP2M1, CYP2K1 and CYP3A27 in rainbow trout (*Oncorhynchus mykiss*). *Toxicology and Applied Pharmacology* 186:91–101.

Choi, J., H. Roche, and T. Caquet. 2000. Effects of physical (hypoxia, hyperoxia) and chemical (potassium dichromate, fenitrothion) stress on antioxidant enzyme activities in *Chironomus riparius* Mg. (Diptera, Chironomidae) larvae: potential biomarkers. *Environmental Toxicology and Chemistry* 19:495–500.

Das, T. and W. B. Stickle. 1994. Detection and avoidance of hypoxic water by juvenile *Callinectes sapidus* and *C. similis. Marine Biology* 120:593–600.

DeFur, P. L., C. P. Mangum, and J. E. Reese. 1996. Respiratory responses of the blue crab *Callinectes sapidus* to long-term hypoxia. *Biological Bulletin* 178:46–54.

Denslow, N. D., I. Knoebl, and P. Larkin. In press. Approaches in proteomics and genomics for ecotoxicology. In *Biochemistry and Molecular Biology of Fishes,* Vol. 6, T. W. Moon and T. P. Mommsen (eds.). Elsevier, Amsterdam.

Dixon, T. J., J. B. Taggart, and S. G. George. 2002. Application of real time PCR determination to assess interanimal variabilities in CYP1A induction in the European flounder (*Platichthys flesus*). *Marine Environmental Research* 54:267–270.

Ermolaeva, O. D. and E. D. Sverdlov. 1996. Subtractive hybridization, a technique for extraction of DNA sequences distinguishing two closely relate genomes: critical analysis. *Genetic Analysis* 13:49–58.

Funkenstein, B. 2001. Developmental expression, tissue distribution and hormonal regulation of fish (*Sparus aurata*) serum retinol-binding protein. *Comparative Biochemistry and Physiology Part B* 129:613–622.

Gracey, A. Y., J. V. Troll, and G. N. Somero. 2001. Hypoxia-induced gene expression profiling in the euryoxic fish *Gillichthys mirabilis. Proceedings of the National Academy of Sciences of the United States of America* 98:1993–1998.

Heid, C. A., J. Stevens, N. J. Livak, and P. M. Williams. 1996. Real time quantitative PCR. *Genome Research* 6:986–994.

Hightower, L. E. 1993. A brief perspective on the heat-shock response and stress proteins. *Marine Environmental Research* 35:79–83.

Hochachka, P. W. and P. L. Lutz. 2001. Mechanism, origin, and evolution of anoxia tolerance in animals. *Comparative Biochemistry and Physiology Part B* 130:435–459.

Hochachka, P. W., L. T. Buck, C. J. Doll, and S. C. Land. 1996. Unifying theory of hypoxia tolerance: molecular/metabolic defense and rescue mechanisms for surviving oxygen lack. *Proceedings of the National Academy of Sciences of the United States of America* 93:9493–9498.

Hogstrand, C., S. Balesaria, and C. N. Glover. 2002. Application of genomics and proteomics for study of the integrated response to zinc exposure in a non-model fish species, the rainbow trout. *Comparative Biochemistry and Physiology Part B* 133:523–535.

Justic, D., N. N. Rabalais, R. E. Turner, and W. J. Wiseman. 1993. Seasonal coupling between riverborne nutrients, net productivity and hypoxia. *Marine Pollution Bulletin* 26:184–189.

Kultz, D. and G. N. Somero. 1997. Differences in protein patterns of gill epithelial cells of the fish *Gillichthys mirabilis* after osmotic and thermal acclimation. *Journal of Comparative Physiology Part B* 166:88–100.

Kumar, R. S., S. Ijiri, and J. M. Trant. 2001. Molecular biology of the channel catfish gonadotropin receptors: 2. Complementary DNA cloning, functional expression, and seasonal gene expression of the follicle-stimulating hormone receptor. *Biology of Reproduction* 65:710–717.

Larkin, P., L. C. Folmar, M. J. Hemmer, A. J. Poston, and N. D. Denslow. 2003a. Expression profiling of estrogenic compounds using a sheepshead minnow cDNA macroarray. *Environmental Health Perspectives Toxicogenomics* 111:839–846.

Larkin, P., I. Knoebl, and N. D. Denslow. 2003b. Differential gene expression analysis in fish exposed to endocrine disrupting compounds. *Comparative Biochemistry and Physiology Part B* 136:149–161.

Lisitsyn, N. and M. Wigler. 1993. Cloning the differences between two complex genomes. *Science* 259:946–951.

McMahon, B. R. 2001. Respiratory and circulatory compensation to hypoxia in crustaceans. *Respiration Physiology* 128:349–364.

Mangum, C. P. 1997. Adaptation of the oxygen transport system to hypoxia in the blue crab, *Callinectes sapidus*. *American Zoologist* 37:604–611.

Manning, C. S, A. L. Schesny, W. E. Hawkins, D. H. Barnes, C. S. Barnes, and W. W. Walker. 1999. Exposure methodologies and systems for long-term chemical carcinogenicity studies with small fish species. *Toxicological Methods* 9:201–217.

May, E. B. 1973. Extensive oxygen depletion in Mobile Bay, Alabama. *Limnology and Oceanography* 18:353–366.

Meiller, J. C. and B. P. Bradley. 2002. Zinc concentrations effect at the organismal, cellular and subcellular levels in the eastern oyster. *Marine Environmental Research* 54:401–404.

Paerl, H. W., J. L. Pinckney, J. M. Fear, and B. L. Peierls. 1998. Ecosystem responses to internal and watershed organic matter loading: consequences for hypoxia in the eutrophying Neuse River estuary, North Carolina, USA. *Marine Ecology Progress Series* 166:17–25.

Renaud, M. L. 1986. Detecting and avoiding oxygen deficient sea water by brown shrimp, *Penaeus aztecus*, and white shrimp, *Penaeus setiferus*. *Journal of Experimental Marine Biology and Ecology* 98:283–292.

Ringwood, A. H. and C. J. Keppler. 2002. Water quality variation and clam growth: is pH really a non-issue in estuaries? *Estuaries* 25:901–907.

Robb, T. and M. V. Abrahams. 2003. Variation in tolerance to hypoxia in a predator and prey species: an ecological advantage of being small? *Journal of Fish Biology* 62:1067–1081.

Rose, K. A. 2000. Why are quantitative relationships between environmental quality and fish populations so elusive? *Ecological Applications* 10:367–385.

Rose, K. A., C. A. Murphy, S. L. Diamond, L. A. Fuiman, and P. Thomas. 2003. Using nested models and laboratory data for predicting population effects of contaminants on fish: a step toward a bottom-up approach for establishing causality in field studies. *Human and Ecological Risk Assessment* 9:231–257.

Russell, W. J., Y. S. Ho, G. Parish, and R. M. Jackson. 1995. Effects of hypoxia on MnSOD expression in mouse lung. *American Journal of Physiology–Lung Cellular and Molecular Physiology* 13:L221–L226.

Schena, M., R. A. Heller, T. P. Theriault, K. Konrad, E. Lachenmeier, and R. W. Davis. 1998. Microarrays: biotechnology's discovery platform for functional genomics. *Trends in Biotechnology* 16:301–306.

Seliger, H. H., J. A. Boggs, and W. H. Biggley. 1985. Catastrophic anoxia in the Chesapeake Bay in 1984. *Science* 228:70–73.

Semenza, G. L., P. H. Roth, H. M. Fang, and G. L. Wang. 1994. Transcriptional regulation of genes encoding glycolytic enzymes by hypoxia-inducible factor 1. *Journal of Biological Chemistry* 269:23757–23763.

Shepard, J. L. and B. P. Bradley. 2000. Protein expression signatures and lysosomal stability in *Mytilus edulis* exposed to graded copper concentrations. *Marine Environmental Research* 50:457–463.

Sokal, R. R. and F. J. Rohlf. 1995. *Biometry,* 3rd ed. W.H Freeman, New York, 887 pp.

Suter, G. W., II, S. B. Norton, and S. M. Cormier. 2002. A methodology for inferring the causes of observed impairments in aquatic ecosystems. *Environmental Toxicology and Chemistry* 21:1101–1111.

Syring, R. A., T. Hoexum Brouwer, and M. Brouwer. 2000. Cloning and sequencing of cDNAs encoding for a novel copper-specific metallothionein and two cadmium-inducible metallothioneins from the blue crab *Callinectes sapidus*. *Comparative Biochemistry and Physiology Part C* 125:325–332.

Ton, C., D. Stamatiou, V. J. Dzau, and C. C. Liew. 2002. Construction of a zebrafish cDNA microarray: gene expression profiling of the zebrafish during development. *Biochemical and Biophysical Research Communications* 296:1134–1142.

Trant, J. M., S. Gavasso, J. Ackers, B. C. Chung, and A. R. Place. 2001. Developmental expression of cytochrome P450 aromatase genes (CYP19a and CYP19b) in zebrafish fry (*Danio rerio*). *Journal of Experimental Zoology* 290:475–483.

Turner, R. E. and N. N. Rabalais. 1994. Coastal eutrophication near the Mississippi delta. *Nature* 368:619–621.

U.S. EPA. 1999. Ecological Condition of Estuaries in the Gulf of Mexico. EPA 620-R-98-004. U.S. Environmental Protection Agency, Office of Research and Development, National Health and Environmental Effects Research Laboratory, Gulf Ecology Division, Gulf Breeze, FL.

Wannamaker, C. M. and J. A. Rice. 2000. Effects of hypoxia on movements and behavior of selected estuarine organisms from the southeastern United States. *Journal of Experimental Marine Biology and Ecology* 249:145–163.

Williams, T. D., K. Gensberg, S. D. Mincin, and J. K. Chipman. 2003. A DNA expression array to detect toxic stress response in European flounder (*Platichthys flesus*). *Aquatic Toxicology* 65:141–157.

Winn, R. N. and D. M. Knott. 1992. An evaluation of the survival of experimental populations exposed to hypoxia in the Savannah River estuary. *Marine Ecology Progress Series* 88:161–179.

Wu, R. S. S., P. K. S. Lam, and K. L. Wan. 2002. Tolerance to, and avoidance of, hypoxia by the penaeid shrimp (*Metapanaeus ensis*). *Environmental Pollution* 118:351–355.

Zimmerman, R. J. and J. M. Nance. 2001. Effects of hypoxia on the shrimp fishery of Louisiana and Texas. In *Coastal Estuarine Studies. Coastal Hypoxia: Consequences for Living Resources and Ecosystems*, N. N. Rabalais and E. Turner (eds.). American Geophysical Union, Washington, D.C., pp. 293–310.

18

Spionid Polychaetes as Environmental Indicators: An Example from Tampa Bay, Florida

Thomas L. Dix, David J. Karlen, Stephen A. Grabe, Barbara K. Goetting, Christina M. Holden, and Sara E. Markham

CONTENTS

Introduction

The polychaete family Spionidae is one of most ubiquitous benthic taxa (Fauchald, 1977; Johnson, 1984; Blake, 1996). This family is dominated by many cosmopolitan species (Foster, 1971; Johnson, 1984; Herrando-Perez et al., 2001) that tolerate a wide salinity range from tidal freshwater to hypersaline marine conditions (Foster, 1971; Taylor, 1971).

Spionids inhabit all types of substrata in estuarine habitats (Dauer, 1985) although most genera are present in soft-bottom sediments (Hartman, 1951; Foster, 1971; Taylor, 1971; Fauchald, 1977; Dauer et al., 1981; Johnson, 1984; Flint and Kalke, 1986; Blake, 1996; Frouin et al., 1998; Sato-Okoshi, 2000; Zajac et al., 2000). All spionids construct mucus-lined burrows or tubes (Day, 1967; Sato-Okoshi, 1999). Some species in the genera *Polydora* and *Boccardia* burrow or attach to hard substrates such as rock, limestone, dead coral, mollusk shells, wood, and sponges (Day, 1967; Blake, 1971; Fauchald, 1977; Rice, 1978; Johnson, 1984; Blake, 1996; Martin, 1996; Sato-Okoshi and Okoshi, 1997; Warner, 1997; Sato-Okoshi, 1999).

Day (1967) and Fauchald and Jumars (1979) classified spionids as surface deposit feeders, but they have also been considered suspension feeders or as both deposit and suspension feeders (Myers, 1977; Dauer et al., 1981; Dauer, 1985; Gaston and Nasci, 1988; Gaston et al., 1988; Dauer, 1991; Dauer and Ewing, 1991; Blake, 1996; Frouin et al., 1998; Garcia-Arberas and Rallo, 2002). Spionids use palps to gather food particles from the substratum (Day, 1967) and from the water column (Dauer, 1985; Dauer and Ewing, 1991).

A few species are opportunistic and occupy areas that are disturbed or organically enriched (Boesch et al., 1976; Steimle and Radosh, 1979; Thistle, 1981; Weston, 1990; Grizzle and Penniman, 1991; Gaston and Young, 1992; Dauer, 1993; Zmarzly et al., 1994; Posey et al., 1996; Bachelet et al., 2000; Morris and Keough, 2002; Posey and Alphin, 2002). Opportunistic species demonstrate plasticity in reproduction and development, which enable them to dominate in coastal habitats (McCall, 1977; Rice, 1978; Levin, 1984; Dauer, 1993; Lardicci et al., 1997; Mackay and Gibson, 1999). They can also tolerate oxygen-stressed areas (Lamont and Gage, 2000; Rosenberg et al., 2001; Rabalais et al., 2002) and can potentially act as indicator species for degraded habitats (Grizzle and Penniman, 1991; Rizzo and Amaral, 2001; Morris and Keough, 2002).

A variety of polychaete distribution and ecology studies have been done in Florida, especially among estuaries in the eastern Gulf of Mexico. Taylor (1971) wrote a systematic and ecological account of polychaete worms at 363 sites in Tampa Bay, Florida from 1963–1964 and 1969. Taylor's (1971) study, however, did not include the tributaries in the Tampa Bay watershed. Santos and Simon (1974) studied the distribution and abundance of polychaetes from vegetated and nonvegetated zones at Lassing Park in St. Petersburg. Intraspecific variations among *Polydora cornuta* (as *P. ligni*) populations were studied from Ballast Point, Courtney Campbell Causeway, Ft. DeSoto, and Cockroach Bay (Rice, 1978; Rice and Simon, 1980). A new spionid species, *Streblospio gynobranchiata*, was described from the Courtney Campbell Causeway in Old Tampa Bay (Rice and Levin, 1998; Schulze et al., 2000). These studies described the polychaetes of Tampa Bay, community structures, or populations, but did not use spionids as environmental indicators. This chapter looks at the utility of using spionids as indicators of estuarine habitat type in the Tampa Bay watershed.

Materials and Methods

The sediment contaminant and benthic-monitoring program adopted for Tampa Bay employed a stratified (by bay segment), random probability-based design (Coastal Environmental, 1994). A hexagonal grid was overlaid on Tampa Bay and within each hexagon the sampling coordinates were randomly determined, with a known probability of inclusion. The same sites were sampled during both 1993 and 1994; thereafter, new sampling coordinates were generated annually. Sample sites in the tributaries were randomly generated by computer, but did not use the hexagonal grid system. All samples were collected during the late summer/early fall.

In all, 1532 stations were sampled from 1993 to 2001 (Figure 18.1). Station location was determined in the field using differential GPS (global positioning system). Water column measurements of temperature, dissolved oxygen, salinity, and pH were made using a Hydrolab® Surveyor 3. Measurements were recorded every meter during both the descent and ascent; beginning in 1998 only surface and bottom depths were sampled for water quality variables.

Sediments were sampled using a stainless steel, Young-modified Van Veen grab sampler (0.04 m²). Sediment contaminant samples were collected by removing the upper 2 cm from one to three grab samples. A separate grab sample was collected for benthic community analysis. A 60-ml core subsample was removed and the depth to the apparent redox potential discontinuity layer (RPD) (Gray, 1981) was measured to the nearest millimeter. The sediment subsample was retained for later analysis of the silt/clay content (%SC). Determination of %SC was after Folk (1968). During 1993 and 1994 the faunal sample was sieved immediately through a 0.5-mm mesh sieve and then fixed in a 10% solution of borax-buffered Formalin with Rose Bengal stain. Beginning in 1995, a solution of magnesium sulfate (Epsom Salt) was added to the sample as a relaxant and the sample was then placed on ice. All samples were sieved and preserved at the conclusion of the day's sampling. Within 2 weeks of collection, samples were transferred

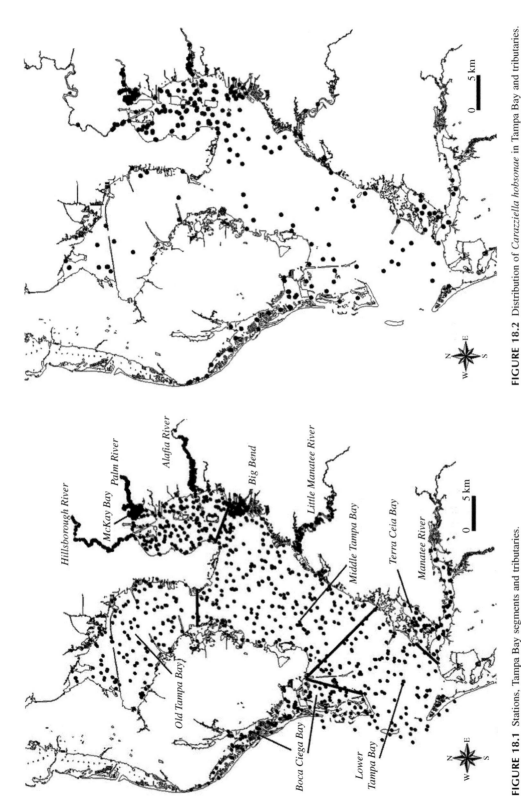

FIGURE 18.2 Distribution of *Carazziella hobsonae* in Tampa Bay and tributaries.

FIGURE 18.1 Stations, Tampa Bay segments and tributaries.

to 70% isopropanol with Rose Bengal for preservation. Benthic samples were sorted using a dissecting microscope, and organisms were identified to the lowest practical taxonomic level.

The total number of species and abundance for all 1532 stations were analyzed as a group and for each bay and tributary segment. The median and range for environmental variables were calculated for each species. Distributional maps were created for the dominant species using ArcView® Version 3.2.

Results

Six spionid species accounted for 90% of the total familial abundance (Table 18.1): *Carazziella hobsonae* (25.8%), *Prionospio perkinsi* (23.1%), *Paraprionospio pinnata* (14.9%), *Streblospio gynobranchiata* (14.6%), *Prionospio heterobranchia* (6.7%), and *Spio pettiboneae* (4.0%). *Prionospio perkinsi* and *Paraprionospio pinnata* had the highest frequency of occurrence and were found at 38.1% and 37.1% of the sites, respectively.

Most taxa showed a wide range for environmental variables with which they were associated (Table 18.1) but some distinct distributional patterns were apparent. *Carazziella hobsonae* (Figure 18.2) was the dominant spionid in McKay Bay, but was also found in high abundance in all of the major bay segments except Lower Tampa Bay (Table 18.2 and Table 18.3) as well as in most of the tributaries (Table 18.4). *Prionospio perkinsi* (Figure 18.3) was the dominant spionid in Hillsborough Bay, Old Tampa Bay, and Middle Tampa Bay and highly abundant in all other segments and tributaries (Table 18.2 through Table 18.5). *Paraprionospio pinnata* (Figure 18.4) and *Apoprionospio pygmaea* (Figure 18.5) were also widespread throughout Tampa Bay. *Paraprionospio pinnata* was the dominant spionid recorded in Boca Ciega Bay, Terra Ceia Bay, the Manatee River, and at the Big Bend stations (Table 18.3 and Table 18.5). *Apoprionospio pygmaea* occurred in low abundances baywide, and was among the dominant spionid taxa in Lower Tampa Bay, Terra Ceia Bay, and Big Bend (Tables 18.2, 18.3, and 18.5). *Streblospio gynobranchiata* (Figure 18.6) was dominant in all four tributaries (Tables 18.4 and 18.5) and highly abundant in Hillsborough Bay, the Manatee River, McKay Bay, and at Big Bend (Tables 18.2 through 18.5). This species also had the lowest median salinity (Table 18.1). *Prionospio heterobranchia* (Figure 18.7), *Polydora cornuta* (Figure 18.8), and *Dipolydora socialis* (Figure 18.9) had scattered distributions throughout Tampa Bay, with *Prionospio heterobranchia* among the five dominant spionids in the four main bay segments as well as in Terra Ceia Bay, the Little Manatee River, Palm River, McKay Bay, and Big Bend (Tables 18.2 through 18.5). *Polydora cornuta* was particularly abundant in Boca Ciega Bay (Table 18.3) and *Dipolydora socialis* in the Hillsborough, Alafia, and Little Manatee Rivers (Table 18.4). *Scolelepis texana* (Figure 18.10) was found in shallow-water, sandy habitats throughout the bay but was not among the dominants in any segment or tributary. *Spio pettiboneae* (Figure 18.11) and *Prionospio steenstrupi* (Figure 18.12) were distributed mainly in Middle Tampa Bay and Lower Tampa Bay. *Spio pettiboneae* was the dominant spionid in Lower Tampa Bay and was also abundant in Old Tampa Bay, Middle Tampa Bay, and the Manatee River (Tables 18.2 and 18.3) while *P. steenstrupi* was among the top five dominants in Lower Tampa Bay (Table 18.2).

Discussion

The six dominant spionid taxa (*Carazziella hobsonae*, *Prionospio perkinsi*, *Paraprionospio pinnata*, *Streblospio gynobranchiata*, *Prionospio heterobranchia*, and *Spio pettiboneae*) are known to tolerate a wide range of environmental conditions, but several of these taxa showed distinct distributions within Tampa Bay. These six species along with several less abundant taxa that display unique distributional patterns or habitat preferences are reviewed below. The other species of spionids from this study are considered rare taxa: they had a frequency of occurrence less than 5% and relative abundance of less than 0.5%. Generally, these taxa occurred in the lower part of Tampa Bay and no conclusions were made on their ecological preferences.

TABLE 18.1

Dominant Spionid Species and Associated Physical Parameters

	Relative Abundance (%) (n = 36,528)	Cumulative Abundance (%)	Frequency of Occurrence (%) (n = 1,532)	Depth (m) Median (min.–max.)	Temperature (C) Median (min.–max.)	Salinity (‰) Median (min.–max.)	Dissolved O$_2$ (mg/L) Median (min.–max.)	% Silt + Clay (%) Median (min.–max.)
Carazziella hobsonae	25.8	25.8	20.0	2.0 (0.1–11.1)	29.1 (23.8–34.0)	25.9 (1.8–35.9)	4.6 (0.1–11.3)	6.3 (0.0–74.4)
Prionospio perkinsi	23.1	49.0	38.1	2.7 (0.1–11.1)	28.8 (23.9–35.5)	25.7 (7.4–35.9)	5.1 (0.1–11.0)	4.5 (0.0–74.4)
Paraprionospio pinnata	14.9	63.9	37.1	2.0 (0.1–11.1)	29.1 (22.9–33.4)	25.2 (2.1–35.9)	4.7 (0.1–11.0)	5.4 (0.0–86.8)
Streblospio gynobranchiata	14.6	78.5	19.1	1.0 (0.1–11.0)	29.0 (22.9–36.2)	19.9 (0.0–35.9)	4.1 (0.1–14.0)	6.4 (1.0–86.8)
Prionospio heterobranchia	6.7	85.2	17.5	1.0 (0.1–9.0)	28.8 (21.6–35.5)	26.6 (11.5–35.9)	5.6 (1.0–14.0)	3.7 (0.0–82.2)
Spio pettiboneae	4.0	89.2	12.4	3.5 (0.1–9.0)	28.2 (23.4–31.2)	28.3 (11.7–34.8)	5.7 (1.6–9.3)	2.1 (0.0–49.6)
Apoprionospio pygmaea	2.9	92.1	18.3	2.3 (0.1–11.8)	28.6 (21.6–34.0)	26.8 (11.6–35.9)	5.6 (0.0–11.0)	3.1 (0.0–81.1)
Polydora cornuta	2.4	94.5	7.8	0.8 (0.1–9.6)	28.9 (21.6–32.6)	26.8 (6.0–34.4)	5.7 (0.1–14.0)	5.0 (1.2–82.2)
Prionospio steenstrupi	1.6	96.1	5.5	4.0 (0.1–12.5)	28.0 (23.9–31.2)	29.0 (16.0–34.2)	5.7 (3.2–11.3)	2.1 (0.0–19.1)
Prionospio cristata	1.1	97.2	5.7	3.4 (0.1–9.0)	28.8 (23.9–30.9)	28.1 (17.2–35.6)	5.7 (2.7–11.0)	2.9 (0.0–15.3)
Scolelepis texana	1.0	98.2	5.2	1.0 (0.1–8.9)	28.6 (23.0–33.4)	25.1 (10.4–33.7)	5.4 (2.0–10.7)	2.8 (1.0–42.3)
Dipolydora socialis	1.0	99.2	7.5	2.0 (0.1–11.0)	28.1 (21.6–32.2)	28.1 (0.0–35.9)	5.5 (0.0–11.3)	5.2 (0.0–82.2)

TABLE 18.2

Relative Abundance and Frequency of Dominant Taxa in Hillsborough Bay, Old Tampa Bay, Middle Tampa Bay, and Lower Tampa Bay

Abundance/Frequency (%)			
Hillsborough Bay (*n* = 213)	**Old Tampa Bay** (*n* = 142)	**Middle Tampa Bay** (*n* = 197)	**Lower Tampa Bay** (*n* = 152)
Prionospio perkinsi 31.3/56.8	*Prionospio perkinsi* 40.6/66.2	*Prionospio perkinsi* 44.0/60.4	*Spio pettiboneae* 28.9/50.0
Carazziella hobsonae 36.5/39.4	*Paraprionospio pinnata* 9.4/44.4	*Carazziella hobsonae* 12.8/18.3	*Prionospio perkinsi* 10.8/29.6
Paraprionospio pinnata 16.1/49.8	*Prionospio heterobranchia* 13.2/29.6	*Paraprionospio pinnata* 7.9/29.4	*Prionospio steenstrupi* 13.9/22.4
Streblospio gynobranchiata 7.8/18.3	*Carazziella hobsonae* 17.0/12.7	*Spio pettiboneae* 6.5/28.4	*Apoprionospio pygmaea* 6.9/34.9
Prionospio heterobranchia 2.9/16.4	*Spio pettiboneae* 8.4/16.9	*Prionospio heterobranchia* 11.1/14.2	*Prionospio heterobranchia* 8.8/23.0

TABLE 18.3

Relative Abundance and Frequency of Dominant Taxa in Boca Ciega Bay, Terra Ceia Bay, and the Manatee River

Abundance/Frequency (%)		
Boca Ciega Bay (*n* = 125)	**Terra Ceia Bay** (*n* = 61)	**Manatee River** (*n* = 97)
Paraprionospio pinnata 16.5/50.4	*Paraprionospio pinnata* 42.2/65.6	*Paraprionospio pinnata* 35.5/63.9
Prionospio heterobranchia 16.4/39.2	*Apoprionospio pygmaea* 11.1/49.2	*Streblospio gynobranchiata* 33.5/22.7
Carazziella hobsonae 17.2/33.6	*Prionospio perkinsi* 9.1/49.2	*Prionospio perkinsi* 15.1/29.9
Polydora cornuta 16.9/26.4	*Prionospio heterobranchia* 8.2/41.0	*Carazziella hobsonae* 8.7/10.3
Prionospio perkinsi 9.0/32.8	*Carazziella hobsonae* 7.6/26.2	*Spio pettiboneae* 2.0/10.3

TABLE 18.4

Relative Abundance and Frequency of Dominant Taxa in the Hillsborough River, Alafia River, and Little Manatee River

Abundance/Frequency (%)		
Hillsborough River (*n* = 87)	**Alafia River** (*n* = 151)	**Little Manatee River** (*n* = 79)
Streblospio gynobranchiata 29.1/32.2	*Streblospio gynobranchiata* 62.6/39.7	*Streblospio gynobranchiata* 44.7/25.3
Carazziella hobsonae 31.6/12.6	*Prionospio perkinsi* 18.1/17.9	*Prionospio perkinsi* 12.9/8.9
Paraprionospio pinnata 16.2/16.1	*Paraprionospio pinnata* 9.0/23.8	*Prionospio heterobranchia* 10.7/10.1
Prionospio perkinsi 15.4/10.3	*Carazziella hobsonae* 9.3/11.3	*Paraprionospio pinnata* 6.3/10.1
Dipolydora socialis 5.5/6.9	*Dipolydora socialis* 0.5/5.3	*Dipolydora socialis* 8.1/5.1

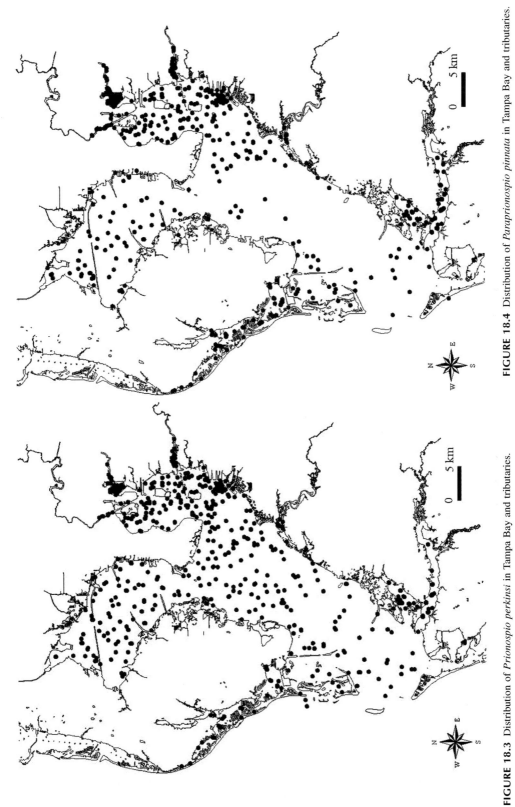

FIGURE 18.4 Distribution of *Paraprionospio pinnata* in Tampa Bay and tributaries.

FIGURE 18.3 Distribution of *Prionospio perkinsi* in Tampa Bay and tributaries.

TABLE 18.5

Relative Abundance and Frequency of Dominant Taxa in the Palm River,
in McKay Bay, and at Big Bend

Abundance/Frequency (%)		
Palm River (*n* = 78)	**McKay Bay** (*n* = 76)	**Big Bend** (*n* = 74)
Streblospio gynobranchiata 72.0/24.4	*Carazziella hobsonae* 80.5/44.7	*Paraprionospio pinnata* 33.5/51.4
Paraprionospio pinnata 21.4/17.9	*Paraprionospio pinnata* 7.3/61.8	*Prionospio perkinsi* 18.0/35.1
Carazziella hobsonae 2.2/7.7	*Prionospio perkinsi* 5.5/40.8	*Streblospio gynobranchiata* 23.9/23.0
Prionospio heterobranchia 3.1/3.8	*Streblospio gynobranchiata* 5.5/39.5	*Prionospio heterobranchia* 15.8/28.4
Prionospio perkinsi 0.9/5.1	*Prionospio heterobranchia* 0.9/17.1	*Apoprionospio pygmaea* 4.3/14.9

Carazziella hobsonae Blake, 1979

Taylor (1971) found *Carazziella hobsonae* (reported as *Pseudopolydora* sp.) in all areas of Tampa Bay except Terra Ceia Bay, but especially abundant in Old Tampa Bay and Hillsborough Bay. In this study, *C. hobsonae* was found in all areas of Tampa Bay and in all of the tributaries, but its highest abundance was in McKay Bay and Hillsborough Bay. *Carazziella hobsonae* has been found in a wide variety of substrates, salinities, and depths (Taylor, 1971; Blake, 1979). The salinity range in this study (2 to 36‰) was greater than in Taylor's (1971) study (18 to 34‰), but both studies had a mean salinity of 26‰.

Prionospio perkinsi Maciolek, 1985

Prionospio perkinsi was recorded as *P. cirrobranchiata* by Taylor (1971) and identified as *Minuspio cirrifera* in several other studies (Foster, 1971; Day, 1973). *Prionospio perkinsi* was found throughout Tampa Bay in a variety of environmental conditions. Taylor (1971) found *P. perkinsi* throughout Tampa Bay, except for Lower Tampa Bay, in wide variety of substrates. *Prionospio perkinsi* is a coastal species that occurs in the intertidal zone, but not greater the 40 m (Foster, 1971; Blake, 1983). The salinity range (7 to 36‰) in this study was greater than in Taylor's (1971) study (21 to 34‰), but both studies had a mean salinity of 26‰. Foster (1971) stated that this species was more common in mud bottoms and associated with areas of disturbance, but in our study it was found more commonly in sandy environments.

Paraprionospio pinnata (Ehlers, 1901)

Paraprionospio pinnata was widespread throughout Tampa Bay, but was least common in Middle Tampa Bay. Taylor (1971) also found *P. pinnata* throughout Tampa Bay, but it was less common in Lower Tampa Bay. *Paraprionospio pinnata* is a cosmopolitan species and common in American waters from less than 3 to 1300 m (Foster, 1971; Blake, 1996). *Paraprionospio pinnata* has been a dominant component of benthic communities in several studies (Stickney and Perlmutter, 1975; Dauer et al., 1981; Mauer et al., 1994; Simboura et al., 1995; Flemer et al., 1997; Delgado-Blas, 2001). The salinity range (2 to 36‰) in this study was greater than in Taylor's (1971) study (16 to 35 ‰), but both studies had similar mean salinities (25 vs. 27‰, respectively). *Paraprionospio pinnata* has been found in a variety of sediment types (Foster, 1971; Taylor, 1971; Dauer et al., 1981; Johnson, 1984). Earlier studies have suggested that *P. pinnata* has a preference for muddy substrates (Johnson, 1984; Flint and Kalke, 1986; Blake, 1996), but in Tampa Bay it was dominant in sandy habitats. Even though they were found chiefly in sandy habitats in Tampa Bay, Dauer (1985) found that *P. pinnata* feeds preferentially on particles that are less than 63 μm in diameter. *Paraprionospio pinnata* can feed at the sediment–water interface on suspended, resuspended, and deposited materials (Dauer, 1985). *Paraprionospio pinnata* has been found

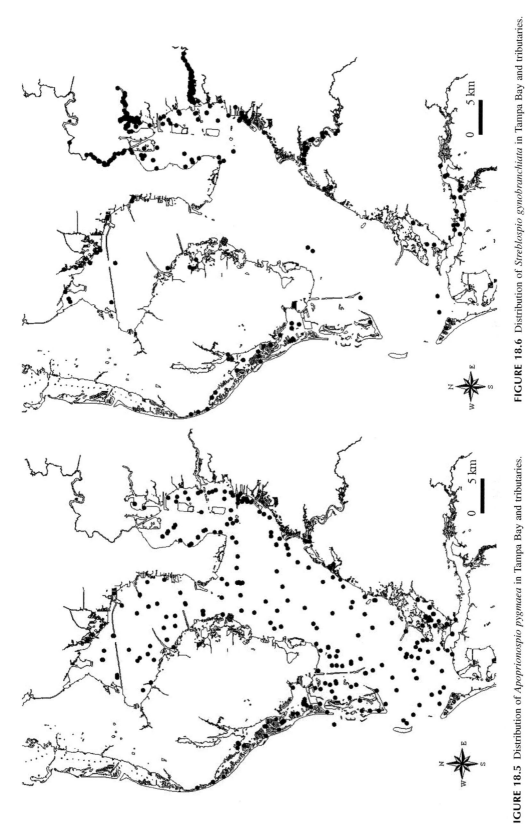

FIGURE 18.6 Distribution of *Streblospio gynobranchiata* in Tampa Bay and tributaries.

FIGURE 18.5 Distribution of *Apoprionospio pygmaea* in Tampa Bay and tributaries.

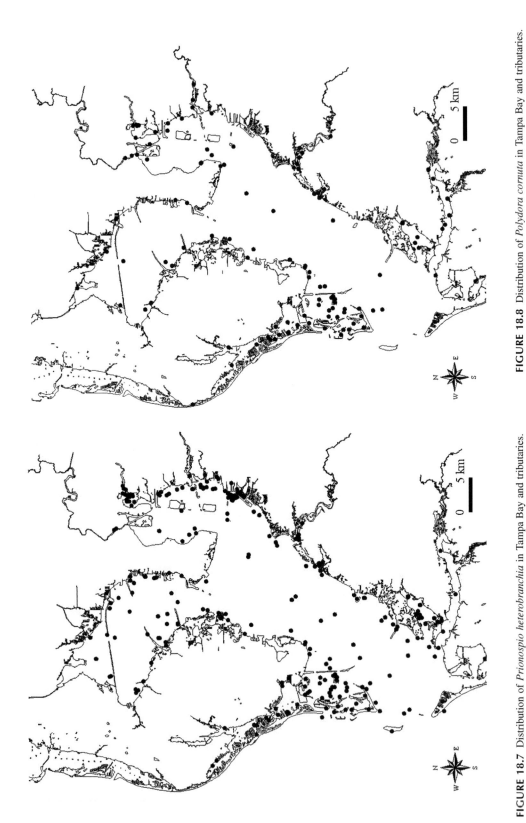

FIGURE 18.8 Distribution of *Polydora cornuta* in Tampa Bay and tributaries.

FIGURE 18.7 Distribution of *Prionospio heterobranchia* in Tampa Bay and tributaries.

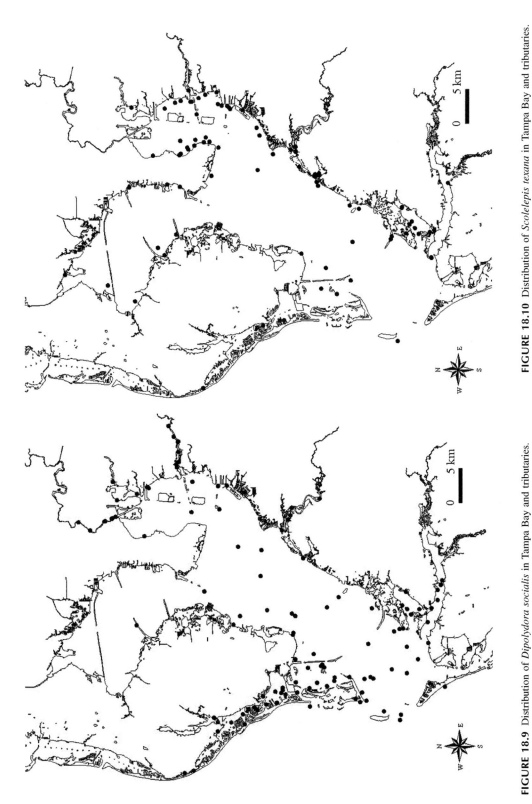

FIGURE 18.10 Distribution of *Scolelepis texana* in Tampa Bay and tributaries.

FIGURE 18.9 Distribution of *Dipolydora socialis* in Tampa Bay and tributaries.

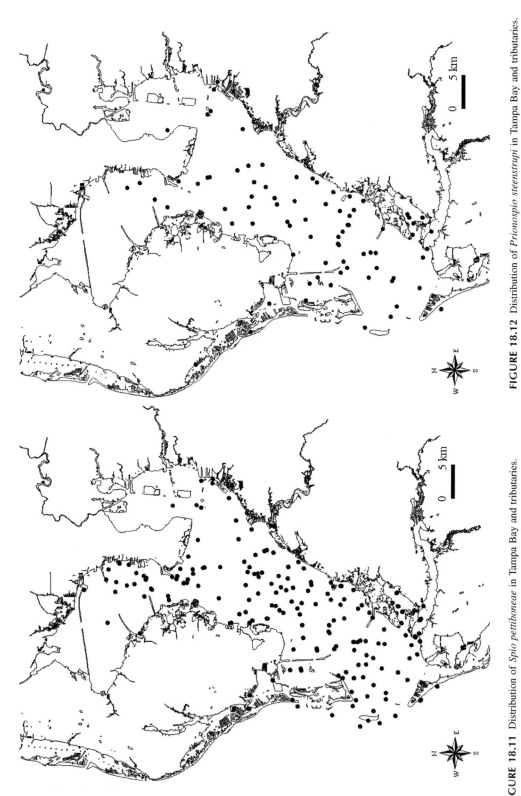

FIGURE 18.12 Distribution of *Prionospio steenstrupi* in Tampa Bay and tributaries.

FIGURE 18.11 Distribution of *Spio pettiboneae* in Tampa Bay and tributaries.

in physically disturbed and polluted areas (Boesch et al., 1976; Dauer, 1985; Blake, 1996), and it exhibits several adaptations to stressed habitats such as the ability to delay metamorphosis at low oxygen concentrations (Rabalais et al., 2002), the ability to produce respiratory currents to meet their respiratory needs (Dauer, 1985), and a high rate of growth and fecundity (Boesch et al., 1976). *Paraprionospio pinnata* also is highly motile compared to other spionids (Dauer et al., 1981), and may be able to move when conditions are stressed. The multiple ways of collecting food, reproduction strategies, and motility may be factors for the widespread abundance in Tampa Bay.

Streblospio gynobranchiata Rice and Levin, 1998

Streblospio gynobranchiata was identified as *S. benedicti* in earlier studies in Tampa Bay. In this study, *S. gynobranchiata* was dominant in the tributaries, but it was also found in Hillsborough Bay and Boca Ciega Bay. Taylor (1971) found this species throughout Tampa Bay, except in Lower Tampa Bay, but the highest abundance was in Hillsborough Bay and Boca Ciega Bay. The salinity range in this study (0 to 36‰) was greater than in Taylor's (1971) study (16 to 35‰); our study also had a lower mean salinity (19‰) than Taylor's (27‰). *Streblospio gynobranchiata* has been found in a wide variety of sediment types (Taylor, 1971; Rice and Levin, 1998) and has been found with high abundances in stressed or disturbed habitats (Rice and Levin, 1998).

Prionospio heterobranchia Moore, 1907

Prionospio heterobranchia has a widespread distribution in the Atlantic and Pacific Oceans (Blake, 1983). *Prionospio heterobranchia* had a scattered distribution throughout Tampa Bay in this study. Taylor (1971) did not find *P. heterobranchia* in Hillsborough Bay, but it was present in Boca Ciega Bay and Lower Tampa Bay. The salinity range (12 to 36‰) in this study was greater than in Taylor's (1971) study (24 to 35‰), but both studies had similar mean salinities (27 vs. 31‰, respectively). *Prionospio heterobranchia* has been found in a variety of sediment types, on sponges, and also associated with vegetation (*Halimeda*, *Thalassia*, and decaying vegetation) (Foster, 1971). Santos and Simon (1974) found that *P. heterobranchia* attained their highest density in the *Thalassia* zone in Tampa Bay. *Prionospio heterobranchia* is a suspension feeder and has a maximum sediment depth of 4 cm (Myers, 1977). Foster (1971) stated that *P. heterobranchia* does not tolerate polluted waters and is highly abundant in unpolluted bays.

Apoprionospio pygmaea (Hartman, 1961)

This study found *Apoprionospio pygmaea* to be widespread in Tampa Bay. Taylor (1971) did not find *A. pygmaea* in Hillsborough Bay, but it was found throughout the remainder of Tampa Bay. The salinity range (11.6 to 35.9‰) in this study was greater than in Taylor's (1971) study (19 to 35‰), but both studies had similar mean salinities (27 vs. 29‰, respectively). *Apoprionospio pygmaea* has been found in a variety of sediment types and in *Thalassia* flats (Foster, 1971; Taylor, 1971; Johnson, 1984; Blake, 1996).

Spio pettiboneae Foster, 1971

Spio pettiboneae was dominant in Lower Tampa Bay, and it had a low frequency of occurrence in Hillsborough Bay. Taylor (1971) did not find *S. pettiboneae* in Hillsborough Bay. The salinity range (12 to 35‰) in this study was greater than in Taylor's (1971) study (22 to 35‰), but both studies had similar mean salinities (28 vs. 30‰). *Spio pettiboneae* was found in variety of sediment types, but predominately it was found in fine to medium sand (Taylor, 1971; Johnson, 1984). *Spio* spp. have been known to form dense colonies on sand banks (Day, 1967). Taylor (1971) found that about 30% of the stations that contain *S. pettiboneae* had vegetation present.

Prionospio steenstrupi Malmgren, 1867

Prionospio steenstrupi is widely distributed in both the Atlantic and Pacific Oceans (Blake, 1983). *Prionospio steenstrupi* distribution in this study was mainly in Middle Tampa Bay and Lower Tampa Bay. Taylor (1971) did not record *P. steenstrupi* in his study. *Prionospio steenstrupi* has been found in a variety of sediment types (Foster, 1971; Johnson, 1984; Blake, 1996) and has been recorded as a surface deposit-feeder or filter-feeder (Zajac et al., 2000). Although it is tubiculous, it also has the ability to be mobile (Zajac et al., 2000).

Scolelepis texana Foster, 1971 and *Scolelepis squamata* (O. F. Müller, 1806)

In this study, *Scolelepis texana* was scattered throughout Tampa Bay in shallow, sandy habitats. *Scolelepis squamata* was also collected, but its frequency of occurrence was less than 0.3%, while *S. texana* had a frequency of occurrence of over 5.0%. *Scolelepis texana* was described only in 1971, and was therefore not reported in Taylor's (1971) study. However, he did report *S. squamata* at 43 (12%) of his stations, and it is possible that some of these may have been *S. texana*. *Scolelepis texana* and *S. squamata* have similar ecological habitats and these species are considered together. *Scolelepis* spp. are typically found in intertidal sandy beaches and subtidal sandy flats (Foster, 1971; Day, 1973; Johnson, 1984; Dauer and Ewing, 1991; Blake, 1996; Frouin et al., 1998; Rizzo and Amaral, 2001; Prezant et al., 2002). *Scolelepis texana* has been found in sandy and silty sediments (Johnson, 1984; Blake, 1996), and it is deeply infaunal (Virnstein, 1979). *Scolelepis* spp. are highly mobile and can quickly retract into their tubes of several tens of centimeters in length. Some species have tubes 60 to 80 cm long (Frouin et al., 1998). *Scolelepis squamata* depends on high sediment permeability because the species cannot reverse the respiratory current into their burrows (Dauer, 1985). *Scolelepis* spp. appear to be suspension feeders (Frouin et al., 1998) and are not selective on particle size. The gut contents of *Scolelepis* spp. have been found to contain sand particles, small invertebrates, and fecal pellets (Dauer and Ewing, 1991; Blake, 1996; Rizzo and Amaral, 2001). *Scolelepis* spp. in organically enriched habitats have been reported to be both suspension and surface deposit feeders (Weston, 1990). The salinity range found in this study for *S. texana* (10 to 34‰) was greater than in Taylor's (1971) study for *S. squamata* (salinity range 17 to 35‰), and our study had a mean salinity of 25‰ while Taylor's had a mean salinity of 29‰. *Scolelepis squamata* is known to tolerate oscillations in salinity, biological contamination, and oil spills (Rizzo and Amaral, 2001). In San Francisco, *S. squamata* is not related to sediment type, but is associated with hydrodynamics and organic enrichment (Rizzo and Amarl, 2001).

Dipolydora socialis (Schmarda, 1861)

Dipolydora socialis (reported as *Polydora socialis* in earlier studies) was scattered throughout Tampa Bay with the highest abundances in the lower part of Tampa Bay and lowest in Old Tampa Bay. This species also occurred in the tributaries. Taylor (1971) found *D. socialis* throughout Tampa Bay, and found that it was least abundant in Old Tampa Bay and Hillsborough Bay. In this study, *D. socialis* in Hillsborough Bay had slightly lower abundance than in the lower part of Tampa Bay. This species has been associated with calcareous substrates as well as soft sediments (Hartman, 1951; Blake, 1971; Foster, 1971; Blake and Evans, 1973; Johnson, 1984; Blake, 1996; Sato-Okoshi, 2000). *Dipolydora socialis* has been known to form grooves on and in the crevices of oysters (living and dead), but it apparently does not bore into the shell structure (Sato-Okoshi, 2000). *Dipolydora socialis* was a dominant component of benthic infaunal communities in some areas of California (Blake, 1996), and has been known to form extensive beds (Blake, 1971). The salinity range (0 to 36‰) in this study was similar to Taylor's (1971) study (4 to 35‰), and both studies had similar mean salinities (28 vs. 30‰, respectively). *Dipolydora socialis* is not a rapid crawler but it has the ability to move (Sato-Okoshi, 2000). *Dipolydora socialis* is considered a surface deposit and suspension feeder (Gaston and Nasci, 1988).

Polydora cornuta Bosc, 1802 and *Polydora websteri* Hartman, 1943

Polydora cornuta was reported as *P. ligni* in earlier studies before Blake and Maciolek (1987) reestablished *P. cornuta* as the senior synonym for this taxon. *Polydora cornuta* has been reported as a dominant component of benthic communities in previous studies (Dauer et al., 1981; Gaston et al., 1988; Gaston and Young, 1992). In our study, *P. cornuta* was scattered throughout Tampa Bay. Taylor (1971) did not record *P. cornuta*, but *P. websteri* was recorded at 73 (20%) of his sites. *Polydora websteri* was also collected in this study, but its frequency of occurrence was less than 0.3%, while *P. cornuta* was found at 5.5% of our sites. It should be noted that the grab sampler used in our study does not adequately sample hard substrates or oyster reefs, while Taylor (1971) employed several sampling methods in his study, so the differences between our studies may be a sampling artifact. Even through *P. cornuta* and *P. websteri* have slightly different ecological habitats, these species are considered together. *Polydora cornuta* inhabits soft sediments and calcareous substrates but is a nonboring species (Blake, 1971; Foster, 1971; Rice, 1978; Dauer et al., 1981; Johnson, 1984; Blake, 1996; Sato-Okoshi and Okoshi, 1997; Radashevsky and Hsieh, 2000; Sato-Okoshi, 2000). *Polydora cornuta* is also found on tunicates, tubes of the onuphid polychaete *Diopatra sugokai*, serpulid polychaete tubes, rocks, wood, shells, wharf pilings, and vegetated sites (Blake, 1971; Foster, 1971; Rice, 1978; Dauer et al., 1981; Radashevsky and Hsieh, 2000; Bolam and Fernandes, 2002). *Polydora websteri* prefers living shell substrata (Sato-Okoshi, 1999) and burrows into the nacreous layers of bivalves forming "mud blisters," while *P. cornuta* forms grooves on and in crevices of living and dead oysters, but does not bore into the shell (Blake, 1971; Foster, 1971; Blake and Evans, 1973; Rice, 1978; Johnson, 1984; Handley, 1995; Blake, 1996; Radashevsky, 1999; Sato-Okoshi, 2000; Prezant et al., 2002). *Polydora cornuta* has been known to form extensive beds that can sometimes cause extensive mortalities to oysters (Blake, 1971; Foster, 1971; Blake and Evans, 1973; Rice, 1978; Johnson, 1984; Blake, 1996). *Polydora websteri* has been reported from ten different bivalve hosts but heavy infestations rarely cause mortality to its host; however, the host may be more susceptible to parasites or disease (Blake, 1996). The salinity range for *P. cornuta* (6 to 34‰) in this study was similar to *P. websteri* in Taylor's (1971) study (4 to 35‰) and both studies had similar mean salinities (27 vs. 30‰, respectively). *Polydora cornuta* is highly motile (Sato-Okoshi, 2000), while *P. websteri* is more sessile (Sato-Okoshi, 1999). *Polydora cornuta* is classified as a surface deposit and suspension feeder (Gaston and Nasci, 1988), and both *P. cornuta* and *P. websteri* are known to selectively sort food particles (Blake, 1996). *Polydora cornuta* has been considered a pollution indicator (Rice, 1978; Rice and Simon, 1980; Radashevsky and Hsieh, 2000) and can tolerate low dissolved oxygen levels (Rice, 1978). The *Polydora* species complex as a whole has been associated with organically enriched environments (Rice, 1978; Sato-Okoshi, 2000; Morris and Keough, 2002), and they are rapid colonizers (Rice, 1978; Rosenberg et al., 2001).

Conclusions

Most spionid taxa showed wide ranges for physical parameters; however, several taxa did show distinctive distributional patterns. *Streblospio gynobranchiata* was dominant in the tributaries. *Spio pettiboneae* and *Prionospio steenstrupi* distributions were mainly in Middle and Lower Tampa Bay. *Carazziella hobsonae*, *P. perkinsi*, *Paraprionospio pinnata*, *Streblospio gynobranchiata*, *Prionospio heterobranchia*, and *Spio pettiboneae* comprised 90% of the relative abundance for the spionids in Tampa Bay and its tributaries. Spionids seem to have some usefulness in characterizing environmental habitats, but there is a wide range of physical parameters for each species. Spionids may be useful in detecting changes in disturbance or recovery in an estuary, for example, changes in the distribution of *P. heterobranchia* and *Apoprionospio pygmaea* in Hillsborough Bay since Taylor's (1971) survey may be an indication that this region of Tampa Bay is showing signs of recovery from historical anthropogenic degradation.

Acknowledgments

Funding for this study was provided by the Tampa Bay Estuary Program (1993–1998), the Environmental Protection Commission of Hillsborough County, Hillsborough County Board of County Commissioners, and the Phosphate Severance Tax. The U.S. EPA/Gulf Breeze provided additional laboratory support for the 1993 and 1997 surveys. Tom Ash, Glenn Lockwood, Richard Boler, and Eric Lesnett assisted with field collections and instrument calibration. Sediment particle size analysis was provided by Manatee County's Environmental Management Department. Laboratory assistance was provided by a plethora of temporary employees over the years. Tom Perkins verified voucher specimens.

References

Bachelet, G., X. de Montaudouin, I. Auby, and P. J. Labourg. 2000. Seasonal changes in macrophyte and macrozoobenthos assemblage in three coastal lagoons under varying degrees of eutrophication. *Journal of Marine Science* 57:1495–1506.

Blake, J. A. 1971. Revision of the genus *Polydora* from the East Coast of North America (Polychaeta: Spionidae). *Smithsonian Contributions to Zoology* 75:1–32.

Blake, J. A. 1979. Four new species of *Carazziella* (Polychaeta: Spionidae) from North and South America, with a redescription of two previously described forms. *Proceedings of the Biological Society of Washington* 92:466–481.

Blake, J. A. 1996. Family Spionidae Grube, 1850. In *Taxonomic Atlas of the Benthic Fauna of the Santa Maria Basin and Western Santa Barbara Channel*, Vol. 6: *The Annelida* Part 3. Santa Barbara Museum of Natural History, Santa Barbara, CA, pp. 81–223.

Blake, J. A. and J. W. Evans. 1973. *Polydora* and related genera as borers in mollusk shells and other calcareous substrates (Polychaeta: Spionidae). *Veliger* 15:235–249.

Blake, J. A. and N. J. Maciolek. 1987. A redescription of *Polydora cornuta* Bosc (Polychaeta:Spionidae) and designation of a neotype. *Biological Society of Washington Bulletin* 7:11–15.

Blake, N. M. 1983. Systematics of Atlantic Spionidae (Annelida: Polychaeta) with Special Reference to Deep-Water Species. Dissertation, Boston University, Boston, 400 pp.

Boesch, D. F., R. J. Diaz, and R. W. Virnstein. 1976. Effects of Tropical Storm Agnes on soft-bottom macrobenthic communities of the James and York estuaries and the Lower Chesapeake Bay. *Chesapeake Science* 17:246–259.

Bolam, S. G. and T. F. Fernandes. 2002. The effects of macroalgal cover on the spatial distribution of macrobenthic invertebrates: the effect of macroalgal morphology. *Hydrobiologia* 475/476:437–448.

Coastal Environmental, Inc. 1994. Monitoring Program to Assess Environmental Changes in Tampa Bay, Florida. Tampa Bay National Estuary Program Tech. Rep. 02-93. Coastal Environmental, Inc., St. Petersburg.

Dauer, D. M. 1985. Functional morphology and feeding behavior of *Paraprionospio pinnata* (Polychaeta: Spionidae). *Marine Biology* 85:143–151.

Dauer, D. M. 1991. Functional morphology and feeding behavior of *Polydora commensalis* (Polychaeta: Spionidae). *Ophelia* Supplement 5:607–614.

Dauer, D. M. 1993. Biological criteria, environmental macrobenthic community structure. Reports. *Marine Pollution Bulletin* 26:249–257.

Dauer, D. M. and R. M. Ewing. 1991. Functional morphology and feeding behavior of *Malacoceros indicus* (Polychaeta: Spionidae). *Bulletin of Marine Science* 48:395–400.

Dauer, D. M., C. A. Maybury, and R. M. Ewing. 1981. Feeding behavior and general ecology of several spionid polychaetes from the Chesapeake Bay. *Journal of Experimental Marine Biology and Ecology* 54:21–38.

Day, J. H. 1967. *A Monograph on the Polychaeta of Southern Africa*. Part 2. Sedentaria. The British Museum, London, 878 pp.

Day, J. H. 1973. New Polychaeta from Beaufort, with a Key to All Species Recorded from North Carolina. NOAA Technical Report NMFs CIRC-375, 140 pp.

Delgado-Blas, V. H. 2001. Distribution especial y temporal de poliquetos (Polychaeta) benticos de la plataforma continental de Tamaulipas, Golfo de Mexico. *Revista de Biologica Tropical* 49:141–147.

Fauchald, K. 1977. *The Polychaete Worms: Definitions and Keys to the Orders, Families and Genera.* Science Series 28. Natural History Museum of Los Angeles County, Los Angeles, CA, 188 pp.

Fauchald, K. and P. A. Jumars. 1979. The diet of worms: a study of polychaete feeding guilds. *Oceanography and Marine Biology Annual Review* 17:193–284.

Flemer, D. A., B. F. Ruth, C. M. Bundrick, and G. R. Gaston. 1997. Macrobenthic community colonization and community development in dredged material disposal habitats off coastal Louisiana. *Environmental Pollution* 96:141–154.

Flint, R. W. and R. D. Kalke. 1986. Niche characterization of dominant estuarine benthic species. *Estuarine, Coastal, and Shelf Science* 22:657–674.

Folk, R. L. 1968. *Petrology of Sedimentary Rocks.* University of Texas, Austin, 170 pp.

Foster, N. M. 1971. Spionidae (Polychaeta) of the Gulf of Mexico and the Caribbean Sea. *Studies on the Fauna of Curacao and Other Caribbean Islands* 36:1–183.

Frouin, P., C. Hily, and P. Hutchings. 1998. Ecology of spionid polychaetes in the swash zone of exposed beaches in Tahiti (French Polynesia). *Comptes Rendus de l'Academie des Sciences de la Vie. Paris* 321:47–54.

Garcia-Arberas, L. and A. Rallo. 2002. The intertidal soft-bottom infaunal marcobenthos in three Basque estuaries (Gulf of Biscay): a feeding guild approach. *Hydrobiologia* 475/476:457–468.

Gaston, G. R. and J. C. Nasci. 1988. Trophic structure of macrobenthic communities in the Calcasieu Estuary, Louisiana. *Estuaries* 11:201–211.

Gaston, G. R. and J. C. Young. 1992. Effects of contaminants of marcobenthic communities in the Upper Calcasieu Estuary, Louisiana. *Bulletin of Environmental Contamination and Toxicology* 49:922–928.

Gaston, G. R., D. L. Lee, and J. C. Nasci. 1988. Estuarine marcobenthos in Calcasieu Lake, Louisiana: community and trophic structure. *Estuaries* 11:192–200.

Gray, J. S. 1981. *The Ecology of Marine Sediments.* Cambridge University Press, New York, 185 pp.

Grizzle, R. E. and C. A. Penniman. 1991. Effects of organic enrichment on estuarine macrofaunal benthos: a comparison of sediment profile imaging and traditional methods. *Marine Ecology Progress Series* 74:249–262.

Handley, S. J. 1995. Spionid polychaetes in Pacific oysters, *Crassostrea gigas* (Thunberg) from Admiralty Bay, Marlborough Sounds, New Zealand. *New Zealand Journal of Marine and Freshwater Research* 29:305–309.

Hartman, O. 1951. The littoral marine annelids of the Gulf of Mexico. *Publications of the Institute of Marine Science* 2:7–124.

Herrando-Perez, S., G. San Martin, and J. Nunez. 2001. Polychaete patterns from an oceanic island in the eastern Central Atlantic: La Gomera (Canary Archipelago). *Cahiers de Biologie Marine* 42:27–287.

Johnson, P. G. 1984. Family Spionidae. In *Taxonomic Guide to the Polychaetes of the Northern Gulf of Mexico,* Vol. 2. Vittor and Associates, Inc., Mobile, AL, pp. 6-1 to 6-69.

Lamont, P. A. and J. D. Gage. 2000. Morphological responses of marcobenthic polychaetes to low oxygen on the Oman continental slope, NW Arabian Sea. *Deep-Sea Research* Part II 47:9–24.

Lardicci, C., G. Ceccherelli, and F. Rossi. 1997. *Streblospio shrubsolii* (Polychaeta: Spionidae): temporal fluctuations in size and reproductive activity. *Cahiers de Biologie Marine* 38:207–214.

Levin, L. A. 1984. Life history and dispersal patterns in a dense infaunal polychaete assemblage: community structure and response to disturbance. *Ecology* 65:1185–1200.

Mackay, J. and G. Gibson. 1999. The influence of nurse eggs on variable larval development in *Polydora cornuta* (polychaeta: spionidae). *Invertebrate Reproduction and Development* 35:167–176.

Martin, D. 1996. A new species of *Polydora* (polychaeta, spionidae) associated with the excavating sponge *Cliona viridis* (porifera, hadromerida) in the northwestern Mediterranean Sea. *Ophelia* 45:159–174.

Mauer, D., G. Robertson, and T. Gerlinger. 1994. Community structure of soft-bottom macrobenthos of Newport Submarine Canyon, California. *Marine Ecology* 16:57–72.

McCall, P. L. 1977. Community patterns and adaptive strategies of the infaunal benthos of Long Island Sound. *Journal of Marine Research* 35:221–266.

Morris, L. and M. J. Keough. 2002. Organic pollution and its effects: a short-term transplant experiment to assess the ability of biological endpoints to detect change in a soft sediment environment. *Marine Ecology Progress Series* 225:109–121.

Myers, A. C. 1977. Sediment processing in a marine subtidal sandy bottom community: II. Biological consequences. *Journal of Marine Research* 35:633–647.

Posey, M. and T. Alphin. 2002. Resilience and stability in an offshore benthic community: responses to sediment borrow activities and hurricane disturbance. *Journal of Coastal Research* 18:685–697.

Posey, M., W. Lindberg, T. Alphin, and F. Vose. 1996. Influence of storm disturbance on an offshore benthic community. *Bulletin of Marine Science* 59:523–529.

Prezant, R. S., R. B. Toll, H. B. Rollins, and E. J. Chapman. 2002. Marine macroinvertebrates diversity of St. Catherine's Island, Georgia. *American Museum Novitate* 3367:1–33.

Rabalais, N. N., R. E. Turner, and W. J. Wiseman, Jr. 2002. Gulf of Mexico hypoxia, a.k.a. "The dead zone." *Annual Review of Ecology and Systematics* 33:235–63.

Radashevsky, V. I. 1999. Description of the proposed lectotype for *Polydora websteri* Hartman in Loosanoff & Engle, 1943 (Polychaeta: Spionidae). *Ophelia* 51:107–113.

Radashevsky, V. I. and H. Hsieh. 2000. *Polydora* (Polychaeta: Spionidae) species from Taiwan. *Zoological Studies* 39:203–217.

Rice, S. A. 1978. Intraspecific Variation in the Opportunistic Polychaete *Polydora ligni* (Spionidae). Dissertation, University of South Florida, Tampa, 203 pp.

Rice, S. A. and L. A. Levin. 1998. *Streblospio gynobranchiata*, a new spionid polychaete species (Annelida: Polychaeta) from Florida and the Gulf of Mexico with an analysis of phylogenetic relationships within the genus *Streblospio*. *Proceedings of the Biological Society of Washington* 111:694–707.

Rice, S. R. and J. L. Simon. 1980. Intraspecific variation in the pollution indicator polychaete *Polydora ligni* (Spionidae). *Ophelia* 19:79–115.

Rizzo, A. E. and A. C. Z. Amaral. 2001. Environmental variables and intertidal beach annelids of Sao Sebastiao Channel (State of Sao Paulo, Brazil). *Revista de Biologica Tropical* 49:849–857.

Rosenberg, R., H. C. Nilsson, and R. J. Diaz. 2001. Response of benthic fauna and changing sediment redox profiles over a hypoxic gradient. *Estuarine, Coastal, and Shelf Science* 53:343–350.

Santos, S. L. and J. L. Simon. 1974. Distribution and abundance of the polychaetous annelids in a South Florida estuary. *Bulletin of Marine Science* 24:669–689.

Sato-Okoshi, W. 1999. Polydorid species (Polychaeta: Spionidae) in Japan, with descriptions of morphology, ecology and burrow structure. 1. Boring species. *Journal of the Marine Biological Association of the United Kingdom* 79:832–848.

Sato-Okoshi, W. 2000. Polydorid species (Polychaeta: Spionidae) in Japan, with descriptions of morphology, ecology and burrow structure. 2. Non-boring species. *Journal of the Marine Biological Association of the United Kingdom* 80:443–456.

Sato-Okoshi, W. and K. Okoshi. 1997. Survey of the genera *Polydora*, *Boccardiella* and *Boccardia* (Polychaeta, spionidae) in Barkley Sound (Vancouver Island, Canada), with special reference to boring activity. *Bulletin of Marine Science* 60:482–493.

Schulze, S. R., S. A. Rice, J. L. Simon, and S. A. Karl. 2000. Evolution of poecilogony and the biogeography of North American populations of the polychaete *Streblospio*. *Evolution* 54:1247–1259.

Simboura, N., A. Zenetos, P. Panayotidis, and A. Makra. 1995. Changes in benthic community structure along an environmental pollution gradient. *Marine Pollution Bulletin* 30:470–474.

Steimle, F. W., Jr. and D. J. Radosh. 1979. Effects on the benthic invertebrate community. In Oxygen Depletion and Associated Benthic Mortalities in New York Bight, R. L. Swanson and Sindermann (eds.). NOAA Prof Paper 11. U.S. Department of Commerce–NOAA, Rockville, MD, pp. 281–292.

Stickney, R. R. and D. Perlmutter. 1975. Impact of intracoastal waterway maintenance dredging on a mud bottom benthos community. *Biological Conservation* 7:211–226.

Taylor, J. L. 1971. Polychaetous Annelids and Benthic Environments in Tampa Bay, Florida. Dissertation, University of Florida, Gainesville, 1332 pp.

Thistle, D. 1981. Natural physical disturbances and communities of marine soft bottoms. *Marine Ecology Progress Series* 6:223–228.

Virnstein, R. W. 1979. Predation on estuarine infauna: response patterns of component species. *Estuaries* 2:69–86.

Warner, G. F. 1997. Occurrence of epifauna on the periwinkle, *Littorina littorea* (L.), and interactions with polychaete *Polydora ciliata* (Johnston). *Hydrobiologia* 355:41–47.

Weston, D. P. 1990. Quantitative examination of macrobenthic community changes along an organic enrichment gradient. *Marine Ecology Progress Series* 61:233–244.

Zajac, R. N., R. S. Lewis, L. T. Poppe, D. C. Twichell, J. Vozarik, and M. L. DiGiacomo-Cohen. 2000. Relationships among sea-floor structure and benthic communities in Long Island Sound at regional and benthoscape scales. *Journal of Coastal Research* 16:627–640.

Zmarzly, D. L., T. D. Stebbins, D. Pasko, R. M. Duggan, and K. L. Barwick. 1994. Spatial patterns and temporal succession in soft-bottom macroinvertebrate assemblages surrounding an ocean outfall on the southern San Diego shelf: relation to anthropogenic and natural events. *Marine Biology* 118:293–307.

19

Trematode Parasites as Estuarine Indicators: Opportunities, Applications, and Comparisons with Conventional Community Approaches

Todd C. Huspeni, Ryan F. Hechinger, and Kevin D. Lafferty

CONTENTS

Introduction

Biomarkers, Bioindicators, and Management

Wetland managers face a difficult task: how best to use resources to acquire information about wetland condition necessary to make appropriate management decisions. It can be difficult to choose from the array of potential approaches, ranging from directly monitoring whole biotic communities to using proxy biomarkers and bioindicators (Adams and Ryon, 1994). Indicators employing population and community measures vary considerably in the scope of the information conveyed, as well as in the costs associated with using them. This leaves us with the two basic and pertinent questions when evaluating an indicator: (1) Does it provide information useful for management? (2) Is it cost-effective relative to other options?

In this chapter, we review and elaborate on the application potential for using larval trematode parasites in snail hosts as biodiversity indicators. As we will demonstrate, larval trematodes in host snails are potentially very sensitive, ecologically relevant, and cost-effective bioindicators that reflect free-living community abundance, diversity, and trophic links. Our goals in this chapter are (1) to review the logical framework and supporting evidence of larval trematode parasites as bioindicators; (2) to list what types of these parasites and hosts are available around the world for use as indicators; (3) to articulate how larval trematode parasites could be included in a monitoring program; and (4) to explore the costs and benefits of using these parasites compared to traditional measures.

Larval Trematodes as Bioindicators

Parasites have great potential as population and community level bioindicators, as they may be highly sensitive to many types of impacts (either directly or indirectly through effects on host populations) (Marcogliese and Cone, 1997; Overstreet, 1997; Lafferty and Holt, 2003). However, parasite responses to environmental impacts are complex and do not yield a simple set of predictions (Lafferty, 1997). Whether a particular impact will increase or decrease parasites depends on whether the impact depresses host resistance (good for parasites), depresses host populations (bad for parasites), or affects parasites more than hosts (bad for parasites) (Lafferty and Holt, 2003). Generally, parasites requiring multiple hosts to complete their life cycles are negatively influenced by disturbance and impacts that could interfere with the efficiency of transmission from one host to the next.

One of the most promising parasite indicators is larval trematodes in snail hosts. These parasitic flatworms are typically hermaphroditic and sexually reproduce in vertebrate final (or "definitive") hosts (Figure 19.1). Trematode eggs pass in the excreta of the final host, and a ciliated swimming stage called a miracidium infects a specific mollusk host, usually a snail. If the appropriate snail species is present, the miracidium penetrates the host and asexually reproduces in this molluscan first intermediate host. Each infection typically persists for the life of the snail unless the trematode is displaced by a superior competitor (Lie et al., 1965; Heyneman and Umathevy, 1968; Lim and Heyneman, 1972). Some snail species (e.g., *Cerithidea californica*) serve as first intermediate host to several different species of trematode (Martin, 1972; Yamaguti, 1975; Kuris and Lafferty, 1994). Tailed swimming stages called

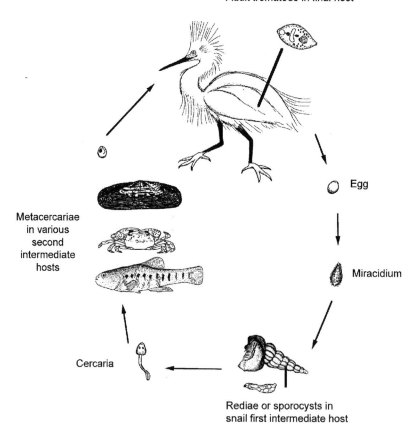

FIGURE 19.1 Generalized life cycle of trematodes using *Cerithidea californica* as a first intermediate host. (Modified from Huspeni and Lafferty, 2004). The exact type of second intermediate host used (e.g., fishes, crustaceans, or mollusks) depends on the species of trematode. *Note:* Not all reported second intermediate hosts for trematodes infecting *C. californica* are depicted in this figure.

cercariae are released from the first intermediate host mollusk and these cercariae may swim for several hours in search of an appropriate second intermediate host (e.g., fish, other mollusks, polychaetes, and crustaceans). Upon contact with an appropriate host, cercariae lose their tails and encyst in or on the second intermediate host. The type of, and specificity for, each second intermediate host depends on the trematode species. Some trematodes are highly host specific for the second intermediate host. For example, the trematode *Euhaplorchis californiensis* encysts only on the brain of the California killifish, *Fundulus parvipinnus* (Martin, 1950). Other trematode species are only specific to higher-level taxa (e.g., some can encyst in several species of fish, others in several species of crabs) and some simply encyst on hard substrates like crab exoskeletons and snail opercula. The trematode life cycle is completed when a final host preys on an infected second intermediate host. Because transmission to the vertebrate final host is by predation, we refer to this process as "trophic transmission" (Lafferty, 1999). This aspect of trematode life cycles permits inferences about which trophic links are functioning in a habitat (Marcogliese, 2001). Parasites in first intermediate host snails, therefore, are *positive* indicators of host communities and functioning trophic links. This runs counter to general impressions of parasites (i.e., that parasites indicate unhealthy conditions) and must be acknowledged and appreciated at the outset.

Hypotheses, Predictions, and Supporting Evidence

The complex life cycles of digenean trematodes (Figure 19.1) permit us to formulate host- and habitat-related hypotheses, and we discuss four of these below. While we acknowledge that in terms of scientific methodology hypotheses may only be falsified and not proved, we assert that the evidence from studies directly contradicts the alternative (i.e., falsifiable) hypotheses in each case presented below, and we therefore present the observed evidence as "support" for the hypotheses outlined below.

> **Hypothesis 1:** *The diversity and abundance of larval trematodes in mollusk first intermediate host populations directly reflects the diversity and abundance of final hosts.*

A diverse and abundant trematode community in first intermediate host snails is impossible without a diverse and abundant final host community. This is because final hosts are the sources of the trematode stages infectious to snails (Figure 19.1), and final hosts vary in what adult trematode communities they harbor. Therefore, more final hosts, and more species of final hosts, will result in more trematodes and more species of trematodes in first intermediate host snail populations.

Support for this hypothesis comes from several lines of evidence. The earliest observation of a correlation between larval trematode infections in first intermediate host snails and the presence of final hosts was made by Hoff (1941) who observed that snails infected with a trematode more frequently occurred near an aggregation of final host gulls. Subsequent to Hoff's study, many workers have noted that trematode infections in snails are higher at locations close to areas of heavier bird use (Robson and Williams, 1970; Matthews et al., 1985; Bustnes and Galaktionov, 1999). Smith (2001) was the first to explicitly relate bird density to trematode prevalence in snails. Working in Florida mangroves, she investigated the causal chain of events that link bird abundance to trematode prevalence in snails. She found significant correlations between bird abundance and number of roosting perches; between number of roosting perches and abundance of bird droppings; and between the abundance of bird droppings and prevalence of trematodes in caged sentinel snails. Finally, we have recently performed studies in a California salt marsh that directly link bird communities to larval trematode communities in snails (Hechinger and Lafferty, unpubl.), and we found positive significant associations between (1) bird abundance and trematode abundance and (2) larval trematode abundance and bird species richness.

> **Hypothesis 2:** *The diversity and abundance of larval trematode infections in first intermediate host snails indirectly reflects the diversity and abundance of prey species that may serve as second intermediate hosts.*

If final hosts disproportionately spend time at sites with abundant food (i.e., fishes and invertebrates), final hosts should transmit more trematodes to snails at sites with more abundant prey. Also, since different final host species prey on different food types, we expect that more final host species will use sites that have greater species richness of fishes and invertebrates. Additionally, local completion of trematode life cycles requires the presence of second intermediate hosts because the trematodes need those hosts to infect final hosts.

Support for the second hypothesis has only recently become available. Huspeni et al. (unpubl.) sampled larval trematodes, fishes, and benthic infauna at 27 sites in three California estuaries (Morro Bay, Carpinteria, and Pt. Mugu). They observed significant positive correlations between larval trematodes infecting snails and the benthic and fish communities at a site. Specifically, larval trematode prevalence in *C. californica* and the density of benthic infauna were positively associated, and trematode species richness was strongly correlated with infauna and fish species richness at a site.

Hypothesis 3: *The abundance of trematodes in first intermediate hosts reflects water quality.*

Miracidia and cercariae, as free-swimming trematode stages (see Figure 19.1), have direct contact with ambient environmental conditions. They are also sensitive to heavy metals and other toxins. Poor water quality, therefore, should impede trematode transmission. Evidence for this interpretation is provided by numerous studies showing the negative effects of toxics on both the molluscan hosts and the free-living stages of trematodes (Pietrock and Marcogliese, 2003). Free-living stages of trematodes are susceptible to environmental stressors such as increased temperature, extreme pH, and salinity shifts, as well as pollutants, particularly organics and heavy metals (Pietrock and Marcogliese, 2003). For example, trace metals kill free-living trematode cercariae and miracidia (Siddall and des Clers, 1994). Additionally, infected snails may be more susceptible to pollution than uninfected snails, and this can reduce the prevalence of infection in the snail population (Guth et al., 1977; Stadnichenko et al., 1995). Lefcort et al. (2002) also reported lower larval trematode diversity in first intermediate host snails at sites contaminated with heavy metals relative to reference sites.

Hypothesis 4: *Larval trematodes in first intermediate hosts are negatively influenced by environmental impacts and are less abundant and less diverse in impacted habitats.*

Because they have complex life cycles and require multiple hosts, perturbations affecting hosts or transmission at any point of the life cycle will lead to a reduction of trematodes measured in first intermediate host populations. Many workers have noted the logic and anecdotal evidence supporting this prediction (Kuris and Lafferty, 1994; MacKenzie et al., 1995; Lafferty, 1997). Cort et al. (1960) were the first to suspect that environmental impacts could result in drops of larval trematode prevalence in first intermediate host snails. They sampled trematode infections in a first intermediate host snails at several sites in a Michigan lake (Cort et al., 1937). More than 20 years later, Cort et al. (1960) resampled one of the sites and found a precipitous drop in trematode prevalence and suggested the cause was habitat loss and degradation of remaining habitat for final host bird populations. More recently, Keas and Blankespoor (1997) resampled three of Cort et al.'s 1937 sites and found that they had also dropped in trematode prevalence in snails. Huspeni and Lafferty (2004) used a salt marsh restoration project as an opportunity to explicitly test this hypothesis by performing a before-after-control-impact study on the larval trematodes in the California horn snail. Prior to restoration, impacted sites had significantly fewer larval trematodes (lower prevalence and species richness) in snails than control sites located in intact salt marsh (Figure 19.2A and B). After being restored, the impacted sites experienced a significant increase in trematode prevalence and species richness (Huspeni and Lafferty, 2004).

Methods: Employing Trematodes in Assessments: Worldwide Opportunities, Application Methodology, and Comparisons with Other Techniques

Opportunities exist for using larval trematodes in different habitats and regions of the world. We identify promising snails from a variety of aquatic habitats around the globe. To accomplish this, the literature was examined for snail first intermediate hosts reported to be parasitized by rich communities of larval trematodes. Because our principal focus for this chapter is estuarine habitats, we largely restricted our evaluation to estuarine snails. However, because trematodes are ubiquitous components of most aquatic and wetland

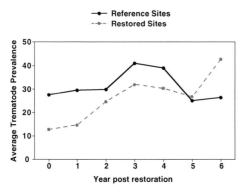 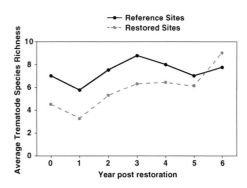

FIGURE 19.2 (A) Average trematode prevalence at control and restored sites in Carpinteria salt marsh. Time 0 represents samples collected before the restoration. Times 1 to 6 represent number of years since the restoration was completed. Year 0 prevalence at control sites was significantly greater than at sites to be restored *(p < 0.05)*. Restored sites were significantly higher in trematode prevalence in years 2, 4, 5, and 6 relative to prevalence at those sites in year 0. (B) Average trematode species richness at control and restored sites in Carpinteria salt marsh. Time 0 represents samples collected before the restoration. Times 1 to 6 represent number of years since the restoration was completed. Year 0 species richness at control sites was significantly greater than at sites to be restored *(p < 0.05)*. Restored sites were significantly higher in trematode species richness in year 6 relative to species richness at those sites in year 0. (Data reproduced from Huspeni and Lafferty, 2004.)

habitats (Dawes, 1946; Yamaguti, 1971; Yamaguti, 1975), some candidate snail hosts from other aquatic systems are also reported.

To date, the specific steps required to include larval trematodes in a monitoring or management program have not been fully articulated. This chapter outlines steps for using larval trematodes as bioindicators of the abundance, richness, and trophic links operating in surrounding communities. We describe how to sample and identify larval trematodes, and construct a methodology to analyze these data to infer information about the abundance, richness, and trophic interactions of surrounding benthic, fish, mammal, and bird communities.

Finally, the costs (i.e., effort) and benefits (i.e., informational content) of various traditional community assessment techniques are compared to using larval trematodes to evaluate ecosystem function. The extent to which the various techniques provide information about spatial and temporal variation in community organization is especially examined.

Results

Larval Trematodes: Worldwide Opportunities for Application

Although trematodes are ubiquitous components of nearly all aquatic and marine habitats, those most tractable for estuarine assessments infect snails from intertidal or shallow subtidal zones. Our review of the literature found several estuarine snails and families that might be suitable for use in monitoring programs using larval trematodes (Table 19.1). Species in the potamidid genus *Cerithidea* have a worldwide tropical and subtropical distribution in estuaries, and are frequently abundant and long-lived (Houbrick, 1984). *Cerithidea* species also host rich larval trematode communities comprising either the same trematode species, or trematode species closely related to those in *C. californica* (Huspeni, 2000). For example, in North America, *C. californica,* which serves as first intermediate host to 18 trematode species, has already been examined, and occurs in many critical habitats in southern California and northern Baja California (Martin, 1972). *Cerithidea mazatlanica* occurs from southern Baja California through the Gulf of California south to Ecuador (Keen, 1971), and shares *C. californica's* 18 trematode species (Huspeni, 2000). *Cerithidea pliculosa* hosts at least 12 trematode species (Wardle, 1974) and is found in estuaries in the Gulf of Mexico, while *C. costata* hosts a similar 12 species and is found in the Gulf of Mexico and throughout the Caribbean (Cable, 1956). Outside North America, few other *Cerithidea* species have been examined for trematodes, but *C. cingulata* in the Persian Gulf, India, and Japan hosts at least 15 trematode species (Mani and Rao, 1993; Abdul-Salam and Sreelatha, 1998) (Table 19.1).

TABLE 19.1

Estuarine Snail Species Reported as Infected with Larval Trematodes[a]

Snail Family and Species	Region	No. of Reported Trematode Species	Second Intermediate Host Types	Final Host Types	Ref.
Potamididae					
Cerithidea californica	W. North America	18	Crustaceans, fishes, and mollusks	Birds and mammals	25, 33
Cerithidea cingulata	S. Asia, Indopacific	15	Crustaceans, fishes, mollusks, and vegetation	Birds and fishes	1, 32
Cerithidea costata	S.W. North America, Caribbean, E. Central America	12	Crustaceans, fishes, and mollusks	Birds and mammals	7
Cerithidea mazatlanica	S.W. North America, N.W. South America	18	Crustaceans, fishes, and mollusks	Birds and mammals	23, 33[b]
Cerithidea pliculosa	Gulf of Mexico to Central America	12	Crustaceans, fishes, and mollusks	Birds and mammals	34, 49
Cerithidea rhizophorarum	Japan	3	Crustaceans and mollusks	Birds	18, 19
Cerithidea scalariformis	S.E. North America	12	Crustaceans, fishes, and mollusks	Birds and mammals	21, 44
Tympanotonus microptera	Japan	3	Hard substrates	Birds	26
Velacumantus australis	Australia	8	Crustaceans, fishes, and mollusks	Birds	3–5, 16, 47, 48
Cerithiidae					
Cerithium mediterraneum	Mediterranean	3	Fishes	Birds	40–42
Cerithium litteratum	Caribbean and Venezuela	3	Fishes	Birds and fishes	8, 35
Cerithium moniliferum	Australia	11	Crustaceans, fishes, and mollusks	Birds and fishes	9, 10
Cerithium muscarum	Gulf of Mexico	3	Fishes and mollusks	Birds	14, 37, 46
Cerithium rupestre	Mediterranean	3	Crustaceans and fishes	Birds	6, 39, 41

Cerithium scabridum	S.W. Asia, Suez Canal	12	Crustaceans, fishes, and mollusks	Birds and fishes	2
Cerithium stercusmuscarum	W. North America	5	Crustaceans, fishes, and mollusks	Birds, mammals, and fishes	15, 24
Cerithium vulgatum	Mediterranean and Black Sea	3	Fishes	Birds	36, 38, 50
Nassariidae					
Ilyanassa obsoleta	E. North America	9	Fishes, crustaceans, cnidarians, polychaetes, turbellarians, and mollusks	Birds and fishes	11, 12, 17, 45
Nassarius orissaensis	South Asia	3	Mollusks	Fishes	30, 31
Batillariidae					
Batillaria cumingi	East Asia	8	Crustaceans, fishes, and mollusks	Birds	19, 20, 43
Batillaria minima	Gulf of Mexico	4	Crustaceans and mollusks	Birds	7, 46
Hydrobiidae					
Hydrobia acuta	Mediterranean	15	Crustaceans, fishes, and mollusks	Birds and fishes	13
Hydrobia ulvae	N. and W. Europe	32	Crustaceans, fishes, and mollusks	Birds and fishes	13, 22
Hydrobia ventrosa	N. and W. Europe, Mediterranean, Black Sea	19	Crustaceans, fishes, and mollusks	Birds and fishes	13, 27–29

[a] Hosts with fewer than three reported trematode species are not listed.

[b] Huspeni (2000) demonstrated that the trematode species present in *Cerithidea californica* are also present in *C. mazatlanica*. Consequently, Martin's 1972 key to the larval trematodes infecting *Cerithidea californica* can be used for trematodes infecting *C. mazatlanica*.

References for Table 19.1

1. Abdul-Salam, J. and B. S. Sreelatha. 1998. A list of larval digenetic trematodes parasitizing some marine invertebrates in Kuwait Bay. *Kuwait Journal of Science and Engineering* 25:409–434.

2. Abdul-Salam, J. and B. S. Sreelatha. 1999. Component community structure of larval trematodes in the snail *Cerithium scabridum* from southern Kuwait Bay. *Current Science* 77:1416–1417.

3. Appleton, C. C. 1983. *Austrobilharzia terrigalensis* (Trematoda: Schistosomatidae) in the Swan Estuary, western Australia: frequency of infection in the intermediate host population. *International Journal of Parasitology* 13:51–60.

4. Appleton, C. C. 1983. Studies on *Austrobilharzia terrigalensis* (Trematoda: Schistosomatidae) in the Swan Estuary, western Australia: observations on the biology of the cercaria. *International Journal of Parasitology* 13:239–248.

5. Appleton, C. C. 1989. Translocation of an estuarine whelk and its trematode parasites in Australia. *Environmental Conservation* 16:172–182.

6. Bartoli, P. and G. Prevot. 1978. Ecological research work on a trematode life cycle in a Provençal lagoon, France. Part 2. The life cycle of *Maritrema misenensis* (Microphallidae). *Annales de Parasitologie Humaine et Comparee* 53:181–194.

7. Cable, R. M. 1956. Marine cercariae of Puerto Rico. In *Scientific Survey of Porto Rico and the Virgin Islands,* Vol. 16, Part 4, R. W. Miner (ed.). New York Academy of Sciences, New York, pp. 491–577.

8. Cable, R. M. and F. M. Nahhas. 1962. *Bivesicula caribbensis* sp. n. (Trematoda: Digenea) and its life history. *Journal of Parasitology* 48:536–538.

9. Cannon, L. R. G. 1978. Marine cercariae from the gastropod *Cerithium moniliferum* at Heron Island, Great Barrier Reef. *Proceedings of the Royal Society of Queensland* 89:45–58.

10. Cannon, L. R. G. 1979. Ecological observations on *Cerithium moniliferum* (Gastropoda: Cerithiidae) and its trematode parasites at Heron Island, Great Barrier Reef, Australia. *Australian Journal of Marine and Freshwater Research* 30:365–374.

11. Ching, H. L. 1991. Lists of larval worms from marine invertebrates of the Pacific Coast of North America. *Journal of the Helminthological Society of Washington* 58:57–68.

12. Curtis, L. A. 1997. *Ilyanassa obsoleta* (Gastropoda) as a host for trematodes in Delaware estuaries. *Journal of Parisitology* 83:793–803.

13. Deblock, S. 1980. Inventaire des trematodes larvaires parasites des mollusques Hydrobia (Prosobranches) des cotes de France. *Parassitologia* 22:1–105.

14. Dennis, E. A. and L. R. Penner. 1971. *Mesostephanus yedeae,* new species (Trematoda: Cyathocotylidae): its life history and descriptions of the developmental stages. *University of Connecticut Occasional Papers Biological Science Series* 2:5–15.

15. Dronen, N. O. and L. R. Penner. 1975. Concerning *Philophthalmus andersoni,* new species (Trematoda: Philophthalmidae): another ocular helminth from birds which develops in a marine gastropod. *University of Connecticut Occasional Papers Biological Science Series* 2:217–223.

16. Ewers, W. H. 1965. The incidence of larval trematodes in adults of *Velacumantus australis* (Quoy and Gaimard) (Gastropoda: Potamididae). *Journal of Helminthology* 39:1–10.

17. Gambino, J. J. 1959. The seasonal incidence of infection of the snail *Nassarius obsoletus* (Say) with larval trematodes. *Journal of Parasitology* 45:440, 456.

18. Harada, M. 1989. *Cercaria shikokuensis,* new species (Trematoda) from littoral gastropods in Kagawa Prefecture, Shikoku, Japan. *Japanese Journal of Parasitology* 38:135–138.

19. Harada, M. and S. Suguri. 1989. Surveys on cercariae in brackish water snails in Kagawa Prefecture, Shikoku, Japan. *Japanese Journal of Parasitology* 38:388–391.

20. Hechinger, R. F. unpublished manuscript.

21. Holliman, R. B. 1961. Larval trematodes from the Appalachee Bay area Florida, with a checklist of the known marine cercariae arranged in a key to their superfamilies. *Tulane Studies in Zoology* 9:2–74.

22. Honer, M. R. 1961. Some observations of the ecology of *Hydrobia stagnorum* (Gmelin) and *H. ulvae* (Pennant), and the relationship ecology–parasitofauna. *Basteria* 25:7–29.

23. Huspeni, T. C. 2000. A Molecular Genetic Analysis of Host Specificity, Continental Geography, and Recruitment Dynamics of a Larval Trematode in a Salt Marsh Snail. Ph.D. dissertation, University of California, Santa Barbara.

24. Huspeni, T. C. Unpublished data.

25. Huspeni, T. C. and K. D. Lafferty. 2004. Using larval trematodes that parasitize snails to evaluate a salt-marsh restoration project. *Ecological Applications* 14:795–804.

26. Ito, J. 1957. Studies on the brackish water cercariae in Japan. III. Three new echinostome cercariae in Tokyo Bay, with a list of Japanese echinostome cercariae. *Japanese Journal of Medical Science and Biology* 10:439–453.

27. Kube, J., S. Kube, and V. Dierschke. 2002. Spatial and temporal variations in the trematode component community of the mudsnail *Hydrobia ventrosa* in relation to the occurrence of waterfowl as definitive hosts. *Journal of Parasitology* 88:1075–1086.

28. Kube, S., J. Kube, and A. Bick. 2002. Component community of larval trematodes in the mudsnail *Hydrobia ventrosa*: temporal variations in prevalence in relation to host life history. *Journal of Parasitology* 88:730–737.

29. Lauckner, G. 1986. Analysis of parasite-host systems in the western Baltic Sea. *Ophelia* 129–138.

30. Madhavi, R. and U. Shameem. 1991. *Cercaria chilkaensis*, new species II. A new zoogonid cercaria from the snail *Nassarius orissaensis* from Chilka Lake, India. *Journal of the Helminthological Society of Washington* 58:31–34.

31. Madhavi, R. and U. Shameem. 1993. Cercariae and metacercariae of *Stephanostomum cloacum* (Trematoda: Acanthocolpidae). *International Journal of Parasitology* 23:341–347.

32. Mani, G. G. and K. H. Rao. 1993. Studies on Indian marine cercariae: two new echinostome cercariae. *Journal of the Helminthological Society of Washington* 60:250–255.

33. Martin, W. E. 1972. An annotated key to the cercariae that develop in the snail *Cerithidea californica*. *Bulletin of the Southern California Academy of Sciences* 71:39–43.

34. McNeff, L. L. 1978. Marine Cercariae from *Cerithidea pliculosa* Menke from Dauphin Island, Alabama; Life Cycles of Heterophyid and Opisthorchiid Digenea from *Cerithidea* Swainson from the Eastern Gulf of Mexico. Ph.D. dissertation, University of Alabama, Tuscaloosa.

35. Nasir, P. 1983. Marine larval trematodes. VI. *Cercaria criollisimae*: *Cercaria criollisima* XI. n.sp., a vivax larva from Venezuela. *Rivista di Parassitologia* 44:71–76.

36. Pearson, J. C. and G. Prevot. 1985. A revision of the subfamily Haplorchinae (Trematoda: Heterophyidae) Iii. genera *Cercarioides* and *Condylocotyla*, new genus. *Systematic Parisitology* 7:169–198.

37. Penner, L. R. and J. J. Trimble. 1970. *Philophthalmus larsoni*, new species, an ocular trematode from birds. *University of Connecticut Occasional Papers Biological Science* 1:265–273.

38. Prevot, G. 1967. Studies on the prosobranch cercariae in the Marseilles region, *Cercaria mirabilicaudata*, new species, (Trematoda: Digenea) Opisthorchioidea of *Cerithium vulgatum*. *Bulletin de la Societe Zoologique de France* 92:515–522.

39. Prevot, G. 1972. Contribution to the study of Microphallidae Travassos 1920: trematoda life cycle of *Megalophallus carcini*, parasite of the herring gull, *Larus argentatus* Michaellis. *Bulletin de la Societe Zoologique de France* 97:157–163.

40. Prevot, G. 1973. Life cycle of *Galactosomum timondavidi* (Trematoda: Heterophyidae): parasite of the herring gull, *Larus argentatus*. *Annales de Parasitologie Humaine et Comparee* 48:457–467.

41. Prevot, G. and P. Bartoli. 1978. Life cycle of *Renicola lari* (Trematoda: Renicolidae). *Annales de Parasitologie Humaine et Comparee* 53:561–576.

42. Prevot, G., P. Bartoli, and S. Deblock. 1976. The life cycle of *Maritrema misenensis*, new combination, (Trematoda: Microphallidae) from French Mediterranean coast. *Annales de Parasitologie Humaine et Comparee* 51:433–446.

43. Shimura, S. and J. Ito. 1980. Two new species of marine cercariae from the Japanese intertidal gastropod, *Batillaria cumingi* (Crosse). *Japanese Journal of Parasitology* 29:369–375.

44. Smith, N. F. 2001. Spatial heterogeneity in recruitment of larval trematodes to snail intermediate hosts. *Oecologia* 127:115–122.

45. Stunkard, H. W. 1983. The marine cercariae of the Woods Hole, Massachusetts region of the USA: a review and a revision. *Biological Bulletin* 164:143–162.

46. Trimble, J. J. and L. R. Penner. 1971. A comparison of the larval stages of *Philophthalmus hegeneri* and *Philophthalmus larsoni* (Trematoda: Philophthalmidae). *Zoologischer Anzeiger* 186:373–379.

47. Walker, J. 1976. Aspects of the host parasite relationship between *Velacumantus australis* and its trematode parasites. *Malacological Review* 9:138.

48. Walker, J. C. 1979. *Austrobilharzia terrigalensis* a schistosome dominant in interspecific interactions in the molluscan host. *International Journal of Parasitology* 9:137–140.

49. Wardle, W. J. 1974. A Survey of the Occurrence, Distribution and Incidence of Infection of Helminth Parasites of Marine and Estuarine Mollusks from Galveston, Texas. Ph.D. dissertation, Texas A&M University, College Station.

50. Zdun, V. I. and S. M. Ignatiev. 1980. Black Sea mollusk *Cerithium vulgatum* (Gastropoda: Cerithiidae): a new intermediate host of trematodes. *Parazitologiya* 14:345–348.

Other estuarine snails reported to host larval trematodes are also listed in Table 19.1. Snail species in the family Nassariidae and in the genera *Cerithium and Batillaria* that have been examined for larval trematodes typically host rich larval trematode faunas. Like *Cerithidea* species, *Cerithium, Batillaria,* and nassariid species are frequently abundant where they occur, but relatively few of the species in these taxa have been thoroughly examined for larval trematode parasites. *Ilyanassa obsoleta* on the East Coast (and introduced to the West Coast) of the United States is ideally suited for trematode assessments and, more broadly, other nassariid species are likely to host rich trematode faunas.

Table 19.1 is not intended to be an exhaustive list of hosts that should have rich communities of larval trematodes. The hosts listed are necessarily limited by both our scope of review (e.g., many snail species that had fewer than three species of larval trematodes reported were excluded), as well as basic scientific exploration of hosts for larval trematodes. There is strong consistency between snail species in the same family regarding whether or not they host rich trematode communities (Ewers, 1964). Snail species are ecologically and taxonomically related to those reported (i.e., intertidal and shallow subtidal species in the same genus or family) will also likely host rich larval trematode communities.

While the principal focus in this chapter is to produce information on promising snails for potential use in larval trematode assessments in estuaries, a more general listing of potential snail hosts in other aquatic habitats is also provided. Specifically, we examined the literature for reported diverse larval trematode communities in snail hosts from shallow freshwater habitats (lakes, ponds, and streams) and intertidal and shallow subtidal marine habitats (rocky and sandy beach). In freshwater habitats, lymnaeids (e.g., Rees, 1932; Anteson, 1970; Brown et al., 1988), physids (e.g., Brown et al., 1988; Snyder and Esch, 1993; Sapp and Esch, 1994), planorbids (e.g., Goater et al., 1989; Fernandez and Esch, 1991a,b), and hydrobiids (e.g., Winterbourn, 1974; Krist et al., 2000) are frequently parasitized by diverse guilds of larval trematodes. The hydrobiids are a very large group of snails that are found in freshwater to marine habitats, and many hydrobiid species are undescribed. In temperate freshwater habitats, many snails often live only 1 year, limiting inferences to a very recent timescale (see below). Muricids (e.g., Ching, 1991) and littorinids (e.g., Matthews et al., 1985) are good candidates in rocky intertidal habitats, while olivellids and buccinids are likely candidates in sandy beach habitats.

Steps Required for Conducting a Larval Trematode Assessment

1. Choose the Snail First Intermediate Host

Trematode communities will vary greatly among snail species and, therefore, comparisons among sites should hold the host snail species constant. Because trematodes are ubiquitous, many snail hosts are available. If present in a habitat, we believe the snail hosts and families listed in Table 19.1 and above are likely to be infected with diverse trematode guilds. Other considerations are important as well, and to be useful as a bioindicator of the diversity and abundance of final hosts and the broader free-living community, a snail host should be abundant and at densities that permit sampling for parasites (e.g., averaging at least one snail per square meter).

2. Control for Habitat

Within an estuary, larval trematode infections in snails may vary between habitat types (e.g., channels vs. mudflats). For example, significant differences were found in prevalence and community composition of larval trematode infections in *Cerithidea californica* across channels, vegetated marsh, mudflat, and pans at Estero de Punta Banda, Baja California (Huspeni et al., unpubl.). To control for this potential source of variation, we recommend limiting comparisons to specific habitat types (e.g., channels or mudflats), or sampling habitat types equally across sites.

3. Control for Snail Age

Each snail is analogous to a datalogger that records an infection event over time. Therefore, trematode prevalence and diversity in samples of first intermediate host snails will increase with the amount of time during which snails have been alive and exposed to infection. Because older snails are likely to be larger, it is important to control for snail size when comparing among samples. If snails vary significantly in size, an initial collection of snails across all sizes present should first be attempted. Focus should be given to larger size classes of snails because, in most systems, prevalence and species richness of larval trematodes increases with snail size (i.e., age) (Kuris, 1990; Sousa, 1993; Kuris and Lafferty, 1994; Lafferty et al., 1994). That is, older snails are more likely to be infected than younger snails. The magnitude of this effect can be demonstrated by plotting prevalence (i.e., percent snails infected) as a function of size in a system under consideration. If a positive relationship exists, it will be necessary to control for size in analyses, and this can be accomplished by restricting subsequent collections to a narrow size class.

Three factors should be considered when choosing the size class for sampling. First, sampling smaller-sized classes will emphasize more recent conditions, while sampling large-sized classes will provide a longer temporal integration. Variation in exposure time with snail size could permit assessment of changes in estuarine community abundance, richness, and trophic functioning over time by comparing changes in the trematode community across snail size classes. This has not been attempted to our knowledge, but could allow the identification of temporal changes in the absence of baseline data. Second, the selected size class should be abundant across sites to ensure a sufficient sample size. Third, the size class should have, on average, an intermediate prevalence of infection in order to increase statistical power when making comparisons (i.e., it is difficult to compare samples for differences when almost none or nearly all the snails are infected). For example, in *C. californica,* the largest adults measure 30 to 35 mm in size. We frequently use 20- to 25-mm snails in our comparisons, as this size class is common and has intermediate prevalences.

4. Choose the Sampling Effort and Scales

Snails move very little, so samples at a particular site will reflect conditions on the order of tens of meters. This makes it possible to assess fine-scale heterogeneity in habitat quality within an estuary. It is important to control for sampling effort and a constant number of snails should be assessed from each site. In our studies, 100 snails has proved to be a good minimum sample size to assess trematode prevalence and species richness at a site because this sample size is sufficient to produce a small variance when estimating prevalence of trematodes in a population of hosts.

Repeated sampling may provide information on year-to-year variation. Because some estuarine fauna have strong seasonal variation in abundance and distribution, it is important to control for sample season when comparing samples. Some snail hosts are only seasonally abundant and, therefore, need to be sampled at particular times of year. Assessments of trematode communities over time or that compare sites within or between wetlands require appropriate levels of replication for statistical tests to be valid (generally five or more sites per wetland). Tests for significant differences in prevalence can be done with simple comparisons of confidence intervals (calculated for proportions) for sites or sites pooled across habitats. Other robust comparisons for prevalence and species richness may be made using resampling statistical procedures (Lafferty et al., 1994; Edgington, 1995). Other significance tests are possible for comparing sites with standard community measures and are outlined in Krebs (1999).

5. Assess Snails for Larval Trematode Infections

Prior to dissecting, snails should be rinsed to remove mud and organic matter, and the length (or width, if appropriate) of each snail should be recorded. Begin a dissection by placing a snail in a shallow container and gently cracking its shell. A hammer or vise is useful for large snails, and pressing down with the bottom of a glass vial is sufficient for smaller ones. Pieces of shell can be removed with fine forceps and the internal organs exposed. Fortunately for the dissector, larval trematode stages typically make up from 30 to 50% of the tissue weight of infected snails (Kuris and Lafferty, 1994). Examine the gonad and digestive gland carefully, as these are common sites of trematode infection. The kidney, heart, pericardial region, and mantle regions are also possible sites of larval trematode infection. In some cases, multiple species may be observed infecting a single snail, particularly when overall percent of infected snails is high. In these cases, it is important to identify and record each species.

6. Identification of Trematode Species

Most basic parasitology texts (e.g., Roberts and Janovy, 2000) provide descriptions of the general types of larval stages of trematodes. Such texts also provide a good overall introduction to digenean trematodes. Schell's classic *How to Know the Trematodes* (1970) is also a great resource for descriptions of the intramolluscan stages. Using this or other keys, it is straightforward to identify the cercaria stage to a digenean family. A determination of trematode infection to this level in the vast majority of cases is sufficient to identify the typical second intermediate host (e.g., mollusk, copepod, fish, etc.) and final host (e.g., fish, amphibian, reptile, bird, or mammal). In the cases of well-studied snail first intermediate hosts (Table 19.1), many trematode life cycles have been completely described, with second intermediate and final hosts identified. In other cases, a worker may simply work with operational taxonomic units (e.g., Bucephalid 1, Heterophyid 1, etc.). At initial screening stages, the focus should be on carefully identifying infections to the lowest operational taxonomic unit.

7. Data Analysis

Several summary measures are useful in comparing larval trematode infections in snails across sites.

a. Prevalence and Abundance

Prevalence is a measure of the percent of hosts with a larval trematode infection. For example, if 100 snails were assessed and 25 trematode infections were observed, then the trematode prevalence at this site would be 0.25 or 25%. At sites with a significant frequency of multiply infected snails (i.e., double or triple infections), a modified prevalence (abundance) may be calculated by dividing the total number of infections observed (e.g., with double infections in a single snail counting as two) and dividing this number by the total number of snails examined. At high prevalence sites, abundance may exceed 1.0 or 100%. It is worth remembering that asexual reproduction of trematode stages takes place within infected snails. Therefore, the appropriate unit of measure is not the number of individual clonal stages in a snail, but the number of independent infections that have occurred. This is estimated by the number of different species infecting an individual snail.

If two trematode species infect the same snail, one often predictably outcompetes the other (Kuris and Lafferty, 1994). Because of this, in cases where infection rates are high, the infection rate of subordinate species may be underestimated. In other words, using the "snail as a datalogger" analogy, infections with dominant species can "overwrite" records of previous subordinate infections. These overwritten data are meaningful because they correspond to the presence of a final host that carried a subordinate trematode species. Fortunately, it is possible to estimate the frequency at which each subordinate species has been replaced by a dominant because analytical techniques have been developed to estimate the pre-interactive prevalence of subordinate species (Lafferty et al., 1994). Doing so requires knowledge of the trematode dominance hierarchy, which can be postulated for most trematode communities using simple rules (Kuris, 1990; Kuris and Lafferty, 1994). Pre-interactive prevalence is the preferred indicator of transmission from final hosts to snails and should be calculated where possible.

b. Species Richness and Other Community Measures

Species richness is simply the number of trematode species (or operational taxonomic units) observed at a site. It is extremely important to control for sampling effort when comparing species richness (i.e., examining an equal number of hosts per site, or using analytical resampling approaches), because more trematode species are likely to be found as more hosts are examined. Other diversity measures (e.g., Simpson's index or the Shannon–Wiener index) can be calculated as well, and instructions and recommendations for these are given in Krebs (1999). Community similarity indices may also be useful for evaluations that compare impacted or restored sites with reference sites.

c. Second Intermediate Host Use

To reveal specific trophic links operating in a habitat, larval trematode communities can be compared on the basis of second intermediate hosts required. To accomplish this, it is necessary to partition the community of larval trematodes according to second intermediate host use. For example, 100 snails assessed at a site may yield 40 infections (e.g., prevalence = 40%), of which 30 are heterophyids (using fishes), five are microphallids (using crustaceans), and five are echinostomes (using mollusks). The proportional breakdown of the larval trematode community permits inferences about the diversity of final hosts and their likely prey choice at a site. Sites and habitats can then be compared on the basis of second intermediate host use (see Huspeni and Lafferty, 2004).

d. Final Host Use

In some cases, infections in first intermediate host snails may be partitioned into groups based on final host use. For example, *I. obsoleta* hosts nine different larval trematode species: four that use birds as final hosts, four that use fishes as final hosts, and one that uses turtles as final hosts (Table 19.1). Partitioning the larval trematode community in snails on the basis of final hosts used permits inferences to be made regarding final host communities present at each site.

Comparison of Larval Trematodes with Other Community Assessment Approaches

Table 19.2 shows a comparison of commonly employed estuarine sampling techniques for community assessments. Estuarine communities frequently assessed by managers include fishes, large and small benthic invertebrates, birds, and plants (Table 19.2). These communities differ markedly in the way they may vary temporally at a site. For example, because of their vagility, fish and bird communities at a site often vary significantly over short time intervals. Large benthic infauna (e.g., those captured on a 3-mm mesh) are less temporally variable; but small benthic infauna (e.g., those captured on a 0.5-mm mesh) often vary seasonally. Plant communities are often the least temporally variable in estuarine systems.

Not surprisingly, conventional sampling techniques vary in their ability to capture temporal variation in the target community. They also differ in the environmental damage that results from sampling. For example, a single seining effort with blocking nets provides a very limited temporal snapshot of the fish community, requires at least several person-hours to accomplish, and is destructive to muddy habitats. Fish traps can be deployed *in situ* for several days and are, therefore, more temporally integrative than seining. However, many fish species do not enter traps, traps vary in efficiency for the fish species they do capture, and traps cannot provide an estimate of absolute density of fishes at a site. While not environmentally damaging, single bird surveys are also limited in their ability to capture the temporal variation in bird communities at a site. More temporally integrated views of the bird community can be achieved through repeated surveys or videotaped observations, but each of these requires substantial effort. Benthic coring and processing for large infauna is laborious in the field and destructive to the habitat. Sampling for the smaller benthic community requires less field labor, but can be incredibly laborious in the laboratory, depending on the taxonomic level of identification desired.

By comparison, assessments of larval trematodes in first intermediate host snails provide significant information about local communities with relatively little field laboratory effort, and minimal environmental disturbance, and provide temporally integrated views of bird, fish, and benthic communities. Furthermore, trematode surveys of snail hosts can provide a temporally integrated view of the final host use of the habitat over the life of the snail host. With the exception of larval trematode surveys, all other methods provide information largely limited to the community of focus. For example, bird surveys, either using standard bird count surveys or videography, provide information largely limited to the bird community (although observations of foraging birds will indirectly indicate benthic invertebrate or fish communities). The sampling techniques that have a high information yield typically require significant field and/or laboratory effort, are usually destructive to the habitat, and sample a narrow range of species.

Discussion

Assessing larval trematode infections in snail hosts can provide a quantitative, comparative, comprehensive, temporally integrative, environmentally safe, and cost-effective approach for inferring community structure and trophic linkages in an estuary. Larval trematodes in first intermediate host snails provide indirect information about vertebrate and invertebrate communities as well as trophic links between second intermediate and final hosts.

It is important to reemphasize here that unlike many parasites (e.g., ciliates on the gills of a fish), larval trematodes in first intermediate host snails should be viewed as positive indicators of broader host communities. If this predicted connection seems tenuous, we submit the following. Several ecological indicators have

TABLE 19.2

Comparison of Estuarine Community Sampling Methodologies

Sampling Method	Community of Interest	Temporal Variation in Community	Ability of Method to Capture Temporal Variation	Primary Gear Required to Sample a Site (field and lab)	Degree of Habitat Destruction Caused by Assessment Method	No. of Visits Required to Sample a Site	Estimated per Site Person-Hours[a]		
							Field	Lab	Total
Seining	Fishes	High	Low	Seine, blocking nets	High	1	4	0	4
Trapping	Fishes	High	Moderate	Fish traps	Low	2	1	0	1
Large cores	Benthic infauna (large)	Low	High	Corers, wide mesh sieves	High	1	6–8	0	6–8
Small cores	Benthic infauna (small)	Moderate	Moderate	Corers, small mesh sieves, microscopes	Low	1	0.5	10–30	11–31
Single bird survey	Birds	High	Low	Binoculars, spotting scope	Low	1	0.5	0	0.5
Monthly bird survey	Birds	High	Moderate		Low	12	0.5	0	0.5
Bird video	Birds	High	Moderate	Video equipment	Low	30[b]	0.5[c]	2	2.5
Plant survey	Plants	Low	High	Transect tapes	Low	1	1	0	1
Larval trematode survey	Birds, benthos, fishes	High	High	Calipers, mesh bags, hammer, microscope	Low	1	0.5	3	3.5

[a] Person-hours equal the number of workers multiplied by number of hours required to sample a site from our protocols for estuary assessment; estimated efforts for each method do not include travel times to and from sites or the effort involved in transporting required equipment to each site.

[b] This is for sampling two seasons, with 15 visits each season.

[c] Field time for video method also includes equipment setup and removal (divided by 30 visits).

been proposed (and are being used) based on the logic that conditions that are favorable for a particular indicator (e.g., spionid polychaetes) might also be good for other species in the community. By comparison, the conditions that are favorable for larval trematodes in snails *are* the host organisms. This is because abundant and rich larval trematode communities are impossible without abundant and diverse host communities. It is also important to consider that diverse and abundant trematode communities, while they reflect abundant hosts, might not always reflect positive environmental conditions. For example, some sites with large host populations (such as dump sites that attract gulls) are considered degraded yet may lead to prevalent trematode infections in snails (Bustnes and Galaktionov, 1999; Bustnes et al., 2000).

The ease of use and applicability of larval trematode assessments will vary by hosts available in a region and habitat, the number of trematode species infecting these hosts, and the work to date on descriptions of larval trematodes from particular hosts. In these respects, we submit that in estuarine habitats in the tropics and subtropics, *Cerithidea* spp. offer good potential. So too do species of *Cerithium* and *Batillaria*. In some cases (e.g., *Cerithium* spp.*)*, the larval trematodes infecting a host snail may encompass two different types of final host (e.g., fishes and birds). In these cases, larval trematode infections give information about very different final hosts. Because trematodes are ubiquitous in many aquatic habitats (Dawes, 1946; Schell, 1970; Yamaguti, 1971; Yamaguti, 1975), larval trematodes are potentially powerful indicators in far more than just estuarine habitats.

The comprehensive information provided by larval trematode assessments comes at a relatively low cost. Collection of hosts and identification of larval trematodes requires less work than many traditional community survey techniques (Table 19.2). We also emphasize that the work associated with identifying species or operational taxonomic units for larval trematodes is no more laborious, and requires no more expertise, than similar identifications of invertebrates, fishes, or plants.

The use of larval trematode communities should not completely supplant traditional taxonomic surveys, especially in cases where particular species are of interest. Rather, larval trematode community assessments can be used to inform these lists and highlight trophic linkages between them. However, since they have a high information yield and low cost, using larval trematode bioindicators might be given high priority when considering how to use limited resources in a monitoring project.

Acknowledgments

The authors acknowledge very useful comments on the manuscript from Armand Kuris and assistance from Vi Mababa in tracking down literature. Special thanks to Steve Bortone for coordinating the workshop for which this chapter is a contribution. This chapter has also benefited from support received from the National Science Foundation through the NIH/NSF Ecology of Infectious Disease Program (DEB-0224565), and a grant from the U.S. Environmental Protection Agency's Science to Achieve Results (STAR) Estuarine and Great Lakes (EaGLe) program through funding to the Pacific Estuarine Ecosystem Indicator Research (PEEIR) Consortium, U.S. EPA Agreement R-882867601. However, the chapter has not been subjected to any EPA review and therefore does not necessarily reflect the views of the Agency, and no official endorsement should be inferred.

References

Abdul-Salam, J. and B. S. Sreelatha. 1998. A list of larval digenetic trematodes parasitizing some marine invertebrates in Kuwait Bay. *Kuwait Journal of Science and Engineering* 25:409–434.

Adams, M. and M. G. Ryon. 1994. A comparison of health assessment approaches for evaluating the effects of contaminant-related stress on fish population. *Journal of Aquatic Ecosystem Health* 3:15–25.

Anteson, R. K. 1970. On the resistance of the snail, *Lymnaea catascopium pallida* (Adams) to concurrent infection with sporocysts of the strigeid trematodes, *Cotylurus flabelliformis* (Faust) and *Diplostomum flexicaudum* (Cort and Brooks). *Annals of Tropical Medicine and Parasitology* 64:101–107.

Brown, K. M., B. K. Leathers, and D. J. Minchella. 1988. Trematode prevalence and the population dynamics of freshwater pond snails. *American Midland Naturalist* 120:289–301.

Bustnes, J. O. and K. Galaktionov. 1999. Anthropogenic influences on the infestation of intertidal gastropods by seabird trematode larvae on the southern Barents Sea coast. *Marine Biology,* 133:449–453.

Bustnes, J. O., K. V. Galaktionov, and S. W. B. Irwin. 2000. Potential threats to littoral biodiversity: is increased parasitism a consequence of human activity? *Oikos* 90:189–190.

Cable, R. M. 1956. Marine cercariae of Puerto Rico. In *Scientific Survey of Porto Rico and the Virgin Islands*, R. W. Miner (ed.), Vol. 16, Part 4. New York Academy of Sciences, New York, pp. 491–577.

Ching, H. L. 1991. Lists of larval worms from marine invertebrates of the Pacific Coast of North America. *Journal of the Helminthological Society of Washington* 58:57–68.

Cort, W. W., D. B. McMullen, and S. Brackett. 1937. Ecological studies on the cercariae in *Stagnicola emarginata angulata* (Sowerby) in the Douglas Lake region, Michigan. *Journal of Parisitology* 23:504–532.

Cort, W. W., K. L. Hussey, and D. J. Ameel. 1960. Seasonal fluctuations in larval trematode infections in *Stagnicola emarginata angulata* from Phragmites flats on Douglas Lake. *Proceedings of the Helminthological Society of Washington* 27:11–12.

Dawes, B. 1946. *The Trematoda.* Cambridge University Press, London.

Edgington, E. S. 1995. *Randomization Tests,* 3rd ed. Marcel Dekker, New York.

Ewers, W. H. 1964. An analysis of the molluscan hosts of the trematodes of birds and mammals and some speculations on host-specificity. *Parasitology* 54:571–578.

Fernandez, J. and G. W. Esch. 1991a. The component community structure of larval trematodes in the pulmonate snail *Helisoma anceps. Journal of Parasitology* 77:540–550.

Fernandez, J. and G. W. Esch. 1991b. Guild structure of larval trematodes in the snail *Helisoma anceps*: patterns and processes at the individual host level. *Journal of Parasitology* 77:528–539.

Goater, T. M., A. W. Shostak, J. A. Williams, and G. W. Esch. 1989. A mark-recapture study of trematode parasitism in overwintered *Helisoma anceps* (Pulmonata) with special reference to *Halipegus occidualis* (Hemiuridae). *Journal of Parasitology* 75:553–560.

Guth, D. J., H. D. Blankespoor, and J. J. Cairns. 1977. Potentiation of zinc stress caused by parasitic infection of snails. *Hydrobiologia* 55:225–230.

Heyneman, D. and T. Umathevy. 1968. Interaction of trematodes by predation within natural double infections in the host snail *Indoplanorbis exustus. Nature* 217:283–285.

Hoff, C. C. 1941. A case of correlation between infection of snail hosts with *Cryptocotyle lingua* and the habits of gulls. *Journal of Parasitology* 27:539.

Houbrick, R. S. 1984. Revision of higher taxa in genus *Cerithidea* Mesogastropoda:Potamididae based on comparative morphology and biological data. *Bulletin of the American Malacological Union* 2:1–20.

Huspeni, T. C. 2000. A Molecular Genetic Analysis of Host Specificity, Continental Geography, and Recruitment Dynamics of a Larval Trematode in a Salt Marsh Snail, Ph.D. dissertation, University of California, Santa Barbara, CA.

Huspeni, T. C. and K. D. Lafferty. 2004. Using larval trematodes that parasitize snails to evaluate a salt-marsh restoration project. *Ecological Applications* 14:795–804.

Keas, B. E. and H. D. Blankespoor. 1997. The prevalence of cercaria from *Stagnicola emarginata* (Lymnaeidae) over 50 years in northern Michigan. *Journal of Parasitology* 83:536–540.

Keen, A. M. 1971. *Sea Shells of Tropical West America Marine Mollusks from Baja California to Peru.* Stanford University Press, Stanford, CA.

Krebs, C. J. 1999. *Ecological Methodology,* 2nd ed. Addison-Wesley Educational Publishers, Menlo Park, CA.

Krist, A. C., C. M. Lively, E. P. Levri, and J. Jokela. 2000. Spatial variation in susceptibility to infection in a snail-trematode interaction. *Parasitology* 121:395–401.

Kuris, A. 1990. Guild structure of larval trematodes in molluscan hosts: prevalence, dominance and significance of competition. In *Parasite Communities: Patterns and Processes*, G. W. Esch, A. O. Bush, and J. M. Aho (eds.). Chapman & Hall, New York, pp. 69–100.

Kuris, A. M. and K. D. Lafferty. 1994. Community structure: larval trematodes in snail hosts. *Annual Review of Ecology and Systematics* 25:189–217.

Lafferty, K. D. 1997. Environmental parasitology: what can parasites tell us about human impacts on the environment? *Parasitology Today* 13:251–255.

Lafferty, K. D. 1999. The evolution of trophic transmission. *Parasitology Today* 15:111–115.

Lafferty, K. D. and R. D. Holt. 2003. How should environmental stress affect the population dynamics of disease? *Ecology Letters* 6:797–802.

Lafferty, K. D., D. T. Sammond, and A. M. Kuris. 1994. Analysis of larval trematode communities. *Ecology* 75:2275–2285.

Lefcort, H., M. Q. Aguon, K. A. Bond, K. R. Chapman, R. Chaquette, J. Clark, P. Kornachuk, B. Z. Lang, and J. C. Martin. 2002. Indirect effects of heavy metals on parasites may cause shifts in snail species compositions. *Archives of Environmental Contamination and Toxicology* 43:34–41.

Lie, K. J., P. F. Basch, and T. Umathevy. 1965. Antagonism between two species of larval trematodes in the same snail. *Nature* 206:422–423.

Lim, H.K. and Heyneman D. 1972. Intramolluscan inter trematode antagonism: a review of factors influencing the host parasite system and its possible role in biological control. *Advances in Parasitology,* 10:191–268.

MacKenzie, K., H. Williams, B. Williams, A. H. McVicar, and R. Siddall. 1995. Parasites as indicators of water quality and the potential use of helminth transmission in marine pollution studies. *Advances in Parasitology* 35:85–144.

Mani, G. G. and K. H. Rao. 1993. Studies on Indian marine cercariae: two new echinostome cercariae. *Journal of the Helminthological Society of Washington* 60:250–255.

Marcogliese, D. J. 2001. Pursuing parasites up the food chain: Implications of food web structure and function on parasite communities in aquatic systems. *Acta Parasitologica* 46:82–93.

Marcogliese, D. J. and D. K. Cone. 1997. Parasite communities as indicators of ecosystem stress. *Parassitologia* 39:227–232.

Martin, W. E. 1950. *Euhaplorchis californiensis n.g.,* n.sp., Heterophyidae, Trematoda, with notes on its life cycle. *Transactions of the American Microscopical Society* 194–209.

Martin, W. E. 1972. An annotated key to the cercariae that develop in the snail *Cerithidea californica. Bulletin of the Southern California Academy of Sciences* 71:39–43.

Matthews, P. M., W. I. Montgomery, and R. E. B. Hanna. 1985. Infestation of littorinids by larval digenea around a small fishing port. *Parasitology* 90:277–288.

Overstreet, R. M. 1997. Parasitological data as monitors of environmental health. *Parassitologia* 39:169–175.

Pietrock, M. and D. J. Marcogliese. 2003. Free-living endohelminth stages: at the mercy of environmental conditions. *Trends in Parasitology* 19:293–299.

Rees, F. G. 1932. An investigation into the occurrence, structure, and life histories of trematode parasites of four species of *Lymnaea* (*L. trunculata* (Mull.), *L. pereger* (Mull.), *L. palustris* (Mull.), and *L. stagnalis* (Linne)), and *Hydrobia jenkinsi* (Smith) in Glamorgan and Monmouth. *Proceedings of the Zoological Society of London* 1932:1–32.

Roberts, L. S. and J. Janovy. 2000. *Foundations of Parasitology,* 6th ed. McGraw-Hill, New York.

Robson, E. M. and I. C. Williams. 1970. Relationships of some species of digenea with the marine prosobranch *Littorina littorea.* Part 1. The occurrence of larval digenea in *Littorina littorea* on the North Yorkshire coast. *Journal of Helminthology* 44:153–168.

Sapp, K. K. and G. W. Esch. 1994. The effects of spatial and temporal heterogeneity as structuring forces for parasite communities in *Helisoma anceps* and *Physa gyrina. American Midlland Naturalist* 132:91–103.

Schell, S. 1970. *How to Know the Trematodes.* Wm. C. Brown, Dubuque, IA.

Siddall, R. and S. des Clers. 1994. Effect of sewage sludge on the miracidium and cercaria of *Zoogonoides viviparus* (Trematoda: Digenea). *Helminthologia* 31:143–153.

Smith, N. F. 2001. Spatial heterogeneity in recruitment of larval trematodes to snail intermediate hosts. *Oecologia* 127:115–122.

Snyder, S. D. and G. W. Esch. 1993. Trematode community structure in the pulmonate snail *Physa gyrina. Journal of Parasitology* 79:205–215.

Sousa, W. P. 1993. Interspecific antagonism and species coexistence in a diverse guild of larval trematode parasites. *Ecological Monographs* 63:103–128.

Stadnichenko, A. P., L. D. Ivanenko, I. S. Gorchenko, O. V. Grabinskaya, L. A. Osadchuk, and S. A. Sergeichuk. 1995. The effect of different concentrations of nickel sulphate on the horn snail (Mollusca: Bulinidae) infected with the trematode *Cotylurus cornutus* (Strigeidae). *Parazitologiy* 29:112–116.

Wardle, W. J. 1974. A Survey of the Occurrence, Distribution and Incidence of Infection of Helminth Parasites of Marine and Estuarine Mollusks from Galveston, Texas, Ph.D. dissertation, Texas A&M University, College Station.

Winterbourn, M. J. 1974. Larval trematoda parasitizing the New Zealand species of *Potamopyrgus* (Gastropoda: Hydrobiidae). *Mauri Ora* 2:17–30.

Yamaguti, S. 1971. *Synopsis of Digenetic Trematodes of Vertebrates.* Keigaku, Tokyo.

Yamaguti, S. 1975. *A Synoptical Review of Life Histories of Digenetic Trematodes of Vertebrates with Special Reference to the Morphology of Their Larval Forms.* Keigaku, Tokyo.

20

Macrobenthic Process-Indicators of Estuarine Condition

Chet F. Rakocinski and Glenn A. Zapfe

CONTENTS

Introduction

Background and Need

Healthy estuaries within the Gulf of Mexico of the United States contribute high fisheries production, ample biodiversity, recreational benefits, transportation, and water supply (U.S. EPA, 1999). In response to mounting pressures on national estuarine resources, however, the U.S. Environmental Protection Agency (EPA) Environmental Monitoring and Assessment Program–Estuaries (EMAP–E) was implemented to quantify the status and condition of estuaries (Summers et al., 1991). The objective of the EMAP-E multiyear monitoring program was to assess the ecological condition and biotic integrity of entire coastal regions, or provinces. Macrobenthic organisms provide reliable indicators of biotic integrity; and the benthic index was one of the most useful measures of estuarine condition developed by the U.S. EPA EMAP-E Program. However, there are various disadvantages of existing benthic indices: (1) they represent a static expression of ecological condition; (2) they are not explicitly linked to changes in ecological function; (3) they may not be specific with respect to different kinds of stressors; (4) they are subject to underlying taxonomic changes across estuarine gradients; (5) they can be labor intensive (e.g., sorting specimens and taxonomic identification); and (6) they are not applied consistently across biogeographic provinces.

Excessive nutrient loading is one of the primary causes of degradation of coastal estuaries across the northern Gulf of Mexico (U.S. EPA, 1999). The magnitude of nutrient loading via riverine discharge and runoff into estuaries of the northern Gulf of Mexico has doubled in just a few decades, resulting in the elevated production of algae and marked increases in hypoxia (Turner and Rabalais, 1999). Adverse ecosystem effects of severe eutrophication include excessive production of phytoplankton as well as epiphytic and benthic algae, shifts in phytoplankton species, changes in macrophyte production, increased turbidity, oxygen depletion and attendant shifts in microbial processing, and ultimately, the reduction of biodiversity and fisheries production (Carpenter et al., 1998). Excessive phytoplankton production leads to the deposition and buildup of organic material within sediments. Consequent degraded sediment conditions include organic enrichment, hypoxia, and upward shifts in the location of the redox discontinuity layer. Complications also ensue when organic enrichment and low dissolved oxygen (DO) effects coincide and interact with other environmental stresses, such as sediment contamination.

The benthic environment plays a pivotal role in the regeneration of nutrients through various benthic–pelagic coupling mechanisms involving both physical and biotic processes (Twilley et al., 1999). Macrobenthic communities mediate trophic functioning of the estuarine ecosystem in ways that affect rates, directions, and pathways of exchange and transformations of energy and materials, including nutrients, between the water column and the sediment (Hansen and Kristensen, 1997). Clearly, changes in trophic function induced by nutrient loading that involve interactions with the macrobenthos need to be understood and monitored in estuaries. Thus, there is a pressing need to develop broadly applicable indicators of macrobenthic processes related to ecosystem function; and to validate and transfer these broad indicators to regional monitoring programs.

Definition of Macrobenthic Process Indicators

Effective ecological indicators can be recognized by how well they meet certain criteria; they should (1) be straightforward and easy to measure; (2) reflect changes in ecosystem integrity, function, or resilience; (3) convey information about ecological processes; (4) foster examination of responses and linkages at various levels of organization and spatiotemporal scales; (5) be amenable to validation of their reliability through quantitative assessments of sensitivity and background variability (*sensu* Cottingham and Carpenter, 1998). Various macrobenthic measures potentially convey information about ecosystem function, including secondary production rates, annual production:biomass (P:B) ratios, biomass-size spectra, and trophic structure. These dynamic measures can be estimated in a straightforward manner from standing macrobenthic samples. A suite of macrobenthic process indicators potentially could be used to measure changes in benthic ecosystem function associated with nutrient loading and other attendant anthropogenic stresses. Moreover, to realize their full potential, such macrobenthic indicators require integration with other indicators of ecosystem function derived at various levels of organization and spatiotemporal scales.

Standing macrobenthic biomass has been used as an ecological indicator, usually in the context of concerted species-abundance-biomass (SAB) relationships involving species richness, total abundance, and total biomass in response to organic enrichment or sediment contamination (Pearson and Rosenberg, 1978; Rakocinski et al., 2000). The classic SAB model depicts how these three indicators covary along an organic pollution gradient as the macrobenthic community becomes dominated by fewer opportunistic small-bodied taxa with increasing organic enrichment. As useful as this model is for showing how benthic communities respond to environmental stress, these three indicators are static measures of macrobenthic condition. However, secondary production is a dynamic measure of condition, or a process-indicator, that can convey useful information about ecological function. For example, because metabolic rate varies allometrically with body-size (Edgar, 1990; Edgar et al., 1994), two sites with the same standing biomass can have radically different production rates. Changes in secondary production will integrate and reflect shifts in size distributions, abundances, and the diversity of benthic organisms; thus, estimates of macrobenthic secondary production should provide an important indicator of ecological function that can be readily compared across habitats, sites, gradients, or regions.

Cohort analysis of the dominant members of the benthic fauna represents the traditional means of estimating secondary production, but this approach can be cumbersome and deficient. A more practical means of estimating daily production was employed by Edgar (1990), based on the application of allometric scaling relationships linking body-size-dependent metabolic rates with body-size distributions. Since it is impractical to measure body sizes from many large samples of small invertebrates, Edgar estimated macrobenthic production by partitioning size distributions of organisms via a series of nested sieves of standard geometrically spaced mesh sizes. Based on many other published studies, he determined relationships between body weight, water temperature, and production rate for major groups of macroinvertebrates. The dependence of the daily production rate (P) on organism ash-free dry weight (AFDW) or biomass (B) and water temperature (T) was evaluated using the allometric equation: $P = a \times B^b \times T^c$, where the coefficients a, b, and c were calculated from the logarithm (base 10) transformed regression equation. Equations relating faunal AFDW (μg) and sieve mesh size (mm) also were developed.

An additional process indicator that results from the relationship between secondary production and standing biomass is the P:B ratio, which reflects the average lifespan of macrobenthic organisms (Gray, 1981), and thus the stability of the macrobenthic community. Furthermore, P:B values provide practical measures of faunal turnover rates when expressed on an annual scale. Various factors may influence P:B values, including temperature, organism growth rates, predation rates, population growth rates, and population size structure. The last two factors are particularly important. P:B values would be expected to increase in connection with environmental stress due to a shift in the macrobenthic community toward small-bodied, short-lived, opportunistic species.

Biomass-size spectra provide an aggregate allometric expression of biogeochemical processes, trophic organization, and ecosystem function (Strayer, 1986; Boudreau et al., 1991; Rasmussen, 1993; Ramsay et al., 1997; Kerr and Dickie, 2001). Moreover, biomass-size spectra can be readily compared across habitats, because they are independent of taxonomic composition. This property may be especially useful in estuaries, where strong environmental gradients elicit distinct transitions in community structure (Rakocinski et al., 1997; 2000). Indeed, characteristic shapes of macrobenthic biomass-size spectra can be conservative across estuarine habitats in healthy aquatic ecosystems, such as sediment types or salinity gradients (Warwick, 1984; Duplisea and Drgas, 1999; Parry et al., 1999). However, Rasmussen (1993) demonstrated shifts in macrofaunal biomass-size spectra in relation to different types of submerged macrophytes and, moreover, with respect to increasing primary productivity. Indeed, parameters of biomass-size spectra may vary in relation to various forms of stress, including organic enrichment, sediment contamination, high primary productivity, and disturbance. Such shifts in macrobenthic size distributions as reflected by parameters of the biomass-size spectrum should represent a reliable process indicator (Warwick, 1993).

Macrobenthic trophic structure consists of information about the distribution of organisms among various feeding types; and conveys information about benthic/pelagic coupling. Changes in trophic structure imply attendant shifts in ecosystem function (Rakocinski et al., 1997). Shifts in macrobenthic trophic structure have been associated with sediment contamination in the northern Gulf of Mexico (Brown et al., 2000). Such changes in macrobenthic trophic function can affect rates of decomposition, nutrient cycling, and energy transfer (Gaston et al., 1998). Thus, a direct connection between nutrient loading and macrobenthic trophic function also might also be expected. Trophic structure can be characterized by the distribution of organisms among a suite of trophic categories, which convey their feeding behavior and type of food. For example, various macrobenthic trophic categories can be distinguished, including: filter feeders, subsurface deposit feeders, carnivores, etc. (Gaston and Nasci, 1988). A more dynamic and accurate picture of trophic structure than can be obtained based on abundances of organisms might be based on distributions of biomass or production rates among the trophic categories. Viewed in terms of production rates, trophic structure could yield further insights into effects of stress on trophic pathways and changes in ecosystem function due to nutrient loading. Knowledge about variability in production might prove more useful if macrobenthic production could be partitioned into the various trophic categories. More focused hypotheses about changes in trophic dynamics and other ecosystem processes in response to nutrient loading might then be possible. For example, rates of nutrient recycling might be very different for two systems with the same total macrobenthic production, depending on how the production is distributed among trophic categories. Additional insights might also be gained by considering production with respect to a combination of trophic category and body size.

Hypothesis

Given the kinds of changes that are known to occur in abundances and body-size distributions of benthic organisms with increasing organic enrichment (Pearson and Rosenberg, 1978), one might hypothesize a nonlinear response in macrobenthic function along a gradient of increasing eutrophication (Figure 20.1). More specifically, a hypothetical response might be defined by an initial linear increase in macrobenthic function up to peak levels, followed by an eventual exponential decrease in function as levels of nutrient loading intensify. Such a decline would reflect the impaired capacity of the ecosystem to process available resources at such high stress levels (Xu et al., 1999). For example, relatively low rates of production should occur at low nutrient levels due to nutrient limitation, followed by enhancement of production at moderate levels of nutrient enrichment, whereas production should decline at high nutrient levels due to stress imposed by eutrophication. Stressors include various detrimental effects associated with excessive nutrient loading and hypoxia.

Objectives

Our goal is to develop practical methods for characterizing macrobenthic function through the use of process indicators derived from standard macrobenthic samples. The general objectives of this chapter are twofold: (1) to present detailed procedures for processing macrobenthic samples to obtain macrobenthic process indicators, including secondary production, P:B, and biomass-size spectra; and (2) to

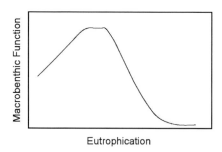

FIGURE 20.1 Hypothetical nonlinear response in macrobenthic function along a gradient of increasing eutrophication.

present a case study of spatial variability in macrobenthic function conducted within the Grand Bay National Estuarine Research Reserve (GBNERR) located in Mississippi.

Methods

Standard Operating Procedures

An important prerequisite for the development of reliable ecological indicators was the establishment of standard operating procedures (SOP). Codified procedures were needed for every phase of indicator development, including field methods and laboratory methods, as well as for data management, indicator application, and interpretation.

Field Methods

SOP for characterizing new indicators followed existing procedures as much as possible. Field procedures paralleled the U.S. EPA-sponsored Mississippi National Coastal Assessment (NCA) protocol to a large extent, except three pairs of grabs were taken per site, three for macrofauna (rather than one) and three for sediments. Sediments were not sampled for trace metals; however, sediments were sampled for pore water nutrients.

Site-Event Characterization

Site events were characterized prior to taking samples. Predetermined latitude and longitude coordinates were located using a Garmin® GPSMAP 76S enabled with Wide Area Augmentation System (WAAS). Associated information was recorded, including the unique field identifier, date, time, tide, depth, weather, and seas.

Water Column Profile

The water column was characterized for each site event using a YSI® 6600 multiparameter sonde. Prior to using the sonde, actual and Secchi depths were measured. The protocol used for generating a water column profile differed somewhat from the Mississippi NCA SOP. First, the depth of the sonde was recorded at the bottom before creating the profile so the relative position of the sonde within the water column could be determined. Hence, the sonde depth sensor was used to locate sample points within the water column. Measurements of parameters were always taken from 0.2 m below the surface and above the bottom. If the depth was less than 3 m, the profile measurements were recorded at 0.5 m increments; or measurements were made every 1.0 m for depths equal to or greater than 3 m. Variables measured include salinity, temperature, DO, pH, turbidity, and chlorophyll. Both downcast and upcast profiles were made for each site event, following the Mississippi NCA protocol.

Macrofaunal Samples

Following site-event characterization and water-column profiles, three successive pairs of benthic grabs were taken using a 0.0413 m² modified Van Veen grab. One member of each pair was processed for macrofauna, and the other member was used to characterize sediment properties. Macrofaunal samples were passed through a 0.5-mm-mesh standard sieve to remove fines in the field. Labeled macrofaunal samples were preserved in buffered 10% Formalin and returned to the laboratory for processing.

Sediment Samples

For each site event, subsamples from the three grabs for sediment analysis were pooled and kept on ice in a stainless-steel bowl during sampling. The upper 2-cm layer of sediment was removed from each grab using a large stainless-steel spoon. The sediment comprising the three samples was homogenized before taking aliquots for sediment characterization. Using a stainless-steel utensil, aliquots of 500, 300, and 50 ml of sediment were placed into labeled wide-mouth NALGENE® bottles for later determinations of pore water nutrients, sediment composition, and grain-size and total organic carbon (TOC), respectively. These samples were stored on ice and returned to the laboratory for further processing. Finally, sediment oxidation-reduction potential (ORP) was routinely measured at 2.5 cm below the intact sediment surface on the second of the three sediment grabs using an Orion® QuiKcheK Model 108 ORP Pocket Meter.

Laboratory Methods

Although SOP for laboratory characterization of macrobenthic process indicators followed existing procedures wherever possible, considerable development of new protocols was required for this project. New procedures that were implemented involved size fractionation of macrofaunal samples and the use of image analysis along with calibrated squash plates to quantify organism volumes.

Processing Macrofaunal Samples

Macrobenthic sample sorting followed established quality assurance–quality control (QA-QC) procedures. A small amount of Rose Bengal was added to the samples before sorting. Unsorted sample material was fractionated using a set of nested sieves with 63 μm as the final mesh size, and sieve fractions were sorted separately. All macrobenthic organisms for each benthic grab sample were sorted from remaining material in a gridded petri dish, placed into one vial and preserved in 5% buffered Formalin. Alcohol was not used as preservative to avoid shrinkage and the leaching of lipids from the organisms. QA-QC procedures for sorting efficiency followed established EPA-EMAP protocol; 10% of the samples were resorted and examined for 90% accuracy.

SOP were developed for completing three progressive stages of laboratory processing of sorted macrobenthic organisms: size fractionation, taxonomic identification, and volumetric determinations. A sequence of steps was adopted for size-fractionating macrofaunal samples, which involved the subdivision of organisms into nine possible size fractions. All organisms from each sorted sample were sifted through a series of nine nested sieves of 8.0, 5.6, 4.0, 2.8, 2.0, 1.4, 1.0, 0.71, and 0.50 mm mesh sizes (Edgar, 1990). Each set of two sequential sieves was progressively employed, starting with the largest two sieves. Sets of sieves were placed in a tub filled with enough water to reach the bottom of the top sieve and then gently agitated to separate organisms into size fractions. Pieces of organisms within smaller size fractions were accordingly recombined with the appropriate larger size fractions. Organism remnants that passed through the finest sieve were also recombined with the smallest size fraction, if appropriate. Macrofaunal size fractions were transferred to taxonomic experts for identification of the organisms, usually to species. Organisms were assigned a taxonomic code and counted, resulting in the breakdown of size fractions into taxonomic categories. This level of detail allowed calculations of conventional indicators as well as macrobenthic process indicators.

Sets of squash plates (Hellawell and Abel, 1971) were constructed using microscope slides and coverslips. Several sets of different squash plates were constructed for use at finer or coarser volumetric scales of resolution. Eppendorf® micropipettes were used to deliver known volumes of KOH-glycerol solution onto the squash plates for calibration. For each set of plates, either seven or nine different volume levels were used. Three replicate measurements were made for every volume level for each set of plates. From these data, standard area–volume curves for use at a particular magnification were generated for each set of squash plates.

Once identified to lowest possible taxon, sample taxon-size fractions were ready for volumetric determinations using a Nikon® image analysis system consisting of a DMX 1200 digital camera attached to an SMZ 1500 stereomicroscope. MetaVue® 5.0 imaging software was used to estimate volumes from two-dimensional areas of macrobenthic organisms compressed to a uniform thickness using calibrated squash plates. The outlines of squashed organisms within a given taxonomic size fraction were traced two times to obtain a mean area. A WACOM® digital tablet was used to readily trace area outlines from printed images of squashed organisms. The printed images were scaled proportionately for a one-to-one correspondence with computer screen images. This method was very precise; areas of individual macrobenthic organisms could readily be measured. Volumes were then estimated from the specific calibration curve associated with the particular set of squash plates used. Only the soft tissue was measured from mollusks. Occasionally, organisms too large to be handled practically by the squash plate method were blotted and weighed to the nearest 10^{-5} g using an Ohaus® Analytical Plus microbalance.

Processing Sediment Samples

Corresponding sediment samples were processed for pore water nutrients, sediment composition, grain size analysis, and TOC. Pore water was extracted using vacuum aspiration, stored at $-70°C$, and later examined. Total nitrate-nitrite, total Keldahl nitrogen, ammonia, orthophosphate, and total phosphate levels were determined following EPA standard methods for each specific analyte using a Bran+Luebbe® AutoAnalyzer 3. Sediment composition and grain-size parameters were determined following Folk (1974) and Plumb (1981). Sediments were characterized in terms of silt/clay and sand content through granulometric analysis, and grain-size percentages were measured using standard procedures. TOC was determined in duplicate by combustion of 2 to 3 g subsamples of acidified and dried sediment. Dried and ground sediment was combusted in a microprocessor-based LECO® C-200 Carbon Determinator Analyzer at 2500°C for about 60 s. During combustion all carbon was converted to CO_2, and the carbon content was then determined by infrared absorption (IR).

Results

Process-Indicator Development

Although the necessary concepts for developing macrobenthic process indicators are available, the development of efficient methods for estimating process indicators poses a challenge. Practical methods are prerequisite for implementation in monitoring programs. A combination of image analysis, data conversions, and allometric scaling was implemented for estimating process indicators for each site event.

Volume Conversions

All calibration curve regressions assumed a zero intercept, and explained more than 99% of the variation in volume. Using calibrated squash plates, volume (in μL) could be extrapolated directly from the area of compressed organisms (in mm^{-2}) based on one conversion factor (i.e., the slope of the calibration curve). Depending on the particular set of squash plates and the magnification level, conversion factors generally ranged from 0.25 to 0.52.

Mass Conversions

Ash-free dry mass values were required for estimating daily production rates of macrobenthic organisms. Wet mass values were obtained from volumes of squashed organisms using the conversion factor, 1.13 g cm^{-3} (i.e., specific gravity of 1.13) (Wieser, 1960; Gerlach et al., 1985). A value of 0.16 was used for the conversion of wet soft-mass to ash-free dry mass values based on Ricciardi and Bourget (1998). Wet soft-mass values for highly calcified organisms that needed to be directly weighed, such as sand dollars and brittle stars, were obtained as 0.45 of the wet mass, before obtaining the ash-free dry mass (i.e., 0.072 of wet mass). Future wet mass conversions of calcified organisms will follow Ricciardi and Bourget (1998).

Daily Production Estimates

Daily production estimates were obtained from dry mass values based on the general allometric equation presented by Edgar (1990). However, rather than projecting daily production from entire sieve fractions, daily production was calculated by aggregating estimated production values for individual organisms. This is a more accurate approach given the nature of the macrobenthic samples, which contained various small, elongate organisms that may not sort well enough into fractions corresponding with empirical relationships provided for sieve fractions by Edgar (1990). Nevertheless, resulting taxon-size fractions contained uniform body sizes, facilitating accurate production estimates. Edgar (1990) concluded that the influences of taxonomic and functional groups on the estimation of daily production were not very great. Thus, the general allometric equation of Edgar (1990) was used to convert dry mass into estimated production per day per individual for organisms within each taxon-size fraction:

$$P_{ind} = 0.00489778 \times B_{ind}^{0.8} \times T^{0.89}$$

where P_{ind} = daily individual production in µg AFDW d^{-1}, B_{ind} = individual µg AFDW, and T = °C water temperature. The P_{ind} value was then multiplied by the number of individuals within the taxon-size fraction.

Aggregation to Macrobenthic Production

Daily production was aggregated from the taxon-size fraction level for all three benthic grabs per site event. Finally, the resulting daily site production value was scaled up to a square meter basis for comparability.

Estimating the P:B Ratio

A proxy annual P:B ratio was also calculated as the quotient between the total dry mass and the 365-fold site production value. The P:B value reflects the turnover rate of the macrobenthic community resulting from the lifespan and body-size composition of member organisms. As it does not incorporate real seasonal changes, it is best taken as an index of the community turnover rate. The reciprocal of P:B approximates the community turnover rate in months (i.e., P:B = 6 reflects a 2-month turnover rate).

Standardized Biomass-Size Spectra

Biomass-size spectra were standardized in their linear form (Sprules and Munawar, 1986; Hanson, 1990; Rasmussen, 1993; reviewed in Kerr and Dickie, 2001). Spectra were generated for each site event by distributing macrobenthic biomass along an octave scale of size classes defined by body-mass doubling increments (Schwinghamer, 1988), starting with the smallest size class of 150 µg dry weight. Linear biomass-size spectra were obtained by standardizing total biomass per size class with respect to the size-class range (i.e., difference between the beginning and ending mass within each size class), and then taking the logarithms (base 10) of the resulting values. These values were fitted against the logarithms (base 10) of the midpoints of the associated size classes to define decreasing linear relationships, from which parameters (i.e., slope and intercept) could be used to characterize site events.

Case Study — Grand Bay National Estuarine Research Reserve

We conducted a case study in the Grand Bay National Estuarine Research Reserve (GBNERR) to develop and assess the use of macrobenthic process indicators. Four specific objectives of this study were (1) to apply procedures to effectively characterize macrobenthic process indicators; (2) to examine spatial variation in macrobenthic process indicators within two parallel bayou systems; (3) to determine whether the macrobenthic process indicators were sensitive to moderate differences in levels of presumed nutrient loading; and (4) to evaluate the hypothetical nonlinear macrobenthic response to eutrophication.

Study Design

Fieldwork for the GBNERR case study was conducted from 18 to 20 July 2002, during the summer index period of the Mississippi NCA. A 7.5-km transect was established within each of two bayou systems, Bayou Heron and Bayou Cumbest (Figure 20.2). Five sites were located along each transect, and sites were placed at distance octaves proceeding from the upper bayous to the adjoining bays (i.e., 0.5, 1.0, 2.0, and 4 km between sites). Sites were spaced closer in the upper regions of the systems, where organic loading and DO stress were more likely to occur. Bayou Heron is in a relatively pristine

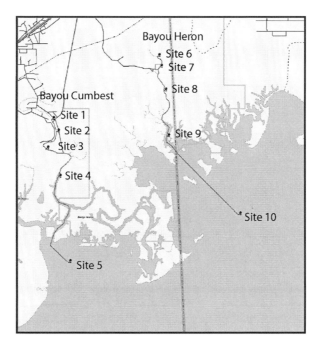

FIGURE 20.2 Map of the GBNERR study area showing transects as defined by five sites within each of the two bayou systems, Bayou Cumbest and Bayou Heron.

area, but this system exhibits relatively low DO in the uppermost portion. Bayou Cumbest is thought to be affected by moderate residential wastewater runoff due to poor septic facilities (MSU-CREC, 2000). The latter system is also subject to considerable land use as evidenced by altered shorelines in its upper reaches. Bayou Cumbest is also a larger, less dendritic system with higher flow rates than Bayou Heron. Combined differences in both land use and geomorphology should contribute to generally higher levels of nutrient loading in Bayou Cumbest relative to Bayou Heron. Three pairs of benthic grabs were taken at each site for macrofauna and sediments; and water quality profiles as well as other information were also obtained, as described above.

Spatial Patterns in Process Indicators

Three macrobenthic process indicators were examined in the case study: macrobenthic production, P:B values, and standardized biomass-size spectra. Interesting spatial variation in these three indicators was evident within the GBNERR system (Table 20.1; Figure 20.3); and both longitudinal and cross-system patterns were apparent. Overall, macrobenthic production increased from upestuary to downestuary sites; while P:B values were higher at upestuary sites, reflecting the tendency for downestuary sites to contain macrobenthic communities consisting of larger and longer-lived organisms. Moreover, macrobenthic production was clearly higher within Bayou Cumbest than in Bayou Heron: values ranged over one order of magnitude, from 8,248 to 85,308 μg m^{-2} d^{-1} in the Bayou Cumbest system; and only from 95 to 15,192 μg m^{-2} d^{-1} in the Bayou Heron system. This difference was congruent with the moderate nutrient enrichment hypothesis. Ironically, the lowest production value occurred at the uppermost site in Bayou Heron, which was located near the upper limit of the main channel.

Linear relationships were significant for eight of the ten standardized biomass-size spectra from the GBNERR sites; and reasonable biomass-size spectrum parameters were readily obtained for the other two sites. The intercept for the uppermost Bayou Heron site was estimated by assuming the same slope as that observed for the corresponding uppermost site in Bayou Cumbest. Fitted linear relationships had negative slopes, reflecting the overall effect of diminishing biomass, when scaled to the magnitude of the size category, with increasing body size. As observed for macrobenthic production, longitudinal and cross-system differences in biomass-size spectra were evident. Slopes and intercepts reflected underlying

TABLE 20.1

Spatial Variation in Daily Secondary Production, Annual P:B, and Numbers of Organisms Sampled within Three Benthic Grabs Totaling 0.124 m^{-2} at Each of Ten Sites within the GBNERR Study Area

	BC Site 1	BC Site 2	BC Site 3	BC Site 4	BC Site 5
Daily prod	21,955	8,248	29,327	22,140	85,308
Annual P:B	11.11	13.13	8.02	9.00	5.21
Number Org	485	199	403	347	342

	BH Site 6	BH Site 7	BH Site 8	BH Site 9	BH Site 10
Daily prod	95	4,890	4,073	10,277	15,192
Annual P:B	14.14	13.32	11.79	9.05	6.23
Number Org	3	125	64	86	72

Note: Daily prod = daily production in μg m^{-2} d^{-1}, annual P:B values = the number of inferred faunal turnover periods per year based on the daily production rate. Longitudinal and cross-system variation is evident, as discussed in the text.

properties of the biomass-size spectra: higher biomass, especially of small organisms, raised the intercepts, whereas the presence of larger organisms lowered the slopes. The longitudinal pattern in biomass-size spectra reflected shifts in abundances of organisms and in the body-size composition of the macrobenthic community, ranging from high biomass of small sizes and narrow size distributions at upper sites to lower biomass of small sizes and broad size distributions at downestuary sites. Biomass-size spectra at Bayou Cumbest sites tended to have higher intercepts and steeper declining slopes than those at Bayou Heron sites.

Linking Macrobenthic Process Indicators

Although linkages among the three macrobenthic process indicators were not readily apparent, relationships among these indicators should exist. Furthermore, interpreting biomass-size spectra can be difficult, because their slopes and intercepts are interrelated and inherently variable. However, when the intercepts of the standardized biomass-size spectra were regressed against the corresponding slopes for the ten GBNERR sites, a significant general relationship was defined ($r = -0.83$; $P < 0.001$) (Figure 20.4). Moreover, for each site the relative deviation of the intercept from the general relationship, or the standardized residual, reflected the degree to which biomass was higher or lower than expected for a biomass-size spectrum of a known slope. Notably, standardized residuals from the general biomass-size spectrum parameter relationship were positively related to logarithm (base 10) production ($r = 0.95$; $P < 0.0005$), reflecting a connection between biomass-size spectra and production. Furthermore, when plotted within a third dimension, P:B values varied inversely in relation to production and residual biomass-size spectrum parameter values. Presumably, the three process indicators reflected different facets of macrobenthic function; thus, a single composite factor was derived from a principal components analysis (PCA) of the three indicators, which all loaded at 0.9 or higher on the same component, that is, 0.95 logarithm (base 10) production; –0.906 P:B; 0.983 spectrum residual. This single PCA component accounted for 89.46% of the overall variation in the three indicators.

Relating Macrobenthic Process Indicators and Ecosystem Function

When characterized as a composite variable based on the three process indicators, site scores for the first PCA component provided a succinct measure of macrobenthic function. Site PCA scores in turn were significantly related to several functional environmental variables, including concentrations of pore water ammonia and pore water total phosphorus, as well as surface chlorophyll and bottom DO (Figure 20.5). Although the relationship with DO was relatively weak, the link was surprising given that the DO values were only point measures of a labile variable measured at different times of the day. Overall, these results suggest that the macrobenthic process indicators did reflect ecosystem function.

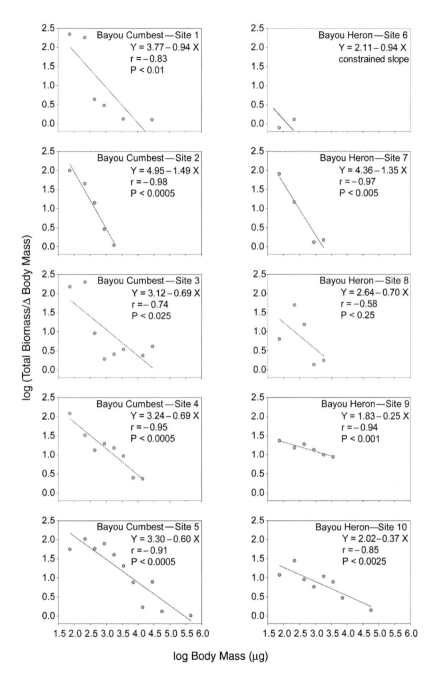

FIGURE 20.3 Spatial variation in standardized biomass-size (log base 10) spectra in the form of linear relationships for each of the ten sites within the GBNERR study area. Longitudinal and cross-system variation is evident, as discussed in the text.

Discussion

Discernable spatial variation in the process indicators probably tracks variability in the trophic condition of the ecosystem. For example, downestuary macrobenthic communities appear to be relatively stable compared to communities at upper sites, which may be subject to more direct effects of nutrient loading

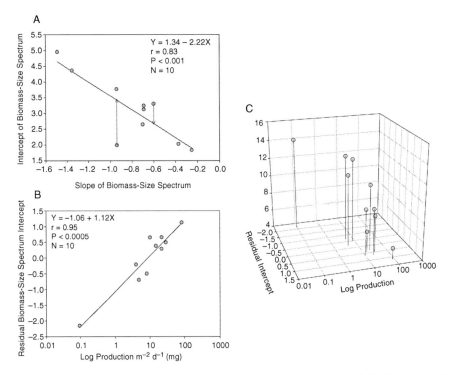

FIGURE 20.4 Linkages among the three macrobenthic process indicators. (A) General relationship between intercepts and slopes of standardized size spectra for the ten GBNERR sites. (B) Significant relationship between standardized residuals from A and log (base 10) macrobenthic production estimates. (C) P:B values align well in the third dimension relative to the two variables in B.

and hypoxia (Gonzalez-Oreja and Saiz-Salinas, 1999). Although such a longitudinal trophic shift might occur naturally within estuarine systems, the general relationship between intercepts and slopes of standardized biomass-size spectra proved to be a useful baseline for identifying departures in trophic condition, including those that may be caused by anthropogenic stressors. However, the site that showed the greatest impairment in macrobenthic function was located near the upper limit of the main channel within the relatively "natural" Bayou Heron system. This site was also characterized by relatively high chlorophyll and low DO, as well as relatively high pore water ammonia and total phosphorus.

Importantly, the macrobenthic process indicators appear sensitive enough to detect moderate levels of nutrient loading, as indicated by relatively high secondary production in Bayou Cumbest relative to Bayou Heron. Furthermore, the results from this study were congruent with the hypothetical nonlinear macrobenthic response to eutrophication, in that macrobenthic function within Bayou Cumbest was generally elevated. Moreover, particularly low macrobenthic function at the uppermost Bayou Heron site occurred where chlorophyll levels were highest and bottom DO values were lowest. However, further study is needed to more fully address whether the macrobenthic function hypothesis is robust.

Various potential sources of bias may affect estimates of macrobenthic process indicators. The approach presented in this study is intended to be straightforward and practical enough for use in monitoring programs; but these same useful qualities may introduce some inaccuracies. For example, the use of a 0.5-mm mesh sieve to process benthic grabs in the field was chosen because: (1) 0.5 mm is the standard mesh size typically used as a cutoff for distinguishing macrofauna from meiofauna; and (2) 0.5-mm mesh is also practical for processing sediments. However, some early stages of macrobenthic organisms undoubtedly can pass through this mesh size, leading to underestimates of biomass and production, especially for the smaller macrofaunal organisms. Another potential source of error is gear bias resulting from the limited spatial scale in terms of area and depth, which might underestimate large and deep burrowing organisms. Nevertheless, fairly large organisms were routinely collected in the GBNERR, including brittle stars, sand dollars, and large gastropods. It was expedient to use general

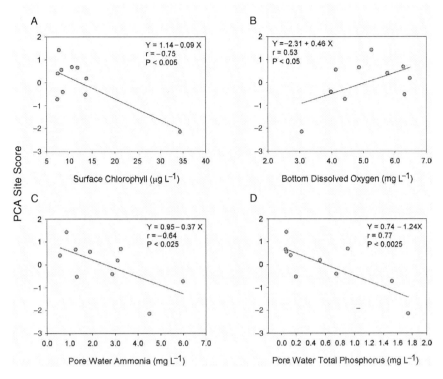

FIGURE 20.5 Relationships between macrobenthic function as characterized by a composite PCA variable based on the three process indicators and four functional environmental variables. Significant relationships suggest that the macrobenthic process indicators reflected ecosystem function.

conversion and scaling factors rather than detailed conversions for particular taxonomic groups. However, production-scaling factors specific for broad taxonomic groups are similar to the general production-scaling factor (Edgar, 1990), although some biases may be introduced by using the general conversion. Also, the use of conversion factors is indirect and approximate compared to directly obtaining such data. On the other hand, systematic biases should not be too consequential when comparing relative differences between sites. Inexactness due to natural variability and sampling variability probably greatly exceeds such systematic errors. An advantage of using image analysis to estimate volumes rather than weights of preserved material is that biases due to preservation effects should be minimized. Organisms are also retained in Formalin to minimize shrinkage effects of ethanol preservation. The interpretation of standardized biomass-size spectra should be made cautiously when well-defined peaks and troughs in biomass are present across the size spectrum, due to autocorrelation of serial values that can bias the parameters of the relationship (Rasmussen, 1993). Another potential source of inaccuracy arises from the extrapolation of the snapshot perspective to a longer-term perspective, for example, in the calculation of proxy annual P:B values. An accurate estimate of annual P:B would require sampling at multiple times of the year to integrate seasonal changes in the community composition and water temperature. Nevertheless, the use of the proxy P:B values reflected the faunal turnover period and provided values that were comparable with other studies. Although some of the P:B values obtained in the present study were fairly high, the observed range of P:B from 5.21 to 14.14 generally agrees with macrobenthic P:B values obtained in other estuarine studies (reviewed in Schwinghamer et al., 1986; Maurer et al., 1992; Edgar et al., 1994; Smith and Kukert, 1996). Perhaps a more realistic way to scale P:B data would be by calculating faunal turnover time in terms of days, as represented by 365/(P:B) (Schwinghamer et al., 1986). The production values obtained in the case study were also reasonable when compared with estimates of secondary production from other estuarine studies using various methods (reviewed in Edgar, 1990; Maurer et al., 1992; Edgar et al., 1994; Tumbiolo and Downing, 1994; Smith and Kukert, 1996).

Additional process-indicator information in addition to that used in the GBNERR case study might be useful for assessing estuarine condition. For example, slopes of standardized biomass-size spectra relative to a value of –1, or scale regularity, convey whether biomass is increasing or decreasing with body size (Ramussen, 1993). Furthermore, variability in slopes of biomass-size spectra may reflect shifts in primary productivity. The presence and positions of modes within biomass-size spectra may also be diagnostic of ecological condition. Indeed, both slopes and modes of biomass-size spectra may reflect the trophic organization of the underlying community (Borgmann, 1987; Boudreau et al., 1991; Rasmussen, 1993). Other scaling relationships may also be applied to size structure data to obtain further insights into trophic function. For example, various studies have employed allometric relationships to examine metabolic rates for benthic communities from size-structured data (Gerlach et al., 1985; Schwinghamer et al., 1986; Morin and Nadon, 1991). Metabolic information would facilitate investigations of production efficiency and ecosystem production capacity. Finally, other valuable information not utilized in the case study should consider taxonomic diversity and trophic structure.

Relevant spatial and temporal scales for the macrobenthic process indicators still need to be resolved. Macrobenthic communities integrate ecosystem effects that are manifested on a temporal scale that is reflected by community turnover rates, which in the present study were inferred to range between 26 and 70 days, or from 1 to 2 months. The present study was intended to show landscape-level patterns through sampling of discrete areas separated by distances between 0.5 and 4.0 km. However, the effects of habitat heterogeneity on the macrobenthic indicators are still largely unknown. For example, although biomass size spectra appear to be conservative with respect to certain habitat features, such as substrate type; other habitat-specific differences in biomass size-spectra have been shown, such as between different types of submerged macrophytes (Rasmussen, 1993). Clearly, the degree to which macrobenthic indicators are sensitive to heterogeneity on various spatial scales needs to be closely examined.

Conclusions and Recommendations

Several noteworthy conclusions are evident from the GBNERR case study. Macrobenthic function was reflected through the combined use of three process indicators, and a composite variable incorporating three process indicators was significantly related to several environmental variables associated with trophic status, including pore water nutrient and surface chlorophyll concentrations. This implies that the macrobenthic process indicators do reflect ecosystem function.

Macrobenthic process indicators would reflect variability in ecosystem function, regardless of the cause. Thus, distinguishing anthropogenic effects from natural effects on such indicators presents a formidable challenge. However, anthropogenic effects can potentially be distinguished from natural causes of ecosystem stress by (1) defining baselines through general relationships; (2) relating macrobenthic indicator data from areas of known anthropogenic stress to designated reference areas; and (3) integrating of macrobenthic indicators with a suite of coherent biological indicators representing multiple levels of biological organization. For example, it was useful to define a general trend between intercepts and slopes of standardized biomass-size spectra for the GBNERR case study to identify sites that departed greatly from the trend. Further studies should determine how robust such trends are across estuarine systems. Boundary conditions as defined by known reference sites vs. known stressed sites should also help distinguish natural from altered areas. Finally, the importance of taking a multidisciplinary approach to the development and application of ecological indicators is evident from the need to distinguish anthropogenic from natural causes.

Acknowledgments

This research was supported by a grant from the U.S. Environmental Protection Agency's Science to Achieve Results (STAR) Estuarine and Great Lakes (EaGLe) program through funding to the Consortium for Estuarine Ecoindicator Research for the Gulf of Mexico (CEER-GOM), U.S. EPA Agreement

R-82945801-0. Although the research described in this chapter was funded by the U.S. Environmental Protection Agency, it has not been subjected to the agency's required peer and policy review and therefore does not necessarily reflect the views of the agency and no official endorsement should be inferred. Additional thanks are due to those individuals who helped to produce the GBNERR macrobenthic data set, including Jerry McLelland, Kathy VanderKooy, Deborah Vivian, Sara Turner, and Patsy Tussy. Sediment composition and TOC data were provided by the Geology and Environmental Chemistry Sections of the University of Southern Mississippi Gulf Coast Research Laboratory, and sediment nutrient data were provided by the Center for Environmental Diagnostics and Bioremediation (CEDB) of the University of West Florida. Finally, we thank the personnel of the Grand Bay National Estuarine Research Reserve for permission to work in the reserve.

References

Borgmann, U. 1987. Models on the slope of, and biomass flow up, the biomass size spectrum. *Canadian Journal of Fisheries and Aquatic Science* 44:136–140.

Boudreau, P. R., L. M. Dickie, and S. R. Kerr. 1991. Body-size spectra of production and biomass as system-level indicators of ecological dynamics. *Journal of Theoretical Biology* 152:329–339.

Brown, S. S., G. R. Gaston, C. F. Rakocinski, and R. W. Heard. 2000. Macrobenthic trophic structure responses to environmental factors and sediment contaminants in northern Gulf of Mexico estuaries. *Estuaries* 23:411–424.

Carpenter, S. R., N. F. Caraco, D. L. Correll, R. W. Howarth, A. N. Sharpley, and V.H. Smith. 1998. Nonpoint pollution of surface waters with phosphorus and nitrogen. *Ecological Applications* 8:559–568.

Cottingham, K. L. and S. R. Carpenter. 1998. Population, community, and ecosystem variates as ecological indicators: phytoplankton responses to whole-lake enrichment. *Ecological Applications* 8:508–530.

Duplisea, D. E. and A. Drgas. 1999. Sensitivity of a benthic, metazoan, biomass size spectrum to differences in sediment granulometry. *Marine Ecology Progress Series* 177:73–81.

Edgar, G. J. 1990. The use of size structure of benthic macrofaunal communities to estimate faunal biomass and secondary production. *Journal of Marine Biology and Ecology* 137:195–214.

Edgar, G. J., C. Shaw, G. F. Watson, and L. S. Hammond. 1994. Comparisons of species richness, size-structure and production of benthos in vegetated and unvegetated habitats in Western Port, Victoria. *Journal Marine Biology and Ecology* 176:201–226.

Folk, R. L. 1974. *Petrology of Sedimentary Rocks*. Hemphill, Austin, TX, 170 pp.

Gaston, G. R. and J. C. Nasci. 1988. Trophic structure of macrobenthic communities in the Calcasieu Estuary, Louisiana. *Estuaries* 11:201–221.

Gaston, G. R., C. Rakocinski, S. S. Brown, and C. M. Cleveland. 1998. Trophic structure in estuaries: response of macrobenthos to natural and contaminant gradients. *Marine and Freshwater Research* 49:833–846.

Gerlach, S. A., A. E. Hahn, and M. Schrage. 1985. Size spectra of benthic biomass and metabolism. *Marine Ecology Progress Series* 26:161–173.

Gonzalez-Oreja, J. A. and J. I. Saiz-Salinas. 1999. Loss of heterotrophic biomass structure in an extreme estuarine environment. *Estuarine, Coastal, and Shelf Science* 48(3):391–399.

Gray, J. S. 1981. *The Ecology of Marine Sediments*. Cambridge University Press, Cambridge, U.K., 185 pp.

Hansen, K. and E. Kristensen. 1997. Impact of macrofaunal recolonization on benthic metabolism and nutrient fluxes in a shallow marine sediment previously overgrown with macroalgal mats. *Estuarine, Coastal, and Shelf Science* 45:613–628.

Hanson, J. M. 1990. Macroinvertebrate size-distributions of two contrasting freshwater macrophyte communities. *Freshwater Biology* 24:481–491.

Hellawell, J. M. and R. Abel. 1971. A rapid volumetric method for the analysis of the food of fishes. *Journal of Fish Biology* 3:29–37.

Kerr, S. R. and L. M. Dickie. 2001. *The Biomass Spectrum: A Predator-Prey Theory of Aquatic Production.* Columbia University Press, New York, 320 pp.

Maurer, D., S. Howe, and W. Leathem. 1992. Secondary production of macrobenthic invertebrates from Delaware Bay and coastal waters. *Internationale Revue der gesamten Hydrobiologie* 77:187–201.

Mississippi State University–Coastal Research and Extension Center. 2000. Fecal Coliform TMDL for Bayou Cumbest/Bangs Lake Watershed Coastal Streams Basin, Jackson County, Mississippi. Mississippi Department of Environmental Quality, Office of Pollution Control, TMDL/WLA Section of the Water Quality Assessment Branch, Jackson, MS.

Morin, A. and D. Nadon. 1991. Size distribution of epilithic lotic invertebrates and implications for community metabolism. *Journal of the North American Benthological Society* 10:300–308.

Parry, D. M., M. A. Kendall, A. A. Rowden, and S. Widdicombe. 1999. Species body size distribution patterns of marine benthic macrofauna assemblages from contrasting sediment types. *Journal of the Marine Biological Association of the United Kingdom* 79:793–801.

Pearson, T. H. and R. Rosenberg. 1978. Macrobenthic succession in relation to organic enrichment and pollution of the marine environment. *Oceanography and Marine Biology Annual Review* 16:229–311.

Plumb, R. H., Jr. 1981. Procedures for Handling and Chemical Analysis of Sediment and Water Samples. U.S. Environmental Protection Agency/U.S. Army Corps of Engineers Technical Committee on Criteria for Dredged and Fill Material, 478 pp.

Rakocinski, C. F., S. S. Brown, G. R. Gaston, R. W. Heard, W. W. Walker, and J. K. Summers. 1997. Macrobenthic responses to natural and contaminant-related gradients in northern Gulf of Mexico estuaries. *Ecological Applications* 7:1278–1298.

Rakocinski, C. F., S. S. Brown, G. R. Gaston, R. W. Heard, W. W. Walker, and J. K. Summers. 2000. Species-abundance-biomass responses to sediment chemical contamination. *Journal of Aquatic Ecosystem Stress and Recovery* 7:201–214.

Ramsay, P. M., S. D. Rundle, M. J. Attrill, M. G. Uttley, P. R. Williams, P. S. Elsmere, and A. Abada. 1997. A rapid method for estimating biomass size spectra of benthic metazoan communities. *Canadian Journal of Fisheries and Aquatic Science* 54:1716–1724.

Rasmussen. J. B. 1993. Patterns in the size structure of litoral zone macroinvertebrate communities. *Canadian Journal of Fisheries and Aquatic Science* 50:2192–2207.

Ricciardi, A. and E. Bourget. 1998. Weight-to-weight conversion factors for marine benthic macroinvertebrates. *Marine Ecology Progress Series* 163:245–251.

Schwinghamer, P. 1988. Influence of pollution along a natural gradient and in a mesocosm experiment on biomass-size spectra of benthic communities. *Marine Ecology Progress Series* 46:199–206.

Schwinghamer, P., B. Hargrave, D. Peer, and C. M. Hawkins. 1986. Partitioning of production and respiration among size groups of organisms in an intertidal benthic community. *Marine Ecology Progress Series* 31:131–142.

Smith, C. R. and H. Kukert. 1996. Macrobenthic community structure, secondary production, and rates of bioturbation and sedimentation at the Knāne'ohe Bay Lagoon Floor. *Pacific Science* 50:211–229.

Sprules, W. G. and M. Munawar. 1986. Plankton size spectra in relation to system productivity, size, and perturbation. *Canadian Journal of Fisheries and Aquatic Science* 43:1789–1794.

Strayer, D. 1986. The size structure of a lacustrine zoobenthic community. *Oecologia* 69:513–516.

Summers, J. K., J. M. Macauley, and P. T. Heitmuller. 1991. Implementation Plan for Monitoring the Estuarine Waters of the Louisianian Province–1991. EPA/600/05-91-228. U.S. Environmental Protection Agency, Environmental Research Laboratory, Gulf Breeze, FL.

Tumbiolo, M. L. and J. A. Downing. 1994. An empirical model for the prediction of secondary production in marine benthic invertebrate populations. *Marine Ecology Progress Series* 114:165–174.

Turner, R. E. and N. N. Rabalais 1999. Suspended particulate and dissolved nutrient loadings to Gulf of Mexico estuaries. In *Biogeochemistry of Gulf of Mexico Estuaries,* T. S. Bianchi, J. R. Pennock, and R. R. Twilley (eds.). John Wiley & Sons, New York, pp. 89–107.

Twilley, R. R., J. Cowan, T. Miller-Way, P. A. Montagna, and B. Mortaavi. 1999. Benthic nutrient fluxes in selected estuaries in the Gulf of Mexico. In *Biogeochemistry of Gulf of Mexico Estuaries,* T. S. Bianchi, J. R. Pennock, and R. R. Twilley (eds.). John Wiley & Sons, New York, pp. 163–209.

U.S. EPA. 1999. The Ecological Condition of Estuaries in the Gulf of Mexico. EPA 620-R-98-004. U.S. Environmental Protection Agency, Office of Research and Development, National Health and Environmental Effects Research Laboratory, Gulf Ecology Division, Gulf Breeze, FL, 71 pp.

Warwick, R. R. 1984. Species size distributions in marine benthic communities. *Oecologia* 61:32–41.

Warwick, R. R. 1993. Environmental impact studies on marine communities: pragmatical considerations. *Australian Journal of Ecology* 18:63–80.

Wieser, W. 1960. Benthic studies in Buzzards Bay. II. The meiofauna. *Limnology and Oceanography* 5:121–137.

Xu, F.-L., S. E. Jørgensen, and S. Tao. 1999. Ecological indicators for assessing freshwater ecosystem health. *Ecological Modelling* 116:77–106.

21

Using Macroinvertebrates to Document the Effects of a Storm Water–Induced Nutrient Gradient on a Subtropical Estuary

Gregory Graves, Mark Thompson, Gitta Schmitt, Dana Fike, Carrie Kelly, and Jillian Tyrrell

CONTENTS

Introduction

Biscayne Bay is a subtropical estuary on the southeastern coast of Florida, located in Miami-Dade and Monroe Counties, extending from Dumfounding Bay in the north to Barnes Sound in the south (about 88 km). Depths are shallow, generally ranging between 0.5 and 3 m. Much of the bay, including the entire southern portion, is classified as an Outstanding Florida Water (OFW) and is subject to Florida's most protective water quality regulations.

Historically, fresh water entered the system via overland flow from the Everglades through bay-fringing estuarine coastal wetlands and through artesian upwelling in amounts sufficient to create freshwater "boils" in Biscayne Bay. The water quality in the late 1800s was low in nutrients, low in turbidity, and

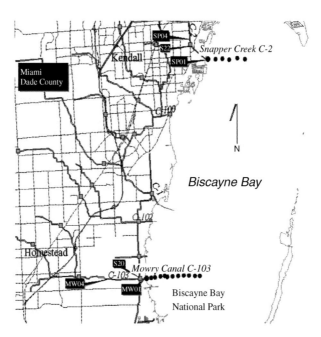

FIGURE 21.1 Lower Biscayne Bay showing the location of sampling sites (dots), canals, and salinity structures (squares).

high in light transmittance — conditions that promoted luxuriant seagrass meadows (U.S. Army Corps of Engineers, or USACOE, and South Florida Water Management District, or SFWMD, 1999). The bay was in a relatively natural condition when the City of Miami was founded in 1896. As development of the watershed progressed, canals replaced creeks and wetlands. These canals are managed to provide drainage and flood protection, to provide irrigation supply, and to prevent aquifer saltwater intrusion. Freshwater flow is now dominated by pulse-released direct canal discharges, resulting in a dramatic change in the timing, distribution, quantity, and quality of freshwater inflows in comparison to historical conditions (USACOE and SFWMD, 2002).

The Biscayne Bay watershed is located next to the most densely populated area in Florida and contains a mix of urban and agricultural lands underlain by the unconfined surficial Biscayne Aquifer. Despite being situated in this densely populated area, most of southern Biscayne Bay remains a viable ecosystem and contains the 700 km² Biscayne National Park/Aquatic Preserve. Southern Biscayne Bay benthic habitats include lush seagrass beds and large areas of hard bottom and corals supporting a diversity of marine life. However, habitat along the western shore of the bay has been affected by lowered water tables and the conversion of free-flowing streams to controlled canal discharges (USACOE and SFWMD, 2002).

SFWMD Canal C-103 (Mowry Canal) discharges directly into Biscayne National Park through SFWMD control structure S-20F, located 900 m north of the Biscayne National Park headquarters (Figure 21.1). The S-20F structure is located about 650 m upstream from the mouth of C-103 at the bay. Major land uses within the 105 km² watershed are agriculture (50%) and urban (39%) (1995 SFWMD land use cover). SFWMD Canal C-102 (Princeton Canal) discharges into the bay through control structure S-21A about 4 km north of the mouth of C-103. The S-21A control structure is located about 690 m upstream of its mouth at the bay. The canal drains a 66 km² watershed composed primarily of agriculture (60%) and urban (25%) land uses. SFWMD Canal C-2 (Snapper Creek Canal) is located in eastern Miami-Dade County in the cities of Sweetwater and Kendall (Figure 21.1) and discharges to Biscayne Bay through salinity control structure S-22. This control structure is located approximately 500 m upstream of Biscayne Bay. The canal drains a 120 km² basin composed primarily of urban development (80%) with a small proportion being agriculture (5%).

Predominant soil types in the C-103 and C-102 have high permeability and low water-holding capacity. These soil conditions result in low use efficiency of fertilizers and potential leaching of nutrients into groundwater (Li, 2001; Li and Zhang, 2002). Nitrate nitrogen is highly soluble, mobile, and perhaps the most widespread contaminant of groundwater in the world (Hallberg and Keeney, 1993). Studies have documented the relationship between groundwater nitrate concentrations and the percentage of land area in fertilized crops (Embleton, 1986). Nitrate in groundwater can also originate in urban areas where overwatering and overfertilization may occur (Morton et al., 1988).

In the continental United States, South Florida's coastal waters are the only place where hermatypic (reef-forming) coral communities are found (Lapointe et al., 2002). These communities are highly diverse and thrive in clear, warm, nutrient-poor tropical water. A widespread decline in coral cover and in abundances of characteristic reef fish in the Florida Reef Tract has been attributed to cumulative land-based nutrient discharges from upland watersheds (Lapointe et al., 2002). It is generally understood that periodic loading of carbon, nitrogen, and phosphorus from land runoff coupled with a general shallow geomorphology drives and limits estuarine productivity. Changes in primary production induced by increased nutrient loading can result in changes in species abundance, distribution, and diversity, leading to alteration of the estuarine community structure (Livingston, 2001). Nutrient enrichment can adversely affect seagrass beds by stimulating growth of epiphytic and planktonic algae that inhibit transmittance of light (Duarte, 1995; Valiela et al., 1997). Changes in phytoplankton and epiphytic algal growth rates also affect the type and availability of food for each level of trophic inhabitant within the estuarine ecosystem.

In recent years several studies have been initiated to assess perceived impacts of canal and groundwater discharges to southern Biscayne Bay (Szmant, 1987; Byrne and Meeder, 1999; Lietz, 1999; Langevin, 2000). Findings suggest nitrogen-enriched groundwater enters Biscayne Bay through canals and underwater springs. A Biscayne Bay National Park research management report documented a change in seagrass community composition near the mouth of C-102 and C-103 and found lower salinities and higher water column nutrient concentrations in the affected areas (Szmant, 1987). It was speculated that seagrass communities were affected by C-102 and C-103 Canal discharges.

Despite the input of large amounts of nitrogen from freshwater discharges, the bay has low water column chlorophyll concentrations (Brand, 2002). Typically, nitrogen is the principal limiting nutrient to primary production in estuaries (Kennish, 1999); however, in Biscayne Bay, as well as in many other tropical carbonate shallow waters, phosphorus is chemically scavenged from the water by calcium carbonate and becomes the limiting nutrient (Brand, 1988). Relatively elevated chlorophyll concentrations in southern Biscayne Bay have been related to discharges from canals (including C-103 and C-102) in association with dry-season phosphorus buildup in soils and groundwater flushing into the bay at the onset of the rainy season (Brand, 1988, 2002; Brand et al., 1991).

Benthic fauna are primarily sedentary and have limited escape mechanisms to avoid disturbances (Bilyard, 1987). Measuring changes in species composition and abundance can provide a record of short- and long-term environmental changes and perturbations. Benthic macroinvertebrates have been used extensively as indicators of impacts of both pollution and natural fluctuations in estuarine environments (Reish, 1986; Bilyard, 1987; Holland et al., 1987). Shifts in primary productivity from macrophytic dominance toward phytoplankton and epiphytic algae will be reflected in changes in benthic macroinvertebrate community functional feeding group composition, i.e., higher percentage scrapers and grazers (Duarte, 1995). Also, changes in substrate characteristics resulting from greater organic deposition caused by nutrient enrichment and increased decomposition activity will affect benthic community structure by changing environmental conditions, e.g., finer particles, lower dissolved oxygen (DO), etc. (Gray, 1981). Utilization of macroinvertebrate community data in environmental investigations relies on the premise that similar ecological conditions give rise to similar species assemblages, and that sites that share common macroinvertebrate abundance characteristics are assumed to share similar water quality and ecological conditions (Clarke and Ainsworth, 1993). The primary objective of this study is to characterize and attribute differences in benthic macroinvertebrate community structure to conditions resulting from episodic discharges of nutrient-laden urban and agricultural storm water from the SFWMD C-103 Canal into Biscayne National Park and to extend those findings to the C-102 Canal, where comparable conditions and impacts would be expected.

Methods

General

Water quality and flow conditions were evaluated using 10 years of monthly SFWMD data from monitoring sites in Biscayne Bay and tributary canals. Differences between C-103 and C-2 basin water quality were examined by comparing data from stations upstream of the control structures (MW04 and SP04, respectively) with data from stations at the mouths of each canal (MW01 and SP01, respectively). Monitoring stations MW01 and SP01 are located 650 and 500 m, respectively, downstream (east) of the C-103 and C-2 Canal flow control structures.

Two approaches were employed to evaluate potential impacts associated with canal discharges. The first was to evaluate the benthic macroinvertebrate community structure along a transect extending west–east away from the C-103 Canal discharge. Benthic macroinvertebrates would be exposed to a gradient of decreasing nutrient concentration moving eastward along this transect. The second approach was to evaluate differences between benthic communities exposed to similar freshwater inputs, but at much lower nitrogen loading rates, i.e., along a similar east–west transect originating near the mouth of the C-2 Canal. Sample sites were chosen along each 4000-m transect using a stratified random design. Three strata were defined within each transect: nearshore, intermediate, and far (offshore sites).

At each station, benthic macroinvertebrate samples were collected using a Young-modified Van Veen sampler (0.04 m²). Material obtained by the grab was washed through a 0.5-mm mesh sieve and the retained material was preserved with 10% buffered Formalin containing Rose Bengal stain (FDEP, 2000a). The fauna contained within each sample was separated from the sediment in the laboratory (FDEP, 1999a, 2000b) and stored in 80% ethanol (FDEP, 1999b). Identifications were conducted to the lowest practical taxonomic level (FDEP, 2003) using standard quality assurance procedures (FDEP, 2000c).

At each sample site, DO, pH, temperature, and conductivity were measured using a YSI model 6920 datasonde (FDEP, 2002a). Estimates of submerged aquatic vegetation (SAV) and macroalgae coverage were obtained using Braun Blanquet techniques (FDEP, 1995) with a 1 m², 10×10-cm subunits randomly deployed three times at each sampling site. Sediment samples were obtained at each site for the purpose of estimating organic content based on a modified loss-on-ignition (LOI) procedure, which corrects for error caused by salt content in pore water (Fishman and Friedman, 1989; R.L. Miller, USGS, pers. commun.).

Because a great amount of inorganic nitrogen is delivered via C-103 and C-102 to a phosphorus-limited environment, inorganic nitrogen was assumed to be conservative in the study area. To test this assumption, during a release from the C-103 structure on 29 July 2003, conductivity measurements in the discharge plume were used to collect 14 nutrient samples at locations within a spectrum of fresh to marine conditions to verify a linear relationship between inorganic nitrogen concentration and conductivity (Figure 21.2). A two-dimensional hydrodynamic salinity model of Biscayne Bay (G. Brown, USACOE, pers. commun.) was used to develop a salinity/distance curve based upon a 5950 L/s discharge from the C-103 Canal. The 10-year (1991–2000) mean daily flows from the C-103 (6230 L/s) and C-2 (5240 L/s) Canals were both comparable to this discharge rate. Using the linear relationship between salinity and nitrate/nitrite concentrations, estimates of inorganic nitrogen and total phosphorus concentration exposures at each site were determined by inserting 10-year mean nutrient values from monitoring stations MW01 and SP01 into the model and developing concentration–distance relationships (Figure 21.3).

Benthic Community Differences

Species-abundance data used in these analyses were reduced to include only those species that comprise at least 3% of any one sample to prevent spurious domination of patterns by a large number of rare species as suggested by Field et al. (1982). Bray–Curtis similarity coefficient matrices were developed with square-root-transformed species abundance data, since square-root-transformed data produced the lowest stress values in two-dimensional, non-metric, multidimensional scaling (MDS) ordinations. MDS is a multivariate procedure that maps objects so that their relative positions in reduced space reflect the degree of similarity between the objects (Hair et al., 1992; Clarke and Warwick, 1994). The

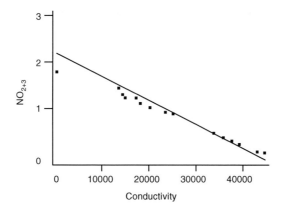

FIGURE 21.2 Linear regression results for NO_{2+3} vs. conductivity for samples taken during a storm event discharge from C-103 in July.

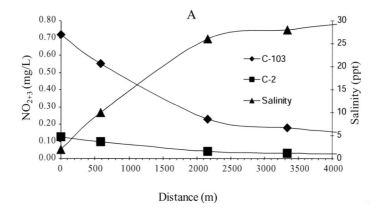

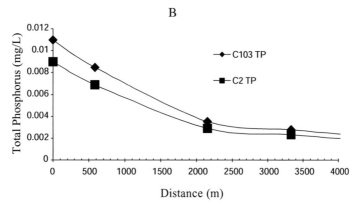

FIGURE 21.3 Curves used to estimate nutrient concentrations at sample sites. These curves were developed using the salinity vs. distance results from the Biscayne Bay two-dimensional hydrodynamic model used by the USACOE. A linear relation was verified between nitrate-nitrite concentrations and salinity. Mean nutrient concentrations were established at the mouth of each canal using a 10-year period of SFWMD data. Nutrient concentrations were determined by using the same rate of decay as salinity vs. distance. (A) Curve developed showing salinity vs. distance from shore and showing estimates of nitrate-nitrite concentration vs. distance from shore. (B) Curve developed that shows total phosphorus concentration vs. distance from shore.

PRIMER® Analysis of Similarity (ANOSIM) program was utilized to detect differences in similarity matrices between benthic communities among the strata. Hierarchical clustering analysis was used on Bray–Curtis similarity coefficient matrices to further strengthen the validity of observed groupings of similar samples as recommended by Clarke and Warwick (1994).

Characterization of Community Difference

The FDEP Biological Database software program was used to determine species abundance matrices, diversity indices, and functional feeding group proportions. Mean species sensitivity indices were calculated for each station utilizing the guidelines developed by Farrell (1992). Geometric abundance plots were produced using PRIMER software and were evaluated to determine which samples had a majority of their abundance concentrated within a relatively few taxa.

Differences in Environmental Factors

Comparisons of nutrient concentrations and flow rates among sites were made utilizing paired t-tests. Relationships between different types of measurements (e.g., concentration and loads) were evaluated using Spearman rank correlation. Minitab® (Minitab, Inc.) software was used to perform most statistical analyses on water quality data and biological indices. Differences among biological descriptors were evaluated using pairwise Wilcoxon's signed rank and Mann–Whitney tests. Unless otherwise noted, the level of significance for all tests was $\alpha = 0.05$.

Correlation of Degraded Community Structure to Environmental Factors

Principal components analysis (PCA) was used to find groupings of similar sites based on environmental characteristics. Six environmental variables were investigated using PCA, namely, total phosphorus and nitrate-nitrite concentrations, total volatile solids, depth, seagrass coverage, and conductivity. The PRIMER BIO-ENV procedure (Carr, 1996) was subsequently used to identify water quality factors that had the highest weighted Spearman rank correlation with the biotic similarity matrix using the pooled data set. To evaluate any change in correlation of abiotic to biological data within transects, the C-103 and C-2 data were separately analyzed using BIO-ENV.

Results

General

Based on a 10-year (1991–2000) mean, concentrations of nitrate-nitrite (paired t-test, $t = 16.2$, $P < 0.001$) and inorganic nitrogen (paired t-test, $t = 15.7$, $P < 0.001$) were significantly higher at the C-103 station upstream of structure S-20F (MW04) than at the station located downstream of the structure at the canal mouth (MW01) (Table 21.1). Conversely, total phosphorus (paired t-test, $t = 2.6$, $P = 0.013$) and ammonia (paired t-test, $t = 5.9$, $P < 0.001$) were lower upstream at MW04 than concentrations downstream at station MW01. Based on a mean annual flow of 193.7 million m^3/year, the estimated total phosphorus and inorganic nitrogen loads from C-103 were 1,590 and 446,000 kg/year, respectively (Table 21.2). More than 98% of the inorganic load was in the nitrate-nitrite form, and the average nitrogen-to-phosphorus (N:P) ratio within the C-103 Canal was typically greater than 600:1. Concentrations of total phosphorus (Spearman $r_s = 0.344$, $P = 0.003$) and inorganic nitrogen (Spearman $r_s = 0.471$, $P < 0.001$) at the mouth of C-103 (MW01) were positively correlated to the inorganic nitrogen load calculated at the discharge structure S-20F. Primary productivity in the bay at MW01 was phosphorus limited given the calculated mean N:P ratio of 144:1. Episodes of higher inorganic nitrogen concentrations observed in the bay were a result of increases in nitrate-nitrite concentration in C-103 since the ammonia nitrogen concentration was unaffected by flow (Figure 21.4). The mean N:P ratio for the nearby C-102 Canal was greater than 1000:1, and the inorganic nitrogen load was similar to that of C-103 (Table 21.2).

TABLE 21.1

Ten-Year (1991–2000) Mean Monitoring Values for Specific Parameters Evaluated in This Study

Station	DO (mg/L)	Ammonia (mg/L)	Nitrate–Nitrite (mg/L)	Total Phosphorus (mg/L)
C103 u/s S20F (MW04)	6.5	0.030	2.30	0.007
C103 d/s S20F (MW01)	5.4	0.057	0.66	0.011
C102 u/s S21A (PR03)	5.5	0.045	3.99	0.007
C102 u/s S21A (PR01)	5.4	0.140	1.52	0.011
C2 u/s S22 (SP04)	4.0	0.085	0.15	0.010
C2 d/s S22 (SP01)	4.2	0.071	0.06	0.009
Biscayne Bay (BB41)	6.4	0.061	0.04	0.002

Note: Station identifiers are in parentheses. Stations are located upstream (u/s) and downstream (d/s) of control structures.

TABLE 21.2

Ten-Year (1991–2000) Mean Discharge Volumes, Nutrient Loads, and Concentrations of Select Canals Discharging into Southern Biscayne Bay

Canal	Flow (m³/yr)	Inorganic Nitrogen (kg/yr)	Inorganic Nitrogen (mg/L)	Total Phosphorus (kg/yr)	Total Phosphorus (mg/L)
C-103	193.7	446,000	2.300	1359	0.007
C-102	105.9	424,219	3.990	700	0.007
C-1	173.7	60,103	0.345	1254	0.007
C-2	165.3	43,297	0.261	1112	0.009
Military	20.3	14,148	0.694	227	0.011
C-100	39.3	6,153	0.156	242	0.006
C-3	14.2	8,247	0.580	693	0.049

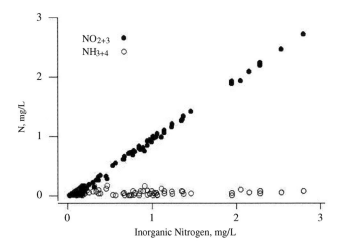

FIGURE 21.4 Comparison of nitrate–nitrite component of inorganic nitrogen load to the ammonia nitrogen load. Note that the ammonia nitrogen remains constant and the nitrate–nitrite component accounts for the entire change in concentration.

Monitoring station BB41 is located in Biscayne Bay 5500 m east of the mouth of the C-103 Canal. Inorganic nitrogen concentrations at BB41 were significantly correlated (Spearman $r_s = 0.294$, $P = 0.017$) to loads discharged from the C-103 Canal when the concentration data was lagged by 1 month. The mean N:P ratio at BB41 was 111:1, indicating primary productivity was phosphorus limited.

Monitoring station SP04 is located on the C-2 Canal upstream of the S-22 structure and station SP01 is located downstream of the structure at the mouth of the canal. Based on 10-year mean values (1991–2000), the concentration of nitrate-nitrite at station SP01 was significantly lower (paired t-test, $t = 9.5$, $P < 0.001$) than station SP04, while total phosphorus and ammonia were not significantly different (Table 21.1). Based on a mean flow of 165.3 million m^3/year, the estimated total phosphorus and inorganic nitrogen loads from C-2 were 2,229 and 43,297 kg/year, respectively (Table 21.2), the latter being about one tenth that from C-103. The mean N:P ratios at both C-2 stations indicate a phosphorus-limited environment (SP04 = 43:1, SP01 = 31:1).

Differences in Environmental Factors

Inorganic nitrogen concentrations (paired t-test, $t = 7.44$, $P < 0.001$) at the mouth of the C-2 Canal were significantly lower than those observed at the mouth of C-103, while total phosphorus and ammonia concentrations were not significantly different.

Seagrass coverage at sample sites ranged from zero to 98% (Table 21.3). Generally, *Thallasia testudinum* was the dominant seagrass, except at sites near the C-103 and C-102 Canal mouths where *Halodule wrightii* occurred in significant quantities. Water depths at C-2 sites were generally deeper than C-103 sites (Table 21.3).

As a group, samples in the C-103 had significantly higher percent volatile solids than the C-2 strata (pairwise Wilcoxon's signed rank test, $P = 0.0156$). The nearshore C-103 stratum had significantly higher volatile solids than strata farther offshore (Mann–Whitney test, $P < 0.03$). Percent volatile solids correlated inversely with increasing distance from the C-103 Canal mouth (Spearman $r_s = 0.656$, $P = 0.021$). No association could be found between percent volatile solids and increasing distance along the C-2 transect.

Benthic Community Differences

Hierarchical clustering showed samples nearest the mouth of C-103 to be closely related to one another and distinct from all other samples in community composition (Figure 21.5). The two-dimensional MDS ordination (stress = 0.18) also inferred that sites nearest the mouth of C-103 were different from the remaining sites (Figure 21.6). ANOSIM tests of Bray–Curtis similarity matrices found significant differences in community structure between the C-103 nearshore and both the C-103 intermediate and offshore strata. Significant differences were also found between C-103 nearshore and all of the C-2 strata. All other comparisons between strata produced no significant differences.

Characterization of Community Difference

Taxonomic diversity (Shannon index) was significantly greater at C-2 sites when compared to the C-103 site data (pairwise Wilcoxon's signed rank test, $P = 0.016$). Farrell's species sensitivity scores were marginally greater at the C-2 sites than at the C-103 sites (pairwise Wilcoxon's signed rank test, $P = 0.10$). No significant difference was seen between taxa density for the April C-103 samples compared to the July C-103 samples (pairwise Wilcoxon's signed rank, $P = 0.395$). Taxa densities were marginally higher at the C-2 stations when compared to the pooled C-103 stations (pairwise Wilcoxon's signed rank test, $P = 0.10$). Plots of geometric abundances showed that the nearshore C-103 stratum had more individuals concentrated within fewer taxa than C-103 strata farther offshore or any C-2 strata (Figure 21.7). Evaluating community composition based on functional feeding group, the nearshore C-103 sample sites had significantly greater percentages of grazers, scrapers, and scavengers than the intermediate and offshore C-103 strata, or any of the C-2 strata (Mann–Whitney test, $P < 0.05$).

TABLE 21.3

Environmental Data Used in Comparisons and Correlations between Sample Sites

Sample Site	Conductivity (µS/cm)	Percent Seagrass Coverage	Depth (m)	Percent Volatile Solids	Inorganic Nitrogen (mg/L)	Total Phosphorus (mg/L)
C103-1	23,942	8	1.2	5.77	0.95	0.007
C103-2	24,951	91	1.2	4.64	0.90	0.007
C103-3	25,612	60	1.2	3.20	0.86	0.007
C103-4	35,728	5	1.2	1.63	0.34	0.006
C103-5	37,029	1	1.6	2.18	0.27	0.006
C103-6	38,252	0	1.4	2.24	0.21	0.005
C103-7	39,476	0	1.4	1.54	0.15	0.004
C103-8	41,922	0	1.9	2.42	0.02	0.003
C103-9	42,447	2	1.6	1.38	0	0.003
C103-10	42,447	1	1.6	2.28	0	0.003
C103-11	42,447	5	1.8	1.91	0	0.003
C103-12	42,447	35	2.2	2.40	0	0.003
C103-2.1	23,942	85	1.2	3.89	0.95	0.017
C103-2.2	24,951	58	1.2	5.16	0.90	0.008
C103-2.3	25,612	48	1.2	2.03	0.86	0.007
C103-2.4	35,728	13	1.2	1.46	0.34	0.006
C103-2.5	37,029	28	1.6	1.20	0.27	0.005
C103-2.6	38,252	1	1.4	1.89	0.21	0.004
C103-2.7	39,476	0	1.4	2.21	0.15	0.004
C103-2.8	41,922	0	1.9	2.13	0.03	0.003
C103-2.9	42,447	55	1.6	2.02	0	0.003
C103-2.10	42,447	55	1.6	1.63	0	0.003
C103-2.11	42,447	27	1.8	1.76	0	0.003
C103-2.12	42,447	15	2.2	2.45	0	0.003
C2-1	23,942	10	2.1	0.10	0.06	0.009
C2-3	25,612	27	2.3	0.43	0.06	0.006
C2-5	37,029	97	3.2	0.22	0.02	0.005
C2-7	39,476	50	3.5	0.90	0	0.003
C2-9	42,447	98	3.6	0	0	0.003
C2-11	42,447	89	3.6	0.98	0	0.002

Note: C-103 samples 1 to 12 collected April 2003 and samples 2.1 to 2.12 collected July 2003. C-2 samples collected July 2003.

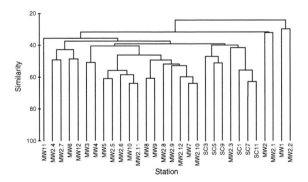

FIGURE 21.5 Cluster analysis results showing the distinct grouping of the C-103 nearshore samples (MW1, MW2, MW2.1, MW2.2). Samples labeled MW refer to sites offshore of C-103 Canal. Samples labeled SC refer to sites offshore of C-2 Canal. Samples farther offshore were assigned a progressively larger numerical site number.

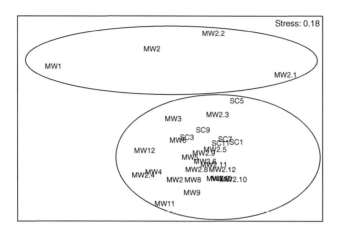

FIGURE 21.6 Results of MDS showing apparent grouping of nearshore C-103 samples (MW1, MW2, MW2.1, MW2.2) separately from the remainder of samples. The groupings are based upon Bray–Curtis similarity calculations for each sample site. Samples labeled MW refer to sites offshore of C-103 Canal. Samples labeled SC refer to sites offshore of C-2 Canal. Samples farther offshore were assigned a progressively larger numerical site number.

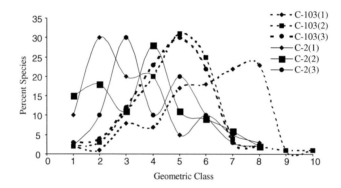

FIGURE 21.7 Geometric abundance plot for each stratum showing the percentage of species within the sample that had specific abundance of individuals. Geometric classes are proportioned as: class 1 = 1 individual, class 2 = 2–3 individuals, class 3 = 4–7 individuals, class 4 = 8–15 individuals, etc. The nearshore stratum for C-103 is designated C-103(1), the intermediate stratum is designated C-103(2), and offshore stratum is designated C-103(3). The plot shows that the nearshore C-103 stratum shows a greater abundance of individuals concentrated within fewer species (skewed toward right of graph).

Correlation of Degraded Community Structure to Environmental Factors

A PCA ordination of environmental data showed samples were similar according to transect, i.e., C-103 vs. C-2, and distance from shore, i.e., nearshore vs. offshore (Figure 21.8). BIO-ENV identified percent volatile solids, total phosphorus, and nitrate-nitrite concentration as the subset of environmental data that best correlated with the ordination of benthic community data (Table 21.4). The BIO-ENV analysis further indicated that the addition of conductivity, water depth, and/or seagrass coverage only served to weaken the correlation between environmental variables and biological data groupings.

Discussion

General

Several of Florida's estuarine systems are classified as potentially impaired due to nutrient enrichment (FDEP, 2002b) and exhibit well-documented adverse impacts to oyster populations, seagrass beds, water

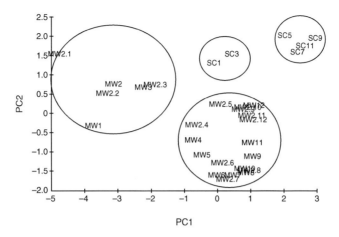

FIGURE 21.8 PCA performed on six environmental variables for all sample sites produced these groupings of samples. Samples labeled MW refer to sites offshore of C-103 Canal. Samples labeled SC refer to sites offshore of C-2 Canal. Samples farther offshore were assigned a progressively larger numerical site number.

TABLE 21.4

Results of the Multivariate Spearman Rank Correlation of Environmental Data to Benthic Community Data Using BIO-ENV Showing the Ten Best Combinations of Variables

Rank	Environmental Variables Correlated	Correlation
1 (best)	TP, % Volatile solids, IN	0.656
2	TP, % Volatile solids	0.647
3	TP, % Volatile solids, Conductivity	0.644
4	TP, % Volatile solids, IN, Conductivity	0.636
5	% Volatile solids, IN	0.623
6	% Volatile solids, IN, Conductivity	0.621
7	% Volatile solids, Conductivity	0.611
8	TP, % Seagrass, % VS, IN, Conductivity	0.595
9	TP, % Seagrass, % Volatile solids, Conductivity	0.588
10	TP, % Seagrass, % Volatile solids, IN	0.588

Note: The correlation value can be used only as a relative measure of correlation when comparing the effects of variable combinations. The match of environmental variables that best explain the separation of biological data is ranked first in this table. TN = total nitrogen; IN = inorganic nitrogen.

clarity, oxygen regime, species diversity, and sedimentation rates (Livingston, 2001; FDEP, 2002c). Because Biscayne Bay has a wide and direct connection to the Atlantic Ocean, tidal flushing might be assumed to reduce impacts from contaminants entering the bay via dilution and quick transport of contaminants away from sources of input; however, recent U.S. ACOE modeling results have shown that nearshore retention rates of canal inflows are on the order of several weeks (G. Brown, U.S. ACOE, pers. commun.).

The annual input of inorganic nitrogen to Biscayne Bay from the C-103 and C-102 Canals is disproportionately high compared to inputs from most other tributaries discharging into Florida estuaries (Table 21.5). Both C-103 and C-102 mean inorganic nitrogen concentrations were greater than the 95th percentile of mean concentrations for all streams in Florida (J. Hand, FDEP, pers. commun.). In the C-103 Canal, mean inorganic nitrogen concentration was 2.3 mg/L, which is 16 times greater than the median value for all streams in Florida. The mean concentration for the C-102 Canal was 4.0 mg/L, or 29 times greater than the median value for all streams in Florida. Loading rates from the C-2 Canal were typical (Table 21.5) of other tributaries in Florida. The mean inorganic nitrogen concentration was 0.3 mg/L.

TABLE 21.5

A Comparison of Inorganic Nitrogen Loadings from Major Tributaries Discharging
to Florida Estuaries

Tributary	Estuary	Annual Inorganic Nitrogen Loading (kg/yr)
Apalachicola	Apalachicola Bay	7,422,317
C-43	Caloosahatchee/San Carlos Bay	470,387
C-103	Biscayne Bay	446,000
C-102	Biscayne Bay	424,219
Eleven Mile Creek	Perdido Bay	372,000
Hillsborough River	Tampa Bay	325,868
Alafia River	Tampa Bay	313,167
Perdido River	Perdido Bay	237,250
Little Manatee River	Tampa Bay	140,851
C-44	St. Lucie Estuary	138,674
C-51	Lake Worth Lagoon	115,468
Deep Creek	St. Johns	89,000
Ortega River	St. Johns	84,700
C-2	Biscayne Bay	43,297
Black Creek	St. Johns	49,700
C-23	St. Lucie Estuary	41,451
Trout River	St. Johns	37,200
C-24	St. Lucie Estuary	35,375
C-25	Indian River Lagoon	34,313
Six Mile Creek	St. Johns	27,300
Peace River	Charlotte Harbor	23,293
Myakka River	Charlotte Harbor	10,394

Source: Based on FDEP Basin Status Reports and SFWMD data (1991–2000).

Because benthic macroinvertebrate community composition should reflect environmental impacts (Gibson et al., 2000), and high nutrient inputs (as found in the C-103 and C-102 Canals) can affect ecosystem function (Livingston, 2001), changes in invertebrate benthic community should be expected as a result of C-102 and C-103 discharges. Changes should be greatest near the mouths of the canals where nutrient concentrations are highest and should decrease with seaward distance from the mouths. In contrast, impacts associated with the C-2 Canal should reflect the much lower load of nutrients. However, detection of effects due to pollution stress may be confounded by salinity gradients around canal mouths (Livingston, 2001). Since the annual freshwater discharge from C-2 was comparable to that from C-103, and each of these canals discharges into areas with similar coastal morphologies, salinity regimes should also be comparable leaving nutrient exposure as a determinate variable.

The study aimed to resolve the following questions: Was there a measurable difference in benthic community structure between samples collected in the vicinity of the C-103 Canal and those collected in the vicinity of C-2 Canal or those farther away from the C-103 mouth? If there was a difference, was the difference unfavorable? If biological community differences were measurable, what environmental factors were significantly different between sites? If there were differences in environmental variables, what variables correlated to the unfavorable biotic differences? If correlated, could the relationships thus identified be interpreted within a justifiable cause-and-effect framework?

Benthic Community Differences

multidimensional scaling (MDS) and hierarchical clustering were used to illustrate differences in biological community structures. The degree to which the final plot agrees with the structure of the source data is given by the "stress" value, where a smaller value of stress indicates a better degree of fit (see Clarke and Warwick, 2001). When interpreting an MDS ordination, Clarke and Warwick (2001) recommended that a hierarchical cluster analysis be employed to validate the interpretation when stress is above a level of 0.1. Because the stress level of the MDS ordination in this study was above the

TABLE 21.6

A Comparison of C-103 and C-2 Canal Data Including Watershed Land Use and Soil Types

Attribute	C-103	C-2
Mean annual discharge (m³/yr)	193.7	165.3
Mean salinity at mouth (ppt)	27.2	28.1
Mean depth nearshore strata (m)	1.2	2.1
Mean inorganic nitrogen at mouth (mg/L)	0.719	0.127
Mean total phosphorus at mouth (mg/L)	0.011	0.009
Mean % volatile solids nearshore strata	3.47	0.265
Mean % seagrass coverage N/S strata	46	21
Mean Shannon index N/S strata	3.90	4.15
Mean Farrell sensitivity score N/S strata	2.19 (more tolerant)	3.05
Mean taxon density nearshore strata (taxa/m²)	860	1225
Mean % grazer species N/S strata	9.1	1.7
Mean % scraper species N/S strata	5.4	2.8
Land use	Agriculture 50%, urban 39%	Agriculture 5%, urban 80%
Soil type	Gravely loam, gravely marl, limestone	Urban fill, limestone

threshold level, hierarchical clustering was performed to support MDS results. Both of these analyses showed that samples within the nearshore stratum of C-103 were distinct compared to samples analyzed in the remainder of the strata (Figure 21.5 and Figure 21.6). ANOSIM showed significant differences in macroinvertebrate community structure between C-103 nearshore samples and C-103 intermediate and offshore samples. Significant biotic differences were also apparent between samples in the stratum near the mouth of the C-103 and all C-2 strata. No differences in community structure were apparent between nearshore C-2 strata and intermediate or offshore C-2 strata, or between offshore C-103 and any C-2 strata. These results indicate a difference exists between C-103 nearshore samples and all others, indicating that some unique environmental factor was influencing the nearshore C-103 faunal composition.

Characterization of Community Difference

Values of Shannon diversity indices were significantly less for the C-103 nearshore stratum compared to all C-2 sites (Table 21.6). In addition, taxonomic density was diminished in the nearshore C-103 stratum compared to C-103 intermediate and offshore strata. Pairwise comparisons of strata using Farrell's sensitivity scores showed that benthos animals in the C-103 samples were marginally more tolerant than in C-2 samples. Gray and Pearson (1982) recommend plotting the number of species in geometric abundance classes as a means of detecting effects of pollution stress on communities. In unpolluted situations, the mode of the plot is well to the left. In polluted situations there are fewer rare species and more abundant species so that the higher abundance classes shift the mode to the right. Plots of geometric abundance for all sample strata show that the mode of the C-103 nearshore stratum is skewed to the right of the modes of all other strata (Figure 21.7). The decrease in diversity and taxonomic density, the increase in tolerant species, and the concentration of abundance within a few taxonomic groups all indicate the C-103 nearshore stratum was degraded compared to other strata in the study.

Differences in Environmental Factors

A correlation of nitrogen and phosphorus concentrations in Biscayne Bay at the mouth of C-103 to upstream loading rates indicates canal discharges create measurable changes in water quality in nearshore areas. The correlation of inorganic nitrogen concentrations at the mid-bay station (BB41), located 5500 m east of the mouth of C-103, with inorganic nitrogen concentrations being discharged from the mouth of C-103, and presumably C-102, indicates that measurable exposure to nutrient loadings from these canals is far reaching. Inorganic nitrogen concentrations and percent sediment organics were found to be significantly greater in the nearshore stratum of C-103 compared to nearshore C-2 and to C-103 strata farther offshore, while no significant difference was found comparing total phosphorus, depth, seagrass coverage, and ammonia nitrogen. Differences in land use types appear to explain the observed differences in inorganic nitrogen concentrations between the C-103 and C-2 basins (Table 21.6). Land uses in the C-103 basin are primarily agriculture atop a thin layer of permeable soil. In contrast, land use in the C-2 basin is primarily urban atop a shallow, moderately permeable fill. The predominance of agricultural lands in the C-103 basin is the probable source of the much greater concentrations of nitrate-nitrites than found in the C-2 basin. Byrne and Meeder (1999) found that the concentration of inorganic nitrogen in these canals was similar to that present in the basin's groundwater and storm water runoff.

Small changes in nutrient loading may have great impacts on primary production in a nutrient-limited environment (Brand, 2002). Higher chlorophyll concentrations near the mouths of C-103 and C-102 Canals suggest that although the entire western coast of Biscayne Bay may be phosphorus limited, sources of additional nitrogen contribute to increased primary production. In fact, the propagation of some plant and plankton species may be encouraged by higher N:P ratios (L. Brand, University of Miami, pers. commun.). Phosphorus undergoes no oxidation-reduction transformation during cycling, while nitrogen undergoes complex transformations involving several oxidation states; thus, the rate of phosphorus recycling can be orders of magnitude faster than that of nitrogen (Kadlec and Knight, 1996). It follows that a large pool of readily available inorganic nitrogen such as provided by C-103 and C-102 discharges would exacerbate conditions otherwise limited by phosphorus availability.

The higher percentage of volatile solids in the C-103 nearshore stratum compared to intermediate and offshore strata, and to all C-2 samples, is attributed to increased phytoplankton production in the nutrient-enriched plume near the mouth of the C-103 Canal. A similar effect could be assumed present adjacent to the C-102 Canal. Previous studies have shown that planktonic activity is correlated to the organic content of sediments (Gray, 1981), whereby increases in phytoplankton, epiphyton, and/or epibenthic production result in a corresponding increase in the rate of accumulation of organic matter. Brand (2002) detected high chlorophyll concentrations associated with discharges near the mouths of C-103/C-102 Canals, and Szmant (1987) found significantly greater amounts of epiphytic growth (calcareous algae, foraminifera, hydrozoans, etc.) on artificial grass blades deployed in areas affected by the C-103 and C-102 discharges.

Increased planktonic biomass reduces water clarity and can induce a shift in primary production from a seagrass-based system toward a phytoplankton-based system (Duarte, 1995). Szmant (1987) reported that historically lush beds of shade-sensitive *Thallasia testudinum* seagrass were being replaced by low-light-tolerant *Halodule wrightii* in the vicinity of the C-103 and C-102 discharges. Additionally, increased phytoplankton concentrations at nearshore C-103 sites could provide an expanded food source for grazers while increased epiphytic growth could support a greater abundance of scrapers. A significant increase in the proportion of each of these feeding groups was found in the nearshore C-103 stratum (Table 21.6). These findings provide corroborating evidence that C-103/C-102 discharges are causing nutrient-induced imbalances in the phytoplankton, epiphytic algae, and macrophyte communities, which in turn has altered the makeup of the benthic macroinvertebrate community. These changes are now evident in the vicinity of C-103 and C-102 Canals in Biscayne Bay.

Correlation of Degraded Community Structure to Environmental Factors

Environmental, hydrologic, and/or morphologic factors that could result in observable differences in biotic community structure include water depth, salinity, flow rates, circulation patterns, seagrass coverage, sediment depth and composition, sedimentation rate, and water quality. The BIO-ENV multivariate correlation procedure indicated nutrient concentrations and organic composition of the substrate best explained observed benthic community differences. When C-103 samples were analyzed separately from C-2 samples using BIO-ENV, the combination of percent volatile solids, nitrate-nitrite, and total phosphorus concentrations was again identified as best correlating the biotic (community structure) and abiotic (environmental) factors. This suggests that influences of nitrogen, phosphorus, and sediment organics were greater in the C-103 nearshore stratum compared to other strata in the study. BIO-ENV also indicated that salinity was a lesser influence on community differences than nutrient exposure and/or sediment organic content.

Conclusions

This study sought to attribute undesirable differences in macroinvertebrate community composition to elevated nutrient concentrations, and to correlate these elevated nutrient concentrations directly to canal discharges from the C-103, and by inference to discharges from the C-102 Canal, which had similar nutrient loading rates. The nearshore stratum of C-103 was exposed to significantly greater concentrations of nitrate-nitrite than C-2 samples or the C-103 samples farther offshore although both canals had comparable discharge rates (Table 21.6). Multivariate analyses including hierarchical clustering, multi-dimensional scaling, and analysis of similarity showed that significant differences in benthic community structure existed between the C-103 nearshore stratum and all other strata. Diversity indices, taxonomic densities, and sensitive species scores were significantly lower in the nearshore C-103 stratum, indicating unfavorable changes had occurred. Multivariate correlation analyses were able to tie these unfavorable faunal differences to sediment organic content and water column nutrient concentrations. The higher organic composition of sediment in this stratum is a probable consequence of enhanced epiphytic and planktonic algal growth induced by nutrient discharges. A higher percentage of algae-consuming grazer and scraper organisms was documented as a result of these nutrient-induced changes.

Under Florida's Impaired Waters Rule, a water body may be listed as impaired based on "other information indicating an imbalance in flora and fauna due to nutrient enrichment" (Chapter 62-303 Florida Administrative Code). This study verifies that an imbalance in flora and fauna due to anthropogenic nutrient enrichment exists in the area adjacent to the C-103 and C-102 Canals. Even though this study documented unfavorable biological community impacts only in the stratum nearest nutrient inputs, the finding that water column inorganic nitrogen concentrations 5500 m offshore correlated with canal discharges suggests that actual impacts may be more widespread.

References

Bilyard, G. R. 1987. The value of benthic infauna in marine pollution monitoring studies. *Marine Pollution Bulletin* 18:581–585.

Brand, L. 1988. Assessment of plankton resources and their environmental interaction in Biscayne Bay, Florida. Dade Environmental Resource Management Technical Report 88-1, 79 pp. and appendices.

Brand, L. 2002. The transport of terrestrial nutrients to South Florida coastal waters. In *The Everglades, Florida Bay, and Coral Reefs of the Florida Keys: An Ecosystem Sourcebook*, J. W. Porter and K. G. Porter (eds.). CRC Press, Boca Raton, FL, pp. 361–413.

Brand, L. E., M. D. Gottfried, C. C. Baylon, and N. S. Romer. 1991. Spatial and temporal distribution of phytoplankton in Biscayne Bay, Florida. *Bulletin of Marine Science* 49:599–613.

Brown, G. 2003. Biscayne Bay Feasibility Study. Hydrodynamic Model Scenario Run. U.S. Army Corps of Engineers (USACOE), Engineer Research and Development Center, Vicksburg, MI.

Byrne, M. and J. Meeder. 1999. Groundwater discharge and nutrient loading to Biscayne Bay. U.S. Geological Survey (USGS). Available at http://sofia.usgs.gov/projects/grndwtr_disch/grndwtrdisabfb1999.html.

Carr, M. R. 1996. PRIMER® User Manual. Plymouth Routes in Multivariate Ecological Research. Plymouth Marine Laboratory, Plymouth, U.K.

Clarke, K. R. and M. Ainsworth. 1993. A method of linking environmental community structure to environmental variables. *Marine Ecology Progress Series* 92:205–219.

Clarke, K. R. and R. M. Warwick. 1994. *Change in Marine Communities: An Approach to Statistical Analysis and Interpretation*. Plymouth Marine Laboratory, Plymouth, U.K.

Clarke, K. R. and R. M. Warwick. 2001. *Change in Marine Communities: An Approach to Statistical Analysis and Interpretation*, 2nd ed. PRIMER-E, Plymouth, U.K.

Duarte, C. M. 1995. Submerged aquatic vegetation in relation to different nutrient regimes. *Ophelia* 4:87–112.

Embleton, T. W., M. Matsumura, L. H. Stolzy, D. A. Devitt, W. W. Jones, R. El-Motaium, and L. L. Summers. 1986. Citrus nitrogen fertilizer management, groundwater pollution, soil salinity, and nitrogen balance. *Applied Agricultural Research* 1:57–64.

Farrell, D. H. 1992. A community based metric for marine benthos. Florida Department of Environmental Regulation SW District Office, Tampa, FL. Unpublished report. 15 pp.

Field, J. G., K. R. Clarke, and R. M. Warwick. 1982. A practical strategy for analyzing multispecies distribution patterns. *Marine Ecology Progress Series* 8:37–52.

Fishman, M. J. and L. C. Friedman. 1989. Methods for Determination of Inorganic Substances in Water and Fluvial Sediments, 3rd ed. Techniques of Water-Resources Investigations of the USGS, Book 5, Laboratory Analysis, Open-File Report 85-495.

Florida Department of Environmental Protection (FDEP). 1995. Submerged Lands and Environmental Resource Program Operations and Procedures Manual. SLER 0856. Tallahassee, FL.

FDEP. 1999a. SOP IZ05 Invertebrate Grab/Dredge Sample Preparation. (Modified from standard methods 10500C) FDEP Central Laboratories, Tallahassee, FL. Reviewed 2003.

FDEP. 1999b. SOP IZ09 Preparation of 80% Ethanol. FDEP Central Laboratories, Tallahassee, FL. Reviewed 2002.

FDEP. 2000a. SOP IZ06 Invertebrate Enumeration and Taxonomic Analysis. (Modified from standard methods 10500C) FDEP Central Laboratories, Tallahassee, FL. Reviewed 2003.

FDEP. 2000b. SOP IZ12 Invertebrate Sorting Quality Control. FDEP Central Laboratories, Tallahassee, FL. Reviewed 2002.

FDEP. 2000c. SOP IZ13 Invertebrate Taxonomic Identification Quality Control. FDEP Central Laboratories, Tallahassee, FL. Reviewed 2002.

FDEP. 2002a. SOP IZ37 Calibration and Use of YSI 610D Data Logger and 600XL Sonde. DEP, Central Laboratories, Tallahassee, FL. Reviewed 2003.

FDEP. 2002b. Florida's Water Quality Assessment 2002 305(b) Report. Water Resource Management Program. Tallahassee, FL.

FDEP. 2002c. Saint Lucie Estuary Evidence of Impairment. July 2002. Southeast District Water Quality Section, Port St. Lucie, FL.

Gibson, G. R., M. L. Bowman, J. Gerritsen, and B. D. Snyder. 2000. Estuarine and Coastal Marine Waters: Bioassessment and Biocriteria Technical Guidance. EPA 822-B-00-024. U.S. Environmental Protection Agency, Office of Water, Washington, D.C.

Gray, J. S. 1981. *The Ecology of Marine Sediments.* Cambridge University Press, Cambridge, U.K.

Gray, J. S. and T. H. Pearson. 1982. Objective selection of sensitive species indicative of pollution-induced change in benthic communities. I. Comparative methodology. *Marine Ecology Progress Series* 9:111–119.

Hair, J. F., L. E. Anderson, L. L. Tatham, and W. C. Black. 1992. *Multivariate Data Analysis with Readings.* Macmillan, New York.

Hallberg, G. R. and D. R. Keeney. 1993. Nitrate. In *Regional Ground-Water Quality,* W. M. Alley (ed.). Van Nostrand Reinhold, New York.

Holland, A. F., A. T. Shaughnessy, and M. H. Hiegel. 1987. Long-term variation in mesohaline Chesapeake Bay macrobenthos: spatial and temporal patterns. *Estuaries* 10: 227–245.

Kadlec, R. H. and R. L. Knight. 1996. *Treatment Wetlands.* CRC Press, Boca Raton, FL.

Kennish, M. J. 1999. *Estuary Restoration and Maintenance.* CRC Press, Boca Raton, FL.

Langevin, C. 2000. Ground-Water Discharge to Biscayne Bay. USGS. October 2000, Greater Everglades Ecosystem Restoration Conference Presentation.

Lapointe, B., W. Matzie, and P. Barile. 2002. Biotic phase-shifts in Florida Bay and fore reef communities of the Florida Keys: linkages with historical freshwater flows and nitrogen loading from Everglades runoff. In *The Everglades, Florida Bay, and Coral Reefs of the Florida Keys: An Ecosystem Sourcebook,* J. W. Porter and K. G. Porter (eds.). CRC Press, Boca Raton, FL, pp. 629–648.

Li, Y. 2001. Calcareous Soils in Miami-Dade County. July 2001. Fact Sheet SL 183, Soil and Water Science Department, Florida Cooperative Extension Service, Institute of Food and Agricultural Sciences, University of Florida, Gainesville.

Li, Y. and M. Zhang. 2002. Major problems affecting agriculture in Miami-Dade: natural resources and crop management (fertilizer). In Miami-Dade County Agricultural Land Retention Study: Final Report, R. L. Degner, T. J. Stevens III, and K. L. Morgan (eds.). Florida Department of Agricultural and Consumer Services Contract No. 5218. Florida Agricultural Market Research Center, Institute of Food and Agricultural Sciences, University of Florida, Gainesville.

Lietz, A. C. 1999. Nutrient transport to Biscayne Bay and water-quality trends at selected sites in Southern Florida. October 2000, USGS Water Quality Workshop.

Livingston, R. J. 2001. *Eutrophication Processes in Coastal Systems.* CRC Press, Boca Raton, FL.

Morton, T. G., A. J. Gold, and W. M. Sullivan. 1988. Influence of overwatering and fertilization on nitrogen losses from home lawns. *Journal of Environmental Quality* 17:124–30.

Reish, D. J. 1986. Benthic invertebrates as indicators of marine pollution: 35 years of study. *Oceans* 86(3):885–888.

Szmant, A. M. 1987. Biological Investigations of the Black Creek Vicinity. Biscayne National Park. Research Resources Management Report SER-87. U.S. Department of Interior, National Park Service.

U.S. ACOE and South Florida Water Management District (SFWMD). 1999. April 1999, Central and Southern Florida Project Comprehensive Review Study, Final Feasibility Report and Programmatic Environmental Impact Statement.

U.S. ACOE and SFWMD 2002. Central and Southern Florida Project, Comprehensive Everglades Restoration Plan, Project Management Plan: Biscayne Bay Coastal Wetlands, Jacksonville, FL.

U.S. Department of Agriculture. 1996. Soil Survey of Dade County Area, Florida. Natural Resource Conservation Service, Miami.

U.S. EPA. 2001. Environmental Monitoring and Assessment Program: National Coastal Assessment Quality Assurance Project Plan 2001–2004. EPA-620/R-01-002. Office of Research and Development, National Health and Environmental Effects Research Laboratory, Gulf Ecology Division, Gulf Breeze, FL.

USGS. 1999. Methodology for estimating nutrient loads discharged from the east coast canals to Biscayne Bay, Miami-Dade County, Florida. USGS Water Resources Investigations Report 99-4094 as part of the USGS South Florida Ecosystem Program.

Valiela, L., J. McClelland, J. Hauxwell, P. J. Behr, D. Hersh, and K. Foreman. 1997. Macroalgal blooms in shallow estuaries: controls and ecophysiolgical and ecosystem consequences. *Limnology and Oceanography* 42:1105–1118.

22

Nekton Species Composition as a Biological Indicator of Altered Freshwater Inflow into Estuaries

Michael Shirley, Patrick O'Donnell, Vicki McGee, and Tim Jones

CONTENTS

Introduction

Estuaries, by nature, are habitats with fluctuating salinity due to changes in freshwater inflow and modification by oceanic processes. Natural patterns in the quantity and timing of freshwater inflow create a spatially and temporally heterogeneous environment. This dynamic condition supports a diverse assemblage of species and some of the most biologically productive habitats on Earth. Freshwater inflow directly and indirectly influences factors that control nekton species composition, including larval recruitment, reproduction, food availability, predation, competition, and physiological constraints related to osmoregulation and hypoxia (Gilmore et al., 1983; Stanley and Nixon, 1992; Cartaxana, 1994; Van Den Avyle and Maynard, 1994; Whitfield, 1999; Riera et al., 2000).

Freshwater inflow can also influence primary productivity, zooplankton biomass, and nekton abundance by influencing nutrient concentrations (Grange et al., 2000). Organic matter flowing into an estuary via freshwater inflow has been shown to contribute significantly to the food available for detritivores, such as shrimp (Riera et al., 2000). Several studies have documented that brackish water estuaries serve as refuges from predation for vulnerable species (Rozas and Hackney, 1984; Shirley et al., 1990). Turbidity gradients or olfactory signals associated with freshwater inflow may also serve as cues for fish to orient toward a specific nursery area (Blaber et al., 1989; Whitfield, 1999). Anthropogenic alterations in the quantity, quality, and timing of freshwater inflow can have wide-ranging consequences, thereby altering species composition (Montague and Ley, 1993; Sklar and Browder, 1998; Grange et al., 2000; Meng and Matern, 2001).

The natural freshwater inflow to most estuaries within the Rookery Bay National Estuarine Research Reserve has been altered by channelization of wetland sheetflow. As a result, the salinity of these estuaries can be influenced more by canal management than by tides and rainfall. Altered freshwater inflow has been identified as the most important threat to the natural biodiversity of this protected area (Shirley et al., 1997).

Unfortunately, no nekton data are available for the reserve prior to watershed channelization. Yokel (1975a,b) conducted the earliest nekton monitoring of this reserve. Yokel's 2-year otter trawl study of Henderson Creek, an estuary with altered freshwater inflow, was initiated in June 1970 and concluded in July 1972. In this study, 728 trawl samples collected 25,790 fish (34.4 CPUE, catch per unit effort) representing 64 species. The most abundant species were pinfish (*Lagodon rhomboides*), silver jenny (*Eucinostomus gula*), pigfish (*Orthopristis chrysoptera*), silver perch (*Bairdiella chrysoura*), spotfin mojarra (*E. argenteus*), and lane snapper (*Lutjanus synagris*). Fish and pink shrimp (*Penaeus duorarum*) abundances peaked during the summer, the onset of the region's rainy season.

Carter et al. (1973) combined the results of seines, surface trawls, and otter trawls to compare fish populations in a freshwater inflow–altered estuary, Faka Union Bay, with Fakahatchee Bay, a nearby estuary with a more natural freshwater inflow. This study collected 273,270 fish representing 96 species. The top six species, in order of abundance, were anchovies (*Anchoa* spp., primarily *A. mitchilli*), yellow fin menhaden (*Brevoortia smithi*), scaled sardine (*Harengula jaguana*), pinfish, silver perch, and silver jenny. Fakahatchee Bay had a higher species diversity and nekton abundance than Faka Union Bay. Both estuaries are now managed by the Rookery Bay Reserve.

Colby et al. (1985) conducted surface and bottom trawls in Faka Union Bay as well as in four bays to the east (including Fakahatchee Bay). These collections were made from August 1982 to August 1983. Species composition, size, relative abundance, and food consumed were related to salinity, sediment type, and aquatic vegetation. The top six species collected were anchovy, spotfin mojarra, silver jenny, black-cheeked tonguefish (*Symphurus plagiusa*), lined sole (*Achirus lineatus*), and sand seatrout (*Cynoscion arenarius*). Anchovy, menhaden, silversides (family Atherinidae), and needlefish (family Belonidae) were predominately collected with surface trawls, while pinfish, pigfish, sand seatrout, and silver perch were more common in otter trawls. Similar to Carter et al. (1973) the numbers and biomass of fish and the numbers of certain macroinvertebrates were substantially less in Faka Union Bay than in the other sampling locations. Bighead sea robin (*Prionotus tribulus*), tidewater silversides (*Menidia beryllina*), Gulf toadfish (*Opsanus beta*), and mullet (*Mugil* spp.) were collected in all locations except Faka Union Bay. Spotfin mojarra were collected in disproportionately greater numbers in Faka Union Bay relative to the other sampling locations. Fish densities were reduced with the onset of the rainy season. This reduction was greater in Faka Union Bay, where fish abundance was reduced by 83%, as opposed to 70% in the bays to the east and 50% in the bays to the west.

Browder et al. (1986) used surface and otter trawls to collect fishes and macroinvertebrates during a monthly study conducted from July 1982 to June 1984. Stations included Fakahatchee Bay and Faka Union Bay and a freshwater-deprived estuary to the west, Pumpkin Bay. Surface and bottom trawl data were combined. A total of 85,561 fish representing 83 species were collected. Included in this data set were 441 otter trawl samples that collected 18,252 fish (41.4 CPUE) representing 71 species. Fakahatchee Bay had higher species diversity than Faka Union Bay, but fish abundance was seldom greater in Fakahatchee Bay than in Faka Union Bay. The total number of species sampled in Faka Union Bay and Fakahatchee Bay were 55 and 64, respectively. The top six most abundant species were the same as those reported by Carter et al. (1973): bay anchovy (primarily *Anchoa mitchilli*), yellow fin menhaden, scaled sardine, pinfish, silver perch, and silver jenny. Blue crabs (*Callinectes sapidus*) and pink shrimp were most abundant in Pumpkin Bay. During the dry season, pink shrimp were collected in equal amounts in Pumpkin and Faka Union Bay and less in Fakahatchee Bay. Blue crab abundance peaked in the dry season, and pink shrimp abundance peaked in the wet season.

These previous studies were focused primarily on species abundance. The results indicate that altered freshwater inflow has adversely affected species abundance. The primary goal of the research described in this chapter was to examine nekton species composition patterns relative to water quality for two estuaries with altered freshwater inflow, Faka Union Bay and Henderson Creek, relative to a reference estuary, Fakahatchee Bay. This study was designed to test the hypothesis that altered freshwater inflow influences nekton species composition. In addition to testing this hypothesis, we examined the relationship of inter- and intra-annual water quality conditions with nekton species composition. Another goal of this study was to identify the species contributing most to differences in species composition between the freshwater inflow–altered estuaries and the reference estuary. This approach is used to develop performance targets for two Southwest Florida Everglades Restoration Projects and to guide the management of freshwater inflow to other estuaries in the region. The applicability of this approach to the conservation of worldwide estuarine biodiversity is also discussed.

Methods

Study Sites

The study sites, Fakahatchee Bay, Faka Union Bay, and Henderson Creek, are located in a subtropical biogeographic region within the Rookery Bay National Estuarine Research Reserve, Southwest Florida (Figure 22.1). This region has an average annual rainfall of 127 to 140 cm (50 to 55 in.). The heaviest average monthly rainfall, 20 to 23 cm (8 to 9 in.), occurs from June through September. Lowest average rainfall, 2 to 5 cm (1 to 2 in.), occurs from November through March. Approximately 66% of the total yearly rainfall occurs between the months of June and October. Salinity patterns follow rainfall patterns with a short time lag. Based on his review of monthly salinity data collected from 1986 to 1992, Christenson (1998) reported that it was not uncommon to find hypersaline conditions in these estuaries, particularly in March and April. In general, minimum salinity occurs in September and the maximum salinity occurs in May (Christenson, 1998). Seasonal rainfall patterns with corresponding estuarine salinity conditions can be categorized as early dry season (December through February), late dry season (March through May), early wet season (June through August), and late wet season (September through November).

Fakahatchee Bay's fresh water flows from a 48,305-ha watershed, primarily as natural overland sheetflow through a watershed that is mostly protected by wildlife conservation areas. Faka Union Bay receives the most freshwater inflow of the three estuaries. This estuary currently receives fresh water through 222 km of canals that drain a sparsely developed watershed of 57,800 ha. The primary purpose

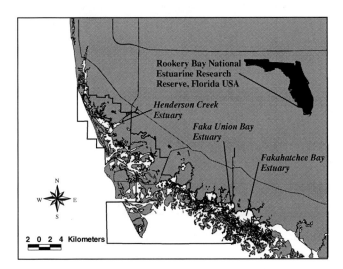

FIGURE 22.1 Location of three estuaries used for this study within the Rookery Bay National Estuarine Research Reserve, Florida (outlined). The Fakahatchee Bay estuary serves as a reference site because of its more natural freshwater inflow pattern.

of the canal system that empties into Faka Union Bay is to quickly drain the watershed and prevent upstream flooding of residential areas. Henderson Creek, the third study area, has a smaller watershed (14,385 ha) and the most closely managed freshwater inflows. Henderson Creek receives most of its fresh water from a canal system that drains a watershed area that is approximately 50% developed. There is a weir at the headwaters of this estuary that retains water during the dry season and early wet season and releases water during the late wet season.

The shorelines of the Fakahatchee Bay, Faka Union Bay, and Henderson Creek estuaries consist of red mangrove forests. Attached submerged vegetation is rarely observed. Sediments range from silty-sand to sand with occasional oyster shell. Tides range from 0 to 2.76 m (average 1.06 m). Salinities range from 0 to 42 ppt.

Water Quality Monitoring

During this study the reserve maintained automated water quality monitoring stations and also conducted monthly water quality monitoring in Fakahatchee Bay, Faka Union Bay, and Henderson Creek. A YSI® 6600 datalogger was deployed at a fixed location within each of the three estuaries and operated according to the protocol established by the National Estuarine Research Reserve System-Wide Monitoring Program (SWMP). All three automated water quality stations recorded temperature, turbidity, salinity, pH, dissolved oxygen, and depth at 1/2 h intervals. In addition, monthly water quality monitoring was conducted during each nekton trawl using a YSI handheld model 85 water quality meter measuring temperature, salinity, and dissolved oxygen.

Dataloggers were deployed for 2 weeks at a time. After 2 weeks each datalogger was replaced with a newly calibrated datalogger to allow for continuous, uninterrupted water quality monitoring. Each datalogger was deployed inside a 10-cm diameter PVC pipe attached to an "aid to navigation" piling. The pipe was oriented vertically and attached to the piling with stainless-steel hose clamps. The end of the pipe was open and located about 15 cm above the substrate, and 30 5-cm diameter holes were drilled around the lower half of the pipe to allow adequate water flow around the sensors. The PVC housing for the sensors was cleaned monthly and replaced biannually.

Nekton Collection

The reserve staff conducted four bottom trawls per estuary each month from December 2000 through November 2002. The collection sites were randomly selected within each estuary from a 10×10 m grid drawn on a nautical chart. Each collection used a 6-m otter trawl with a 3-mm mesh liner pulled for 5 min (approximately 0.1 km). The collections for each month were scheduled within the same 4-day period during a slack high tide. All fish, pink shrimp, and blue crabs were identified in the field to species, when possible, and then counted and measured (cm).

Statistical Analyses

CPUE was calculated as the number of fish, shrimp, and crabs collected per trawl. For statistical analysis, CPUEs were log-transformed by $\log_{10} (x + 1)$ to control heteroscedasticity and to conform to normality assumptions (Sokal and Rohlf, 1995). An analysis of variance (ANOVA) followed by a Tukey–Kramer multiple comparison procedure was used to test for the effects of location, season, and year on CPUE. In addition, patterns in species composition as influenced by year, season, and location and the relative influence of water quality on these patterns were analyzed using nonparametric analyses (PRIMER® version 5.29, PRIMER-E LTD, 2002).

For these latter procedures, the fish CPUE was transformed (fourth root) to emphasize the contribution of less abundant species and a Bray–Curtis similarity matrix was calculated for each year, estuary, and season. Group average sorting from these matrices was used to produce dendrograms. Non-metric multidimensional scaling (MDS) analyses were used to convert the similarity matrices into two-dimensional ordinations. An analysis of similarity (ANOSIM) was then used to test for significant differences in species composition for each season and location. For each year, the five species contributing most to significant seasonal and spatial differences in nekton species composition were identified using the PRIMER SIMPER procedure. Finally, the PRIMER BIOENV (Spearman Rank correlation) procedure was used to relate the abiotic and biotic data. The abiotic data used in this analysis were the means, maximum and standard deviation of salinity (ppt), salinity change (ppt/h), temperature (°C), dissolved oxygen (ppm), depth (m), pH (units), and turbidity (NTU), summarized over each season, year, and location from the automated water quality monitoring data set. In addition, the PRIMER BIOENV analysis was conducted with the means, maximum, and standard deviation of salinity (ppt), temperature (°C), dissolved oxygen (ppm), and depth, summarized over each season, year, and location from the manually collected monthly trawl water quality data set. P values for the resultant correlation coefficients were obtained using the probability calculator function of NCSS® 2004 (Hintze, 2001).

Results

Water Quality

As previously observed by Christenson (1998), the salinity of the estuaries in this region followed seasonal rainfall patterns. Highest salinities for all three estuaries were recorded during the late dry season (March to May) and lowest salinities were recorded during the late wet season (September to November) (Figure 22.2). The early wet season (June to August) and early dry season (December to February) exhibited intermediate salinities to these extremes. For most of the 2-year period the Faka-hatchee Bay estuary exhibited salinities below the Henderson Creek estuary and higher than the Faka Union Bay estuary. The use of near-continuous data allowed the mean change in salinity, as ppt/h, to be calculated for each estuary by season (Figure 22.3). An extreme difference in salinity fluctuations was seen at the Faka Union Bay site vs. the other two study sites. Combined, these two figures illustrate that Henderson Creek is an estuary with restricted freshwater inflow and perhaps unnaturally low salinity fluctuations. A more detailed examination of the raw data set revealed that Henderson Creek did indeed exhibit reduced fluctuations in salinity, especially during the dry season (Figure 22.4) but this estuary also had more extreme salinity fluctuations during the late rainy season than did the reference site, Fakahatchee Bay (Figure 22.5).

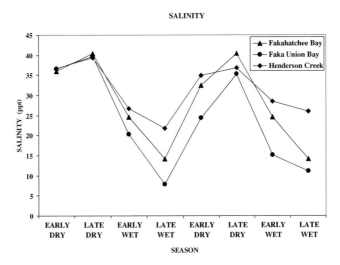

FIGURE 22.2 Mean seasonal salinity (ppt) recorded in each estuary by season over a 2-year period (December 2000 to November 2002). Each data point represents approximately 4000 datalogger readings.

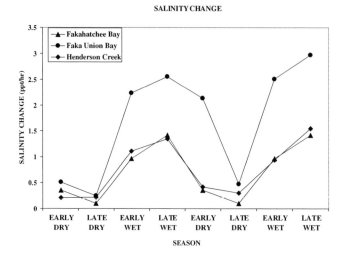

FIGURE 22.3 The change in salinity (ppt/h) recorded in each estuary by season over a 2-year period (December 2000 to November 2002). Each data point represents approximately 4000 datalogger readings.

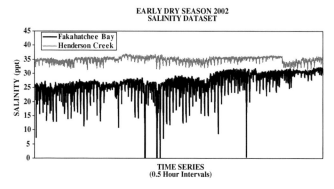

FIGURE 22.4 Near-continuous salinity measurement for Henderson Creek and Fakahatchee Bay during the early dry season (December to February 2002).

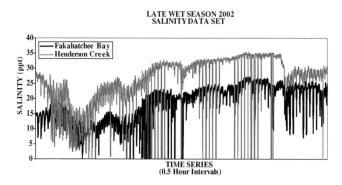

FIGURE 22.5 Near-continuous salinity measurement for Henderson Creek and Fakahatchee Bay during the late wet season (September to November, 2002).

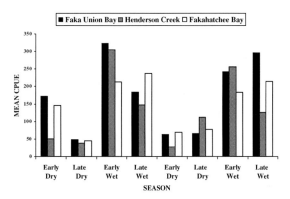

FIGURE 22.6 Mean seasonal CPUE for Fakahatchee Bay, Faka Union Bay, and Henderson Creek for December 2000 through November 2002.

Nekton Species Composition

In both years, three taxa accounted for over 75% of the total catch. *Eucinostomus argenteus* (spotfin mojarra), *E. gula* (silver jenny), and *Penaeus duorarum* (pink shrimp) dominated the catch of all estuaries during both years. The next five most abundance species were *Etropus crossotus* (fringed flounder), pigfish, anchovies, *Symphurus plagiusa* (blackcheek tonguefish), and pinfish. The results of the ANOVA on the log-transformed CPUE data set indicated highly significant effects of location ($p = 0.008$) and season ($p < 0.001$). The interaction effect of location $\times$ season was also significant ($p = 0.01$) and the year $\times$ season term was marginally significant ($p = 0.054$). Year effects ($p = 0.61$), the interaction effects of location $\times$ year ($p = 0.68$), and year $\times$ season $\times$ location ($p = 0.47$) were not statistically significant.

A Tukey–Kramer multiple-comparison test of the mean log-transformed CPUE values indicated that, for both years combined, nekton CPUE was less in Henderson Creek than in Fakahatchee Bay or Faka Union Bay and that wet season CPUEs were greater ($\alpha = 0.05$) than dry season CPUEs. No significant difference ($\alpha = 0.05$) was found in CPUEs during the same season across years. Comparisons of CPUE for Henderson Creek or Faka Union Bay vs. Fakahatchee Bay during the same season indicated only a significant ($\alpha = 0.05$) reduction in CPUE for Henderson Creek during the early dry season (nontransformed data presented in Figure 22.6).

Multidimensional scaling ordination analyses of the similarity in species composition for Fakahatchee Bay revealed distinct seasonal patterns (Figure 22.7). This analysis identified the late wet season and the late dry season as extremes separated by the "transitional" early dry and early wet seasons. As with

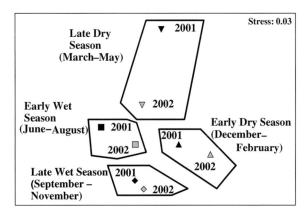

FIGURE 22.7 Results of MDS ordination of species composition similarity for Fakahatchee Bay by season and year.

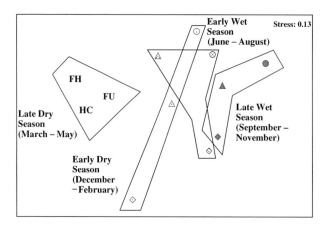

FIGURE 22.8 Results of MDS ordination of species composition similarity for Fakahatchee Bay (FH), Faka Union Bay (FU), and Henderson Creek (HC) in 2001 by season and location.

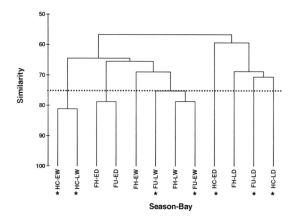

FIGURE 22.9 Dendrogram of species composition similarity for Fakahatchee Bay (FH), Faka Union Bay (FU), and Henderson Creek (HC) during the early dry (ED), late dry (LD), early wet (EW), and late wet (LW) season of 2001. The asterisks indicate a significant ($p < 0.05$) difference from Fakahatchee Bay during the same season. The dashed line indicates a Bray–Curtis similarity value of 75.

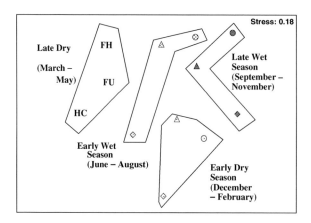

FIGURE 22.10 Results of MDS ordination of species composition similarity for Fakahatchee Bay (FH), Faka Union Bay (FU), and Henderson Creek (HC) in 2002 by season and location.

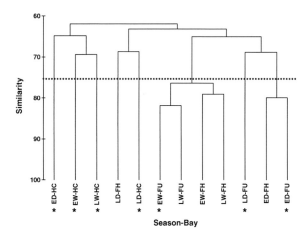

FIGURE 22.11 Dendrogram of species composition similarity for Fakahatchee Bay (FH), Faka Union Bay (FU), and Henderson Creek (HC) during the early dry (ED), late dry (LD), early wet (EW), and late wet (LW) season. The asterisks indicate a significant ($p < 0.05$) difference from Fakahatchee Bay during the same season. The dashed line indicates a Bray–Curtis similarity value of 75.

the ANOVA result above, the results of the ANOSIM procedure indicated a significant ($p < 0.05$) influence of season but not year on this pattern. The ANOSIM procedure also indicated a significant interaction effect of season and year with the species composition significantly ($p < 0.05$) different across each year during the same season except for the late wet season.

Multidimensional scaling ordination of the species composition similarities of all three estuaries in 2001 revealed a similar pattern to Fakahatchee Bay but with less distinction between the transitional seasons (Figure 22.8). The ANOSIM procedure indicated that Fakahatchee Bay species composition was significantly different from the other two estuaries during all seasons with the exception of Faka Union Bay during the early dry season (Figure 22.9).

Multidimensional scaling ordination analyses of the species composition similarities of all three estuaries in 2002 revealed a similar overall pattern to the 2001 MDS analyses (Figure 22.10). During 2002, the ANOSIM procedure indicated that Fakahatchee Bay species composition was significantly different from the two other estuaries during all seasons with the exception of Faka Union Bay during the late wet season (Figure 22.11).

The five species contributing most to the dry season vs. wet season differences in species composition for the reference estuary, Fakahatchee Bay, varied from year to year, with the exception of pigfish (Table 22.1). In 2001, pink shrimp were one of the five species contributing most to the species composition difference between late dry season and late wet season. This species was collected in greater abundance during the late dry season. Pigfish, lined sole, blackcheek tonguefish, and *Eucinostomus* spp. were also among the five species contributing most to this seasonal species composition difference but these species were collected in greater abundance during the late wet season.

TABLE 22.1

The Five Species Contributing Most to the Difference in Nekton Species Composition in Fakahatchee Bay Estuary in 2001 and 2002 for the Late Wet vs. the Late Dry Season

2001		2002	
Species	**Common Name**	**Species**	**Common Name**
Symphurus plagiusa	Blackcheek tonguefish	**Orthopristis chrysoptera*	Pigfish
Achirus lineatus	Lined sole	*Anchoa* spp.	Anchovy
Orthopristis chrysoptera	Pigfish	*Chloroscombrus chrysurus*	Atlantic bumper
*Penaeus duorarum	Pink shrimp	*Prionotus tribulus*	Bighead searobin
Eucinostomus spp.	Mojarra	**Prionotus scitulus*	Leopard searobin

Note: * = species that were collected in greater abundance in the late dry season.

TABLE 22.2

The Five Species Contributing Most to the difference in Nekton Species Composition in the Fakahatchee Bay Estuary vs. Faka Union Bay Estuary in 2001 and 2002 for the Seasons Found to Have Significantly Different Species Composition

2001			
Early Dry	**Late Dry**	**Early Wet**	**Late Wet**
NS	*Achirus lineatus*	**Floridichthys carpio*	*Diapterus auratus*
	*Mycteroperca microlepis	**Etropus crossotus*	*Lucania parva*
	*Prionotus scitulus	**Arius felis*	**Prionotus tribulus*
	*Orthopristis chrysoptera	**Chilomycterus schoepfi*	**Bagre marinus*
	Lutjanus griseus	**Orthopristis chrysoptera*	**Gobiosoma robustum*

2002			
Early Dry	**Late Dry**	**Early Wet**	**Late Wet**
*Lutjanus synagris	**Orthopristis chrysoptera*	**Lagodon rhomboides*	NS
Anchoa spp.	**Etropus crossotus*	*Anchoa* spp.	
*Orthopristis chrysoptera	**Cynoscion nebulosus*	**Orthopristis chrysoptera*	
Syngnathus louisianae	**Lagodon rhomboides*	**Chloroscombrus chrysurus*	
*Arius felis	**Opsanus beta*	*Gobionellus smaragdus*	

Note: * = species that were collected in greater abundance in Fakahatchee Bay. NS = species composition not found to be significantly different (ANOSIM, $p > 0.05$).

In 2002, pigfish and leopard searobin (*Prionotus scitulus*) were two of the five most important contributors to the observed seasonal species composition difference in Fakahatchee Bay. These two species were collected in greater abundance during the late dry season. The other three species in this ranking were *Anchoa* spp., Atlantic bumper (*Chloroscombrus chrysurus*), and the bighead searobin (*Prionotus tribulus*); all three were collected in greater abundance during the late wet season.

Similarly, for each season, the five species contributing most to the species composition differences between Fakahatchee Bay and Faka Union Bay, with the exception of pigfish, varied during both years (Table 22.2). For the comparison of Fakahatchee Bay to Henderson Creek, the five species contributing most to the species composition differences between corresponding seasons varied each year with three exceptions (Table 22.3). In both 2001 and in 2002 the code goby (*Gobiosoma robustum*), blackcheek tonguefish, and the leopard searobin were among the five species contributing most to the difference in species composition between these two bays during the early dry season, late dry season, and the late wet season, respectively. Overall, 34 species ranked within the five species contributing most to the species composition differences between the reference estuary and the estuaries with altered freshwater inflow. More than 75% of these species were in greater abundance in the reference estuary, Fakahatchee Bay (Table 22.2 and Table 22.3).

Linking Abiotic and Biotic Patterns

Using the water quality data collected with the automated datalogger stations, the PRIMER BIOENV (Spearman Rank correlation) procedure indicated that, for both years, minimum salinity (i.e., maximum freshwater inflow), $\rho_s = 0.65$ for 2001; $\rho_s = 0.45$ for 2002) was the best-correlated single abiotic variable to the observed species composition patterns. In 2001, the best-correlated group of abiotic variables was average temperature, minimum salinity, average salinity, and maximum salinity change ($\rho_s = 0.65$). In 2002, the best-correlated group of abiotic variables were maximum temperature, minimum salinity, average salinity, average salinity change, and minimum pH ($\rho_s = 0.55$). These correlations were all significant at the $p = 0.05$ level.

When the water quality data collected during the monthly trawls were used in this same analysis, maximum temperature ($\rho_s = 0.55$) in 2001 and the standard deviation in salinity ($\rho_s = 0.05$) in 2002 were identified as the two most-correlated single abiotic variables. The best-correlated group of variables were mean salinity, minimum salinity, minimum temperature, mean depth, and maximum depth ($\rho_s = 0.58$)

TABLE 22.3

The Five Species Contributing Most to the Difference in Nekton Species Composition in the Fakahatchee Bay Estuary vs. the Henderson Creek Estuary in 2001 and 2002 for the Seasons Found to Have Significantly Different Species Composition

2001			
Early Dry	**Late Dry**	**Early Wet**	**Late Wet**
*Syngnathus scovelli	Anchoa spp.	*Lagodon rhomboides	*Gobiosoma robustum
*Penaeus duorarum	*Opsanus beta	*Syngnathus scovelli	*Symphurus plagiusa
*Anchoa spp.	Ogcocephalus radiatus	Microgobius thalassinus	*Lagodon rhomboides
*Gobiosoma robustum	Microgobius thalassinus	Arius felis	Gobionellus smaragdus
*Achirus lineatus	*Prionotus scitulus	*Floridichthys carpio	*Bagre marinus

2002			
Early Dry	**Late Dry**	**Early Wet**	**Late Wet**
*Archosargus probatocephalus	*Prionotus scitulus	*Bairdiella chrysoura	*Symphurus plagiusa
*Gobiosoma robustum	*Cynoscion nebulosus	*Gobiosoma robustum	*Anchoa spp.
Ogcocephalus radiatus	*Paralichthys albigutta	*Opsanus beta	*Syngnathus louisianae
*Eucinostomus spp.	*Achirus lineatus	*Trinectes maculatus	*Syngnathus scovelli
*Microgobius gulosus	Menidia menidia	Eucinostomus spp.	*Trinectes maculatus

Note: * = species that were collected in greater abundance in Fakahatchee Bay.

in 2001 and the standard deviation in salinity and the standard deviation in dissolved oxygen ($\rho_s = 0.06$) in 2002. The 2001 correlations between the monthly water quality monitoring and the trawl species composition patterns were significant ($p < 0.05$), but the 2002 correlations were not ($p > 0.05$).

Discussion

A striking difference between Faka Union Bay and the reference estuary, Fakahatchee Bay, was the extreme salinity fluctuation observed during May through February in Faka Union Bay. The management strategy of quickly draining the watershed to prevent upstream flooding of residential areas delivers unnaturally large volumes of fresh water into this estuary and extends the influence of the wet season into the early dry season. In 2001 and 2002, these estuaries had significantly different species compositions, and the species contributing most to this difference were in greater abundance in Fakahatchee Bay. Despite these differences, the overall CPUE for these estuaries for the same season during each year was not statistically different.

A comparison of Henderson Creek's salinity patterns with the patterns in Fakahatchee Bay is more complex. The management of the freshwater inflow into Henderson Creek is a delicate balance between flood control, aquifer recharge, and the prevention of saltwater intrusion. The headwaters of this estuary are dammed by a water control structure with gates that can be set to different levels depending on the water management needs of the watershed. This management scenario causes Henderson Creek to have higher average salinities most of the year but also results in periods of higher salinity fluctuations followed by periods of lower salinity fluctuations during the rainy season relative to Fakahatchee Bay. Nevertheless, as with the comparison of species composition differences between Fakahatchee Bay and Faka Union Bay, there was a significant difference in species composition between Henderson Creek and Fakahatchee Bay. For both years during each season there were statistically significant species composition differences between these estuaries. In addition, for each year, more than 70% of the five species contributing most to these seasonal species composition differences were in greater abundance in Fakahatchee Bay. In contrast, the overall CPUE comparisons of Henderson Creek and Fakahatchee Bay indicated greater nekton abundance only during the early dry season in Fakahatchee Bay.

As with Yokel (1975a,b), we found increased nekton abundances during the wet season. Similar results in other regions of the world indicate that nekton catch rates can be positively correlated to increases in primary productivity triggered by freshwater inflow at the onset of the rainy season (Whitfield, 1999; Grange et al., 2000). Yokel (1975a,b) reported that the five most abundant fish species collected were *Eucinostomus* spp., pinfish, pigfish, silver perch, and lane snapper. In the present study, the most abundant fish species were *Eucinostomus* spp., pinfish, pigfish, and *Anchoa* spp. Despite similarities in dominant species, the total number of species collected and collection techniques, our results produced an average CPUE for Henderson Creek of 132, substantially greater than the 34.4 CPUE reported by Yokel (1975a,b). Browder et al. (1986) collected 18,252 fish (41.4 CPUE) using otter trawl to sample Fakahatchee Bay, Faka Union Bay, and Pumpkin Bay. Again, this CPUE is nearly fourfold less than the CPUE we obtained from sampling these same estuaries in 2001 and 2002.

An examination of the results of other past studies from this region reveals different patterns in seasonal nekton abundance. Colby et al. (1985) found the numbers of fish to be substantially less in Faka Union Bay in the wet season than in the other sampling locations. We found no statistical difference between the seasonal nekton abundance of these two estuaries. Also, Colby et al. (1985) reported a reduction in fish densities with the onset of the rainy season that was more evident in Faka Union Bay, where numbers were reduced by 83%, as opposed to 70% in the Fakahatchee Bay. In contrast, and similar to Yokel (1975a,b), Carter et al. (1973), and Browder et al. (1986), we found wet season nekton abundances to increase in Fakahatchee Bay and in Faka Union Bay relative to dry season abundances. This difference in the CPUE-based results from present and past nekton studies within the same region highlights the complexity of the relationship between species abundance and freshwater inflow and the limitations of using this metric to establish baseline conditions or set restoration targets.

For these past studies and our present study, an analysis of CPUE data sets using standard statistical techniques identified 11 species that contribute to the six most abundant species collected for any given study. In contrast, species composition analyses indicated 34 species ranked within the five species contributing most to the interannual seasonal species composition differences between Fakahatchee Bay, Henderson Creek, and Faka Union Bay. Furthermore, more than 75% of species contributing most to statistically significant seasonal species composition differences were collected in greater abundance in the reference estuary, Fakahatchee Bay.

Clearly, species composition analysis provides an approach consistent with multispecies management that can be used to assess the influence of anthropogenic alterations on natural biodiversity. This approach shifts the focus from the most abundant species to those most sensitive to habitat alterations. We found that the total number and types of species collected within each estuary were similar, but the seasonal patterns in composition of these species for a reference estuary vs. two estuaries with altered freshwater inflow estuaries were different. Even within our reference estuary, Fakahatchee Bay, species composition can vary seasonally within a year and can be dissimilar from one year to next. Despite this variability, a consistent pattern of species composition emerged, with the late wet and late dry seasons being most dissimilar and the early wet and early dry season as a transitional period. These results underscore the value of comparing species composition patterns of reference vs. altered estuaries using data collected within the same year and season and argues against using historic data sets to set restoration targets.

Making comparisons of species composition within the same year and season, we found that Faka Union Bay and Henderson Creek had significant differences in species composition relative to Faka-hatchee Bay. Of these two estuaries, Henderson Creek's species composition had the least in common with Fakahatchee Bay's species composition. This result indicates that the tightly controlled inflow of fresh water into Henderson Creek may be more disruptive to maintaining natural species composition than the freshwater inflow management approach affecting Faka Union Bay.

As with studies in other estuaries with extreme seasonally dependent freshwater inflow (Kanandjembo et al., 2001), the present study found significant differences in species composition correlated to seasonal changes in environmental salinity. Noteworthy, the present study produced higher correlation coefficients between abiotic factors and species composition using similar statistical techniques (0.64 for 2001 and 0.55 for 2002) relative to those of 0.34 reported by Kanandjembo et al. (2001). Correlations between near-continuous water quality data and nekton species composition were always better than the correlations between species composition and the water quality data obtained from once-per-month samples made during each trawl (0.58 and 0.059 for 2001 and 2002, respectively). Indeed, during 2002, in contrast to the near continuous data set, our monthly water quality data were not significantly correlated with the observed species composition patterns. This outcome emphasizes the importance of using near-continuous water quality monitoring data to characterize estuarine habitats.

Conclusions

The primary management goal of the Rookery Bay National Estuarine Research Reserve is to conserve natural biodiversity not to manage freshwater inflow to optimize conditions for an individual species or a group of species. We recommend using species similarity coefficients derived from comparing altered estuaries to a reference site, as a target for restoration. We further recommend that an adaptive management approach, with the species similarity target being increased or decreased as future research augments the abiotic water quality data set with more complete habitat characterization information. This step is important to account for natural differences in factors affecting species composition differences between reference and altered estuaries. Based on the present study, three additional recommendations can be made: (1) nekton species composition patterns are a useful, more sensitive and consistent indicator of altered freshwater inflow than nekton abundance, (2) water quality monitoring using near-continuous sampling is preferred over sampling at larger timescales, and (3) comparisons of the reference and altered estuaries should be made within the same season and year as opposed to relying on historic data sets to provide baseline conditions.

For the conservation of worldwide estuarine biodiversity it is vital that specific estuaries, and their watersheds, be identified to serve as reference sites for each biogeographic region to guide local coastal management decisions. The conservation of natural conditions, including the quality, quantity, and timing of freshwater inflow, of these estuarine protected areas should be given top priority by the responsible governments.

Acknowledgments

The authors thank past and present staff and volunteers of the Rookery Bay National Estuarine Research Reserve for their invaluable contribution to this ongoing research project. We also greatly appreciate the constructive comments of an anonymous reviewer and the book's editor Dr. Bortone. The first author would also like to express his sincere gratitude to Kathy Shirley, who contributed significantly to the completion of this chapter through editorial comments and her patient tolerance of the many hours required to prepare the manuscript. This work would not be possible without the continued support of staff of the Florida Department of Environmental Protection's Office of Coastal and Aquatic Managed Areas, the South Florida Water Management District's Big Cypress Basin Office, and the National Oceanic and Atmospheric Administration Estuarine Research Reserve Division.

References

Blaber, S. J. M., D. T. Brewer, and J. P. Salini. 1989. Species composition and biomasses of fishes in different habitats of a tropical northern Australian estuary: their occurrence in adjoining sea and estuarine dependence. *Estuarine, Coastal, and Shelf Science* 29:509–531.

Browder, J. A., A. Dragovich, J. Tashiro, E. Coleman-Duffie, C. Foltz, and J. Zweifel. 1986. A Comparison of Biological Abundances in Three Adjacent Bay Systems Downstream from the Golden Gate Estates Canal System. Final Report to the U.S. Army Corps of Engineers, Jacksonville, FL, from the Miami, FL, Laboratory of the Southeast Fisheries Center, National Marine Fisheries Service, NOAA.

Cartaxana, A. 1994. Distribution and migrations of the prawn *Palaemon longirostris* in the Mira River Estuary (Southwest Portugal). *Estuaries* 17(3):685–694.

Carter, M. R., L. A. Burns, T. R. Cavender, K. R. Dugger, P. L. Fore, D. B. Hicks, H. L. Revells, and T. W. Schmidt. 1973. Ecosystem Analysis of the Big Cypress Swamp and Estuaries. U.S. Environmental Protection Agency, Region IV, Atlanta, GA.

Christenson, T. 1998. Mesoscale Spatial and Temporal Water Quality Trends in the Rookery Bay Estuary. Master's thesis, University of South Florida, Tampa, 157 pp.

Colby, D., G. Thayer, W. Hettler, and D. Peters. 1985. A Comparison of Forage Fish Community in Relation to Habitat Parameters in Faka Union Bay, Florida, and Eight Collateral Bays during the Wet season. NOAA Technical Report NMFS SEFC-162. Southeast Fisheries Center, Beaufort Laboratory, Beaufort, NC, 87 pp.

Gilmore, R. G., C. J. Donohoe, and D. W. Cooke. 1983. Observations on the distribution and biology of East-Central Florida populations of the common snook, *Centropomus undecimalis* (Bloch). *Florida Scientist* 46(3/4):313–336.

Grange, N., A. K. Whitfield, C. J. De Villiers, and B. R. Allanson. 2000. The response of two South African east coast estuaries to altered river flow. *Marine and Freshwater Ecosystems* 10(3):155–177.

Hintze, J. 2001. NCSS, Number Cruncher Statistical System. Kaysville, UT.

Kanandjembo, A. N., I. C. Potter, and M. E. Platell. 2001. Abrupt shifts in fish community of the hydrologically variable upper estuary of the Swan River. *Hydrological Processing* 15:2503–2517.

Meng, L. and S. A. Matern. 2001. Native and introduced larval fishes of the Suisun Marsh, California: effects of freshwater flow. *Transactions of the American Fisheries Society* 130:750–765.

Montague, C. L. and J. A. Ley. 1993. A possible effect of salinity fluctuation on abundance of benthic vegetation and associated fauna in northeastern Florida bay. *Estuaries* 6(4):703–717.

Popowski, R., J. Browder, M. Shirley, and M. Savarese. 2003. Hydrological and Ecological Performance Measures and Targets for the Faka Union Canal and Bay. Final Draft Performance Measures: Faka Union Canal. U.S. Fish and Wildlife Service Technical Document, August 2003.

PRIMER 5. 2002. Version 5.29, PRIMER-E LTD, Plymouth, U.K.

Riera, P., P. A. Montagna, R. D. Kalke, and P. Richard. 2000. Utilization of estuarine organic matter during growth and migration by juvenile brown shrimp *Penaeus aztecus* in a South Texas estuary. *Marine Ecology Progress Series* 199:205–216.

Rozas, L. P. and C. T. Hackney. 1984. Use of oligohaline marshes by fish and macrofaunal crustaceans in North Carolina. *Estuaries* 7:213–224.

Shirley, M. A., A. H. Hines, and T. G. Wolcott. 1990. Adaptive significance of habitat selection by molting adult blue crabs *Callinectes sapidus* (Rathbun) within a subestuary of central Chesapeake Bay. *Journal of Experimental Marine Biology and Ecology* 140:107–119.

Shirley, M., J. Haner, H. Stoffel, and H. Flanagan. 1997. Estuarine Habitat Assessment: Rookery Bay National Estuarine Research Reserve and the Ten Thousand Islands Aquatic Preserve, Naples, FL. Report to the Florida Coastal Zone Management Program, 41 pp.

Sklar, F. and J. Browder. 1998. Coastal environmental impacts brought about by alteration to freshwater flow in the Gulf of Mexico. *Environmental Management* 22(4):547–562.

Sokal, R. R. and F. J. Rohlf. 1995. *Biometry,* 3rd ed. W. H. Freeman, New York.

Stanley, D. W. and S. W. Nixon. 1992. Stratification and bottom-water hypoxia in the Pamlico River Estuary. *Estuaries* 15(3):270–281.

Van Den Avyle, M. J. and M. A. Maynard. 1994. Effects of saltwater intrusion and flow diversion on reproductive success of striped bass in the Savannah River Estuary. *Transactions of the American Fisheries Society* 123:886–903.

Whitfield, A. K. 1999. Ichthyofaunal assemblages in estuaries: a South African case study. *Review of Fish Biology and Fisheries* 9:151–186.

Yokel, B. J. 1975a. Estuarine Biology. Conservation Foundation. Rookery Bay Land Use Studies: Environmental Planning Strategies for the Development of a Mangrove Shoreline. Conservation Foundation, Washington, D.C.

Yokel, B. J. 1975b. A Comparison of Animal Abundance and Distribution in Similar Habitats in Rookery Bay, Marco Island and Fakahatchee on the Southwest Coast of Florida. Preliminary Report. Conservation Foundation, Washington, D.C.

23

Evaluating Nearshore Communities as Indicators of Ecosystem Health

Donna Marie Bilkovic, Carl H. Hershner, Marcia R. Berman, Kirk J. Havens, and David M. Stanhope

CONTENTS

Introduction

As human populations and coastal development increase, pressures on aquatic resources multiply. Maintaining the desired benefits derived from aquatic systems requires informed management of anthropogenic impacts. Given that coastal systems are exceptionally complex, few individuals or agencies have the resources to undertake comprehensive analysis of every potential impact. Efficiency in decision making requires distillation of complex interactions and data into a few information-rich metrics that can provide guidance to those with management authority. The effort to generate these metrics is embodied in much of the ongoing research on environmental indicators.

Most of the extant work on indicators is aimed at detection of condition, diagnosis of causes of condition, and/or communication of condition to nontechnical audiences. In estuarine systems a common focus of indicator development is on ecosystem health or integrity. Ecosystem health and integrity can represent two related but fundamentally different concepts (Karr and Chu, 1999; Lackey, 2001). Ecosystem integrity is often defined as the unimpaired condition of a self-sustaining ecosystem. The implicit assumption is little or no anthropogenic impact. In contrast, ecosystem health is the optimal state of a system modified by human activity (e.g., the best possible environmental condition in an agricultural or urban landscape; Karr and Chu, 1999). Regier (1993) maintains that ecosystem health and integrity are interchangeable concepts when one accepts the existence of multiple benchmarks for ecosystem integrity based on societal choices. Both approaches recognize that systems in which human use is an integral component will differ significantly from pristine conditions.

With this recognition, the single benchmark of condition based on an absence of human impact can be replaced by multiple reference conditions. Each reference is defined by an attainable optimum, given existing management goals. For example, the optimum condition attainable in a highly developed harbor would differ from the condition attainable in a rural agricultural setting. The advantage of multiple references lies in the precision of guidance that can be developed for managers. Information on conditions in an area is developed relative to an attainable outcome with the difference supposedly amenable to management.

Measures of ecosystem condition include the use of physical, chemical, and biological indicators. Biological indicators are significant because they represent the integration of aquatic conditions. Thus, they can be used to assess both abiotic and biotic condition as well as cumulative effects. In fact, the use of individual species to indicate pollution degradation has a long history (Bain et al., 2000). Subsequent analyses have evolved community-level, ecosystem, and landscape approaches, including a growing number of studies that incorporate multiple aspects of ecosystem structure and function into indices of ecosystem condition (Deegan et al., 1997; Jordan and Vaas, 2000; Hughes et al., 2002).

Fish community characteristics have been used since the early 1900s to measure relative ecosystem health (Fausch et al., 1990). Within the last 20 years, advances stem from the development of integrative measures of ecological condition, such as the index of biotic integrity (IBI), which relates fish communities to abiotic and biotic conditions of the ecosystem. Fish community IBIs were first developed for use in freshwater, midwestern U.S. streams, and were subsequently modified for application in Great Lakes bays, reservoirs, streams, and large rivers throughout the United States and other countries. Fish community IBIs have now expanded to multiple versions applicable in a variety of environments. The common thread that connects the various IBIs is a multimetric approach, which describes biotic community structure and function and relates it to the ecosystem or habitat.

Estuarine systems are arguably some of the most complex aquatic systems. Their natural variability compounds the problems of detecting anthropogenic impacts. Until now, use of fish community IBIs in estuarine systems has been limited, with varying degrees of success (Carmichael et al., 1992; Deegan et al., 1997; Jordan and Vaas, 2000; Hughes et al., 2002; Meng et al., 2002). With growing recognition that effective management of estuarine systems can occur only at ecosystem levels, the need for further development of these metrics is widely accepted.

The complexity of estuarine systems argues for integrated measures of condition, as well as a suite of stratified indicators of habitat condition and anthropogenic impacts. The broad gradients of salinity and the more localized gradients of physical forces interact to create a diversity of habitats within which biota are sorted based on preference and tolerance. Patterns of human use modify these habitats and add another layer of diversity.

Shallow-water habitat provides critical nursery and spawning areas, protection from predators, and foraging opportunities for numerous fish species. The cumulative impact of shoreline armoring has been demonstrated to drastically reduce available shallow-water habitat structure and associated fish communities (Beauchamp et al., 1994; Jennings et al., 1999). For example, over the past 10 years in Virginia, it is estimated that 342 km of tidal shoreline have been altered with riprap (stone revetments) and retaining walls (bulkheads) (Center for Coastal Resources Management, Tidal Wetlands Impacts data; www.vims.edu/rmap/wetlands). As efforts to manage fisheries evolve toward an ecosystem approach, information on the habitat quality of the nearshore and riparian zones is essential. Evaluation of in-stream habitat and shoreline condition in conjunction with descriptions of biological communities may establish links between landscape and the biota.

To address the need for multiple indicators of aquatic ecosystem health in coastal plain estuarine systems, we are developing an estuarine shallow-water index (ESI) that incorporates fish community metrics, habitat condition variables (shoreline, in-stream habitat, and land use), prey community metrics (e.g., the benthic macroinvertebrate index of biotic integrity), and water quality parameters, for example, dissolved oxygen, total suspended solids (TSS), and salinity. Simon and Lyons (1995) recommend the inclusion of appropriate habitat information with biological communities assessments. We propose that the use of integrated measures of health (physical, chemical, biological, and socioeconomic parameters) will most accurately characterize ecosystem health within estuarine systems.

The Chesapeake Bay is the largest estuary in the contiguous United States and because of the varied land use offers an excellent study area to develop measures of aquatic health based on social choice.

The population of the bay region is projected to increase at a rate of 300 people per day approaching 18 million by 2020 (U.S. EPA, 1982; Chesapeake Bay Program, 1999). Watersheds that were predominantly based on forestry or agricultural product are shifting to suburban and urban. Forests are being converted at a rate of 40.5 ha/day (Scientific and Technical Advisory Committee, 2003). The capacity of watershed ecosystems to maintain water quality parameters at a minimal level to sustain aquatic resources is a measurement of vital importance to resource managers. Local communities need the ability to gauge the health of their aquatic resources in relation to the social and economic conditions they have chosen for their watershed. For example, if a community has chosen to retain its agriculture-based economy, what conditions within the landscape must be conserved or adjusted to improve or maintain appropriate water quality standards? Conversely, if a community has chosen to move toward a predominantly residential suburban watershed, what pattern of land use will allow for continued aquatic health?

As a first step to the development of an ESI, fish community metrics were developed and tested in the Chesapeake Bay, and the relationships between habitat condition and nekton communities were evaluated.

Methods

Watershed Selection

For the study, 14 watersheds (14-digit hydrologic unit codes) were selected throughout the Chesapeake Bay based on several criteria: land use classifications, availability of long-term fish community data, salinity regime, and accessibility. Each watershed was broadly placed into one of three land use categories based on principal land use percentages throughout the watershed (Figure 23.1; Table 23.1). The categories were identified as forested, agricultural (includes mixed-agricultural), and developed (includes mixed-developed). Three or more representative watersheds in the oligo-mesohaline salinity regime were sampled from each of the three principal land use categories.

Guild Development

As a first step in the calculation of metrics, fish species were placed into several guilds based on their documented life histories. Guilds were constructed based on reproductive strategy, trophic level, primary life history, habitat preference, and origin. Primary sources of life history information included Lippson and Moran (1974), Hardy (1978), Jenkins and Burkhead (1994), and Murdy et al. (1997). Categorization is based on the predominant behavior of each species at the life stage typically observed in nearshore estuarine waters from July to September. The reproductive strategy guild categorizes species by spawning location. Within the trophic level guild, species are classed as omnivores, carnivores, or benthivores based on their primary prey items. Categorization of the primary life history guild is based on how each species relates to the estuarine system; for example, nonresident species that have estuarine-dependent larval or juvenile stages are placed in the estuarine-dependent nursery category. The habitat guild broadly classifies species by typical position in the water column (i.e., pelagic or benthic). The origin guild separates species into estuarine residents (present year-round) and nonresidents (Table 23.2).

Metric Selection

In the manner of Karr et al. (1986), eight metrics were assessed for consistency as indicators of aquatic ecosystem health based on fish community structure and function. Metrics were chosen that represent key aspects of fish community integrity, as well as the elements of life history that are dependent on estuarine condition. Several metrics were extracted from current literature that addressed similar estuarine environments. Metrics were placed into four broad categories: taxonomic richness and diversity, abundance, trophic composition, and nursery function (Table 23.3). For each site, individual metric values were calculated based on observed species composition and abundance in 2002.

FIGURE 23.1 Sampling locations with associated watershed land use categories in the Chesapeake Bay, 2002.

TABLE 23.1

Land Use Classification Criteria for Watershed Categorization

Land Use Category	Criteria
Forested	Greater than 60% total forest cover (forest, mixed, forest wetland) and <10% urban
Agricultural	Greater than 50% total agricultural cover (pasture, crop)
Developed	Greater than 50% total urban cover (low and high residential and industrial areas)
Mixed-agriculture	20–50% total agricultural cover
Mixed-developed	20–50% total urban cover

Data Sources and Field Sampling

Within each watershed, five sites were examined. Data used in metric calculation were obtained from two sources: (1) long-term data sets from a beach-seine, finfish monitoring program and (2) nearshore seine sampling in July 2002. In each selected watershed, every attempt was made to use long-term survey locations that included samples from 2002. In those instances where fewer than five long-term sites were present in a selected watershed, additional sites were sampled in 2002 based on a stratified random design. Data from July 2002 were extracted from long-term monitoring surveys at each station, and combined with values obtained from auxiliary 2002 sampling sites for metric compilation.

TABLE 23.2

Fish Guild Categories Used in the Development of Metrics

Fish Guilds	Categories
1. Reproductive	Marine spawner
	Anadromous
	Freshwater spawner
	Estuarine spawner
2. Trophic level	Carnivore
	Planktivore
	Benthivore
3. Primary life history	Marine
	Estuarine
	Fresh water
	Diadromous
	Estuarine-dependent nursery
4. Habitat	Pelagic
	Benthic
5. Origin	Estuarine resident
	Estuarine nonresident

Note: Categorization is based on the predominant behavior of each species at the respective life stage typically observed within the nearshore estuarine waters from July to September. The reproductive guild categorizes the location of spawning of each species. Species are placed in respective trophic level categories based on their primary prey items. The primary life history guild describes a critical aspect or primary ecosystem on which the species success depends. The habitat guild broadly classifies the position in the water column where each species spends the majority of its time. The origin guild separates species that are year-round estuarine species.

TABLE 23.3

Fish Community Metrics Assessed for Use in a Multimetric Index and Associated Source

Fish Community Metrics	Ref.
Species richness/diversity measures	
Species richness (SR = no. of species – 1/log(no. of individuals))	This chapter
Proportion of benthic-associated species (no. benthic-associated species/total no. of species)	Deegan et al., 1997
No. of dominant species (no. of species that make up 90% of total abundance)	Deegan et al., 1997
No. of resident species	Deegan et al., 1997
Fish abundance	
Ln abundance	Deegan et al., 1997
Trophic composition	
Trophic index (relative proportions of three broadly defined trophic guilds: piscivores, planktivores, and benthivores (scaled to 5)	Jordan and Vaas, 2000
Nursery function	
No. of estuarine spawning species	Deegan et al., 1997
No. of estuarine nursery species	Deegan et al., 1997

Fish abundance data were acquired from two long-term monitoring, finfish data sets: (1) the Maryland Estuarine Juvenile Finfish Survey (Durell and Weedon, 2002; http://www.dnr.state.md.us/fisheries/juvin-dex/index.html), and (2) the Virginia Institute of Marine Science, Juvenile Striped Bass Beach Seine Survey (http://www.fisheries.vims.edu/trawlseine/sbmain.htm). Both were beach seine surveys conducted from July to September in the tributaries of the Chesapeake Bay. The seine survey protocols were slightly different between programs. At each seine station, Maryland conducted three rounds of two seine hauls per round for a total of six hauls per year, while Virginia conducted five rounds of two seine hauls per round for a total of ten hauls per year. To normalize the data between seine surveys, only stations with complete data for July 2002 were included in the analyses. For both surveys, abundance and species counts were extracted for the first round (July) of sampling with data from the replicate seine hauls combined for metric calculations.

Auxiliary seine sampling protocols were designed to correspond with protocols used in long-term finfish surveys. Two replicate seine hauls (30.5×1.22-m bagless beach seine of 6.4-mm bar mesh) were conducted at each site during July 2002. One end of the seine was held on shore. The other was fully stretched perpendicular to the beach and swept with the current over a quarter circle quadrant. Ideally, the area swept was equivalent to a 729-m^2 quadrant. When depths of 1.22 m or greater were encountered, the offshore end was deployed along this depth contour. An estimate of distance from the beach to this depth was recorded. Replicate hauls were combined and counts and total lengths recorded for each finfish species (or a subsample of at least 25 individuals); selected crustacean species were also enumerated.

Metric Analyses

Metric distributions were normalized with natural logarithms or arc-sin transformations when necessary (only in two instances). The metrics abundance and trophic index required transformations (natural logarithms and arc-sin, respectively), but all other metrics had normal distributions and were not transformed. Individual metrics were standardized based on each metric distribution and aggregated, without weighting, into a fish community index (FCI) score. For example, each species richness metric value was divided by the largest observed richness measure to standardize values (0–1) based on existing conditions for the year (no reference condition was considered); standardized metrics were then added to obtain the aggregate FCI.

The applicability of metrics and variability of the FCI and metrics were assessed by calculating correlation coefficients for metric scores, graphing relationships between individual metrics, and the FCI, and examining principal components analysis (PCA) coefficients of the metrics. By plotting the FCI vs. individual metrics, the variability of the FCI can be visually assessed. The precision of the FCI can be estimated based on the proximity of points to a 45° line when relating individual metrics to the aggregate FCI. PCA was applied to individual fish community metrics to evaluate the usefulness of the multimetric index (FCI) as a descriptor of ecosystem integrity. Those metrics that are supported in a multimetric index should exhibit similar associations. Metrics that exhibited similar trends in correlation (high and positive) with the aggregate FCI of all eight tested metrics were combined into a final FCI by summing standardized individual metric values.

Habitat Assessment

Within approximately 150 m of each site, several habitat parameters were assessed: (1) riparian land use; (2) shoreline alteration; (3) in-stream habitat structure (e.g., submerged aquatic vegetation, or SAV; woody debris; shell); and (4) amount of fringing marsh and beach. The assessments were based in part on quantitative, continuous shoreline surveys that combine onsite GPS (global positioning system) surveys with GIS (geographic information system) technology, underway in Virginia and Maryland (Center of Coastal Resources Management; http://ccrm.vims.edu/gis/gisdata.html). For the purposes of this study two habitat variables were used: shoreline alteration and in-stream habitat structure. Onsite visual evaluation and field photographs were used to estimate percentages of the specified habitat. Percentages were scored on a sliding scale (0 to 20) similar to rapid bioassessment protocols developed by the U.S. Environmental Protection Agency (Barbour et al., 1999).

Fish Community Indices and Habitat Measures

Final FCI scores were independently compared with shoreline condition, in-stream habitat, and watershed land use. Shoreline condition and in-stream habitat scores were ranked based on data distributions and placed into three groups that represented gradients of anthropogenic alterations (25 and 75% quartile cutoffs were used to establish the three groups as scores ranked <25%, 25 to 75%, or >75%). Shoreline condition was represented as highly, moderately, or minimally altered. In-stream habitat was categorized as minimal, moderate, or abundant. Land use patterns were used as surrogates for human disturbance levels and categorized as developed, agriculture, or forested. Correlation analyses were used to explore trends in the data, and ANOVA (analysis of variance) analyses further examined relationships between categorized habitat values and FCI scores. k-dominance curves depicted cumulative ranked abundances plotted against species rank to examine differences in communities at grades of shoreline condition and watershed land use. Curves with relatively shallow slopes are indicative of communities dominated by one or a few species and are thought to be representative of impacted sites (Attrill, 2002).

Relationships between habitat measures were also assessed. Associations between shoreline alteration and available in-stream structure to fish communities were predicted and examined. Additionally, we tested linkages between watershed and riparian land use by assessing correlations between percentages of each land use type (developed, agricultural, and forested) in a watershed, and the corresponding riparian land use category for a subset of 13 watersheds in the Chesapeake Bay.

Results

All but one of the examined fish community metrics were positively and highly correlated ($r \geq 0.5$) with the summed metrics (FCI). The majority of correlations among metrics were positive. Total number of individuals (transformed into natural logarithms) had low, nonsignificant correlations with the FCI and negative correlations with other individual metrics. Similarly, plots of FCI and individual metrics indicated strong linear relationships in all but one metric (i.e., abundance, natural logarithm transformed) (Figure 23.2). PCA of individual fish community metrics supported the use of all but two of the metrics (i.e., abundance, natural logarithm transformed, and proportion of benthic-associated species) in a composite FCI. The first and second principal components accounted for 74% of the variance in the data set (Table 23.4). All metrics were positively associated with PC1, with lower loading values for proportion of benthic-associated species and total abundance. When considering correlation patterns and PCA analyses, the use of all the metrics, with the exception of total abundance, was supported for the development of a nearshore FCI in coastal plain estuarine ecosystems.

Links among habitat conditions were supported in the relationships between in-stream and shoreline condition, as well as shoreline and adjacent watershed land use. Shoreline condition and in-stream habitat measures were significantly correlated ($p < 0.0001$; $r = 0.58$) indicating a negative association between shoreline alterations and available in-stream structural habitat, such as SAV and woody debris (Figure 23.3). Dominant watershed land use was reflected in shoreline land use conditions for all three of the categories (developed: $p = 0.05$, $r = 0.54$; agricultural: $p = 0.04$, $r = 0.58$; forested: $p = 0.02$, $r = 0.65$) (Figure 23.4). Correlations are noted cautiously for agricultural and developed dominant land use, however, because of the lack of sampled watersheds with high values of each land use (i.e., low number of data points with dominant land use > 40%).

Biotic responses were correlated with habitat condition in the nearshore, riparian zone, and watershed. As FCI values increased, sites typically consisted of adjacent shorelines with minimal alterations and abundant in-stream habitat ($p = 0.01$ and $p < 0.0001$, respectively) (Figure 23.5). Differences in FCI values were not significant among shoreline condition categories (ANOVA; $p = 0.09$). However, further examination by Tukey's pairwise comparisons illustrated differences between FCI scores in highly and minimally altered conditions. FCI scores were significantly different among all three categories of available in-stream habitat (minimal, moderate, and abundant habitat) (ANOVA; $p < 0.0001$). Likewise, FCI scores in developed watersheds were lower than, but not significantly different from, agricultural

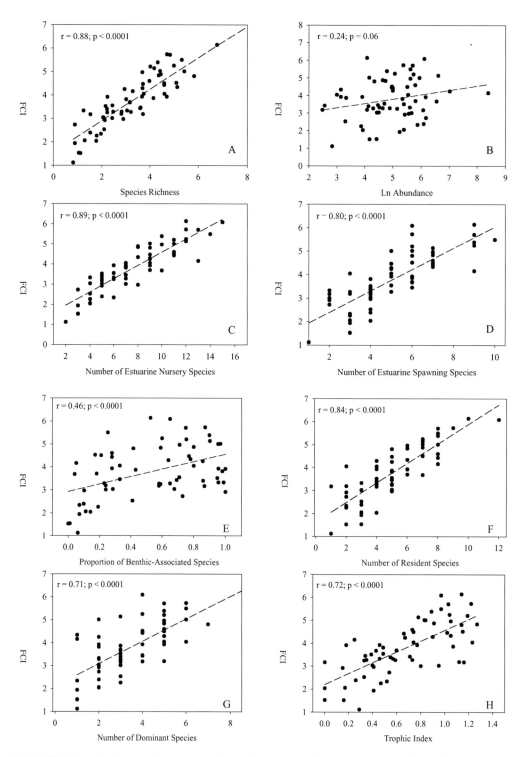

FIGURE 23.2 Individual raw metrics scores: (A) species richness, (B) abundance, natural logarithm transformed, (C) number of estuarine nursery species, (D) number of estuarine spawning species, (E) proportion of benthic-associated species, (F) number of resident species, (G) number of dominant species, and (H) trophic index vs. an aggregate FCI of all eight metrics. Metrics showed constantly increasing responses to increasing FCI scores, with the exception of abundance, natural logarithm transformed (B).

TABLE 23.4

Eigenvectors and Accountable Variances of the First Two Principal Components
Based on Individual Fish Community Metrics

PC1	PC2	Variable
0.44	0.16	Species richness
0.12	0.18	Proportion of benthic-associated species
0.35	0.36	No. of dominant species
0.42	0.25	No. of resident species
0.10	0.71	Ln total abundance
0.34	0.40	Trophic index
0.41	0.20	No. of estuarine spawning species
0.45	0.22	No. of estuarine nursery species
55%	19%	Accountable variance

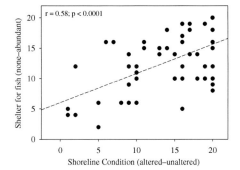

FIGURE 23.3 Comparison between available in-stream habitat (scaled from none to abundant habitat) and shoreline condition (scaled from highly altered to unaltered states) per site.

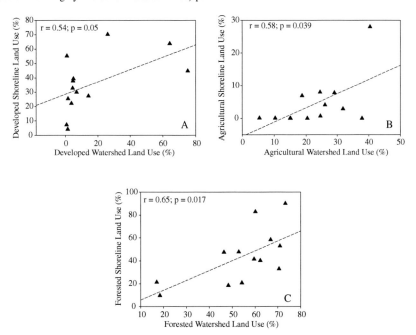

FIGURE 23.4 Comparison between percentages of each land use type in a watershed, and the corresponding riparian land use category: (A) developed, (B) agricultural, or (C) forested. Data were extracted from a subset of 13 watersheds in the Chesapeake Bay: Back, Battle, Breton Bay, Chickahominy, Elizabeth, Lower Rappahannock, Lower James, Pagan, Piankatank, Severn, St. Clements, St. Mary's, and Totuskey.

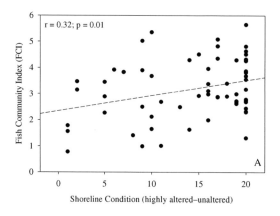

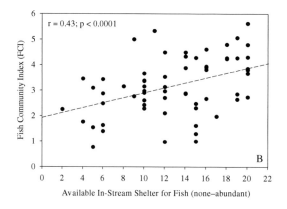

FIGURE 23.5 FCI scores (an aggregate of all consistently correlated metrics; Ln abundance was excluded) in relation to (A) shoreline condition (scaled from highly altered to unaltered), and (B) in-stream habitat (scaled from none to abundant).

or forested watersheds (ANOVA; $p = 0.08$) (Figure 23.6). Percentages of cumulative dominance by species rank indicated that watersheds with developed land use and highly altered shoreline conditions had fewer dominant species than agricultural or forested land uses and moderate or minimally altered shoreline, respectively (Figure 23.7).

Discussion

Biotic community responses to habitat condition were measured with metrics that represented key aspects of fish community integrity, as well as elements of life history dependent on estuarine condition. In this regard, the metrics chosen to evaluate estuarine condition follow the traditional IBI premise that compelling indicators should encompass critical components of fish assemblages by depicting elements of structure, composition, and function of a given community (Karr, 1981; Cairns et al., 1993; Hughes et al., 1998). Because estuarine fish communities differ markedly from stream fish communities, modified metrics had to be tested for effectiveness in this new system. All but one tested metric was consistently associated with fish community structure and function in the sampled estuarine systems. Thus, the derived FCI may be a useful indicator of shallow-water estuarine condition.

Attempts were made to diminish the effects of some factors such as salinity by limiting the sites to oligo-mesohaline estuarine segments, and selecting metrics that are independent of salinity influences (e.g., trophic index), but slight negative correlations between FCI and salinity existed. Further data limitations include temporal variability inherent in estuarine fish populations, which places constraints

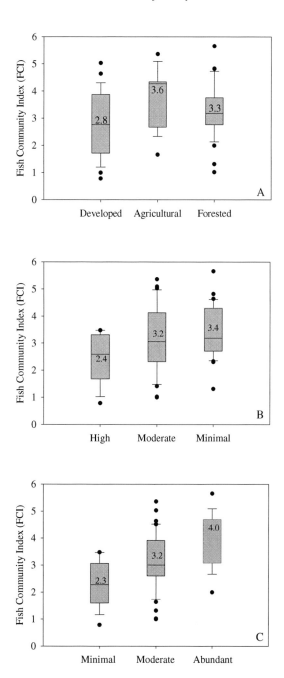

FIGURE 23.6 FCI variability by categorized habitat: (A) watershed land use, (B) shoreline condition, and (C) in-stream habitat abundance. Shoreline and in-stream habitat were categorized based on data distribution thresholds of 25 and 75% quartiles. Land use was categorized broadly as developed, agricultural, and forested based on dominant land use patterns. Average FCI values and 95th and 5th percentiles and outliers per habitat category are depicted on each associated bar graph. One-way ANOVA analyses and Tukey's pairwise multiple comparisons were used to assess differences in condition. (A) Developed watersheds had lower, but not significantly different FCI average values in relation to agricultural and forested watersheds (one-way ANOVA, $p = 0.09$), Tukey's multiple comparisons indicted differences between FCI scores in developed and agricultural conditions; (B) Highly altered shoreline had lower, but not significantly different associated FCI average values in relation to moderately or minimally altered shoreline (one-way ANOVA; $p = 0.09$), and Tukey's multiple comparisons indicted differences between FCI scores in highly altered and unaltered conditions; (C) FCI scores were significantly different among all three in-stream habitat conditions (minimal, moderate, and abundant available habitat in the nearshore) (one-way ANOVA; $p < 0.0001$; Tukey's multiple comparisons).

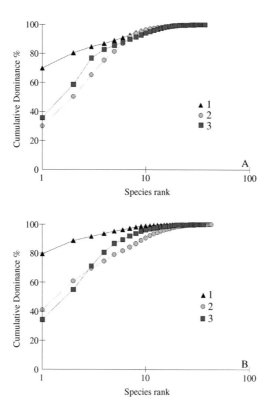

FIGURE 23.7 *k*-dominance curves depict cumulative ranked abundances of fish plotted against species rank to examine differences in communities at grades of (A) watershed land use and (B) shoreline condition. Curves with relatively shallow slopes are indicative of communities dominated by one or a few species and are considered to be representative of impacted sites. Categories of habitat condition were (A) watershed land use (1 = developed; 2 = agricultural; 3 = forested), and (B) shoreline condition (1 = highly altered; 2 = moderately altered; 3 = minimally altered).

on the interpretation of results. While we strove to use long-term survey data, historic data were unavailable for most of the developed watersheds; thus only data from 2002 were available for comparison with auxiliary sites. As sampling took place during a single year, statements can be made about the relative condition of the estuaries during that time, but only tentative conclusions may be made regarding the long-term condition.

Temporal trends in fish community data are being examined to assess the ability of the FCI to track estuarine health across years. Preliminary analysis of historic data (1989 to 2000) indicates correlation between mean and variance in FCI; in other words, as FCI scores decrease, their variances increase. While several studies reported similar trends in variability across condition (Karr, 1987; Steedman, 1988; Fore et al., 1994; Yoder and Rankin, 1995), some have noted higher variability with high IBI scores as opposed to low scores (Hugueny et al., 1996) or highest variation at intermediate values (Hughes et al., 1998). This association may be a real reflection of increases in variability in disturbed systems, a statistical artifact, or a regional phenomenon. The possibility of using variability as an indicator may be pursued if the trend proves to be consistent across systems.

Correlative trends between biotic responses and habitat condition were most evident for the in-stream habitat, and to a lesser extent for shoreline condition and watershed land use. This result is intuitive since fish may not be responding directly to riparian or watershed alterations, but to changes in nearshore habitat. In fact, several recent studies have addressed the association between incremental shoreline alteration and nearshore and littoral habitat condition, as well as aquatic community impacts (Bryan and Scarnecchia, 1992; Beauchamp et al., 1994; Ward et al., 1994; Christensen et al., 1996; Jennings et al., 1999). If linkages exist between in-stream habitat and shoreline condition, as indicated by correlation analyses, then shoreline condition may be considered an indirect indicator of shallow-water ecosystem health in the Chesapeake Bay.

Biotic responses to intense watershed and riparian alterations in fresh water, and to a lesser extent in estuarine systems, have typically been characterized by lower species diversity, less trophic complexity, altered food webs, altered community composition, and reduced habitat heterogeneity (Angermeier and Karr, 1984; Howarth et al., 1991; Schlosser, 1991; Everett and Ruiz, 1993; Hoss and Thayer, 1993; Roth el al., 1996). In support, this study observed evidence of fish community structural and functional changes in relation to extreme habitat alterations.

Although declines in fish community indices could be discriminated where watersheds and shorelines had the highest levels of human impact, moderate alteration of habitat could not be determined using community metrics alone. One explanation is that in this study agricultural watersheds often had high percentages of forested riparian zones. This fact may account for the failure to detect differences in biotic indices between forested and agricultural watersheds. From the fish community perspective, there may be no difference in the two landscapes, or the habitat impacts associated with a buffered agricultural landscape may not exceed the response threshold in fish communities.

The problem of resolving impacts on condition at moderate levels of disturbance is universal in ecological indicator development. In estuarine systems this challenge is particularly acute. The FCI developed in this study can resolve good system conditions and poor system conditions, but discrimination of intermediate states is difficult. In freshwater lake systems, lower fish community diversities have been observed at sites with bulkheads than at ones with riprap (Jennings et al., 1999). It is hypothesized that riprap may offer more habitat heterogeneity, and bulkheads may act to reduce available shallow water habitat and woody debris inputs. But in estuarine systems, the consequences of these differences may be lost in fish community metrics simply because the community varies in response to other more potent natural signals in the system.

The absence of a finely resolved biotic response along the entire gradient of human impact in estuarine systems may reflect a variable resistance in biotic communities, particularly fish. It is probable that fish continually respond to an aggregation of multiple stressors that do not co-vary. The result would be an apparent reaction to any one stressor that is quite variable. For this reason, biological indicators may not be suitable as sole indicators of condition along the entire stressor gradient. Additional metrics may be needed to improve indicator resolution and accuracy.

In this study we have attempted to develop initial relationships between easily observable habitat conditions and the FCI. A limited amount of variance in the data could be explained with habitat condition alone. In addition, the integrative descriptors, watershed land use and shoreline condition, appear to be effective estuarine condition indicators at the regional scale. With more detailed analysis, shoreline condition may indicate system condition at multiple spatial scales. Additional biological communities (sedentary, benthic macroinvertebrates) are being examined in association with habitat condition, to assess biotic responses in disparate assemblages, and ascertain if stronger signals are apparent with different target organisms. These are first steps in identifying other system metrics that will be considered as elements in a multimetric indicator. The objective is to find useful indicators for every point along the stressor gradient, and to combine them in a simple synthesized metric (estuarine shallow-water index, or ESI) that integrates a lot of detailed data about a complex system. Where the FCI can describe condition, and the shoreline condition index can identify potential causative factors, the ESI is intended to both indicate and communicate condition relative to attainable goals.

Acknowledgments

This research has been supported by a grant from the U.S. Environmental Protection Agency Science to Achieve Results (STAR) Estuarine and Great Lakes (EaGLe) program through funding to the Atlantic Slope Consortium, Virginia Institute of Marine Science, U.S. EPA Agreement (R-82868401). We thank Eric Durell with the Maryland Department of Natural Resources and Chris Bonzek and Herb Austin with the Virginia Institute of Marine Science for providing beach seine survey data. Many thanks to those who contributed to the field and laboratory components of this project, including Kory Angstadt,

Harry Berquist, Liz Herman, Sharon Killeen, Walter Priest III, Tamia Rudnicky, Dan Schatt, and David Weiss. Although the research described in this article has been funded wholly or in part by the United States Environmental Protection Agency through cooperative agreement R-82868401 to Atlantic Slope Consortium, Virginia Institute of Marine Science, it has not been subjected to the Agency's required peer and policy review and therefore does not necessarily reflect the views of the Agency and no official endorsement should be inferred. This chapter is contribution No. 2168 of the Virginia Institute of Marine Science, The College of William and Mary.

References

Angermeier, P. L. and J. R. Karr 1984. Relationships between woody debris and fish habitat in a small warmwater stream. *Transactions of the American Fisheries Society* 113(6):716–726.

Attrill, M.J. 2002. Community-level indicators of stress in aquatic ecosystems. In *Biological Indicators of Aquatic Ecosystem Stress*, S. M. Adams (ed.). American Fisheries Society, Bethesda, MD, pp. 473–508.

Bain, M. B., A. L. Harig, D. P. Loucks, R. R. Goforth, and K. E. Mills. 2000. Aquatic ecosystem protection and restoration: advances in methods for assessment and evaluation. *Environmental Science and Policy* 3:S89–S98.

Barbour, M. T., J. Gerritsen, B. D. Snyder, and J. B. Stribling. 1999. Rapid Bioassessment Protocols for Use in Streams and Wadeable Rivers: Periphyton, Benthic Macroinvertebrates and Fish, 2nd ed. EPA 841-B-99-002. U.S. Environmental Protection Agency, Office of Water, Washington, D.C.

Beauchamp, D. A., E. R. Byron, and W. A. Wurtsbaugh. 1994. Summer habitat use by littoral-zone fishes in Lake Tahoe and the effects of shoreline structures. *North American Journal of Fisheries Management* 14(2):385–394.

Bryan, M. D. and D. L. Scarnecchia. 1992. Species richness, composition, and abundance of fish larvae and juveniles inhabiting natural and developed shorelines of a glacial Iowa lake. *Environmental Biology of Fishes* 35(4):329–341.

Cairns, J., P. V. McCormick, and B. R. Niederlehner. 1993. A proposed framework for developing indicators of ecosystem health. *Hydrobiologia* 263(1):1–44.

Carmichael, J. B., B. Richardson, M. Roberts, and S. J. Jordan. 1992. Fish Sampling in Eight Chesapeake Bay Tributaries. Technical Report, Chesapeake Bay Research and Monitoring Division, Maryland Department of Natural Resources, Annapolis.

Chesapeake Bay Program. 1999. The State of the Chesapeake Bay. EPA 903-R99-013, CBP/TRS 222/108, U.S. Environmental Protection Agency, Washington, D.C., 59 pp.

Christensen, D. L., B. J. Herwig, D. E. Schindler, and S. R. Carpenter. 1996. Impacts of lakeshore residential development of coarse woody debris in north temperate lakes. *Ecological Applications* 6(4):1143–1149.

Deegan, L. A., J. T. Finn, S. G. Ayvazian, C. A. Ryder-Kieffer, and J. Buonaccorsi. 1997. Development and validation of an Estuarine Biotic Integrity Index. *Estuaries* 20(3):601–617.

Durrell, E. Q. and C. Weedon. 2002. Striped bass seine survey juvenile index webpage. http://www.drr.state.md.us/fisheries/juvindex/index.html.

Everett, R. A. and G. M. Ruiz. 1993. Coarse woody debris as a refuge from predation in aquatic communities. *Oecologia* 93(4):475–486.

Fausch, K. D., J. Lyons, J. R. Karr, and P. L Angermeier. 1990. Fish communities as indicators of environmental degradation. In *Biological Indicators of Stress in Fish,* S. M. Adams (ed.). American Fisheries Society Symposium 8, Bethesda, MD, pp. 123–144.

Fore, L. S., J. R. Karr, and L. L. Conquest. 1994. Statistical properties of an index of biological integrity used to evaluate water resources. *Canadian Journal of Fisheries and Aquatic Science* 51(5):1077–1087.

Hardy, J. D., Jr. (ed.). 1978. *Development of Fishes of the Mid-Atlantic Bight: An Atlas of the Egg, Larval and Juvenile Stages. Vol. III. Aphredoderidae through Rachycentridae.* U.S. Fish and Wildlife Service Biological Services Program, FWS/OBS-78/12.

Hoss, D. E. and G. W. Thayer. 1993. The importance of habitat to the early life history of estuarine dependent fishes. *American Fisheries Society Symposium* 14:147–158.

Howarth, R. W., J. R. Fruci, and D. Sherman. 1991. Inputs of sediment and carbon to an estuarine ecosystem: influence of land use. *Ecological Applications* 1(1):27–39.

Hughes, R. M., P. R. Kaufmann, A. T. Herlihy, T. M. Kincaid, L. Reynolds, and D. P. Larsen. 1998. A process for developing and evaluating indices of fish assemblage integrity. *Canadian Journal of Fisheries and Aquatic Science* 55(7):1618–1631.

Hughes, J. E., L. A. Deegan, M. J. Weaver, and J. E. Costa. 2002. Regional application of an index of estuarine biotic integrity based on fish communities. *Estuaries* 25(2):250–263.

Hugueny, B., S. Camara, B. Samoura, and M. Magassouba. 1996. Applying an index of biotic integrity based on fish assemblages in a West African river. *Hydrobiologia* 331(1–3):71–78.

Jenkins, R. E. and N. M. Burkhead. 1994. *Freshwater Fishes of Virginia.* American Fisheries Society, Bethesda, MD.

Jennings, M. J., M. A. Bozek, G. R. Hatzenbeler, E. E. Emmons, and M. D. Staggs. 1999. Cumulative effects of incremental shoreline habitat modification on fish assemblages in north temperate lakes. *North American Journal of Fisheries Management* 19(1):18–27.

Jordan, S. J. and P. A. Vaas. 2000. An index of ecosystem integrity for Northern Chesapeake Bay. *Environmental Science and Policy* 3:S59–88.

Karr, J. R. 1981. Assessment of biotic integrity using fish communities. *Fisheries* 6(6): 21–27.

Karr, J. R. 1987. Biological monitoring and environmental assessment: a conceptual framework. *Environmental Management* 11(2):249–256.

Karr, J. R. and E. W. Chu. 1999. *Restoring Life in Running Waters: Better Biological Monitoring.* Island Press, Washington, D.C.

Karr, J. R., K. D. Fausch, P. L Angermeier, P. R. Yant, and I. J. Schlosser. 1986. Assessing biological integrity in running waters, a method and its rationale. Illinois Natural History Survey Special Publication, 5:1–28.

Lackey, R. T. 2001. Values, policy, and ecosystem health. *Bioscience* 51(6):437–443.

Lippson, A. J. and R. L. Moran. 1974. Manual for Identification of Early Developmental Stages of Fishes of the Potomac River Estuary. Prepared for Maryland Department of Natural Resources, Power Plant Siting Program. PPSP-MP-13.

Meng, L., C. D. Orphanides, and J. C. Powell. 2002. Use of a fish index to assess habitat quality in Narragansett Bay, Rhode Island. *Transactions of the American Fisheries Society* 131(4):731–742.

Murdy, E. O., R. S. Birdsong, and J. A. Musick. 1997. *Fishes of Chesapeake Bay.* Smithsonian Institution Press, Washington, D.C.

Regier, H. A. 1993. The notion of natural and cultural integrity. In *Ecological Integrity and the Management of Ecosystems,* S. Woodley, and G. Francis (eds.). St. Lucie Press, Delray Beach, FL, pp. 3–18.

Roth, N. E., J. D. Allan, and D. L. Erickson. 1996. Landscape influences on stream biotic integrity assessed at multiple spatial scales. *Landscape Ecology* 11(3):141–156.

Schlosser, I. J. 1991. Stream fish ecology: a landscape perspective. *Bioscience* 41(10):704–712.

Scientific and Technical Advisory Committee. 2003. Chesapeake Futures: Choices for the 21st Century. STAC Publication Number 03-001, Chesapeake Research Consortium, Inc., Edgewater, MD.

Simon, T. P. and J. Lyons. 1995. Application of the Index of Biotic Integrity to evaluate water resource integrity in freshwater ecosystems. In *Biological Assessment and Criteria — Tools for Water Resource Planning and Decision Making,* W. S. Davis and T. P. Simon (eds.). Lewis Publishers, Boca Raton, FL, pp. 245–262.

Steedman, R. J. 1988. Modification and assessment of an index of biotic integrity to quantify stream quality in southern Ontario. *Canadian Journal of Fisheries and Aquatic Science* 45(3):492–501.

U.S. EPA. 1982. Chesapeake Bay: Introduction to an Ecosystem. U.S. Environmental Protection Agency, Washington, D.C.

Ward, D. L., A. A. Nigro, R. A. Farr, and C. J. Knutsen. 1994. Influence of waterway development on migrational characteristics of juvenile salmonids in the Lower Willamette River, Oregon. *North American Journal of Fisheries Management* 14(2):362–371.

Yoder, C. O. and E. T. Rankin. 1995. Biological response signatures and the area of degradation value: new tools for interpreting multimetric data. In *Biological Assessment and Criteria-Tools for Water Resource Planning and Decision Making,* W. S. Davis and T. P. Simon (eds.). Lewis Publishers, Boca Raton, FL, pp. 263–286.

24

Fishes as Estuarine Indicators

Stephen A. Bortone, William A. Dunson, and Jaime M. Greenawalt

CONTENTS

Introduction

The goal of developing effective estuarine bioindicators is to determine the health of the ecosystem and causes of any detrimental changes. Adverse changes in aquatic habitats are often manifested in stress on fishes (see Adams, 1990, for an inclusive presentation). Studies on the impact of such stress are well documented in freshwater lentic and lotic environments, but are conspicuously rare for estuarine situations (Simon, 1999). It is important to note that our objective here is to use quantifiable indices of fish condition to assess the scale and direction of changes in the estuarine environment. A primary aspect of estuarine health is the response its biological components make to changes in abiotic and biotic factors of natural and anthropogenic origin. Separation of these two major causative agents poses a great problem, but the effort is crucial in assessing whether observed changes in the estuary are perceived as adverse or benign. The solution to this conundrum lies with a precise application of the scientific method across the entire spectrum of experimental control of variables and the realism of the conditions of study. This requires both field and laboratory observations and experiments, and includes experimental simulations extending from mathematical models, to highly controlled laboratory experiments, to mesocosms, to studies in the most complex but least controlled of natural circumstances. Any one level of study, no matter how elegantly conducted, will be inadequate to provide a full-scale explanation of natural events and their separation from anthropogenic influences. Some fish are ideally suited for this modern approach as they lend themselves well to multiple levels of inquiry.

0-8493-2822-5/05/$0.00+$1.50
© 2005 by CRC Press

Species Choices

There are a plethora of species that have the potential to serve as indicators of estuarine condition. Below we examine many of the categorical descriptors that help to organize the decision-making process when considering using fishes as estuarine indicators. The consideration is neither hierarchical nor inclusive but is demonstrative of the multiplicity of features that should be considered.

Taxonomic Group

Briefly, fishes are considered to be those "aquatic vertebrates that have gills throughout life and limbs, if any in the shape of fins" (Nelson, 1994, p. 2). Fish diversity includes groups such as hagfishes (Myxiniformes); lampreys (Petromyzontiformes); sharks, skates and rays (Chondrichthyes); and a variety of Osteichthyes (bony fishes) that includes gars (Semionotiformes), sturgeons (Acipenseriformes), eels (Elopiformes), shad and herrings (Clupeiformes), Osteriophysines (including catfishes and carp), salmon (Salmoniformes), cods and haddock (Gadiformes), killifishes (Cyprinodontiformes), and the Percomorpha (including the perches, groupers, and flatfishes among others). This diversity is extraordinary among vertebrates (but not so impressive to our colleagues who study invertebrates!) in that fishes are represented by more than 482 families and at least 24,600 species (Nelson, 1994). While it is difficult to accurately estimate the number of estuarine-dependent fish species, nevertheless, the number is at least in the thousands worldwide.

Knowledge of the ecological affinities and physiological tolerances of the phylogenetic group being studied is of great importance in the choice of estuarine fish bioindicators. For example, many species of gobies (Gobiidae) are found strictly in fresh water, but their phylogenetic lineage clearly has an affinity to the marine environment (Nelson, 1994). Differential responses to salinity conditions for some species might, therefore, be explained in large part to the phylogenetic linkages rather than to purely environmental adaptations.

Life Stage

Life stages in estuarine fishes can offer advantages and disadvantages where developing effective estuarine indicators. Dovel (1971) summarized the egg and larval distribution of many temperate North American fish species, indicating the dynamic relationship that many eggs and larvae have with the spatial and temporal salinity regimes of estuaries.

Certain species, for example, are known to occur in the estuary but have affinities to spawn and/or undergo juvenile development in the lower salinity (even up to and including freshwater) portions of the estuary and even upstream (e.g., striped bass, *Morone saxatilis,* and hogchoker, *Trinectes maculatus*). Many species are known to spawn in the mid-salinity areas of the estuary. For example, the tidewater silverside (*Menidia beryllina*) spawns in shallow areas, its eggs attaching to the emergent vegetation, while the bay anchovy (*Anchoa mitchilli*) spawns near the surface in open portions of the estuary. Last, there are species that are more often found in marine waters and they tend to spawn near the downstream portions of the estuary. Fish such as the Atlantic croaker (*Micropogonias undulatus*) and Atlantic menhaden (*Brevoortia tyrannus*) are examples of these latter species.

Thus, it is important to consider the preferred spawning salinity and preferred habitat for eggs and larvae when developing an estuarine indicator fish species study. Choices regarding salinity and season for exposures can greatly influence study results.

Level of Organization

As with any group of organisms, the level of organization examined in fishes can serve as an effective indicator of estuarine conditions. Studies on fish have been conducted at virtually every level of organization: molecular, physiological, population, and community. Other authors in this book (e.g., Adams, Chapter 2) have indicated the relationship between the level of organization studied and the time-response

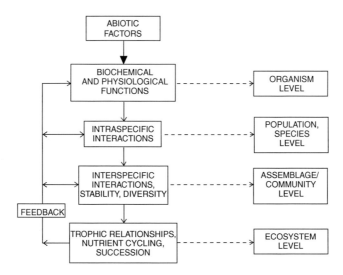

FIGURE 24.1 Flow diagram of the interaction among levels of biological organization (right side) and factor-responses.

scale with sensitivity and ecological significance. It is sufficient to indicate here that the measured response variables in fish should be chosen to adequately answer the questions being asked with regard to sensitivity and significance to the estuarine change being monitored. Figure 24.1 illustrates the relationship between levels of organization and the respective hierarchical biological features.

Trophic Status

Estuarine fishes can occupy several trophic levels. While only a few estuarine fish species can be considered facultative herbivores (such as the sheepshead minnow, *Cyprinodon variegatus*); mullet (*Mugil* spp.) can be considered detritivores; anchovies (*Anchoa* spp.) are planktivores; gobies and blennies (*Gobiosoma* spp. and *Chasmodes* spp.) are low-level invertebrate pickers; croakers and drums (Sciaenidae) consume larger invertebrates such as crabs, shrimp, and mollusks; red drum (*Sciaenops ocellatus*) and flounder (*Paralichthys* spp.) can be considered piscivores, along with the large sharks. Thus, fishes present the investigator with a broad array of trophic level consumers. It should be noted, however, that some fish species occupy a broad range of trophic levels in estuaries, but a narrow range at different stages of their lives. In addition, fish species often change or alter their trophic level membership ontogenetically. For example, smaller spotted seatrout (*Cynoscion nebulosus*) may consume predominantly smaller crustaceans while the largest spotted seatrout may consume mostly fishes (Baltz et al., 2003). While changes in a species' trophic parameters may be reflective of changes in the trophic level it occupies, these changes may also be indicative of larger-scale conditions in the entire ecosystem.

Estuarine Dependency

Estuarine dependency is an important qualifier when considering the development of a fish species as an estuarine indicator because the length of time spent in the estuary could affect the sensitivity of the response to adverse impacts. Phylogenetic history and life stage clearly can dictate their level of estuarine dependency. The diversity of fishes includes the diadromous fishes (i.e., migratory species that move between fresh and saltwater at specific stages of their life cycle) and includes a number of anadromous (freshwater spawning marine fishes such as salmon), catadromous (marine spawning freshwater fishes such as the American eel), and amphidromous fishes (such as gobies that move back and forth between fresh and saltwater as mature adults) (Helfman et al., 1997). Life-stage dependency varies ontogenetically. This can occur not only at the earliest life stage, for example, at the egg or larval stage (Dovel, 1971), but also at the juvenile life stage (Baltz et al., 2003) and even at the adult stage as fishes alternate between preferred spawning areas. Importantly, the more the fish species moves around, the less resolution one will be able to obtain from the species as an indicator. Consequently, indication of local estuarine conditions can only be provided by fish that remain in the local estuary.

Sensitivity

Estuarine fishes are noted for their tolerance to the wide range of changes they might encounter, especially with regard to salinity, temperature, and oxygen. Nevertheless, each species shows varying degrees of sensitivity. For example, Dunson et al. (1998) showed that the estuarine-dependent sheepshead minnow (*Cyprinodon variegatus*) responded to extremely low salinities with lower growth and reproductive rates. Some of these responses may be due to long-term evolutionary adaptations of the species, their short-term physiological adaptations, or even the density and community present (Rowe and Dunson, 1995). An indicator species should be chosen that will have a biological response with the sensitivity that accommodates detection of variation in the variable being examined.

Prior Database

The availability of a prior database of information from which to compare the status and trends of biological responses to estuarine conditions is, in some circumstances, an overriding factor in the decision process to determine the most appropriate species. This should not deter the development of "new" species as potential estuarine indicators. However, a prior database of both laboratory and field data on a species allows the researcher the "luxury" of a more efficient study design and certainly reduces overall associated indicator development costs. Also important are the temporal comparisons that are permitted when such prior databases exist.

Importance to Stakeholders

Even though this last consideration is not strictly biological in nature, in many circumstances it may prove to be one of the most important. Too often biologists make their indicator choices purely on biological appropriateness. Nevertheless, being able to relate study results to the general public is becoming paramount in the consideration of estuarine indicator choices. It is easy to argue that "all things being equal" one should choose a species that facilitates interpretation and dialog with the general public. However, when one considers that scientific study results often serve research managers as "just another opinion" in the larger scale of decision making, it is of even greater importance to consider the relevance that a particular group (i.e., taxonomic group, species, or stock) has to stakeholders. On one extreme, selecting an obscure species risks not gaining attention from both the public and resource managers. Oppositely, choosing a species that is of paramount economic or social interest among the general public is likely to garner substantial interest among the diversity of stakeholders.

Laboratory vs. Field

Studies on fishes (and other organisms as well) can include the examination of biological responses to specific controlled factors and the ability to replicate environmental conditions to verify the observed response. These examinations are most often laboratory studies that offer a low level of realism. In contrast, field studies offer realism with a concomitant reduction in the ability to control factors and produce precise replication. Clearly, opportunities are available for studies that make use of mesocosms (either enclosed or exposed) that may be intermediate for control and realism. Figure 24.2 illustrates the conundrum of situational realism vs. the ability to control and effectively replicate the response fishes may make to a stressor or stressors. Studies should opt to examine the effects of any factor on at least two parts of the continuum.

The killifishes (Cyprinodontidae of the genera *Cyprinodon* and *Fundulus*) have been commonly used as estuarine bioindicators because a few species are widespread in eastern North America in coastal

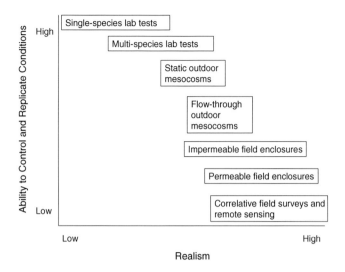

FIGURE 24.2 Diagram indicating the relationship between the level of realism in studies with the level of control and replication.

areas, are conveniently small for experimental purposes, and will reproduce well under controlled conditions. There is a very large body of literature based almost entirely on laboratory studies of toxins on these fishes. However, the usefulness of killifish could be far greater if more attention was paid to comparisons of field and laboratory results to evaluate the impacts of potential stress on estuarine systems. An extremely important variable in such studies would be the interaction between salinity and xenobiotic stress. For example, Loftus and Kushlan (1987) indicated from their field studies that the estuarine *C. variegatus* was excluded from typical freshwater systems of lower specific conductance. Dunson et al. (1998) studied *C. variegatus* in outdoor mesocosms where they found that low salinity conditions negatively affected growth and reproduction — features that could impart a competitive disadvantage to sheepshead minnows in lower salinity areas. Fish on the edges of their tolerance range for salinity would likely be more susceptible to xenobiotic stresses. Thus, it is important for studies of estuarine stress to adequately simulate the natural background levels of salinity. Salinity variation may itself affect the toxicity of metals to estuarine fish by changing the levels of free metal ions, which are the most toxic form (see Lin and Dunson, 1993, for an example using cadmium). A killifish (*Rivulus marmoratus*) with unique genetic characteristics (i.e., self-fertilizing hermaphrodite) has many advantages for use as a tropical estuarine bioindicator. It produces fertilized eggs of nearly identical genetic composition and can be maintained in small culture dishes in large numbers. This enables the investigator to conduct very complex tests in the laboratory such as the interactive effects of temperature, salinity, and food on growth and fecundity (Lin and Dunson, 1995). The tolerances to metals (cadmium, copper, zinc) are similar for *Rivulus* and *Fundulus* (Lin and Dunson, 1993) and *R. marmoratus* can be experimentally manipulated in the estuary by placement in small-screened enclosures that allow for periodic assessment of the effects of environmental stress (Dunson and Dunson, 1999). This species is a classic example of a "pure" bioindicator, which has no economic value itself, in marked contrast to the next case to be discussed.

A Case Study: A Plan and an Example to Establish a Fish Species as an Estuarine Indicator

During the course of trying to establish a fish species as an effective indicator of estuarine conditions along the Gulf of Mexico, a search began to identify a *fish* species that would serve the broadest aspects of being able to assess a specific estuarine condition while being able to detect changes within and between estuaries. In addition, fish estuarine indicators should have many of the features outlined above.

The species should be broadly distributed within an estuary and should be found in a variety of estuaries. Populations should be genetically similar between estuaries yet distinct enough to indicate a lack of movement of individuals between estuaries.

The choices of potential species were limited by the above restrictions. Few species had sufficient data available to make the determinations about genetic structure or, when the data were available, there were few species that met the other criteria. Among these, the criteria of prior database, estuarine dependency, and importance to stakeholders even further limited the selection of potential species in the Gulf of Mexico.

Colleagues suggested that the spotted seatrout, *Cynoscion nebulosus* (a sciaenid fish related to the croakers and drums), was a logical candidate. Preliminary suspicions were that this species satisfied most of the criteria but further refinement and examination was required before the species could be adopted for assessment purposes.

Method

Specialists from the broadest range of experience and expertise were invited to participate in the task of bringing together the most recent information available on the biology of the spotted seatrout. The result was an edited volume (Bortone, 2003a) that summarized existing information or, in some instances, presented previously unpublished information. Topics included the widest range of life history features, including taxonomy, genetics, age-and-growth, reproduction, juvenile and adult habitat preferences, sound production, diseases and parasites, management, and potential for population modeling. Each chapter author was asked to present data that allowed an examination as to the appropriateness of each of these biological features to serve as a metric for the species' responsiveness to estuarine change. Below is a brief summary of the authors' efforts.

Results

Several features of the spotted seatrout make it a reasonable candidate for use as an estuarine indicator species. The spotted seatrout is a broadly distributed species that occurs from Nova Scotia to Florida along the Atlantic Coast of the United States, and in the Gulf of Mexico from Florida to Mexico (Chao, 2003). Thus, the species has the potential to serve as an indicator for a large number of estuaries throughout its range. Data presented by both Gold et al. (2003) and Wiley and Chapman (2003) using genetic information garnered from nuclear microsatellites confirm that the species consists of a series of overlapping stocks. Gold et al. (2003) examined variation in the genetic composition off Texas and determined that individual stocks are centered on "home" estuaries but there is sufficient interchange and isolation that makes the stocks recognizable while preventing significant genetic divergence. Wiley and Chapman (2003) came to similar conclusions regarding subpopulations of spotted seatrout off the southeastern Atlantic Coast of the United States. Further, they suggested that the subpopulations responded to classic zoogeographic barriers indicating that historical events and residential behavior helped explain genetic population structure in the species. Thus, the spotted seatrout displays an almost idyllic population structure: fishes can be ascribed to home, natal estuaries, but physical distance between estuaries is an indication of genetic distance. The species maintains estuarine-specific genetic identity while retaining a "nearest neighbor" zoogeographic similarity between contiguous subpopulations.

Age and growth studies perhaps provided the most useful biological information on the spotted seatrout relative to the task of allowing inter- and intra-estuarine comparisons. Murphy and McMichael (2003, p. 41) presented some general information that the growth rates in the species could serve as a "good indicator of subtle changes in the biotic and abiotic conditions within individual estuaries." This was chiefly because, for the most part, spotted seatrout are lifetime residents of a single estuary and have a high degree of flexibility in their growth relative to environmental features in the home estuary.

Taking this type of analysis a step further, Bedee et al. (2003) compared the age and growth of spotted seatrout from six estuaries along the northern Gulf of Mexico (Figure 24.3), analyzing for males and females separately. Estimating age from annular rings on otoliths, they noted that female spotted seatrout generally grew larger and faster than males. More importantly for our overall objective of determining

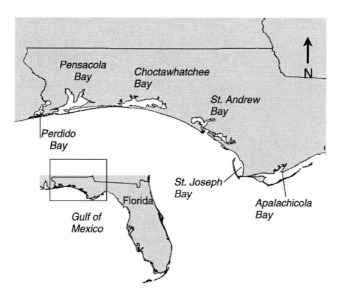

FIGURE 24.3 Estuaries along the northern Gulf of Mexico where spotted seatrout were sampled. (From DeVries, D. A. et al. 2003. In *Biology of the Spotted Seatrout,* S. A. Bortone (ed.). CRC Press, Boca Raton, FL, pp. 79–98. With permission.)

the utility of the use of growth rates as estuarine indicators, they noted significant estuarine specific growth characteristics throughout this area of northwest Florida (Figure 24.4 and Figure 24.5). Fish (regardless of sex) grew faster and attained a larger size from St. Joseph Bay while growth rates among fish from Apalachicola Bay were slower (as indicated by the slopes in Figure 24.4 and Figure 24.5). It is interesting that the two bays closest in geographic proximity (i.e., St. Joseph and Apalachicola Bays) displayed the greatest difference in growth rates. This observation supports the contention that biological responses were more a function of estuarine condition rather than genetic affinity.

Brown-Peterson (2003) presented the case that reproductive features varied geographically with regard to size at sexual maturity (most probably a growth-related phenomenon). DeVries et al. (2003) built upon the postulates presented by Brown-Peterson (2003) and examined inter-estuarine differences in reproductive characters between the same six estuaries (Figure 24.3) studied by Bedee et al. (2003). DeVries et al. (2003) noted that spawning season varied among estuaries. Other reproductive variables they examined also indicated inter-estuarine differences. For example, the gonadosomatic index (GSI, an index of reproductive condition) was bimodal in two bays (i.e., St. Andrew and Apalachicola Bays) for both sexes — a feature not observed in fish from the other four estuaries examined.

DeVries et al. (2003) also determined estuarine specific differences in total instantaneous mortality. An examination of these rates in Figure 24.6 indicates that the longitudinal/spatial relationship of the estuaries was not related to the measured instantaneous mortality rates calculated for each estuary and each sex. This indicates that there was no geographically related, clinal variation in this biological variable, further indicating the utility that these measures in spotted seatrout may have in assessing estuarine-specific conditions.

One of the features, mentioned above, that makes a fish species a good indicator is sensitivity to various environmental variables. Holt and Holt (2003) presented evidence from both field and laboratory studies that spotted seatrout respond to varying salinity conditions, with juveniles especially sensitive to changes in salinity. Interestingly, adult female spotted seatrout spawn in a variety of salinity regimes, apparently accommodating the changes in salinity by producing eggs with different tolerances to osmotic conditions. Under laboratory conditions, however, spawning success is apparently constrained by salinity (Holt and Holt, 2003).

Baltz et al. (2003) presented habitat preference data for both adult and juvenile spotted seatrout for a variety of environmental variables. They noted an ontogenetic habitat shift as juveniles are found in shallower portions of the estuary but move into progressively deeper waters as they mature. Thus, studies examining the habitat preferences of spotted seatrout should take into account their broad preference

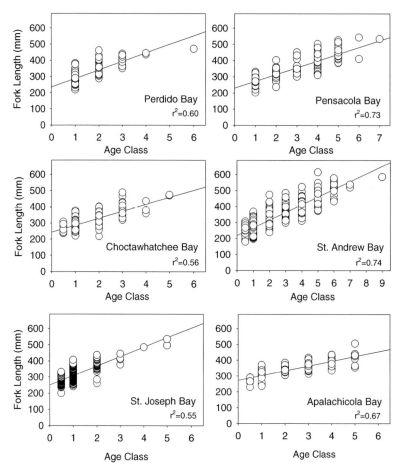

FIGURE 24.4 Plot of length–age regressions for male spotted seatrout from Perdido Bay, Pensacola Bay, Choctawhatchee Bay, St. Andrew Bay, St. Joseph Bay, and Apalachicola Bay. (From Bedee, C. D. et al. 2003. In *Biology of the Spotted Seatrout,* S. A. Bortone (ed.). CRC Press, Boca Raton, FL, pp. 57–77. With permission.)

range for a variety of environmental features but must also account for the changes in preference these fish display as they mature.

Sound production (and detection) is a feature of spotted seatrout that should serve researchers well in the future when it comes to monitoring their abundance, distribution, and reproductive potential. Gilmore (2003) presented a strong argument for developing our level of sophistication in being able to use sound in future studies on their basic biology. The ability to detect these populational features while virtually eliminating actual contact should prove effective in obtaining quality life history data that is estuarine specific.

Huspeni et al. (Chapter 19 of this volume) appropriately indicated the utility of examining estuarine fish parasites to explain estuarine conditions. Blaylock and Overstreet (2003) used this same approach to assess the environmental influences of estuaries by detailing the potential disease and parasite relationships with spotted seatrout. Their study can serve as a reference guide to future studies as a database is developed that allows assessment of changes in parasite faunas from a background level of intensity and incidence of infestation. Changes in these variables can be used as monitors to changes in estuarine conditions surrounding the host species — in this case, the spotted seatrout.

Another attribute that helps determine if a particular fish species has priority when establishing a fish-based estuarine indicator is the interest of the stakeholder. Baltz et al. (2003) and VanderKooy and Muller (2003) present summaries of commercial and recreational landings data on spotted seatrout. Clearly, this species has the interest of a broad range of the fishing public. Recreationally, many local sportsfishing

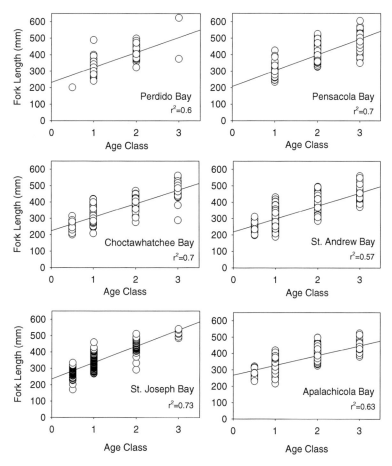

FIGURE 24.5 Plot of length–age regressions for female spotted seatrout from Perdido Bay, Pensacola Bay, Choctawhatchee Bay, St. Andrew Bay, St. Joseph Bay, and Apalachicola Bay (From Bedee, C. D. et al. 2003. In *Biology of the Spotted Seatrout,* S. A. Bortone (ed.). CRC Press, Boca Raton, FL, pp. 57–77. With permission.)

clubs include spotted seatrout as a target species in fishing tournaments. Commercially, the fishery had been well developed but more recently, many states in the United States have passed laws restricting the use of entangling gears in nearshore areas, consequently reducing the inclusion of spotted seatrout in commercial catches. Nevertheless, information on spotted seatrout with regard to its status, trends, and responsiveness to local environmental conditions has the attention of the general public. Thus, the species satisfies the criteria of having an interest among stakeholders.

Modeling is a natural extension of the research database in environmental studies. Models allow predictions with regard to the biological responses species make to anticipated estuarine conditions. There are sufficient historical and current data available on the spotted seatrout that allow modeling efforts. Two of these (Ault et al., 2003; Clark et al., 2003) were recently constructed, demonstrating the utility of the spotted seatrout biological database in developing two very different modeling approaches to spotted seatrout. It is anticipated that these models will become more refined as model verification efforts help improve performance. Coupled with ongoing assessment studies, these models will serve fisheries managers well in being able to forecast responses that spotted seatrout will make to changes in their estuarine-specific environmental conditions.

A summary of the arguments for the utility of spotted seatrout as a biological indicator of estuarine conditions was presented by Bortone (2003b). After the extensive examination presented in the complete volume and its inclusive chapters (Bortone, 2003a) it is even more evident that a species such as the

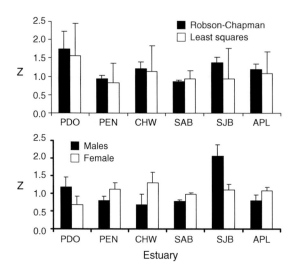

FIGURE 24.6 Estimates of total instantaneous mortality (Z) of spotted seatrout by estuary. Top panel: estimates of Z calculated using the Robson–Chapman least squares methods. (Bottom) Robson–Chapman estimates of Z by sex. Error bars = 95% confidence limits. PDO = Perdido Bay, PEN = Pensacola Bay, CHW = Choctawhatchee Bay, SAB = St. Andrew Bay, SJB = St. Joseph Bay, and APL = Apalachicola Bay. (From DeVries, D. A. et al. 2003. In *Biology of the Spotted Seatrout,* S. A. Bortone (ed.). CRC Press, Boca Raton, FL, pp. 79–98. With permission.)

spotted seatrout satisfies most, if not all, of the qualifiers suggested here to direct the decision-making process of selecting a fish species to serve as an estuarine indicator. This case study should serve as an example of the data-gathering and decision-making process that researchers might want to consider when making choices regarding the most appropriate fish species to use as an estuarine indicator. Above all, it is important for researchers to remember that study results on a small number of fish species can never fully reflect effects at the ecosystem level. Although one species is never sufficient, the spotted seatrout is one of the most useful model species available.

Conclusions

When considering a fish species to serve as a potential indicator of estuarine condition, researchers should be aware that interactions between individual species and community ecology, given the plethora of environmental variables, are exceedingly complex. This is especially significant when investigating ecotoxicology. To make the studies on fishes meaningful, it is important to select fishes that are representatives of food webs or trophic levels in the affected habitat. Study designs should try to incorporate all reasonable permutations and interactions of potentially significant abiotic and biotic factors. Field studies, coupled with laboratory studies, can help scientists approach a level of investigation that varies control of environmental conditions with the realism found in nature. This makes study results meaningful and more easily understandable by the general public.

References

Adams, S. M. (ed.). 1990. *Biological Indicators of Stress in Fish.* American Fisheries Symposium 8. American Fisheries Society, Bethesda, MD.

Ault, J. S., J. Luo, and J. D. Wang. 2003. A spatial ecosystem model to assess spotted seatrout population risks from exploitation and environmental changes. In S. A. Bortone (ed.). *Biology of the Spotted Seatrout,* CRC Press, Boca Raton, FL, pp. 267–296.

Baltz, D. M., R. G. Thomas, and E. J. Chesney. 2003. Spotted seatrout habitat affinities in Louisiana. In *Biology of the Spotted Seatrout,* S. A. Bortone (ed.). CRC Press, Boca Raton, FL, pp. 147–175.

Bedee, C. D., D. A. DeVries, S. A. Bortone, and C. L. Palmer. 2003. Estuary-specific age and growth of spotted seatrout in the northern Gulf of Mexico. In *Biology of the Spotted Seatrout,* S. A. Bortone (ed.). CRC Press, Boca Raton, FL, pp. 57–77.

Blaylock, R. B. and R. M. Overstreet. 2003. Diseases and parasites of the spotted seatrout. In *Biology of the Spotted Seatrout,* S. A. Bortone (ed.). CRC Press, Boca Raton, FL, pp. 197–225.

Bortone, S. A. (ed.). 2003a. *Biology of the Spotted Seatrout.* CRC Press, Boca Raton, FL.

Bortone, S. A. 2003b. Spotted seatrout as a potential indicator of estuarine conditions. In *Biology of the Spotted Seatrout,* S. A. Bortone (ed.). CRC Press, Boca Raton, FL, pp. 297–300.

Brown-Peterson, N. J. 2003. The reproductive biology of the spotted seatrout. In *Biology of the Spotted Seatrout,* S. A. Bortone (ed.). CRC Press, Boca Raton, FL, pp. 99–133.

Chao, N. L. 2003. Taxonomy of the seatrout, genus *Cynoscion* (Pisces, Sciaenidae), with artificial keys to the species. In *Biology of the Spotted Seatrout,* S. A. Bortone (ed.). CRC Press, Boca Raton, FL, pp. 4–15.

Clark, R. D., W. Morrison, J. D. Christensen, M. E. Monaco, and M. S. Coyne. 2003. Modeling the distribution and abundance of spotted seatrout: integration of ecology and GIS technology to support management needs. In *Biology of the Spotted Seatrout,* S. A. Bortone (ed.). CRC Press, Boca Raton, FL, pp. 247–265.

DeVries, D. A., C. D. Bedee, C. L. Palmer, and S. A. Bortone. 2003. The demographics and reproductive biology of spotted seatrout, *Cynoscion nebulosus,* in six northwest Florida estuaries. In *Biology of the Spotted Seatrout,* S. A. Bortone (ed.). CRC Press, Boca Raton, FL, pp. 79–98.

Dovel, W. L. 1971. Fish Eggs and Larvae of the Upper Chesapeake Bay. NRI Special Report 4. Natural Resources Institute, University of Maryland.

Dunson, W. A. and D. B. Dunson. 1999. Factors influencing growth and survival of the killifish, *Rivulus marmoratus,* held inside enclosures in mangrove swamps. *Copeia* 1999(3):661–668.

Dunson, W. A., C. J. Paradise, and D. B. Dunson. 1998. Inhibitory effect of low salinity on growth and reproduction of the estuarine sheepshead, *Cyprinodon variegatus. Copeia* 1998(1):235–239.

Gilmore, R. G., Jr. 2003. Sound production and communication in the spotted seatrout. In *Biology of the Spotted Seatrout,* S. A. Bortone (ed.). CRC Press, Boca Raton, FL, pp. 177–195.

Gold, J. R., L. B. Stewart, and R. Ward. 2003. Population structure of spotted seatrout (*Cynoscion nebulosus*) along the Texas Gulf Coast, as revealed by genetic analysis. In *Biology of the Spotted Seatrout,* S. A. Bortone (ed.). CRC Press, Boca Raton, FL, pp. 17–29.

Helfman, G. S., B. B. Collette, and D. E. Facey. 1997. *The Diversity of Fishes.* Blackwell Science, Malden, MA.

Holt, G. J. and S. A. Holt. 2003. Effects of salinity on reproduction and early life stages of spotted seatrout. In *Biology of the Spotted Seatrout,* S. A. Bortone (ed.). CRC Press, Boca Raton, FL, pp. 135–145.

Lin, H.-C. and W. A. Dunson. 1993. The effect of salinity on the acute toxicity of cadmium to the tropical estuarine, hermaphroditic fish, *Rivulus marmoratus:* a comparison of Cd, Cu, and Zn tolerance with *Fundulus heteroclitus. Archives of Environmental Contamination and Toxicology* 25:41–47.

Lin, H.-C. and W. A. Dunson. 1995. An explanation of the high strain diversity of a self-fertilizing hermaphroditic fish. *Ecology* 76(2):593–605.

Loftus, W. F. and J. A. Kushlan. 1987. Freshwater fishes of southern Florida. *Bulletin of the Florida State Museum, Biological Sciences* 31:1–344.

Murphy, M. D. and R. H. McMichael, Jr. 2003. Age determination and growth of spotted seatrout, *Cynoscion nebulosus* (Pisces: Sciaenidae). In *Biology of the Spotted Seatrout,* S. A. Bortone (ed.). CRC Press, Boca Raton, FL, pp. 41–56.

Nelson, J. S. 1994. *Fishes of the World.* John Wiley & Sons, New York.

Rowe, C. L. and W. A. Dunson. 1995. Individual and interactive effects of salinity and initial fish density on a salt marsh assemblage. *Marine Ecology Progress Series* 128:271–278.

Simon, T. P. (ed.). 1999. *Assessing the Sustainability and Biological Integrity of Water Resources Using Fish Communities.* CRC Press, Boca Raton, FL.

VanderKooy, S. J. and R. G. Muller. 2003. Management of spotted seatrout and fishery participants in the U.S. In *Biology of the Spotted Seatrout,* S. A. Bortone (ed.). CRC Press, Boca Raton, FL, pp. 227–246.

Wiley, B. A. and R. W. Chapman. 2003. Population structure of spotted seatrout, *Cynoscion nebulosus,* along the Atlantic Coast of the U.S. In *Biology of the Spotted Seatrout,* S. A. Bortone (ed.). CRC Press, Boca Raton, FL, pp. 31–40.

25

Habitat Affinities of Juvenile Goliath Grouper to Assess Estuarine Conditions

Anne-Marie Eklund

CONTENTS

Introduction

The overall goal in managing and monitoring an estuarine ecosystem should be a "healthy" system. Ecosystem health and ecosystem integrity have been variously defined as synonyms of each other and as synonyms of stability, sustainability, resilience, balance, and productivity (Simberloff, 1998; Jordan and Smith, Chapter 30, this volume). A healthy ecosystem is one that also supports reasonable human uses through the long term (Simberloff, 1998). To maintain a healthy estuarine system or to restore a degraded system to a stable and functioning state, it is essential to be able to measure that system's health. When managing or restoring an estuary to a better condition, it is important to understand its present state, what the current trends are in the ecosystem's status and how long it will take to achieve a healthy system (Simberloff, 1998).

In Chapter 30, Jordan and Smith describe two different paths in assessing ecosystem health. One method is to describe effects of stressors on the sediment, habitat, nutrients, etc. and then define the relationships between abiotic and biotic components of an estuary, by means of a complex conceptual model. The model would then have outputs as response variables that describe ecosystem health. Another method to assess the status of an ecosystem is to define direct relationships between stressors and responses. Using the second method does not require a complete understanding of a conceptual model and all of the interrelationships and components of a complex food web (Jordan and Smith, Chapter 30, this volume). It does require the use of a suitable indicator that integrates the system's responses to stressors and predicts changes in ecosystem status.

As outlined by Jordan and Smith, indicators of ecological integrity can be from population, community, or ecosystem levels. They can be species groups, broad taxa, or indices that reflect an attribute of a community, such as species composition, diversity, evenness, or richness. Often, however, a single species may be designated an indicator of ecosystem function, particularly as it may be more practical and possible to monitor and manage a single species, rather than to monitor and manage a suite of species or many components of an ecosystem.

Of course, using indicator species in conservation biology is not without problems. First, it is difficult to choose one and hard to understand just what it is supposed to indicate. Species are chosen for monitoring and management if their presence or abundance gives us a better understanding of the system. Often a flagship or high-profile species is chosen for its political "palatability," but it may not be a good indicator of the status of a community or ecosystem (Zacharias and Roff, 2000).

If one is to use a single species as an indicator of ecosystem health, restoration success, or change, it is imperative to choose an effective one. First, the indicator species needs to be abundant in the area that is being studied and it must be relatively easy to catch or observe, to ensure some success at monitoring. Of course, the species must be measurably affected by changes that are occurring and there must be some understanding of the mechanisms affecting the species and causing the observed changes.

When assessing the effects of anthropogenic or natural changes in an estuary, species or species groups from lower trophic levels may be used as indicators, because they are more abundant than those from the higher trophic groups. Furthermore, using a higher tropic-level species as an indicator can be problematic. Effects on higher trophic levels can be more difficult to understand, because there are many steps in the food web that link to the top-trophic levels. Abiotic effects may be mitigated (or exacerbated) by intermediate steps in the food chain. In addition, highly mobile organisms may respond quickly to suboptimal conditions by leaving those environments. Another complicating factor is that bottom-up effects (such as water quality and pollution) may or may not be as important as top-down effects such as fishing and other human-extractive activities. For example, in Chapter 30, Jordan and Smith present information on the striped bass in Chesapeake Bay and the fact that the top-down forces of fishery management affected the species to the extent that any bottom-up relationship between the fish and habitat restoration was not measurable.

Despite these difficulties, it is vitally important to try to understand the relationships between the natural and anthropogenic perturbations of the system and the effects of such perturbations on fish and other higher trophic-level organisms, particularly because there is often an inherent interest in them by managers and the general public. Furthermore, the ultimate success in restoration is to revive the natural system, so that all trophic levels — including the highest levels — benefit from restoration activities.

Using fish as indicators of estuarine ecosystem health may turn out to be ecologically important to other areas as well. Many predatory fish and other species that are the target of fisheries use estuaries as nursery areas. Their success in the nursery will affect the ecology of other systems as they grow and recruit to their adult habitats (e.g., coral reefs). Many commercially and recreationally valuable species depend on estuaries for some part of their life history. In the southeastern United States, more than 95% of the fish that are landed commercially and the majority of the recreational species caught are dependent on estuaries for some part of their life cycle (Nakamura et al., 1980). Estuaries benefit juveniles, in particular, because they provide rich food resources with fewer predators and less competition from adults of the same species (Colby et al., 1985).

The juvenile goliath grouper, *Epinephelus itajara* (formerly referred to as jewfish before a proposed common name change by Nelson et al., 2001) thrives in the estuary of the Ten Thousand Islands of southwest Florida. It is an important species in the estuarine ecosystem as a top-level predator, and the system is important to the goliath grouper as its primary nursery habitat. Goliath grouper can be easily caught, tagged, measured, and recaptured (Eklund and Schull, 2001) in the same general area, and their abundance is affected by certain variables that are changing with habitat restoration. Because the only known predators of goliath grouper (even as juveniles) are sharks and humans and because there currently is no harvest of goliath grouper allowed, total mortality is probably quite low and top-down effects on the species are minimal to nonexistent.

In this chapter I explain how goliath grouper can be used as an indicator of estuarine condition. I describe what we can learn about the restoration of an ecosystem and the recovery of a protected species and how the two are related.

Goliath Grouper

The goliath grouper is the largest grouper in the western North Atlantic. As adults they inhabit shallow reefs, wrecks, canals, seawalls, bridges, and piers, although they are also found on offshore wrecks and reefs down to at least 70 m (Sadovy and Eklund, 1999). The larvae settle in the fall in the estuaries of Florida (Bullock and Smith, 1991), including the Ten Thousand Islands (personal observation). They grow up in the estuary and are found along mangrove-lined creeks and mangrove islands in tidal passes from settlement size up to about 1 m, and ages 0 to 6 or 7 years (Bullock et al., 1992). The fish are site faithful (Eklund and Schull, 2001) based on mark–recapture data, making it possible to assess habitat preferences and describe microhabitats.

According to Bullock and Smith (1991), the species' center of abundance along Florida's west coast is in the Ten Thousand Islands estuary, due to the extensive habitat of mangrove swamps for juveniles. The juveniles have been collected in poorly oxygenated canals (Lindall et al., 1975) and in mangrove swamps with tidal currents that are strong enough to scour holes in the bottom and undercuts in the ledges (Bullock and Smith, 1991). Goliath grouper appear to tolerate a large range of salinity (Sadovy and Eklund, 1999), but are susceptible to cold water-induced mortality (Gilmore et al., 1978) and red tide (Smith, 1976).

Goliath grouper are europhagic carnivores, but they are more likely to consume crustaceans and slow-moving benthic fishes (Odum et al., 1982; Bullock and Smith, 1991; Sadovy and Eklund, 1999; personal observation). Even as juveniles, goliath grouper are top predators in the estuarine system, because they grow to a size greater than most other fish in the area, within their first 2 years of life (Bullock and Smith, 1991).

The U.S. fishery for goliath grouper expanded rapidly in the 1980s, until the populations were overexploited to the point of economic extinction (Sadovy and Eklund, 1999). In the early 1990s, the Gulf of Mexico, South Atlantic, and Caribbean Fishery Management Councils passed amendments to prohibit retention of goliath grouper in U.S. waters. Their stocks may be recovering due to fishing prohibitions that have been in place since that time (Porch et al., 2003); however, it is not clear how variable year-class strengths are and how vulnerable juvenile goliath grouper are to environmental perturbations in their nursery habitats.

In the Ten Thousand Islands of southwest Florida, the proximity of natural riverine habitat to that of dredged canals with altered freshwater flow patterns enables one to compare the abundance of juveniles in altered and unaltered habitats. This comparison has set the stage for long-term monitoring of goliath grouper abundance in order to indicate the health of the Ten Thousand Islands estuary and the success of the restoration of that system. These rivers and canals link the upstream freshwater system of the Big Cypress Basin to the system of bays, which empty into the Gulf of Mexico through a series of channels around mangrove islands. Most of the area is completely undeveloped and protected, yet it is downstream from areas in the Big Cypress Basin that have been subjected to massive changes in water delivery, timing, and quantity over the years. The entire area is included in the Comprehensive Everglades Restoration Project (CERP) (U.S. Army Corps of Engineers/South Florida Water Management District, 2000), which is an attempt to restore the system, as much as possible, to historical conditions.

Ten Thousand Islands Estuary

The Ten Thousand Islands is one of the largest estuaries in the United States (Browder et al., 1986) and is composed of a series of shallow bays that are separated from the Gulf of Mexico by thousands of small mangrove islands stretching approximately 30 km from Goodland to Chokoloskee, Florida (Figure 25.1). Several tidally influenced rivers empty into the bays, and each bay is connected to the Gulf of Mexico through convoluted passes.

The natural, undeveloped freshwater system of southwest Florida was one in which fresh water slowly flowed as a broad sheet over gently sloping prairies and eventually into the estuary, with a lag of several months between upstream rainfall and inflow into the bays (Browder et al., 1986). Although the area is under the protection of the Ten Thousand Islands National Wildlife Refuge, Rookery Bay Estuarine Research Reserve, and Everglades National Park, it has been, and continues

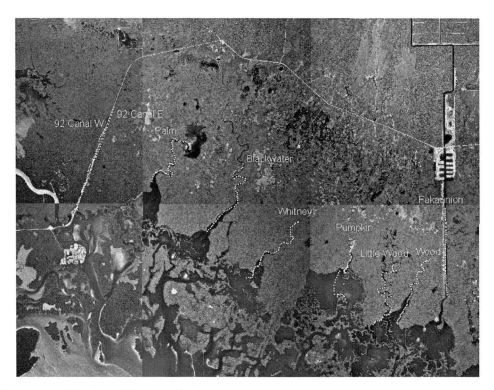

FIGURE 25.1 (Color figure follows p. 266.) Six natural, tidally influenced rivers and three canals in the Ten Thousand Islands of southwest Florida. Canals on either side of U.S. Highway 92 are labeled 92 Canal West and 92 Canal East. The dots on the river and canal transects designate locations where fish traps and crab traps were placed to catch juvenile goliath grouper, *Epinephelus itajara*, from 1999–2000.

to be, adversely affected by upstream water management practices. The natural slough system of freshwater sheet flow is no longer active, due to the vast network of canals that are actually deeper than the groundwater table (Popowski et al., 2003). The most drastic drainage activity in the area was that of the Southern Golden Gate Estates (SGGE) Canal system, which includes more than 294 canals that were built for a housing project that never materialized (Browder et al., 1986). As a result of the canal network, more than 600 km² of wetlands are drained into Faka Union Bay (U.S. Army Corps of Engineers/South Florida Water Management District, 2000). Pre-drainage, the flow would occur over land much more broadly and would drain into a larger area of the estuary. As a result, Faka Union Bay receives about five times more water annually than it did historically, but the nearby bays receive much less water because water is lost from surface flow and from the groundwater as well.

The excessive drainage of the SGGE has made the area a target for restoration. The Golden Gates Estates Feasibility Study and the CERP include immediate plans for restoring the system by disassembling the canals, removing roads, and adding spreader canals and pumps (U.S. Army Corps of Engineers/South Florida Water Management District, 2000). If canals are plugged, sheet flow restored, and upland water storage increased, then estuarine systems are expected to improve. Discharges into Faka Union Bay should decrease in the wet season, the base flow to the entire system should increase in the dry season, and a more natural salinity gradient should be reestablished (U.S. Army Corps of Engineers/South Florida Water Management District, 2000; Popowski et al., 2003).

Browder et al. (1989) and Sklar and Browder (1998) reviewed the few studies that had been conducted in the bays of the Ten Thousand Islands and found that every study that compared animal abundances among or between bays found lower numbers of the study organisms in the Faka Union system than in adjacent systems (Carter et al., 1973; Colby et al., 1985; Browder et al., 1986). Reasons for the lower numbers in Faka Union Bay could be less area of suitable salinity for many organisms or the difficulty in larval transport against high canal flow rates. These studies found depressed numbers of fishes and invertebrates throughout the year, indicating that the large, wet-season discharges had long-term effects.

None of the above-mentioned studies detected any changes in species composition, however. Thus, using species composition alone as an indicator of ecosystem health would fail, in this case, to detect any changes in the altered system.

The tidal streams and rivers of the Ten Thousand Islands have received considerably less attention and little research activity. Colby et al. (1985) is the only published study that covers the entire area of embayments of the Ten Thousand Islands. However, neither that study nor any of the others included an investigation of the tidal rivers. Those rivers are primarily fringed with red mangroves (*Rhizophora mangle*) and connect the freshwater marshes with the shallow bays.

Mangrove communities, in general, are characterized by turbid surface water with low dissolved oxygen (DO), low concentrations of macronutrients (mainly phosphorus), and extreme ranges in salinity from 0 to 40 ppt (or above) (Odum et al., 1982). Typically, DO concentrations are between 2 and 4 ppm and often approach zero when waters are stagnant or after heavy storm runoff (Odum et al., 1982).

Mangrove swamps provide habitat for many organisms through the tree canopy, the aerial roots, and the associated muddy substrates in the adjacent creeks and embayments. The riverine mangrove forest system of southwest Florida supports a dense and speciose fish assemblage, with 47 to 60 species per river system (Odum et al., 1982). The mangrove shorelines include vast undercuts of eroded banks that provide shelter for many species of invertebrates and fishes. Personal observations include goliath grouper, gag grouper (*Mycteroperca microlepis*), snook (*Centropomis undecimalis*), and gray snapper (*Lutjanus griseus*) co-occurring in high densities under the mangrove overhangs. Invertebrate species diversity is moderately high and includes such organisms as spiny lobsters, barnacles, sponges, polychaetes, gastropods, oysters, mussels, isopods, amphipods, mysids, crabs, shrimp, copepods, ostracods, coelenterates, nematodes, insects, bryozoans, and tunicates (Odum et al., 1982). The leaf litter forms the basis of a detrital food web.

The fish assemblages of mangrove communities have not been studied extensively as a result of inherent gear limitations (Serafy et al., 2003). Comparisons of fish abundance inside and outside of mangrove habitats are rare, and those comparisons are often problematic due to the difficulty in sampling in the mangroves (Beck et al., 2001). Often different gears are used in and out of mangrove habitats, making comparisons difficult. Most studies have collected fish adjacent to the mangrove forests, not actually within the flooded forest (Beck et al., 2001).

With Everglades restoration efforts under way, water quality, quantity, and timing of water delivery will soon be altered due to restoration. The SGGE project has already begun. If we can predict the response of a top-level predator to the changes in the water quality of the system, then we may be able to successfully monitor, manage, and shape decisions about future restoration activities. The objectives of this study were to estimate the abundance, size distribution, site fidelity, and movement patterns of juvenile goliath grouper in altered and unaltered rivers and canals in the Ten Thousand Islands of southwest Florida and to ascertain whether that species could be used as an indicator of ecosystem restoration.

Methods

For a pre-restoration "baseline" data assessment, in 1999 and 2000, the abundance of juvenile goliath grouper in natural tidal passes, or rivers, was compared to their abundance in channelized canals (Figure 25.1). The natural rivers should provide optimal microhabitat for the juvenile goliath grouper, including mangrove overhangs along eroded shorelines (Figure 25.2) and rocky depressions in tidal passes. Oppositely, canals tend to have straightened shorelines with little to no eroded banks and mangrove overhangs. Because they are dredged, canals are also of relatively uniform bathymetry, lacking the natural depressions that rivers contain. The hypothesis was that the physical features of the two habitat types would differ and that the goliath grouper would be more abundant in the natural rivers.

In each river and canal, 40 crab traps and 10 fish traps were placed every 92.6 m (0.05 nautical miles) along a linear transect. Two rivers and one canal were sampled concurrently for 3 weeks with the traps

FIGURE 25.2 (Color figure follows p. 266.) Photographs of typical eroded shorelines in the Ten Thousand Islands of southwest Florida. The erosion along the mangrove shorelines provides for underwater habitat underneath the mangrove overhangs.

inspected and sampled weekly within the 3-week period. At the end of 3 weeks, traps were moved to new locations. At the end of 9 weeks, all locations had been sampled and the first sites were sampled again. In the second year, YSI® datasondes were deployed to continuously measure temperature, salinity, DO, and depth. A datasonde was secured to a fish trap in each river, so that the water quality parameters measured would reflect the water quality adjacent to and inside the traps that were currently fishing. One datasonde was placed in a fish trap either in the upper, middle, or lower part of a river/canal and remained deployed for 1 week. When the traps were inspected, the water quality data were downloaded, the datasonde was calibrated, if necessary, and subsequently moved to another part of the river/canal. At the end of the 3-week sampling period, the datasonde would have acquired water quality information at all three sections of the river/canal.

The amount of eroded (vs. depositional or straight) shoreline was measured by taking Global Positioning System (GPS) waypoints at the beginning and end of each section of eroded shoreline and measuring the distance between the two points using Geographic Information System (GIS) ArcView® software. The heterogeneity of the bottom (i.e., the presence/absence of rocky depressions and other obstructions) was estimated by taking a depth reading every 185.2 m (0.1 nautical miles) along each side of the river/canal. The change in depth from each reading was then calculated and averaged for the entire river/canal.

The duration of hypoxic events was determined for each datasonde deployment by calculating the percent of time that the datasonde recordings (made every 15 min) were below 2 parts per million (ppm). The percentage was calculated for each datasonde deployment and averaged for each river for the year.

Because goliath grouper are found at a broad range of salinity and appear to tolerate even fresh water to a certain degree, it was appropriate to look at the rate of salinity change rather than the absolute value of salinity. Thus, the change in salinity from one reading to the next (15 min between readings) was calculated. This difference between each reading was averaged for each deployment. The average change in salinity for all the deployments in each river was then calculated for a grand mean for each river.

A multilinear regression analysis was used with catch-per-unit-effort (CPUE) as the dependent variable, and two physical habitat variables (meters of eroded shoreline and bathymetric complexity) and two chemical variables (percent of time that DO was below 2 ppm and mean change in salinity) were the independent variables. A Pearson's rank correlation coefficient was calculated between CPUE and the four above-listed habitat variables.

In addition to the analysis made on an annual and entire river basis, analysis of CPUE was also divided into parts of the river/canal (upper, middle, lower), so that a comparison of CPUE and water quality could be made for each sampling period and each river/canal section.

Results

A total of 687 juvenile goliath grouper were caught in nine rivers and canals of the Ten Thousand Islands from 1999 to 2000. Many of these fish were recaptured at least once, with previously tagged fish comprising 38% of the total catch. Fish demonstrated movement within rivers but not among river/canal systems. In only a few cases ($n < 5$) were marked fish recaptured outside their original river or canal. Goliath grouper CPUE and total catch were highest in Little Wood River, Palm River, and Blackwater River, and few goliath grouper were caught in the Wood, Pumpkin, and Whitney Rivers (Table 25.1 and Figure 25.3A). While two of the canals, Faka Union and 92 West Canal, had lower CPUE and total catch of goliath grouper than in several rivers, 92 East Canal had higher CPUE and total catch than many of the natural rivers (Figure 25.3A).

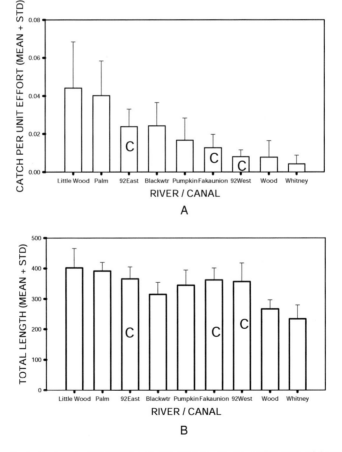

FIGURE 25.3 Mean + standard error (A) CPUE and (B) total length (in millimeters) of juvenile goliath grouper, *Epinephelus itajara*, caught in fish traps and crab traps in six tidal rivers and three canals (designated with a C) of the Ten Thousand Islands of southwest Florida from 1999–2000.

TABLE 25.1

CPUE and Total Catch of Goliath Grouper, *Epinephelus itajara*, and Dead Bycatch from Crab and Fish Traps Set in Rivers and Canals of the Ten Thousand Islands of Southwest Florida[a]

River/Canal	Mean CPUE		Total Catch		Dead Organisms		Eroded Shoreline		Depth Change (per 0.1 nmi)	Time Hypoxic % of time	Salinity Change per 15 min	Temperature (C)		Salinity (ppt)		DO (ppm)	
	1999	2000	1999	2000	1999	2000	(m)	%				Mean	Range	Mean	Range	Mean	Range
92 Canal W.	0.006	0.009	11	15	0	3	2.6	42	0.61	29.41	0.60	25.69	16.4–32.5	24.79	0.1–38.0	3.63	0.17–10.22
92 Canal E.	0.024	0.020	43	20	3	1	4.6	59	0.47	32.63	0.35	27.00	17.1–32.6	26.67	2.1–36.7	2.90	–0.33–7.88
Palm R.	0.040	0.040	81	65	1	2	3.4	51	3.45	8.56	0.29	27.59	21.1–34.0	25.34	0.1–35.0	4.19	0.77–10.24
Blackwater R.	0.026	0.020	59	36	10	6	1.3	20	1.98	25.49	0.29	27.66	23.4–25.6	28.46	8.8–36.6	3.44	0.13–8.49
Whitney R.	0.003	0.006	9	13	78	29	3.0	48	2.16	49.84	0.19	26.73	15.0–33.1	27.81	12.1–37.4	2.21	0.03–7.85
Pumpkin R.	0.013	0.015	31	20	73	11	0.7	22	1.28	31.76	0.16	26.17	14.9–33.4	29.00	0.55–39.2	3.97	–0.31–13.03
Little Wood R.	0.045	0.043	119	87	1	6	4.7	70	4.31	16.52	0.10	27.09	16.6–33.0	25.68	0.64–33.8	3.44	0.21–8.75
Wood R.	no data	0.007	no data	18	no data	34	0.6	10	1.08	23.76	0.12	25.89	17.1–33.5	22.34	0.40–36.7	4.03	0.29–9.68
Faka Union	0.011	0.015	27	33	6	0	0	0	0.63	4.05	0.55	27.78	22.5–33.0	17.05	0.40–37.1	4.16	0.92–7.99

[a] Along with measurements of length of eroded shoreline, percent of shoreline with eroded banks, mean change in depth (per 185 m or 0.1 nautical mile along the length of the river/canal, percent of time that DO concentration was below 2 ppm, mean change in salinity per 15-min period, and the mean, minimum, and maximum temperature, salinity, and DO measured in each river and canal from June to December 2000.

Within each river and canal, certain sections were more productive, in terms of goliath grouper CPUE. In general, the upper sections of the Little Wood River and Palm River, the lower sections of the Whitney and Wood, and the middle section of the Pumpkin River/Bay area were more productive than the other sections of the respective rivers. The Blackwater River and the canals were more variable in where the highest CPUE occurred (see Eklund et al., 2002, for details on each section of river and canal).

The goliath grouper caught in the Ten Thousand Islands ranged in length from 133 to 903 mm total length (TL), practically the entire range of the juvenile life-history stage (Sadovy and Eklund, 1999). Consistently, the largest fish were caught in the Palm and Little Wood Rivers, and only small fish were caught in the Whitney and Wood Rivers (Figure 25.3B). The canals and Blackwater and Pumpkin Rivers were more temporally varied in the sizes caught. There did not seem to be a consistent pattern with size distribution along sections of the rivers, and there was no indication of ontogenetic migration upstream or downstream.

Goliath Grouper Habitat Description

The amount of erosion along the shorelines of the meandering rivers should be a good measure of the amount of suitable or optimal habitat for goliath grouper, as the erosion provides for a mangrove undercut area where the fish can reside (Figure 25.2). The Little Wood River and the 92 East Canal have the greatest length of eroded shoreline (more than 4.5 km), comprising 70 and 59% of those systems, respectively (Table 25.1). Very little of the shorelines of the Blackwater, Pumpkin, and Wood Rivers is undercut (1.3, 0.7, 0.6 km, respectively; less than 25% of the system) (Table 25.1), and Faka Union Canal is a completely straight canal with no meandering and resultant erosion/deposition along the shorelines (Table 25.1).

All three canals have uniform depths, with a change of 0 to 1 m between readings (data were recorded 185.2 m apart) (Table 25.1). Little Wood River and Palm River had the greatest variation in depth, with an overall mean change in depth readings equal to 4.31 and 3.45 m, respectively. The Whitney, Blackwater, and Pumpkin Rivers were intermediate in depth variation, and the Wood River was more similar to the canals with a mean change just slightly greater than 1 (Table 25.1).

Water depth varied with time, due to tidal changes and upstream flow. The rivers and canals appeared to experience similar changes in depth over time, with all nine systems having a mean depth change between 1.0 and 1.3 m, within the week's sampling period.

While currents were not directly measured, it was possible to gather a relative description of overall flow, based on movement of the traps. The Palm and Little Wood Rivers and parts of the Blackwater River had the strongest flow, based on the fact that the traps had to be secured to trees along the riverbanks to prevent their loss. The Highway 92 Canals also had high water flow at times, probably due to pulses of upstream water releases. Thus, traps had to be secured to the banks of those two canals as well. Faka Union Canal also received upstream water pulses, but that canal is very wide with the overall flow dampened somewhat across the stream. The other rivers received such little flow that the traps did not move appreciably, unless there was a storm event.

The water temperature range in the Ten Thousand Islands rivers and canals was from 15 to 34°C, with mean water temperatures similar among rivers and canals, between 26 and 28°C (Table 25.1). The rivers and canals that were sampled concurrently yielded almost the exact mean temperatures, meaning that water temperature changes were reflective of greater environmental conditions and not of the individual river/canal systems.

There was a lot of fluctuation in salinity readings in the Ten Thousand Islands (Table 25.1). The lowest overall mean salinity was found in Faka Union Canal. The Palm River and all three canals often experienced a large range of salinities, at times the readings went from completely fresh water (less than 5 ppt) to almost salt water (greater than 30 ppt) within 1 week. The lower section of the 92 East Canal and the lower Blackwater River, on the other hand, maintained higher salinity with minimal variation. Overall, most of the rivers experienced a 10 ppt change in salinity within a week's period.

Perhaps more germane to the survival or habitat preference of goliath groupers and other organisms in the area is the rate of salinity change during the week. In general, the canals experienced more rapid changes in salinity over short time periods (Table 25.1), with Faka Union Canal and 92 West Canal having much faster rates of change than 92 East Canal. Blackwater River and Palm River also experienced relatively high rates of salinity changes. The other rivers had much lower rates of change, particularly the Wood and Little Wood Rivers (Table 25.1).

Although the rivers differed in their patterns of DO concentration, their overall means were similar (Table 25.1), except for the Whitney River and 92 East Canal, whose means were less than 3.0 ppm. The Whitney River had the lowest overall mean DO concentration; all sections of that river had minimum DO less than 0.30 throughout the year, except for the lower Whitney in midsummer, which had a minimum DO of 1.01. The upper Pumpkin and Wood Rivers always had minimum DO less than 0.35 and, until toward the end of the wet season, the middle sections also had minimums less than 1.0. The only parts of the Pumpkin and Wood Rivers that consistently had high DO were the lower sections, which were really part of the bay systems and less riverine in their physical nature (Figure 25.1). The 92 East Canal also had low DO levels (actually becoming anoxic) in the upper and middle sections during the middle of the summer, but those low levels did not persist.

More important to sustaining most life in the rivers is the length of time that hypoxic conditions persisted. The datasonde in Whitney River measured DO concentrations below 2 ppm over 49% of the time that the probes were in the water. The Wood, Blackwater, and Pumpkin Rivers and both of the Highway 92 Canals were hypoxic one fourth to one third of the time that they were sampled. The Little Wood River, Palm River, and Faka Union Canal had fewer periods of hypoxic conditions (Table 25.1).

Perhaps indicative of anoxic conditions, the traps from the Wood, Whitney, and Pumpkin Rivers often contained dead blue crabs (*Callinectes sapidus*), hardhead catfish (*Arius felis*), *Tilapia* spp., and various other fish. Dead crabs or fish were a rare occurrence in the other rivers and canals (Table 25.1).

No significant relationships were found between any abiotic variables and CPUE when examined by specific river section or time period. However, much stronger relationships were revealed when CPUE and abiotic factors were averaged for each river for the entire year of sampling (Figure 25.4). A Pearson product-moment correlation coefficient indicated a significant ($\alpha < 0.05$) positive correlation between bathymetric complexity and CPUE. The multilinear regression had an $r^2 = 0.92$ when all four factors were used in the analysis:

$$\text{CPUE} = 0.0218 + (0.00367 \times \text{meters of eroded shoreline}) + (0.00364 \times \text{bathymetric complexity})$$
$$- (0.000591 \times \text{percentage of time hypoxic}) - (0.00938 \times \text{salinity change})$$

Salinity change had the lowest r^2 (0.058; Figure 25.4A), and the least effect on the regression when it was removed from the equation. Bathymetric complexity had the strongest relationship with CPUE ($r^2 = 0.639$; Figure 25.4B), explaining more than half the variation among rivers. Percent of hypoxic conditions and length of eroded shoreline each explained about one third of the variation in CPUE ($r^2 = 0.313$ and 0.312, respectively; Figure 25.4C and D).

Discussion

Goliath grouper catch was variable among the rivers in the Ten Thousand Islands, making direct comparisons of rivers and canals more difficult than anticipated. These differences, however, illuminated differences in physical-chemical habitat and underscored how restoration success could be indicated by the abundance of these juvenile fish. Goliath grouper were most abundant in the Little Wood and Palm Rivers, and those rivers also had the greatest amount of bathymetric heterogeneity and eroded shoreline, and neither river experienced many periods of hypoxia. In addition, the Little Wood River had less variation in salinity than that of the other rivers. The presence of both bathymetric complexity and eroded shoreline are indications of good physical habitat for these fish. Rocky holes and mangrove undercuts provide optimal habitat for goliath grouper in the form of shelter from current and an ideal location for ambush predator activities. In addition, the lack of hypoxic events and extreme salinity changes helped maintain a quality habitat for the fish.

The Little Wood and Palm Rivers also had the largest goliath grouper caught, another indication of optimal habitat. The two rivers that had the lowest goliath grouper catch, the Wood and Whitney Rivers, also had the smallest fish caught. In some instances, catching more small fish could be an

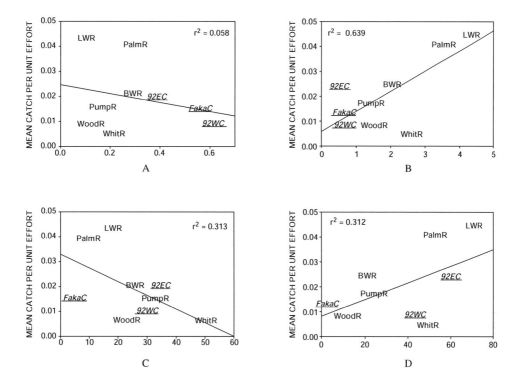

FIGURE 25.4 The relationship between juvenile goliath grouper CPUE and (A) mean change in salinity between each 15-min datasonde recording period; (B) bathymetric complexity (as determined by averaging the change in depth between each 0.1 nautical mile of river/canal bank); (C) the percentage of time that hypoxic conditions (<2 ppm DO concentration) were recorded; and (D) the percentage of the river/canal shoreline that was eroded in each river and canal in the Ten Thousand Islands. LWR = Little Wood River, PalmR = Palm River, BWR = Blackwater River, PumpR = Pumpkin River, WoodR = Wood River, WhitR = Whitney River, *92EC* = 92 Canal East, *92WC* = 92 West Canal, and *FakaC* = Faka Union Canal.

artifact of low sample size, since younger fish are usually more abundant than older fish due to natural mortality. However, in this case, the trend is real, since the blue crab traps do not select for the smallest size classes. Rather, the blue crab traps used in this study selected for fish between 300 and 400 mm TL. The fact that the few fish caught at Whitney and Wood Rivers were smaller than the selected size range indicates that the fish in those rivers really were smaller. The goliath grouper were likely settling throughout the system, and then either emigrating or suffering mortality in those areas of suboptimal habitat. In Chapter 2 of this volume, Adams accentuates the need to notice and measure physiological affects of animals as an alarm before there is a response to population levels. However, highly mobile animals will emigrate as a response to environmental change, before a physiological response can be detected. Thus, a change in abundance is what we need to measure, not a change in physiology.

The hypothesis that natural rivers would provide better and more habitat than canals for juvenile goliath grouper was based on the assumption that the shorelines and bathymetry would be vastly different between rivers and canals. Our visual assessment of Faka Union Canal bore out that hypothesis, as the canal is completely straight with no eroded shorelines and a uniform bathymetry. The two other canals, however, are not completely straight and the slight meanders have resulted in some shoreline erosion, providing undercut habitat for goliath grouper and other fish (Table 25.1). In fact, the 92 East Canal had more eroded shoreline than all of the natural rivers, except the Little Wood River, with large tracts of erosion along its eastern bank, away from the roadbed. The Faka Union Canal and the Wood, Pumpkin, and Blackwater Rivers had shorelines with little erosion. None of the canals had heterogeneous bathymetry, as expected, because the canals are created by dredging activities. Also as expected, the canals all received much greater fluctuations in salinity (but the 92 East Canal did not fluctuate as much as the other two); yet these results contradict the assumption that salinity would be the main driving affect on fish distribution (Browder et al., 1989; Popowski et al., 2003).

The Whitney River is the biggest outlier in the relationships between physical habitat characteristics and CPUE, but the low CPUE is explained by the extremely low DO concentrations that occurred in the Whitney River almost half the time it was measured. The Pumpkin River was intermediate to low in physical and chemical habitat values and contained a low abundance of goliath grouper as well. In addition, our qualitative observations revealed that there was much less water flow in the Pumpkin, Wood, and Whitney Rivers than there was in the Little Wood, Palm, and Blackwater Rivers and in all the canals.

The rivers with consistently low DO (Pumpkin, Wood, and Whitney) had consistently low CPUE of goliath grouper. The datasonde measured the conditions on the bottom of the rivers and canals, and it is likely that the systems are stratified in oxygen concentration. Mobile animals, such as fish and crabs, can move easily when the water becomes anoxic. Thus, systems such as the 92 East Canal and the Little Wood River, which experienced anoxia or hypoxia for short time periods, could still sustain healthy populations of goliath grouper. Systems that appeared to maintain an extremely low DO throughout the sampling period were the rivers that had low catch of goliath grouper. The number of dead fish and crabs is another indication of the unsuitability of these rivers. It is apparent that catfish and crabs were living in the rivers, but they were caught in the traps during bad water quality time periods and subsequently died. The low number of dead (or live) goliath grouper caught with the dead organisms indicates that the grouper either move out very quickly during low DO events, or more likely, they do not inhabit the areas at all.

It is important to realize that the correlations between the abiotic variables and CPUE were not significant when viewed on a small spatial (river section) and temporal (specific sampling run) scale; yet the physical habitat variables and the degree of hypoxia exhibited stronger relationships with CPUE when viewed from the scale of an entire river or canal and whole sampling season. These results are indicative of the integrative properties of a mobile, predatory fish. The effect of antecedent conditions are demonstrated by the fact that abiotic conditions for a particular time and space do not indicate what the abundance and distribution of goliath grouper will be. Although each river was subject to great variations in abiotic factors, it is the average value that has greater meaning with regard to fish abundance. As DO, salinity, water depth, and temperature vary, so does the fish's behavioral response and resulting fish abundance, but it is the integration of these responses throughout the year that result in the establishment of a home site for the goliath grouper and in the effectiveness of goliath grouper as an ecological indicator.

What is most surprising about this study is that the biggest differences in goliath grouper abundance were not between the canals and the rivers, considering the huge impact of the canals (particularly the Faka Union Canal) on freshwater input into the system. Rather, there was wide variation among the natural rivers, which leads to the question of why one river would provide more suitable habitat than an adjacent river. All the rivers in the study are downstream of the same Big Cypress Basin, but aerial photographs of the area help explain the differences among the rivers (see Figure 25.1). The rivers with the highest goliath grouper catch rate (Little Wood, Palm, and Blackwater) obviously separate from the rivers of low catch rate (Pumpkin, Wood, and Whitney). Little Wood, Palm, and Blackwater Rivers are all connected to natural bodies of water upstream. Mud bay connects Blackwater and Palm Rivers to each other and the upper Blackwater River is navigable almost all the way to U.S. Highway 41 (Figure 25.1). The upstream source of the Little Wood River is a rich labyrinth of streams and ponds within the mangrove forest, providing a vehicle for greater overall flow of oxygen and nutrients (Figure 25.1). These upstream areas may also be providing more habitat for goliath grouper to settle and/or grow over time. In addition, the productivity, while not measured in this study, is probably much higher in these upstream areas, yielding a rich food web that can better support juvenile goliath grouper and other fishes. The Pumpkin, Wood, and Whitney Rivers, on the other hand, all appear to lead to upstream dead ends, with little water flow coming into the rivers (Figure 25.1).

The CERP and specifically the SGGE restoration projects will have direct impacts on the habitat of the juvenile goliath grouper, as well as other juvenile fishes, in the Ten Thousand Islands. Precious little is known concerning the animals in the natural rivers and canals in the area, with the bulk of the research activities taking place in upland and freshwater areas. However, the restoration activities will surely affect areas downstream.

As the CERP and SGGE projects proceed, there will be an increase in the freshwater flow and somewhat of a return to the sheet flow of the natural system (Popowski et al., 2003). As canals are plugged and the freshwater flow is increased over time and space, there will be an immediate change in the patterns of salinity and DO. It is the increase in flow, providing more oxygen-rich waters to the rivers, that may result in an increase in goliath grouper. As the juveniles appear to settle throughout the system (personal observation) and then either emigrate from or die in the suboptimal habitats, there is potential for immediate effects of an improvement of water quality (DO and flow).

An increase in flow should also have longer-term effects, by changing the physical nature of the tidal rivers and providing new physical habitats (undercuts and rocky depressions). These physical changes will occur very slowly, however, as evidenced by the Whitney River that still retains its undercuts and holes, even though flow has been absent there for a long time. Possible medium-term effects may occur with an increase in upstream linkages throughout the system, as more water is allowed to fill dried-up depressions and creek beds. Eventually, if sheet flow is restored somewhat, other rivers, besides the Little Wood River, will have more extensive and productive upstream linkages.

Conclusions and Research Needs

There is a direct relationship between goliath grouper abundance and certain habitat and water quality variables that are integrated over time. As the Ten Thousand Islands estuary is restored somewhat to a more natural system, more freshwater flow should provide more oxygen, more productivity, and eventually the creation of more physical habitat for goliath grouper and other fishes. Although top-level predators may be at times difficult to use as indicators, the juvenile life history stage of this species is showing a direct response to water management. The fish is probably not limited by food resources but is limited by the quality of available habitat; therefore, changes in habitat should be reflected in changes in distribution and abundance of this species. It is easy to measure this effect because abundance alone can serve as the metric, and it is not necessary to measure physiological stresses. Also, by using a single species as an indicator of restoration, cause–effect relationships can be predicted and measured without using a complex food web or conceptual model.

It may be unexpected that a top-level predator's abundance would reflect the differences in attributes of these rivers, but it is reasonable to believe that when conditions are good enough for goliath grouper to thrive in an area, they are good enough for a number of yet-unstudied estuarine species to thrive also. Thus, the abundance of juvenile goliath grouper can be used as a performance measure of the Big Cypress Basin restoration work.

Much more information is needed, however, to understand restoration effects on goliath grouper and other co-occurring species. Baseline (pre-restoration) conditions must be quantified, particularly flow patterns in the rivers. Currently, only one of the rivers and one of the canals are being monitored for flow (Popowski et al., 2003). This study has demonstrated that each river is unique and that all have undoubtedly been affected by upstream water management. The entire system is "altered," not just the canals, and we cannot adequately compare one or two altered canals with one or two "natural" rivers without understanding whether the natural river is healthy or how it compares to the other rivers.

The study on goliath grouper in the tidal rivers and canals of the Ten Thousand Islands has provided some insight on how increased freshwater flow may affect the estuary's biota. This baseline information can be used to structure hypotheses to be tested as restoration activities proceed, even without a complete understanding of the complexity and variability of the system. Indeed, we will never comprehend all of the intricacies of a complex ecosystem. However, knowledge of the relationships between abiotic variables and the abundance of an important top-level predator, the juvenile goliath grouper, is a huge step forward in the ability to assess restoration success.

This study has demonstrated that discovering a few direct relationships between stressors (DO, bathymetry, erosion) and response variables (fish abundance) has given us information to predict responses to habitat change and ecosystem restoration. Although an estuary is a complex ecosystem with a large food web of interconnecting organisms, it is possible to find direct links between a predatory fish species and its nursery habitat. As a result, we can predict changes to fish distribution based on habitat changes. Fish are integrators of environmental change on different spatial and temporal scales, and while such integration may appear to cloud the situation, it actually provides us a more accurate depiction of the biological effects of habitat alteration and restoration.

Acknowledgments

I thank my colleagues, Jennifer Schull, Matt Finn, Chris Koenig, Felicia Coleman, and Todd Bevis, whose hard work and dedication in goliath grouper research have helped me draw insights into this subject matter. I particularly acknowledge Steve Wong who has produced detailed, georeferenced, aerial maps of the Ten Thousand Islands region. I also thank Larry Demere, Gary Weeks, Jeff White, and Beth White who helped find the juvenile goliath grouper in the first place, and Don DeMaria whose unwavering interest and passion about the goliath grouper has fueled interest and determination to learn more about this protected species. This research was funded by NOAA-Fisheries, Southeast Fisheries Science Center's Marine Fisheries Initiative and its Essential Fish Habitat program, and by the NOAA-Fisheries Office of Protected Resources and a grant from the National Fish and Wildlife Foundation. I dedicate this work to Peter and Mary Gladding, who serve as fine examples of sincere, passionate defenders of the oceans and estuaries.

References

Beck, M. W., K. L. Heck, K. W. Able, D. L Childers, D. B. Eggleston, B. M. Gillanders, B. Halpern, C. G. Hays, K. Hoshino, T. J. Minello, R. J. Orth, P. F. Sheridan, and M. Weinstein. 2001. The identification, conservation, and management of estuarine and marine nurseries for fish and invertebrates. *Bioscience* 51(8):633–641.

Browder, J. A, A. Dragovich, J. Tashiro, E. Coleman-Duffie, C. Foltz, and J. Zweifel. 1986. A Comparison of Biological Abundances in Three Adjacent Bay Systems Downstream from the Golden Gate Estates Canal System. NOAA Technical Memorandum NMFS-SEFC-185.

Browder, J. A, J. D. Wang, J. Tashiro, E. Coleman-Duffie, and A. Rosenthal. 1989. Documenting estuarine impacts of freshwater flow alterations and evaluating proposed remedies. In *Proceedings of International Symposium on Wetlands and River Corridor Management*, 5–9 July 1989, Charleston, SC.

Bullock, L. H. and G. B. Smith. 1991. *Seabassses (Pisces: Serranidae)*. *Memoirs of the Hourglass Cruises*. Florida Marine Research Institute, St. Petersburg, 243 pp.

Bullock, L. H., M. D. Murphy, M. F. Godcharles, and M. E. Mitchell. 1992. Age, growth, and reproduction of jewfish, *Epinephelus itajara*, in the eastern Gulf of Mexico. *Fishery Bulletin* 90:243–249.

Carter, M. R., L. A. Burns, T. R. Cavender, K. R. Dugger, P. L. Fore, D. B. Hicks, H. L. Revells, and T. W. Schmidt. 1973. Ecosystem Analysis of the Big Cypress Swamp and Estuaries. U.S. Environmental Protection Agency, Ecological Report DI-SFEP-74-5. EPA, Region IV, Atlanta, GA.

Colby, D., G. Thayer, W. Hettler, and D. Peters. 1985. A comparison of forage fish community in relation to habitat parameters in FU Bay and eight collateral bays during the wet season. NOAA Technical Report NMFS SEFC-162, 87 pp.

Eklund, A. M. and J. Schull. 2001. A stepwise approach to investigating the movement patterns and habitat utilization of goliath grouper, *Epinephelus itajara*, using conventional tagging, acoustic telemetry and satellite tracking. In *Electronic Tagging and Tracking in Marine Fisheries*, J. R. Sibert and J. L. Nielsen (eds.), Kluwer Academic Publishers, Dordrecht, pp. 189–216.

Eklund, A. M., S. Wong, J. Schull, and M. Finn. 2002. Nassau Grouper and Jewfish Habitat. A Final Report to the National Fish and Wildlife Foundation. Grant 99-35, 149 pp.

Gilmore, R. G., L. H. Bullock, and F. H. Berry. 1978. Hypothermal mortality in marine fishes of south-central Florida, January 1977. *Northeast Gulf Science* 2(2):77–97.

Lindall, W. N., Jr., W. A. Fable, Jr., and L. A. Collins. 1975. Additional studies of the fishes, macroinvertebrates, and hydrological conditions of upland canals in Tampa Bay, Florida. *Fishery Bulletin* 73(1):81–85.

Nakamura, E. L., J. R. Taylor, and I. K. Workman. 1980. The occurrence of life stages of some recreational marine fishes in estuaries of the GMEX. NOAA Technical Memorandum NMFS-SEFC-45.

Nelson, J. S., E. J. Crossman, H. Espinosa-Perez, H. Findley, C. R. Gilbert, R. N. Lea, and J. D. Williams. 2001. Recommended change in the common name for a marine fish: goliath grouper to replace jewfish (*Epinephelus itajara*). *Fisheries* May:31.

Odum, W. E., C. C. McIvor, and T. J. Smith III. 1982. The Ecology of the Mangroves of South Florida: A Community Profile. U.S. Fish and Wildlife Service Biological Service. FWS OBS-81/24, 144 pp.

Popowski, R., J. Browder, M. Shirley, and M. Savarese. 2003. Hydrological and Ecological Performance Measures and Targets for the Faka Union Canal and Bay. Report to the South Florida Water Management District. 23 pp.

Porch, C. E., A. M. Eklund, and G. P. Scott. 2003. An Assessment of Rebuilding Times for Goliath Grouper. NOAA-Fisheries, Southeast Fisheries Science Center, Sustainable Fisheries Division Contribution SFD-2003-0018. 16 pp.

Sadovy, Y. and A. M. Eklund. 1999. Synopsis of Biological Information on the Nassau Grouper, *Epinephelus striatus* (Bloch 1792), and the Jewfish, *E. itajara* (Lichtenstein 1822). NOAA Technical Report, NMFS 146, and FAO Fisheries Synopsis 157, 65 pp.

Serafy, J. E., C. H. Faunce, and J. J. Lorenz. 2003. Mangrove shoreline fishes of Biscayne Bay, Florida. *Bulletin of Marine Science* 72:161–180.

Simberloff, D. 1998. Flagships, umbrellas, and keystones: is single-species management passé in the landscape era? *Biological Conservation* 83:247–257.

Sklar, F. and J. Browder. 1998. Coastal environmental impacts brought about by alterations to freshwater flow in the Gulf of Mexico. *Environmental Management* 22:547–562.

Smith, G. B. 1976. Ecology and Distribution of Eastern Gulf of Mexico Reef Fishes. Florida Marine Research Institute Publication 19, 78 pp.

U.S. Army Corps of Engineers and South Florida Water Management District. 2000. Southern Golden Gate Estates Hydrologic Restoration Project. Final Project Management Plan. March 2001.

Zacharias, M. A. and J. C. Roff. 2000. Use of focal species in marine conservation and management: a review and critique. *Conservation Biology* 14:1–8.

26

Using Waterbirds as Indicators in Estuarine Systems: Successes and Perils

Eric D. Stolen, David R. Breininger, and Peter C. Frederick

CONTENTS

Introduction

Estuarine resource managers need reliable information about the state of the ecosystem and how it is changing over time due to natural or anthropogenic perturbations. Animal indicators are often used for this purpose, as they are highly visible, reactive, easily measured, and of intuitive value to the wider public (Morrison, 1986; Landres et al., 1988; Kushlan, 1993; Frederick and Ogden, 2003). Various attributes of animal populations or communities (e.g., population size, reproductive success, habitat use, species composition) may provide information about other ecosystem attributes that are more difficult to measure (e.g., trophic structure, hydrology, contamination). When one or a small number of species are used to provide information about other members or attributes of an ecosystem, they are referred to as "surrogate species" (Caro and O'Doherty, 1999). The role of surrogate species may fall into one of the following categories: health indicator, biodiversity indicator, umbrella species, keystone species, or flagship species. There is much literature on the use of surrogate species; Table 26.1 provides definitions of terms commonly used when referring to vertebrate indicator species, and Table 26.2 lists attributes that must be considered when choosing an appropriate indicator species within a specific ecological system.

    The term *waterbirds* refers to birds that spend all or key parts of their lifetimes in wetlands, including the avian orders Ciconiiformes, Charadriiformes, Gaviiformes, Podicipediformes, Procellariiformes, Pelecaniformes, Anseriformes, and Gruiformes. Although a highly diverse group of taxa, many waterbird species have similar life-history characteristics, many of which are useful for monitoring. Thus, it seems natural that any program of monitoring the integrity and function of estuaries would include waterbird populations as indicators of ecosystem processes, attributes, and biodiversity (e.g., Gawlik et al., 1998; Frederick and Ogden, 2003). However, it should be noted that investigations of the application, utility, and reliability of waterbirds as indicators are at least partly lacking.

TABLE 26.1

Definitions of Terms Used in Discussions of Vertebrate Indicator Species

Term	Definition	Ref.[a]
Indicator	An attribute of an ecosystem that is used to access the current state or change in state of other features of that ecosystem	4
Indicator species	An organism whose characteristics are used to measure changes in the attributes of other organisms, structures, or functions within an ecosystem	1, 2, 3
Health indicator	An indicator that measures changes in the structural or functional attributes of an ecosystem	1, 2, 3
Population indicator	An indicator that measures changes in populations of other organisms	1, 2
Biodiversity indicator	An indicator that measures the level of biodiversity within an ecological system of interest	1, 2
Umbrella species	A species whose broad habitat requirements are used to represent the composite requirements of a community of organisms	1, 3, 5
Keystone species	A species with a unique and functionally important ecological role within an ecological system, upon which a large part of the biotic community depends	1, 3, 5
Flagship species	A species with broad appeal to the public, which makes it useful as a representative of an ecological system, often one that is threatened by anthropogenic disturbance	1, 3, 5

[a] *References:* (1) Caro and O'Doherty, 1999; (2) Landres et al., 1988; (3) Simberloff, 1998; (4) Kushlan, 1993; (5) Noss, 1990.

TABLE 26.2

Attributes of Vertebrate Indicator Species That Should Be Considered When Designing Monitoring for a Specific Ecological System

Term	Criteria	Ref.[a]
Generation time/life span	Shorter generation times usually mean shorter reaction time to perturbation (but longer lifespan may be useful in some circumstances)	1, 2
Reproductive rate	Faster rates allow more specificity and less time lag but slower rates integrate more information	1, 2, 3
Individual variability	Lower individual variability desirable (reduces noise in signal, easier statistical estimation of parameters)	1, 2
Sensitivity (to disturbance)	High sensitivity to ecosystem attributes desirable; low sensitivity to other disturbances reduces noise	1, 2
Home range	Larger home range allows more integration of information; smaller home range allows more specificity	1, 2, 3, 4
Distribution	More widespread species allows better detection of local changes	3
Diet (generalist/specialist)	Generalists are not as useful as specialists (but may be cases where generalist desirable)	2, 3
Trophic level	Top carnivores most useful (but certain diet specialists may be useful)	1, 2, 3
Habitat specificity	Habitat specialists are the most useful	1, 2, 3
Adaptability	Highly adaptable species not as useful (adaptations mask responses to perturbations)	2

[a] *References:* (1) Caro and O'Doherty, 1999; (2) Landres et al., 1988; (3) Kushlan, 1993; (4) Noss, 1990.

Several reviews of birds as indicators conclude that, more often than not, bird populations are not robust indicators in biological systems (e.g., Morrison, 1986; Temple and Wiens, 1989; Niemi et al., 1997). This is often due to logistical problems, such as the inability or failure to collect data that are repeatable and amenable to statistical analysis, or to poor definition of variables or goals. Many avian monitoring studies have not been long enough or included corollary information about the ecosystem to allow an unbiased test of the indicator's utility. There is a growing recognition that bioindicators in general are only useful for monitoring specific attributes of an ecosystem and should be used only in conjunction with other indicators to make general statements about ecosystem status and function. Ecological relationships are likely to differ widely among species and ecosystems, requiring specific information that may take years to collect. For these reasons, it is not surprising that early attempts to use birds as indicators have met with mixed success. However, there are numerous examples of birds that have been successfully used as bioindicators, and the lessons from these examples are becoming clearer.

There are three components to developing a scientifically credible wildlife indicator. First, one must clearly define what attributes of the estuarine system are to be indicated. Examples of ecosystem attributes include biodiversity, energy flow, contamination, habitat structure, or community structure. In contrast, terms such as *wetland health* or *wetland barometer* may convey an intuitive and comfortable sense of completeness, but in practice are too general or complex to be defined as parameters. At this stage, it is important to distinguish between the need to monitor waterbirds for their own sake as a component of the biota and the desire to use some attribute of birds as an ecological indicator. The former is a perfectly sound reason for monitoring birds, but the reasoning and expectations behind the monitoring should be explicitly stated.

Next, a model of how the proposed indicator will respond to perturbations must be defined. This model can be stated qualitatively (i.e., as a narrative) or may be explicitly defined including quantitative predictions of responses to perturbation. For example, if a manager is interested in monitoring changes to forage fish density due to a disruption of wetland hydrology, then a prediction of reduced foraging success by wading birds might be made. Finally, links between ecosystem attributes and the changes in bird indicators must be validated either through literature review or using quantitative methods in pilot studies (Temple and Wiens, 1989; Kushlan, 1993). Establishing such links may take considerable time and effort, and links and relationships vary among ecosystems, so pilot studies of some depth and duration are an almost inevitable part of the process.

A bioindicator might be useful in the absence of a clear understanding of mechanisms, if a statistical association between the bioindicator and the ecological parameter of interest exists (Erwin and Custer, 2000). However, use of such an indicator relies on the assumption that the statistical relationship remains in effect during the period of interest and under the conditions observed. Relying on such untested assumptions is risky. Another alternative approach, advocated by many conservation biologists (e.g., Noss, 1990), is to use surrogate species such as umbrella or flagship species for purposes other than strictly monitoring ecosystem changes (see Table 26.1). For example, a natural resource manager might select one (or a suite of) waterbird species to monitor, under the assumption that when conditions were favorable for that species, they would also be favorable for most others in the system. In this way, the species would serve as an "umbrella" protecting the biodiversity within an estuarine system. The manager might later be interested in determining in more detail the sources of stress to the ecosystem, but initially the surrogate species is used to provide general information about the integrity and functioning of the ecosystem.

Methodological Approach

The scope of this chapter precludes a thorough review of avian indicators, so the goal here is to indicate both the potential benefits and the potential pitfalls of using waterbirds as monitors of estuarine systems; much of the information will also be useful for other wetlands. First, the advantages and disadvantages of using waterbirds as indicators are reviewed. Next, two case studies are used to illustrate the use of waterbirds as indicators in estuarine systems: long-term monitoring of wading bird nesting populations in the Everglades/Florida Bay Ecosystem, and long-term monitoring of wading bird foraging habitat use in salt marshes of the northern Indian River Lagoon system. Finally, recommendations are given for ways to improve the practice of using waterbird populations as indicators in estuarine systems.

Results

Benefits of Using Waterbirds as Indicators in Estuarine Systems

There are many attributes of waterbird populations that could potentially be used as indicators in estuarine systems (Table 26.3). In general, waterbirds are often an abundant, conspicuous, and functionally

TABLE 26.3

Attributes of Waterbird Populations That Could Be Used for Monitoring and Attributes of Estuarine Systems To Be Monitored

Category	Attributes
Waterbird Population Attributes	
Demographic	Population size (nesting, wintering, adults, juveniles), adult survival, juvenile survival, fecundity, philopatry (number returning to nest in colony), colony site dynamics
Foraging ecology	Foraging behavior (individual, group), foraging success (capture rate, total intake, efficiency), diet (prey selection), habitat use, habitat selection
Organismal	Parasite load, chemical contaminant load, development of young
Ecosystem Attributes	
Structural	Trophic structure, habitat availability, habitat suitability, vegetation structure, chemical contamination
Functional	Hydrology, energy flow, nutrient cycling

important component of estuarine systems (e.g., Kushlan, 1976; Christy et al., 1981; Berrutti, 1983; Erwin, 1985; Montague and Weigert, 1990; Frederick, 2001). Waterbirds are highly mobile and may respond quickly to environmental change. As apex predators in aquatic food webs they may be good monitors of bioaccumulative toxins and diseases (Spalding et al., 1993; Erwin and Custer, 2000). They are conspicuous and (compared to soil, vegetation, or small aquatic organisms) relatively easy and inexpensive to quantify. To the public, they are charismatic and thus often serve as emblems of wetland conservation. Finally, their basic ecology, habitat preferences, and systematics are often well established, and there is an abundance of specimens including skins, eggshells, and feathers collected over many decades.

One of the greatest strengths of using waterbird populations as indicators is the ease of monitoring them. They are often numerous and conspicuous components of their ecosystems, making them both energetically important and also well suited to estimation of numbers over time. Many species congregate at feeding, roosting, or nesting sites during particular times making it easy to survey their populations within focal areas (Bibby et al., 2000; Erwin and Custer, 2000; Williams et al., 2001). Similarly, demographic information is often obtained by marking nests and individuals in breeding colonies (Erwin and Custer, 1982; Erwin et al., 1996; Cezilly, 1997; Bibby et al., 2000). Waterbirds also have great appeal to the general public, making it easier to obtain resources necessary for monitoring, than it might be for other, less well-known components of estuarine systems (Frederick and Ogden, 2003). Some waterbird species are protected by laws, which often mandate monitoring; for example, the wood stork (*Mycteria americana*), snail kite (*Rostrhamus sociabilis*), and piping plover (*Chararius melodus*) are protected by the Endangered Species Act in the United States.

For many species of waterbirds, their relatively large body size and position near the top of food chains enable them to integrate information about the trophic structure of the ecosystem. For example, a waterbird population or community may be a good indicator of an estuarine system's prey base, which is itself of ecological importance. Waterbird populations are often key components of ecosystems through vital functions such as predation effects (Kushlan, 1976; Hafner and Britton, 1983; Hafner et al., 1993), nutrient deposition at nesting or roosting colonies (Onuf et al., 1977; Bildstein et al., 1991; Frederick and Powell, 1994), and soil perturbation during foraging (Bildstein, 1993). Many waterbirds are opportunistic in their resource use, shifting their foraging habitat or prey preferences between seasons (Butler, 1993; Stolen et al., 2002), making foraging locations reliable indicators of prey location. Similarly, through large-scale movements, waterbirds serve as integrators, providing information about average conditions across large wetland or coastal ecosystems.

When several different species co-occur within an estuarine system, monitoring the waterbird community allows greater specificity for indicating changes within the estuary. Although many are termed "generalists" because of their large movements and use of widely different habitats, waterbird species can be divided into more specific foraging niches, especially within taxonomic groups. For example, the various waterfowl, shorebird, and wading bird taxa all include an array of foraging specialists such as species specializing on fish or invertebrate prey. This allows flexibility in aligning an indicator with a specific attribute within an ecosystem. Monitoring multiple species within the same estuarine system also provides information about multiple levels of ecosystem function. Finally, because the overall level of biodiversity is an important component of an estuarine system, monitoring a diverse waterbird community may be useful in itself (Caro and O'Doherty, 1999).

Pitfalls of Using Waterbirds as Indicators in Estuarine Systems

The most basic problem in the use of waterbirds as indicators is determining which species, attribute, or population is to be monitored within a given ecosystem, especially given the diversity of life histories, spatial habitat use patterns, ecological interactions, and behaviors. It is crucial to understand a species and its role in the ecosystem prior to attempting to use it as an indicator. It is also important to consider how to define the study population and the factors that influence its dynamics, as home ranges of waterbirds range from endemic flightless rails to highly mobile nomadic waterfowl, ibises, and storks (Frederick and Ogden, 1997). An indicator species may be present in the estuary year-round, or may only occur during certain seasons (e.g., wintering or breeding populations). If a nonresident population is selected, factors outside of the estuarine system may affect its performance as an indicator. Another complication occurs when an indicator species has two or more overlapping populations present at various times, as when breeding populations are supplemented by wintering birds from more northern breeding populations during part of the year (Mikuska et al., 1998). In some cases, highly mobile species may not be suitable as indicators.

In addition to regular seasonal patterns of movements, many populations of waterbirds are highly nomadic. For these, it is important to consider the potential environmental attractants and repellents in other parts of their range when interpreting local fluctuations. For example, declines in numbers of white ibis (*Eudocimus albus*) in Florida were observed during the late 1980s (Runde, 1991) and could have been interpreted as an indication of changes in habitat conditions in the region. However, it was later learned that a large increase in white ibis was occurring in Louisiana where the birds were attracted by an increasing crayfish aquaculture industry (Fleury and Sherry, 1995; Frederick et al., 1996; Frederick and Ogden, 1997).

Surveys of individual birds or their nests within some prescribed area produce an estimate of the total number of individuals comprising a population (Bibby et al., 2000). Examples include systematic aerial surveys conducted over large areas, ground counts of successive waves of migrating shorebirds at a stopover point, or repeated boat surveys along transects through a coastal marsh or creek system. Although these methods of estimating population size are the least expensive and easiest to use, they may have many unstated or unmeasured biases that lead to interpretation problems. For example, nearly all attempts by humans to count or estimate numbers of birds larger than ten tend to be underestimates (Caughley, 1974; Erwin, 1982; Frederick et al., 2003) and the degree of bias is highly variable among investigators (Frederick et al., 2003) and may have little relationship to training, experience, or age, making it difficult to estimate the degree of bias.

Many surveys must also deal with visibility interference, in which some unknown number of targets is blocked by vegetation, diving patterns, or are easily confused with other species (Caughley, 1974; Pollock and Kendall, 1987; Frederick et al., 2003). These problems may be generally addressed by measuring the degree of error within and between observers or techniques (Erwin, 1982), and adjusting to obtain a less-biased estimate of population size (Dolbeer et al., 1997). It is often valuable to use complementary techniques with different strengths. For example, systematic aerial surveys at colony or foraging locations can yield spatially reliable information over a large area, while site visits can yield accurate information about species composition, status and behavior, population size, and biological interactions. The best advice is to measure and understand the limitations of the survey technique prior to widespread use, and especially prior to interpretation of data. The use of surveys in any ecosystem should be iterative, involving pilot studies, accuracy measurement, modeling of bias, and field validation.

Surveys may be inappropriate for species that are cryptic or especially where individual-specific information is needed (e.g., demographic studies). In this case, mark–recapture/resighting and radiotelemetry may be used. Mark–recapture/resighting studies are conducted by capturing a portion of a waterbird population and applying marks such as conspicuous wing tags or leg bands, which allow individuals to be identified later in the field using observation or capture (Williams et al., 2002). This may yield information on location, movements, population size, and population parameters (e.g., survival). Radio or satellite transmitters can also be used to follow movements and habitat use of individuals, and can be used as a means of estimating survival (Cezilly, 1997).

Case Study 1: Long-Term Monitoring of Nesting Wading-Bird Populations in Everglades/Florida Bay Ecosystem

The Everglades is a large, flat wetland (>4000 km^2) located at the southern end of the Florida peninsula that includes a mosaic of freshwater and estuarine components (Gunderson and Loftus, 1993). Most rainfall occurs May to October, with a distinct dry season November to April; variation in dry season rainfall typically exceeds 80% of the long-term mean (Gunderson and Loftus, 1993). Wading birds are dominant predators in this ecosystem in terms of position in the food web and biomass (Frederick and Powell, 1994; Ogden, 1994), and both foraging and reproduction are directly affected by estuarine conditions.

Reproduction by wading birds is in general strongly affected by the availability of prey (Frederick, 2001), which in the Everglades is largely controlled by the timing and degree of seasonal drying. During the dry season, surface water levels can recede by 2 to 8 mm/day, forcing fishes and macroinvertebrates into shallow ponds and pools where they become available to the birds. At the interface of the freshwater marsh and tidal mangrove regions, aquatic fauna are concentrated by numerous streams and short rivers (Lorenz, 1999, 2000). Fish and macroinvertebrate densities can increase during the dry season by up to one, and biomass by up to two orders of magnitude (Loftus et al., 1986; Loftus and Eklund, 1994; Lorenz, 2000). Historically, most of the large colonies of wading birds were concentrated in the estuarine regions of the Everglades (Pierce, 1962; Frohring et al., 1988; Ogden, 1994). The colonies were probably located there for a variety of reasons, including protection from mammalian predators and production and availability of prey due to the close proximity of a variety of foraging habitats with a range of hydrological conditions (McIvor et al., 1994; Ogden, 1994; Gawlik, 2002).

Several important inferences have emerged from historic records, which go back discontinuously for nearly 100 years (Pierce, 1962; Robertson and Kushlan, 1974; Frohring et al., 1988; Ogden, 1994). First, historical nesting patterns suggest very high interannual variability, ranging from 0 to 100,000 pairs, sometimes in consecutive years (Ogden, 1994). Second, pairs of nesting wading birds have decreased by 80 to 90% compared with averages and peaks established in the 1930s (Ogden, 1994; Frederick and Ogden, 2003). Third, timing of nest initiation by wood storks changed from November to December in the period prior to the 1970s, to January to March in the recent period. Fourth, the success of nesting by at least one species (wood stork) was documented to have declined markedly after the mid-1960s (Ogden, 1994). Finally, the location of nesting changed dramatically, with coastal nesting having been largely abandoned by the mid-1980s, and 80 to 90% of nesting moved to freshwater marshes farther inland (Frederick and Ogden, 2003).

The population decline, change in timing and success of nesting, and change in location of nesting for wading birds occurred during a period of dramatic, anthropogenically induced hydrological change during the latter half of the 20th century (Ogden, 1994). During this period the flow of water through the Everglades was interrupted by a series of canals that lowered water levels and reduced flow to the coastal regions, thereby altering their estuarine nature (Light and Dineen, 1994). The most intensive period of alteration was between 1950 and 1970, when the freshwater portion of the Everglades was divided by levees into a series of large pools. The net effect of these hydrological alterations was to reduce water flow to estuarine systems by more than 60%. In addition, because water flow was subsequently managed for both flood control (wet season) and water supply (dry season), the timing of surface flows was also dramatically changed.

The alterations in wading bird population size as well as the timing and success of nesting were greatest during the period of most rapid and intense hydrological change — between the 1950s and 1970s. There are mechanistic links between reduced water flow to the coast and nesting. Both survey and experimental studies have demonstrated that the densities of small forage fishes are strongly affected by hydroperiod and freshwater flows, in both freshwater marshes (Loftus and Eklund, 1994; Trexler et al., 2003) and, particularly, in coastal mangrove areas (Lorenz, 1999). Coupled with experimental work demonstrating that wading birds give up foraging at specific densities of prey (Gawlik, 2002), this is strong support for the hypothesis that wading birds stopped nesting in the region because reduced flows resulted in substandard prey production or availability.

Ornithologists first noted changes to wading bird populations during the 1960s (Ogden, 1994), but it took nearly 20 additional years to uncover the specific mechanisms by which secondary productivity of the coastal zone had undergone dramatic collapse (Browder, 1985; McIvor et al., 1994; Lorenz, 1999; Frederick and Ogden, 2003). Thus wading birds were the first biological indicator of ecological problems, by some 20 years when compared with other animal populations. The data demonstrating the estuarine indicator function of Everglades' wading bird populations have been sound enough to allow the formulation of hypotheses for restoration of the ecosystem. A central goal of the Comprehensive Everglades Restoration Plan is to restore freshwater flows to the coastal regions as a means to restore wading bird breeding. Both volume of freshwater flow and the size of the wading bird nesting population also serve as metrics for the determination of restoration success (Frederick and Ogden, 2003).

Wading bird populations in the Everglades have also served as indicators of contaminants. The Everglades ecosystem is highly contaminated with mercury (Hg) from local waste incineration (Dvonch et al., 1999; Frederick, 2000). Although the problem of mercury contamination was initially discovered through monitoring of fish flesh for human health standards (Ware et al., 1990), the history and trends of mercury contamination have been successfully monitored using wading birds (Spalding et al., 1994; Sunlof et al., 1994; Frederick et al., 1999, 2002, 2004; Sepulveda et al., 1999). Annual monitoring of mercury in the feather tissue of nestling great egrets (*Ardea albus*) between 1993 and 2000 has shown that mercury contamination declined by over 73%, a change not detectable through analysis of water and air samples (Frederick et al., 2002). An analysis of wading bird museum skins has similarly yielded a record of mercury profiles in Everglades' biota extending back for nearly a century (Frederick et al., 2004).

Case Study 2: Using Wading Birds to Monitor Estuarine Habitat Restoration

Located on the Atlantic Coast of Florida, the Kennedy Space Center/Merritt Island National Wildlife Refuge encompasses the northern quarter of the 250-km-long subtropical estuary known as the Indian River Lagoon System. Historically, the eastern shore of the Indian River Lagoon System was extensively vegetated with irregularly flooded salt marsh habitat (Schmalzer, 1995). Human impacts on saline and brackish marshes of this system culminated in the impoundment of nearly all of the fringing marshes by 1970. Permanent flooding of these impoundments resulted in a profound change in vegetative and animal communities. The isolation of wetland habitat from the Indian River Lagoon System is believed to have reduced ecological benefits of the system, and efforts are under way to reconnect over three quarters of all impounded wetlands in the estuary (Brockmeyer et al., 1997). The Indian River Lagoon System supports abundant wading bird populations that utilize freshwater and salt marsh habitats for feeding, roosting, and nesting (Schikorr and Swain, 1995: Sewell et al., 1995; Smith and Breininger, 1995; Mikuska et al., 1998; Stolen et al., 2002).

To assess the impact of space vehicle launch operations on impounded wetland habitat, 13 focal impoundments were selected from among 75 impoundments on the Kennedy Space Center/Merritt Island National Wildlife Refuge. These impoundments contained roughly one fifth of the nearly 11,000 ha of impounded marsh habitat. Monthly wading bird habitat use surveys were conducted between April 1987 and April 2002 using a helicopter flying at an altitude of 60 m, and a speed of 110 km/h. Impoundments were flown systematically such that all area within was observed, and all individuals visible within the impoundment were counted. Between 1987 and 2002, 6 of the 13 impoundments were reconnected to the estuary through the installation of culverts, 1 was restored by completely removing the perimeter dike, and 5 remained isolated from the estuary. One of the 13 impoundments was restored prior to the study period and represented the probable condition of salt marsh in the system prior to impounding.

Although this study was not designed to monitor the effects of reconnection, wading bird foraging habitat use may be used to illustrate the use of waterbirds to detect changes in estuarine systems. In this case, the specific hypothesis tested was that reconnection would strongly affect wading bird use of foraging habitat by altering productivity and availability of prey within impoundments. To focus the analysis on the resident populations of wading birds, the mean monthly density of individuals observed in impoundments between April and August was used as a metric of habitat suitability. In this analysis wading birds were divided into piscivores (great blue heron, *Ardea herodius;* great egret; snowy egret, *Egretta thula;* tricolored heron, *E. tricolor;* and reddish egret, *E. rufescens*) and probers (white ibis, and glossy ibis, *Pelegadis falcinellus*). An investigation of visibility bias (E. Stolen and G. Carter, unpubl. data) showed that detectability of white-plumaged wading birds within impoundments was nearly 0.9 (i.e., one in ten missed), while dark-plumaged birds were more likely to be missed (detectability around 0.7).

Over the study period, the mean density of piscivores was 0.19 individuals/ha and the mean density of probers was 0.12 individuals/ha. Snowy egret made up 51% of all piscivores and white ibis made up 79% of all probers. The annual mean density of piscivorous wading birds observed foraging within the 13 impoundments varied by a factor of 30, ranging from 0.06 to 1.8 individuals/ha, demonstrating the extreme variability in the use of foraging habitat by wading birds in the system. Although some of the impoundments showed an increase in the density of foraging piscivorous wading birds following reconnection with the estuary, the patterns were inconsistent and the effect size was similar to variation in the impoundments that remained unconnected with the estuary (Figure 26.1A and B). The variation in monthly density observed within impoundments was very high, making it difficult to interpret differences between years. Regression models were used to test the hypothesis that restoration had an effect on the density of foraging wading birds when the effect of year was controlled. Only two of the six restored impoundments showed any evidence of such an effect (Figure 26.1A). Similarly, the annual mean density of probing wading birds observed foraging within the 13 impoundments varied by a factor of 155, ranging from 0.009 to 1.400 individuals/ha. As was the case with piscivores, the density of probers within impoundments following reconnection with the estuary did not differ from the density before, and monthly variation within the impoundments was high (results not given due to space limitations).

As a result of the high variation between monthly estimates of abundance, wading bird use of foraging habitat was not a sensitive indicator of the effect of reconnection on impounded salt marsh habitat at the Kennedy Space Center/Merritt Island National Wildlife Refuge. Although it may be that reconnection did not alter food supply or availability enough to affect wading birds, it is not possible to make any conclusions from the data available. There are factors that made it difficult to use this monitoring data to indicate change within this system. First, using impoundments as a sampling unit may not provide a good metric for the way the birds view habitat. Foraging wading birds commonly move between foraging locations within and between impoundments over time periods spanning minutes to hours. A better approach would have been to measure wading bird foraging success within impoundments, before and after reconnection. This could also have been coupled with measurement of prey density within the impoundments. In addition, information regarding other factors such as hydrology and management history that may have affected wading bird use of foraging habitat more than reconnection status was not available. Such information might be obtained, even after a disturbance event occurred, by using aerial photography and hydrological modeling incorporating historical water level data; such analysis is currently under way for this system (E. Stolen and D. Breininger, in prep.).

This example illustrates the importance of critically evaluating the usefulness of waterbird populations as indictors prior to investing heavily in their use. Had this study been designed to indicate environmental change within this system, it would not have succeeded in that goal. In this case, a pilot study of the use of wading bird populations would have quickly revealed the high variability in monthly densities and the need for habitat mapping and collection of management and hydrologic data. It might also have been decided that investigations that focus on the relative magnitude of intrinsic vs. extrinsic factors influencing wading bird use of foraging habitat would improve the usefulness of wading birds as indicators in this system (i.e., monitoring movements of birds at a larger spatial scale).

Conclusions

Many problems with the current use of waterbirds as indicators are due to a lack of information about the structure and demography of their populations. In the majority of situations, managers do not have adequate information about factors affecting waterbird populations that are only resident in the system for part of the time. The Everglades/Florida Bay ecosystem is quite large and this has allowed a good understanding of waterbird populations within. By contrast, the northern Indian River Lagoon site is smaller, and is a part of a larger ecosystem. In this case, information about factors affecting waterbird populations outside of the system was lacking and this made use of waterbird populations as an indicator difficult. These examples highlight the need for a thorough understanding of waterbird populations before their use as indicators in estuaries. However, the development of this type of information may be beyond the ability of individual managers to obtain. In many cases, collaboration between adjacent areas and within regions will be needed to gain such knowledge about the dynamics of waterbird populations.

Monitoring waterbirds has often been justified based on their value as indicators of environmental changes, and there have been several examples that have proved this to be true (e.g., brown pelican, *Pelecanus occidentalis*; osprey, *Pandion haliatus*). But there will be many cases in which waterbirds will not be particularly efficient indicators of changes in estuarine systems. In such cases, it is important to remember that waterbirds are important elements of biological diversity that often depend on estuarine systems, even if they are sometimes not the best indicators of particular environmental changes within those systems. Time lags in population responses can mask the consequences of habitat change for long periods (Nagelkerke et al., 2002), and there is ample evidence that many populations of waterbirds are declining (e.g., Kushlan and Hafner, 2000). For these reasons, continued research and monitoring of waterbird populations as important elements of biological diversity are justified.

The most useful aspects of waterbird populations as indicators within estuaries include the large spatial extent sampled, often numerous species from which to target an indicator, the ability to tailor the indicator to a specific ecosystem attribute, their sensitivity to contaminants, and the rapidity with which populations can respond to perturbations. Although the most powerful applications of indicators include detailed validation of links between ecosystem changes and indicator response, some useful knowledge can be gained from even the most preliminary investigations incorporating waterbird indicators. For example, most current knowledge of wading birds populations came from monitoring studies that were initially blind (or mostly blind) to ecological relationships. Therefore, we encourage the continued use of waterbird populations as bioindicators within estuaries, and we challenge investigators to improve on the reliability of the information obtained from such studies.

Acknowledgments

Work on the Kennedy Space Center/Merritt Island National Wildlife refuge was funded as part of the NASA Life Sciences Support Contract at the Kennedy Space Center. We thank B. Summer Field, K. Gorman, R. Smith, and the staff of the Merritt Island National Wildlife Refuge. Work in the Everglades was funded by the U.S. Army Crops of Engineers and the Florida Department of Environmental Protection.

A

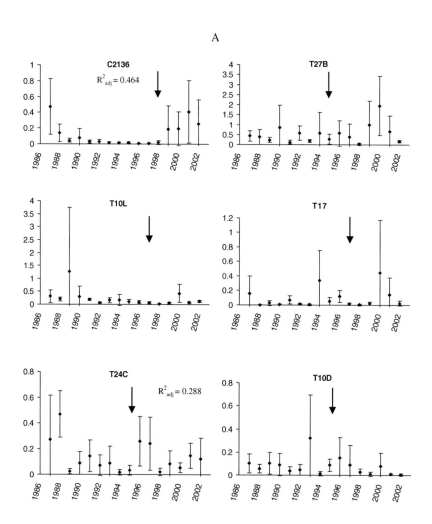

FIGURE 26.1 (A) Yearly mean density of foraging wading birds within impoundments that were reconnected with the estuary at Kennedy Space Center/Merritt Island National Wildlife Refuge showed inconsistent evidence of positive ecological effects. Shown is yearly mean density (individuals/ha) of monthly aerial foraging habitat use surveys (April to August) within each of six impoundments that were reconnected to the estuary by culverts; arrows indicate year of reconnection. Error bars are 95% confidence intervals of mean. R^2_{adj} values are given for sites with evidence of significant difference between years before and after restoration (other sites not significant at $\alpha = 0.05$ level).

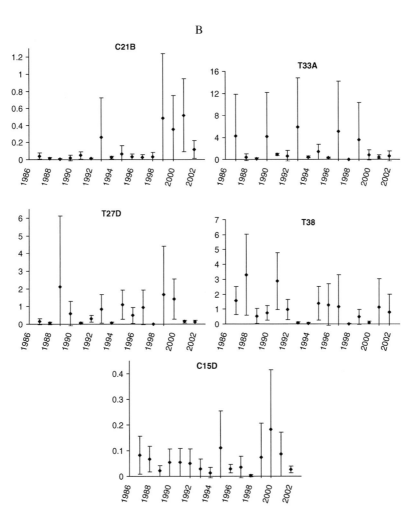

FIGURE 26.1 (CONTINUED) (B) Yearly mean density of foraging wading birds within impoundments that remained isolated from the estuary at Kennedy Space Center/Merritt Island National Wildlife Refuge showed similar patterns as did those that were reconnected. Shown is yearly mean density of monthly aerial foraging habitat use surveys (April to August) within each of five impoundments that were not reconnected to the estuary.

References

Berruti, A. 1983. The biomass energy consumption and breeding of waterbirds relative to hydrological conditions at Lake St. Lucia. *Ostrich* 54:65–82.

Bibby, C. T., N. D. Burgess, D. A. Hill, and S. H. Mustoe. 2000. *Bird Census Techniques.* Academic Press, San Diego, CA.

Bildstein, K. L. 1993. *White Ibis: Wetland Wanderer.* Smithsonian Institution Press, Washington, D.C.

Bildstein, K. L., L. Blood, and P. C. Frederick. 1991. The relative importance of biotic and abiotic vectors in nutrient processing in a South Carolina salt-marsh ecosystem. *Estuaries* 15:147–57.

Brockmeyer, R. E., J. R. Rey, R. W. Virnstein, R. G. Gilmore, and L. Earnest. 1997. Rehabilitation of impounded estuarine wetlands by hydrologic reconnection to the Indian River Lagoon, Florida (USA). *Wetlands Ecology and Management* 4:93–109.

Browder, J. 1985. Relationship between pink shrimp production on the Tortugas and water flow patterns in the Florida Everglades. *Bulletin of Marine Science* 37:839–856.

Butler, R. W. 1993. Time of breeding in relation to food availability of female great blue herons (*Ardea herodias*). *Auk* 110:693–701.

Caro, T. M. and G. O'Doherty. 1999. On the use of surrogate species in conservation biology. *Conservation Biology* 13:805–881.

Caughley, G. 1974. Bias in aerial survey. *Journal of Wildlife Management* 38:921–933.

Cezilly, F. 1997. Demographic studies of wading birds: an overview. *Colonial Waterbirds* 20:121–128.

Christy, R. L., K. L. Bildstein, and P. DeCoursey. 1981. A preliminary analysis of energy flow in a South Carolina salt marsh. *Colonial Waterbirds* 4:96–103.

Dolbeer, R. A., J. L. Belant, and G. E. Bernhardt. 1997. Aerial photography techniques to estimate populations of laughing gull nests in Jamaica Bay, New York, 1992–1995. *Waterbirds* 20:8–13.

Dvonch, J. T., J. R. Graney, G. J. Keeler, and R. K. Stevens. 1999. Use of elemental tracers to source apportion mercury in south Florida precipitation. *Environmental Science and Technology* 33:4522–4527.

Erwin, R. M. 1982. Observer variability in estimating numbers: an experiment. *Journal of Field Ornithology* 53:159–167.

Erwin, R. M. 1985. Foraging decisions, patch use, and seasonality in egrets (Aves: Ciconiiformes). *Ecology* 66:837–844.

Erwin, R. M. and T. W. Custer. 1982. Estimating reproductive success in colonial waterbirds: an evaluation. *Colonial Waterbirds* 5:49–56.

Erwin, R. M. and T. W. Custer. 2000. Conservation and management of herons: herons as indicators. In *Status and Conservation of Herons*, J. A. Kushlan, and H. Hafner (eds.). Academic Press, New York, pp. 311–330.

Erwin, R. M., J. G. Haig, D. B. Stotts, and J. S. Hatfield. 1996. Dispersal and habitat use by post-fledging juvenile snowy egrets and black-crowned night herons. *Wilson Bulletin* 108:342–356.

Fleury, B. E., and T. W. Sherry. 1995. Long-term population trends of colonial wading birds in the southeastern United States: the impact of crayfish aquaculture on Louisiana populations. *Auk* 112:613–632.

Frederick, P. C. 2000. Mercury contamination and its effects in the Everglades ecosystem. *Reviews in Toxicology* 3:213–255.

Frederick, P. C. 2001. Wading birds in the marine environment. In *Biology of Seabirds*, B. A. Schreiber and J. Burger (eds.). CRC Press, Boca Raton, FL, pp. 617–655.

Frederick, P. C. and J. C. Ogden. 1997. Philopatry and nomadism: contrasting long-term movement behavior and population dynamics of white ibis and wood storks. *Colonial Waterbirds* 20:316–323.

Frederick, P. C. and J. C. Ogden. 2003. Monitoring ecosystems using avian populations: seventy years of surveys in the Everglades. In *Interdisciplinary Approaches for Evaluating Ecoregional Initiatives*, D. Bush and J. Trexler (eds.). Island Press, Washington, D.C., pp. 321–350.

Frederick, P. C. and G. V. N. Powell. 1994. Nutrient transport by wading birds in the Everglades. In *Everglades: The Ecosystem and Its Restoration*, S. M. Davis and J. C. Ogden (eds.). St. Lucie Press, Delray Beach, FL, pp. 571–584.

Frederick, P. C., K. L. Bildstein, B. Fleury, and J. C. Ogden. 1996. Conservation of large, nomadic populations of white ibis (*Eudocimus albus*) in the United States. *Conservation Biology* 10:203–216.

Frederick, P. C., M. G. Spalding, M. S. Sepulveda, G. E. Williams, Jr., L. Nico, and R. Robbins. 1999. Exposure of great egret nestlings to mercury through diet in the Everglades of Florida. *Environmental Toxicology and Chemistry* 18:1940–1947.

Frederick, P. C., M. G. Spalding, and R. Dusek. 2002. Wading birds as bioindicators of mercury contamination in Florida: annual and geographic variation. *Environmental Toxicology and Chemistry* 21:163–167.

Frederick, P. C., B. Hylton, J. A. Heath, and M. Ruane. 2003. Accuracy and variation in estimates of large numbers of birds by individual observers using an aerial survey simulator. *Journal of Field Ornithology* 74:281–287.

Frederick, P. C., B. Hylton, J. A. Heath, and M. G. Spalding. 2004. A historical record of mercury contamination in southern Florida as inferred from avian feather tissue. *Environmental Toxicology and Chemistry* 23: 1474–1478.

Frohring, P. C., D. P. Voorhees, and J. A. Kushlan. 1988. History of wading bird populations in the Florida Everglades: a lesson in the use of historical information. *Colonial Waterbirds* 11:328–335.

Gawlik, D. E. 2002. The effects of prey availability on the numerical response of wading birds. *Ecological Monographs* 72:329–346.

Gawlik, D. E., R. D. Slack, J. A. Thomas, and D. N. Harpole. 1998. Long-term trends in population and community measures of colonial-nesting waterbirds in Galveston Bay Estuary. *Colonial Waterbirds* 21:143–151.

Gunderson, L. H. and W. F. Loftus. 1993. The Everglades. In *Biotic Communities of the Southeastern United States*, W. H. Martin, S. G. Boyce, and A. C. Echternacht (eds.). John Wiley & Sons, New York.

Hafner, H. and R. H. Britton. 1983. Changes of foraging sites by nesting little egrets (*Egretta garzetta* L.) in relation to food supply. *Colonial Waterbirds* 6:24–30.

Hafner, H., P. J. Dugan, M. Kersten, O. Pineau, and J. P. Wallace. 1993. Flock feeding and food intake in little egrets *Egretta grazetta* and their effects on food provisioning and reproductive success. *Ibis* 135:25–32.

Kushlan, J. A. 1976. Feeding behavior of North American herons. *Auk* 93:86–94.

Kushlan, J. A. 1993. Colonial waterbirds as bioindicators of environmental change. *Colonial Waterbirds* 16:223–251.

Landres, P. B., J. Verner, and J. W. Thomas. 1988. Ecological uses of vertebrate indicator species: a critique. *Conservation Biology* 2:316–328.

Light, S. S. and J. W. Dineen. 1994. Water control in the Everglades: a historical perspective. In *Everglades: The Ecosystem and Its Restoration,* S. M. Davis and J. C. Ogden (eds.). St. Lucie Press, Delray Beach, FL, pp. 47–84.

Loftus, W. F. and A. M. Eklund. 1994. Long-term dynamics of an Everglades small-fish assemblage. In *Everglades: The Ecosystem and Its Restoration,* S. M. Davis and J. C. Ogden (eds.). St. Lucie Press, Delray Beach, FL, pp. 461–484.

Loftus, W. F., J. K. Chapman, and R. Conrow. 1986. Hydroperiod effects on Everglades marsh food webs, with relation to marsh restoration efforts. *Conference on Science in the National Parks* 6:1–22.

Lorenz, J. J. 1999. The response of fishes to physicochemical changes in the mangroves of Northeast Florida Bay. *Estuaries* 22:500–517.

Lorenz, J. J. 2000. Impacts of Water Management on Roseate Spoonbills and Their Piscine Prey in the Coastal Wetlands of Florida Bay. Ph.D. dissertation, University of Miami, Coral Gables, FL.

McIvor, C. C., J. A. Ley, and R. D. Bjork. 1994. Changes in freshwater inflow from the Everglades to Florida Bay including effects on biota and biotic processes: a review. In *Everglades: The Ecosystem and Its Restoration,* S. M. Davis and J. C. Ogden (eds.). St. Lucie Press, Delray Beach, FL, pp. 117–148.

Mikuska, T., J. A. Kushlan, and S. Hartley. 1998. Key areas for wintering North American herons. *Colonial Waterbirds* 21:125–34.

Montague, C. L. and R. G. Weigert. 1990. Salt marshes. In *Ecosystems of Florida,* R. L. Myers and J. J. Ewel (eds.). University of Central Florida Press, Orlando, FL, pp. 481–516.

Morrison, M. L. 1986. Bird populations as indicators of environmental change. *Current Ornithology* 3:429–451.

Nagelkerke, K., J. Verboom, F. v. D. Bosch, and K. V. D. Wolfshaar. 2002. Time lags in metapopulation responses to landscape change. In *Applying Landscape Ecology in Biological Conservation,* K. J. Gutzwiller (ed.). Springer, New York, pp. 330–354.

Niemi, G. J., J. M. Hanowski, A. R. Lima, T. Nicholls, and N. Weiland. 1997. A critical analysis on the use of indicator species in management. *Journal of Wildlife Management* 61:1240–1252.

Noss, R. F. 1990. Indicators for monitoring biodiversity: a hierarchical approach. *Conservation Biology* 4:355–363.

Ogden, J. C. 1994. A comparison of wading bird nesting colony dynamics (1931–1946 and 1974–1989) as an indication of ecosystem conditions in the southern Everglades. In *Everglades: The Ecosystem and Its Restoration,* S. M. Davis and J. C. Ogden (eds.). St. Lucie Press, Delray Beach, FL, pp. 533–570.

Onuf, C. P., J. M. Teal, and I. Valiela. 1977. Interactions of nutrients, plant growth and herbivory in a mangrove ecosystem. *Ecology* 58:514–26.

Pierce, C. W. 1962. The cruise of the *Bonton. Tequesta* 22:3–63.

Pollock, K. H. and W. L. Kendall. 1987. Visibility bias in aerial surveys: a review of estimation procedures. *Journal of Wildlife Management* 51:502–510.

Robertson, W. B. and J. A. Kushlan. 1974. The southern Florida avifauna. *Miami Geological Society Memoirs* 2:414–452.

Runde, D. E. 1991. Trends in Wading Bird Nesting Populations in Florida 1976–1978 and 1986–1989. Nongame Wildlife Section, Division of Wildlife, Florida Game and Fresh Water Fish Commission, Tallahassee, FL.

Schikorr, K. E. and H. M. Swain. 1995. Wading birds — barometer of management strategies in the Indian River Lagoon. *Bulletin of Marine Science* 57:215–229.

Schmalzer, P. A. 1995. Biodiversity of saline and brackish marshes of the Indian River Lagoon: historic and current patterns. *Bulletin of Marine Science* 57:37–48.

Sepulveda, M. S., P. C. Frederick, M. G. Spalding, and G. E. Williams, Jr. 1999. Mercury contamination in free-ranging great egret (*Ardea albus*) nestlings from southern Florida. *Environmental Contamination and Toxicology* 18:985–992.

Sewell, C. E., N. D. Joiner, M. S. Robson, and M. J. Weise. 1995. Wading bird colony protection in the Indian River Lagoon. *Bulletin of Marine Science* 57:237–241.

Simberloff, D. 1998. Flagships, umbrellas, and keystones: is single-species management passé in the landscape era? *Biological Conservation* 83:247–257.

Smith, R. B. and D. R. Breininger. 1995. Wading bird populations of the Kennedy Space Center. *Bulletin of Marine Science* 57:230–236.

Spalding, M. G., G. T. Bancroft, and D. J. Forester. 1993. The epizootiology of eustrongylidosis in wading birds (Ciconiiformes) in Florida. *Journal of Wildlife Diseases* 29:237–249.

Spalding, M. G., R. D. Bjork, G. V. N. Powell, and S. F. Sundlof. 1994. Mercury and cause of death in great white herons. *Journal of Wildlife Management* 58:735–739.

Stolen, E. D., R. B. Smith, and D. R. Breininger. 2002. Analysis of Wading Bird Use of Impounded Wetland Habitat on the Kennedy Space Center/Merritt Island National Wildlife Refuge: 1987–1998. NASA Technical Memorandum 211173.

Temple, S. A. and J. A. Wiens. 1989. Bird populations and environmental changes: can birds be bio-indicators? *American Birds* 43:260–270.

Trexler, J. C. W. F. Loftus, and J. H. Chick. 2003. Setting and monitoring restoration goals in the absence of historical data: the case of fishes in the Florida Everglades. In *Interdisciplinary Approaches for Evaluating Ecoregional Initiatives,* D. E. Bisch and J. C. Trexler (eds.). Island Press, Washington, D.C., pp. 351–376.

Ware, F. J., H. Royals, and T. Lange. 1990. Mercury contamination in Florida largemouth bass. *Proceedings of the Annual Conference of Southeast Association of Fish and Wildlife Agencies* 44:5–12.

Williams, B. K., J. D. Nichols, and M. J. Conroy. 2002. *Analysis and Management of Animal Populations.* Academic Press, San Diego, CA.

27

Using Spatial Analysis to Assess Bottlenose Dolphins as an Indicator of Healthy Fish Habitat

Leigh G. Torres and Dean Urban

CONTENTS

Introduction

Florida Bay, which lies at the southern tip of the Florida peninsula (Figure 27.1), is the terminus of the largest ecosystem restoration project ever attempted in the United States. The Comprehensive Everglades Restoration Project (CERP), with the ambitious goal to improve the quality, quantity, and timing of freshwater inputs into the South Florida ecosystem, covers over 29,000 km² (18,000 square miles) and will require 30 years to complete (1999–2029). Quantifying the spatial and temporal changes in species distributions, abundance, and diversity in Florida Bay is essential to gauge the success of this ambitious restoration project. This task will require developing tangible quantitative metrics to assess the abiotic and biotic changes taking place in response to the CERP management.

Seabirds and cetaceans are ideal focal organisms for the study of ecosystem-level changes in marine systems because they are numerous and conspicuous predators, with large energetic requirements (Croxall, 1989; Moore and DeMaster, 1998; Baumgartner et al., 2000; D'Amico et al., 2003). Furthermore, because they forage on fish and squid, prey that are often difficult to sample by conventional means, these predators can be used to monitor ecosystem structure (Furness and Camphuysen, 1997; Griffin, 1999; Moore et al., 2002). Bottlenose dolphins (*Tursiops truncatus*) are prominent in Florida Bay and, like the spotted owl in old growth forests of the Pacific Northwest (Simberloff, 1987; Lamberson et al., 1992; Caro and O'Doherty, 1999), their distribution may be indicative of important habitats. Aerial surveys for dolphins in Florida Bay are a relatively efficient management tool, both financially and logistically; they can be completed in less than a day and require little labor. Dolphin foraging habitat can quickly be identified as productive fish communities, allowing managers an easy tool to spatially and temporally quantify habitat quality and, through long-term observational data collection, monitor spatial migration of productive fish habitats. This approach will help managers

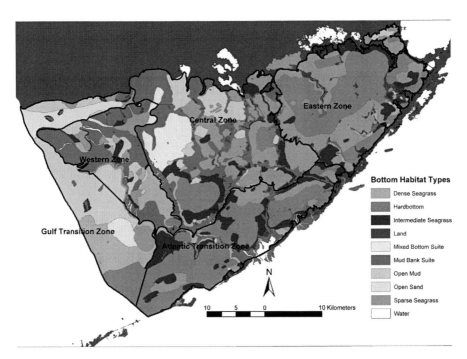

FIGURE 27.1 The location and description of the Florida Bay study area: Benthic habitat types and zones. Note patchiness of habitat types throughout the bay. (Benthic habitat type map provided by Robert Halley and Ellen Prager. 1997. Florida Bay Bottom Types Map: USGS Open-File Reports OFR 97-526, U.S. Geological Survey, Reston, VA.)

mitigate the effects of human development on the Florida Bay ecosystem by identifying and protecting important habitat.

As the restoration project continues, and the ecosystem of Florida Bay begins to change, resource managers will need to identify habitat alterations, determine the cause, predict the impact, and, in some cases, mitigate anthropogenic activities to conserve ecosystem viability. The working hypothesis underlying this work is that the spatially heterogeneous biotic and physical attributes of Florida Bay influence the distribution and composition of the fish community and, subsequently, determine the distribution and habitat use of dolphins throughout the bay. In other words, top predators follow their prey distributions, and therefore distribute themselves relative to biotic (e.g., chlorophyll) and abiotic (e.g., substrate type, water salinity) properties, which predict prey distributions. The two objectives are (1) to determine which environmental variables, or combination of variables, influence fish community distribution and composition throughout Florida Bay, and (2) to quantify the links between productive fish habitats and dolphin distributions and habitat use. The immediate research question asked in this study is "Can bottlenose dolphin distribution in Florida Bay be used to predict the distribution and structure of fish communities?"

In addition to documenting the biophysical linkages supporting upper trophic predators in this heterogeneous environment, our research examines the suitability of the bottlenose dolphin as an indicator species in Florida Bay. (A management indicator species, or MIS, is defined as any species, group of species, or species habitat elements selected to focus management attention for the purposes of resource production, population recovery, or maintenance of population viability or ecosystem diversity; U.S. Forest Service, 2000.) More specifically, we determined the response of dolphin distributions to spatial and temporal variation in water quality, prey density, and habitat type. Therefore, we established a link between the various habitat variables, prey distributions, and dolphin habitat use patterns, in order to evaluate the feasibility of using bottlenose dolphins as an indicator of important fish habitats in Florida Bay.

Background

Historically, Florida Bay was an extremely productive ecosystem supporting high densities of fish, birds, sea turtles, and marine mammals in a complex biological and physical oceanographic environment (Sogard et al., 1989). Since 1881 the watershed of Florida Bay has been highly managed to support agriculture, control floods, and provide water to the growing population of South Florida (Light and Dineen, 1994; McPherson and Halley, 1996). Currently, up to 70% of the freshwater flow through the Everglades is diverted for human consumption (Smith et al., 1989), causing salinity levels in Florida Bay to increase more than 50 ppt during drought conditions (Schmidt, 1979; Fourqurean et al., 1992; McIvor et al., 1994). In addition to the drastic decrease in freshwater discharge into Florida Bay, water quality has also deteriorated as a result of high phosphates and dissolved and particulate matter inputs from runoff due to agricultural practices within the watershed (Rudnick et al., 1999). Increased salinity and elevated nutrient levels from anthropogenic sources have stimulated algal blooms, resulting in large-scale die-offs of seagrasses, sponges, and mangroves (Robblee et al., 1991; Boesch et al., 1993; Butler et al., 1995). In turn, this habitat degradation has resulted in reductions in the standing stocks and changes in the composition of fish communities (Matheson et al., 1999). In short, as a result of these anthropogenic disturbances, fish diversity in Florida Bay has greatly diminished (Tilmant, 1989). However, little is known about the impact of these broad habitat changes on the upper trophic predators.

The Florida Everglades drains directly into Florida Bay, which is a large (2200 km^2), shallow ecosystem that lies within the boundaries of the Everglades National Park and the Florida Keys National Marine Sanctuary. Mudbanks and mangrove islands divide the bay into a series of shallow, semi-isolated basins that restrict circulation (Zieman et al., 1989). The watershed of Florida Bay is approximately 12,000 km^2 of wetlands extending south of Lake Okeechobee; the marine influence in the bay comes primarily from the Gulf of Mexico and, secondarily, through passes to the Atlantic (Fourqurean and Robblee, 1999). A complex gradient between fresh and salt water provides a unique habitat for many protected upper trophic level species including manatees, alligators, crocodiles, sea turtles, and bottlenose dolphins (Holmquist et al., 1989). Additionally, Florida Bay links the mangroves and the coral reef systems in South Florida, acting as an important nursery for many reef and commercial fish (Rutherford et al., 1989; Tabb and Roessler, 1989).

Florida Bay is an ideal field laboratory for assessing the relationship between habitat variability and the distribution of upper trophic predators and their prey. Florida Bay is a complex mosaic of heterogeneous habitats characterized by striking physical and biotic gradients, patchy bottom substrate types, and variable water quality. On the basis of similar physical characteristics (e.g., salinity, temperature, turbidity) and distributions of flora and fauna (e.g., bottom substrate, seagrass coverage, fish assemblage composition), researchers have identified five distinct zones within Florida Bay: Eastern Zone, Central Zone, Western Zone, Gulf Transition Zone, and Atlantic Transition Zone (Figure 27.1) (Sogard et al., 1989; Thayer and Chester, 1989; Zieman et al., 1989; Matheson et al., 1999; Thayer et al., 1999). It is our contention that this physical and biological variability of Florida Bay structures influences fish distribution and consequently influences dolphin habitat use.

Methods

Previous research focused on studying dolphin habitat use patterns has used behavioral sampling techniques (Waples, 1995; Allen and Read, 2000). However, these studies did not fully assess habitat quality, which can be inferred by physical attributes, including temperature, salinity, dissolved oxygen concentration, current velocity, water depth, and bottom substrate (Gibson, 1994; Raven et al., 1998; Fraser et al., 1999), as well as biological parameters, such as the density of predators, prey, and competitors (Gibson, 1994). These indirect metrics were employed in this study to assess spatially and temporally variable habitat quality.

Data Collection/Field Methods

The field methods used to collect the data in Florida Bay during the summers of 2002 and 2003 included boat-based surveys and randomly stratified trawls and gillnet sets. During each summer season, Florida Bay was surveyed twice for the presence/absence of bottlenose dolphins. Surveys were conducted in a small, outboard-powered vessel using standardized techniques (Buckland et al., 1993). To ensure standardized "sightability" across areas and days, surveys were limited to sea state conditions of Beaufort 3 or less and conducted with a minimum of three observers. Global Positioning System (GPS; longitude and latitude) positions were downloaded every 2 min to record the survey route. Habitat variables, water temperature, salinity, depth, clarity (using a Secchi depth measurement), sea state condition, and benthic habitat type were measured at 30-min intervals during surveys. At each dolphin sighting, these same environmental variables were measured, in addition to recording relevant survey (GPS location and the perpendicular distance, in meters, to the sighting) and ecological (group size, behavior, composition) attributes. The standardized behavioral state categories, adapted from the Sarasota Dolphin Research Program (Urian and Wells, 1996), are travel, social, rest, forage, unknown, play, and interacting with the research vessel (e.g., bow riding).

Fish distributions in Florida Bay were sampled with preselected stations with 3-min bottom trawls and 30-min gillnet sets. In 2002, 99 randomly stratified trawls were conducted, and 94 randomly stratified trawls were performed in 2003. In 2002, 24 gillnet sets were conducted using a 50-m-long, 7.6-cm mesh. A 50-m panel of 8.3-cm mesh was added to the gillnet in 2003, and 50 randomly stratified gillnet sets were conducted. Site selection for bottom trawls and gillnet sets were randomly stratified using Geographic Information System (GIS) software (ArcGIS®) to adequately sample the different habitat types within each zone of the bay. Although both trawls and gillnet sets catch potential dolphin prey, the trawl catches were often of a smaller size than the dolphin prey size class. To obtain a more representative sample of dolphin prey availability, gillnet sets were used to size-select the catch. The catch efficiency of dolphin prey with trawls was limited; nevertheless, abundance, species richness, and diversity are considered indicative of habitat quality. Thus, both trawling and gillnetting samples were used to describe the fish community and dolphin prey availability throughout the bay. Using these fishing techniques, prey distributions were related to dolphin behavior by comparing fish community composition at sites where dolphins were observed feeding, present but not feeding, and absent.

All captured fish were placed into an aerated bucket of water and later removed for identification and measurement before being released alive. During this fieldwork in Florida Bay, 98% of fish caught were released alive. The standardized abundance (catch-per-unit-effort, CPUE), Margalef's richness index (($S - 1$)/log N), Simpson's diversity index (p_i^2), and CPUE of dolphin prey (based on fish species and size) were calculated for each trawl and gillnet set.

Data Analysis

The central theme of the analysis was to quantify the spatial structure of Florida Bay and its effect on habitat quality and fish and dolphin distributions. Because inherent correlations exist between environmental variables and space, as well as among environmental variables themselves, the objective here was to "tease apart" the relative influence of each individual variable on fish and dolphin distribution. These patterns were analyzed using geostatistical techniques specially suited to integrate disparate types of autocorrelated data such as the Mantel test and Classification and Regression Trees (CARTs), and then visualized with a GIS. Mantel tests (1967) are multivariate statistics that can explicitly include space as an explanatory variable. Moreover, Mantel tests account for the effects of spatial autocorrelation on dependent and independent variables, thus allowing for the rigorous testing of spatially explicit relationships between dolphin and fish distribution and environmental variables. Subsequently, fish and dolphin foraging habitats were modeled using CARTs (Venables and Ripley, 1997). CARTs recursively partition data using an algorithm that splits observations into groups based on a single best predictor variable, until all points are classified. Essentially, CARTs are a set of nested "if" statements that are used to define the relationship of each response variable to predictor variables. Categorical data cannot be included in a Mantel test. Therefore, habitat type and zone were addressed only with the CART analysis. Figure 27.2 depicts the generalized model tested in this chapter.

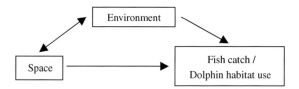

FIGURE 27.2 The general tested model of this chapter: effect of environmental variables on space and the subsequent effect on fish catch and dolphin habitat use; effect of space on environmental variables and the subsequent effect on fish catch and dolphin habitat use; the direct effect of space on fish catch and habitat use; the direct effect of environmental variables on fish catch and dolphin habitat use.

The Mantel tests were used to relate the four indices (CPUE, species richness, species diversity, and dolphin prey) of each data set (2002 and 2003 trawls, and 2002 and 2003 gillnet sets) to the following environmental variables for the 2002 data: temperature, salinity, depth, water clarity, habitat type, zone, chlorophyll, percent saturated dissolved oxygen, turbidity, distance from land, and distance from mudbanks. (Chlorophyll, percent saturated dissolved oxygen, and turbidity data were provided by the SERC-FIU water quality monitoring network. These data points were interpolated in GIS using a Gaussian kriging technique and sampled for each trawl and gillnet location.) Chlorophyll, percent saturated dissolved oxygen, and turbidity data were not available for 2003, so trawl and gillnet catch from 2003 was not tested against these variables.

CART models of CPUE from 2002 and 2003 trawls were built for each of the five zones, using the same environmental data, with the addition of habitat type. The resulting five "zonal" CART models predicting abundant fish habitats in Florida Bay were spatially mapped out in GIS. Additionally, a CART model was preformed to classify the habitat of foraging vs. nonforaging dolphins during the 2002 and 2003 summer field seasons. The predicted dolphin foraging habitat from these CART models were also mapped out spatially using GIS and overlaid on the predicted fish habitats for comparison.

Results

Dolphins were sighted throughout the bay, with foraging behavior observed, during both the 2002 and 2003 summer field seasons (Table 27.1).

Mantel Tests

A multitude of Mantel tests were performed to determine the significance of each variable on each of the four catch indices from the two fish sampling methods. For all tests, space was highly correlated with environmental variables ($p < 0.001$). Moreover, the global Mantel tests revealed that fish catch was related to space, even when the effect of the environmental variables on space was removed ($p < 0.034$ or less). This was true for each index, for each fish catch data set, except gillnet sets from 2003. This result suggested that zone could be an important factor influencing the Florida Bay fish community.

Additionally, the Mantel tests determined that when the effect of spatial autocorrelation is removed, no single tested environmental variable had a consistent significant effect on any of the fish catch indices. Various pure partial Mantel tests were conducted to test the individual effect of each variable, on each index, for the four fish catch data sets. However, only Secchi depth (water clarity), depth, and chlorophyll were occasionally significant (Table 27.2).

Five environmental variables were strongly correlated with space in every test except with the 2003 gillnet data set: depth ($p < 0.05$), salinity ($p < 0.001$), Secchi depth ($p < 0.004$), distance from land and mudbanks ($p < 0.04$), and distance from mudbanks ($p < 0.005$). A Pearson product–moment correlation matrix was also performed for all environmental variables for each fish catch data set to determine relationships among variables. The environmental data from the 2002 trawl and gillnet data sets showed very strong relationships between chlorophyll and turbidity (trawl = 0.80; gillnet = 0.78). Additionally, Secchi depth had a relatively strong inverse relationship to chlorophyll (–0.41) and turbidity (–0.43). Percent-saturated dissolved oxygen had a fairly strong inverse relationship to salinity (–0.54).

TABLE 27.1

Number of Dolphin Sightings and Foraging Observations from Survey Effort
in Florida Bay during the Summers of 2002 and 2003

2002	Sightings	Dolphins	2003	Sightings	Dolphins
Sightings	57	356	Sightings	68	520
Foraging	26 (46%)	190 (53%)	Foraging	24 (35%)	194 (37%)
Nonforaging	31 (54%)	166 (47%)	Nonforaging	44 (65%)	326 (63%)

TABLE 27.2

P Values from the Pure Partial Mantel Tests That Produced Significant or Marginally
Insignificant Results

Data Set and Index	Secchi Depth	Depth	Chlorophyll *a*
Trawls 2002			
CPUE	0.113	0.055	NS
Dolphin prey	0.047	0.051	NS
Simpson's Diversity	0.005	NS	NS
Richness	NS	0.114	NS
Trawls 2003			
Richness	0.054	NS	NS
Gillnet Sets 2002			
CPUE	NS	0.102	0.097
Dolphin prey	0.131	0.116	0.095
Simpson's Diversity	NS	NS	0.056
Richness	NS	NS	0.122
Gillnet Sets 2003			
Simpson's Diversity	0.048	NS	NS
Richness	0.02	NS	NS

Note: A pure partial Mantel test removes the confounding effect of all other variables,
including space, from the test. NS = not significant.

TABLE 27.3

Residual Mean Deviance from CART Models for Each Zone of Fish CPUE from 2002 and 2003 Trawls

	Eastern	Central	Western	Atlantic Transition	Gulf Transition
2002	0.001	0.002	0.464	0.003	0.050
2003	0.230	0.001	NA	0.013	0.026

Note: A higher deviance value indicates that it was more difficult for the model to split the response variable (level of
CPUE) into homogeneous groups based on the available predictor variables. No value is associated with the 2003
Western Zone because the CART model found no significant predictor variable for the response data.

CARTs

A benefit of CARTs is the ability to incorporate both continuous and categorical data into the model. Therefore, CART models were run with all the environmental variables, including habitat type and zone, for each index of each fish catch data set. While each resulting tree looked different, one consistent trend was evident: habitat type and zone were always primary explanatory variables used to split the response data, with Secchi depth often used further down the tree. However, the divisions created by these variables were not consistent. For example, habitat type was not always partitioned so that dense seagrass had a greater fish catch; nor was higher water clarity always indicative of better fish habitat. Figure 27.3, based on dolphin prey subset from 2003 trawls, is an example of a typical CART result from this analysis. The combination of this result and the result from the Mantel tests indicating the significance of space regardless of environmental variables suggested that the effect of zone is significant. Therefore, individual CARTs were performed for each zone on CPUE from the 2002 and 2003 trawl data. The residual mean deviance (a measure of misclassification by the CART) of these models ranged from 0.001 to 0.464 in 2002, and between 0.001 and 0.230 in 2003 (Table 27.3). A higher deviance value indicates that it was more difficult for the model to split the response variable (level of CPUE) into homogeneous groups based on the available predictor variables.

Dolphin foraging habitat was also modeled using CARTs. Environmental data from dolphin sightings for each year were used as predictor variables of behavior state (foraging or nonforaging) during the sighting. Figure 27.4 and Figure 27.5 depict the resulting trees for 2002 and 2003, respectively. In 2002, the CART model used zone, habitat type, and Secchi depth as the predictor variables of dolphin foraging habitat. In 2003, only habitat type and Secchi depth were used by the CART model as explanatory variables.

Abundant fish habitats of each zone were mapped using GIS for both 2002 and 2003 based on the "zonal" CART models. Additionally, the predicted dolphin foraging habitats for 2002 and 2003 were mapped. These maps were overlaid, by year, to compare the areas predicted spatially (Figure 27.6 and Figure 27.7).

There is generally good spatial overlap between those areas predicted as abundant fish habitats and the locations where dolphins forage. In 2002 (Figure 27.6), no area of the Eastern or Western Zones was modeled as dolphin foraging habitat because no dolphins were observed foraging in these zones during this summer. In the other zones of Florida Bay, overlap between models in 2002 is good, especially the Gulf transition zone and the northern and southern portions of the Central Zone. There were only three sightings of foraging dolphins in the Atlantic Transition Zone in 2002. This small sample size made trend analysis and extrapolation of habitat preference to the entire zone difficult. However, both the fish and dolphin foraging CARTs correctly modeled the locations of these three foraging sightings.

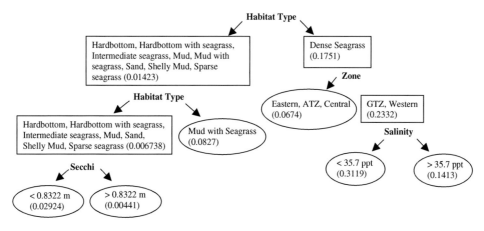

FIGURE 27.3 CART of dolphin prey subset of CPUE from 2003 trawls. Circles denote terminal nodes. Numbers in parentheses are mean CPUE for the specified group. Residual mean deviance = 0.0044. Note that the first three breaks are habitat type and zone.

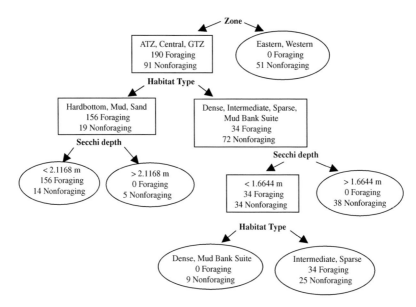

FIGURE 27.4 CART of dolphin foraging habitat for 2002 based on zone, habitat type and Secchi depth (water clarity). Circles denote terminal nodes. Misclassification error rate: 0.1175 = 39/332.

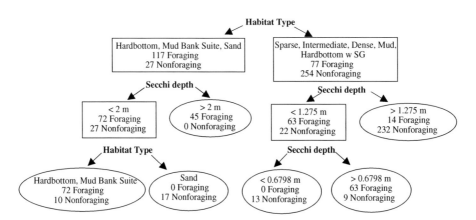

FIGURE 27.5 CART of dolphin foraging habitat for 2003 based on habitat type and zone. Circles denote terminal nodes. Misclassification error rate: 0.069 = 33/475.

In 2003 (Figure 27.7), the CART model of fish habitat for the Western Zone did not find any significant predictor variable. Therefore, the entire Western Zone is classified as abundant fish habitat because all trawls had a high CPUE. Hence, all observations of foraging animals in the Western Zone overlapped with predicted abundant fish habitat. Although more area was predicted as dolphin foraging habitat in the Eastern Zone in 2003 than in 2002, fewer areas were modeled as fish habitat. However, the two sightings of foraging dolphins did overlap with small areas predicted as abundant fish habitat. In the Central Zone, less area was predicted as abundant fish habitat in 2003 than in 2002, but much of the same area in the northern portion of the zone was identified in both years. Only one observation of foraging dolphins occurred in this area in 2003 and did not overlap with predicted fish habitat. Conversely, essentially opposite habitats were predicted as abundant fish habitat in the Gulf Transition Zone between 2002 and 2003. But, in 2003, two sightings of foraging dolphins did overlap, or closely overlap, with predicted fish habitat. In the Atlantic Transition Zone, two large areas of predicted habitat overlapped, but no location of foraging dolphins were captured by both models.

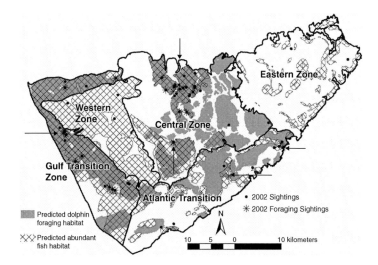

FIGURE 27.6 Overlap of predicted abundant fish habitat (based on zonal CARTs of CPUE from 2002 trawls) and predicted dolphin foraging habitat (based on CART of sightings from 2002) during the summer of 2002. Arrows point to observations of foraging dolphins that both models identified as habitat; these areas are discussed in the text.

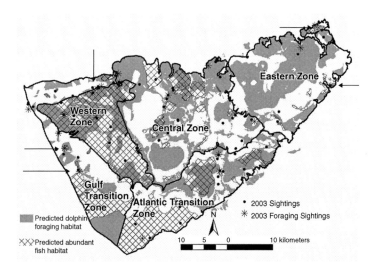

FIGURE 27.7 Overlap of predicted abundant fish habitat (based on zonal CARTs of CPUE from 2003 trawls) and predicted dolphin foraging habitat (based on CART of sightings from 2003) during the summer of 2003. Arrows point to observations of foraging dolphins that both models identified as habitat; these areas are discussed in the text.

Discussion

This chapter addressed two critical research priorities. The first priority was to answer the question, "What is the relationship between environmental and habitat change and the upper trophic levels in Florida Bay?" Population declines of some species, such as wading birds, indicate that the system's carrying capacity has diminished and/or that important habitats have been degraded (Deegan et al., 1998). The second priority, as identified by the Florida Bay Integrated Science Plan, was to examine the hypothesis that bottlenose dolphins are good indicators of the distribution of forage fish and quality of fish habitat in Florida Bay (Dynamic Higher Trophic Levels Science Plan, 2001). Despite considerable prior efforts to understand the ecosystem dynamics of Florida Bay, very little research has focused on a dominant upper trophic level predator, the bottlenose dolphin. Bottlenose dolphins range throughout Florida Bay and their patterns of distribution reflect oceanographic variability, changes in marine food webs, and anthropogenic impacts (Woodley and Gaskin, 1996; Baumgartner, 1997; Kasamatsu et al., 2000). Understanding dolphin distributions, foraging ecology, and habitat use can help resource managers to monitor ecosystem-level changes and to enhance the understanding of biogeographic and ecological processes in this heterogeneous and dynamic ecosystem.

The CART models, and subsequent GIS mapping, of both fish and dolphin foraging habitats, were able to spatially capture the patchy quality of the bay in a realistic manner. In particular, in 2002 and 2003, both models impressively mapped habitats around locations of foraging dolphin sightings, showing fine-scale accuracy and overlap of predicted habitats (denoted by arrows in Figure 27.6 and Figure 27.7). While some areas of predicted habitat in 2002 and 2003 do not coincide, these models can be fine-tuned using additional habitat variables, an improved bottom habitat types map (in progress), and a larger data set. Based on these preliminary results, however, it appears that dolphins can be used to predict the distribution and structure of fish communities throughout Florida Bay. This connection can be very useful for mitigation purposes because as the restoration project continues managers can use this link between foraging dolphins and fish to identify healthy habitats supporting upper trophic level species. Moreover, an understanding of which habitat parameters are important to the health of bottlenose dolphin and fish populations within Florida Bay will allow managers to concentrate restoration efforts on degraded habitats while simultaneously protecting habitats that continue to support a healthy ecosystem.

For simplicity, the "zonal" CART analysis and spatial modeling was done only with the CPUE index from 2002 and 2003 trawls, but these models could have been based on species diversity, species richness, or the dolphin prey subset. The index chosen to determine "healthy fish habitats" is a management question. Management based on CPUE will identify areas of abundant fish, regardless of size or species. However, using species diversity or species richness as the management metric can lead to management of marginal habitats. This is because areas of high species diversity or richness may be intermediate or marginal habitats, supporting a variety of species but not providing optimal habitat for any of the organisms.

Results from this analysis underscore the extremely heterogeneous nature of Florida Bay. No single environmental variable was identified as the determining factor of productive fish habitat. Instead, it appears from both the Mantel and CART results that combinations of variables interact to provide patchy habitats throughout the bay. Moreover, the correlations between certain environmental variables (chlorophyll, turbidity, and Secchi depth; percent saturated dissolved oxygen and salinity) also indicate that the relative effect of each variable is not easily detectable, but rather it is the interplay between variables that creates complexly defined habitat quality. However, certain environmental variables were identified through this analysis as being primary predictor variables of abundant fish communities: zone, habitat type, water clarity, chlorophyll, and salinity.

The complexity of Florida Bay is further confounded by the dynamics of habitat quality across the different zones of the bay. For example, in the hard-bottom communities of the Atlantic Transition Zone, the fish community is chiefly composed of reef-dwelling animals dependent on habitats supported by good water clarity. However, less than 25 km away, in the Central Zone of Florida Bay, the fish community is relatively very abundant and associated with habitat with less water clarity (murkier). Therefore, it is not just the significance of different variables that changes with spatial location (zone) in the bay, but also the magnitude or direction of individual variables.

Finally, this exercise showed that Mantel tests and CARTs are applicable and useful when incorporating space into the analysis of fish and dolphin habitats. One objective of this ongoing study is to establish geospatial statistics essential to detect physical and biological changes in this highly heterogeneous habitat. This is especially important where habitat variables interact to predict habitat use patterns. Here, data were integrated into a predictive habitat-model using spatial analysis, which can help resource managers and stakeholders quantify changes in the unique South Florida ecosystem.

As field research and analysis continues, including stepwise generalized linear models, logistic regressions, and discriminant functions, the models described above will be improved. Our goals are to quantify fine-scale (on the order of tens of meters) habitat quality variation spatially and to relate this variability to upper trophic level predator use. Greater sample sizes and multiple years of data will provide the stronger statistical power needed to elucidate interannual variation. These methods will allow local spatial structure of habitat quality in Florida Bay to be teased out by discriminating between suitable and nonsuitable habitat for the various upper trophic level users considered in this study.

The South Florida human population continues to grow at an astounding rate, adding pressure to the already stressed water management system of the greater Everglades region. These intensive pressures of human development have changed the structure and function of coastal systems throughout South Florida (U.S. Army Corps of Engineers, 1999). In light of the current habitat restoration efforts, research that leads to appropriate management decisions and practices is essential to enhance the preservation of the biological diversity of Florida Bay. This research took an ecosystem-level approach to resource management by synthesizing studies of water quality, habitat heterogeneity, fish distribution, and dolphin behavior to give managers a reliable mitigation tool: the distribution of bottlenose dolphins as a management indicator species of fish habitats. In summary, this research characterized the spatially variable habitat quality supporting abundant fish populations in Florida Bay and demonstrated a link to dolphin foraging behavior, enabling mangers to better understand, conserve, and restore this heterogeneous ecosystem.

Acknowledgments

Fish were sampled in Florida Bay under research permits from the Everglades National Park (EVER-2002-SCI-0049 and EVER-2003-SCI-0042), the Florida Keys National Marine Sanctuary, and the Florida Fish and Wildlife Conservation Commission (02R-451). Additionally, level B harassment, as described under the Marine Mammal Protection Act (1972), of bottlenose dolphins was conducted under a general authorization permit (911-1466) from the U.S. federal government. Chlorophyll, percent saturated dissolved oxygen, and turbidity data were provided by the SERC-FIU water quality monitoring network, which is supported by SFWMD/SERC Cooperative Agreements C-10244 and C-13178 as well as EPA Agreement X994621-94-0. Furthermore, this research could not have been possible without the support of the Dolphin Ecology Project, The Taylor Foundation, and the National Marine Fisheries Service. A great deal of credit must be given to the field assistants for enduring long, hot, muggy Florida summer days: Gretchen Lovewell, Michele Barbieri, Liz Touhy-Sheen, Kate Freeman, and Anne Starling. A special thanks to Andrew Read, David Hyrenbach, Andrew Westgate, Danielle Waples, and Pat Halpin for their support and advice. Finally, we thank the Everglades National Park and the Florida Keys National Marine Sanctuary.

References

Allen, M. C. and A. J. Read. 2000. Habitat selection of foraging bottlenose dolphins in relation to boat density near Clearwater, Florida. *Marine Mammal Science* 16(4):815–824.

Baumgartner, M. F. 1997. The distribution of Risso's dolphin (*Grampus griseus*) with respect to the physiography of the northern Gulf of Mexico. *Marine Mammal Science* 13(4):614–638.

Baumgartner, M. F., K. D. Mullin, L. N. May, and T. D. Leming. 2000. Cetacean habitats in the northern Gulf of Mexico. *Fishery Bulletin* 99(2):219–239.

Boesch, D. F., N. E. Armstrong, C. F. D'Elia, N. G. Maynard, H. W. Paerl, and S. L. Williams. 1993. Deterioration of the Florida Bay Ecosystem: An Evaluation of the Scientific Evidence. Report to the Interagency Working Group on Florida Bay.

Buckland, S. T., D. R. Anderson, K. P. Burnham, and J. L. Laake. 1993. *Distance Sampling: Estimating Abundance of Biological Populations.* Chapman & Hall, London.

Butler, M. J., IV, J. H. Hunt, W. F. Herrnkind, M. J. Childress, R. Bertelsen, W. Sharp, T. Matthews, J. M. Field, and H. G. Marshall. 1995. Cascading disturbances in Florida Bay, USA: *Cyanobacteria* blooms, sponge mortality, and implications for juvenile spiny lobster *Panulirus argus. Marine Ecology Progress Series* 129:119–125.

Caro, T. M. and G. O'Doherty. 1999. On the use of surrogate species in conservation biology. *Conservation Biology* 13(4):805–814.

Croxall, J. P. 1989. Use of Indices of Predator Status and Performance in CCAMLR Fishery Management. Selected Scientific Papers Commission for the Conservation of Antarctic Living Resources 1989 (SC-CAMLR-SSP/6), pp. 353–365.

D'Amico, A., A. Bergamasco, P. Zanasca, S. Carniel, E. Nacini, N. Portunato, V. Teloni, C. Mori, and R. Barbanti. 2003. Qualitative correlation of marine mammals with physical and biological parameters in the Ligurian Sea. *IEEE Journal of Oceanic Engineering* 28(1):29–43.

Deegan, L. A., S. Holt, J. S. Nowlis, and S. M. Sogard. 1998. Higher trophic level initiative for the Florida Bay program. Perspectives on the November 4–5 Workshop from the Florida Bay Oversight Panel *ad Hoc* Committee on Higher Trophic Levels.

Dynamic Higher Trophic Levels Science Plan, 2001. Dynamic Florida Bay Higher Trophic Levels Integrated Science Plan: Linking Florida Bay and Coastal Ecosystems and Their Watershed. Available at http://www.aoml.noaa.gov/fbay/DraftHTLPlan_2001Mar21.html.

Fourqurean, J. W. and M. B. Robblee. 1999. Florida Bay: a history of recent ecological changes. *Estuaries* 22:345–357.

Fourqurean, J. W., J. C. Zieman, and G. V. N. Powell. 1992. Phosphorus limitation of primary production in Florida Bay: evidence from the C:N:P ratios of the dominant seagrass *Thalassia testudinum*. *Limnology and Oceanography* 37:162–171.

Fraser, D. F., J. F. Gilliam, M. P. MacGowan, C. M. Arcaro, and P. H. Guillozet. 1999. Habitat quality in a hostile river corridor. *Ecology* 80(2):597–607.

Furness, R. W. and C. J. Camphuysen. 1997. Seabirds as monitors of the marine environment. *ICES Journal of Marine Science* 54:726–737.

Gibson, R. N. 1994. Impact of habitat quality and quantity on the recruitment of juvenile flatfishes. *Netherlands Journal of Sea Research* 32(2):191–206.

Griffin, R. B. 1999. Sperm whale distributions and community ecology associated with a warm-core ring off Georges Bank. *Marine Mammal Science* 15(1):33–51.

Holmquist, J. G., G. V. N. Powell, and S. M. Sogard. 1989. Sediment, water level and water temperature characteristics of Florida Bay's grass covered mud banks. *Bulletin of Marine Science* 44:348–364.

Kasamatsu, F., P. Ensor, G. G. Joyce, and N. Kimura. 2000. Distribution of minke whales in the Bellingshausen and Amundsen Seas (60°W–120°W), with special reference to environmental/physiographic variables. *Fisheries Oceanography* 9(3):214–223.

Lamberson, R. H., R. McKelvey, B. R. Noon, and C. Voss. 1992. A dynamic analysis of northern spotted owl viability in a fragmented forest landscape. *Conservation Biology* 6:505–511.

Light, S. S. and J. W. Dineen. 1994. Water control in the Everglades: a historical perspective. In *Everglades: The Ecosystem and Its Restoration*, S. M. Davis and J. C. Ogden (eds.). St. Lucie Press, Delray Beach, FL, pp. 47–84.

Mantel, N. 1967. The detection of disease clustering and a generalized regression approach. *Cancer Research* 27:209–220.

Matheson, R. E., Jr., D. K. Camp, S. M. Sogard, and K. A. Bjorgo. 1999. Changes in seagrass-associated fish and crustacean communities on Florida Bay mud banks: the effects of recent ecosystem changes? *Estuaries* 22(2B):534–551.

McIvor, C. C., J. A. Ley, and R. D. Bjork. 1994. Changes in freshwater inflow from the Everglades to Florida Bay including effects on biota and biotic processes: a review. In *Everglades: The Ecosystem and Its Restoration,* S.M. Davis and J.C. Ogden (eds.). St. Lucie Press, Delray Beach, FL, pp. 117–146.

McPherson, B. E. and R. Halley. 1996. The South Florida Environment — A Region under Stress. Circular 1134. U.S. Geological Survey, Reston, VA.

Moore, S. E. and D. P. DeMaster. 1998. Cetacean habitats in the Alaskan arctic. *Journal of Northwest Atlantic Fisheries Science* 22:55–69.

Moore, S. E., W. A. Watkins, M. A. Daher, J. R. Davies, and M. E. Dahlheim. 2002. Blue whale habitat associations in the Northwest Pacific: analysis of remotely-sensed data using a Geographic Information System. *Oceanography* 15(3):20–25.

Raven, P. J., N. T. H. Holmes, F. H. Dawson, and M. Everard. 1998. Quality assessment using river habitat survey data. *Aquatic Conservation: Marine and Freshwater Ecosystems* 8:477–499.

Robblee, M. B., T. R. Barber, P. R. Carlson, Jr., M. J. Durako, J. W. Fourqurean, L. K. Muehlstein, D. Porter, L. A. Yarbro, R. T. Zieman, and J. C. Zieman. 1991. Mass mortality of tropical seagrass *Thalassia thestudinum* in Florida Bay (USA). *Marine Ecology Progress Series* 71:297–299.

Rudnick, D. T., Z. Chen, D. Childers, J. Boyer, and T. Fontaine. 1999. Phosphorus and nitrogen inputs to Florida Bay: the importance of the Everglades watershed. *Estuaries* 22(2B):398–416.

Rutherford, E. S., T. W. Schmidt, and J. T. Tilmant. 1989. Early life history of spotted seatrout (*Cynoscion nedulosus*) and gray snapper (*Lutjanus griseus*) in Florida Bay, Everglades National Park, Florida. *Bulletin of Marine Science* 44:49–64.

Schmidt, T. W. 1979. Ecological Study of Fishes and the Water Quality Characteristics of Florida Bay, Everglades National Park, Florida. Project Report, RSP-EVER N-36 South Florida Research Center, FL, 145 pp.

Simberloff, D. 1987. The spotted owl fracas: mixing academic, applied, and political ecology. *Ecology* 68(4):766–772, 778–779.

Smith, T. J., III, H. H. Hudson, M. B. Robblee, G. V. N. Powell, and P. J. Isdale. 1989. Freshwater flow from the Everglades to Florida Bay: a historical reconstruction based on fluorescent banding in the coral *Solenastrea bournoni*. *Bulletin of Marine Science* 44:274–282.

Sogard, S. M., G. V. N. Powell, and J. G. Holmquist. 1989. Spatial distribution and trends in abundance of fishes residing in seagrass meadows on Florida Bay mudbanks. *Bulletin of Marine Science* 44(1):179–199.

Tabb, D. C. and M. A. Roessler. 1989. History of studies on juvenile fishes of coastal waters of Everglades National Park. *Bulletin of Marine Science* 44:23–34.

Thayer, G. W. and A. J. Chester. 1989. Distribution and abundance of fishes among basins and channels in Florida Bay. *Bulletin of Marine Science* 44(1):200–219.

Thayer, G. W., A. B. Powell, and D. E. Hoss. 1999. Composition of larval, juvenile, and small adult fishes relative to changes in environmental conditions in Florida Bay. *Estuaries* 22(2B):518–533.

Tilmant, J. T. 1989. A history and overview of recent trends in the fisheries of Florida Bay. *Bulletin of Marine Science* 44(1):3–33.

Urian, K. W. and R. S. Wells. 1996. Bottlenose Dolphin Photo-identification Workshop. Final Report to the Southeast Fisheries Science Center, National Marine Fisheries Service, Charleston Laboratory. NOAA Technical Memorandum. NMFS-SEFSC-393, 92 pp.

U.S. Army Corps of Engineers. 1999. Central and Southern Florida Project Comprehensive Review Study: Final Integrated Feasibility Report and Programmatic Environmental Impact Statement.

U.S. Forest Service. 2000. Management Indicator Species: Population and Habitat Trends.

Venables, W. N. and B. D. Ripley. 1997. *Modern Applied Statistics with S-PLUS*, 2nd ed. Springer, New York.

Waples, D. M. 1995. Activity Budgets of Free-Ranging Bottlenose Dolphins (*Tursiops truncatus*) in Sarasota Bay, Florida. M.Sc. thesis, University of California, Santa Cruz, 61 pp.

Woodley, T. H. and D. E. Gaskin. 1996. Environmental characteristics of North Atlantic right and fin whale habitat in the lower Bay of Fundy, Canada. *Canadian Journal of Zoology* 74:75–84.

Zieman, J. C., J. W. Fourqurean, and R. L. Iverson. 1989. Distribution, abundance and productivity of seagrasses and macroalgae in Florida Bay. *Bulletin of Marine Science* 44(1):292–311.

28

A Process for Selecting Indicators for Estuarine Projects with Broad Ecological Goals

Louis A. Toth

CONTENTS

Introduction

Goals for preserving, enhancing, or restoring environmental resources are often broadly defined and expressed in vague terms (e.g., ecosystem health and integrity) that hinder assessment of success (Rapport, 1989; Hartig and Zarull, 1992; Karr, 1999). The goal for restoration of ecological integrity of the Kissimmee River and floodplain (Toth, 1995), for example, requires "reestablishment of an ecosystem that is capable of supporting and maintaining a balanced, integrated, adaptive community of organisms having a species composition, diversity, and functional organization comparable to the natural habitat of the region" (*sensu* Karr and Dudley, 1981). This goal has a powerful ecological context but requires explicit expectations for response and recovery to be effectively evaluated and interpreted (Pastorok et al., 1997; Toth and Anderson, 1998). Evolution of the Kissimmee River evaluation program provides a case study of a process for selecting indicators to evaluate ecological changes associated with management and restoration objectives relating to the health and integrity of natural systems. A similar strategy has been recommended for evaluating the planned restoration of the San Francisco Bay–Delta River System (Levy et al., 1996) and indicates the potential applicability of this process to estuarine systems.

Kissimmee River Restoration Initiative

Between 1962 and 1971 the natural hydrogeomorphic configuration and functionality of the Kissimmee River basin was transformed by the Central and Southern Florida Flood Control Project (U.S. Army

Corps of Engineers, 1956) into a compartmentalized system of levees, canals, and water control structures that are used to manage water levels and flows for flood control. More than 160 km of the meandering river and its 1.5- to 3-km-wide floodplain were channelized into a 90-km-long, 9-m-deep, and 64- to 105-m-wide canal that was divided by levees and water control structures into a series of five impoundments with stabilized water surface profiles. Lakes within the river's 4230 km^2 headwater basin were connected by canals and divided into water storage reservoirs with flood control regulation schedules that reduced the range of water level fluctuation in the lakes and greatly modified discharge regimes to the channelized river (Toth et al., 1997).

Associated losses of wetland and fish and wildlife resources (Toth, 1993; Toth et al., 1995) provided the impetus for a grassroots restoration movement (Woody, 1993), and the basis for adoption of the ecological integrity goal. This goal emanated from recognition that the broad array of lost values and functions could only be restored through reestablishment of the physical, chemical, and biological characteristics, processes, and interactions that governed the ecology and evolution of the historic ecosystem. Accordingly, the ecological integrity goal shifted the focus of restoration planning from independent objectives involving discrete taxonomic components or ecological functions to the organizational determinants and self-sustaining properties of river/floodplain ecosystems (Toth and Aumen, 1994).

The restoration plan for the Kissimmee River was authorized for implementation by the 1992 Water Resources Development Act (Public Law 102-580) and consists of two integrated components. The lower basin component will restore the physical form of the river and floodplain by refilling the flood control canal and secondary drainage canals on the floodplain, removal of water control structures, and degradation of floodplain levees and berms. The upper basin component will modify flood control regulation schedules and operation rules for the headwater lakes to reestablish discharge regimes that are needed to restore the hydrologic determinants of ecological integrity. Reconstruction of the lower basin began in 1999 and is scheduled for completion by 2012.

Objectives of the Restoration Evaluation Plan

The authorized restoration plan includes an evaluation program to measure the success of the project in reestablishing ecological integrity. The evaluation program has been designed to address four related objectives:

1. Document effects of restoration that are of ecological and social importance.

A suite of indicators representing a range of physical, chemical, biological, and functional components must be measured to evaluate ecological integrity (Karr, 1993), but ecological responses to restoration are only meaningful relative to societal values (Rapport, 1989; Kelly and Harwell, 1990). Thus, the restoration evaluation program must account for linkages between physical, chemical, and biological conditions and ecosystem services, including recreational and economic benefits and functions such as water quality. A critical aspect of this objective will be to disseminate results and other restoration-related information to both public and scientific audiences through publications, presentations, and other forums for information exchange.

2. Show cause-and-effect relationships between restoration of the physical components of the ecosystem and quantifiable ecological responses.

A basic premise of the restoration project is that the reestablishment of historical (prechannelization) hydrologic and physical characteristics of the river ecosystem will provide the driving force and hydrogeomorphic habitat template for the restoration of ecological integrity. Definitively establishing this link is required to validate the success of the restoration project and the pioneering principles employed in the project planning and design. Use of ecosystem-level hydrologic and geomorphic criteria and natural processes to affect ecosystem restoration is a simpler conceptual approach than the individual species criteria (e.g., Habitat Evaluation Procedures) that commonly have been used in previous restoration and management efforts (Feather and Capan, 1995).

Because ecosystems involve complex interactions and are dynamic over a range of spatial and temporal scales, establishing cause–effect relationships requires wise selection of sensitive indicators, and careful, well-thought-out studies with appropriate sampling methodologies and statistical designs, including observational and hypothesis-driven experimental approaches. With adequate control (site and statistical), restoration measures are tantamount to experimental manipulations and can be evaluated using the before-after-control-impact (BACI) analytical methodology (Stewart-Oaten et al., 1992; Smith et al., 1993). Restoration of ecological integrity of the Kissimmee River is being evaluated by comparing hypothesized responses to restored hydrology in de-channelized sections (i.e., baseline and post-restoration measurements) to comparable temporal data from remaining channelized reaches.

3. Provide a thorough understanding of the restoration process.

The restoration evaluation program will not be successful if it simply documents change but does not show how or explain why change occurred or, alternatively, does not provide information on why change did not occur when it was expected. Effective repair of any complex system begins with accurate diagnosis of the cause(s) and nature of the problems, which requires an understanding of the difference between the composition and functioning of the healthy and impaired systems. Although the nature and sources of degradation in the Kissimmee River ecosystem are well documented (Toth, 1993), ecosystem response to restoration will be mediated by complex physical, chemical, and biological mechanisms, including potential synergistic and antagonistic effects, which will operate over a continuum of spatial and temporal scales (Dahm et al., 1995). Recognition and understanding of these mechanisms and interactions will provide critical information for adaptive adjustments of the restoration project (National Research Council, 1992). Integrative models can be useful (Minns, 1992) but require insightful predictions and subsequent verification of measured ecological responses to changes in state variables.

4. Provide for continual, scientifically informed fine-tuning of the planning and implementation of the restoration project, and for adaptive management (Holling, 1978) of the recovering and restored ecosystem.

A primary objective of the restoration evaluation program is to ensure that the project's physical form requirements and hydrologic criteria are achieved. Related studies need to provide information on the stability of reconstructed features of the river and floodplain, while verifying the reestablishment of hydrodynamic and geomorphic processes that will enable the reconstructed system to be self-sustaining. These studies will determine any necessity for protective measures (e.g., armoring, vegetation plantings) for precluding gross and/or chronic instability of disturbed areas such as the backfilled canal, degraded spoil banks, and reconstructed river channel, while recognizing that primary geomorphic features are expected to undergo continual change and adjustment (dynamic equilibrium).

The hydraulic performance of the reconstructed system needs to be closely monitored to maintain required levels of flood protection in the basin. Within the context of this flood control constraint, hydraulic monitoring will provide the necessary information for defining the upper limits of the restoration project (i.e., canal backfilling) and for developing optimal operational rules for water management in the headwaters of the reconstructed system.

The restoration evaluation program also will ensure that the reconstruction process is conducted in an environmentally sensitive manner and that appropriate action is taken to minimize any impacts to the system's resources. In addition, evaluation program studies will provide information for facilitating and, if possible, expediting system recovery. This information eventually will serve as a database for sound management of the restored ecosystem and for wise use of its resources.

End Points

The scope of the ecological integrity goal requires a multilayered evaluation program for measuring and judging response and recovery of ecosystem dynamics and resources. Initial and long-term responses must be tracked by a suite of indicators representing physical, chemical, biological, and functional properties at multiple spatiotemporal scales of observation (National Research Council, 1992). Thus, multiple end points (*sensu* Kelly and Harwell, 1990) are needed to reflect the nature of degradation and degree of impacts within the channelized system and to represent the breadth of the ecological integrity goal (Cairns and McCormick, 1991; Cairns et al., 1993).

Response indicators for the Kissimmee River restoration evaluation program were developed within an organizational framework of three interrelated categories (i.e., impact assessment, sociopolitical, and restoration) of end points (Table 28.1). Each of these categories provides for evaluation measures that collectively contribute to the primary objectives of measuring the success of the restoration project and providing for scientifically informed fine-tuning of the physical reconstruction and adaptive management of the recovering and restored ecosystem. The relative importance of each end point was based on the extent to which the ecological issues and values that they represent were affected by channelization and the flood control project, a somewhat subjective judgment of general public perspectives of their significance, and for some impact assessment end points, the potential that the restoration project could have adverse effects.

Impact Assessment End Points

Impact assessment end points reflect concerns that the reconstruction of the river/floodplain ecosystem could result in unintentional degradation of environmental attributes and/or loss of socioeconomic services. For example, the conversion of pasture to wetlands could result in loss of foraging habitat for Audubon's crested caracara (U.S. Fish and Wildlife Service, 1991) and thereby affect the reproductive success of a threatened species that has colonized the drained floodplain. Although potential adverse impacts were thoroughly addressed during the restoration planning process, incorporation of these end points in the evaluation program will ensure that continual efforts are made to alleviate even minor incidental impacts. In addition to reproductive success of Audubon's crested caracara, aspects of the Kissimmee River restoration project have the potential to adversely affect recreational navigation, flood control, and water quality parameters.

The greatest potential for impacts to recreational navigation will likely occur in new river channels that will be constructed at locations where the former river channel was destroyed by excavation of the flood control canal or by deposition of spoil. Shoaling may occur as the bed and banks of these reconstructed channels adjust and equilibrate to river discharges. Natural formation of sandbars within the restored river also presents potential impediments to navigation.

The principal water quality variables of concern are turbidity, suspended solids, dissolved oxygen, nutrient concentrations, and mercury. The greatest potential for generation of high turbidity and/or suspended solids likely will occur during canal backfilling, when the spoil from the original canal excavation is redeposited in the canal. In addition to the backfill material, resuspension of organic deposits on the canal bottom could provide a source of turbidity and suspended solids. Although standard methods (e.g., turbidity screens) will be employed to contain construction-generated turbidity, evaluation of the spatial and temporal extent of this water quality impact could indicate the need for adaptive modifications to the reconstruction process. Subsequently, turbidity and suspended solids could be generated by flow over the unvegetated soils of the backfilled canal and degraded spoil mounds, and as a result of flushing of accumulated organic deposits in remnant river channels. By reducing light penetration, construction-generated turbidity could limit photosynthetic production by phytoplankton and depress dissolved oxygen concentrations within surface strata of the river water column. Similarly, there is some potential for liberation and downstream transport of nutrients as bottom organic deposits are resuspended during canal backfilling and flushed from remnant river channels with reestablished flow.

TABLE 28.1

End Points for Restoration of the Kissimmee River Ecosystem

Impact Assessment End Points
- Water quality — 3
- Bioaccumulation of mercury — 4
- Threatened and endangered species — 5
- Navigation — 4
- Flood control — 1

Ecological Integrity End Points

Sociopolitical End Points
- Wetland area — 1
- Numbers of birds, particularly waterfowl and wading birds — 1
- Diversity of birds — 2
- Numbers of game fish — 1
- Threatened and endangered species — 2
- Nuisance (exotic) species — 3
- Nutrient loads (transport) — 2
- Navigation — 3
- Aesthetic values — 3

Restoration End Points
- Physical Integrity
  - Structure/Composition
    - Hydrology — 1
    - Topography — 2
    - Floodplain soil characteristics — 5
    - River channel substrate characteristics — 3
    - Water quality characteristics — 5
  - Processes
    - Hydrogeomorphic processes — 1
    - Disturbance (e.g., fire frequency) — 3
- Chemical Integrity
  - Structure/Composition
    - Surface water quality — 3
  - Processes
    - Dissolved oxygen dynamics — 2
    - Soil geochemistry — 4
- Biological Integrity
  - Structure/Composition
    - Community structure (species composition and relative abundance, guild structure) — 2
    - Population structure (age structure, sex ratios) — 4
    - Population densities — 3
    - Food web structure (ecosystem scale) — 3
    - Biodiversity (ecosystem scale) — 4
  - Processes
    - Productivity — 3
    - Colonization rates — 3
    - Growth rates — 4
    - Reproductive success/recruitment — 1
    - Survivorship/mortality rates — 3
    - Persistence — 1
- System Functional Integrity
  - Habitat quality, diversity, use, and persistence — 1
  - River/floodplain interactions — 1
  - Energy flow dynamics — 3
  - Nutrient cycling dynamics — 3

Note: Numerical values reflect the relative importance of each end point. End points with values of 1 are most important to restoration of ecological integrity.

Concerns about mercury stem from studies that have shown that soil disturbance, wetland creation, and periodic drying and flooding of wetlands and croplands can mobilize mercury and from recent detection of high concentrations of mercury in predatory fish in Florida, including the Everglades and headwater lakes of the Kissimmee River (Lange et al., 1993, Lange et al., 1994). Concentrations of mercury can be magnified at progressively higher levels of food chains (e.g., wading birds and predatory fish) and have potential human health implications if present in edible tissues of game fish species.

Maintenance of existing levels of flood protection on privately owned lands within the basin is the principal constraint of the Kissimmee River restoration project and will depend on the hydraulic performance of the recovering and restored hydrosystem after dechannelization.

Ecological Integrity End Points

Sociopolitical and restoration end points represent ecosystem attributes and services that have been impacted (i.e., reduced or lost) by channelization and the flood control project. Sociopolitical end points involve measures of ecosystem health that are of widely recognized importance or concern to the public. Although these classes of end points can be subjective and transitory, they reflect current societal values and therefore strongly influence decisions and actions that affect environmental resources (e.g., implementation of restoration and management efforts). Indicators of sociopolitical end points will provide "high-profile" measures of success. Restoration end points represent physical, chemical, biological, and functional attributes of the river/floodplain ecosystem that must be reestablished to restore and maintain ecological integrity, including those measures of ecosystem health reflected by the sociopolitical end points. Many of these end points, particularly restoration end points, require multiple indicator metrics to represent the broad range of spatial, temporal, and ecological (population, community, ecosystem, and landscape) scales incorporated in the Kissimmee River restoration project.

Sociopolitical End Points

Sociopolitical end points for the Kissimmee River restoration project include wetland area, bird densities and diversity, sportfish populations, endangered species, nuisance/exotic species, nutrient loads, and related ecosystem services such as recreational, aesthetic, economic, and natural heritage values. Preservation and restoration of wetlands is an established national goal (e.g., "no net loss"). The restoration of 11,000 ha of functional floodplain wetlands is expected to lead to reduced nutrient loading and transport from the Kissimmee River basin and thereby contribute to regional efforts to prevent accelerated eutrophication of Lake Okeechobee (Toth and Aumen, 1994).

Waterbirds, sportfish, and endangered species represent the "charismatic megafauna" that are expected to benefit from restoration of the Kissimmee River and floodplain. Restoration of functional floodplain wetlands will reestablish habitat for wading birds, waterfowl, and other wetland-dependent species, including the endangered snail kite (*Rostrhamus sociabilis*), wood stork (*Mycteria americana*), and bald eagle (*Haliaeetus leucocephalus*). Recovery of populations of these threatened and endangered species is consistent with established legal mandates for restoring our natural heritage. Restoration of river and floodplain habitat also is expected to reestablish historical populations of gamefish, including a largemouth bass fishery for which the river was nationally renowned. An enhanced sport fishery and recovery of avian populations will increase the river's recreational, aesthetic, natural heritage, and economic value, including a stimulus for the local and regional ecotourism industry. These values also will be enhanced by restoration of a more scenic riverine vista.

Reestablishment of flow and reflooding of the floodplain is expected to eliminate the major nuisance exotic plants that have invaded the drained floodplain (i.e., Brazilian pepper, *Schinus terebinthifolius;* guava, *Psidium guajava;* and cogon grass, *Imperata cylindrica*) and to reduce the occurrence of water hyacinth (*Eichornia crassipes*) and water lettuce (*Pistia stratiotes*) within the river channel. In the Kissimmee system and throughout south Florida, costly weed control programs are implemented to prevent these nuisance plant species from interfering with flood control and navigation, and to preserve and protect natural habitats, flora, and fauna. Thus, contributions of the restoration project toward control or elimination of exotic species have economic implications, as well as natural heritage value.

Reestablishing flow and a natural, longitudinal water surface profile should eliminate navigation impediments that have arisen from growth of aquatic vegetation (both native and exotic species) into the center of the remnant river channel, and from lowered and stabilized water levels within the channelized system. Increased navigation potential will have recreational and economic value.

Restoration End Points

Restoration end points comprise structure and process end points, and are linked by interactions that provide indicators of the functional integrity of the system. Although structure end points may provide more direct and intuitive descriptors of restoration of ecological integrity, functional end points include process-level indicators that can set the stage for response by structural attributes and could ultimately determine if or when full recovery will occur.

Physical integrity end points are based on the fundamental assumption that the reconstruction and reestablishment of hydrology will recreate the hydrogeomorphic habitat template and provide the driving forces for restoration of the river/floodplain ecosystem. Thus, physical integrity end points represent the most proximate indicators of restoration of ecological integrity. Backfilling of canals and degradation of spoil mounds, berms, and levees will reestablish floodplain topography and soil characteristics, and restore the lateral and longitudinal connectivity of hydrogeomorphic features. Hydrologic criteria for restoration of the Kissimmee River require reestablishment of historical stage and flow characteristics, including continuous inflows from the headwater lakes to sustain prolonged floodplain hydroperiods (Toth et al., 1995; Toth et al., 2002). Restoration of flow will reestablish hydrogeomorphic processes and recreate channel morphology (e.g., depth diversity, point bars) and bed characteristics (i.e., shifting sand substrate) that have been obscured by thick accumulations of organic deposits since channelization. Reestablished flow also will eliminate thermal stratification that occurs in the stagnant channels. Restored inflow regimes and associated disturbance processes (i.e., flood, drought, fire) that occurred within the natural river/floodplain system will again be regulated by climatic conditions.

Chemical integrity end points reflect the impacts of channelization and the flood control project on water quality and biogeochemical characteristics. Low dissolved oxygen concentrations are the principal water quality problem and limiting factor for fish and invertebrate communities within the channelized system. Restoration of flow through the river channel is expected to lead to improved dissolved oxygen regimes and associated habitat for river biota. Other chemical integrity end points relate to restoration of hydric soil characteristics on the floodplain, which will provide for the reestablishment of wetland vegetation and associated nutrient filtration and absorption processes.

The biological portion of the ecological integrity goal requires reestablishment of adaptive biological communities with a species composition, diversity, and functional organization that is comparable to that of the historic river and floodplain. Biological integrity end points reflect indicators at population, community, and ecosystem levels of organization and involve multiple taxonomic groups (e.g., plants, invertebrates, fish, birds) because no single taxon will provide an adequate indicator of the ecosystem scale of the restoration goal. However, because impacts of the flood control project on the full array of associated indicators are not known (and measuring everything is not practical), the relative importance of these end points reflects a focus on characteristics that will provide the most reliable indicators of restoration of ecological integrity.

The community structure end point reflects a restoration objective to reestablish more natural species composition and relative abundances of plant, invertebrate, fish, amphibian, reptile, and bird assemblages. This focus on community characteristics provides for evaluation of ancillary population end points, including colonization dynamics, densities, and related variables (e.g., reproductive success, growth rates, survivorship). Population characteristics of sportfish (e.g., largemouth bass), waterfowl such as the endangered wood stork, and indicator plant species (*Pontederia cordata, Sagittaria lancifolia, Cephalanthus occidentalis*) are of particular relevance to the restoration of ecological integrity. The ecosystem scale of this goal is addressed by end points for food web structure and persistence of biodiversity.

End points for restoration of functional integrity reflect the loss and degradation of ecosystem-level attributes that resulted from transformation of the riverine system into reservoir-like impoundments.

Reestablishment of the historic quality, diversity, and persistence of river and floodplain habitats will be indicated by the structure and dynamics of biological communities that use these habitats. A related end point addresses the need to reestablish lateral ecological connectivity between the river channel and floodplain, which has been precluded by the lowering and stabilization of water levels within the channelized system. Key interactions include use of floodplain wetlands by riverine fish species for feeding, reproductive, and nursery habitat during the increasing leg of the flood pulse, and export of invertebrates and organic matter (i.e., fuel for riverine food chains) from the floodplain to river channel during stage recession periods. Other functional integrity end points include energy flow dynamics of the river/floodplain food web and reestablishment of nutrient filtration and absorption processes that were lost with the drainage of floodplain wetlands.

Prioritization of Indicators

The challenge of choosing an appropriate set of indicators that represent all end points, that will be sensitive to the physical restoration of the system, that produce reliable short- and long-term responses, and that provide information for adaptive management of the project and recovering ecosystem was addressed initially through development of a suite of conceptual models for structural and functional attributes of the river/floodplain ecosystem (Dahm et al., 1995; Harris et al., 1995; Toth et al., 1995; Trexler, 1995; Weller, 1995). These models assisted in the selection of indicators by establishing a conceptual basis and explicit constructs for predicting and understanding expected responses by ecosystem components (Pastorok et al., 1997). The models describe the geomorphic habitat template and identify the operative ecological variables and processes that will drive the restoration of ecological integrity.

Based on this conceptual understanding and available data on the ecology of the natural and channelized river and floodplain ecosystem, more than 140 potential indicators representing a range of taxonomic components and all categories of end points were compiled. The final selection of indicators was completed by prioritization using six criteria developed by Kelly and Harwell (1990) to rank their relative utility (Table 28.2). The criteria reflect desirable properties of indicators, viewed in the context of the restoration project and objectives of the evaluation program. One criterion was based on the expected sensitivity of an indicator to the physical changes that will occur as a result of the restoration project. Based on this criterion, the best indicators will have a high expectation of detectable response that will not be confounded by the metric's natural spatial and temporal variability (Karr and Chu, 1999). A related criterion, the reliability of response, was used to evaluate the specificity of potential indicators by judging the likelihood that any responses would clearly be attributable to the restoration project and not some unrelated factor. The rapidity of response criterion favored metrics that would serve as early indicators of success or, alternatively, point to a need for mid-course corrections or fine-tuning of the project, including possible implementation of supplemental restoration or management measures. Response indicators with high potential for providing information for adaptive management of the project and/or restored system were prioritized by the feedback to management criterion. The ease or economy of monitoring criterion was used to ensure practicality by eliminating potential indicators that would be extremely difficult or prohibitively expensive to measure (Caughlan and Oakley, 2001).

Two of the criteria were based on associated end points, the relevance of the indicator to a defined end point, and the relative importance of that end point. The relevance of different taxonomically based indicators to biotic integrity end points differed according to sociopolitical values and available data on ecological degradation within the channelized system. Fish and wading birds, for example, were considered more relevant to the community structure end point than microbes. Indicators can be direct measures of end points or linked to an end point by a string of ecological connections. For example, largemouth bass population densities would be an intrinsic measure of an important sociopolitical end point (i.e., numbers of game fish). Other indicators, such as aquatic invertebrate densities and river channel substrate characteristics, are indirectly related to this end point because they represent a food source and spawning habitat for game fish species. Both intrinsic and indirect indicators of end points are needed because

TABLE 28.2

Criteria Used for Prioritization of Indicators of Restoration of Ecological Integrity of the Kissimmee River Ecosystem

Criteria	Basis for Relative Scoring	
Sensitivity to restoration/intrinsic variability	1 High signal, low noise	5 Low signal, high noise
Rapidity of response	1 Within 1–2 years	5 Greater than 5 years
Reliability of response	1 High specificity	5 Low specificity
Ease/economy of monitoring	1 Easy, inexpensive	5 Difficult, expensive
Relevance to end point	1 Intrinsic indicator	5 Indirectly related
Importance of end point	1 Critical to restoration, high societal value	5 Minor import to restoration, low societal significance
Feedback to management	1 High potential information	5 Low potential information

Note: Potential indicators were scored on a scale of 1 to 5 for each criterion.
Source: Adapted from Kelly and Harwell (1990).

reliance on intrinsic indicator metrics could limit the potential for project modifications and adaptive management, and might not provide adequate information for assessing the long-term status of an end point. Density estimates, for example, need to be supplemented with studies of reproductive success to accurately predict the status of biological populations (Munkittrick and Dixon, 1989). The importance of the end point (see Table 28.1) was based on the relative significance of that end point to the restoration of ecological integrity and a somewhat subjective judgment of general public perspectives of the importance of represented ecological issues and values.

Indicators were prioritized according to their total score for all evaluation criteria. The metric prioritization process was conducted by core members of the restoration evaluation team with solicited input from other scientists, including the project's independent Scientific Advisory Panel. The core team approach fostered consistency, and the collective experience and knowledge of all contributors provided a broad base of insight for evaluating the list of potential response indicators. However, a lack of equivalent information on all potential components, particularly lower trophic level constituents such as microbes and algae, and ecosystem-level processes such as rates of production, decomposition, and herbivory, introduced some unavoidable bias. As indicated above, some subjectivity was inherent in one of the evaluation criteria (i.e., importance of the end point).

Results of the prioritization process (Table 28.3) are in accord with the fundamental premise that the restoration of ecological integrity will be accomplished through the reestablishment of the physical form and hydrologic characteristics of the ecosystem. The highest-priority indicators (discharge, hydroperiod, channel velocity, channel morphology, substrate characteristics) are measures of hydrogeomorphic end points that relate to the physical factors and forces that will lead to the restoration and persistence of habitat structure (spatiotemporal distribution of floodplain vegetation communities, plant succession), water quality (dissolved oxygen regimes), and key aspects of the functional integrity of the system such as river/floodplain interactions and habitat use by various taxonomic groups.

The prioritization also clearly reflects societal values and concerns. Impact assessment measures (suspended solids, turbidity, flow resistance, and bioaccumulation of mercury), nuisance (exotic) vegetation, nutrient dynamics and indicators relating to end points that will track the reestablishment of wetlands (spatiotemporal distribution of floodplain vegetation communities), and the recovery of charismatic megafauna (fish, wading birds, waterfowl, threatened and endangered species) were among the highest-ranked indicators. Fish, wading birds, and waterfowl also ranked as high-priority indicators of community structure, reproductive success/recruitment, and population density end points.

TABLE 28.3

Highest Ranked Indicators and Associated End Points for Restoration of Ecological
Integrity of the Kissimmee River Ecosystem

Indicator	End Point
Discharge regimes	Physical integrity
Floodplain hydroperiods	Physical integrity
Channel velocity	Physical integrity
Channel morphology	Physical integrity
Distribution of floodplain plant communities	Functional integrity
Channel substrate characteristics	Physical integrity
Wading bird reproductive success	Biological integrity
Plant successional processes	Biological integrity
Flow resistance	Impact assessment
Habitat use by waterfowl	Functional integrity
Turbidity	Impact assessment
Suspended solids	Impact assessment
Threatened/endangered birds	Sociopolitical
Floodplain fish community structure	Biological integrity
Floodplain habitat use by fish	Functional integrity
Nuisance plant species	Sociopolitical
Floodplain topography	Physical integrity
Nutrient budgets	Sociopolitical
Dissolved oxygen regimes	Chemical integrity
Sport fish population dynamics	Sociopolitical
Distribution of river plant communities	Functional integrity
Export of invertebrates from the floodplain to river channel	Functional integrity
River channel habitat use by fish	Functional integrity
Habitat use by wading birds	Functional integrity
Forage fish population dynamics	Biological integrity
Floodplain invertebrate community structure	Biological integrity
Fish movements between the river channel and floodplain	Functional integrity

Expectations

The prioritization scheme provided a directional blueprint for focusing effort and distributing available resources among end point–based evaluation program components. Final refinement of the evaluation program involved the establishment of benchmark indicators of ecological integrity for which reference values are available to provide for explicit predictions of expected responses by altered or degraded attributes of the ecosystem (Pastorok et al., 1997). This subset of prioritized metrics provided clearly defined "expectations" (Toth and Anderson, 1998) for evaluating restoration success and for assessing the potential need for developing and implementing adaptive adjustments to the restoration project.

Because restoration of ecological integrity involves the return of the structure and function of the Kissimmee River ecosystem to a condition that existed prior to channelization, expectations for gauging restoration success should be based on historical information from the river and floodplain system and/or reference data from an ecologically similar but undisturbed system (White and Walker, 1997). However, available predisturbance data for the altered Kissimmee River ecosystem are limited primarily to hydrology, spatial distribution of wetland plant communities, avian nesting activity and winter abundance of waterfowl, and replicate reference systems with similar hydrology and biogeographical distributions of local flora and fauna do not exist. Thus, potential expectations for restoration of ecological integrity were limited by availability, quality, and precision of reference conditions for altered ecological attributes.

In addition to establishing (predicting) post-restoration values for altered or degraded attributes, expectations include the mechanism by which attributes will change from the pre-restoration or

TABLE 28.4

Examples of Restoration Expectations for Indicators of Reestablished Ecological Integrity of the Kissimmee River and Floodplain Ecosystem

Indicator	Restoration Expectation
Discharge regimes	Continuous flows with intra-annual monthly means that reflect historic seasonal patterns and interannual variability <1.0
Floodplain plant communities	Broadleaf marsh covering >50% of the floodplain with a distribution overlapping that of prechannelization by 90%
Channel substrate	A ≥65% decrease in mean thickness of organic deposits and ≥165% increase in percent of samples without depositions
Wading bird reproductive success	Reestablishment of at least three breeding colonies with ≥5 wading bird species and an average of >400 nesting pairs
Habitat use by waterfowl	Significant increase in mean monthly abundance of migratory waterfowl during November–March
Threatened and endangered birds	Mean monthly densities of ≥1 wood storks/km^2 on the floodplain
Floodplain fish community structure	Mean annual relative abundance consisting of >30% young of the year or juvenile centrarchids
Nuisance plant species	A decrease in mean proportional cover of floating and mat-forming species to <5%
Dissolved oxygen regimes	Mean daytime dissolved oxygen concentrations of 3–6 mg/L during June–November and 5–7 mg/L during December–May
River channel habitat use by fish	Mean annual relative abundance consisting of ≤1% bowfin, ≤3% gar and golden shiner, ≥9% spotted sunfish, ≥16% redbreast sunfish, ≥21% bluegill, and ≥58% other centrarchids
Forage fish population dynamics	Mean annual densities of ≥18 small (<10 cm total length) fishes/m^2
Floodplain Invertebrate community structure	Broadleaf marsh with macroinvertebrate species richness >125 and species diversity ≥3.0

baseline condition and the projected time frame for the predicted response. Description and verification of a hypothesized mechanism for achievement of expectations integrates predicted responses in an ecosystem context by identifying linkages with changes to state variables. Proposed mechanisms were derived from conceptual models that identified driving factors that controlled ecological structure and function in the natural and channelized river (Harris et al., 1995; Toth et al., 1995; Trexler 1995; Weller, 1995).

Dissection of the mechanisms by which restoration is expected to occur helped establish a projected time frame for achievement of expectations, which will vary among ecosystem components (Dahm et al., 1995). In addition to establishing the timetable for gauging restoration success, temporal response trajectories for expectations will indicate if and when project adjustments may be needed to facilitate recovery and ensure achievement of the goal. Both short- and long-term responses are required to address the scale of the ecosystem restoration goal. Restoration expectations that will be achieved within a relatively short time period (e.g., less than 5 years) are more valuable for adjusting or fine-tuning the restoration project, but many expected responses by fish and wildlife will have prerequisite and sequential dependencies (e.g., restoration of habitat and food base) and will likely require many years to accomplish. Because hydrology will be the driving force in restoration, temporal response trajectories will be dependent on climatic conditions following the reconstruction. Achievement of expectations within projected time frames will be facilitated by normal or wet conditions, but impeded by drought.

In all, 42 expectations have been established for evaluating restoration of ecological integrity of the Kissimmee River ecosystem and provide performance measures for most (23 of 28) of the highest ranked indicators (Table 28.4). Although multiple expectations could lead to an ambiguous interpretation of restoration success if all are not achieved, collective assessment and integration is appropriate because all expectations are indicative of ecological integrity. Collective evaluation of expectations will be enhanced by ranking criteria based on the relative importance and reliability of individual expectations.

Conclusions

Restoration and management goals for ecosystem health and integrity reflect a realization that societal concerns for select functions or resources, such as water quality, endangered species, or recreational values, are best addressed with a holistic perspective. However, successful achievement of these broad goals must be viewed in an operational context whereby relevant end points are identified and used to select the most appropriate indicators and expectations for measuring progress. The prioritization of indicators and establishment of expectations for restoration of ecological integrity of the channelized Kissimmee River ecosystem provides a conceptual approach and process for developing success criteria for evaluating broad goals for restoration or management of natural resources.

Acknowledgments

Expectations for restoration of ecological integrity (Table 28.4) were developed by Joanne Chamberlain (discharge regimes), Laura Carnal (floodplain plant communities), Pat Davis, Don Frei, and David Anderson (channel substrate), Stefani Melvin (wading bird reproductive success and wood stork density), Bruce Dugger and Stefani Melvin (waterfowl), Lawrence Glenn (fish), Dave Colangelo and Brad Jones (dissolved oxygen), Steve Bousquin and Caroline Hovey (nuisance plant species), and Joe Koebel (invertebrates). Karl Havens, Joel Trexler, and Sharon Trost provided constructive reviews of drafts of the chapter.

References

Cairns, J., Jr. and P. V. McCormick. 1991. The use of community- and ecosystem-level end points in environmental hazard assessment: a scientific and regulatory evaluation. *Environmental Auditor* 2:239–248.

Cairns, J., Jr., P. V. McCormick, and B. R. Niederlehner. 1993. A proposed framework for developing indicators of ecosystem health. *Hydrobiologia* 263:1–44.

Caughlan, L. and K. L. Oakley. 2001. Cost considerations for long-term ecological monitoring. *Ecological Indicators* 1:123–134.

Dahm, C. N., K. W. Cummins, H. M. Valett, and R. L. Coleman. 1995. An ecosystem view of the restoration of the Kissimmee River. *Restoration Ecology* 3:225–238.

Feather, T. D. and D. T. Capan. 1995. Compilation and Review of Completed Restoration and Mitigation Studies in Developing an Evaluation Framework for Environmental Resources. Vol. 1. U.S. Army Corps of Engineers IWR Report 95-R-4. Vicksburg, MS.

Harris, S. C., T. H. Martin, and K. W. Cummins. 1995. A model for aquatic invertebrate response to Kissimmee River restoration. *Restoration Ecology* 3:181–194.

Hartig, J. H. and M. A. Zarull. 1992. Towards defining aquatic ecosystem health for the Great Lakes. *Journal of Aquatic Ecosystem Health* 1:97–107.

Holling, C. S. 1978. *Adaptive Environmental Assessment and Management.* John Wiley & Sons, London.

Karr, J. R. 1993. Defining and assessing ecological integrity: beyond water quality. *Environmental Toxicology and Chemistry* 12:1521–1531.

Karr, J. R. 1999. Defining and measuring river health. *Freshwater Biology* 41:221–234.

Karr, J. R. and E. W. Chu. 1999. *Restoring Life in Running Waters: Better Biological Monitoring.* Island Press, Washington, D.C.

Karr, J. R. and D. R. Dudley. 1981. Ecological perspective on water quality goals. *Environmental Management* 5:55–68.

Kelly, J. R. and M. A. Harwell. 1990. Indicators of ecosystem recovery. *Environmental Management* 14:527–545.

Lange, T. R., H. E. Royals, and L. L. Connor. 1993. Influence of water chemistry on mercury concentration in largemouth bass from Florida lakes. *Transactions of the American Fisheries Society* 122:74–84.

Lange, T. R., H. E. Royals, and L. L. Connor. 1994. Mercury accumulation in largemouth bass (*Micropterus salmoides*) in a Florida lake. *Archives of Environmental Contamination and Toxicology* 27:466–471.

Levy, K., T. F. Young, R. M. Fujita, and W. Alevizon. 1996. *Restoration of the San Francisco Bay-Delta-River System: Choosing Indicators of Ecological Integrity.* Center for Sustainable Resource Development, University of California, Berkeley, CA.

Minns, C. K. 1992. Use of models for integrated assessment of ecosystem health. *Journal of Aquatic Ecosystem Health* 1:109–118.

Munkittrick, K. R. and D. G. Dixon. 1989. A holistic approach to ecosystem health using fish population characteristics. *Hydrobiologia* 188/189:123–135.

National Research Council. 1992. Restoring Aquatic Ecosystems: Science, Technology, and Public Policy. National Academy Press, Washington, D.C.

Pastorok, R. A., A. MacDonald, J. R. Sampson, P. Wilber, D. J. Yozzo, and J. P. Titre. 1997. An ecological decision framework for environmental restoration projects. *Ecological Engineering* 9:89–107.

Rapport, D. J. 1989. What constitutes ecosystem health? *Perspectives in Biology and Medicine* 33:120–132.

Smith, E. P., D. R. Orvos, and J. Cairns, Jr. 1993. Impact assessment using the before-after-control-impact (BACI) model: concerns and comments. *Canadian Journal of Fisheries and Aquatic Sciences* 50:627–637.

Stewart-Oaten, A., J. R. Bence, and C. W. Osenberg. 1992. Assessing effects of unreplicated perturbations: no simple solutions. *Ecology* 73:1396–1404.

Toth, L. A. 1993. The ecological basis of the Kissimmee River restoration plan. *Florida Scientist* 56:25–51.

Toth, L. A. 1995. Principles and guidelines for restoration of river/floodplain ecosystems — Kissimmee River, Florida. In *Rehabilitating Damaged Ecosystems*, 2nd ed., J. Cairns, Jr. (ed.). Lewis Publishers/CRC Press, Boca Raton, FL, pp. 49–73.

Toth, L. A. and D. H. Anderson. 1998. Developing expectations for ecosystem restoration. In *Transactions 63rd North American Wildlife and Natural Resources Conference*, K. G. Wadsworth (ed.). Wildlife Management Institute, Washington, D.C., pp. 122–134.

Toth, L. A. and N. G. Aumen. 1994. Integration of multiple issues in environmental restoration and resource enhancement projects in southcentral Florida. In *Implementing Integrated Environmental Management*, J. Cairns, Jr., T. V. Crawford, and H. Salwasser (eds.). Virginia Polytechnic Institute & State University, Blacksburg, VA, pp. 61–78.

Toth, L. A., D. A. Arrington, M. A. Brady, and D. A. Muszick. 1995. Conceptual evaluation of factors potentially affecting restoration of habitat structure within the channelized Kissimmee River ecosystem. *Restoration Ecology* 3:160–180.

Toth, L. A., D. A. Arrington, and G. Begue. 1997. Headwater restoration and reestablishment of natural flow regimes: Kissimmee River of Florida. In *Watershed Restoration: Principles and Practices*, J. E. Williams, C. A. Wood, and M. P. Dombeck (eds.). American Fisheries Society, Bethesda, MD, pp. 425–442.

Toth, L. A., J. W. Koebel, Jr., A. G. Warne, and J. Chamberlain. 2002. Implications of reestablishing prolonged flood pulse characteristics of the Kissimmee River and floodplain ecosystem. In *Flood Pulsing in Wetlands: Restoring the Natural Hydrological Balance*, B. A. Middleton (ed.). John Wiley & Sons, New York, pp. 191–221.

Trexler, J. 1995. Restoration of the Kissimmee River: a conceptual model of past and present fish communities and its consequences for evaluating restoration success. *Restoration Ecology* 3:195–210.

U.S. Army Corps of Engineers. 1956. Central and Southern Florida, Kissimmee River Basin and Related Areas. Supplement 5: General Design Memorandum, Kissimmee River Basin. U.S. Army Corps of Engineers, Jacksonville District, Jacksonville, FL.

U.S. Fish and Wildlife Service. 1991. Kissimmee River Restoration Project. Fish and Wildlife Coordination Act Report. U.S. Fish and Wildlife Service, Vero Beach, FL.

Weller, M.W. 1995. Use of two waterbird guilds as evaluation tools for the Kissimmee River restoration. *Restoration Ecology* 3:211–224.

White, P. S. and J. L. Walker. 1997. Approximating nature's variation: selecting and using reference information in restoration ecology. *Restoration Ecology* 5:338–349.

Woody, T. 1993. Grassroots in action: the Sierra Club's role in the campaign to restore the Kissimmee River. *Journal of the North American Benthological Society* 12:201–205.

29

Environmental Indicators as Performance Measures for Improving Estuarine Environmental Quality

M. Jawed Hameedi

CONTENTS

Introduction

The rationale for developing an environmental indicator or index is to simplify, and to strive toward a parsimonious description of environmental measures that still retains meaning. Conceptually, environmental indicators are expected to function like economic indicators, i.e., relating to the economy as a whole. Some well-known examples of economic indicators include the jobless rate, hourly earning, interest rates, etc. The Consumer Price Index, based on a weighted average of retail prices of over 300 goods and services in selected localities around the country, is used to compare the cost of living in different geographical areas or time periods. Similarly, environmental indicators can be used to relay credible information about the overall condition of the environment whether it is used for scientific analyses or for policy decisions. This chapter provides background and a framework for using environmental indicators to manage estuarine resources and amenities, with a few illustrative examples of research that has been carried out in response to some pioneering legislation in the United States.

Paraphrasing Hammond et al. (1995), an indicator may be described as a measure that provides a clue to a matter of larger significance or makes perceptible a trend or phenomenon that is not immediately or distinctly discernible. For example, the presence of lead in a child's blood above a certain threshold level means impending metabolic disorders and impaired learning abilities later in life. Similarly, the appearance of blue green alga or cyanobacteria in a lake signifies impending eutrophication; and increased frequency of DNA damage in cells could be a precursor of impaired fitness parameters and genotoxicity.

The terms "environmental indicator" and "environmental index" have generally been used synonymously in the scientific literature. However, sometimes a distinction is made between an environmental indicator and an environmental index. The term environmental indicator generally refers to some environmental attribute, for example, the number of days when a shellfish bed is closed for harvesting or, conversely, the number of restored shellfish beds. Indicators can be presented singly (number of red tide days in a year) or they may be listed simultaneously (e.g., nutrient load, dissolved oxygen content, chemical contaminants, and incidence of fish disease) for a given water body as an "environmental quality profile." The term environmental index is often used to represent a single number derived from two or more indicators and is viewed as a composite measure of the overall ecological condition, environmental quality, or biological productivity.

Historical Perspective

Indices of environmental quality have existed for nearly 100 years. One of the earliest such indices, the Saprobic Index, was designed to detect the influence of oxygen-demanding wastes in freshwater streams in Germany (Kolkwitz and Marsson, 1909). In the United States, environmental quality indices were developed following recommendations of the Environmental Pollution Panel of the President's Science Advisory Committee in 1965. The recommendations included "assigning a numerical index of chemical pollution to water samples" that would be sensitive to chemical pollutants and roughly proportional to the "unfavorable effects of the pollution on man or aquatic life." Most notable among the early environmental indicators are Horton's Water Quality Index (Horton, 1965) and an index to initiate source control actions during air pollution episodes (Green, 1966). Horton's index, based on a weighted sum of eight variables and two coefficients, was designed to rate different water bodies (rivers and streams), evaluate pollution abatement programs, and communicate information to the public at large. Interestingly, the index did not consider toxic substances because the author did not believe that substances injurious to humans, animals, or aquatic life should be in rivers and streams.

A profound impetus and much wider interest in the development of indices were generated in the United States following the passage of the National Environmental Policy Act (NEPA) in 1969. One of the NEPA goals was to attain "the widest range of beneficial uses of the environment without degradation, risk to health or safety, or other undesirable and unintended consequences." The act directed federal agencies to "identify and develop methods and procedures … which will ensure that presently unquantified environmental amenities and values may be given appropriate consideration in decision-making along with economic and technical considerations." The NEPA requirements were later elaborated to signify the meaning and importance of environmental indices: "policy-making neither can nor should become totally scientific … but we must strive to make the maximum use of scientific evidence available to us, and development of environmental indices is one important way of doing this" (Train, 1972). Later, the National Academy of Sciences published a report in response to a request by the President's Council on Environmental Quality (NAS, 1975). The report called for a coordinated multiagency program for the development and application of indicators, monitoring to evaluate the efficacy of indices, and international efforts for developing globally compatible systems. Another impetus for developing environmental indicators was in the early 1980s as a result of a court decision that required relevant agencies to evaluate whether ocean dumping of sewage sludge caused "unreasonable degradation" of human health, welfare, and amenities in the context of all relevant environmental, social, and economic factors. The term "unreasonable degradation" was specifically mentioned in Section 102 of the Marine Protection, Research, and Sanctuaries Act (MPRSA) for issuing of permits for dumping of wastes in the ocean. In the aftermath of this court decision, 11 indices of coastal estuarine degradation were proposed (O'Connor and Dewling, 1986).

As a general observation, air quality indices are more consistent, simpler in their composition, and easier to interpret than are the water quality indices. This is due to more uniform national standards (National Ambient Air Quality Standards), as well as delineation of wide geographical regions that face similar air pollution problems, for example, the South Coast Air Basin in California. As a result of its topography, the South Coast Air Basin, encompassing some 30,000 km^2, is noted for frequent thermal inversions. When coupled with bright sunshine and warm temperatures, emissions from 9 million motor vehicles and hundreds of industrial installations are accentuated under such inversions, resulting in frequent occurrences of heavy smog and unhealthy conditions. However, using emissions of carbon

TABLE 29.1

Examples of Biomarkers Studied under the NOAA National Status and Trends Program

Cellular Integrity and Cytogenetic Damage

Lysosomal destabilization
DNA adducts
DNA strand breakage

Stress Proteins/Detoxification Response

Phase I enzymes (CYP1A, BPH)
Phase II enzymes (GST)
Multi-xenobiotic resistance proteins (MXRs)
Stress proteins (hsp70, hsp76, chaperonin)
Metallothioneins
Antioxidants

Impaired Reproduction

Gonadotropins
Steroids (plasma estradiol, testosterone)
Vitellogenin

Impaired Immune System

Hemocyte numbers and types
Killing index/phagocytic index
Serum lysozymes

Wellness and Condition

Darwinian "fitness parameters"
Atrophied organs and connective tissue
Parasitic infection
Disease and abnormalities
Lesions and tumors

---

monoxide as an indicator, the number of days with air quality problems in the Los Angeles area has declined from 100 to 10 per year during the past 20 years. This reduction is largely due to fairly stringent vehicle emission controls. On the other hand, water clarity in Lake Tahoe has been declining since 1968, despite efforts to improve the lake's water quality (California Resource Agency, 2002).

In the case of coastal and marine environments, the National Oceanic and Atmospheric Administration (NOAA) and the U.S. Environmental Protection Agency (EPA) have provided leadership in developing and encouraging the use of environmental indicators (NOAA, 1990, 1991; McKenzie et al., 1992; Hartwell, 1998). The NOAA National Status and Trends Program, parts of which have continued since 1984, includes development of diagnostic and predictive methodologies and tools for describing coastal environmental conditions. In terms of indicators, the program has focused on both ecological indicators and biomarkers: ecological indicators are those where measurements are made principally at the population, community, or ecosystem level; in the case of biomarkers, measurements are made principally at the individual, cellular, or subcellular level. Species diversity is an example of ecological indictors, and induction of the cytochrome P450 enzyme system is an example of biomarkers (Table 29.1).

Pressure-State-Response Framework

The need for sustainable development of coastal areas, marine environmental protection, and use of indicators in decision making was given strong support at the United Nation's Conference on Environment and Development (UNCED), held in Rio de Janeiro in 1992. Chapter 40 of Agenda 21 called for development of the concept of environmental indictors and promotion of their use in the sustainable

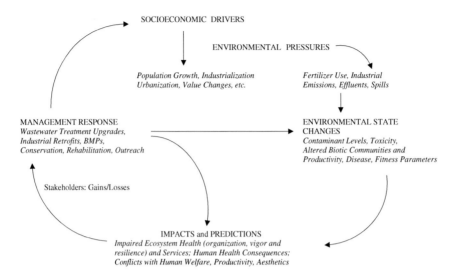

FIGURE 29.1 The PSIR framework for indicators related to environmental contamination.

development of Earth's resources (Robinson, 1993). Agenda 21 reinforced efforts to develop environmental indicators that were already under way in a number of countries, including Canada, the Netherlands, and the United States. Earlier, U.S. EPA had categorized environmental indicators into four categories (Hunsaker and Carpenter, 1990). They included stressor indicators (to identify sources of pollution or other stresses), exposure indicators (to define measurements that define degraded conditions), habitat indicators (parameters that define basic characteristics of the environment), and response indicators (to quantify the response of ecological components to individual or multiple stressors). There was no indicator to suggest social response or management action to close the feedback process.

The pressure-state-response (PSR) framework to categorize environmental indicators was already in place at the time of the Rio Conference. It was based largely on work done at the Organization for Economic Cooperation and Development following a G-7 Economic Summit in 1989 (OECD, 1991). The framework follows the "cause-consequences-societal response" logic and embodies the concept of causality: human activities exert pressure on the environment and its resources that could affect environmental quality and other aspects of the quantity and quality of natural resources (or state of the environment), requiring regulatory or policy changes or other remedial actions (or societal response). For example, in the case of environmentally persistent, bioaccumulative, and toxic organochlorine compounds in coastal waters, it can be surmised that indicators of pressure could include the manufacture, use, and discharge of organochlorines; indicators of state could include levels of organochlorines in tissues of regional fauna, physiological dysfunction, incidence of histopathology, and prevalence of genotoxic conditions in affected species; indicators of response could include development and use of new chemicals or other pest control practices, cleanup of degraded environments, more effective pollution control strategies, and public education and awareness campaigns. Similar sets of indicators can be developed for nutrient overenrichment and eutrophication in coastal waters, excessive harvest of species, biodiversity, ozone depletion, etc. The PSR framework has recently been appended to include assessment of negative effects (impacts) accruing to the system, and added consideration is given to drivers, which are socioeconomic forces or actions resulting in increased pressures on the environment (Figure 29.1). In its new form, it is known as the pressure-state-impact-response (PSIR) framework (Turner et al., 1998).

Although integrated coastal zone management and the use of environmental indicators have received considerable prominence in resource management decision making in the United States and elsewhere, the PSIR framework has not been used as a blueprint for such decision making (Hameedi, 1997). For the most part, this is due to agency mandates and institutional tendencies that are largely sectoral and fragmented. Integration among agencies, collaboration between public and private sectors in making resource use decisions, and multidisciplinary approaches are not a common occurrence in the management of coastal and marine resources.

TABLE 29.2

The U.S. EPA Water Quality Objectives and Indicators

Objective I. Conserve and Enhance Public Health

1. Water supply systems in violation of health-based requirements
2. Surface water systems at risk from microbial pollution
3. Population served by drinking water systems exceeding lead action levels
4. Drinking water systems using groundwater with programs to protect them from pollution
5. Fish consumption advisories
6. Shellfish growing waters with approval for harvest for human consumption

Objective II. Conserve and Enhance Aquatic Systems

7. Biological integrity
8. Species at risk
9. Wetland acreage

Objective III. Support Designated Usage of Water by the States and Tribes

10. Water quality standards to support designated use for the water body (drinking water supply, fish and shellfish consumption, recreation, and propagation of aquatic life) and criteria to assure such use

Objective IV. Conserve and Improve Ambient Conditions

11. Groundwater pollutants
12. Surface water pollutants
13. Concentration levels of selected pollutants in oysters and bivalves
14. Estuarine eutrophication conditions
15. Contaminated sediment

Objective V. Reduce or Prevent Pollutant Loadings and Other Stressors

16. Pollutant loadings from point sources to surface water and groundwater
17. Pollutant loadings to surface water from non-point sources
18. Marine debris

Source: U.S. EPA (1996).

The U.S. EPA, using the PSR framework, identified 18 indicators to accomplish its water quality objectives along a series of milestones (Table 29.2). These indicators, described as understandable measures, were designed to provide information not only on water quality but also on the agency's progress toward its water quality goals, objective, and milestones. The Mid-Atlantic Integrated Assessment (MAIA)–Estuaries is an attempt to integrate environmental assessment, monitoring, and related research activities across agencies to provide a holistic description of current conditions of the region's estuaries, their natural resources, anthropogenic stressors, and vulnerabilities. This project, prompted by the President's Committee on Environment and Natural Resources, National Science and Technology Council, is viewed as a "proof of concept" for integrated regional monitoring and assessment rather than agency-specific, resource-specific, or site-specific assessments and monitoring (Kiddon et al., 2003).

Attempts at restoration of seagrass beds in Tampa Bay, Florida provide a good example where a reasonably well-articulated resource management goal, coupled with collaborative efforts between public and private sectors to implement a nitrogen control strategy, have proved useful in reversing eutrophic conditions in the bay (Johansson and Greening, 2000). Even though the PSIR framework was not explicitly used, various scientific activities and management decisions could be construed within that framework. In this case, high nutrient discharges and runoff and shoreline modifications (pressure) resulted in high primary productivity, including seasonally dense blooms of planktonic cyanobacteria (state) that increased turbidity, lowered dissolved oxygen levels, and resulted in large losses of seagrass beds (impact). The management goal of achieving seagrass cover equal to 95% of that present in 1950 required reduction of nitrogen levels in effluents from municipal wastewater treatment plants, storm water upgrades, agricultural best-management practices, and emission reductions from power plants (response). At present, this strategy is achieving water quality conditions necessary to restore seagrass beds, and nitrogen reduction targets are being exceeded in most parts of the bay. As a result, the seagrass beds are steadily increasing in size each year. It should be noted that the primary seagrass species in the bay, turtle grass (*Thalassia testudinum*), took about 8 years to respond to nutrient control measures (chlorophyll levels in the water column decreased more quickly). According to estimates, it might take another 20 years to achieve the management target of seagrass recovery in the bay.

Unlike restoration of seagrass, a strategy for reducing chemical contaminant load at highly polluted sites or "hot spots" in Tampa Bay has not been effective. This is despite contaminant monitoring and a number of biological effects studies, including some by NOAA that have been carried out in the bay since 1986. This is due in large part to a lack of reasonable and agreed-upon numerical targets (or management goals) to signify restoration of contaminated areas. A wide spectrum of adverse biological effects that are associated with contaminant exposure has made the issue even more tenuous.

PSIR Framework and Environmental Monitoring

The PSIR framework is also suitable for establishing environmental monitoring programs linked to coastal resource management goals and objectives. Inclusion of a scientific construct, restatement of environmental issues as testable null hypotheses, and an objective field sampling strategy are key ingredients to acquiring credible monitoring data (Hameedi, 1997). The resulting data and information products from such a monitoring program are likely to be highly pertinent for developing diagnostic parameters, analytical models, or risk assessment procedures for making resource management decisions. Given high variability in environmental parameters and uncertainty in measurements, considerable attention should also be placed on the sampling design for environmental monitoring, including site selection, sampling frequency, and the overall cost of detecting change of a certain magnitude. Figure 29.2 provides a schematic arrangement of the PSIR components in an integrated monitoring framework that was proposed by an *ad hoc* Coastal Environmental Monitoring Committee in the NOAA National Ocean Service (CEMC, 2000).

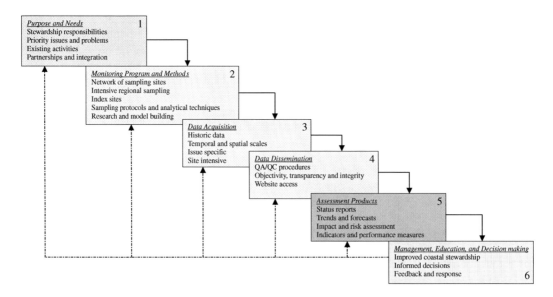

FIGURE 29.2 Environmental monitoring strategy in the context of the PSIR framework. From left to right, the rectangles indicate pressure (rectangle 1), major components of an observational program to define the state of the system, including data dissemination (next 3 rectangles), assessment products (rectangle 5), and management response (rectangle 6). Note the feedback loops and iterative nature of the strategy.

Continuing Challenges

Largely as a result of government-wide and mandatory efforts, there is an increasing emphasis for disseminating environmental data and for maximizing the quality, objectivity, integrity, and utility of the disseminated data. As a result, detailed data and derived information products can be accessed and downloaded from various data portals of government agencies and private sources. Many nongovernmental organizations, such as the Heinz Center for the Environment, the Pew Oceans Commission, and the Ocean Conservancy, are also playing important roles in gathering, analyzing, and reporting environmental data. In addition, these groups comment on environmental policy objectives that are disseminated widely, from the public to lawmakers. Thus, it is entirely reasonable for the public to ask the question: What might be the cost to society of environmental pollution in U.S. estuaries and coastal areas that may have rendered 44% of the tested estuarine areas "unfit" for uses such as swimming, fishing, or supporting aquatic life (The Ocean Conservancy, 2002)? Furthermore, the public may want to know about the government programs to abate pollution, and how to judge the success of such programs. Such questions pose formidable challenges, not only to resource managers but also to the scientific community, including educators.

A More Comprehensive Scientific Approach

Volumes have been written about the need for more comprehensive and holistic scientific approaches, including modeling that accounts for uncertainty, in resolving resource use conflicts and other environmental issues, for example, pollution control (NAS, 2001), protection of coral reefs (Risk et al., 2001), and overenrichment of nutrients (Howarth et al., 2003). However, in practice, a sectoral approach is more often the case (i.e., addressing a particular environmental matrix, a subregion, or a subset of the contaminants of concern). This is the result of a lack of planning or involvement of the scientific community, fragmented data sources, and exigencies resulting from a court order or some other external requirement. In a recent study to restore the "fishable" designated use of the Delaware Estuary, the total maximum daily load (TMDL) calculations are being made to develop new, stricter water quality criteria for polychlorinated biphenyls (PCBs). In the United States, the TMDL defines the amount of a specific pollutant that a water body can receive and still meet water quality standards; often such calculations are made to restore impaired water bodies. The Delaware Estuary has had fish consumption advisories from the coastal states since the 1980s, based on PCB levels in fish exceeding 24 to 60 parts per billion (ppb). The observed concentrations of PCBs in white fish and channel catfish over the past 15 years have remained much higher, in some years exceeding 1000 ppb, with no obvious declining trend. The study involved review and analysis of PCB discharge data from various industrial and municipal sources and use of hydrodynamic and water quality models to derive TMDL values for PCBs in four separate segments of the estuary. The calculated TMDL values ranged from 18 to 257 mg/day (DRBC, 2003). However, the model did not consider PCB contributions from the adjacent bay waters, major tributaries of the estuary, or the atmosphere. Instead, it was assumed that input of PCBs from those sources were at levels that would ensure that the water quality criterion of 8 pg/l (parts per quadrillion) is achieved. In actuality, PCB concentrations in the major tributaries and adjacent bay water exceed this criterion by orders of magnitude. Neither did the study consider a large reservoir of PCBs in the estuary's sediment. Based on results of a recently concluded study, that reservoir could amount to 1 to 2 metric tons in the TMDL study area (Hartwell et al., 2001). The sediment-bound PCBs probably represent a major source of PCB exposure and subsequent accumulation in the estuary's fish. It can be argued that the proposed waste load allocations, based on the TMDL calculations, may place a disproportionate burden of water quality improvement on point source dischargers or set criteria that may not achieve their desired objective, i.e., reducing PCB residue in fish to acceptable levels.

The Delaware Estuary TMDL study is not the only case of its kind; it is outlined here only as an example of a well-meaning study carried out with constraints of time and a limited scientific approach. It was necessitated by and scheduled under a consent decree and settlement agreement following a lawsuit filed in 1997 for adoption of TMDLs in Delaware Estuary by 15 December 2003. Its results may not produce the desired outcome for improving the estuary's water quality. Further, this example exacerbates the need for broader scientific perspective and participation in setting up environmental indicators and attainable criteria. To be effective, such participation ought not to be limited to conducting a certain study or performing a particular policy analysis; it should also include defining the issue, devising hypotheses and alternatives, thoughtful and objective review, and careful and even-handed synthesis of information that may appeal to societal sectors with disparate foci but similar interests (Platt, 1964; Suchman, 1995).

Bridging a Widening Scientific Divide

In the United States, environmental indicators have been formulated over the past 25 years along two distinct lines: one with a foundation in U.S. federal environmental statutes and societal values, and the other on perceived ecosystem properties that would respond to stress (Table 29.3). Very considerable effort has been expended in some landmark legislation to define and elucidate phrases, such as "health of the ocean" and "health of the marine environment" (both found in Title II of the Marine Protection, Research and Sanctuaries Act), "biological integrity of the estuary" and "balanced indigenous populations" (both found in Section 302 of the Clean Water Act), and "health of the Nation's coastal ecosystems" (Title V of the Marine Protection, Research and Sanctuaries Act). It is speculated that such health-related phrases reflected pollution and anoxic conditions in many rivers, portions of the North American Great Lakes, and some estuaries during the late 1960s and early 1970s. For example, in 1969, an oil slick on the Cuyahoga River, Ohio, caught fire and the river was described as being under "critical condition." Lake Erie, one of Great Lakes, was described as "dead" due to very low dissolved oxygen content. In the northeastern United States, thousands of lakes were described as "dead" or "near death" due to effects of acid rain and other pollutants.

The obvious advantage of health-related terms is that they provide a focus for maintaining and restoring the beneficial uses of the coastal environment and ecosystems. They also attract the public's attention to the importance of continued propagation of fish and wildlife resources, and seafood that is safe to eat, minimizing resource-use conflicts, and developing early warning indicators, similar to those for human health. The word "health" elicits little disagreement (value conflict) about whether ecosystem health or environmental health is a good idea. Impaired "ecosystem health" has also been interpreted to mean declining ecosystem properties and services that have human health implications, such as the presence of *Vibrio cholerae* (from ballast water discharge), consumption of fish contaminated from point and non-point sources of pollution, and algal toxins (associated with harmful algal blooms). However, this is much too narrow a focus with which to examine effects of stressors on coastal ecosystems (National Research Council, 1999; Boesch and Paul, 2001). In either case, one is expected to develop normative

TABLE 29.3

Examples of Environmental Indicators with Roots in Federal Environmental Legislation (Panel A) or a Scientific Construct (Panel B) Based on Information Theory, Probability Functions, Phenotypic Deviations, etc.

Panel A	Panel B
Fishable and swimmable waters	Biological productivity
Balanced indigenous populations	Species diversity
Index of Biotic Integrity	Stability
Benthic Index	Resilience
Farrell Epifaunal Index	Resistance
Estuarine Health Index	Ascendancy
Index of Unreasonable Degradation	Developmental instability
	Vitality

Note: Key references are Pielou, 1975; Pimm, 1984; Rapport, 1989; Ulanowicz, 1992; Farrell, 1993; Sherman, 1994; Cooper et al., 1994; Alados, et al., 1998; Costanza and Mageau, 1999; Anderson, 2000.

criteria or indicators for healthy environments and ecosystems. If a case is to be made for an environmental indicator as something that makes "perceptible a phenomenon or trend that is not immediately detectable," then continued research and analysis will be needed. This will not be an easy task. Even in the case of a selected few biochemical and physiological biomarkers, where a cause–effect relationship is well established, there is a general lack of data on their normal range, limiting their use as operational monitoring parameters (Ringwood et al., 1999).

The term "health" has seldom been featured in ecosystem literature, which for a long time has been preoccupied with organizational properties of ecosystems — with origins in information theory and network topology — and energy flow. As a result, "ecosystem health" definitions are varied and, quite often, convoluted and complex; their meaning or measures are not easily communicated to the informed public. Karr et al. (1986) defined ecological health as the "condition when a system's inherent potential is realized, its condition is stable, its capacity for self-repair, when perturbed, is preserved, and minimal external support for management is needed." Ulanowicz (1992, p. 191) defined it as an ecosystem whose "trajectory toward a climax is relatively unimpeded and whose configuration is homeostatic to influences that would displace it back to earlier successional stages" and described it in the context of species richness, niche specialization, recycling and feedback loops, and overall activity. Costanza (1992) and Costanza and Mageau (1999) defined an ecosystem health index comprising three parameters: system organization (ecosystem structure), vigor (productivity), and resilience to stress. Sherman (2000) described ecosystem health on the basis of diversity, productivity, biomass yield, resilience, and stability. These examples demonstrate a degree of similarity in concept but also considerable incongruity that makes their use in coastal resource management decision making tenuous at best. Further, due to the abstract nature of some of their constituent terms, such indices are difficult to quantify, except categorically, and to communicate to resource managers. Very seldom have they been used by themselves to document ecosystem changes due to anthropogenic stressors, such as pollution. However, several simpler and monometric environmental parameters have found wider and more successful applications, including population changes in a certain keystone or valued species, extent and severity of contamination, loss of wetland acreage, curtailed nutrient delivery, etc. Such indicators describe environmental status, and perhaps improvements, but have little value in the forecasting of environmental response to status quo or any specific pollution abatement strategy.

Another notion deriving its impetus from verbiage in the Clean Water Act is "biological integrity." It was defined by Karr (1991) as "a balanced, integrated, adaptive community of organisms having a species composition, diversity, and functional organization comparable to that of a natural habitat of that region" and interpreted by Nielsen (1999) as a community having developed without human disturbance. As in the case of ecosystem health, biological integrity is easily understood and relatable but harder to define in terms of its constituent parts or by a holistic measure.

The Index of Biotic Integrity (IBI), originally developed for freshwater streams, is based on the scoring of a number of individual environmental metrics (as poor, good, or excellent, or numerically as 1, 3, or 5) affecting, for example, fish assemblages in a stream (Karr et al., 1986). A reference site, or the best available site, provides a standard for comparison between numerical scores. The principles behind the IBI have since been applied to benthic invertebrate communities in coastal waters and estuaries (Weisberg et al., 1997; Engle and Summers, 1999; Van Dolah et al., 1999).

Benthic indices have served a useful heuristic purpose; they are usually derived from specific data sets from a specific estuary, such as summer field sampling in a particular year. They have generally been successful in distinguishing heavily degraded sites from reference or "clean" sites. A benthic IBI for estuaries in the southeastern United States was reported to correctly classify 95% of the stations sampled in the developmental stage of the index, and 75% of the stations in the index verification stage (Van Dolah et al., 1999). Such a performance of the index is encouraging. However, indices developed so far lack generality, i.e., a new index needs to be formulated for each application (Engle and Summers, 1999). They have not been used to elucidate environmental factors that determine benthic faunal distribution at the majority of sites in an estuary, sites that are neither clean nor heavily degraded. In some analyses, data from such sites are ignored to facilitate identification of discriminating factors responsible for the observed distribution of benthic fauna. Perhaps it is for such reasons that Suter (1993) described ecosystem health or biotic integrity indices in a tautological sense and stated that "[indices] are justified on the basis of field studies rather than any theory of ecosystem health or any societal or ecological value of the index or its components. That is, an ecosystem's health is bad because the index is low, and a low index value indicates bad health because the index is low for unhealthy ecosystems." Nonetheless, considerable progress is being made in explaining benthic faunal composition in relation to habitat degradation, whether from anthropogenic or natural factors, on the continental shelf and estuaries, and to further clarify the concept of "response sequence" to different levels of stress (Smith et al., 2001; Preston and Shackelford, 2002; Freeman and Rogers, 2003).

It should also be noted that in comparison with freshwater streams, for which the IBI was originally designed, it is much more difficult to define impacted and non-impacted sites in estuaries. Further, the guidelines or criteria often used to define reference and degraded sites, such as the widely used sediment quality guidelines (Long et al., 1995; MacDonald et al., 1996) have a questionable value in this regard (O'Connor and Paul, 2000) and need to be further improved (Wenning and Ingersoll, 2002). On occasions, results of the Microtox test are used to define degraded conditions; this test is quite sensitive and found responsive to a large number of chemicals, but, by themselves, its results have no ecological significance (Harmon et al., 2003). It has also been shown that the simultaneous occurrence of several stressors at a site and large variability in their levels often preclude well-defined bounds of reference conditions, i.e., a "reference envelope," in an estuary (Smith, 1995). Defining reference conditions or a reference envelope also remains a problematic topic (Anderson et al., 1997; Weisberg et al., 1997; Hyland et al., 1999).

Governance Indicators

As noted earlier, the PSIR framework offers the means to gather data as well as a platform to discuss and integrate information on environmental management and policy issues in which resource managers, stakeholders, and the public have key roles to play. An important feature of the framework is that it incorporates indicators or targets for formulating and evaluating coastal resource management scenarios and options. The framework can be used to ask the proverbial question: How do we know that the minister has kept his promise? (Mahmood, 2000). For this reason, the governance-related indicators, like other elements of the framework, should be relevant, measurable, and amenable to monitoring. Enactment of new legislation, development of an action plan, hiring and training of staff, and even expenditure of funds are not sufficient to gauge progress without causal linkages to the issue at hand and measurable milestones to demonstrate progress and achievements at various stages of project implementation.

For contaminated sediments, a frequently used management option is the "provisional control option," which addresses the problem through resource-use controls (e.g., fish consumption advisories) and increased outreach toward stakeholders. The option does not necessarily include discussion of other options. If a broader suite of options, including those with socioeconomic measures, is considered in the PSIR framework, the process may lead to a wider support of the management option at hand in view of overall cost to society.

Time Frame for Environmental Response

All living systems, from cells to ecosystems, have the capability to tolerate and compensate for endogenous and exogenous stresses for a certain length of time; injury or disruption occurs when this capacity is exceeded. Their recovery also takes time, depending largely on the nature and severity of pollution and the effectiveness of the remediation efforts. It has been shown that recovery of oxygenated waters is relatively rapid if the principal sources of the biochemical oxygen demand (BOD) in a water body are curtailed. In the Houston Ship Channel, the most remarkable improvement in environmental quality from management response (improved wastewater treatment, minimized sewage overflows, produced water management, and control of point source discharge of contaminants) was a reduction in BOD values in the upper reaches of the channel from 200,000 kg/day to less than 9,000 kg/day (GBNEP, 1994). The latter value is considered to be lower than the BOD input prior to the rapid population growth and industrialization that took place in the region in the 1930s (Sibley, 1968). In such a case, environmental state response to management action was direct and presumably linear. In other instances, there might be a lag time before the state indicators show a response. For hypothetical nitrate reduction scenarios in New Jersey, it has been shown that surface waters would show a declining nitrogen concentration more quickly than the water supply wells; the lag time could be over a period of decades (Kauffman et al., 2001). For environmentally persistent and biochemically recalcitrant chemicals, such as PCBs, the recovery period may be much longer. It is already evident from the 18-year record of the NOAA Mussel Watch Project data that after an initial rapid decline in PCB residues in bivalves, the declining trends have slowed and are apparently no longer linear with time. It should be pointed out that despite a ban on the manufacture and new use of PCBs for more than 25 years, they are still found in many municipal and industrial discharges. For example, in the Delaware Estuary, the annual input of PCBs from point sources is estimated at 4 kg (DRBC, 2003); in San Francisco Bay, external loading is estimated as high as 30 kg/year. Using a simple mass balance model for PCBs in San Francisco Bay, Davis (2003) showed that even if sustained loading of PCBs were eliminated completely, the PCB mass in the bay would be reduced to one half of the current value in 20 years and to one quarter in 40 years.

These results demonstrate long residence times of PCBs in the environment and that environmental restoration could be a costly endeavor over the long term, longer than many decision makers are accustomed to. "Ecosystem-based" management becomes an even more challenging issue largely because of problems in defining ecosystem boundaries, evaluating decadal or longer-scale changes in ecosystem structure and dynamics in the context of current changes, and making decisions with a sparse and fragmented information base that is so typical for most estuaries and coastal waters. Practical application of ecosystem-based management, now a familiar phrase in resource management agencies, would require a long-term commitment far beyond a fiscal year or some other political time frame; it would require time frames that transcend human lifetimes (Christensen et al., 1996).

Conclusion

Environmental indicators have long been used to communicate information about the state of streams, lakes, coastal bays, and estuaries as well as the socioeconomic benefits derived from these resources. In recent years, indicators have also been crafted as performance measures that are becoming an integral part of budgeting and review of government programs to conserve and better manage the environmental resources and amenities, with a broader aim of their sustainability over the long term. Additionally, indicators are becoming a focal point for assessing the changing state of large marine ecosystems and climate-induced decadal changes (Robertson and Hameedi, 1999; Sherman, 2000). Examples of such measures include the number of shoreline miles that have been characterized in terms of relative vulnerability to oil spill impacts, reduction in the spatial extent and severity of contamination in lakes and estuaries, number of depleted fish stocks that have recovered as a result of conservation efforts, and change in species assemblages over climate cycles, to name a few. There is now a more pronounced need for "usable science": science that can influence environmental policy and resource management decisions and meet the criteria of being adequate, effective, valuable, and legitimate (Clark and Majone, 1985; Hameedi, 1986).

The PSIR framework is important in this regard. The framework is based on the premise that in any given area there is a distribution of socioeconomic activities and related land uses (urban development, industrialization, fisheries, agriculture/forestry/aquaculture, commerce, and transportation). That distribution translates into demands (*drivers*) for a variety of goods and services within the defined area and from outside the area. As a consequence, (1) environmental *pressures* build up via the socioeconomic driving forces causing changes in the *state* of the environmental systems; (2) changes in the environmental state affect human and nonhuman receptors in a number of perceived ways (benefits and costs); and (3) such effects or *impacts*, in turn, provide a stimulus for management *response* (Robinson, 1993).

The PSIR framework is well suited for an integrative approach and a formal link between natural sciences, different dimensions of socioeconomic values, and resource management (Russel, 1995). It is now used to address a range of environmental issues including air and aquatic pollution (California Resources Agency, 2002), management of estuary resources (DEP, 2001), coral reef protection and management (Castro, 2001), and offshore wind power (Elliott, 2002). However, this is not the only framework available for environmental management purposes. The World Bank has long used its "objective-input-output-outcome-impact" framework to track progress in its projects (Segnestam, 1999). More recently, Olsen (2003) proposed the "four Orders of Outcomes" framework for assessing progress in integrated coastal zone management. However, the PSIR framework has an attractive feature of portraying the environment (through a set of indicators of state) in a context where stakeholders and resource managers can participate in defining what needs to be done and what is the outcome of actions previously taken to improve the situation.

Since the framework is for indicators, it is important to select indicators that have certain desirable features and are useful for meeting specific needs. The International Joint Commission for the Great Lakes identified 16 criteria for selecting or developing indicators of "ecosystem health" (IJC, 1991). It is surmised here that the selected indicators should at least be SMART: specific (with clearly stated objective), measurable (both in time and quantity), achievable (within available resources and intellectual capital), relevant (to elucidate the issue at hand), and trackable (amenable to evaluation and determining progress). Such attributes are critical if environmental indicators are to serve their principal role of communicating environmental information to broad and varied audiences.

The apparent dichotomy of argument about the desirability of ecological indicators as performance measures and the problem of defining some of them is reflective of the nascent nature of their application in a broad societal context, such as the PSIR framework. The relevance of some of the ecosystem-based or multimetric indices (Table 29.3) needs to be communicated better both to scientists and resource managers. There are encouraging examples of the use of such a framework, for example, seagrass restoration strategy in Tampa Bay. However, it is also important to remain cognizant of certain drawbacks and limitations of indicators. Edicts and prescriptions stifle innovation, and it is entirely common to focus on indicators that are easily measurable but address trivial issues. Indicators should be viewed as tools in our scientific toolbox; new tools will always be needed.

Acknowledgments

This chapter is based, in part, on a presentation made by the author at the Third Joint Meeting of the Coastal Environmental Science and Technology Panel of the United States–Japan Cooperative Program in Natural Resources that was held in Yokusuka, Japan in July 2002. The chapter has benefited from many fruitful discussions with colleagues, both within and outside government; however, the views expressed are entirely those of the author. Mention of any trade name or commercial product does not constitute endorsement or recommendation for use.

References

Alados, C. L., J. M. Emlen, B. Wachocki, and D. C. Freeman. 1998. Instability of development and fractal structure in dryland plants as an index of grazing pressure. *Journal of Arid Environments* 36:63–76.

Anderson, B., J. Hunt, S. Tudor, J. Newman, R. Tjeerdema, R. Fairey, J. Oakden, C. Bretz, C. J. Wilson, F. LaCaro, G. Kapahi, M. Stephenson, M. Puckett, J. Anderson, E. R. Long, T. Fleming, and K. Summers. 1997. Chemistry, Toxicity and Benthic Community Conditions in Sediments of Selected Southern California Bays and Estuaries. National Oceanic and Atmospheric Administration, Special Report. Silver Spring, MD, 146 pp. plus appendices.

Anderson, J. J. 2000. A vitality-based model relating stressors and environmental properties to organism survival. *Ecological Monographs* 70:445–470.

Boesch, D. F. and J. F. Paul. 2001. An overview of coastal environmental health indicators. *Human and Ecological Risk Assessment* 7:1409–1418.

California Resource Agency. 2002. Environmental Protection Indicators for California (Summary Report). State of California, California Environmental Protection Agency, Sacramento, 28 pp.

Castro, N. G. 2001. Monitoring ecological and socioeconomic indicators for coral reef management in Colombia. *Bulletin of Marine Science* 69:847–859.

CEMC. 2000. Current and Planned NOS Environmental Monitoring Activities. National Oceanic and Atmospheric Administration, National Ocean Service, Special Report. Silver Spring, MD, 63 pp.

Christensen, N. L., A. M. Bartuska, J. H. Brown, S. Carpenter, C. D'Antonio, R. Francis, J. F. Franklin, J. A. MacMahon, R. F. Noss, D. J. Parsons, C. H. Peterson, M. G. Turner, and R. G. Woodmansee. 1996. The report of the Ecological Society of America Committee on the scientific basis for ecosystem management. *Ecological Applications* 6:665–691.

Clark, W. C. and G. Majone. 1985. The critical appraisal of scientific inquiries with policy implications. *Science, Technology, and Human Values* 10:6–19.

Cooper, J. A. G., A. E. L. Ramm, and T. D. Harrison. 1994. The Estuarine Health Index: a new approach to scientific information transfer. *Ocean and Coastal Management* 25:103–141.

Costanza, R. 1992. Toward an operational definition of health. In *Ecosystem Health: New Goals for Environmental Management,* R. Costanza, B. G. Norton, and B. D. Haskell (eds.). Island Press. Washington, D.C., pp. 239–256.

Costanza, R. and M. Mageau. 1999. What is a healthy ecosystem? *Aquatic Ecology* 33:105–115.

Davis, J. A. 2003. The Long Term Fate of PCBs in San Francisco Bay. San Francisco Estuary Institute, Technical Report. Oakland, CA, 56 pp.

DEP. 2001. Delaware Estuary: Environmental Indicators. Delaware Estuary Program, West Trenton, NJ, 25 pp.

DRBC. 2003. Total Maximum Daily Loads for Polychlorinated Biphenyls (PCBs) for Zone 2–5 of the Tidal Delaware River. Delaware River Basin Commission, West Trenton, NJ,. 48 pp. plus appendices. Available at http://www.state.nj.us/drbc/TMDL/Sept2003.htm.

Elliott, M. 2002. The role of the DPSIR approach and conceptual models in marine environmental management: an example for offshore wind power (editorial). *Marine Pollution Bulletin* 44:iii–vii.

Engle, V. D. and J. K. Summers. 1999. Refinement, validation, and application of a benthic condition index for northern Gulf of Mexico estuaries. *Estuaries* 22:624–635.

Farrell, D. H. 1993. Bioassessment in Florida. In Proceedings of the Estuarine and Near Coastal Bioassessment and Biocriteria Workshop, Annapolis, MD. U.S. Environmental Protection Agency, Office of Science and Technology, Washington, D.C., pp. 17–26.

Freeman, S. M. and S. I. Rogers. 2003. A new analytical approach to the characterization of the macro-epibenthic habitats: linking species to the environment. *Estuarine, Coastal, and Shelf Science* 56:749–764.

GBNEP. 1994. The State of the Bay: A Characterization of the Galveston Bay Ecosystem. Galveston Bay National Estuary Program, Report GBNEP-44. Webster, TX, 232 pp.

Green, M. H. 1966. An air pollution index based on sulfur dioxide and smoke shades. *Journal of the Air Pollution Control Association* 11:703–706.

Hameedi, M. J. 1986. Management needs of scientific data. In The Gulf of Alaska, D. W. Hood and S. T. Zimmerman (eds.). U.S. Government Printing Office, Washington, D.C., pp. 597–617.

Hameedi, M. J. 1997. Strategy for monitoring the environment in the coastal zone. In *Coastal Zone Management Imperative for Maritime Developing Nations,* B. U. Haq, S. M. Haq, G. Kullenberg, and J. H. Stel (eds.). Kluwer Academic, Dordrecht, the Netherlands, pp. 111–142.

Hammond, A., A. Adriaanse, E. Rodenburg, D. Bryant, and R. Woodward. 1995. Environmental Indicators: A Systematic Approach to Measuring and Reporting on Environmental Policy Performance in the Context of Sustainable Development. World Resources Institute, Washington, D.C., 43 pp.

Harmon, M. R., A. S. Pait, and M. J. Hameedi. 2003. Sediment Contamination, Toxicity and Macroinvertebrate Infaunal Community in Galveston Bay. National Oceanic and Atmospheric Administration, Technical Memorandum NOS NCCOS CCMA 122. Silver Spring, MD, 60 pp. plus appendices.

Hartwell, S. I. (ed.). 1998. Biological Habitat Quality Indicators for Essential Fish Habitat Workshop Proceedings, National Oceanic and Atmospheric Administration, Technical Memorandum NMFS-F/SPO-32. Silver Spring, MD, 125 pp.

Hartwell, S. I., J. Hameedi, and M. Harmon. 2001. Magnitude and Extent of Contaminated Sediment and Toxicity in Delaware Bay. National Oceanic and Atmospheric Administration, Technical Memorandum NOS ORCA 148. Silver Spring, MD, 107 pp. plus appendices.

Horton, R. K. 1965. An index-number system for rating water quality. *Journal of the Water Pollution Control Federation* 37:300–306.

Howarth, R. R., R. Marino, and D. Scavia. 2003. Nutrient Pollution in Coastal Waters: Priority Topics for an Integrated National Research Program for the United States. National Oceanic and Atmospheric Administration, Special Report. Silver Spring, MD, 21 pp.

Hunsaker, C. T. and D. E. Carpenter.1990. Environmental Monitoring and Assessment Program: Ecological Indicators. U.S. Environmental Protection Agency, Office of Research and Development, Report EPA 600/3-90-060. Research Triangle Park, NC.

Hyland, J. L., R. F. Van Dolah, and T. R. Snoots. 1999. Predicting stress in benthic communities of southeastern U.S. estuaries in relation to chemical contamination of sediments. *Environmental Toxicology and Chemistry* 18:2557–2564.

IJC. 1991. A Proposed Framework for Developing Indicators of Ecosystem Health for the Great Lakes Region. International Joint Commission, Report of the Council of Great Lakes Managers to the International Joint Commission. Washington, D.C./Ottawa, Canada, 47 pp.

Johansson, J. O. R. and H. S. Greening. 2000. Seagrass restoration in Tampa: a resource-based approach to estuarine management. In *Seagrasses: Monitoring, Ecology, Physiology, and Management,* S. A. Bortone (ed.). CRC Press, Boca Raton, FL, pp. 279–294.

Karr, J. R. 1991. Biological integrity: a long-neglected aspect of water resource management. *Ecological Applications* 1:66–84.

Karr, J. R., D. Fausch, P. L. Angermeier, P. R. Yant, and I. J. Scholsser. 1986. Assessing Biological Integrity in Running Waters: A Method and Its Rationale. Illinois Natural History Survey, Special Publication 5. Champaign, IL.

Kauffman, L. J., A. L. Baehr, M. A. Ayers, and R. E. Stackelberg. 2001. Effects of Land Use and Travel Time on the Distribution of Nitrate in the Kirkwood-Cohansey Aquifer System in Southern New Jersey. U.S. Geological Survey, National Water Quality Assessment Program, USGS Water-Resources Investigations Report 01-4117, West Trenton, NJ, 49 pp.

Kiddon, J. A., J. F. Paul, H. W. Buffum, C. S. Strobel, S. S. Hale, D. Cobb, and B. S. Brown. 2003. Ecological conditions of US Mid-Atlantic estuaries, 1997–1998. *Marine Pollution Bulletin* 46:1224–1244.

Kolkwitz, R. and M. Marsson. 1909. Ecology of the animal saprobes: contributions to the theory of the biological waters evaluation. *Internationale Revue der gesamten Hydrobiologie und Hydrographie* 2:126–152 [in German].

Long, E. R., D. D. MacDonald, S. L. Smith, and F. D. Calder. 1995. Incidence of adverse biological effects within ranges of chemical concentrations in marine and estuarine sediments. *Environmental Management* 19:81–97.

MacDonald, D. D., R. S. Carr, F. D. Calder, E. R. Long, and C. G. Ingersoll. 1996. Development and evaluation of sediment quality guidelines for Florida coastal waters. *Ecotoxicology* 5:253–278.

Mahmood, S. A. 2000. How do we know that the minister has kept his promise? Presentation at EB2000: Linking Individuals, Regions, Ideas and Actions, 4 pp. Available at www.eb2000.org/short_note_13.htm.

McKenzie, D. H., D. E. Hyatt, and V. J. McDonald. 1992. *Ecological Indicators.* Elsevier Applied Science, New York, 1567 pp. (2 vol.).

NAS. 1975. Planning for Environmental Indices. National Academy of Sciences, Report of the Planning Committee on Environmental Indices, Washington, D.C., 47 pp.

NAS. 2001. Assessing the TMDL Approach to Water Quality Management. National Academy of Sciences, Report of the Commission on Geosciences, Environment, and Resources, Washington, D.C., 122 pp.

National Research Council. 1999. From Monsoons to Microbes: Understanding the Ocean's Role in Human Health. National Academy Press, Washington, D.C., 132 pp.

Nielsen, N. O. 1999. The meaning of health. *Ecosystem Health* 5:65–66.

NOAA. 1990. Selected Environmental Indicators of the United States and the Global Environment. National Oceanic and Atmospheric Administration, Office of the Chief Scientist, NOAA Environmental Digest, Rockville, MD, 66 pp.

NOAA. 1991. Selected Environmental Indicators of the United States and the Global Environment. National Oceanic and Atmospheric Administration, Office of the Chief Scientist, NOAA Environmental Digest, Rockville, MD, 134 pp. plus appendices.

Ocean Conservancy. 2002. Health of the Ocean Report, 2002. The Ocean Conservancy (formerly the Center of Marine Conservation). Washington, D.C., 80 pp.

O'Connor, J. S. and R. T. Dewling. 1986. Indices of marine degradation: their utility. *Environmental Management* 10:335–343.

O'Connor, T. P. and J. F. Paul. 2000. Misfit between sediment toxicity and chemistry. *Marine Pollution Bulletin* 40:59–64.

OECD. 1991. Environmental Indicators: A Preliminary Set. Organisation for Economic Co-Operation and Development, OECD Publications Services, Paris, France, 77 pp.

Olsen, S.B. 2003. Frameworks and indicators for assessing progress in integrated coastal management initiatives. *Ocean and Coastal Management* 46:347–361.

Pielou, E. C. 1975. *Ecological Diversity.* Wiley-Interscience, New York, 165 pp.

Pimm, S. L. 1984. The complexity and stability of ecosystems. *Nature* 307:321–326.

Platt, J. R. 1964. Strong inference. *Science* 146:347–353.

Preston, B. L. and J. Shackelford. 2002. Multiple stressor effects on benthic biodiversity of Chesapeake Bay: implications for ecological risk assessment. *Ecotoxicology* 11:85–99.

Rapport, D. J. 1989. What constitutes ecosystem health? *Perspectives in Biology and Medicine* 33:120–132.

Ringwood, A. H., M. J. Hameedi, R. F. Lee, R. Brouwer, C. Peters, G. I. Scott, S. N. Luoma, and R. T. Digiulio. 1999. Bivalve Biomarker Workshop: overview and discussion group summaries. *Biomarkers* 4:391–399.

Risk, M. J., J. M. Heikoop, E. N. Edinger, and M. V. Erdman. 2001. The assessment "tool box": community-based reef evaluation methods coupled with geochemical techniques to identify sources of stress. *Bulletin of Marine Science* 69:443–458.

Robertson, A. and J. Hameedi. 1999. A pollution monitoring module for assessing changing state of Large Marine Ecosystem. In *The Gulf of Mexico Large Marine Ecosystem*, H. Kumpf, K. Steidinger, and K. Sherman (eds.). Blackwell Science, Malden, MA, pp. 431–437.

Robinson, N. A. (ed.). 1993. *Agenda 21: Earth's Action Plan.* Oceana Publications, New York, 683 pp.

Russel, C. S. 1995. Old lessons and new contexts in economic-ecological modeling. In *Integrating Economic and Ecological Indicators,* J. W. Milon and J. F. Shogren (eds.). Praeger, Westport, CT, pp. 9–25.

Segnestam, L. 1999. Environmental Performance Indicators. The World Bank, Environmental Department Paper 71. Washington, D.C., 38 pp.

Sherman, K. 1994. Sustainability, biomass yields, and health of coastal ecosystems: an ecological perspective. *Marine Ecology Progress Series* 112:277–301.

Sherman, K. 2000. Why regional coastal monitoring for assessment of ecosystem health: an ecological perspective. *Ecosystem Health* 6:205–216.

Sibley, M. M. 1968. *The Port of Houston, a History.* University of Texas Press, Austin, 246 pp.

Smith, R. W. 1995. The Reference Envelope Approach to Impact Monitoring. U.S. Environmental Protection Agency, Report by EcoAnalysis to U.S. Environmental Protection Agency, Region 9, San Francisco, CA, 67 pp.

Smith, R. W., M. Bergen, S. B. Weisberg, D. Cadien, A. Dalkey, D. Montagne, J. K. Stull, and R. G. Velarde. 2001. Benthic response index for assessing infaunal communities on the southern California mainland shelf. *Ecological Applications* 11:1073–1087.

Suchman, M. C. 1995. Managing legitimacy: strategic and institutional approaches. *Academy of Management Review* 20:571–610.

Suter, G. W. 1993. A critique of ecosystem health concepts and indexes. *Environmental Toxicology and Chemistry* 12:1533–1539.

Train, R. E. 1972. The quest for environmental indices. *Science* 178:121.

Turner, R. K., W. N. Adger, and I. Lorenzoni. 1998. Towards Integrated Modelling and Analysis in Coastal Zones: Principles and Practices. LOICZ Reports and Studies No. 11, LOICZ IPO, Texel, the Netherlands, 122 pp.

Ulanowicz, R. E. 1992. Ecosystem health and trophic flow network. In *Ecosystem Health: New Goals for Environmental Management*, R. Costanza, B. G. Norton, and B. D. Haskell (eds.). Island Press, Washington, D.C., pp. 190–206.

U.S. EPA. 1996. Environmental Indicators of Water Quality in the United States. U.S. Environmental Protection Agency, Office of Water, Report EPA 841-R-96-002. Washington, D.C., 25 pp.

Van Dolah, R. F., J. L. Hyland, A. F. Holland, J. S. Rosen, and T. R. Snoots.1999. A benthic index of biological integrity for assessing habitat quality in estuaries of the southeastern USA. *Marine Environmental Research* 48:269–283.

Weisberg, S. B., J. A. Ranasinghe, D. M. Dauer, L. C. Schaffner, R. J. Diaz, and J. B. Frithsen. 1997. An estuarine Benthic Index of Biotic Integrity (B-IBI) for Chesapeake Bay. *Estuaries* 20:149–158

Wenning, R. J. and C. G. Ingersoll. 2002. Use of Sediment Quality Guidelines and Related Tools for the Assessment of Contaminated Sediments: Executive Summary. SETAC Press, Pensacola, FL, 44 pp.

30

Indicators of Ecosystem Integrity for Estuaries

Stephen J. Jordan and Lisa M. Smith

CONTENTS

Introduction

Why do we need indicators of ecosystem integrity for estuaries, what are they, and how would one or more of these indicators inform and contribute to management of these diverse, complex systems? The importance of estuaries was expressed nicely by Welsh (1984, p. xiii): "Estuaries … are one of the most heavily utilized and most productive zones in our planet. Their integrative processes … weave a web of complexity far out of proportion to their occupation of less than 1% of the planet's surface." Citizens groups, environmental managers, and elected officials want to know the status of estuarine ecosystems, locally, regionally, nationally, and globally. Especially where there have been large public investments in pollution controls and other preventive and restorative measures, people want to know if their money has been well spent (Jordan and Vaas, 2000). Thus, there is a clear need for indicators that will provide comprehensive answers to these questions at appropriate intervals. Murawski (2000, p. 655) recommended "simple, robust indices of ecosystem state that gauge … production, diversity, and variability," and emphasized that indicators should have the capacity to predict the results of management. Although Murawski was writing in the context of fisheries and fisheries management, these principles apply more broadly to the integrity of ecosystems in general.

The problem, then, is to formulate indicators that are simple in presentation and interpretation. They should be robust (i.e., not sensitive to small perturbations or irrelevant factors) and predictive, but also grounded in the complexity and variability that are essential properties of the ecosystem. In terms of time and effort, organization of data, and realism in representing the system, such indicators are intermediate between simple descriptive statistics and complex, process-oriented mathematical models. In this chapter we offer several relevant concepts and a few examples of indicators of the large-scale structure and behavior of estuarine ecosystems. We also outline principles for development and application of indicators.

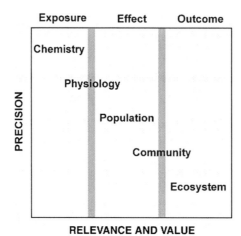

FIGURE 30.1 A hierarchy of indicators.

Terminology and Definitions

Indicators can be constructed at various scales of organization, from single chemical, biochemical, or physiological measurements to highly integrated composites of system attributes (Figure 30.1). This chapter is concerned principally with the integrative, value-oriented indicators located in the lower right quadrant of Figure 30.1. Indicators at molecular, sub-organismal, organismal, population, and community levels of organization can be informative for various purposes, but indicators at the whole-ecosystem level are essential for managing ecosystems and answering the public's most basic questions.

The terms "integrity" and "health" have been used widely to designate desirable states of ecosystems. The goal of the U.S. Clean Water Act is to "restore and maintain the chemical, physical, and biological integrity of the Nation's waters" (U.S. Code 33:26:1251a), but the act does not attempt to define integrity. Without further definition, neither health nor integrity is a useful or measurable descriptor of an ecosystem. A dictionary (Davies, 1976, p. 370) defines integrity as "soundness; completeness; unity" and health as "the state of an organism with respect to functioning, disease, and abnormality at any given time … optimal functioning with freedom from disease and abnormality. … flourishing condition; vitality." Extending these definitions to an ecological context, integrity might imply that the structural properties (state variables) of a system exhibit an expected, undisturbed condition; health might be used to indicate how well the system is functioning (rate variables) with respect to expectations. Karr et al., (1987) defined biological integrity as "the ability to support and maintain … a balanced, integrated, adaptive community of organisms having a species composition, diversity, and functional organization comparable to that of natural habitat of the region." This definition is problematic for estuaries given their complexity, open boundaries, and almost universal lack of "natural habitat." "Condition" is a more general, less value-laden term used in some publications (e.g., U.S. EPA, 2001; Vølstad et al., 2003) in reference to the status of ecosystems. Condition, along with the first definition of health ("the state of an organism … at any given time"), implies a continuum, whereas the other definitions — along with the definition of integrity — are categorical: one is healthy or not; one has integrity or does not. Each of these words can have inappropriate connotations for nonspecialists. Integrity has moral implications. Health can be consciously or subconsciously associated with diseases and toxicity. Condition may suggest negative status unless modified with positive adjectives (good condition, excellent condition).

One of the problems with terminology is the attempt to express several properties in a single word. Chesapeake Bay Program (CBP, 1987) policy-makers wrote, "the entire system must be balanced, healthy, and productive." This statement contains more information than the individual terms discussed above, and reaches the center of what we are trying to express with ecosystem indicators. "Balanced,"

TABLE 30.1

Definitions of the Terms Balanced, Healthy, and Productive, as They Apply to Estuarine Ecosystems

---

Balanced

"Having sufficient populations of prey species to support the species at the top of the food chain, and to limit overabundance at the bottom of the food chain; no major function of the ecosystem dominates the others"

Healthy

"Having diverse populations that fluctuate within acceptable bounds; free from serious impacts of toxic contaminants, parasites, and pathogens; having sufficient habitat to support a diversity of species"

Productive

"Providing sufficient production of harvestable products to serve human needs without depleting predator and grazer populations to the point where internal, functional balance is disrupted"

Source: CBP (1993).

"healthy," and "productive" were specifically defined in the context of the Chesapeake restoration in a later report (Table 30.1). The phrase "the entire system" is also important, suggesting a need for indicators that apply simultaneously and comprehensively to the whole system in all its vastness and complexity.

Apparently, it is a relatively new idea to capture the essence of large, complex ecosystems in a single indicator or small set of indicators. Only in recent years have monitoring programs begun to generate the necessary data, and then only for a select, few systems. We should expect that less ambiguous terminology will arise, but for now, health, integrity, and condition are used as mostly interchangeable descriptors of what we are attempting to quantify with these indicators. In this chapter, we use the term integrity, more for consistency with recent literature than by preference.

Conceptual Models

Conceptual models are fundamental to indicator development (Boyle et al., 2001). Conceptual models of ecosystems can have many forms, ranging from minutely detailed flow diagrams based on energy or materials, to highly aggregated box models, to qualitative descriptions of expectations or values. Typically, conceptual models portray ecosystems as arrangements of interacting parts (molecules, species, trophic guilds, landscape, or seascape mosaics). These concepts lead naturally to indicators based on a suite of interactions, or relationships between inputs and outputs. There is an alternative route to indicators, as portrayed in the upper oval in Figure 30.2. Rather than tracing the effects of multiple stressors through intricate pathways, it is possible to integrate system responses into an indicator of ecosystem integrity. Relationships of magnitude and variation between stressors and the indicator, and its component responses can be used to make causal inferences and form testable hypotheses, if those are our concerns.

A useful type of conceptual model considers the ecosystem as a unitary whole that responds predictably to stress (Figure 30.3) and changes over time (Figure 30.4). In this conceptual mode, we ask not how system components interact, but how best to represent the system as a whole. That is, what indicator or set of indicators accurately tracks changes in the integrity of the ecosystem? How does one quantify the response axis of Figure 30.3 and Figure 30.4? The response variable, an indicator of ecosystem integrity, should answer important, comprehensive questions about ecosystems:

What is the status of the ecosystem with respect to one or more reference points?

What is the current direction of change?

How will the ecosystem change in response to external forces, especially management actions?

How long will it take to reach a desired or stipulated level of integrity?

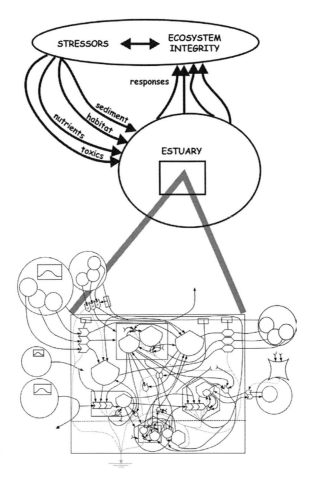

FIGURE 30.2 The contrast between process-oriented models and holistic ecosystem analysis. Indicators of ecosystem integrity are derived from integrated system responses as depicted in the upper part of the diagram. (The flow diagram at the bottom was adapted, with permission, from an ecosystem model developed by Dan Campbell.)

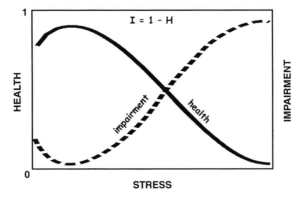

FIGURE 30.3 Conceptual model of the relationship between ecosystem integrity and stress on the ecosystem. On a relative scale, health is the complement of impairment.

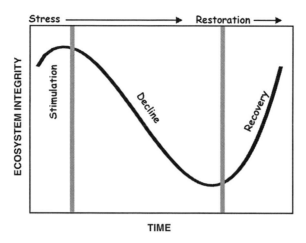

FIGURE 30.4 Integrity of a stressed ecosystem in the time domain (see also Cairns et al., 1992). (From Jordan, S. J. and P. A. Vaas. 2000. *Environmental Science and Policy* 3:S59–S88. With permission.)

Examples of Indicators

The following examples of indicators begin with physical and chemical measurements, and then proceed according to the hierarchy depicted in Figure 30.1, ending with an example at the ecosystem-level of organization. At each level, we discuss some of the merits and concerns associated with the indicators.

Water and Sediment Quality Indicators

Water quality is the most traditional indicator of estuarine condition. The ecological and aesthetic problems associated with anoxia, turbidity, eutrophication, and bacterial pollution are obvious even to nonscientists (dead fish, brown water, excessive algal growth, human illnesses). Even though these problems have been recognized in some estuaries for many decades, the development of water quality criteria, standards, and pollutant load capacities specific to estuaries is in its infancy. The implications of using a suite of water quality indicators for the ecological integrity of estuaries is not always clear. The U.S. Environmental Protection Agency (U.S. EPA) Environmental Monitoring and Assessment Program (EMAP) developed indicators of water quality based on dissolved oxygen, chlorophyll *a*, nutrient concentrations, and water clarity to assess eutrophication. Although EMAP is designed to make ecological assessments over large spatial and temporal scales, the data are used to determine condition for geographic regions rather than ecosystems. Water quality measurements cannot stand alone as indicators of the integrity of an entire ecosystem, because they merely "brush the surface" of biological integrity and ecosystem value. We note that EMAP also samples fish and benthic communities, and employs indicators of biological community structure and health in its assessments.

As with water quality, indicators of sediment quality offer only a piece of the puzzle. Many sediment contaminants are persistent and may be associated with degraded benthic communities, but no criteria for contaminants in estuarine sediments have been established, only guidelines (Long and Morgan, 1990). Contaminants interact with sediment constituents in ways that can greatly affect their biological availability; thus, sediment concentrations may not be directly correlated with toxicity and biological effects (Bayne et al., 1985). Recent sediment research in estuaries has focused not on contaminants, but on organic constituents (e.g., pollen and diatom frustules) in benthic cores as paleological records of ecosystem responses to anthropogenic stressors (Bianchi et al., 2000; Dell'Anno et al., 2002).

The National Coastal Condition Report (U.S. EPA, 2001) combined indicators of water quality, sediment quality, fish tissue contaminants, wetland loss, and benthic communities into a single index to determine the overall condition of the nation's estuaries. This method improved multi-indicator integration, but its lack of a strong biological basis limited its interpretation with respect to ecosystem integrity.

Nevertheless, the application of this index at regional and national scales was an improvement in integrated assessments.

Single-Species (Population-Level) Indicators

Abundance or life history traits (recruitment, growth, mortality) of widely distributed, abundant species, or perhaps less abundant, environmentally sensitive species, could be candidates for indicators of ecosystem health. This idea is particularly attractive when the species is well known to the public, for example, popular sport fishes or important commercial species. Bortone (2003) discussed the potential of spotted seatrout (*Cynoscion nebulosus*) as an indicator species for southeastern estuaries. Because of their popularity and depleted status, striped bass (*Morone saxatilis*) and Eastern oysters (*Crassostrea virginica*) became *de facto*, although scientifically unreliable, indicators of the health of Chesapeake Bay.

Striped bass recruitment in the Chesapeake displayed a long-term pattern of increase, followed by a precipitous decline, followed by dramatic recovery (Figure 30.5). This entire dynamic ultimately could be explained by changes in fisheries and fishery management, but the pattern was also curiously similar to the conceptual model of ecosystem stimulation, decline, and recovery shown in Figure 30.4. A senior manager once asserted that the striped bass population was the only indicator needed to track the Chesapeake restoration — if the fish recovered, the system would recover. Subsequently, the striped bass population recovered within a decade after stringent fishery management controls were initiated, but the ecosystem remained in many ways far from its desired state. For example, the extent of hypoxic and anoxic water did not decline; seagrass coverage, although increasing, was far less than stipulated by restoration goals; and the long-term decline of oyster populations continued (CBP, 2003).

These few of many possible examples of single-species indicators illustrate two points. First, recovery of a "flagship" species can be encouraging, but nevertheless may occur within a system that is far out of balance or far from its desired state. Second, fisheries and fishery management are integral components of estuarine ecosystems; they should not be seen as apart from or irrelevant to concerns about water quality, habitat conditions, and diversity.

Community Indicators

Community-level analysis may be indicative of environmental change caused by single or multiple stressors, and predictive of consequences at the ecosystem level. According to Attrill and Depledge (1997), community-level investigations are ecologically relevant because changes in communities can be extrapolated to the health of the ecosystem through changes in food web structure. Seagrasses (submerged aquatic vegetation [SAV]) and benthic macrofauna are the most prevalent community-based

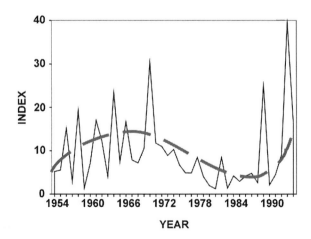

FIGURE 30.5 Maryland index of juvenile striped bass relative abundance 1954–1995. (From Maryland DNR, 2003.)

indicators of estuarine ecosystem health. Long-term changes in seagrass coverage in several estuaries have been strongly associated with nutrient loading and its indirect effects on the availability of light to the plants. This light limitation may result in shifts from SAV-dominated production to proliferation of phytoplankton and macroalgae (McClelland and Valiela, 1998). Long-term losses and gains in seagrasses rank highly as indicators of eutrophication, but also have some drawbacks as comprehensive indicators of ecosystem integrity. They are vulnerable to physical disturbances (boating, dredging, and commercial fishing operations), diseases, major storms, and other climatic extremes; thus, interpretation of their distribution and abundance can be ambiguous.

Infaunal macrobenthic communities have been attractive as indicators largely because of their lack of mobility. This trait makes them reliable indicators of exposure, and susceptible to stressors such as toxic contaminants and severe hypoxia. The structure of these communities, however, is sensitive to factors not directly related to ecosystem integrity such as sediment grain size and organic content. The patchiness of benthic communities over very small spatial scales also can be a drawback in assessing ecosystem integrity over large areas. Ranking and categorical reduction of the data (methods for generalizing almost any indicator) have been applied to minimize the problem of patchiness at any scale (U.S. EPA, 2001).

Indicators based on fish communities or assemblages have received less attention than seagrasses or benthos. The impracticality of complete sampling of estuarine fish communities, along with the migratory behavior of many species are universal difficulties. There have been several attempts to develop estuarine indices of biotic integrity (IBI; Karr et al., 1987) analogous to those used in freshwater systems. Examples can be found in Hughes et al. (2002) and Jordan et al. (1991). Although IBI approaches are feasible, IBIs for estuarine systems tend to be specific to particular habitats, showing less spatial generality and sensitivity to multiple stressors than might be desired.

Vaas and Jordan (1990) used long-term data from seine surveys in the Maryland portion of Chesapeake Bay (Maryland DNR, 2003) to indicate changes in ecosystem integrity. They portrayed graphically 3-year means of relative abundance at intervals of decades. A simple model based on management goals and covariation among species predicted future community structure (Figure 30.6). The tolerance groupings shown in Figure 30.6, developed by Vaas and Jordan (1990) from life history information and cluster analysis of the seine data, showed an interesting and by now familiar pattern when further synthesized for this chapter (Figure 30.7).

Community indicators are generally more robust than single-species indicators, because they integrate responses over broader sectors of the ecosystem and a wider range of environmental influences. Fish community indicators, for example, would be less sensitive to a single fishery management decision than the single-species indicators described above.

Ecosystem Indicators

Indicators at the ecosystem level of organization integrate data over biotic communities and trophic levels, and may include abiotic components (e.g., measures of water quality and physical habitat). Indices such as mean trophic level, system-level trophic transfer efficiency, capacity, ascendancy, and overhead have been used to characterize and compare estuarine ecosystems (e.g., Baird and Ulanowicz, 1989). These types of indices are based on flows of materials (usually carbon) or energy within and through the system. A stressed ecosystem, for example, might exhibit lower values for mean trophic level, transfer efficiency, and ascendancy, along with higher overhead, than an unstressed system. In simpler terms, a stressed system would be less organized and less efficient, while dissipating energy and materials more rapidly (relative to their supply) than an unstressed system. Such indices are valuable tools for gaining understanding about the status and relative functions of ecosystems. Their principal weakness is that they are mathematical abstractions.

Multivariate analysis of ecosystem attributes has been used to organize environmental data. The spatial relationships of these attributes can then be used to develop a relative indication of ecosystem integrity. Jordan and Vaas (2000) used cluster analysis of 12 metrics (selected by screening many candidate metrics) in developing an index of ecosystem integrity for Chesapeake Bay tributaries (Table

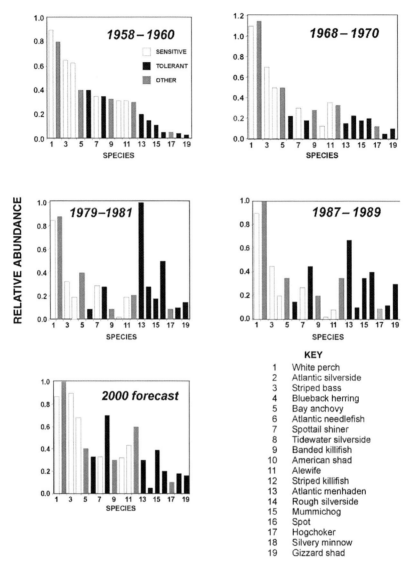

FIGURE 30.6 Observed and predicted changes in relative abundance of 19 species of fish from Maryland Chesapeake Bay seine surveys. Data in the top four bar graphs are 3-year averages. Note the apparent disruption of the community in the two middle graphs, and the predicted partial recovery in 2000. (Adapted from Vaas and Jordan, 1990.)

30.2). The analysis included a broad array of metrics representing communities and trophic levels ranging from phytoplankton and SAV to fish. The index included four water quality metrics in addition to the biotic variables. Six of the metrics were normalized to restoration goals previously established by the Chesapeake Bay Program, so that both societal and ecological values were represented. The final index was constructed by ranking six clusters of observations (spanning 9 years and 40 to 50 sites) into an ordinal scale of ecosystem integrity (Figure 30.8). The index was sensitive to watershed proportions of land cover: urban land predicted low ecosystem integrity, and forested land predicted high ecosystem integrity. The index was further collapsed into three nominal categories of ecosystem integrity by calculating the mean cluster value for each site over 9 years, and assigning "good," "fair," and "poor" designations to the each third of the distribution of means. This procedure produced a simple display of long-term, large-scale geographic patterns (Figure 30.9).

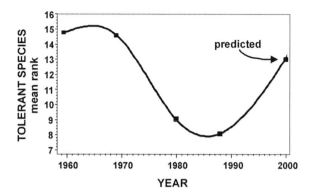

FIGURE 30.7 Mean rank abundance of tolerant species in Maryland Chesapeake Bay seine surveys, 1960–2000. Lower values indicate higher abundance of tolerant species and lower ecosystem integrity.

TABLE 30.2

Metrics Used to Construct an Index of Ecosystem Integrity for Chesapeake Bay

Metric	Ref.
Submersed Aquatic Vegetation Success	
Percentage of potential habitat vegetated (+)	Batiuk et al., 1992
Deviations from Water Quality Goals	
Dissolved inorganic nitrogen ($NO_2 + NO_3 + NH_4$) (−)	Batiuk et al., 1992
Dissolved inorganic phosphorus (−)	Batiuk et al., 1992
Secchi depth (+)	Batiuk et al., 1992
Chlorophyll *a* (−)	Batiuk et al., 1992
Plankton	
Biomass of nuisance algal species (cyanophytes and dinoflagellates) (−)	Jordan and Vaas, 2000
Ratio of mesozooplankton to microzooplankton abundance (+)	Buchanan et al., 1993
Biomass of microzooplankton (−)	Buchanan et al., 1993
Fish	
Trophic index (+)	Jordan and Vaas, 2000
Number of fish species in bottom trawl (+)	Carmichael et al., 1992
Benthos	
Benthic Restoration Goal Index (+)	Ranasinghe et al., 1993
Dissolved Oxygen	
Percentage of observations < 1 mg/L in bottom waters or < 5 mg/L in above pycnocline waters (−)	Waters-Jordan et al., 1992

Note: Plus (+) and minus (−) signs indicate positive or negative relationship to ecosystem integrity.
Source: Adapted from Jordan and Vaas (2000).

Discussion

Ecosystem integrity is a human construct that defies rigorous scientific definition. To assess the integrity of an ecosystem requires reference points defined by humans, who lack perfect knowledge of the system's structure and functions. Because we cannot determine from first principles "what is a good or bad ecosystem," we ask reasonable people what they want from the system. For Chesapeake Bay, the desired

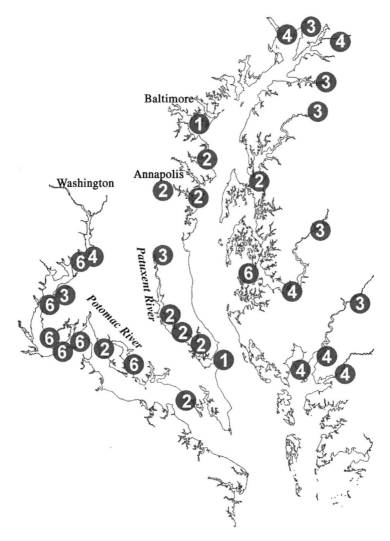

FIGURE 30.8 An index of ecosystem integrity based on cluster analysis of 12 metrics (Table 30.2). Higher values represent higher ecosystem integrity. The year 1991 is shown as an example; the analysis spanned 1986–1994. (From Jordan, S. J. and P. A. Vaas. 2000. *Environmental Science and Policy* 3:S59–S88. With permission.)

states are health, balance, and productivity (CBP, 1987), as defined in the introduction to this chapter. A more recent document (CBP, 2000) explicitly acknowledged the human and conceptual elements in calling for "a shared vision — a system with abundant, diverse populations of living resources," and reiterated the need for the entire system to be healthy and productive.

The examples of indicators presented here all reflect one or more elements of balance, health, or productivity. The striped bass index is a strong indicator of productivity, but tells little about health or balance. In fact, as the Chesapeake striped bass population recovered and became very abundant in the late 1990s and early 2000s, it became less healthy, displaying consistent, abnormally high prevalence of starvation and disease (Jacobs et al., 2002). Perhaps in addition to the obvious health implications, these problems were symptomatic of an unbalanced condition, with overproduction of an important predator.

The Chesapeake Bay fish community indicators include elements of balance (species dominance), productivity (abundance), and, less directly, health (tolerance groups). They lack reference to values or reasoned expectations, however. Instead, they depend on an internal historical reference (the structure of the community ca. 1960). The pristine condition could not be used as a reference because it was

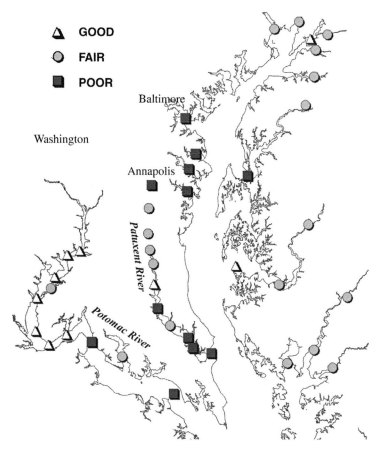

FIGURE 30.9 Long-term ecosystem integrity of northern Chesapeake Bay. Mean cluster values (1–6; see Figure 30.8) for each site over 9 years (1986–1994) were collapsed into three categories of ecosystem integrity. (From Jordan, S. J. and P. A. Vaas. 2000. *Environmental Science and Policy* 3:S59–S88. With permission.)

unknown. Where sufficient data exist, historical references are better than none; they do indeed represent values when society seeks to restore overly stressed ecosystems to former, less-stressed conditions.

The index of ecosystem integrity for Chesapeake Bay (Jordan and Vaas, 2000) includes indicators of balance, health, productivity, and diversity. Moreover, it is based on values, as 6 of 12 component metrics are normalized to specific, numerical restoration goals for water quality and biotic communities. Its weaknesses include problems with missing data and fixed-station sampling designs that lead to difficulties in statistical and geographical inferences. The potential of applying the index in a predictive mode has not been explored fully, although the strong relationship with land cover could be a starting point for predictive analysis.

The scarcity of data for most estuaries is a major impediment to the development and application of ecosystem-level indicators, which require extensive, comprehensive long-term monitoring programs. The efforts of the National Coastal Assessment (U.S. EPA, 2001) to establish consistent monitoring and universal indicators for U.S. estuaries are improving the quantity, quality, and availability of data relevant to ecosystem integrity. The program would provide more support for indicators of ecosystem integrity by monitoring a broader spectrum of biological responses (plankton and aquatic vegetation in addition to fish and benthic communities). Longevity will be crucial. In large estuarine systems, ecosystem-level responses to changes such as contaminant load reductions or land management practices can take years or decades to become detectable. Therefore, monitoring programs must be sustained indefinitely. The fish community indicators described here were possible only because the Maryland seine surveys generated a consistent record extending over several decades. The longevity of a monitoring program is strongly related to the potential for robust indicators and predictive models. Taken together, Figure 30.5 through Figure 30.7 illustrate the relationships between short-term interannual variability, which could only be interpreted as noise, and long-term patterns that contain useful information.

Conclusions

- Whole-system indicators of ecosystem integrity for estuaries can be developed and applied successfully.
- These types of indicators require conceptual models that view the ecosystem as a unit changing with time, rather than a static assemblage of parts.
- Balance, health, productivity, and diversity are important elements of ecosystem-level indicators.
- Indicators should include information about human values and societal goals; numerical goals provide reference points to which data can be normalized.
- Indicators of ecosystem integrity have far lower precision than indicators at lower levels of organization, but can support robust, accurate predictions.
- Long-term, consistent, comprehensive monitoring programs are essential to ecosystem-scale indicators and analysis.

Acknowledgments

We thank the many people and organizations that contributed their work, data, thoughts, and resources to the ideas and examples presented in this chapter. Naming them all would require yet another chapter. The information in this document has been funded in part by the U.S. Environmental Protection Agency. It has been subjected to review by the National Health and Environmental Effects Research Laboratory and approved for publication. Approval does not signify that the contents reflect the views of the agency, nor does mention of trade names of commercial products constitute endorsement or recommendation for use. This is Contribution 1194 of the Gulf Ecology Division.

References

Attrill, M. J. and M. H. Depledge. 1997. Community and population indicators of ecosystem health: targeting links between levels of biological organisation. *Aquatic Toxicology* 38:183–197.

Baird, D. and R. E. Ulanowicz. 1989. The seasonal dynamics of the Chesapeake Bay ecosystem. *Ecological Monographs* 59:329–364.

Batiuk, R. P., R. Heasley, R. Orth, K. Moore, J. Capelli, J. C. Stevenson, W. Dennison, L. Staver, V. Carter, N. Rybicki, R. E. Hickman, S. Kollar, and S. Beiber. 1992. Chesapeake Bay Submerged Aquatic Vegetation Habitat and Restoration Goals: A Technical Synthesis. U.S. Environmental Protection Agency Chesapeake Bay Program Report, Annapolis, MD.

Bayne, B. L., D. A. Brown, K. Burns, D. R. Nixon, A. Ivanovici, D. R. Livingstone, D. M. Lowe, M. N. Moore, A. R. D. Stebbing, and J. Widdows. 1985. *The Effects of Pollution on Marine Animals*. Praeger Scientific, New York, 384 pp.

Bianchi, T., P. Westman, C. Rolff, E. Engelhaupt, T. Andrén, and R. Elmgren. 2000. Cyanobacterial blooms in the Baltic Sea: natural or human-induced? *Limnology and Oceanography* 45:716–726.

Bortone, S. A. (ed). 2003. *Biology of the Spotted Seatrout*. CRC Press, Boca Raton, FL.

Boyle, M., J. J. Kay, and B. Pond. 2001. Monitoring in support of policy: an adaptive ecosystem approach. In *Encyclopedia of Global Environmental Change,* Vol. 4, T. Munn (ed.). John Wiley & Son, New York, pp. 116–137.

Buchanan, C., R. W. Alden III, R. S. Birdsong, F. Jacobs, and K. G. Sellner. 1993. Development of Zooplankton Community Environmental Indicators for Chesapeake Bay: A Report on the Project's Results through June 1993. Interstate Commission on the Potomac River Basin, Rockville, MD.

Cairns, J., J. R. McCormick, V. Paul, and B. R. Niederlehner. 1992. A proposed framework for developing indicators of ecosystem health. *Hydrobiologia* 263:1–44.

Carmichael, J. B., B. Richardson, M. Roberts, and S. J. Jordan. 1992. Fish Sampling in Eight Chesapeake Bay Tributaries. Technical Report, Chesapeake Bay Research and Monitoring Division, Maryland Department of Natural Resources, Annapolis.

CBP. 1987. Chesapeake Bay Agreement. Chesapeake Bay Program, Annapolis, MD. Available at http://www.chesapeakebay.net/pubs/1987ChesapeakeBayAgreement.pdf.

CBP. 1993. Chesapeake Bay Strategy for the Restoration and Protection of Ecologically Valuable Species. Chesapeake Bay Program CBP/TRS 113/94, Annapolis, MD.

CBP. 2000. Chesapeake 2000: A Watershed Partnership. Chesapeake Bay Program, Annapolis, MD. Available at http://www.chesapeakebay.net/agreement.htm.

CBP. 2003. Bay Trends and Indicators. Chesapeake Bay Program Web page: http://www.chesapeakebay.net/indicators.htm.

Davies, P. (ed.) 1976. *The American Heritage Dictionary of the English Language.* Dell, New York.

Dell'Anno, A., M. L. Mei, A. Pusceddu, and R. Danovaro. 2002. Assessing the trophic state and eutrophication of coastal marine systems: a new approach based on the biochemical composition of sediment organic matter. *Marine Pollution Bulletin* 44:611–622.

Hughes, J. E., L. A. Deegan, M. J. Weaver, and J. E. Costa. 2002. Regional application of an index of estuarine biotic integrity based on fish communities. *Estuaries* 25:250–263.

Jacobs, J., M. Matsche, S. Jordan, C. Driscoll, H. Speir, B. Kibler, T. Litwiler, and S. Knowles. 2002. Striped Bass Health. Maryland Department of Natural Resources, Oxford. Available at http://www.dnr.state.md.us/fisheries/oxford/research/fwh/stripedbass/index.html.

Jordan, S. J. and P. A. Vaas. 2000. An index of ecosystem integrity for Northern Chesapeake Bay. *Environmental Science and Policy* 3:S59–S88.

Jordan, S., P. Vaas, and J. Uphoff. 1991. Fish assemblages as indicators of environmental quality in northern Chesapeake Bay. In Biological Criteria: Research and Regulation 1990. Proceedings of a Symposium. U.S. Environmental Protection Agency Office of Water, Washington, D.C.

Jordan, S. J., C. Stenger, M. Olson, K. Mountford, and R. Batiuk. 1992. Dissolved Oxygen Restoration Goals for the Chesapeake Bay Living Resources. U.S. Environmental Protection Agency Chesapeake Bay Program, Annapolis, MD.

Karr, J. R., P. R. Yant, and K. D. Fausch. 1987. Spatial and temporal variability of the index of biotic integrity in three midwestern streams. *Transactions of the American Fisheries Society* 116:1–11.

Long, E. R. and L. G. Morgan. 1990. The Potential for Biological Effects of Sediment-Sorbed Contaminants Tested in the National Status and Trends Program. NOAA Technical Memorandum NOS OMA 52, Rockville, MD.

Maryland DNR. 2003. Striped Bass Seine Survey Juvenile Index Page. Available at http://www.dnr.state.md.us/fisheries/juvindex/index.html.

McClelland, J. W. and I. Valiela. 1998. Changes in food web structure under the influence of increased anthropogenic nitrogen inputs to estuaries. *Marine Ecology Progress Series* 168:259–271.

Murawski, S.A. 2000. Definitions of overfishing from an ecosystem perspective. *ICES Journal of Marine Science* 57:649–658.

Ranasinghe, J. A., S. B. Weisburg, J. B. Frithsen, L. C. Schaffner, R. J. Diaz, and D. M. Dauer. 1993. Chesapeake Bay Benthic Community Restoration Goals. Technical Report, prepared for The Chesapeake Bay Program Office and The Governor's Council on the Chesapeake Bay. Versar, Inc., Annapolis, MD.

Rapport, D. J. 1999. On the transformation from healthy to degraded aquatic ecosystems. *Aquatic Ecosystem Health and Management* 2:97–103.

U.S. EPA. 2001. National Coastal Condition Report. U.S. EPA Office of Research and Development, EPA-620/R-01/005, Washington, D.C. Available at http://www.epa.gov/owow/oceans/NCCR/index.html.

Vaas, P. A. and S. J. Jordan. 1990. Long term trends in abundance indices for 19 species of Chesapeake Bay fishes: reflection of trends in the bay ecosystem. In New Perspectives in the Chesapeake System: A Research and Management Partnership. Proceedings of a Conference. Chesapeake Research Consortium Publication 137.

Vølstad, J. H., N. K. Neerchal, N. E. Roth, and M. T. Southerland. 2003. Combining biological indicators of watershed condition from multiple sampling programs — a case study from Maryland, USA. *Ecological Indicators* 3:13–025.

Welsh, B. L. 1984. Foreword. In *The Estuary as a Filter*, V. S. Kennedy (ed.). Academic Press, New York, 511 pp.

31

Using the Human Disturbance Gradient to Develop Bioassessment Procedures in Estuaries

Ellen McCarron and Russel Frydenborg

CONTENTS

Introduction

Storm water and other non-point sources of pollution are major contributors to the degradation of surface water and groundwater resources (FDEP, 1999). In recognition of this disturbing trend, in the early 1980s the Florida Department of Environmental Protection (FDEP) developed and implemented what is now a nationally recognized program to manage non-point source pollution. One of the main difficulties facing this newly formed program was the lack of appropriate techniques to monitor and assess surface water impairment. The traditional chemistry-based analysis of water samples was inadequate because of the inherently transient and unpredictable nature of non-point pollution.

An alternative approach, developed at the Department in the late 1980s and early 1990s, provided an ecologically based solution to the problem of monitoring non-point sources. The solution was the development and implementation of community-level biological assessment (bioassessment) tools using multiple metrics (a multimetric index). Each of these metrics is an attribute that responds in a predictable manner to human disturbance.

The conceptual approach for developing a multimetric index was proposed by Karr and Chu (1999). Completing this process for streams in Florida has taken well over a decade. Resource limitations, as well as the nature of emerging national guidance from the U.S. Environmental Protection Agency (U.S.

EPA) on bioassessment and biocriteria development, forced the FDEP to build its bioassessment tools slowly (McCarron and Frydenborg, 1997).

Following the national models, the FDEP has made steady progress in developing bioassessment tools for different water body types. At present, bioassessment tools for streams (i.e., the Stream Condition Index and BioReconnaissance, or BioRecon) and lakes (the Lake Condition Index) are completed and are well into the implementation phase in many of the FDEP monitoring programs, while the wetland tools are nearing completion.

Florida's Total Maximum Daily Load Program is one of several high-profile programs in the Department that uses the stream and lake bioassessment tools. Recently adopted rules for this program (Rule 62-303, Florida Administrative Code) set forth objective criteria for determining impairment and are used to determine which waters will be placed on the state's official list of impaired waters that is submitted to the federal government for approval. The rules incorporate the Stream Condition Index, BioRecon, and Lake Condition Index as measures of surface water quality impairment. By including these bioassessment measures in Rule 62-303, the FDEP has greatly enhanced its ability to detect impairment and restore waters in need of remediation, through its Total Maximum Daily Load Program.

The development of a bioassessment tool for estuaries has long been one of the FDEP's goals. Because of the inherent complexities of estuarine systems and the lack of any proven national models, however, an estuarine tool that is applicable statewide has not been developed. The recent use of a Human Disturbance Gradient approach to recalibrate the Stream Condition Index could potentially bring renewed energy to the development of estuarine bioassessment procedures.

The following section on methods describes how the stream bioassessment tool was recalibrated using the Human Disturbance Gradient approach. The discussion and conclusions sections explore how the Human Disturbance Gradient theory underlying this important stream recalibration project could provide a potential framework for developing a bioassessment approach in estuarine systems.

Methods

Following national guidance from the U.S. EPA (Plafkin et al., 1989) in the early 1990s two stream macroinvertebrate multimetric bioassessment tools were developed in Florida. These tools, the Stream Condition Index and BioRecon, were developed, tested, and implemented by comparing the biological condition at sites with minimal human influence (reference sites) and sites with known disturbance (test sites) (Barbour et al., 1996).

In accordance with guidance from the U.S. EPA, a geographic framework was also developed for the Stream Condition Index and BioRecon tools, because regional differences translate into different expectations for the biological communities present. For example, there is naturally much higher taxonomic diversity in northern Florida, compared with central and southern Florida. A first attempt at developing a regional geographic framework used the nationally designated ecoregions in Florida (U.S. EPA, 1989) and partitioned them further into subecoregions, using methods similar to those used in mapping the national ecoregions. This project produced a multiuse map of Florida's subecoregions (U.S. EPA, 1994) (Figure 31.1). Biological data were intentionally not used in the development of subecoregions to avoid circularity in the methodology. A statistical evaluation of macroinvertebrate data against the subecoregional framework led to the identification of the three "bioregions" used for the Stream Condition Index and BioRecon assessment tools, i.e., the Florida panhandle, peninsula, and northeast bioregions (Figure 31.2).

The Stream Condition Index and BioRecon have proved useful in the FDEP's assessment of Florida's biological communities. However, the Human Disturbance Gradient approach, which emerged in the early 1990s, offers a greater level of discrimination between biological condition (y axis) and environmental condition (x axis). As a result, the FDEP has just completed contract work with Leska Fore of Statistical Design, Inc. to recalibrate the Stream Condition Index and BioRecon tools using a Human Disturbance Gradient approach. Fore's report (November 2003) discusses in detail the methods used in the Stream Condition Index recalibration. The following sections summarize the Human Disturbance Gradient theory and how it was applied in recalibrating the stream bioassessment tool.

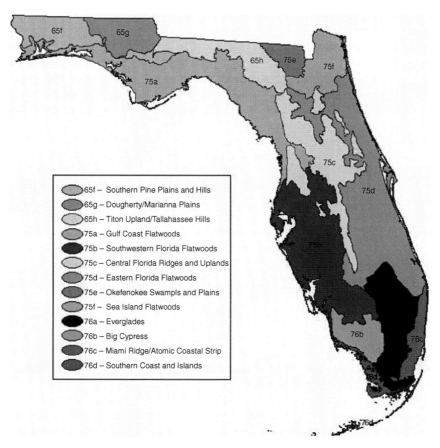

FIGURE 31.1 Subecoregions of Florida.

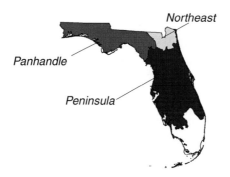

FIGURE 31.2 Bioregions of Florida.

The Human Disturbance Gradient

The Human Disturbance Gradient was first introduced by Karr (1993). This landmark publication built on what Karr and his colleagues described as the five factors that summarize how humans alter and degrade water resources (Karr et al., 1986, 2000): hydrology, physical habitat structure, water quality, energy source, and biological interactions. Data were available in Florida to measure the first four of these five factors, as follows.

Hydrology. A hydrologic scoring system was developed based on FDEP biologists' knowledge of water removal (drainage, consumptive use), patterns of drought, and hydrographs for the specific sites used in the recalibration exercise

Physical habitat structure. Habitat assessment criteria were developed for the FDEP's stream bio-assessments, and a standardized form is always completed for each Stream Condition Index and BioRecon assessment. The habitat assessment score includes substrate condition and availability, water velocity, habitat smothering, channelization, bank stability, and the width and vegetative condition of the riparian zone.

Water quality. Ammonia concentration was used because it had the most complete record of data, has a high correlation with other water quality measures, and is indicative of degradation from a wider range of land uses than other measures.

Energy source. As a direct measure of energy types and quantities was not available, a surrogate was selected that is an index of nonrenewable energy flow in the surrounding catchment, or geographic area. This surrogate, the Land Development Intensity Index, was developed by Dr. Mark Brown and colleagues at the University of Florida (Brown and Vivas, in press). It is calculated as the percentage area within the catchment of particular types of land uses, multiplied by a coefficient of energy associated with that land use, and then summed over all land use types.

Selection of Biological Metrics for the Aggregate Stream Condition Index

To determine the variety of attributes that is reliably associated with human disturbance, 36 candidate biological metrics were evaluated using the following criteria: (1) meaningful measure of ecological structure or function; (2) strong correlation with the Human Disturbance Gradient; (3) statistically robust, with low measurement error; (4) reflecting multiple categories of biological organization; (5) cost-effective to measure; and (6) provide information that is not redundant with other metrics.

The final metrics selected had all of these characteristics. They were then normalized into unitless scores, using the 95th or 5th percentile values for each metric. Each metric value was divided by its range and multiplied by 10. For metrics that decreased with an increasing Human Disturbance Gradient, the values were scored 0 to 10, and for metrics that increased with a Human Disturbance Gradient, the values were scored 10 to 0, so that in either case, the best metric score was a 10.

As described in Fore (2003), this process was carried out separately for each of the three Florida bioregions, and calibration factors were assigned to the metrics as needed to ensure similar responses for all metrics in all regions. The unitless metric scores were then summed into a single Stream Condition Index value for each site. Finally, statistical tests were applied to the data set to determine the number of categories that the Stream Condition Index aggregate index could detect.

Results

Ten metrics were selected for inclusion in the aggregate Stream Condition Index, and most of them required regional scoring calibration (Table 31.1). Also, the FDEP's Human Disturbance Gradient was found to be a reliable predictor of biological condition, as it was highly correlated with the Stream Condition Index (Figure 31.3).

Using the statistical tests described in Fore (2003), there are approximately five assessment categories for two visits at the same site and four categories when only one visit at a site is obtained (Table 31.2). The FDEP's current rules (Rule 62-303, Florida Administrative Code) require two visits to a site to make a definitive assessment using the Stream Condition Index.

Discussion: Developing Bioassessment Criteria for Estuaries

Using the procedure followed for streams as a template, the following steps can be taken to develop effective bioassessment criteria for estuaries: (1) based on knowledge of a region's estuaries, properly

TABLE 31.1

List of Ten Metrics Showing Regional Thresholds

SCI Metric	Northeast	Panhandle	Peninsula
Total taxa	16–42	16–49	16–41
Ephemeroptera taxa	0–3.5	0–6	0–5
Trichoptera taxa	0–6.5	0–7	0–7
% Filterer	1–42	1–45	1–40
Long-lived taxa	0–3	0–5	0–4
Clinger taxa	0–9	0–15.5	0–8
% Dominance	54–10	43–10	54–10
% Tanytarsini	0–26	0–26	0–26
Sensitive taxa	0–11	0–19	0–9
% Very tolerant	78–0	36–0	59–0

Source: Fore (2003).

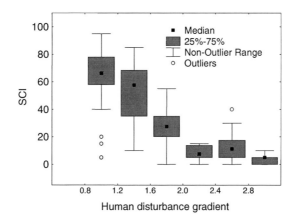

FIGURE 31.3 Stream Condition Index vs. Human Disturbance Gradient. (From Fore, 2003.)

classify these estuaries into meaningful units; (2) develop criteria for a Human Disturbance Gradient for estuaries; (3) establish the reference condition, or the biological expectations within each classified unit, based on the Human Disturbance Gradient; (4) sample the biota at sites representing conditions across the Human Disturbance Gradient, using the same methods and taxonomic resolution; (5) examine biological attributes associated with estuarine community structure, composition, and life history; (6) determine a variety of attributes that is reliably associated with human disturbance (see the earlier discussion on the selection of biological metrics for the Stream Condition Index); and (7) use this information to assess the quality of an estuarine site in a manner that is easily communicated to stakeholders and policy makers (e.g., the multimetric index approach).

Classifying Estuaries — Developing Reasonable Expectations and Making Valid Comparisons

The appropriate classification of estuarine sites is extremely important for establishing biological community expectations. Following the Karr and Chu (1999) model, geographic setting should be the first element in the classification scheme. Florida contains three marine provinces: the Louisianan, West Indian, and Carolinian. Climate, prevailing currents, the availability of specific habitats, and organism recruitment broadly influence the composition of biological communities in these provinces.

Habitat type is the next logical element in developing a classification system for bioassessment. Marine habitats are generally separated into floral-based, faunal-based, and mineral-based communities (Figure 31.4). Obviously, the biota inhabiting a bed of submerged aquatic vegetation (a floral-based

TABLE 31.2

Category Names, Ranges of Values for Stream Condition Index (SCI), and Example Descriptions of Biological Conditions Typically Found for That Category

SCI Category	SCI Range	Description
One Sample		
Good	73–100	Similar to natural conditions, up to 10% loss of taxa expected
Fair	46–73	Significantly different from natural conditions; 20–30% loss of Ephemeroptera, Trichoptera, and long-lived taxa; 40% loss of clinger and sensitive taxa; percentage of very tolerant individuals doubles
Poor	19–46	Very different from natural conditions; 30% loss of total taxa; Ephemeroptera, Trichoptera, long-lived, clinger, and sensitive taxa uncommon or rare; filterer and Tanytarsini individuals decline by half; 25% of individuals are very tolerant
Very poor	0–19	Extremely degraded; 50% loss of expected taxa; Ephemeroptera, Trichoptera, long-lived, clinger, and sensitive taxa missing or rare; 60% of individuals are very tolerant
Two Samples		
Excellent	81–100	Proportion and abundance of taxa similar to natural conditions; minimal loss of taxa
Good	62–81	Similar to natural conditions with up to 10% loss of taxa; 25% loss of Ephemeroptera, Trichoptera, clinger, and sensitive taxa
Fair	43–62	25% loss of total taxa; 50% loss of Ephemeroptera, Trichoptera, clinger, and sensitive taxa; 33% loss of long-lived taxa
Poor	24–43	High percentage of individuals present belong to very tolerant taxa; only tolerant Ephemeroptera, Trichoptera, and clinger taxa present; one sensitive or long-lived taxon may be present
Very poor	0–24	Extremely degraded; 50% loss of expected taxa; Ephemeroptera, Trichoptera, long-lived, clinger, and sensitive taxa missing or rare; 60% of individuals are very tolerant

Note: Statistical analysis indicates that there are five assessment categories for two visits at the same site and four categories when only one visit at a site is obtained. Range of SCI values represent 90% confidence intervals. Square brackets indicate a value is included in the range; round brackets indicate a listed value is not included.

Source: Fore (2003).

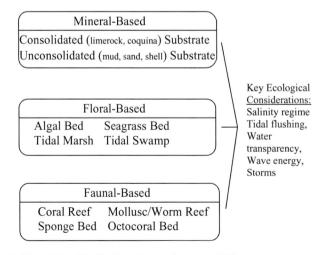

FIGURE 31.4 Marine habitats. (From Florida Natural Areas Inventory, 1990.)

community) differ from those living in a mud flat (a mineral-based community), and so it is important that sampled habitats are controlled to assure valid comparisons. If a mineral-based community such as unconsolidated substrate is the sampling unit, the percentage of organic matter and grain size must be similar when comparing test with reference conditions. Floral- or faunal-based marine communities such as seagrass beds, coral reefs, mollusk/worm reefs, sponge beds, and octocoral beds are more characteristic of fairly transparent waters (e.g., greater than 3-m Secchi depth). Floral-based tidal marsh and tidal swamp communities are usually distributed in shallow, lower wave-energy areas.

Finally, the salinity regime is extremely important in estuarine systems. Nearshore estuarine environments are dynamic, making the process of identifying the typical salinity regime challenging. Salinity is often low during periods of large freshwater inputs and relatively high when offshore water is prevalent. If salinity fluctuates widely at a given site, it may partially mask adverse human influences. Because salinity exerts an overriding influence on the distribution of biological communities, monitoring is critical to establishing expectations for an estuarine system's biological communities. Low-salinity systems are generally more depauperate in terms of species richness than are areas with a higher salinity regime. In addition, the salinity regime preceding a sampling event greatly affects the types of organisms collected. For bioassessment purposes, it is critical to classify reference condition expectations according to the antecedent, or preceding, salinity regime.

Determining the Potential Elements of an Estuarine Human Disturbance Gradient

Because estuaries are complex, with water flowing in variable directions from the effects of tides and winds, careful thought should be given to the potential components of a Human Disturbance Gradient and their application to estuaries. These components include the following:

Land use. A land use–based component, similar to the Landscape Development Intensity Index used in the Florida Stream Condition Index recalibration, may be an effective element if water current patterns in an estuary are known. Some estuarine areas may receive runoff from human land uses, while other, adjacent portions may receive water currents from offshore or undisturbed areas. Therefore, it is important that the Landscape Development Intensity Index value assigned to a particular estuarine zone accurately reflect the associated land use influencing that zone.

Water quality. Water chemistry variables associated with typical human activities are likely components of a human disturbance gradient. Important parameters to include in the evaluation include oxygen-demanding substances, turbidity, nutrients, chlorophyll *a*, and toxic compounds such as metals or pesticides.

Habitat. Habitat quality directly affects biological communities. Estuarine habitat assessment procedures may need to be modified to match a particular habitat type (e.g., transparency may be an element for seagrass meadows, but not for tidal marshes).

Condition of the fishery. "Fishing pressure" is a potentially useful component of a human disturbance gradient. The removal of biomass from estuaries through commercial or sport fishing is widespread. If intensive enough, it could adversely affect predator–prey relationships, with cascading effects throughout trophic levels.

Sediment quality. The FDEP has developed guidelines for assessing sediment quality (MacDonald et al., 2003) for metals and selected organic contaminants. These evaluation techniques would be a logical element of an estuarine human disturbance gradient.

Establishing the reference condition. There are three general ways to establish reference conditions in association with the human disturbance gradient in estuaries, as follows:

1. Using historical data. If historical data are available (i.e., prior to human perturbations), an historical reference condition can be established to define biological expectations for an estuary. This is especially useful when defining systemwide restoration targets. For example, paleolimnologic data indicates that before vast quantities of Everglades marsh water were rerouted away from Florida Bay in the early 1900s, seagrass meadows were rare in portions of the bay, presumably due to naturally lower salinities and reduced light transparency associated with the calcareous sediments (Halley, 2004). During the past 60 years, the bay's seagrasses have expanded, but periodically exhibit catastrophic crashes. Paleolimnologic information does

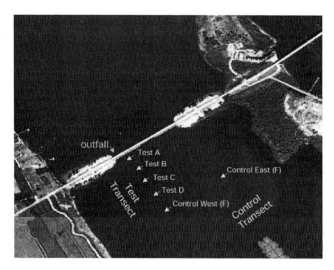

FIGURE 31.5 Potential biological condition disturbance gradient in the Indian River Lagoon.

suggest, however, that the contemporary overall seagrass abundance is higher than conditions prevalent in 1900. Because the current conditions were artificially created, a historically appropriate restoration goal would be to improve the hydroperiod and allow the bay's natural communities to flourish, even if seagrass coverage decreased.

2. Using an unaffected portion of an estuary. In estuaries where the disturbance is confined to a discrete area, the unaffected portions could represent reference conditions. This is especially useful when evaluating the impact of a point source discharge whose effects are confined to a specific area. Figure 31.5 illustrates a potential Human Disturbance Gradient in the Indian River Lagoon system using transects downgradient from a facility outfall that could be used to document biological impairment (Wolfe, 2002). In this example, the FDEP's sediment quality assessment tools were used to ensure that substrate conditions were equivalent in both the test and reference transects, while establishing that a clear chemical disturbance gradient (i.e., contamination from polyaromatic hydrocarbons) was present (Figure 31.6) (Wolfe, 2002).

3. Using an adjacent system. Since many estuaries have been affected by humans, using similar, adjacent estuaries may be useful for establishing the "minimally affected" or reference condition. For example, Florida estuarine researchers have used the Econfina River Estuary as a reference system for comparisons with the Fenholloway system (affected by a point source discharge), have used the Nassau River Estuary to establish expectations for the Amelia (point source discharge) or St. Johns Estuaries (point- and non-point-source inputs), and have used the Matanzas system as a surrogate for the Halifax system (point and non-point source inputs).

Figure 31.7 illustrates how salinity zones in the St. Johns River Estuary could be used to establish parallel reference salinity zones in the adjacent Nassau River Estuary. For bioassessment purposes, it is critical to compare test and reference salinity regimes in adjacent systems appropriately. Using the example in Figure 31.7, test sites with salinity levels of 1 to 5 parts per thousand (ppt) in the St. Johns system should be compared with salinity sites of 1 to 5 ppt in the adjacent "reference" Nassau system. The substrate type must also be similarly measured and compared with an adjacent estuarine system.

Sampling Biota across a Gradient of Human Disturbance

Once the appropriate Human Disturbance Gradient data are available for a variety of sites, it is necessary to sample biological communities from the "best" to "worst" possible condition. This is analogous to establishing a dose–response relationship in toxicity testing (Karr and Chu, 1999). It allows an examination of the community attributes that vary in direct proportion to human disturbance, in the same way

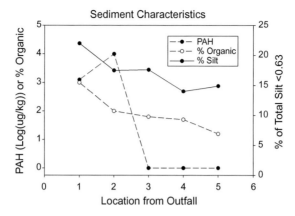

FIGURE 31.6 Documentation of a disturbance gradient (PAH [polyaromatic hydrocarbon] sediment contamination) from a facility outfall in the Indian River Lagoon.

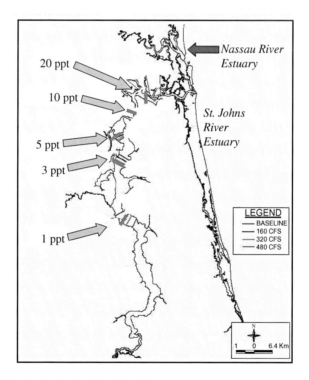

FIGURE 31.7 Salinity isohalines in the St. Johns River Estuary.

that the Stream Condition Index was recalibrated using the Human Disturbance Gradient. For example, in Figure 31.5, Stations A and B, closest to a point source discharge, also have the highest concentration of polyaromatic hydrocarbons in sediment (Figure 31.6). These objective data suggest that the biological differences at these sites, compared with reference conditions, are attributable to human disturbance.

Examining Biological Attributes Associated with Estuarine Community Health

Characteristics of the biological community that should be considered during this process include taxonomic composition (tolerance, keystone species), life history elements (feeding groups, habit),

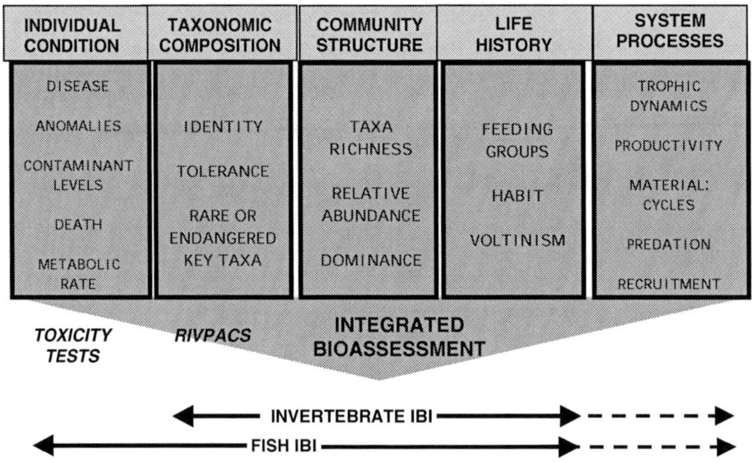

FIGURE 31.8 Attribute groups.

community structure (richness, abundance, dominance), individual organism condition (disease, anomalies), and system processes (productivity, recruitment). Each attribute should be tested against the Human Disturbance Gradient to establish those that effectively distinguish the presence of anthropogenic impairment.

Selecting Community Attributes Associated with Human Disturbance

Figure 31.8 lists categories of elements that provide a variety of nonredundant types of information for assessing a biological community. When formulating an index that summarizes the health of a site, one or two of the strongest metrics from each category should be selected for inclusion in the final index. Other studies have found that the best multimetric biological indexes use 10 to 12 metrics from the above categories (Karr and Chu, 1999).

Communicating the Results Clearly to Stakeholders and Policy Makers

Once a scientifically defensible evaluation of an estuary's condition has been made, restoration activities are unlikely to occur unless both stakeholders and policy makers understand the results and associated ramifications. Reports and presentations should simplify the complexities of ecosystems and clarify the issues into terms appropriate to the target audience (Preston et al., 2004).

Conclusions

The multimetric approach to biological assessment has proved useful in streams, lakes, and wetlands, and it holds great promise for estuaries as well. A number of specific procedures are recommended for developing effective bioassessment tools in estuaries. First, it is necessary to classify estuaries into meaningful units for establishing biological community expectations. The classification criteria should include estuarine geographic province, habitat type (substrate composition), and salinity regime. Next, the development of specific, quantifiable criteria for an estuarine Human Disturbance Gradient is required. Potential elements affected by human activities include land use (the Landscape Development Intensity Index), water quality parameters, habitats, fisheries, and sediment quality. Subsequently, reference conditions, or the biological expectations in each classified estuarine unit, based on the Human Disturbance Gradient, should be established. Frameworks for establishing these reference conditions could include the use of historical data, data from unaffected portions of an estuary, or data from an adjacent, similar estuary. Next, the biota are sampled at sites representing conditions across the Human Disturbance Gradient. This is analogous to establishing a dose–response relationship in toxicity testing, allowing an examination of the biological community attributes that vary in direct proportion to human disturbance. The next task is to analyze a variety of biological attributes in the data that are associated with estuarine community structure, composition, and life history. The strongest measures of human disturbance in each category of community attributes are selected as metrics for the index. The multimetric index is used to summarize the biological condition of an estuary in a manner that is easily communicated to stakeholders and policy makers. This final step is crucial to protecting valuable estuarine resources.

Acknowledgments

The authors express their appreciation to Stephen Bortone, who sponsored and coordinated the 2003 Estuarine Indicators Workshop on Sanibel Island where the material in this chapter was presented; Jim Karr for continuing to mentor the Florida Bioassessment Program and for his lifelong dedication to improving biological monitoring; Leska Fore for her excellent work in recalibrating the Florida Stream Condition Index and BioRecon tools; Eric Livingston for his administrative support in cosponsoring the Sanibel workshop and in making funds available for all of the FDEP's bioassessment research to date; Dave Whiting for helpful discussions and insight; and all of the dedicated FDEP biologists, without whom the Florida Bioassessment Program would not be possible.

References

Barbour, M. T., J. Gerritsen, and J. S. White. 1996. Development of the Stream Condition Index (SCI) for Florida. Florida Department of Environmental Protection Technical Report, Tallahassee, 106 pp.

Brown, M. T. and M. B. Vivas. In press. A landscape development intensity index. *Ecological Monitoring and Assessment.* Available at http://www.dep.state.fl.us/water/surfacewater/docs/nutrient/TAC/tac4_brown-vivas.pdf.

FDEP. 1999. Florida Nonpoint Source Management Program Update. Florida Department of Environmental Protection Bureau of Watershed Management, Tallahassee, 312 pp.

Florida Natural Areas Inventory. 1990. Guide to the Natural Communities of Florida. Florida Natural Areas Inventory and Florida Department of Natural Resources, Tallahassee, 111 pp.

Fore, L. 2003. Draft. Development and Testing of Biomonitoring Tools for Stream Macroinvertebrates in Florida. Florida Department of Environmental Protection Technical Report, Tallahassee, 76 pp.

Halley, R. B. 2004. Florida Bay's Murky Past. U.S. Geological Survey, South Florida Information Access.

Karr, J. R. 1993. Defining and assessing ecological integrity. *Environmental Toxicology and Chemistry* 12:1521–1531.

Karr, J. R. and E. W. Chu. 1999. *Restoring Life in Running Waters: Better Biological Monitoring.* Island Press, Washington, D.C., 206 pp.

Karr, J. R., K. D. Fausch, P. L. Angermeier, P. R. Yant, and I. J. Schlosser. 1986. Assessment of Biological Integrity in Running Water: A Method and Its Rationale. Illinois Natural History Survey Special Publication 5, Champaign, IL.

Karr, J. R., J. D. Allan, and A. C. Benke. 2000. River conservation in the United States and Canada. In *Global Perspectives on River Conservation: Science, Policy, Practice,* P. J. Boon, B. R. Davies, and G. E. Petts (eds.). John Wiley, Chichester, U.K., pp. 3–39.

MacDonald, D. D., C. G. Ingersoll, D. E. Smorong, R. A. Lindskoog, G. Sloane, and T. Biernacki. 2003. Development and Evaluation of Numerical Sediment Quality Assessment Guidelines for Florida Inland Waters. Florida Department of Environmental Protection Technical Report, Tallahassee, 150 pp.

McCarron, E. and R. Frydenborg. 1997. The Florida Bioassessment Program: an agent of change. *Human and Ecological Risk Assessment* 3(6):967–977.

Plafkin, J.L., M. T. Barbour, K. D. Porter, S. K. Gross, and R. M. Hughes. 1989. Rapid Bioassessment Protocols for Use in Streams and Rivers: Benthic Macroinvertebrates and Fish. EPA 440-4-89-001.U.S. Environmental Protection Agency, Office of Water Regulations and Standards, Washington, D.C.

Preston, R. J. Culp, T. DeMoss, R. Frydenborg, E. Kenaga, C. Pittinger, and N. Schofield. 2004. Translating ecological science. In *Ecological Assessment of Aquatic Resources: Linking Science to Decision Making.* SETAC, Pensacola, FL.

U.S. Environmental Protection Agency. 1989. Regionalization as a Tool for Managing Environmental Resources. Publication EPA 600-3-89-060. U.W. EPA, Washington, D.C.

U.S. Environmental Protection Agency. 1994. Unpublished. Florida Regionalization Project Technical Report, 84 pp.

U.S. Environmental Protection Agency. 1999. Rapid Bioassessment Protocols for Use in Wadable Streams and Rivers. Publication EPA 841-B-99-002. U.S. EPA, Washington, D.C.

Wolfe, S. 2002. Personal communication. Florida Department of Environmental Protection.

32

Using Conceptual Models to Select Ecological Indicators for Monitoring, Restoration, and Management of Estuarine Ecosystems

Tomma Barnes and Frank J. Mazzotti

CONTENTS

Introduction

> Indicator: An organism or ecological community so strictly associated with particular environmental conditions that its presence is indicative of those conditions (Merriam-Webster, 1994)

When restoration and management of ecosystems are successful, everyone wins and even failure provides an opportunity to learn how ecosystems work (Ewel, 1987). A frequently asked important question is: How can one determine success? Ecosystem restoration and management seek to repair, improve, or maintain a suite of desired environmental conditions for a specific ecosystem. Ecological monitoring is essential for evaluating ecosystem condition over time. Because it is not feasible to measure, much less monitor, all conditions of an estuarine ecosystem, scientists and managers heavily rely on the concept of ecological indicators to reveal information about their ecosystems (Szaro et al., 1999). As a result, monitoring programs focus on indicators that maximize information on ecosystem patterns and processes while minimizing cost and effort.

After goals and objectives have been established, ecosystem monitoring programs contain three vital components: (1) selection of ecosystem attributes to be used as indicators of overall ecosystem response; (2) design of a sampling program, including methods of data collection and management, quality assurance and control, and statistical protocols; and (3) a method of implementing results of the monitoring program in the decision-making process as desired conditions are met, or not.

The first phase of selecting indicators for monitoring should begin with understanding and describing key components and processes of the ecosystem. Information on these components subsequently allows an assessment of the management actions that may affect them. Conceptual models used in the context of the criteria described above can be used to portray the current status of knowledge about an ecosystem and can be used to determine which components of an ecosystem are most critical for monitoring. This chapter discusses the application of conceptual ecological models in selecting indicators for estuarine ecosystem monitoring programs in southwest Florida.

Conceptual Ecological Models

Conceptual ecological models can be used to identify which biological attributes or indicators should be monitored to best interpret ecosystem conditions, changes, and trends (Rosen et al., 1995; Gentile, 1996; Chow-Fraser, 1998; Twilley, 2000). Conceptual ecological models are simple, nonquantitative models, represented by a diagram that shows a set of relationships between major anthropogenic and natural stressors, biological indicators, and target conditions for the indicators. The development and application of conceptual models may provide benefits to the scientists who create the models and the environmental managers and general public who use them to guide, implement, and understand resource policy (Gentile, 1996; Ogden et al., 1997; Gentile et al., 2001). They are a means of (1) simplifying complex ecological relationships by organizing information and clearly depicting system components and interactions; (2) integrating data to more comprehensively implicit ecosystem dynamics; (3) identifying which species will show ecosystem response; (4) interpreting and tracking changes in targets; and (5) communicating with environmental managers. It has also been suggested that researchers can improve interdisciplinary science through the use of conceptual models as a communication tool (Heemskerk et al., 2003).

Conceptual ecological models have been widely used in other regions of North America in planning several large-scale restoration projects (Rosen et al., 1995; Gentile, 1996; Chow-Fraser, 1998; Twilley, 2000). In South Florida, they are being used to give direction to performance-based ecological monitoring and research plans as part of the Comprehensive Everglades Restoration Plan (CERP) and are currently being created as part of the Southwest Florida Feasibility Study (SWFFS) to identify attributes or indicators for modeling and monitoring.

Methods

The creation of a conceptual ecological model is an interactive and iterative process. A conceptual ecological model can be an effective instrument for developing consensus regarding a set of working hypotheses that explain changes that have occurred within an estuary and identify indicators that are representative of the overall ecological conditions of the ecosystem. Prior to creation of a model, a model coordinator should review existing information on the ecosystem. The next step is to use informal workshops to identify and discuss causal hypotheses that best explain both natural and key anthropogenically driven alterations in the estuary. Workshop attendees should include regional and local experts from multiple disciplines. From these discussions, participants create lists of the appropriate stressors, ecological effects, and attributes (indicators) in the estuarine system. The objective is to identify physical and biological components and linkages in each landscape, that best characterize the changes described by the hypotheses. Preparers (modeling team leaders) use hypotheses and lists of components to lay out an initial draft of the model and to prepare a supporting narrative document to explain the organization of the model and science supporting the hypotheses. Drafts of narratives should be reviewed and revised in subsequent workshops.

The South Florida conceptual ecological models follow a top-down hierarchy of information with the following components (Ogden and Davis, 1999):

Drivers/Sources — Major external driving forces that have large-scale influences on natural systems. Drivers can be natural forces (e.g., sea level rise) or anthropogenic (e.g., regional land use changes).

Stressors — Physical or chemical changes that occur within natural systems. Stressors are brought about by drivers and are directly responsible for significant changes in biological components, patterns, and relationships in natural systems.

Ecological Effects — Biological responses caused by stressors. The links in conceptual ecological models between one or more stressors and ecological effects and attributes are diagrammatic representations of working hypotheses that explain changes that have occurred in ecosystems.

Attributes — Also known as indicators or end points and defined as a frugal subset of all potential biological elements or components of natural systems representative of overall ecological conditions. Attributes typically are populations, species, guilds, communities, or processes. Attributes are selected to represent known or hypothesized affects of stressors (e.g., numbers of nesting wading birds) and elements of systems that have important human values (e.g., endangered species, sport fish). Performance measures and restoration objectives are set for each attribute. Status and trends among attributes are measured by a systemwide monitoring and assessment program as a means of determining success of a program in reducing or eliminating adverse effects of stressors. Measuring both stressors and their attributes also provides a means for evaluating working hypotheses and for better understanding cause-and-effect relationships in natural systems.

Measures — Specific features of each attribute to be monitored to determine how well that attribute is responding to projects designed to correct adverse effects of stressors (i.e., to determine success of the project).

As one connects stressors to indicators, a link is established with a working hypothesis, recognizing that these relationships have different levels of certainty. Knowing this level of certainty allows indicators to be organized into monitoring components and research questions.

A conceptual ecological model should be presented in both graphic and narrative form (Suter, 1996). Each model narrative should include (from Ogden and Davis, 1999):

1. A brief introduction to the dynamics and problems of the landscape
2. Descriptions of specific ecological stressors or external drivers, as well as ecological attributes
3. Ecological effects, including descriptions of major ecological linkages (working hypotheses) affected by stressors, and levels of certainty of the hypotheses
4. Research questions developed from working hypotheses
5. Recommended performance measures and restoration targets for attributes

A good conceptual ecological model will not attempt to explain all possible relationships or include all possible factors influencing the performance measure targets. Instead, it will simplify ecosystem function by containing only information most relevant to ecosystem monitoring goals. Following these criteria, chosen attributes must be measurable, and historical patterns, relationships, and functions well enough understood to interpret their responses (Ogden et al., 2003).

Case Study

The Southwest Florida Feasibility Study is a component of the Comprehensive Everglades Restoration Plan. The Southwest Florida Feasibility Study will result in an independent but integrated implementation plan for Comprehensive Everglades Restoration Plan projects. The Southwest Florida Feasibility Study will provide a framework to address the health and sustainability of aquatic systems. This includes water quantity and quality, flood protection, and ecological integrity. A Southwest Florida Feasibility Study systemwide monitoring and assessment plan will be developed from a minimal set of attributes and performance measures considered by the Southwest Florida Feasibility Study Team as necessary to understand system responses. To identify attributes (indicators) to be measured, the Southwest Florida Feasibility Study Team has developed a set of conceptual ecological models for both inland and coastal systems. The performance measures in the resulting monitoring and assessment plan are arranged into packages describing each performance measure, key uncertainties, related research questions, and monitoring protocol. Three coastal models have been developed. Included here is a conceptual ecological model created for the Ten Thousand Islands region of Southwest Florida (Figure 32.1).

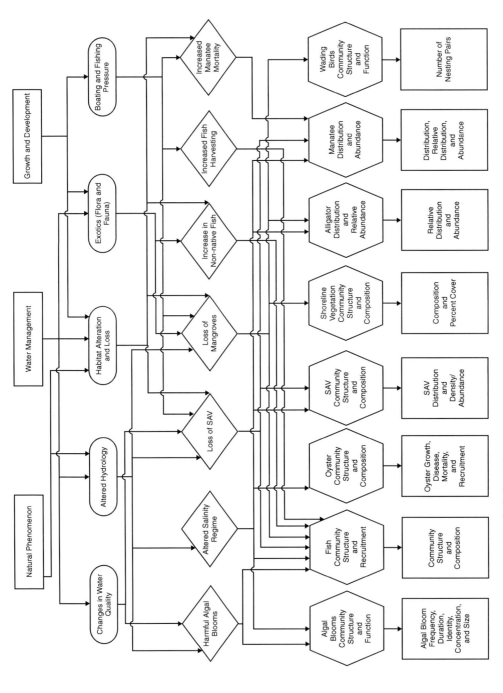

FIGURE 32.1 Ten Thousand Islands conceptual ecological models.

The Ten Thousand Islands region is a shallow, marine-dominated system characterized by mangroves and open-water habitats with areas of coastal strands, saltwater marshes, mudflats, oyster reefs, and seagrasses. The salinity gradient varies spatially with topography and location, and temporally by season (Popowski and Burke, 2003). Because of its location in the downstream portion of the East Collier drainage basin, the Ten Thousand Islands region is affected by seasonal freshwater input and upstream water management practices that alter freshwater flows that drive salinity gradients (Popowski and Burke, 2003).

Altered hydrology, habitat, and changes in water quality are the predominate stressors of the system and have affected distribution and abundance of species historically found within the system. Alterations include conversion of wetlands, dredging of channels and spoil disposal, changes in shoreline, snagging of streams for navigation, and decrease in spatial extent of the estuaries. Increase in nutrients and dissolved organics entering the system as a result of anthropogenic activities in the watershed also have caused changes in water quality (USGS, 1998). Non-native flora and fauna, such as Brazilian pepper (*Schinus terebinthifolius*) and the Mayan cichlid (*Cichlasoma urophthalmus*), and an increase in boating and fishing pressure were identified as stressors on this ecosystem. Use of personal watercraft, airboats, and similar shallow-draft vessels has increased significantly, providing motorized access to previously inaccessible shallow waters. These stressors are the result of three major sources or drivers: water management practices, changes in land use, and natural phenomena (such as hurricanes and sea level rise).

Eight attributes were identified as indicators of ecosystem response for the Ten Thousand Islands (Figure 32.1, Table 32.1). These indicators also were evaluated by other criteria. A general checklist of criteria for evaluating potential indicators was established by the National Research Council (2000). A modified subset applied to this study included the following:

General Importance — Does the indicator provide information about changes in important ecological processes? Does the indicator show us something about major environmental changes that affect significant areas? Is the indicator sensitive to stress? Is it biological or socially relevant?

Reliability — What experiences or other evidence demonstrates the indicator's reliability?

Temporal and Spatial Scales — Does the indicator inform us about ecosystem, community, or species specific conditions, processes, and products? Are the changes measured by the indicator likely to be short term or long term? Can the indicator detect changes at the appropriate temporal and spatial scales without being overwhelmed by variability?

Statistical Properties — Has the indicator been shown to serve its intended purpose in the areas of accuracy, sensitivity, precision, and robustness? Is the indicator sensitive enough to detect important changes but not so sensitive that signals are masked by natural variability? Are its statistical properties understood well enough that changes in its values will have clear and unambiguous meaning?

Data — Are they easily measured? How much and what kinds of data are required for the indicator to show a measurable trend? Are the data integrative?

In addition to a distinct linkage to a stressor, each selected attribute had additional significance, such as being ecologically important for endangered species, or providing habitat for other species (Table 32.1). This final set of attributes allows monitoring to track ecosystem response at different temporal and spatial scales and at different taxonomic levels. Each indicator has at least one performance measure allowing evaluation of changes in response to management manipulations (Table 32.1).

Discussion

Conceptual ecological models can be used as tools for identifying biologically and socially relevant indicators for monitoring (Rosen et al., 1995; Gentile, 1996; Chow-Fraser, 1998; Twilley, 2000). When properly developed, a model can effectively capture important ecosystem processes, and when carefully chosen, indicators can give important information on the overall health of that ecosystem. The model creation process is simple but is often time-consuming because several iterations may be necessary to correctly portray the ecosystem and its subsequent consensus from the working group. Consensus building, through the workshop process, helps assure managers that indicators have been selected using the best professional knowledge and available information on the ecosystem of interest and its significant attributes.

TABLE 32.1

Ten Thousand Islands Conceptual Ecological Model: Summary of Attributes, Linkages, and Performance Measures

Attribute	Criteria, Scale, and Organization Level	Linkages	Performance Measures
Algal blooms	Criteria: Indicator of Water Quality Spatial scale: Landscape Temporal response: Days to weeks Level of organization: Species	Altered hydrology through harmful algal blooms Altered hydrology through altered salinity regime Changes in water quality through harmful algal blooms	Algal bloom frequency, duration, identity, concentration, and size
Fish community structure and function	Criteria: Valued ecosystem component Spatial scale: Ecosystem Temporal response: Less than 1 year Level of organization: Community	Changes in water quality through harmful algal blooms Changes in water quality through loss of SAV Altered hydrology through altered salinity regime Altered hydrology through loss of mangroves Habitat alteration and loss through loss of SAV Habitat alteration and loss through loss of mangroves Habitat alteration and loss through increase in non-native fish Exotics through loss of mangroves Exotics through increase in non-native fish Boating and fishing pressure through loss of SAV Boating and fishing pressure through loss of mangroves Boating and fishing pressure through increased fish harvesting	Juvenile community structure and composition
Oyster structure and function	Criteria: Provides estuarine habitat and structure Spatial scale: Ecosystem Temporal response: Less than 1 year Level of organization: Species	Altered hydrology through altered salinity regime	Oyster growth, disease, mortality, and recruitment
SAV community structure and composition	Criteria: Provides estuarine habitat and structure Spatial scale: Ecosystem Temporal response: Multiple growing seasons Level of organization: Community	Changes in water quality through loss of SAV Altered hydrology through altered salinity regime Altered hydrology through loss of SAV Habitat alteration and loss through loss of SAV Boating and fishing pressure through loss of SAV	SAV distribution, density, and abundance
Shoreline vegetation community structure and composition	Criteria: Provides information on the terrestrial/aquatic transition zone Spatial scale: Ecosystem Temporal response: Years Level of organization: Community	Altered hydrology through loss of mangroves Habitat alteration and loss through loss of mangroves Exotics through loss of mangroves Boating and fishing pressure through loss of mangroves	Composition and percent cover

(continued)

TABLE 32.1 (CONTINUED)

Ten Thousand Islands Conceptual Ecological Model: Summary of Attributes, Linkages, and Performance Measures

Attribute	Criteria, Scale, and Organization Level	Linkages	Performance Measures
Alligator distribution and abundance	Criteria: Keystone species Spatial scale: Ecosystem Temporal response: Years Level of organization: 　Species	Altered hydrology through altered salinity regime Altered hydrology through loss of mangroves Habitat alteration and loss through loss of mangroves Exotics through loss of mangroves Boating and fishing pressure through loss of mangroves	Relative distribution and abundance
Manatee distribution and abundance	Criteria: Endangered species Spatial scale: Ecosystem Temporal response: Years Level of organization: 　Species	Changes in water quality through loss of SAV Altered hydrology through altered salinity regime Habitat alteration and loss through loss of SAV Boating and fishing pressure through loss of SAV Boating and Fishing Pressure through increased manatee mortality	Relative distribution and abundance
Wading bird community structure and function	Criteria: Valued ecosystem component Spatial scale: Ecosystem Temporal response: Multiple nesting seasons Level of organization: 　Community	Altered hydrology through loss of mangroves Habitat alteration and loss through loss of mangroves Exotics through loss of mangroves Boating and fishing pressure through loss of mangroves	Number of nesting pairs

Note: SAV = submerged aquatic vegetation.

Once a model has been created and indicators have been selected, it is very important to take each indicator through an additional checklist of criteria, such as that above, to assure its applicability and usefulness to project specific goals and management needs. It is important to remember throughout the process of identifying indicators that no single indicator can provide a comprehensive picture of the system and that a suite of indicators, at different spatial or temporal scale and biological levels, should be included (Franklin, 1993; Lambeck, 1997). Because ecological monitoring is such a critical factor in understanding ecosystem and landscape response to change, it is very important to select high-quality indicators (Szaro et al., 1999). In South Florida, indicators have been identified through the conceptual ecological process discussed previously and are being incorporated into systemwide and project-specific monitoring plans. In addition, conceptual ecological models have been used to identify the critical linkages between ecosystem stressors, indicators, and performance measures, and the uncertainties associated with the linkages.

The creation of conceptual ecological models has become a crucial part of the adaptive assessment process required for all of the Comprehensive Everglades Restoration Plan projects as part of the Water Resources Development Act of 2000. Adaptive management can be described as a learning process to guide management in the face of uncertainty (Stankey et al., 2003). In Southwest Florida, conceptual ecological models have been used as the foundation for developing an integrated process for adaptive management (Figure 32.2). Regional hydrological and ecological models, such as habitat suitability models or stressor response models, are used to evaluate alternative scenarios and the results used to modify alternatives in the study. These models forecast future conditions and provide an understanding of the potential magnitude of management and restoration alternatives. Models relating output of hydro-

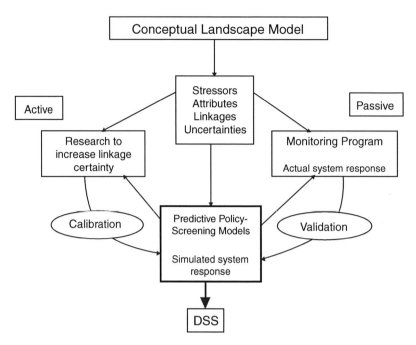

FIGURE 32.2 Conceptual ecological models form the basis for an integrated approach to adaptive ecosystem management and restoration.

logical models and potential changes in the landscape to amount and location of potential habitat for individual species or habitats give both a visual and a quantitative picture of effects of restoration on selected attributes.

The SWFFS conceptual ecological models show how local scientists think the natural areas of southwest Florida have been stressed, and present the working hypotheses used to explain current ecological conditions in these altered systems. These hypotheses are separated by their level of certainty into research or monitoring. Research or experimentation tests the certainty of linkages and provides calibration for models. This provides a process for active adaptive management. Monitoring evaluates the *in situ* system response and evaluates the success or failure of a project. This creates an iterative or passive adaptive management loop. Despite the potential of adaptive management there are serious concerns regarding the ability to incorporate scientific knowledge into the decision-making process (Walters, 1997). Ecosystem management decisions are frequently complex and uncertain, precluding the structure needed for adaptive management to work (Stankey et al., 2003).

Decision support systems are broadly defined as computer-based systems used to aid environmental managers using data and models to solve unstructured problems (Sprague and Carlson, 1982; Rauscher, 1999; Mowrer, 2000). They do not make decisions or set policy; rather they help decision makers to organize, sort, and display decision variables and parameters, and to appreciate impacts of potential policy actions. Decision support systems will provide critical linkage between science and management in the SWFFS.

This approach will enable the SWFFS and other restoration projects to conduct tasks of formulation, evaluation, and modification of alternatives. These alternatives are necessary to develop an effective and feasible plan. In addition, this approach will supply information required by the SWFFS to meet its objectives of conserving and protecting water resources to ensure sustainability of natural resources; improving and protecting quality, heterogeneity, and natural biodiversity in freshwater, upland, estuarine, and marine ecosystems; and protecting and recovering "listed" species.

References

Chow-Fraser, P. 1998. A conceptual ecological model to aid restoration of Cootes Paradise Marsh, a degraded coastal wetland of Lake Ontario, Canada. *Wetlands Ecology and Management* 6:43–57.

Ewel, J. J. 1987. Restoration is the ultimate test of ecological theory. In *Restoration Ecology: A Synthetic Approach to Ecological Research,* W. R. Jordan III, M. E. Gilpin, and J. D. Aber (eds.). Cambridge University Press, Cambridge, U.K., pp. 31–33.

Franklin, J. F. 1993. Preserving biodiversity: species, ecosystems, or landscapes? *Ecological Applications* 3:202–205.

Gentile, J. H. 1996. Workshop on South Florida Ecological Sustainability Criteria. Final Report. University Miami, Center for Marine and Environmental Analysis, Rosenstiel School of Marine and Atmospheric Science, Miami, FL, 54 pp.

Gentile, J. H., M. A. Harwell, W. Cropper, Jr., C. C. Harwell, D. DeAngelis, S. Davis, J. C. Ogden, and D. Lirman. 2001 Ecological conceptual models: a framework and case study on ecosystem management for South Florida sustainability. *Science of the Total Environment* 274:213–253.

Heemskerk, M., K. Wilson, and M. Pavao-Zuckerman. 2003. Conceptual models as tools for communication across disciplines. *Conservation Ecology* 7(3):8. Available at http://www.consecol.org/vol7/iss3/art8.

Lambeck, R. J. 1997. Focal species: a multi-species umbrella for nature conservation. *Conservation Biology* 11:849–856.

Merriam-Webster's Collegiate Dictionary, 10th ed. 1994. Merriam-Webster, Springfield, MA.

Mowrer, H. T. 2000. Uncertainty in natural resource decision making support systems: sources, interpretation, and importance. *Computers and Electronics in Agriculture* 27:139–154.

National Research Council. 2000. Ecological Indicators for the Nation. National Academy Press, Washington, D.C., 180 pp.

Ogden, J. C. and S. M. Davis. 1999. The Use of Conceptual Ecological Landscape Models as Planning Tools for the South Florida Ecosystem Restoration Programs. South Florida Water Management District, West Palm Beach, FL.

Ogden, J. C., S. M. Davis, D. Rudnick, and L. Gulick. 1997. Natural Systems Team Report to the Southern Everglades Restoration Alliance. Final draft. July 1997. South Florida Water Management District, West Palm Beach, FL, 43 pp.

Ogden, J. C., S. M. Davis, and L. A. Brandt. 2003. Science strategy for a regional ecosystem monitoring and assessment program: the Florida Everglades example. In *Monitoring Ecosystems*, D. E. Busch and J. C. Trexler (eds.). Island Press, Washington, D.C.

Popowski, R. and T. Burke. 2003. Ten Thousand Islands Conceptual Ecological Model (Draft manuscript). South Florida Water Management District and U.S. Fish and Wildlife Service.

Rauscher, H. M. 1999. Ecosystem management decision support for federal forests in the United States: a review. *Forest Ecology and Management* 114:173–197.

Rosen, B. H., P. Adamus, and H. Lal. 1995. A conceptual model for the assessment of depressional wetlands in the prairie pothole region. *Wetlands Ecology and Management* 3:195–208.

Sprague, R. H., Jr. and E. D. Carlson, 1982. *Building Effective Decision Support Systems.* Prentice-Hall International, London.

Stankey, G. H., B. T. Bormann, C. M. Ryan, B. Shindler, V. Startevant, R. N. Clark, and C. Philpot. 2003. Adaptive management and the northwest Forest Plan: rhetoric and reality. *Journal of Forestry* 101(1): 40–46.

Suter, G. W., II. 1996. Guide for Developing Conceptual Models for Ecological Risk Assessment. Environmental Risk Assessment Program, Oak Ridge, TN.

Szaro, R., D. Maddox, T. Tolle, and M. McBurney. 1999. Monitoring and evaluation: why monitoring and evaluation are important to ecological stewardship. In *Ecological Stewardship: A Common Reference for Ecosystem Management,* N. C. Johnson, A. J. Malk, R. C. Szaro, and W. T. Sexton (eds.). Elsevier Science, Oxford, pp. 223–230.

Twilley, R. R. 2000. Developing Conceptual Models of Coastal Wetland Restoration: Environmental Drivers of Ecological Succession. Brown Marsh Executive Order Meeting, Baton Rouge, LA.

U.S. Geological Survey. 1998. Water-Quality Assessment of Southern Florida — Wastewater Discharges and Runoff. U.S. Department of the Interior, USGS Fact Sheet FS-032-98.

Walters, C. 1997. Challanges in adaptivve management of riparian and coastal ecosystems. Conservation ecology [online] 1(2):1. http://www.consecol.org/vol1/iss2/art1.

33

Future Directions for Estuarine Indicator Research

S. Marshall Adams and Stephen A. Bortone

CONTENTS

The Current Status of Estuarine Indicators

Estuaries are complex, variable, and diverse; consequently, they present unique challenges relative to understanding the effects of environmental stressors on these systems. Because of these challenges, the main purpose of the Estuarine Indicators Workshop and this subsequent volume was to bring together a diverse group of estuarine scientists to present and discuss the principles, concepts, and practical use and application of indicators in estuarine research and management practices. Thus, the principal objectives of this volume were to (1) identify the major effects that natural and anthropogenic stressors have on estuarine systems, (2) help identify and understand the underlying cause(s) of these effects, and (3) provide guidelines and recommendations for the development and use of indicators in the effective environmental management of estuarine ecosystems.

The complexity and diversity of estuarine systems are reflected in the diversity and variety of presentations in this book. There are myriad ways in which to classify the the chapters in this volume. Current ongoing research can be grouped as structurally related topics, functional-level presentations, and supporting field studies. A preponderance of chapters here are related to structural-level indicators such as descriptions of types of organisms, populations, or communities that occur in particular habitats, relative abundances or occurrences of these biological components, and the species richness or diversity of various biological groups. In several cases, the structurally related presentations describe the occurrence, relative abundance, or diversity of biological components in relation to influential environmental factors of estuaries such as nutrients, habitat availability, food availability, salinity, and other physiochemical factors. Included in this group are chapters focused primarily on such structurally-based indicators of estuarine health as microbes, benthic diatoms, parasites, macrobenthos, fishes, birds, dolphins, seagrasses, and mangroves.

Some of the ongoing research on estuarine indicators can be grouped into a category related to functional-level indicators and processes. These studies focus on understanding the functional processes of various biological components and why organisms and systems respond the way they do to environmental factors within the estuary. Functional or process-based studies are important not only for eluci- dating the mechanistic basis of how environmental stressors affect estuarine biota, but also for providing common links or "currencies" between different ecological components of estuarine systems. Examples of presentations that fit mainly within this category are the molecular indicators of hypoxia, macrobenthic processes, microalgae and nutrients, and seagrasses.

A final category for grouping current estuarine indicator research are those projects supporting ecological studies that are primarily physicochemical in nature. These types of studies, although primarily descriptive and structurally related, are critical to understanding the structural and functional aspects of estuaries as they relate to the status and behavior of estuarine systems. Several studies included here that support structural and functional features of the estuary are salinity dynamics, nutrient dynamics, physical processes and hydrodynamic models, statistical models, and conceptual models and simulations that utilize various physicochemical components to help explain or predict the biological status and trends in estuarine components.

Overriding all the current research on estuaries is the importance of understanding variability within and between estuaries. Indeed, it is the relatively high variability in biotic and abiotic components of estuaries, both predictive and nonpredictive, that contributes to their inherent complexity and, conse- quently, complicates our ability to predict and assess the effects of environmental stressors on estuarine health.

Future Directions

Based on the information represented in the variety of chapters in this volume, some general recom- mendations and guidelines are proposed for future research initiatives and environmental management practices for estuarine ecosystems. At their core, these areas of research and management include the continued development, validation, and application of various types of indicators for assessing the effects of environmental stressors on estuaries and identifying the causes of the effects on estuarine resources. Suggested areas for future research and development and a brief description of each area are provided below.

Understanding of Scaling and Variability

One of the major characteristics of estuaries is their high spatial and temporal variability. Understanding variability in estuaries is important to help account for effects on biota caused by both natural environ- mental factors and those effects resulting from anthropogenic stressors. Estuarine biota can be affected by both chronic, long-term stressors and the shorter, pulsed, or episodic stressors. Subsequently, a suite of indicators should be used that represents a range of different response times, sensitivities, and specificities to stressors. The short-term, sensitive indicators respond relatively rapidly and generally within the same time frame as many of the episodic, pulsed stressors, thereby providing sensitive and early-warning indicators of stress effects. The longer-term indicators, which are not generally sensitive and rapidly responding, are integrative in nature and provide a measure of ecological relevance or significance. Long-term indicators provide the main types of end points used in environmental manage- ment and regulatory decisions.

For truly understanding the impact of scaling and variability in estuarine ecosystems, scientists need to be cognizant of the problems associated with pseudoreplication in their study designs. Pseudoreplication (*sensu* Hurlbert, 1984) is an inevitable consequence of studies predetermined to investigate a single estuary. Because funding agencies are interested in specific problems occurring in specific estuaries, our ability to identify larger-scale trends and the factors that may be driving them has been hindered. Stewart-Oaten and Murdoch (1986) astutely indicated the ramifications of not recognizing problems associated with pseudoreplication, both temporally and spatially. To avoid pseudoreplication and to more fully understand the relationships between environmental stressors and the corresponding ecosystem responses, study designs need to include several estuaries as part of the sampling replicates.

Long-Term Studies

Long term, with regard to the scale of changes in estuaries, can generally be thought of as decadal to centennial in duration. Because of the relatively high spatial and temporal variability in estuaries, the usual study designs for most environmental monitoring and assessments are inadequate. This is because they are unable to separate the effects on estuarine biota caused by natural environmental factors from those that result from anthropogenic stressors. Long-term temporal trends in biota can be associated with changes in large-scale environmental factors such as temperature, eutrophication, salinity, or habitat modification. Thus, trends in long-term data sets filter out or can account for effects resulting from short-term variability.

Understanding of Influential or Causal Environmental Factors

Understanding causal relationships and the mechanistic processes between environmental stressors and effects on biota is important in the effective management and restoration of estuarine ecosystems. In systems experiencing multiple stressors (as is the case for most estuaries), effective management and environmental regulation of estuarine resources depends on (1) determining if an effect has occurred or if resources of concern have been impaired, (2) determining the extent or nature of injury, and (3) identifying the probable cause of such injuries. If a probable cause of injury can be identified, the proper management and regulatory practices can be initiated to restore the resource to some acceptable level. Several possible methods and approaches can be used to assess potential causality including using (1) a suite of indicators at different levels of biological organization to establish relationships between lower-level responses to stressors (rapid, sensitive indicators) and high-level responses to stressors (population, community-level), (2) simulation models that utilize a combination of physicochemical and biological variables to predict the effects of various perturbations on the resource of interest, and (3) multimetric profiling techniques utilizing environmental diagnostic procedures such as development of exposure–effects response profiles for the biota of interest.

Integration of Assessment Approaches and Techniques

Many current approaches for assessing estuarine health utilize a series of standard techniques such as synoptic field surveys for investigating the health, occurrence, or abundance of organisms sometimes in relationship to influential or controlling environmental factors. However, to better understand the relationships between the various physical, chemical, and biological components of estuaries, particularly as they interact and are affected by environmental stressors, a greater range and variety of available assessment techniques should be used. These various approaches include (1) laboratory and field experimental manipulations such as chronic and life cycle tests, microcosms, mesocosms, and caging studies; (2) simulation modeling including both ecosystem and population models; (3) use of multiple indicators representing a range of different levels of biological organization and utilized within a weight-of-evidence approach; and (4) a combination of these techniques. Each of these techniques for assessing estuarine health has its own set of advantages and limitations and the combination of approaches chosen for study can vary depending on the type of system studied, the environmental stressors involved, the study objectives, etc. For example, laboratory studies often lack ecological realism. Moreover, not all potential factors that cause effects can be tested in the laboratory. In addition, these studies usually do not allow prediction of indirect effects, cumulative and synergistic effects are not accounted for, and laboratory studies are usually limited to chemical stressors. Concomitantly, laboratory investigations can be conducted under controlled and replicable conditions, causality can usually be determined for single stressors and effects, and, most importantly, they can be used to validate possible relationships observed in the field from either field or manipulative studies. Simulation models, for example, can be used to help address and mathematically formulate causal mechanisms that control ecosystem, community, or population structure. Predictions generated by simulation models help focus research issues on specific questions concerning controlling mechanisms, which can then be experimentally tested. The limitation of simulation models, however, is that the assumed causal relationships underlying the basic model processes may not be accurate, resulting in unreliable predictions. Also, problems of parameter estimation are one of the main reasons complex simulation models may not yield reliable predictions of system behavior.

Utilization of Indicators within a Holistic Framework

Most indicators, as they are currently applied, are used to assess the health or condition of particular organisms, populations, communities, habitat types, or some functional process such as nutrient dynamics. However, the structural and functional properties of estuaries are not only coupled to each other but these properties are also linked, to varying degrees, to those external factors such as climatic and human-induced stressors, which can also exert controlling influences. Therefore, the next generation of estuarine indicators should include variables that are within a holistic ecosystem context whereby they help link estuarine ecosystem structure to function. In the future, estuarine indicators should be able to identify the coupling of chemical, physical, and biological processes. They will also be useful in making evaluations between habitats instead of only within habitats. For the most part, our current knowledge and understanding is not to the point where these types of holistic indicators can be readily applied but, nevertheless, the development, testing, validation, and application of these types of indicators should be a priority of estuarine scientists in the near future.

New Approaches to Estuarine Indicators

As a result of a variety of factors such as the complexity and variability of estuaries, limitations in funding, etc., it would be difficult to design and conduct a study within an entire estuary whose main goal was to develop and test some of the recommended approaches discussed above. One suggestion would be to choose a relatively localized area of an estuary where effects are known to occur to specific species or communities and then test some of these approaches. To avoid the pitfall of overgeneralizing from improperly replicated study designs, other estuaries should be similarly sampled for localized stressors. For example, such study areas in a diversity of estuaries could be "hot spots" of sediment contamination or where sources of pollution are known to impair organisms, populations, or habitats. A phased, but larger-scale approach for such a test case subsequently could be taken where the decision to proceed to subsequent stages (i.e., more variables, more details, or more analytical sophistication) in the assessment process would be dependent on the results of the preceding phase.

Such a study design will be foiled unless a truly multijurisdictional approach to estuarine research becomes a reality. Local water management districts, states, provinces, and, inevitably, countries have the most to benefit from such spatially extensive collaborations among estuarine scientists. This scale of estuarine examination, however, is mostly a dream. For our future research efforts to be successful, the dream demands the attention of all of us interested in estuaries. In the future, when people ask the inevitable — "How's the estuary doing?" — we will be able to give them a meaningful answer.

References

Hurlbert, S. H. 1984. Pseudoreplication and the design of ecological field experiments. *Ecological Monographs* 54(2):187–211.
Stewart-Oaten, A. and W. M. Murdoch. 1986. Environmental impact assessment: "pseudoreplication" in time. *Ecology* 67(4):929–940.

Index

A

Abdul-Salam and Sreelatha studies, 301
Abel, Hellawell and, studies, 320
Abiotic patterns, 360–361
Abrahams, Robb and, studies, 262
Achnanthes, 138
Acute toxicity, 85, *89–90*
Adams, Titus and, studies, 224
Adams and Bates studies, 195
Adams and Ryon studies, 297
Adams studies, 5–16, 72, 381, 503–506
Adaptive management, 439
Admiraal studies, 127
Aggregate stream condition index, 484
Aggregation, 322
Agusti studies, 194
Ainsworth, Clarke and, studies, 335
Alafia River, Fla., 280
Alberte, Dennison and, studies, 194, 200, 204
Alcoverro studies, 187
Algal communities, *132–133*, 133–137, *135–136*
Alive areas, mangroves, 255–256, *256*
Allen and Read studies, 425
Alphin, Posey and, studies, 278
Altered freshwater flow, *see* Nekton species composition
Amann studies, 100, 112
Amaral, Rizzo and, studies, 278, 290
Americamysis bahia, 82–83, 85, 94
American Public Health Association (APHA), 82–83
American Society for Testing and Materials (ASTM), 69, 81–83
D'Amico studies, 423
Ampelisca abdita, 82, 85, 94
Amphora, 138
Anabaena, 158
Anclote Anchorage, Fla., 93
Anderson, Toth and, studies, 437, 446
Anderson studies, 460
Angermeier and Karr studies, 377
Animal species, sediment habitat assessment, 82–83
Anteson studies, 306
ANZECC studies, 66
Apalachicola Bay, Fla., 387
APHA, *see* American Public Health Association (APHA)
Apoprionospio pygmaea, 289
Araya studies, 113
Army Corps of Engineers, *see* U.S. Army Corps of Engineers (USACOE)

Artemia, 264
Assessment, *see also* Coastal sediments assessment; Sediment habitat assessment
    indicator integration, 505–506
    multiple response bioindicators, 14–16
    trematode parasites, 306–309
ASTM, *see* American Society for Testing and Materials (ASTM)
Atchafalaya River, 54
Atlantic Coast, 67, 69, 386, 415, *see also* specific area
Atlantic Oceanographic and Meteorological Laboratory, 56
Attributes, bioassessment procedures development, 489–490
Attrill and Depledge studies, 8, 472
Attrill studies, 371
Ault studies, 54, 389
Aumen, Toth and, studies, 438, 442
Austin studies, 114
Avicennia, 138, *see also* Mangroves
Van Den Avyle and Maynard studies, 351

B

Bachelet studies, 278
Bacterial communities
    basics, 99–107
    case study, 101–102, *102*
    discussion, 106–107
    materials and methods, 102–103
    results, 103–106, *104–106*
Badylak, Phlips and, studies, 188
Bain studies, 366
Balba and Nedwell studies, 121
Balk studies, 16
Ballast Point, Fla., 278
Baltz studies, 383, 387–388
Bamforth studies, 114
Banana Lake, Fla., 232
Barbour studies, 370, 482
Barciela studies, 148
Barko, Twilley and, studies, 212, 223
Barko studies, 223–224
Barnes Sound, Fla., 333
Barnes studies, 493–500
Barry (tropical storm), 58
Bartak, Rohacek and, studies, 195, 205
Bart and Hartman studies, 11
Bartholomew studies, 7
Bartosch studies, 113
Bartosiewicz studies, 263